AF679765

"Digital Elevation Model Technologies and Applications"
American Society for Photogrammetry and Remote Sensing

Copyright © 2024 by American Society for Photogrammetry and Remote Sensing

All rights reserved.
No part of this book may be reproduced in any form or by any electronic or mechanical means, including information storage and retrieval systems, without written permission from the author, except for the use of brief quotations in a book review.

ISBN: 979-8-89569-011-6

Spines

Digital Elevation Model Technologies and Applications: The DEM Users Manual, 3rd Edition

Edited by David F. Maune, PhD, CP
and Amar Nayegandhi, CP, CMS

American Society for Photogrammetry
and Remote Sensing | Bethesda, MD

FOREWORD

David F. Maune and Amar Nayegandhi

Whenever I autograph a copy of The DEM Users Manual, I write "May all your DEMs come true!" This is an obvious play on words with "May all your dreams come true." When the 2nd edition was published, I had three basic dreams: (1) Development of high-accuracy, affordable elevation technologies for betterment of society; (2) Development and update of DEM technology standards, guidelines and specifications; and (3) Implementation of a nationwide program, such as today's 3D Elevation Program (3DEP), to produce and maintain standardized high-quality DEMs used by all. These three dreams have largely been realized, as documented in this 3rd edition.

My future dreams include: (4) Development of a seamless "3-D Nation" from the tops of the mountains to the depths of the seas, to include inland bathymetry; (5) Use of the latest elevation data to routinely and systematically update the National Hydrography Dataset, flood studies, forest metrics and other datasets that require up-to-date DEMs; and (6) Development of DEM applications to fully support the dozens of business uses and hundreds of mission-critical activities (MCAs) documented in the National Enhanced Elevation Assessment (NEEA) completed in 2012, and the 3-D Nation Requirements and Benefits Study being conducted in 2018-2019. For these future dreams to be fulfilled, it's time for me to "pass the torch" to Amar Nayegandhi and the younger lidar professionals who will lead us into the 3-D Nation of the future. Amar agreed to serve as co-editor for this 3rd edition.

The editors and chapter authors for this 3rd edition recognize that diverse DEM users seek a continuous topographic/bathymetric surface including submerged topography beneath the oceans, rivers, streams, lakes and ponds, as well as a top reflective surface including treetops, rooftops and tops or all other manmade features. Some of these requirements will be satisfied by standard products whereas other requirements will be satisfied by users creating their own derivative products. USGS' 3DEP addresses the traditional topographic surface portion of this vision, and we believe our full vision for a 3-D Nation, including submerged topography, will ultimately be realized worldwide. The editors and authors hope this manual will help to untangle some of the issues involved.

When the 2nd edition was published in 2007, we had no nationally-accepted standards or specifications for lidar, other than the FEMA guidelines and specifications used by the National Flood Insurance Program (NFIP). Since the 2nd edition of this manual was published over a decade ago, much progress has been made toward our current vision of the future:

- In 2010, the U.S. Geological Survey (USGS) published its draft Lidar Guidelines and Base Specifications, V.13, which ultimately became the USGS Lidar Base Specification, V1.0
- In 2012, the National Enhanced Elevation Assessment (NEEA) was published that provided a comprehensive analysis of DEM user requirements and benefits for five Quality Levels (QLs) of topographic DEMs
- In 2012, USGS published its Lidar Base Specification, V1.0
- In 2013, based on the NEEA implementation scenario with the highest return-on-investment, USGS announced the new 3DEP, to deliver QL2 lidar nationwide except for QL5 IfSAR statewide in Alaska
- In 2013, ASPRS published its latest LAS Specification, V1.4
- In 2014, ASPRS published its Positional Accuracy Standards for Digital Geospatial Data that included vertical accuracy classes

- In 2014, USGS published its Lidar Base Specification, V1.2 with detailed specifications for QL0, QL1, and QL2 lidar, consistent with the ASPRS vertical accuracy classes
- In 2015, the U.S. Army Corps of Engineers published a new EM 1110-1-1000, Photogrammetric and Lidar Mapping, endorsing both the ASPRS standards and the USGS Lidar Base Specification
- In 2016, FEMA published its Guidance for Flood Risk Analysis and Mapping, Elevation Guidance, and standardized on QL2 lidar as defined in the USGS Lidar Base Specification, V1.2 for new lidar acquisition, consistent with the goals of the 3DEP, while also aligning with the ASPRS Positional Accuracy Standards for Digital Geospatial Data and the ASPRS LAS Specification V1.4
- In 2017, the National Geodetic Survey (NGS) started teaming with USGS for a NEEA Update and Coastal/Offshore Elevation Requirements and Benefits Study that might establish bathymetric equivalents to the topographic data Quality Levels in the NEEA. This should help to "close the loop" on our vision for a 3-D Nation from the tops of the mountains to the depths of the sea
- In 2018, USGS published its Lidar Base Specification, V1.3, and NOAA and USGS kicked off their 3-D Nation Requirements and Benefits Study

The following summarizes the major focus of each chapter in support of our vision for a 3-D Nation:

- Chapter 1 (Introduction to Digital Elevation Models) establishes the vocabulary and definitions for various forms of elevation raster data, vector data, and point data (including point clouds) so that data users and producers can agree on standardized terminology. It is important that DEM users understand the concepts of mass points, breaklines, and triangulated irregular networks (TINs), for example, and how they are used to produce uniformly-spaced topographic and/or bathymetric DEMs. This chapter also covers topographic and hydrologic surface modeling (e.g., hydro-flattening and hydro-enforcement), elevation derivatives, and 3-D terrain visualization
- Chapter 2 (Vertical Datums) defines a vertical datum as a reference surface representing zero height—whether heights above sea level (with various tidal datums), orthometric heights (elevations) above the geoid with equal gravity potential (with various vertical orthometric datums), or ellipsoid heights above a mathematical ellipsoid that approximates the shape of the Earth as obtained from air-, land-, or sea-based GPS. It is important that DEM users understand geoid undulations, how all U.S. vertical datums of the future will be geoid based vertical datums that result from the National Geodetic Survey's Gravity for the Redefinition of the American Vertical Datum GRAV-D) project, and how to transform from one vertical datum to another
- Chapter 3 (Standards, Guidelines and Specifications) introduces DEM users to the ASPRS Positional Accuracy Standards for Digital Geospatial Data v1.0, the ASPRS LAS Specifications v1.4, the USGS Lidar Base Specifications v1.3, the National Ocean Service (NOS) Hydrographic Survey Specifications based on standards of the International Hydrographic Organization (IHO), and other relevant standards, guidelines and specifications. It is important that DEM users understand these documents and how they must be rigorously enforced to achieve our vision of a seamless, consistent, high-accuracy, high-resolution 3-D Nation, from the tops of the mountains to the depths of the sea
- Chapter 4 (The National Elevation Dataset—NED) provides the background, rationale and history of the legacy NED and how NED data were produced, quality controlled and delivered to the public. It provides information about NED specifications and production processing, accuracy and data quality. The NED was retired when the 3DEP became operational, but USGS is developing a new line of science products known as the Coastal National Elevation Database (CoNED) which integrates recent high resolution coastal lidar data (both topographic and bathymetric) and a temporal component from captures on different dates
- Chapter 5 (The 3D Elevation Program—3DEP) explains USGS' national elevation initiative that forms the elevation layer of The National Map. The 3DEP resulted from analyses and vetting of the NEEA

study which assessed DEM user requirements and benefits and concluded that the highest return on investment would come from QL2 lidar nationwide except for QL5 IfSAR of Alaska. The chapter describes the 3DEP program; the U.S. Interagency Elevation Inventory (USIEI); the USGS Lidar Base Specifications; the Broad Agency Announcement (BAA) process; acquisition trends; 3DEP data quality assurance; 3DEP products, services and data dissemination; current developments and future directions

- Chapter 6 (Photogrammetry) explains airborne and satellite digital imaging systems; project planning considerations; georeferencing and aerotriangulation; photogrammetric data collection methods (softcopy stereoplotters, manual, and automated elevation collection); post processing; data deliverables; enabling technologies; calibration procedures; capabilities and limitations compared with competing/complementary technologies; DEM user applications; cost considerations; and technological advancements cost considerations; and technological advancements including Structure from Motion (SfM). It is important that DEM users understand the capabilities and limitations of photogrammetry compared with lidar and IfSAR, for example
- Chapter 7 (IfSAR) explains how interferometric synthetic aperture radar works, airborne and satellite IFSAR alternatives, how aerial IFSAR is completing the first-ever mapping of Alaska to specified accuracy standards, and how differential IFSAR/InSAR is used to monitor subsidence at the mm level
- Chapter 8 (Airborne Topographic Lidar) explains the basic concepts of topographic lidar scanning and sensors; compares traditional linear-mode lidar with photon-sensitive and Geiger-mode lidar; boresight calibration; airborne lidar project planning; and the status of current lidar sensor technologies from Teledyne Optech, Leica Geosystems, Riegl, and Harris Corp
- Chapter 9 (Lidar Data Processing) explains concepts and approaches to automated filtering of lidar point clouds to include ground and non-ground points, noise, vegetation, structures, and other above-ground features; manual editing of lidar; breakline processing to include area and linear hydrographic features, structures, manual review and editing; elevation assignment to breakline features, to include linear and area hydrographic feature elevation assignment; DEM processing concepts and approaches, processing techniques, incorporating breaklines; DSM processing; and other derivative products including contours
- Chapter 10 (Airborne Lidar Bathymetry) explains the basic concepts of bathymetric lidar scanning and sensors; system design; data processing including system calibration; output formats and deliverables; and the status of current sensors including SHOALS, CZMIL, LADS, Chiroptera II/Hawk Eye III, EAARL, VQ 820/880-G, and Titan; operational and planning considerations; and comparisons with overlapping technologies
- Chapter 11 (Sonar) provides a technology overview and developmental history of acoustic mapping and explains the basic principles used, to include acoustic sources and directional transmit/receive transducers. It explains the different types of sonars (vertical beam, multibeam, side scan, interferometric, focusing, and Doppler); present operating status; platforms and installation; calibration procedures; planning considerations; capabilities and limitations and comparisons with complementary and competing technologies; post processing, quality control, data deliverables, cost considerations, and technology advancements
- Chapter 12 (Enabling Technologies) explains the Global Positioning System (GPS) and other international systems that form the Global Navigation Satellite System (GNSS). It explains GNSS positioning technologies including GNSS single point positioning, differential GNSS, precise phase interferometry positioning, and precise point positioning (PPP). It explains local, regional, and global differential GNSS, Continuously Operating Reference Stations (CORS), sensors and error sources. It explains inertial navigation systems (INS) and GNSS-aided inertial navigation technologies, direct georeferencing systems for airborne DEM generation, boresight calibration, post processing, and motion sensing systems for multibeam sonar bathymetry. These technologies are vital for accurate and cost-effective photogrammetry, IfSAR, lidar and sonar mapping

- Chapter 13 (DEM User Applications) reviews how DEMs are vital for production of digital orthophotos, topographic maps and various other types of maps (soils, geologic, wetlands, forestry, wildlife habitat, cultural resources, urban and regional planning, and flood maps); underwater and coastal mapping applications (Digital Coast, sea level rise viewing, shoreline delineation, coastal management and engineering); transportation applications (land, aviation and marine navigation and safety); military applications; technical applications (national resources conservation, hydrologic and hydraulic modeling, water supply, subsidence, and stormwater management; commercial applications (precision agriculture, mining, renewable energy, oil and gas, telecommunications), and individual applications. It summarized the NEEA's 27 major DEM Business Uses that helped justify the 3DEP and would be relevant also to state or local initiatives
- Chapter 14 (DEM User Requirements and Benefits) explains why DEM users needing lidar data should normally state their requirements for standard QL2 lidar data consistent with the 3DEP so as to receive standard raw and classified point cloud data, standard breaklines, standard metadata, and standard hydro-flattened, bare-earth raster DEMs—all with potential common data upgrades that do not compromise standardization and interoperability. The resulting benefits include: A single authoritative source of high quality and consistent 3DEP data at lower costs for all; standard 3DEP products that use common hardware/software and standard training for data users and producers; easier generation of derivative products from a standard source—all yielding a seamless, consistent elevation surface from the top of the mountains to the depths of the sea
- Chapter 15 (Quality Assessment of Elevation Data) is a 90-page tutorial with images designed to promote consistency in delivery of elevation data acceptable for the 3DEP. This chapter reviews relevant standards, guidelines and specifications; explains procedures for testing and reporting absolute and relative accuracy; and goes into great detail in addressing various forms of qualitative assessments, to include: Source data QA/QC; breakline QA/QC (breakline completeness, variance and topology); macro level reviews of DEM data; and micro level reviews of topographic and topobathymetric DEMs including hydro flattening or enforcement, edge-matching, and bare-earth editing of buildings, bridges and artifacts. This chapter also includes procedures for QA/QC of contours and metadata
- Appendix A is a list of acronyms used in the 3rd edition
- Appendix B is a summary of definitions for technical terms used in the 3rd edition
- Appendix C explains where to download sample elevation datasets
- The Index lists page number where to find discussions of hundreds of technical terms of potential interest

The eBook version has most references hyperlinked for ease of access. All references were accessed between September of 2017 and September of 2018.

I wish to thank all the chapter authors and individual contributors for their professional input and for volunteering their time to help Dewberry and ASPRS publish this important manual. Individual contributors were too numerous to have all their names acknowledged for each chapter.

I hope that this manual will be instrumental to the development of an exciting academic curriculum in which DEM technologies and applications are taught to our future industry leaders. To students and potential DEM users who read this manual and dream of the possibilities of DEM technologies and applications—may all your DEMs come true!

David F. Maune, PhD, CP
Co-Editor

It's been a true pleasure and a privilege working with Dave as co-editor of The DEM Users Manual. This manual is designed to advance the reader's knowledge in the area of 3-D data products and technologies as it relates to mapping the earth's surface, what's above it as well as what's below the water surface in coastal areas. Wide-area topographic and bathymetric data support numerous applications in hydrology/floodplain mapping, forest management, transportation, corridor mapping, urban city planning, natural resources management, archeology, navigation, and coastal resource management. As demonstrated in the NEEA study, the benefits of high-resolution topography and elevation data greatly outweigh the cost to produce these data. This manual is intended to advance the reader's knowledge of the technologies used to produce DEMs, keeping an eye on the future for new and improved technologies as well as giving the reader a full understanding of the standards, guidelines, specifications, and products that are currently available to support various applications.

I am grateful to all the authors and co-authors of the various chapters in this Manual who volunteered their time to support this initiative. This 3rd edition of The DEM Users Manual has lots of new content and information as summarized by Dave above. I am indebted to Dave for his dedication and professional work of countless hours reviewing and organizing the chapters. The graphics team at Dewberry and ASPRS, especially Margarita Girón (Dewberry) and Rae Kelley (ASPRS), who worked closely with Dave and me to ensure that each chapter was carefully formatted for publication. Finally, I support Dave's dreams for the future and hope to expand upon those dreams as DEM technologies and applications continue to evolve.

Amar K. Nayegandhi, CP, CMS
Co-Editor

Special thanks to the Dewberry Board of Directors who agreed to pay for all manuscript preparation, publication, and printing costs for the 3rd edition of this manual. ASPRS is the official publisher and receives the proceeds from all sales of hardcopy and eBook versions of this manual.

TABLE OF CONTENTS

Chapter 1

INTRODUCTION TO DIGITAL ELEVATION MODELS

David F. Maune, Hans Karl Heidemann, Stephen M. Kopp, and Clayton A. Crawford

DIGITAL ELEVATION MODELS (DEMS)

This chapter has three basic goals:

- To explain the basic terminology used in this DEM Users' Manual and simple DEM concepts. See also Appendix A (Acronyms) and Appendix B (Definitions)
- To explain 3-D surface modeling, especially the importance of breaklines, and how elevation data are hydrologically flattened (hydro-flattened), hydrologically enforced (hydro-enforced), and hydrologically conditioned (hydro-conditioned)
- To explain ways in which raster elevation data (uniformly-spaced) are interpolated from irregularly-spaced elevation data

Raster, Vector, and Point Cloud Data

The field of topography is riddled with inconsistencies in vocabulary and definitions. Each given term can mean something subtly or dramatically different depending on the location, workplace, or education of the speaker. Topographic data span all three of the accepted broad geospatial data types: Raster, vector, and point cloud; and they exist as both discrete measurements and as continuous surface constructs. A clarification of common terminology, various meanings, and the definitions used in this volume is therefore necessary.

Most practitioners of modern geography and geographic information systems (GIS) were taught that there are two types of geospatial data: Raster and vector. Many may recall the saying "Raster is faster but vector is correcter," a quaint but often accurate description. Vector data are based on the storage of discrete x, y, and possibly z-values for point, lines, and polygon geometric representations of geographic features. Depending on the application and data system, z-values for lines and polygons may be stored either as a single value for each element (as with a contour line), or as individual values for each vertex in the line (as with a breakline). Because vector elements contain the actual measured position of each feature, vector data are capable of greater precision and accuracy than raster data. The large volume of vector data requires substantially more storage, and because the data are often not maintained in a clearly defined spatial order, they can be slow to process.

Raster data are stored as a set of values in a regular spatial grid consisting of cells (or pixels) arranged in columns and rows. The origin location of the raster, its overall extents (number of columns and rows), and the dimensions of the raster's cells are defined once, in the raster's header or in a companion file. The location of an individual cell is implied through its column/row location, relative to the origin of the overall raster. The body of the raster dataset is composed of the ordered data values corresponding to each cell. For elevation rasters, these are the z-values associated with each cell. Raster data are broadly divided into two sub-types: Thematic or continuous. In a thematic raster, the values assigned to each pixel are derived from a fixed list of values, usually stored as integers, which represent mutually exclusive categories or characteristics of the geography. A continuous raster represents phenomenon or characteristics that do not have distinct boundaries, but change gradually. Continuous data may be stored as integers or real numbers. A raster map of average temperatures across the country would be continuous. A raster map of defined climate zones would be thematic. With the exception of a map of elevation zones, virtually all elevation rasters are continuous.

With the advent of light detection and ranging (lidar) data in the early 2000s, a third fundamental data type, the point cloud, has now been recognized. Although lidar datasets, consisting of millions or billions of xyz point measurements, can technically be considered vector data, the nature of the data and how they are used is fundamentally different from typical point data—enough so that lidar warrants special treatment. With traditional vector point data, each point feature typically stands as an individually meaningful feature. Multiple point features may and often do create a larger more meaningful dataset. For example, a point representing a test well with various chemical concentrations as attributes is individually meaningful to those whose drinking water comes from near that well. A collection of such test wells may be useful for a hydro geologist to map sub-surface water migration across a region.

In a point cloud, individual points also have attributes, but the data of a single point is of marginal interest or meaning. It is only when the point is considered in relationship to its neighbors that it begins to develop meaning. From a practical standpoint, the difference between familiar point vector data and point clouds can be demonstrated by noting that very few traditional GIS tools for the manipulation and analysis of point vector data are—or even can be—applied to point clouds. Moreover, most data structures used for point vector data generally fail outright when used to store the massive point cloud datasets. The differences between point clouds and point vector data are far greater than their similarities.

As each term is defined in this section, it will be described in terms of raster, vector, or point cloud. A second characteristic of each term is whether it refers to discrete data features or a surface. In elevation, discrete data features represent measured elevations at specific locations. Elevations at locations between the measurements are not defined. Discrete point measurements can follow a regular grid, and in some systems these data can be interpreted as a raster surface, as with a lattice. Surface data are realized as an unbroken representation of the elevations within the surface's boundaries. Any location within the extents of the surface can return an elevation value. If locations within the extents are of truly unknown elevation, they are assigned an identifying "NODATA" value, but the key concept is that there is still a value in the dataset for that location. Surface data can be raster or vector.

The distinction between discrete and surface data can be a confusing concept because both types are actually stored on disk as discrete data. The difference lies in how the data are realized within a computer or GIS system, and how the data are used. Still, consistency in usage remains important nonetheless, and to reiterate, there is tremendous diversity across the industry in how various terms are used. The following definitions represent how terminology is used in this volume.

DEM Definitions

Although Digital Elevation Data (DED) would be a more generic term for use in the title of this book, few recognize the acronym "DED," whereas many recognize and use the acronym "DEM," even though they may have different understandings of the meaning of this popular term. The term Digital Elevation Model (DEM) has several different meanings.

As used in the title of this DEM Users Manual, DEM is a generic term for digital topographic and/or bathymetric data, in all their various forms. The term refers to a "model" because computers use such data to digitally model and automatically analyze the Earth's topography and/or bathymetry in three dimensions (3-D), minimizing the need for labor-intensive human interpretation. Unless specifically referenced as a Digital Surface Model (DSM), the generic DEM normally implies elevations of the terrain (bare earth z-values) void of vegetation and manmade features. Some DEMs, those produced from autocorrelation of imagery or from Interferometric Synthetic Aperture Radar (IfSAR), may contain more remnants of above-ground features, due to the way they are collected and produced. DEM elevations of lakes and rivers normally imply the water surface. Bathymetry refers to the depth of the ground below the water surface. All such elevations are referenced to a horizontal and vertical datum. Because of the importance of vertical datums to DEM users, Chapter 2 covers vertical datums in detail, including initiatives to link datasets compiled to different topographic and tidal datums.

For many years, the U.S. Geological Survey (USGS) used the term DEM (pronounced "Dee-Eee-Em") to reference the data and format (.dem) of their digital terrain elevation. As with a DEM, these data represented the elevation of the terrain at regularly spaced intervals in x and y directions, using z-values referenced to a common vertical datum. The USGS DEMs were developed from a variety of sources including original 7.5-minute printed U.S. Topographic Quad Sheets and vary widely in quality, accuracy, and currency. These DEMs were integrated and merged into the USGS National Elevation Dataset (NED), released in 1999, which is now available in various arc-second resolutions in geographic coordinates (one elevation point every 1 or ⅓rd-arc-second, and in some locations 1/9th arc- second, in latitude and longitude). The NED has been continuously updated with newer more accurate data and the original USGS DEMs, while no longer widely distributed, are still used in some circles. The USGS DEM definition is similar to that of Digital Terrain Elevation Data (DTED) and other elevation datasets produced by the National Geospatial-Intelligence Agency (NGA).

More specifically, and as used within this volume, a DEM is a raster (or "gridded") bare-earth surface model, with z-values at regularly spaced intervals in x (Eastings) and y (Northings). Depending on the system used to generate the DEM, the elevation value of each cell may represent the aggregate elevation of the area within the cell, or the elevation at the center of the cell. The resolution (grid spacing; cell or pixel size), datum, coordinate systems, file data format, and other characteristics can vary widely.

Esri's GRID is a familiar example of a raster data structure, as are .TIFs, .IMGs, and all other digital image formats. The cell is a uniform unit that represents a defined area of the Earth, such as a square meter. For most DEMs and DSMs, raster cells are usually square, having the same dimension in width and elevation (easting and northing), and organized in rows and columns of data as in Figure 1.1.

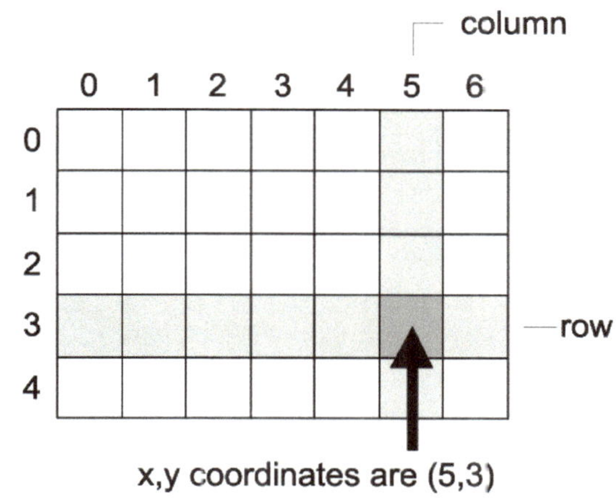

Figure 1.1. Basic structure of a raster.

Raster elevation surfaces are almost always interpolated from other data measurements, either directly from mass points, mass points and breaklines, or from triangulated irregular network (TIN) surfaces. The various techniques are examined in detail later in this Chapter. One notable exception is radar data and its variants Synthetic Aperture Radar (SAR) and Interferometric Synthetic Aperture Radar (IfSAR). These technologies work completely different than traditional collection methods or lidar, and are discussed in more detail in Chapter 7. Another exception are DEMs that are created using autocorrelation of airborne or satellite imagery.

Raster elevation surfaces are stored on disk as discrete data and only become continuous surfaces when realized within a computer application. There are several variations on how this realization is performed, each with its own nuances. Let us first examine how a raster DEM is stored, using a simple generic raster example.

This raster has 12 cells, arranged as 4 rows and 3 columns. The cells are square and 100 by 100 units in size, and the origin of the raster, in this case the lower left, is at the coordinate {100,100}. The first 5 lines are considered header information while the subsequent 4 lines (12 values) are the elevation values. The xy locations of each of these elevation values can be inferred from the header information, as shown in Figure 1.2 on page 4.

Note that in this example, the origin of the raster is NOT the coordinate location of the lower left point, because as the centroid of the cell, the point is offset from the raster's origin by ½ a cell. So while the lower left origin of this example raster is {100,100}, the xy location of the lower left elevation point is {150,150}. When generated, an elevation is interpolated from the source data at the centroid point location of the target cell; that elevation is then accepted as the elevation for the entire area of that cell.

A small number of systems include advanced algorithms that calculate an aggregate elevation for the entire area within each pixel. In either case, this is known as a cell-centered raster and it is the widely accepted standard for elevation rasters

However, some software systems, usually older systems, use a cell corner reference, which would appear as shown in Figure 1.3. In this interpretation, the elevation of the target cell is interpolated from the origin corner of the cell and accepted to apply to the entire cell. This approach is not incorrect, per se, but it can be confusing and introduce error when performing analysis or operations with multiple rasters created using both types of systems. Although this type of reference has become much less common since the early 2000s, users should ensure they are familiar with how their various software systems operate, as well as the lineage of any raster data obtained from outside sources.

ncols	3
nrows	4
xllcorner	100
yllcorner	100
cellsize	100
18 17 15	
16 15 14	
16 16 14	
14 13 12	

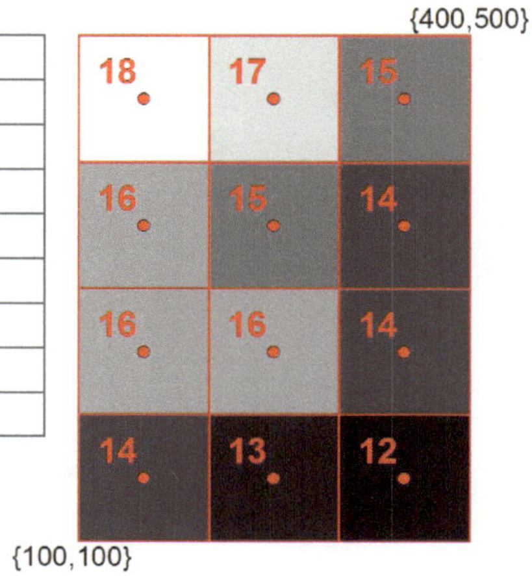

Figure 1.2. Simple cell-centered elevation raster.

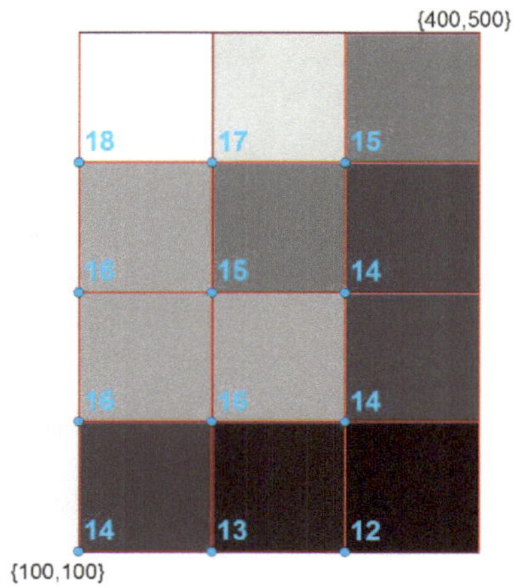

Figure 1.3. Corner referenced elevation raster.

Another type of elevation data model is the lattice, shown in Figure 1.4a-b. This data structure contains a regular grid of elevation points, called mesh points, which can be stored either as vector points or like a raster. Note in the examples, however, that the lattice includes mesh points at all corners of each target cell, not only the lower left. Realizing the lattice as a raster surface requires an additional step to determine the nominal value of each pixel. In practice, the lattice surface is handled more like a TIN surface; the final raster surface is then derived from the TIN surface. Note how the elevation values of the pixels in Figure 1.4b differ from those in Figure 1.3.

In all of the above examples, integer elevation values were chosen for the sake of simplicity and readability. In years past, DEMs often held elevation as integers for storage and processing efficiency. For example, the elevation 123.45 meters would be stored as 12345, with the understanding that all values in the DEM were 100 times the actual elevation. The old USGS DEMs (.dem) stored values this way. With the advances in computer and storage technology over the past decades, working with decimal data (floating point and double precision) has become vastly more efficient, and most DEMs are now stored using decimal data.

The resolution of a raster DEM is frequently called its post spacing. For DEMs stored in Universal Transverse Mercator (UTM), State Plane Coordinate System (SPCS), or other projected Coordinate Reference Systems (CRS), post spacing is defined in linear units (Figure 1.5 with Δx and Δy in feet or meters). In some cases, particularly where the extent of the dataset is so large that a projected CRS will introduce unacceptable

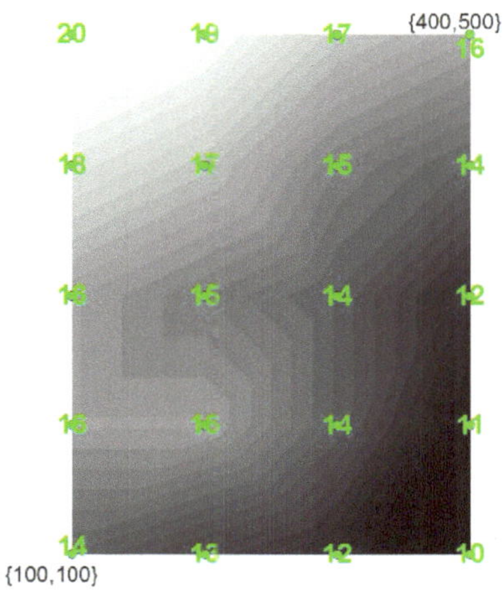

Figure 1.4a. Lattice elevation data, realized as a TIN.

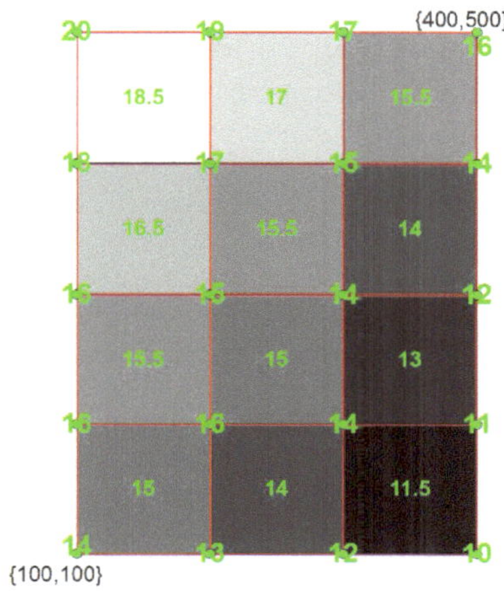

Figure 1.4b. Lattice elevation data, realized as a raster.

distortion, the data can be stored in geographic coordinates, with the post spacing defined in angular units of latitude and longitude (Figure 1.6 with Δx and Δy in arc-seconds). The USGS NED is the most familiar example of this, with its 1, ⅓, and ⅑ arc-second layers. The NED is discussed in greater detail in Chapter 4. Figure 1.5 depicts the schematic arrangement of elevation postings in a legacy USGS DEM (.dem). These files were stored in UTM coordinates, by Zone, but the tiles are cut according to the 7.5 minute latitude/longitude quadrangle boundaries. As a result, the profiles (vertical columns of postings) are vertical but the tile boundary is skewed to the right. In Figure 1.6, the profile lines are skewed to indicate that the linear x distance between points diminishes as the latitude increases, however, the angular distance remains the same. If displayed unprojected, the points in Figure 1.6 would form a regularly spaced grid.

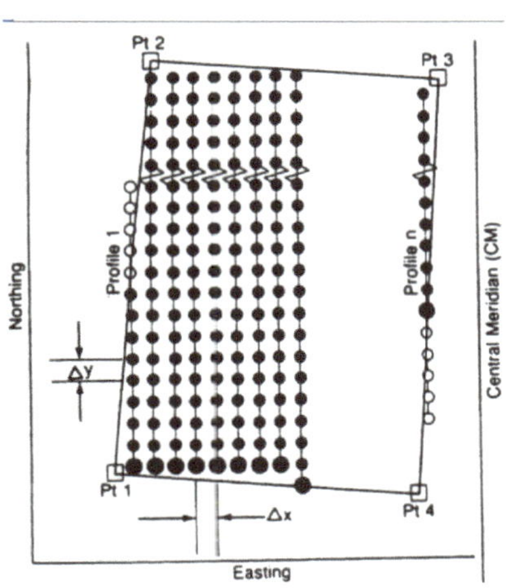

Figure 1.5. DEM post spacing in feet or meters works best for smaller areas where tile boundaries are expressed in UTM or State Plane coordinates.

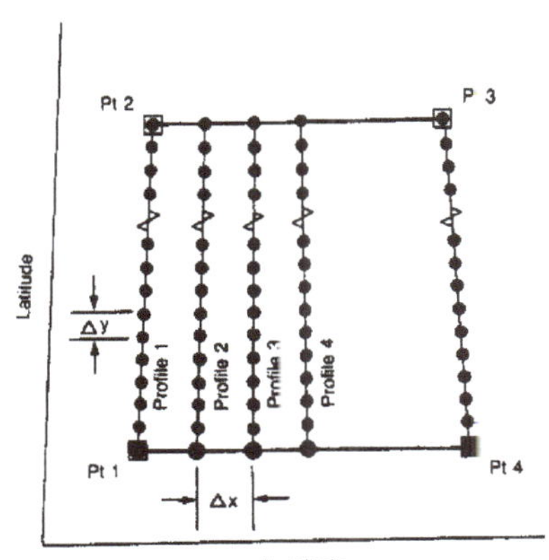

Figure 1.6. DEM post spacing in arc-seconds works best for larger areas where tiles are bounded by meridians of longitude and parallels of latitude or the model extent spans multiple projection zones.

Today's generic DEMs can be stored in any coordinate system but in practice are most often held in projected coordinate systems, the NED, SRTM, and ASTER GDEM being notable exceptions. For the most part, UTM coordinates are currently used for Federal mapping programs that cross state boundaries, for scientific investigations, and for other applications where the more universal metric units are preferred. State Plane coordinates are normally preferred for state, county, and community mapping programs, as their projections are specifically tailored to small areas and therefore have less distortion and greater accuracy.

A uniformly-spaced set of elevation points could be compiled directly with photogrammetric procedures that automatically drive the photogrammetric operator or stereo instrument to pre-calculated xy locations on a regular grid (for example, every 1 meter in Easting and Northing) and then determine the elevations at each of those points using the stereo photogrammetric procedures as explained in Chapter 6. These points could then be directly converted into a raster elevation surface. This approach would be very tedious, slow, and expensive. The more common photogrammetric approach is to first collect breaklines and high-accuracy spot elevations on topographically important features, then to add more widely spaced mass points on a regular, but wider grid. These data comprise a Digital Terrain Model (DTM) from which uniformly-spaced elevation points can be interpolated to create a raster DEM. This process is demonstrated later in this Chapter for both topographic surface modeling and hydrologic surface modeling. Chapter 6 explains a concept known as Semi-Global Matching (SGM) where the computer automatically generates a raster elevation model, but that model is essentially a raster version of a Digital Surface Model (DSM) rather than a raster version of a Digital Terrain Model (DTM), explained below.

Because lidar elevation data are so much denser than data produced photogrammetrically, suitable DEMs can often be generated without the addition of breaklines, unless the application and use of the DEM requires the improved edge detail. When breaklines are required, they can be derived directly from the lidar point cloud data rather than from other photogrammetric sources. There are two common methods to accomplish this: Digitize features from imagery then convert the two-dimensional (2-D) lines into 3-D breaklines by conflating elevations from the lidar to the lines; directly compile 3-D breaklines using lidargrammetry (a type of synthetic stereo photogrammetry that uses lidar elevations and intensity images). These methods are explained in greater detail in Chapter 9.

As with imagery, the resolution of the raster must be appropriate to the application for which it is used. As a general rule, the raster resolution must be twice as small as the smallest feature the user is interested in modeling. Therefore, if an application requires modeling features 10 meters wide, the user should acquire a DEM with a resolution no coarser than 5 meters. At the same time, creating or using a DEM with 1 meter resolution (post spacing) would be more costly and inefficient. Provided that the resolution of the DEM and the density and accuracy of the source data are appropriate for the application, and the loss of precise detail vector data is not critical, raster DEMs can be used with great success in applications from small-scale continental analysis down to large-scale detailed local modeling.

The raster format is a simple, fast and efficient data structure for both storage and analysis algorithms and, because of this, DEMs are the primary surface representation for analysis and visualization. In addition, there are many more analysis tools for raster DEMs than for the other surface data structures. Raster elevation models are widely available at low cost. DEMs produced by the USGS, described in Chapter 4, are examples of raster surface data.

The disadvantages of the raster surface is that if the DEM is of too coarse a resolution for the application, small surface discontinuities such as ridges and stream centerlines can lose their definition, and exact locations for features such as peaks can be lost in the raster interpolation. For these reasons, vector data models are preferred for representing man-made objects such as road cuts, buildings, etc. Raster surfaces, however, have proven their value for elevation models. In areas where elevation values are unchanged from cell to cell, raster DEMs can be an inefficient storage method, as rasters store a value at every location, even if all adjacent cells have the same or nearly the same value. Most modern systems utilize various compression techniques to overcome this disadvantage.

Additional Related Definitions

Digital Terrain Models (DTMs) and Digital Surface Models (DSMs)

A Digital Terrain Model (DTM) also has multiple definitions. In some countries, the term DTM is synonymous with DEM, that is, a raster bare-earth surface. In photogrammetry, a DTM is regarded as the discrete vector mass points, breaklines, and spot elevations compiled from stereo imagery. In this sense, DTMs are the discrete data elements from which TIN surfaces are generated. A DTM can also mean the TIN surface. So the term DTM can mean discrete vector data, a vector surface, or a raster surface, depending on the speaker and context.

As used herein, DTMs may be similar to DEMs, but they frequently incorporate the elevation of important topographic features on the land, via irregularly spaced mass points, spot elevations, and breaklines. The net result of DTMs is that the distinctive terrain features are more clearly defined and precisely located, and contours generated from DTMs more closely approximate the real shape of the terrain. DTMs are therefore preferred for analysis, especially in engineering applications, and they are also commonly used in the production of high quality raster DEMs because the breaklines and surface-specific spot elevations present in the DTMs influence and improve the results of the interpolation process. The surface in the bottom of Figure 1.7 is an example of a DTM.

A Digital Surface Model (DSM) is similar to a DEM or DTM, except that a DSM depicts the elevations of the top surfaces of buildings, trees, towers, and other features elevated above the bare earth. When made from lidar, this is typically referred to as the top or first reflective surface. It is worth noting that measurements from remote sensing technologies such as lidar and IfSAR often do not capture the absolute top of a forest canopy; the elevations usually underestimate the true canopy.

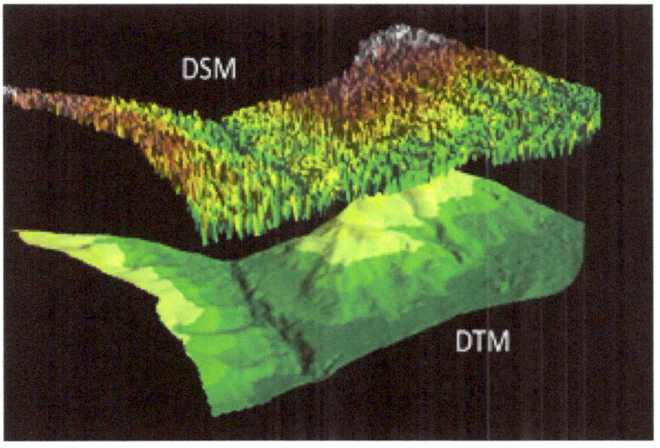

Figure 1.7. DSM of top reflective surface and DTM of bare-earth terrain beneath the vegetation.

DSMs are especially relevant for telecommunications management, forest management, air safety, and 3-D simulation and "fly-throughs." The surface depicted at the top of Figure 1.7 is an example of a DSM showing the detailed tree canopy. Although a DSM can be either raster or vector, within the science and mapping community they are usually raster. Vector DSMs are more commonly used in the animation and 3-D modeling community.

Once realized as a surface (either raster or vector), complete raw elevation data from photogrammetry, lidar or IfSAR, could all be considered DSMs. In practice, a trivial amount of effort, such as filtering on first returns from lidar, can reduce the amount of data needed to produce a DSM. DTMs, on the other hand, require a substantial amount of additional manual or semi-automated post-processing to separate ground from non-ground elevation points, as shown in Figures 1.8 through 1.10 on page 8.

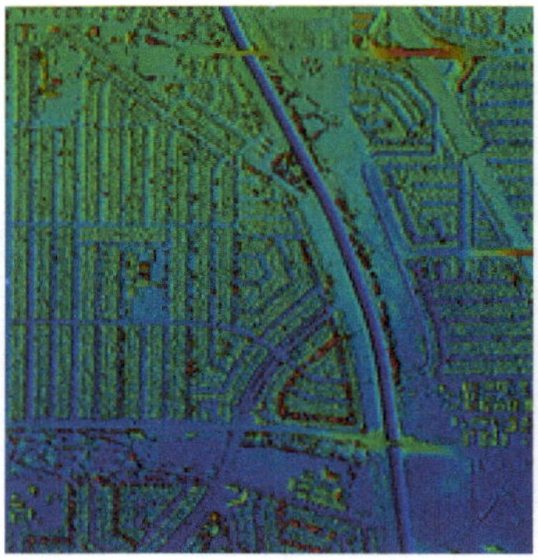

Figure 1.8. Digital Surface Model (DSM) from unclassified lidar data showing the top reflective surface.

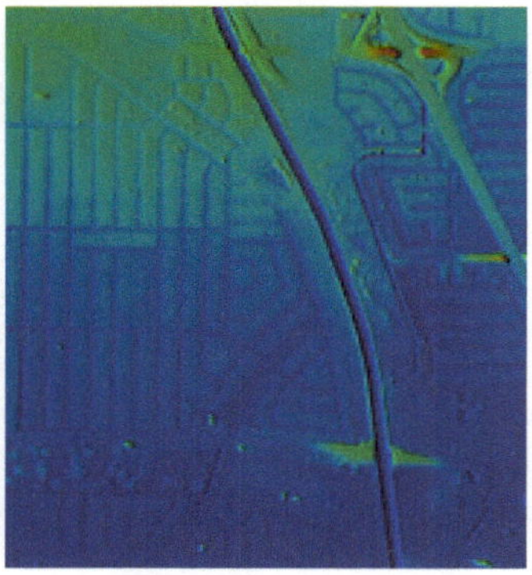

Figure 1.9. Digital Terrain Model (DTM) from classified lidar data showing the bare-earth terrain.

Figure 1.10. Lidar data points on trees, buildings, etc., classified other than ground terrain points.

Mass Points, Breaklines, and Spot Elevations

Mass points and breaklines are the fundamental building blocks of many elevation data and elevation surfaces. Mass points are 3-D point measurements of the ground elevation and in a broad sense include photogrammetric measurements, lidar points, and surveyed elevations. Spot elevations are special high accuracy points to establish elevations of key features. Breaklines are 3-D lines that represent linear and polygonal landscape features. These terms have their origins in photogrammetry and as mentioned in the previous section, are often collectively referred to as a DTM, and are typically (though not always) used to construct a TIN, as explained below.

Mass points are usually irregularly spaced points, each with an xy location and a measured z-value of the ground or reflected surface. They are normally generated by automated methods, e.g., by lidar scanners, semi-automated stereo compilation, or photogrammetric autocorrelation techniques. When generated by automated methods, mass point spacing and pattern depend on characteristics of the technologies used to acquire the data. Some automated methods will create mass points in a regular grid pattern, which can appear to be a lattice or DEM; the important distinction is that mass points are true measurements; they are not interpolated from another elevation surface. In lidar data, the points are extremely dense, irregular in distribution, and discrete measurements. A set of lidar mass points is shown in Figure 1.11. Several items warrant clarification regarding the mass points shown in Figure 1.11.

- The lidar mass points appear to be in north-south strips because the lidar acquisition aircraft was flying east-west while the laser was rapidly scanning in a zigzag pattern to the left and right of the aircraft, i.e., scanning to the north and south
- The areas of the stream, river, and reservoir are shown in white because the points that were collected there are classified as non-ground. For reasons explained in Chapter 8, lidar does not consistently receive returns from water, and the returns that are received are rarely used because they are unreliable and inaccurate
- The white, nearly-rectangular area perpendicular to the river is the elevated deck of a bridge. Because this portion of the bridge is above ground, the lidar points on it are not considered part of a bare-earth elevation dataset. The highest points shown in red are for the north and south ramps leading up to the elevated bridge deck, stopping at the bridge abutments.

Spot elevations are individual points, each with an xy location and a z-value, the same as mass points. However, spot elevations are normally generated manually and are deliberately placed to depict elevations of prominent features representing the highest or lowest elevations in an area, or elevations of prominent features, e.g., dams, levees and road intersections. They may also be chosen to depict the most significant variations in the slope or aspect of TIN triangles. According to National Map Accuracy Standards (Bureau of the Budget, 1947) and ASPRS Accuracy Standards for Large-Scale Maps (ASPRS, 1990), spot elevations are twice as accurate as mass points and contours. Because spot elevations are both more accurate and specifically located they are often depicted directly on topographic contour maps. Spot elevations, therefore, may be either a vector data source, or a cartographic element, or both, again depending on the context. Although arguably incorrect, the term "spot height" is widely used synonymously.

Breaklines are used to model linear features in a surface. They usually describe changes in surface smoothness or continuity. A soft breakline ensures that known z-values along a linear feature are maintained (e.g., elevations along a pipeline, road centerline or drainage ditch), and ensures that linear features and polygon edges are maintained in a TIN surface model, by enforcing the breaklines as TIN edges. Soft breaklines are generally synonymous with 3-D breaklines because they are depicted with series of xyz coordinates. A hard breakline defines interruptions in surface smoothness, e.g., to define streams, shorelines, dams, ridges, building footprints, retaining walls, and other locations with abrupt breaks in slope. Although some hard breaklines are 3-D breaklines, they are often depicted as 2-D breaklines because features such as shorelines and building footprints are normally depicted with a series of xy coordinates only, often digitized from digital orthophotos that include no elevation data. Breaklines may be created in 2-D or 3-D, depending on the method of data capture, but they ultimately become 3-D when incorporated into a surface model. If breaklines were not collected with elevation originally, the elevations must come from attribution (e.g., average lake elevation) or be inferred in some way from surrounding measurements that have elevations. Some systems make a distinction between soft and hard breaklines. The distinctions between soft and hard breaklines only have meaning in the presence of a smooth interpolator; soft lines will not influence smoothness while hard lines will. This will influence how contour generation behaves near and at the breakline. Some systems also use polygonal breaklines to define flat, void, or other special area types. Examples of hydro breaklines are shown in Figure 1.12. Several items warrant clarification regarding the breaklines shown in Figure 1.12.

- The smaller stream feeding from the upper right is represented with a single-line breakline. Single-line hydro breaklines are normally used for streams narrower than the nominal width indicated in project specifications, e.g., nominal width narrower than 100 feet in this example
- The larger river flowing right to left is represented with double lines. The double-line hydro breaklines are normally used for streams wider than the nominal width indicated in project specifications. These

Figure 1.11. Lidar bare-earth mass points, color-coded by elevation. Note data gaps for bridge and water areas.

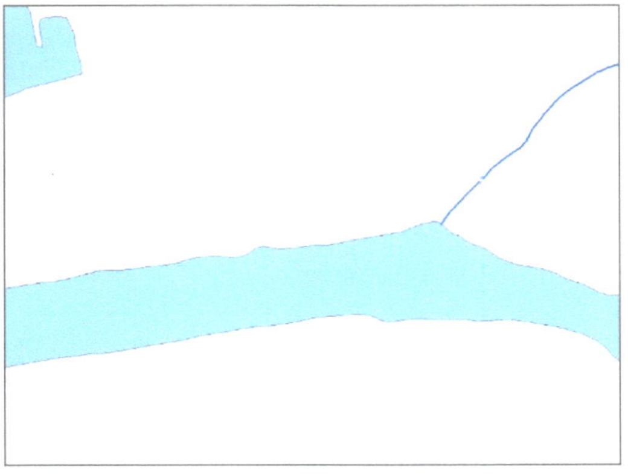

Figure 1.12. Breaklines for single-line stream plus water edges for dual-line stream and water reservoir polygons.

features may be represented as lines or as polygons, depending on the specification, as well as the system and processes being used

- The breakline around the reservoir in Figure 1.12 encircles a closed water body feature for which all breakline xy coordinates should have the same z-values so that the water body is flat from bank to bank. This is common also for ponds and lakes larger than a specified area

Breaklines can be quite complex and difficult to create correctly. Although not depicted in Figure 1.12, there are numerous details regarding breakline generation, including the following.

- For wider double-line streams and rivers, project specifications normally indicate rules for minimizing a "sausage effect" when streams vary in width wider or narrower than the width indicated in the project specifications
- Water surface elevations of large, long reservoirs can and often do slope downhill, much like a river. In these cases, although the feature appears to be a lake, it is represented with breaklines the same as would be used for a river
- Whether single-line or dual-line, hydro breaklines for streams are normally monotonic 3-D breaklines, meaning water surface elevations consistently decrease in the downstream direction. However, adjoining topographic surfaces next to streams normally undulate up and down because of natural variations in the terrain
- Hydrographic breaklines are sometimes captured initially as 2-D breaklines (from digital orthophotos, for example) and subsequently assigned a uniform elevation based on surrounding elevation data
- Breaklines for coastal shorelines are especially difficult because each photographic or lidar flight line is likely acquired during a different stage of the tide, making it sometimes impossible to blend coastal shorelines seamlessly. In cases where this is particularly critical, the collections can be tide controlled, but this can add substantial costs to a project
- In addition to hydro breaklines, other types of breaklines may be collected for road edge-of-pavement; bridge or overpass edge-of-pavement; underground culverts; soft breakline features such as ridges, valleys, or tops of banks; or low confidence areas where lidar bare-earth point density may be inadequate to reliably model the DTM (comparable to areas where dashed contours are used with photogrammetric mapping in densely vegetated areas)
- Because breaklines are generated manually, they are relatively expensive to compile. Higher density lidar datasets often reduce the need for some breaklines, e.g., tops of stream banks for hydrologic and hydraulic modeling, or identification of surface edges where slopes change
- Automated procedures for generation of accurate breaklines from randomly-spaced mass points are still evolving, although some procedures yield reasonable approximations of those breaklines (Poppenga et al., 2010). Nevertheless, it usually takes a human operator to recognize when drainage patterns pass under the visible surface because of bridges and culverts

Triangulated Irregular Networks (TINs)

A Triangulated Irregular Network is a computation of adjacent, non-overlapping triangles computed from irregularly spaced points with xy coordinates and z-values. TINs are the most common form of vector-based surfaces and serve as the foundation for some other data structure types. Mass points, spot elevations, breaklines, and polygons are the core data elements of the TIN. The TIN's data structure stores the topological relationship between the calculated triangles and their adjacent neighbors. TINs are excellent for calculation of slope, aspect, surface area and length; for volumetric and cut-fill analysis; for generation of contours; and for interpolation of surface z-values. The TIN model is preferable to a DEM when it is critical to preserve the precise location of narrow or small surface features such as ditch or stream centerlines, levees, isolated peaks or pits in the data model. TINs are often used as an intermediate data structure when creating raster DEMs from discrete vector source data such as lidar or photogrammetric DTMs.

The simplest interpretation of an elevation model, in terms of defining a continuous surface, is a linear approach and is based on the principal that a plane can be fit to any three (3) non-collinear points (Carter, 1988). Thus, areas of consistent slope and aspect are represented in the TIN by individual triangles. Areas can be considered uniform if changes in slope or aspect are within a given tolerance level. A tolerance allowing for high levels of error will result in a generalized model of the surface. Slope and aspect can be computed rapidly because only individual triangles need to be examined, rather than the relationships between them.

A TIN is a digital terrain model that is based on an irregular array of points that form a set of non-overlapping contiguous triangle facets as shown in Figure 1.13. A TIN is a vector model that supports the incorporation of point-, line-, and area-based features to capture and represent surface morphology, often color-coded by elevation as shown in Figure 1.14 for ease in visualization.

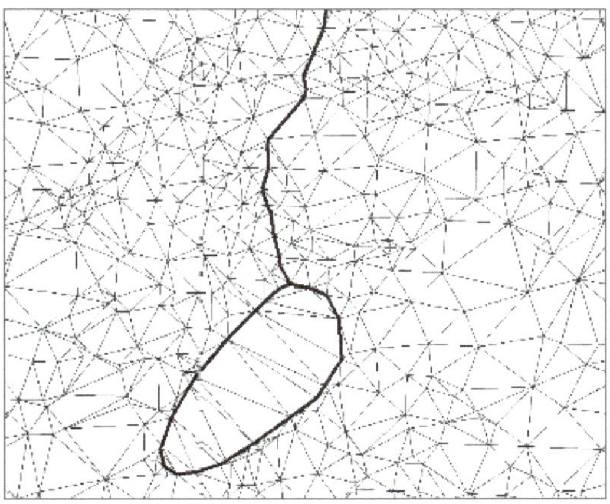

Figure 1.13. A geometric view of a TIN produced from mass points and breaklines where the terrain sample points are connected into a set of non-overlapping triangles. Each node has xy coordinates and a z-value. The pond is not yet flattened.

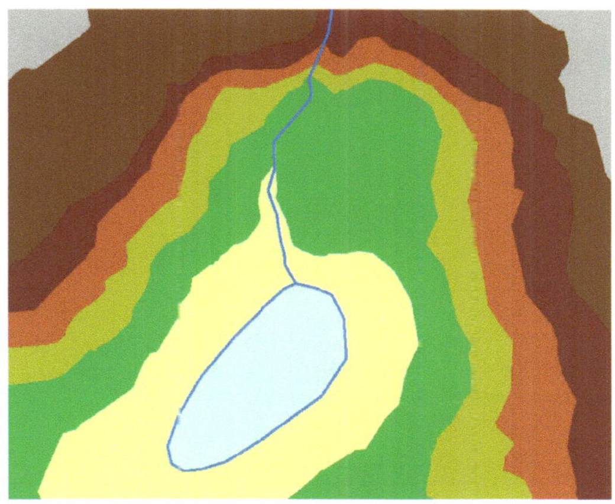

Figure 1.14. The breakline around the pond has been flattened. TINs are used to produce profiles, to interpolate raster DEMs, and to generate derivative products such as contours, hillshades, slope, aspect, and curvature maps.

The points used to construct a TIN come from individually sampled locations (e.g., lidar points) or from the vertices of linear features (breaklines). The points include both a general sampling of the overall terrain, and measurements of critical locations on the surface (peaks, pits, and high accuracy survey). The breaklines define breaks in slope such as ridges, drainages, walls, etc.

Input data locations and elevation values are preserved in the model. No transformation to an intermediate data structure takes place. Analysis is performed on these source locations with the neighborhood relationships, the topology, being calculated and used where necessary to optimize certain operations.

Topological data structure is an important characteristic of the TIN. Once topology for a TIN has been constructed, the neighborhood of a sample point can be reconstructed without resorting to searching (Peuker, et al., 1978). Various approaches to topology exist, each with its own advantages and shortcomings. All will store the xyz coordinates of each point and assign to each a unique index number or node identification (ID). Each triangle is similarly assigned a unique triangle ID. Node-based systems store the three node IDs for each triangle in the TIN; topology can be established by recording the IDs for its three bounding triangles, in an established counter-clockwise order relative to the nodes. Through identification of how nodes are shared, adjacency of the triangles can be established. Alternately, edge-based systems store each edge in the TIN, referencing its two nodes and the triangles on either side of it. Figure 1.15a depicts TIN components graphically; Figures 1.15b,

1.15c, and 1.15d show tables containing the geometry and topology. These examples represent only two common approaches to TIN storage and topology; others exist as well. Every software application uses the storage and topology approach chosen by the developer to meet specific design and performance requirements. Most systems only use the approach for which they were designed. Because of this, TINs are not often used as an interchange or deliverable format; they are primarily used for "in-house" analysis and exploitation.

Regardless of implementation, topology is extremely useful for certain functions such as spatial searching, but the additional disk space required makes it expensive to store and maintain static TIN

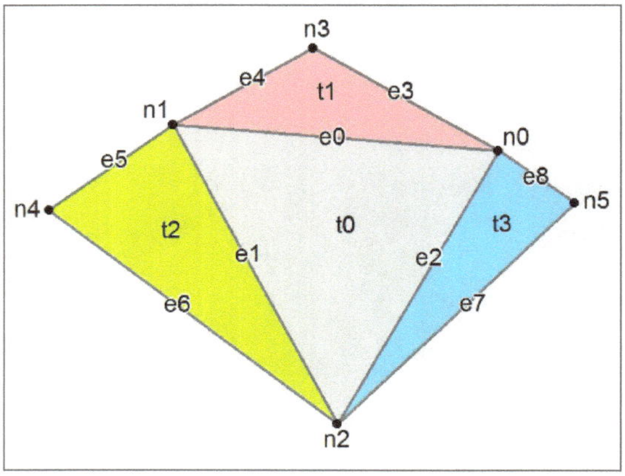

Figure 1.15a. TIN triangle topology. Nodes and edges are referenced in either clockwise or counter-clockwise direction depending on system. In this example all orientation is counter-clockwise.

Geometry			
ID	**X**	**Y**	**Z**
n0	9	6	3
n1	3	6	4
n2	6	1	7
n3	6	8	8
n4	1	5	5
n5	10	5	6

Figure 1.15b. TIN geometry. Each node in the TIN is assigned a unique ID which references its xyz coordinates. In non-topographic 2-D applications, the z-value could represent any other numeric attribute.

Node Topology			
ID	**Node1**	**Node2**	**Node3**
t0	n0	n1	n2
t1	n1	n0	n3
t2	n2	n1	n4
t3	n0	n2	n5

Figure 1.15c. Node-based TIN topology. Each triangle is defined by identifying its three corner nodes, ordered in a system-specific orientation (counter-clockwise in this example). The relationship between the shared nodes establishes the topology.

Edge Topology				
ID	**Node1**	**Node2**	**Left**	**Right**
e0	n0	n1	t1	t0
e1	n1	n2	t2	t0
e2	n2	n0	t3	t0
e3	n0	n3	t1	--
e4	n3	n1	t1	--
e5	n1	n4	t2	--
e6	n4	n2	t2	--
e7	n2	n5	t3	--
e8	n5	n0	t3	--

Figure 1.15d. Edge-based TIN topology. Triangle edges are defined by their component nodes. Triangles and their spatial relationships can be established by referencing the edges associated with each triangle ID in the table. Note how each triangle ID will appear only three times, total, in the Left and Right fields.

topology. Furthermore, making any edits to a TIN requires subsequent rebuilding of the overall topology, adding additional costs in computation time. For small TINs these costs may not be important, but the widespread use of large lidar point clouds is making the use of static topology less and less practical. To overcome this issue, different systems choose to save minimal levels of topology and building detailed topology only when and where necessary at runtime. With the advent of the very large point databases now becoming readily available through such sources as lidar, research is focusing on implicit triangulations that store no topology but information that permits rapid retrieval of points with all topology being constructed on-the-fly (Kidner, et al., 2000).

An accurate, well-designed TIN maintains consistency with the degree of variation in surface elevations found in the terrain. As the terrain becomes more complex the density of the samples used to build the TIN should increase accordingly (Weibel, et al., 1991). In computations, such as query point elevation estimation, elevation values are interpolated for a given location based on the triangle in which it falls or a neighborhood relative to that triangle. Because triangles in the TIN are more densely packed in rough terrain, and the original data values are used as triangle nodes, relatively precise interpolations can be performed.

Once a TIN is available (Figure 1.16), a raster DEM (Figure 1.17) can be produced by interpolating the elevations at the xy coordinates of the centroids of each cell of the DEM. (Figure 1.17 shows exaggerated pixelation for emphasis). This interpolation process yields a DEM surface that is less precise than the dataset used for the interpolation; however, the DEM is normally easier to store in a GIS database. Whenever elevations are estimated by interpolation from surrounding points, the interpolated elevations are less certain. As explained below, the section on 3-D surface modeling explains the rationale for using different forms of interpolation so as to reduce those uncertainties. Explanations of inverse distance weighted interpolation, natural neighbor interpolation, splines and kriging, for example, may appear to be superfluous, but those are valuable concepts for understanding the processes used in interpolation of DEM elevation values.

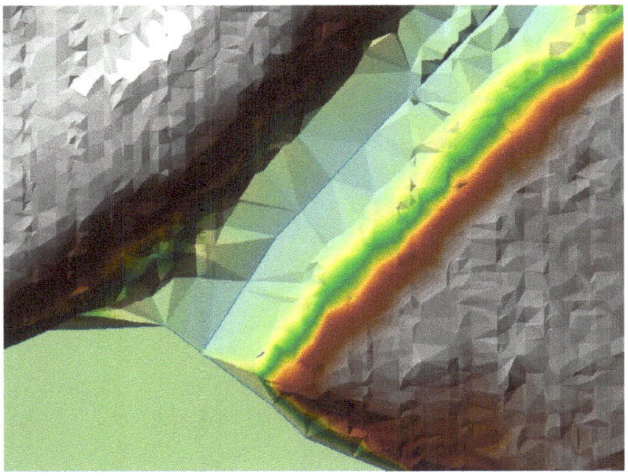

Figure 1.16. TIN from zoomed-in portion of Figure 1.26, showing triangulation artifacts across the single-line stream, too narrow for a hydro-flattened dual-line stream.

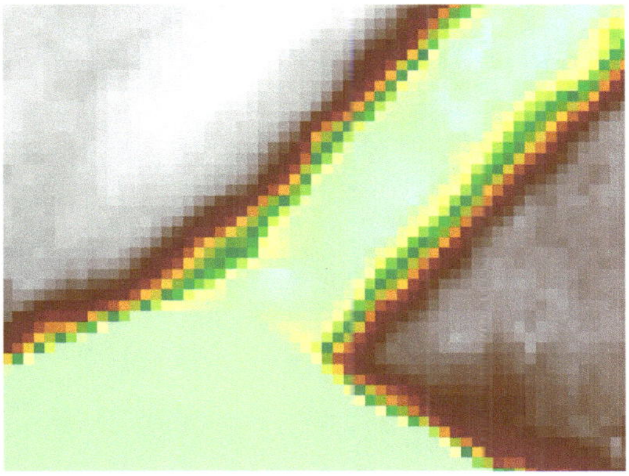

Figure 1.17. Raster DEM (1-meter cell size) from zoomed-in portion of Figure 1.27. This DEM is less detailed and slightly less accurate than the TIN but much more efficient to store and use.

TIN-based surface models have several advantages over other surface representations. One advantage is that TIN models adapt to the variable complexity of terrain. Where there is little variation in the surface only a small amount of data is stored, and where there is much variation, more data are stored. The vector-based structure supports the incorporation of point-, line-, and polygon-based features (Chen, 1988). This enables the original source measurements to be embedded directly in the triangulation. Therefore, the resulting analysis honors the source data exactly. TINs are also the native structure used for 3-D graphics rendering and therefore can be used to improve performance of interactive 3-D visualization or virtual reality.

TINs also have several disadvantages. Good quality data used for their construction, typically photogrammetric- or lidar-based, can be expensive to obtain. Lidar scan density, which is insensitive to terrain complexity, also leads to oversampling in areas of constant slope. This can produce large, sub-optimal TINs that lack the potential "use samples only where needed" benefit the data structure has to offer. As a consequence, extra work is often employed to filter away, or thin, the unnecessary samples. Additionally, there are not as many tools to process TINs as with raster surfaces, and some operations are not as efficient because the data structure is more complex.

A frequently asked question is: "What is the difference between a DTM and a TIN?" In some circles the terms may be synonymous, but the accepted usage here is that a DTM is a collection of mass points and breaklines, and is a set of discrete data. A TIN is a realized surface built from the DTM. A TIN contains a topological data structure by which points and breakline vertices are connected through a series of triangles that connect to create a continuous surface. Each TIN triangle has its own slope and aspect and creates a unique facet in the ground surface model. As a surface, the TIN is extremely useful for visualizing and analyzing slope, aspect, cut and fill of the terrain, and it is used for interpolation of raster DEMs and contours. These tasks cannot be accomplished with the discrete point and line data of the DTM.

From a practical perspective, a TIN's file size is usually 2 to 4 times larger than a DTM covering the same area with the same mass points and breaklines. The growing use of lidar for mass points is substantially increasing the size of even the DTM datasets, and often making TIN storage unmanageable. Moreover, TIN data do not lend themselves to storage as tiled datasets. The introduction of tile boundaries into the TIN invariably causes mismatched triangles along the tile edges; the irregularities these create cannot easily be "edge-matched" out. It is becoming increasingly common to store only the DTM schema and build TINs as needed, on demand in random access memory RAM (Peng, et al., 2004). Unlike many raster data formats, TIN data structures are often proprietary and to date no open TIN file specification has emerged as favored across the industry. Sharing of TIN data can therefore be difficult unless users have the same software system.

Esri Terrain Datasets and LAS Datasets

An Esri Terrain Dataset is a dynamic multi-resolution, TIN-based elevation surface that resides in an Esri geodatabase feature dataset and is built from feature classes stored within the same feature dataset. As shown in Figure 1.18, a Terrain Dataset constructs TINs based on map-scale range and area of interest queries. For each map-scale range, a level of detail (i.e., z-resolution) and pyramid level are defined. The Terrain Dataset establishes a set of user-defined viewing pyramid levels, each having fewer participating source points as the user zooms to smaller scales. Unlike Esri GRIDs or other raster DEMs, Terrain Datasets are generated by utilizing the actual surface points rather than interpolated elevation values for a cell in a raster file. This data storage and visualization method enables faster viewing of large area Terrains at small scales while preserving the most critical information from the source measurements. Figure 1.19 is a schematic representation of an Esri Terrain Dataset. These are surface models that reside in a geodatabase. They

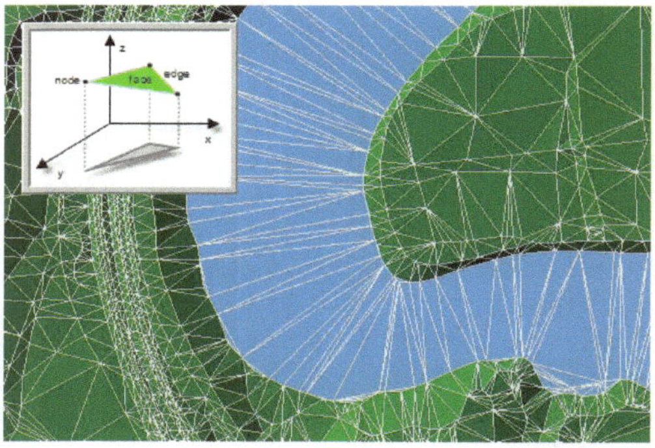

Figure 1.18. Example of a Terrain Dataset showing TIN triangles.

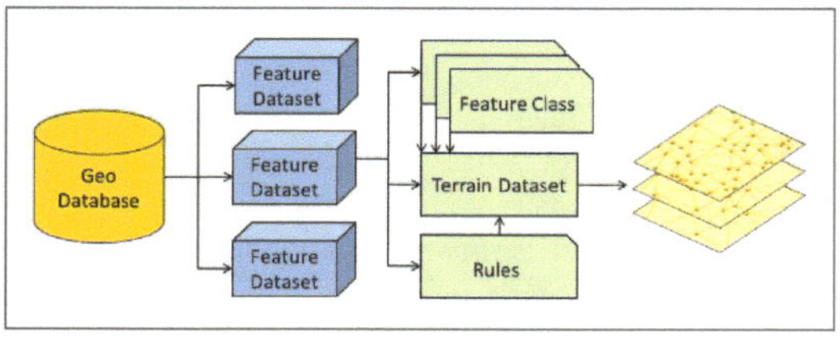

Figure 1.19. Schematic representation of an Esri Terrain Dataset.

reference measurements (e.g., lidar points, breaklines) stored in feature classes and they are assigned rules that control how each feature class is used to define a surface. The actual surfaces are TINs generated by the Terrains on-the-fly based on level of detail and area of interest queries. While Esri Terrain Datasets have become more common and prevalent within the geospatial industry, other software may have their own versions of terrain surfaces. Often these surfaces are created through a TIN-based structure, but will not have associated multi-resolution pyramid levels and may not be compatible with geodatabase storage.

Static, single resolution TINs and DEMs can be exported from the dynamic, multi-resolution Terrain Dataset. Figure 1.20, 1.21 and 1.22 from NRCS show examples of a DEM with 2-meter post spacing/cell size, and when "intelligently-thinned" at two user-defined Terrain Dataset pyramid levels.

Figure 1.20. DEM displayed with full 2-meter DEM post spacing/cell size.

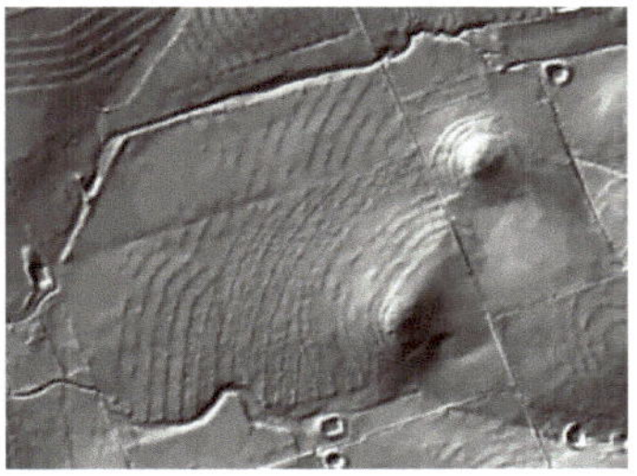

Figure 1.21. Terrain Dataset pyramid Level 1, 5-meter cell size, ¼ meter z tolerance.

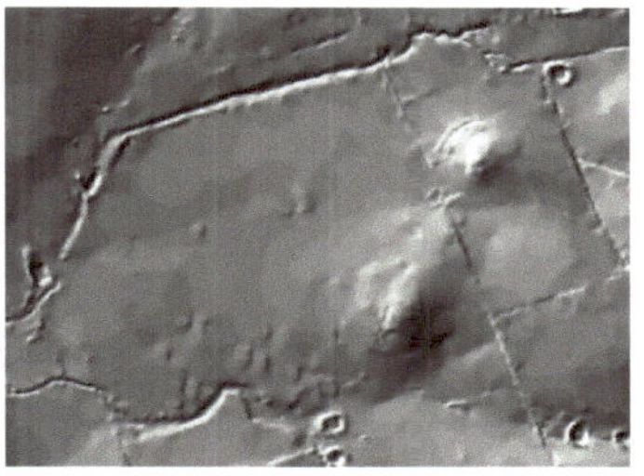

Figure 1.22. Terrain Dataset pyramid Level 2, 5-meter cell size, ½ meter z tolerance.

The Esri LAS Dataset is a related type of surface model. This dataset is similar to terrains in that it builds TIN-based surfaces on the fly from referenced measurements. The primary difference is that a LAS Dataset references lidar points stored in LAS format files rather than a geodatabase. LAS is an industry standard format for lidar defined by ASPRS and is where Esri's LAS Dataset takes its name. Since lidar is usually distributed in LAS format, the ability to use those files directly, without loading them into a geodatabase, can simplify most workflows.

Terrain Datasets and LAS Datasets represent a trend where what is stored as a surface model is actually the schema; the relationships and rules controlling what measurements are used and how. The realized surfaces that are used for analysis and display are made by the models on demand.

Contours

Contour lines are isolines that depict equal elevation on a surface. They are a cartographic construct, as are all isolines, and they are primarily intended and used for 2-D human interpretation of the 3-D terrain surface. Contours are an excellent way to represent the elevations and variations of a surface on a two-dimensional map because they have high visual information content and can easily be understood. With a little training, a map reader can recognize many things from contours such as ridges, valleys, peaks, relative slope, aspect, stream-flow direction, and other characteristics.

In the earliest days of topographic mapping, contours served as both the source data and the cartographic product. Surveyors in the field measured and mapped spot elevations and lines of continuous elevation; in time, technology allowed this to be accomplished from stereo imagery. This practice is referred to as "pulling contours," and it relies as much on the cartographer's artistry as it does the surveyor's skill. With further advances in technology, the rise of computer mapping, and GIS, directly pulling contours has become obsolete. Contours are now, almost exclusively, a secondary or tertiary derivative product, generated from elevation surfaces (TINs and DEMs), which are in turn created from other primary source data (DTMs, lidar). Because contours are themselves a derivative from a surface, they should not be used to re-create a new surface. In addition, contours by their very nature sacrifice accuracy and detail by recording and depicting arbitrary continuous elevations rather than specific topographic features. Moreover, as cartographic constructs, they have usually been edited and modified to make them more readable and interpretable by human eyes, another trade-off of accuracy for cartographic aesthetic.

Unlike 2-D maps, often on paper or Mylar, computer-based modeling and visualization of elevation offers users access to data of higher accuracy and detail, and numerous ways to customize visualization of the 3-D surface. Mass points, breaklines, TINs, DEMs, DTMs, and DSMs provide superior data, and are better for computer-based modeling of the 3-D surface. Visualization of the topography using elevation hillshades, rendered 3-D surfaces, profiles, animations, and other computer-generated visualizations of a TIN or raster provide even greater interpretation opportunities than contours and are becoming increasingly widespread. In spite of these advances, the familiarity with and long history of contours keeps them in constant demand, and virtually all elevation software packages will produce them.

When creating contours from a surface, producers and users must consider the purpose of the contours, accuracy of the source surface, the intended map scale, and the nature of the terrain (complexity, slope, etc.). These factors will guide producers and users in selecting an appropriate contour interval, type of contour, and generation method. There are two types of contours commonly referred to as engineering contours and cartographic contours. Engineering contours are a direct, accurate representation of the TIN or DEM they were derived from; they have little or no smoothing or generalization and as a result they can appear unnaturally jagged and angular. Engineering contours are desirable if users intend to do further visual or computer analysis of the contours, or if they are an archival form of the surface to be used for unspecified future purposes. Cartographic contours are what users are accustomed to seeing on a map. They are smoother, with more rounded angles, and may even be modified to properly point upstream in a valley or appropriately model a bridge overpass or road cut in more detail than the source surface would allow. Traditionally, contours were drawn by hand, and the cartographer would use his/her knowledge and artistry to create a smoother, more representational line for the appropriate map scale. Cartographic contours should be used exclusively for mapping purposes where humans will analyze the contours; for their increased costs, cartographic contours have little additional value for computer analyses of the surface.

To provide additional information to the map reader, it is common to add spot elevations and color in addition to the contours. The spot elevations are most often used to mark the elevation of prominent hills or peaks. Figure 1.23 is an example of cartographic contours with a 2-foot contour interval. This example includes a spot elevation (marked by an "x" with annotation of 689 feet for the elevation of this prominent hilltop, and it also includes index contours (shown here with 10-foot intervals), intermediate contours (2-foot intervals), and depression contours (surrounded by terrain of higher elevation).

Because contours are most often derived from a TIN or DEM surface, and analysis should be performed on the original surface, there are few analytical tools that use contours. A common use of contour data for analysis is the creation of profiles and cross sections that can also be calculated from TIN or raster models. Some software also allows the calculation of areas between contour ranges, which can also be performed on a raster model. When using contour data as input to analysis or input to surface creation, it is important to understand the origin of the data. If the contours were generated from a DEM or TIN, the surface produced from the contours will be no more accurate, and most likely far less accurate, than the surface from which the contours were originally created.

It is critically important that producers and users of contours understand the meaning and importance of the contour interval. Using standard GIS software, contours could (inappropriately) be generated from a DEM or TIN with any contour interval selected. However, it is unprofessional (and potentially unethical) to produce maps for public use that have a contour interval smaller than warranted by the underlying accuracy of the DEM or TIN data from which the contours are generated. In Chapter 3, the ASPRS Positional Accuracy Standards for Digital Geospatial Data explain how the appropriate contour interval must be at least three times larger than the vertical root-mean-square-error (RMSEz). Statistically, the RMSEz must be 0.67-foot or less in order to be accurate enough to support 2-foot contours. If one had a digital elevation dataset that tested with an RMSEz of 1-foot, for example, GIS software could easily enable maps to be drawn with 2-foot contours, but it would be unethical, unprofessional and misleading to do so because such accuracy would only support a 3-foot contour interval or larger, and a misleading accuracy assessment could lead to a potential loss of life, property or money.

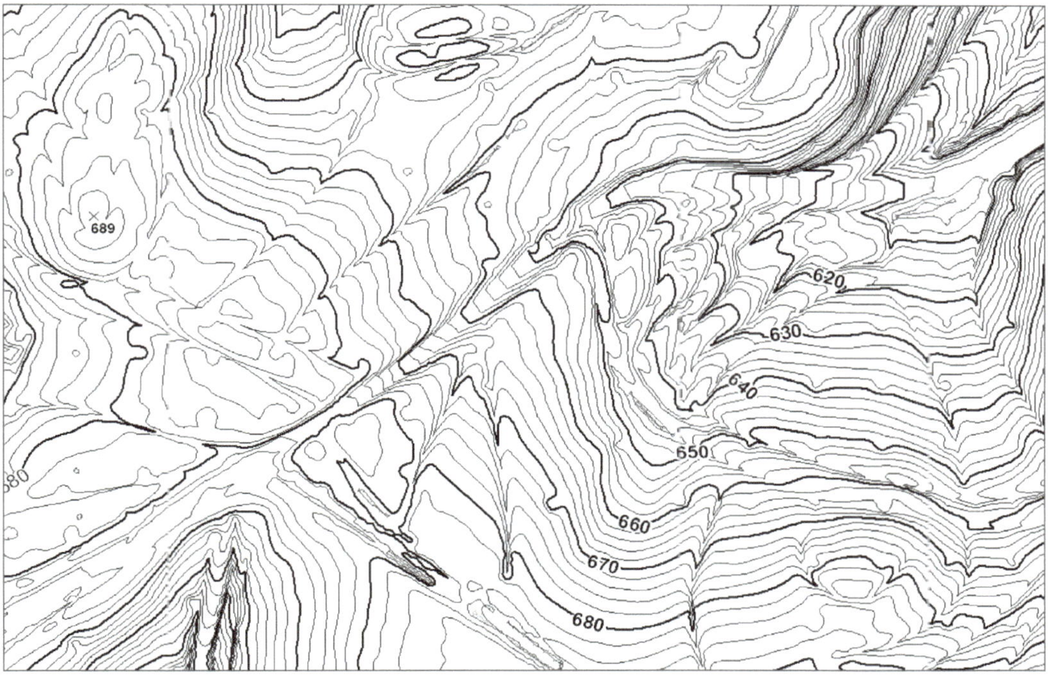

Figure 1.23. Example of 2-foot contours, produced from lidar and breaklines, that depict a road and drainage features. Image courtesy of Dewberry, Fairfax, VA.

3-D SURFACE MODELING

Topographic Surface Modeling

A topographic surface can be considered to be a mathematically continuous surface. No matter how small an area, an infinite number of points may be selected to represent the topography. In practice, only a sample can be utilized. Interpolation is used to construct a digital model of the surface by estimating elevations at any location based on sample elevations.

A digital model used to represent a topographic surface should contain adequate elevation and planimetric measurements compatible in number and distribution with the terrain being modeled, so that the elevation of any location can be interpolated with enough accuracy for the intended applications (Ayeni, 1982). New technologies have emerged recently that have made it easier and less costly to collect large amounts of accurate sample data. Various techniques for collecting elevation data are discussed in Chapters 6 through 11.

Digital terrain data are collected to represent the bare earth, without tree canopy or buildings. Although there is some variability depending on the users and their applications, most topographic surfaces will include linear features such as roads, even where they pass over culverts. This is because road beds over culverts rest on the earth and are considered to be part of the topography. Conversely, because bridges are elevated structures above the earth, they are not part of the topography and are removed.

Topographic surface modeling is traditionally performed by collecting bare-earth elevation mass points (Figure 1.11 on page 9) and breaklines (Figure 1.12 on page 9) described previously. Mass points and breaklines are collectively called a DTM (Figure 1.24). The mass points depicted in Figures 1.11 and 1.24 are bare-earth lidar points collected from an aircraft flying east-west with scan lines in the north-south direction. The mass points are <1 meter apart. The breaklines in this example were generated using lidargrammetry, explained in Chapter 9. Although the term DTM is used in some circles to refer to a topographic surface, this is not technically correct. A DTM is the discrete data from which surfaces are created.

If only mass points are available, a raster surface can be created through direct interpolation of the points. This approach can be processor intensive, slow, and in some situations produce unexpected and unusable results. As a result, it is preferable to generate raster surfaces via intermediate TINs. If TINs are produced from mass points only, the topographic surface model typically includes aesthetically unappealing triangulation artifacts across water bodies (Figure 1.25). Most users prefer to correct this by incorporating monotonic hydrologic breaklines in the TIN generation, creating the bank-to-bank hydro-flattened water bodies seen in Figure 1.26. Monotonicity and hydro-flattening are explained in the section, below, on Hydrologic Surface Modeling.

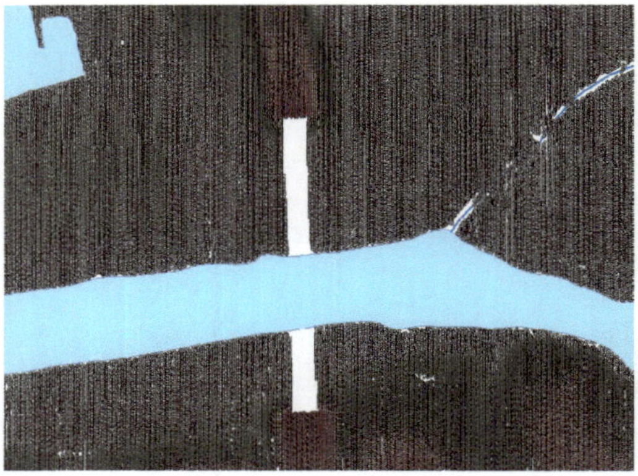

Figure 1.24. Digital Terrain Model (DTM) schema, consisting of mass points and breaklines.

Figure 1.25. This TIN surface with no breaklines shows typical triangulation artifacts across areas with no mass points.

Figure 1.26. This TIN surface with breaklines shows the dual-line stream and reservoir hydro-flattened from bank-to-bank. Figure 1.16 provided a zoomed-in view of this TIN.

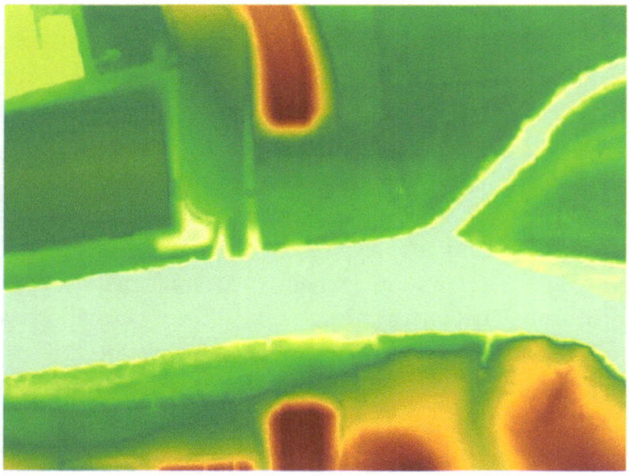

Figure 1.27. Raster Digital Elevation Model (DEM) interpolated from the hydro-flattened Terrain. Figure 1.17 provided a zoomed-in view of this DEM.

From the TIN surface, raster DEMs can then be interpolated using various standard techniques. This process provides the operator with greater input and control, allows the inclusion of a broader array of source data, and ultimately produces more accurate raster DEMs (Figure 1.27), smoother topographic contour lines, and other elevation derivative products described below.

Figure 1.25 and Figure 1.26 show surfaces with hillshade effects that include sun angle. Figure 1.27 does not include hillshading. The color ramps for Figures 1.24, 1.25, 1.26, and 1.27 all vary from light blue (lowest elevations) to beige, green, yellow and red (highest elevations). These datasets were all derived from the mass points and breaklines previously shown in Figures 1.11 and 1.12.

Hydrologic Surface Modeling

The following definitions are introduced here for understanding of hydrologic surface modeling.

Hydrologic modeling is the computer modeling of precipitation and the effects of land cover, soil conditions, and terrain slope to estimate rainfall runoff into streams and lakes. Digital elevation data are used as a vital part of hydrologic modeling.

Hydraulic modeling is the use of digital elevation data, runoff estimates from hydrologic models, stream channel information such as the shape of the stream bed, its surface roughness data, and information on hydraulic structures (e.g., bridges, culverts, dams, weirs, sewers) to predict water depths, estimate flood levels and manage water resources. Hydraulic models are based on computations involving liquids under pressure; note that there are many other definitions of hydraulic modeling that have nothing to do with terrain elevations, e.g., modeling of hydraulic lines in aircraft and automobiles. When performed together, hydrologic and hydraulic modeling is called H&H modeling.

Hydrologic flattening (hydro flattening) is the processing of an elevation surface (DEM or TIN) so that mapped water bodies, rivers, reservoirs, and other cartographically polygonal water surfaces are flat and, where appropriate, level from bank to bank. Surfaces of rivers and long reservoirs demonstrate a gradient change in elevation along their length, consistent with their natural behavior and the surrounding topography. In traditional photogrammetric mapping, this process is accomplished automatically through the inclusion of measured breaklines in the DTM. However, because lidar does not inherently include breaklines, a DEM or TIN derived solely from lidar points will depict water surfaces with unsightly and unnatural artifacts of triangulation. The process of hydro-flattening typically involves the addition of breaklines along the banks of specified water bodies, rivers, etc., which establish elevations for the water surfaces that are consistent with the surrounding topography, and produce aesthetically acceptable water surfaces in the final derived DEM or TIN. Hydro-flattening is not driven by any H&H modeling requirements but solely by cartographic mapping needs. Figure 1.28 and Figure 1.29 on page 20 show two approved methods for creating smooth and stair-stepped gradients while flattening the river from bank to bank. The lower half of each image shows the gradient along the profile lines depicted in the upper half of each image.

Hydrologic enforcement (hydro enforcement) is the processing of mapped water bodies so that lakes and reservoirs are level and streams flow downhill. For example, a DEM, TIN or topographic contour dataset with elevations removed from the tops of selected drainage structures (bridges and culverts) so as to depict the continuous flow of water, as shown in Figure 1.28. Hydro-enforcement enables H&H models to depict water flowing under these structures, rather than appearing in the computer model to be dammed by them because of road deck elevations higher than the water levels. Hydro-enforced TINs also utilize breaklines along shorelines and stream centerlines, for example, where these breaklines form the edges of TIN triangles along the alignment of drainage features. Shore breaklines for streams would be 3-D breaklines with elevations that decrease monotonically as the stream flows downstream, as in Figures 1.28 and 1.29; however, shore breaklines for lakes or reservoirs would have the same elevation for the entire shoreline if the water surface is known or assumed to be level throughout.

With lidar, which captures every irregular feature along a shoreline, it is likely that shorelines will undulate up and down. It is also likely that uniformly-spaced DEM points will "jump" across a small stream and fail to show that a stream passes through. The absence of breaklines in a DEM, by definition, makes them less accurate

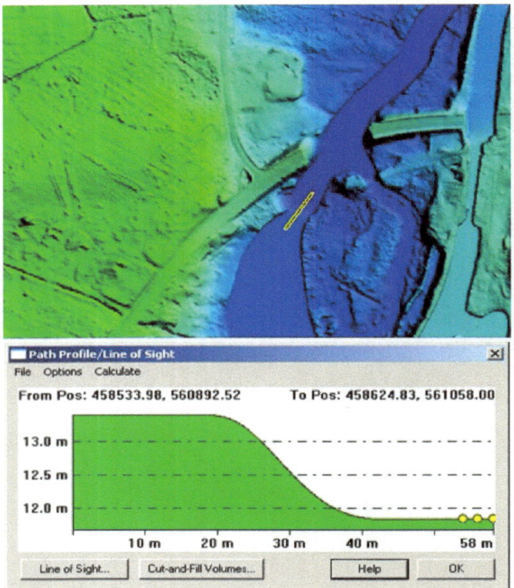

Figure 1.28. Hydro-flattening and hydro-enforcement with smooth gradient.

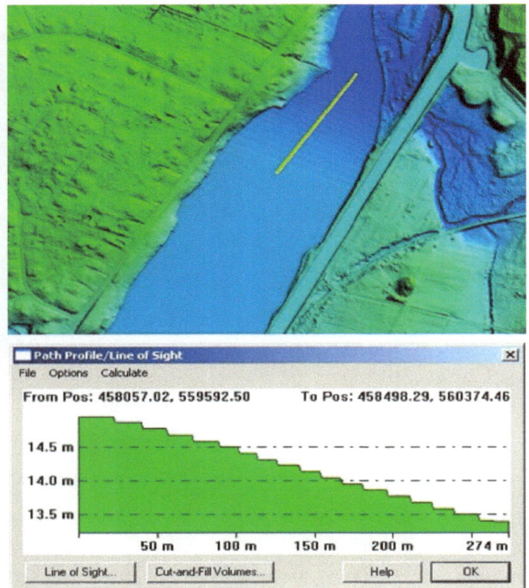

Figure 1.29. Hydro-flattening and hydro-enforcement with stair-stepped gradient.

in depicting the true 3-D shape of the terrain, especially drainage features used for H&H modeling. For this reason, hydro-enforced DEMs are produced by those who need drainage features to be reflected in DEMs and DTMs and where it is necessary for the elevation data to depict the downward flow of water. In most cases, it is necessary for breaklines to be used to enforce the desired depiction of drainage features and water bodies.

An excellent example of hydro-enforcement of lidar data is shown in Figures 1.30 through 1.32. The initial lidar dataset (i.e., mass points) is irregularly spaced and becomes even more irregular after post processing to remove points that do not represent the bare-earth terrain.

Figure 1.30 shows an example of lidar mass points. The mass points are shown in red and yellow; those outside the stream's floodplain were omitted for simplicity. Areas covered by water create natural voids in lidar datasets (between the green lines). Other voids (areas within white lines) were created during post-processing where lidar failed to penetrate dense vegetation. Points shown in yellow represent bare-earth elevations, near shorelines, that are higher than surrounding points, perhaps from rocks or boulders in the stream.

Figure 1.31 shows the TIN when only the lidar mass points are used to create the 3-D surface model. Because of natural undulations along the shorelines, this TIN appears as though water will not flow through areas that are modeled with higher elevations shown with yellow, orange, and red polygons. It is normal for terrain along shorelines to undulate up and down because of natural variations in the topography, including rocks and boulders in the stream.

Figure 1.32 shows how this same TIN changes when breaklines are added for the shorelines and assumed stream centerline. These breaklines were originally 2-D breaklines where xy coordinates were digitized from digital orthophotos. These breaklines were then converted to 3-D breaklines by assigning z-values to the upstream and downstream ends of the 2-D breaklines using elevation data from field-surveyed cross sections at upstream and downstream bridges, then interpolating remaining z-values between upstream and downstream values to make the stream flow downstream.

Lakes and reservoirs are routinely hydro-enforced by digitizing perimeter shorelines in 2-D and then adding a single z-value for an assumed water surface elevation for the entire breakline.

Bridges and major culverts also need to be hydro-enforced. Elevations on the tops of bridges and culverts will appear identical to a dam, so a hydraulic model would erroneously indicate that water cannot pass through this dam. A hydro-enforced DEM or DTM insures that visible surfaces of bridges and culverts are cut by stream breaklines so that H&H computer models will accurately represent stream flow.

Hydro-enforcement almost always requires human intervention to "cut" breaklines through bridges so that all contours "hug" the shoreline under the bridge, and so that computer models will allow the water to flow under the bridge, rather than appearing to be dammed by the bridge when not hydro-enforced. This concept of hydro-enforcement applies, whether dealing with contours, TINs or DEMs.

Hydrologic conditioning (hydro conditioning) is the processing of a DEM or TIN so that the flow of water is continuous across the entire terrain surface, including the removal of all spurious sinks

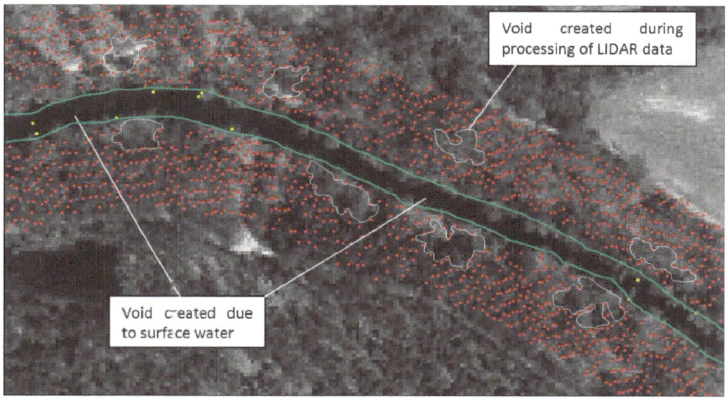

Figure 1.30. Post-processed bare-earth lidar mass points along a North Carolina stream.

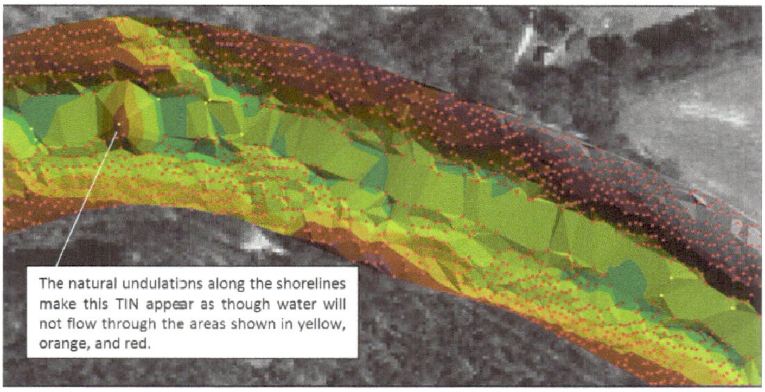

Figure 1.31. Geometric view of a TIN, hillshaded by elevation, created without breaklines.

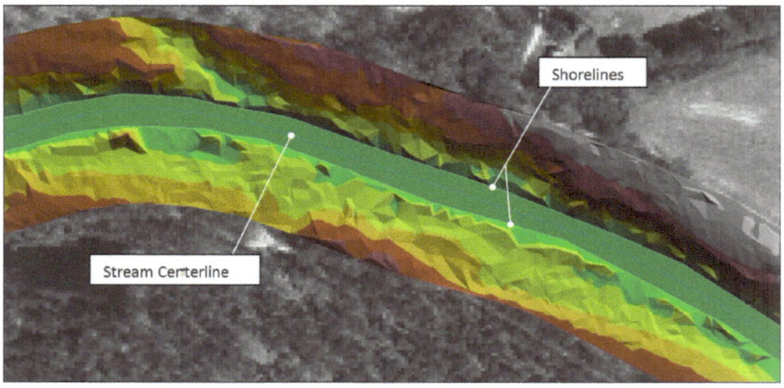

Figure 1.32. The same TIN, hillshaded by elevation, after breaklines are added for hydro-enforcement. Images courtesy of the North Carolina Floodplain Mapping Program (NCFMP), Raleigh, NC.

or pits surrounded by land of higher elevations. The only sinks that are retained are the real ones on the landscape. Whereas "hydrologically-enforced" is relevant to drainage features that are generally mapped, "hydrologically-conditioned" is relevant to the entire land surface and is done so that water flow is continuous across the surface, whether that flow is in a stream channel or not. The purpose for continuous flow is so that relationships/links among basins/catchments can be known for large areas. The physical terrain includes intricate drainage patterns, but the DEM interpolation process creates these artificial sinks. Hydro-conditioning could either fill the sinks, or preferably drain them by using other available information to determine drain paths, e.g., small culverts beneath roads in order to drain a sink, on one side of the road, to an outlet on the other side of the road. Without hydro-conditioning, water appears to flow into these sinks with no apparent outlet. In the DEM creation process, filling sinks is quick and easy, but developing the flow paths to allow sinks to drain is time-consuming and expensive.

Topographic and Hydrologic Surface Examples

Figures 1.33 through 1.38, from USGS, demonstrate how topographic and hydrologic surfaces are modeled differently for diverse user applications.

Photogrammetric DTM

Figure 1.33 shows an example of a traditional DTM compiled from stereo photogrammetry, built from mass points and breaklines, usable for contour mapping. Although hard to distinguish at this scale, the DTM has a much coarser resolution than a DTM produced from lidar. Figure 1.33 demonstrates the familiar and usually expected character of a topographic DEM, most notably, the flat water surfaces. This can be considered the reference image to be emulated with lidar data intended for the 3DEP.

Lidar DTM without Hydro Breaklines

Figure 1.34 shows a basic DTM produced from lidar, using bare-earth lidar points only, without breaklines. The water surface contains extensive triangulation artifacts caused by the absence of lidar returns from water and the absence of breaklines that would define buildings, water, and other features as in the stereo DTM in Figure 1.33. This lidar dataset would be aesthetically and cartographically unacceptable to most users.

Hydro-Flattened Lidar DTM

Figure 1.35 shows a hydro-flattened lidar DTM, achieved with hydro breaklines, acceptable for the 3DEP. This approach removes the most offensive lidar artifacts in the water. Water surfaces of wide streams are flattened from bank to bank, with constant elevations for reservoirs, lakes and ponds. Because water surfaces are developed for aesthetic/cartographic purposes only, there is no implicit or explicit accuracy with regards to the represented water surface elevations. Building voids are not corrected due to high costs.

Lidar DTM with Enhanced Breaklines

Figure 1.36 shows a possible enhanced hydro-flattened lidar DTM. Additional breaklines are used to remove artifacts from building voids and to delineate roads, single-line drainages, ridges, bridge crossings, etc. This approach requires the development of a large number of additional detailed breaklines, yielding a higher quality topographic surface, but is substantially more expensive. Because of the much higher cost, such enhanced lidar DTMs are not cost effective for specifying data for inclusion in the 3DEP.

Hydro-Enforced Lidar DTM

Figure 1.37 shows a hydro-enforced lidar DTM used by engineers for hydrologic & hydraulic (H&H) modeling. It is similar to a hydro-flattened DTM with the addition of single line breaklines for pipelines, culverts, underground streams, etc. The terrain is then cut away at culverts to model hydrologic surfaces.

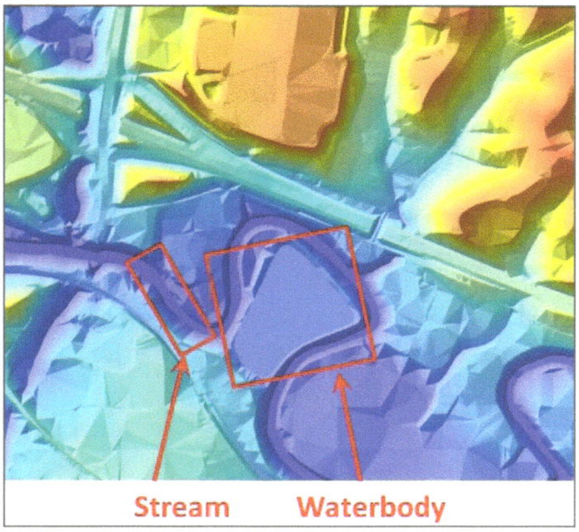

Figure 1.33. Traditional photogrammetric DTM.

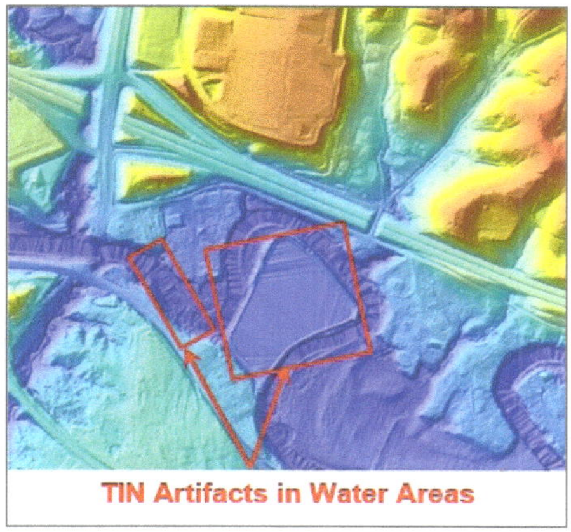

Figure 1.34. Pure lidar DTM without breaklines.

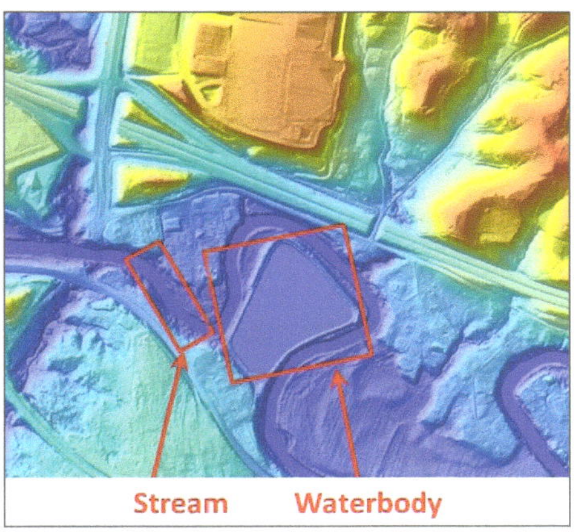

Figure 1.35. Hydro-flattened lidar DTM.

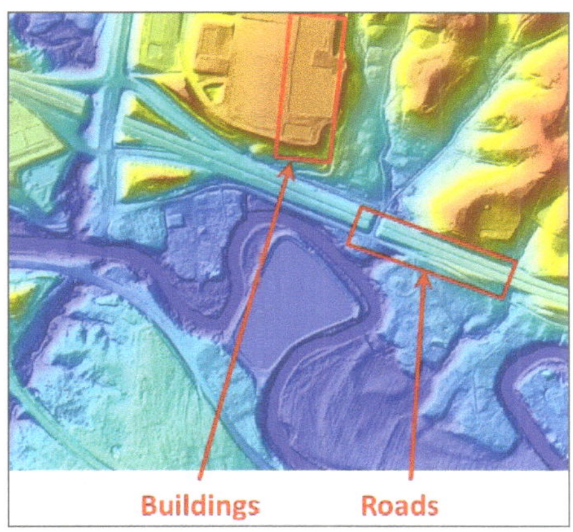

Figure 1.36. Lidar DTM with more breaklines.

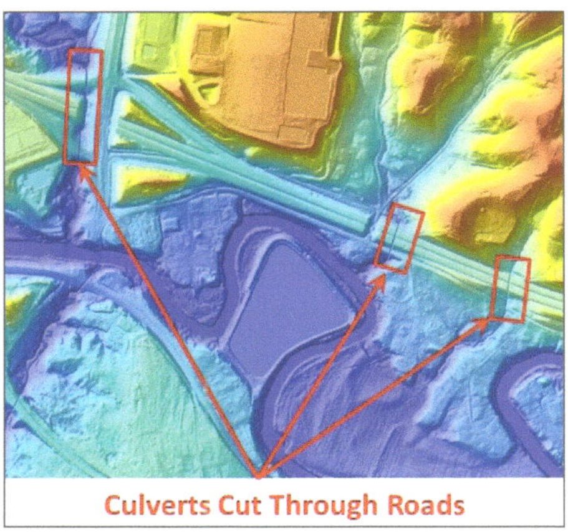

Figure 1.37. Hydro-enforced hydrologic surface.

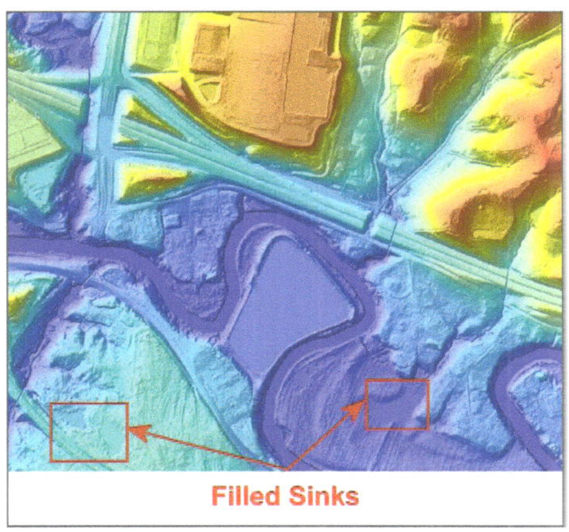

Figure 1.38. Hydro-conditioned hydrologic surface.

Water surface elevations (WSELs) are often set to known values (surveyed or historical). To avoid contour lines passing through underground culverts, hydro-enforced DTMs are not to be used for contours. Data for the 3DEP are not hydro-enforced, but hydro-flattened.

Hydro-Conditioned Lidar DTM

Whereas Figure 1.37 on page 23 shows how hydro-enforced DTMs are used to model hydrologic surfaces of water bodies, Figure 1.38 shows how hydro-conditioned DTMs model the flow of water over the land, outside of water bodies. Also used for H&H modeling, the hydro-conditioned DTM is similar to the hydro-enforced DTM, but with sinks filled. Water flow is continuous across the entire terrain surface with no areas of unconnected internal drainage. This is often achieved with GIS software such as the Arc Hydro tools for ArcGIS or GRASS. Data for the 3DEP are not hydro-conditioned but hydro-flattened.

Elevation Derivatives

As the name implies, elevation derivatives are products derived from 3-D source data. With lidar, the point cloud data are the source from which elevation derivatives are produced, including DSMs, DTMs, raster DEMs, contour lines, hillshades, slope and aspect maps, curvature maps, height-above-ground maps (also called normalized DSM, nDSM, or elevation difference maps), profiles/cross sections, viewsheds, watersheds, etc., including breaklines stereo-compiled using lidargrammetry or photogrammetry.

However, in common usage, users generally refer to elevation derivatives as those elevation products other than the common DSMs, DTMs, DEMs, and contours, which some use as the source for additional derivatives. The authors of this chapter believe that there are hundreds if not thousands of other general and specialized elevation derivatives used by various users for their own, often-unique, functional activities.

To ensure consistent results, elevation derivatives should begin from the same certified source, where possible, e.g., lidar point cloud or photogrammetric stereo model. One notable exception pertains to height-above-ground maps that are most easily produced by subtracting DEM elevations from DSM elevations of an area, rather than attempting to directly use the original lidar point cloud or stereo images to determine elevation differences for thousands or millions of points; but even for this exception, the DEM and DSM should both begin with the same certified elevation source.

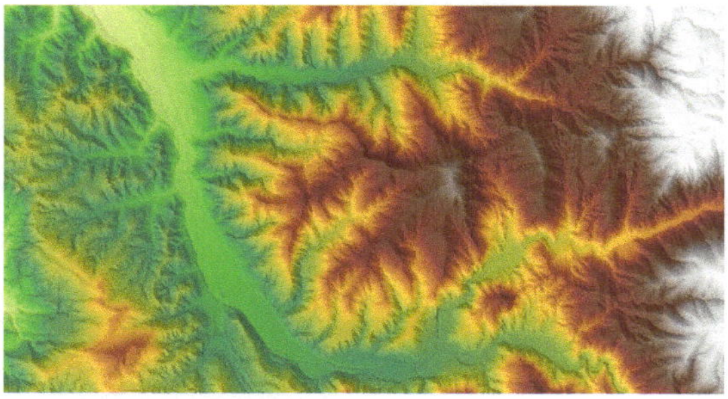

Figure 1.39. Hillshade map example.

Hillshades

Hillshades, or shaded relief maps, with either grey-scale or colored elevation ramps are used to create an illuminated representation of the surface, using a hypothetical light source to enhance terrain visualization effects. In Figure 1.39, the hypothetical light is from the southwest (lower left corner). For many users, hillshades have replaced contour lines as the best way to visualize 3-D topographic surfaces. Hillshade viewing angles and sun angles can be varied to maximize visual interpretability of terrain.

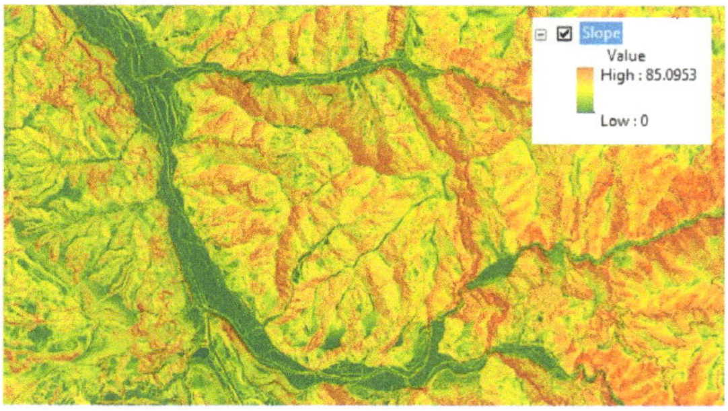

Figure 1.40. Slope map example.

Slope Maps

Slope maps are used for many environmental and engineering applications, including hydrologic and hydraulic modeling, erosion control, and soils mapping. Slope is the measure of change in elevation over distance, expressed either in degrees or as a percent. For example, a rise of 4 meters over a distance of 100 meters describes a slope of 2.3° or 4%. The maximum slope in the map in Figure 1.40 is about 85%.

Aspect Maps

Also used for many environmental and engineering applications, aspect is the slope direction of steepest gradient with respect to points of a compass, measured clockwise in degrees from due north. The value of each location in an aspect dataset indicates the direction the surface slope faces. The example in Figure 1.41 is of the same terrain as shown in Figures 1.39 and 1.40.

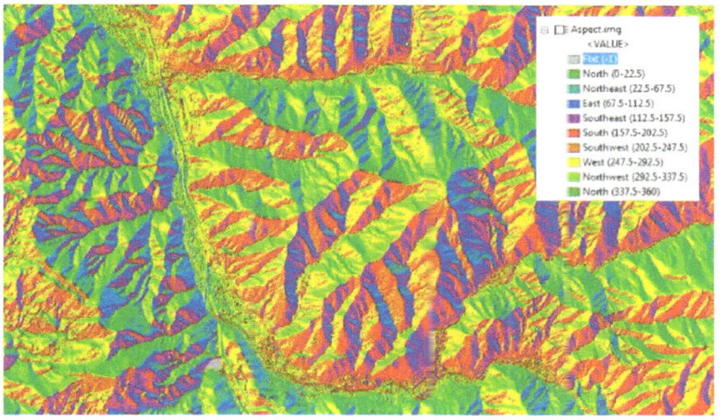

Figure 1.41. Aspect map example.

Curvature Maps

Usually derived from DEMs, curvature maps are key to numerous conservation activities, including soil and wetland mapping, because they model surface flow paths (see Figure 1.42) and soil wetness, for example. There are many types of curvature maps, including plan form curvature, profile curvature and tangential curvature.

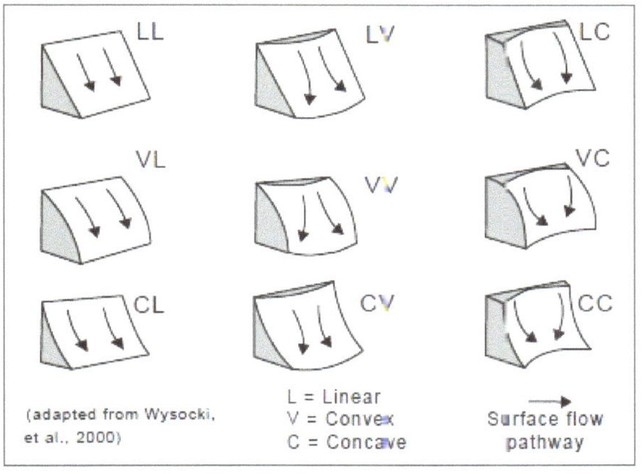

Figure 1.42. Types of curvature surface flow paths.

Profiles/Cross-Sections

Profiles are commonly "cut" through lidar point clouds or elevation surfaces in order to visualize or measure heights or elevation differences along the path of the profile. Figure 1.43 shows an example of a lidar profile cut across a river, to include elevation data on a bridge as well as elevations of bare-earth terrain, vegetation and buildings on either side of the river. Cross sections are profiles that are normally generated perpendicular to linear features such as roads and streams.

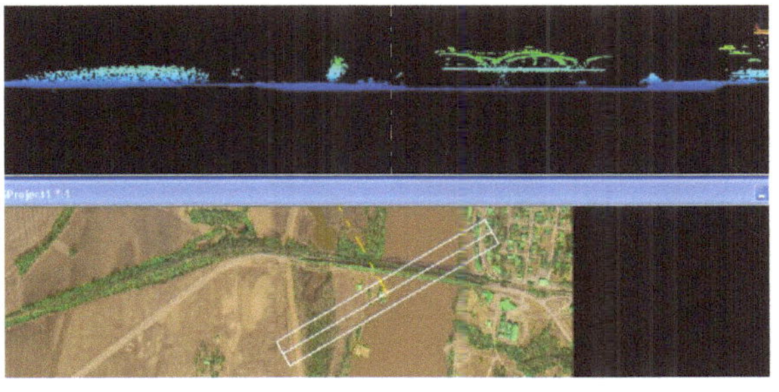

Figure 1.43. Lidar profile (top) showing elevations along the transect (bottom) that crosses the river and bridge.

Height-Above-Ground Maps

Commonly used for determining timber metrics, height-above-ground maps, also known as elevation difference maps and normalized DSM (or nDSM), are produced by subtracting the DEM from the DSM. In Figure 1.44, the original DEM is much higher on the right of this image, including a building on the hill top, but this elevation difference map normalizes the surfaces to determine which trees are taller when considering the terrain to be totally flat.

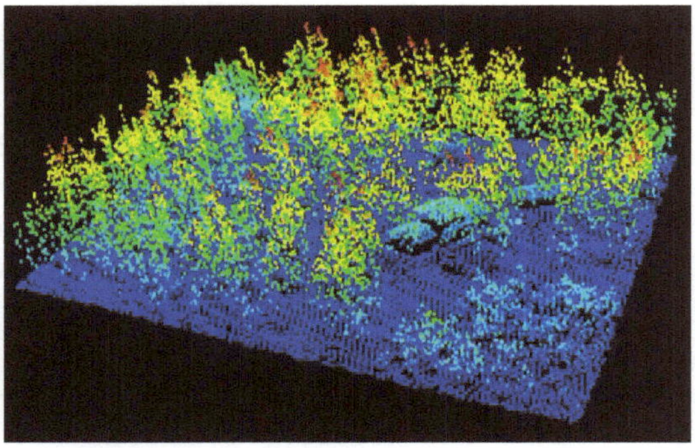

Figure 1.44. Height-above-ground map where heights of trees (and building) are normalized. Image courtesy of USFS.

Viewshed Maps

Viewshed maps, also known as intervisibility or line-of-sight maps, are popularly used by military terrain analysts to determine what can be seen, or not seen, from friendly or enemy positions. Figure 1.45 (Image courtesy of USGS) shows areas in direct line-of-site from the top of a 9.5-foot antenna and areas obscured from such intervisibility. The height of the tower can be virtually elevated to see changing results. Architects use viewshed maps for site selection and design, and real estate agents use them to advertise scenic views.

Figure 1.45. Viewshed map showing what can be seen (red) and not seen (green) from an antenna 9.5 feet tall.

Watershed Maps

Elevation models are the key input to performing hydrologic analysis. The elevation provides the input data for calculating flow direction across the terrain, which is subsequently used for the creation of stream networks and watersheds (Jenson and Domingue, 1988; Tarboton, 1997), as shown in Figure 1.46. Using these tools it is possible to find the upslope contributing area and downslope flow path for any location. These derived data are then used to estimate rainfall runoff, predict flood levels, and manage water resources.

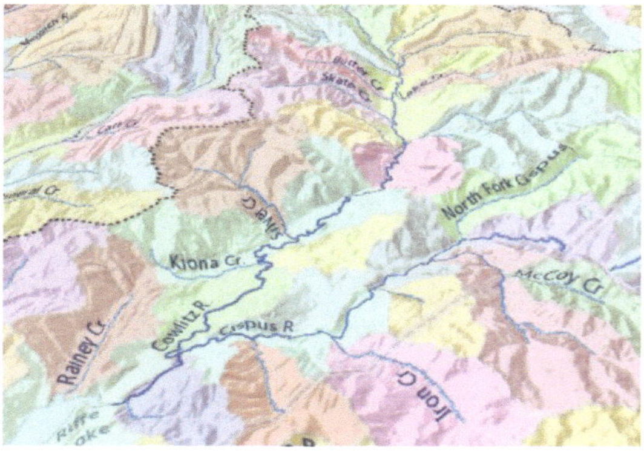

Figure 1.46. Watersheds and stream networks delineated from an elevation model. Hillshading has been added to show the underlying terrain.

Creating Raster Surfaces (Interpolation Methods)

Creating a raster surface from sample data requires the estimation of values for location based on sample data. There are many techniques for creating raster surfaces from sample data, generally referred to as interpolation. As this book is focused on terrain, the discussion of interpolation will focus on the basics, and on methods most appropriate for terrain interpolation. The reason for many techniques is because the input sample data and the surface being modeled may have many different characteristics. As a result, no single interpolation technique is the best for all situations, and no method can be regarded as singularly correct. All figures in this section are courtesy of Esri.

A basic principle shared by all interpolation methods is that things that are closer together tend to be more alike than things that are farther apart. When trying to build an elevation surface, you can assume that the sample values closest to the prediction location will be similar. As you move further away from the prediction location, the influence of the points will decrease. All interpolation methods apply weights to the influence of a sample point based on its proximity to the location being predicted.

There are two classes of interpolation, deterministic and probabalistic. Deterministic methods are based directly on the surrounding measured values and/or mathematical formulas applied to those values. Geostatistical methods are probabalistic statistical models that include autocorrelation, which is the strength of similarity between measured samples accounting for distance and direction. The goal of geostatistical interpolation techniques, such as kriging, is to create a surface while minimizing the error between the predicted values and the statistical model of the surface. All the interpolation techniques discussed here are deterministic techniques, except for kriging. Each example uses a common collection of mass points, shown in Figure 1.47a and 1.47b.

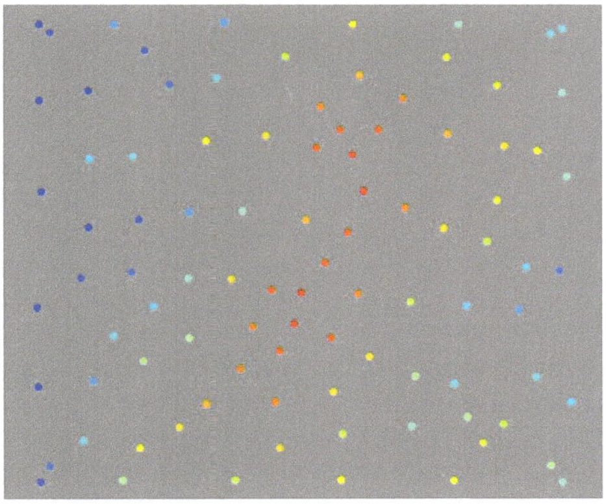

Figure 1.47a. Input sample mass points.

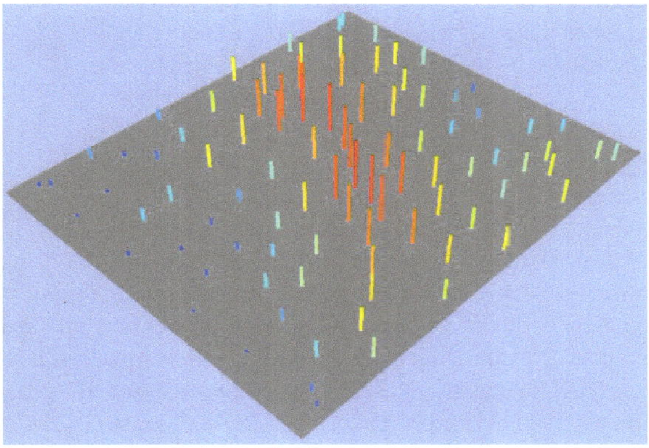

Figure 1.47b. Input sample mass points extruded by height.

Inverse Distance Weighted Interpolation

Inverse distance weighted interpolation (IDW) determines new cell values using a linearly weighted combination of values from nearby points. The weight is a function of inverse distance. Most implementations provide a weighting factor the user can specify to control the significance of input points based upon their distance from the location being interpolated. This method is one of the simplest methods of interpolation and is presented here only as an introduction to interpolation. IDW interpolation is useful when the variable being interpolated decreases in influence with distance from the sample location, such as consumer purchasing power. This assumption is not true for terrain data, and IDW should not be used for terrain. The cross section in Figure 1.48a on page 28 illustrates this effect, which can also be seen in the contours in Figure 1.48b on page 28. The surface has a dimpled effect at the sample point locations.

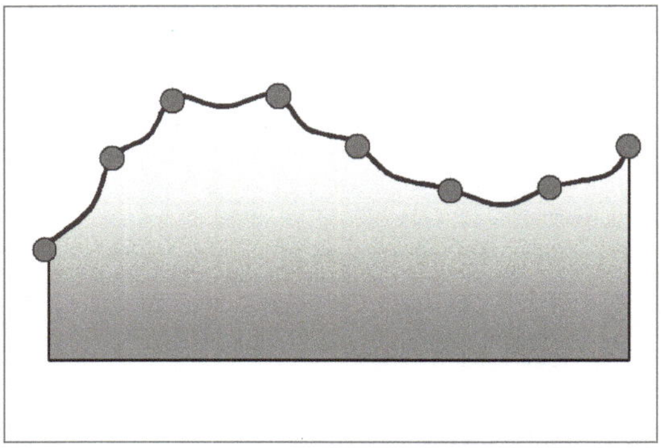

Figure 1.48a. Cross section of inverse distance weighted interpolation.

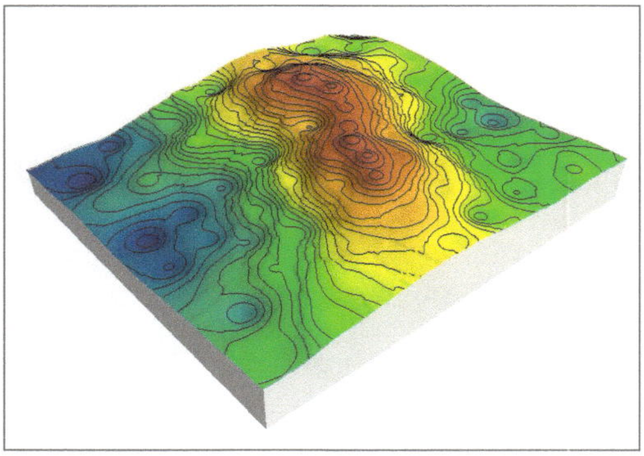

Figure 1.48b. IDW interpolated surface.

Linear Interpolation

Linear interpolation is the most commonly used TIN-based interpolator. This approach interprets every triangle in the TIN as a bounded plane. To estimate height for a query point, a 2-D spatial search, often sped up through the use of topology or indexing, is used to find the encompassing triangle. Once the triangle is known, a point on plane equation can be used to solve for height. This is often performed, for the sake of computational efficiency, using barycentric coordinates that provide a linear weighting scheme between the query point and three samples that define the encompassing triangle.

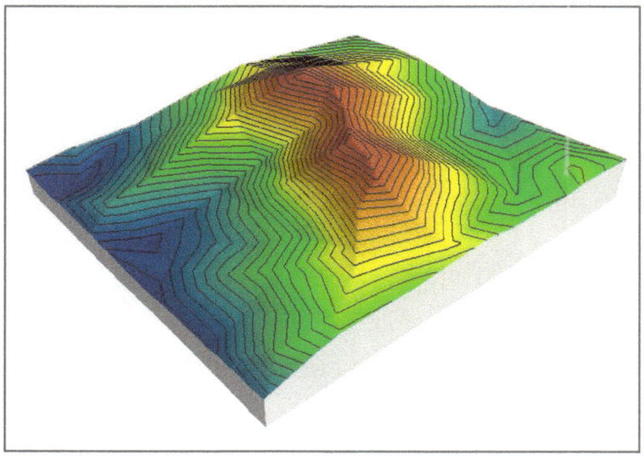

Figure 1.49. Surface derived through linear interpolation.

Linear interpolation is conservative and does not extrapolate heights outside the range of the input data. Peaks and valleys must be included in the input sample data to appear in the surface. As shown in Figure 1.49, the surface is not smooth and can exhibit sharp discontinuities across triangle edges. As a result, this approach is often avoided for cartographic purposes. It is also common for false ridges and valleys to form along triangle edges, a consequence of the approach. In spite of the drawbacks, this form of interpolation is a standard used in terrain modeling because it is fast, simple, good for many forms of analysis, and the results are generally consistent and reproducible between software packages. Additionally, since it is a TIN-based approach, there is full support for breaklines, which is a requirement for some applications.

Natural Neighbor Interpolation

Natural neighbor interpolation, also known as Sibson or "area-stealing" interpolation, finds the closest subset of input samples to a query point and applies weights to them based on proportionate areas in order to interpolate a value (Sibson, 1981). This approach works equally well with regular or irregularly distributed data, produces a smooth yet conservative surface, and completes in a time proportional to the size of the input data (Watson, 1992).

As the name implies, the key strength of this approach is in finding appropriate input samples (neighbors) to use when interpolating a height for a given point. Based on the Delaunay triangulation and its dual, the voronoi (Thiessen) diagram, this algorithm finds the closest points in all directions. It does so by determining a query point's voronoi neighbors (see Figures 1.50a and 1.50b). These are the samples the query point would form triangle edges with if inserted into the Delaunay triangulation of the samples. It therefore provides good representation from the samples in both proximity and direction, with the interpolated value depending only on a local subset of data. Since the neighborhood automatically adjusts itself to fit the local data configuration, there is no need for parameters like search radius or sample quotas that have drawbacks when the input data distribution is unevenly sampled or clustered. Breaklines are fully supported because the underlying data structure is a TIN.

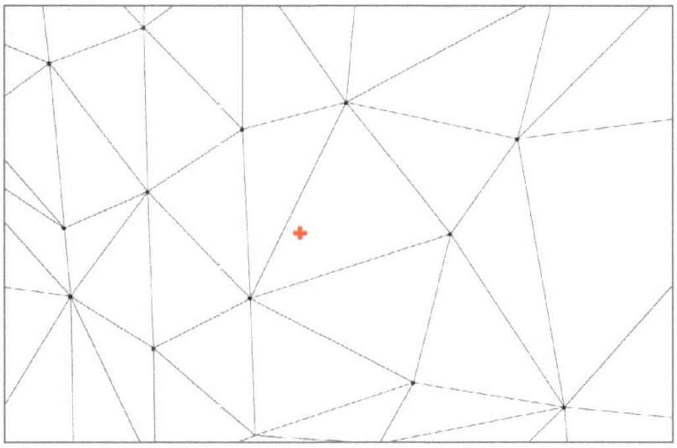

Figure 1.50a. The location of a query point that needs an interpolated height relative to the position of TIN elements.

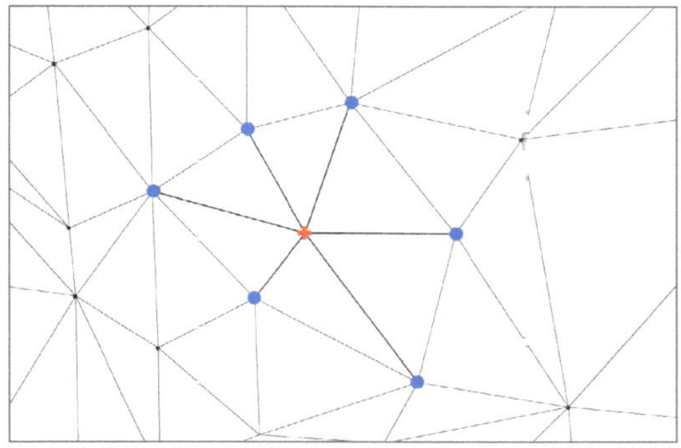

Figure 1.50b. The interpolant has found the closest surrounding nodes in all directions to the query point (the natural neighbors) and establishes a relationship to them for use in height estimation.

Another important aspect of this interpolant is its weighting scheme. Weights are not based on distance but on the amount of area that would be 'stolen' from the voronoi (Thiessen) polygons representing input samples if the query point were inserted into the diagram. As shown at Figure 1.51, this scheme exhibits the important property that the weight of each neighbor drops to zero by the time it ceases to be part of the neighborhood (Gold, 1989).

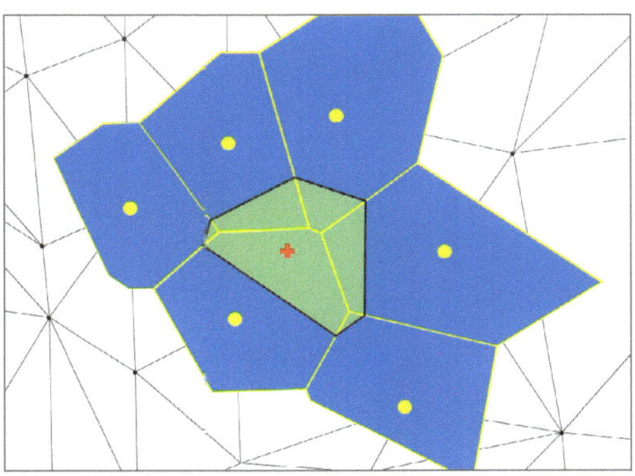

Figure 1.51. The weighting scheme used by natural neighbor interpolation. The point symbolized by a red cross in the center is a query point requiring height estimation. The shaded polygon represents its voronoi region if it were to be inserted in the triangulation. The solid blue polygons are the voronoi regions for the surrounding nodes in the triangulation. The weight of each node is based relative to the area of overlap between its voronoi region, the query point's region, and the areas of overlap for the other nodes. In this example, the natural neighbor to the southwest will be given the most weight.

The natural neighbor surface is smooth everywhere except at the data locations (see Figures 1.52a and 1.52b on page 30). The surface is similar to triangle-based linear interpolation but there is no break in slope along triangle edges. The exception to this is for edges representing hard breaklines in which case the interpolator is adjusted to accommodate the intentional discontinuity.

The natural neighbor interpolator is a conservative interpolator that will not manufacture data. The resultant surface is a true representation of just the input data. This interpolator will honor the input data

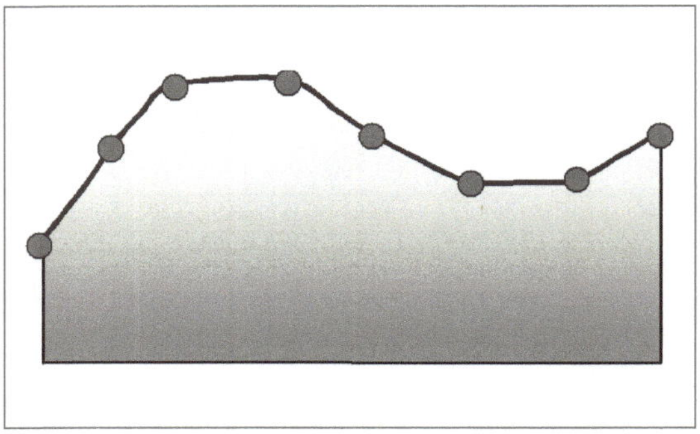

Figure 1.52a. Cross section of natural neighbors interpolated surface.

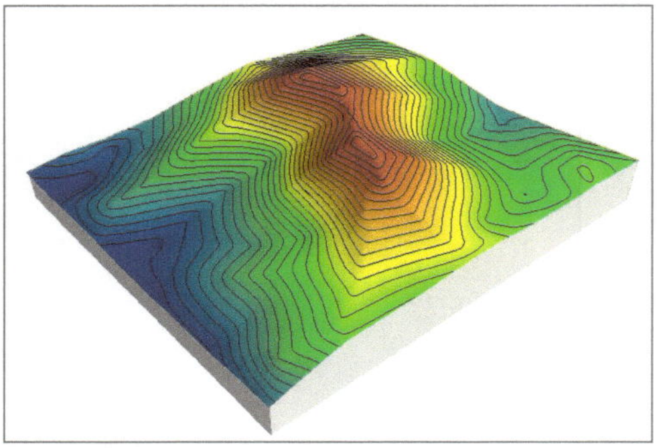

Figure 1.52b. Natural neighbor interpolation.

points and be constrained to the input data range. Therefore peaks and valleys must be included in the input sample data to appear in the surface.

Spline

Splines are a common form of interpolators known as radial basis functions. These interpolation techniques use a mathematical formula to create a surface that minimizes the overall surface curvatures, resulting in a smooth surface that passes through the input points.

Conceptually a spline is like bending a sheet of rubber through a set of points. It fits a mathematical function to a specified number of neighboring input points to determine the value at each location. There are different types of splines and parameters to specify for them that make the surface more tightly conform to the input points or perform more smoothing (Mitas and Mitasova 1988; Franke, 1982).

This method is useful for creating elevation models of areas with gently varying terrain with smooth slope transitions. This method is sensitive to, and not suitable for sharp changes in value over short distances such as steep cliffs or man-made features. In such situations, this interpolator has a tendency to over-exaggerate the value of neighboring cells.

Splines are exact interpolators; they honor the input points. They also extrapolate beyond the input data range so that unsampled peaks or valleys can be represented as demonstrated in Figure 1.53a.

Splined surfaces, because of their smoother characteristic, are often the preferred interpolation method when the end goal is to create cartographic quality contours. Notice the smoothness of the contours in Figure 1.53b. Mathematically smooth continuous surfaces can also be created with other interpolators (Gribov and Krivoruchko, 2004).

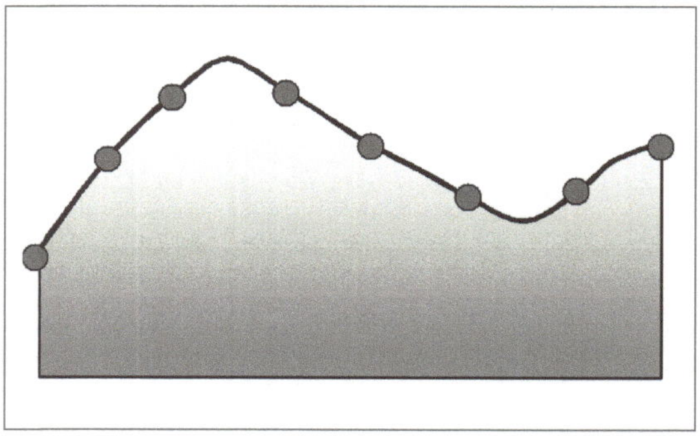

Figure 1.53a. Spline.

Figure 1.53b. Spline interpolated surface.

Local Polynomial

Local polynomial interpolation is another category of common deterministic techniques that are suitable for interpolating elevation data. This approach uses a local moving window to fit a polynomial surface of a user-defined order, where input points near the prediction location are weighted more heavily than those further away (Gribov and Krivoruchko, 2011). This interpolator does not use any global characteristics of the data. Local polynomials are not an exact interpolator; the resulting surface is not guaranteed to pass through the input points and is therefore a particularly useful interpolator when modeling smoothly changing data and knowing there may be some locational or measurement error in the data. Due to the local nature of the interpolator, it is particularly sensitive to uneven spatial distributions of input points, large areas with few or no input points. Figures 1.54a and 1.54b illustrate the smooth behavior of the local polynomial interpolation surface.

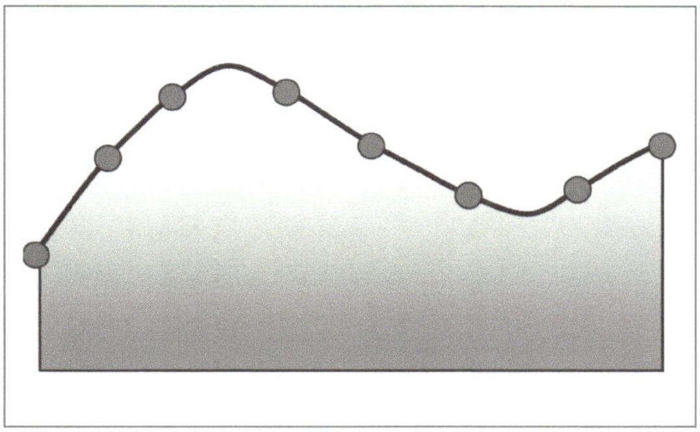

Figure 1.54a. Local Polynomial.

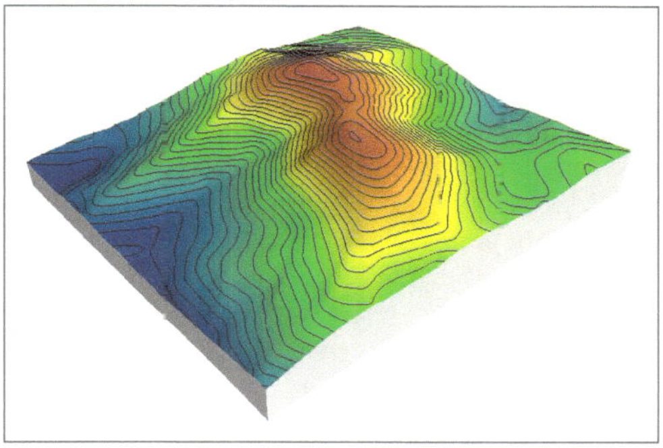

Figure 1.54b. Local Polynomial surface.

Geostatistical Techniques (Kriging)

Kriging is a geostatistical interpolation technique, similar to deterministic interpolation techniques in that it weights the surrounding measured values to derive a prediction for each location. However, the weights are based not only on the distance between the measured points and the prediction location but also on the strength of the overall correlation among the measured points. With kriging, the data are used to define the spatial correlation model to determine the weights for neighboring samples that are used to fit a surface to the points, whereas, with deterministic techniques, the model is based only on user defined input.

These weights are derived through an examination of the data and fitting of a model. The data are examined as a semivariogram, which quantifies the correlation between pairs to points based upon how far apart they are. The semivariogram is constructed by graphing the squared differences in z-value versus map distance between sample point pairs. This builds off the previous assumption that things close together are more alike than things farther apart. Fitting the model involves selecting the appropriate parameters that fit the best line through the semivariogram. There is more to this process than can be sufficiently explained in this text.

A specific common form of kriging known as ordinary kriging is appropriate for interpolation of terrain data. Ordinary kriging can be shown to be statistically optimal if the data have certain properties, such as being normally distributed with a mean and variance that do not change across the terrain. For further discussion of variogram modeling and different kriging options, refer to (Krivoruchko, 2012; Issaks and Srivastava, 1989).

In most cases, kriging is used as an exact interpolator, the surface being created passes exactly through the input points. However, there are situations in which data accuracy may be questionable or there is a known measurement error, such as a measurement device with a known accuracy of 1 foot. In these cases, kriging provides an ability to pay more attention to the local trend of the surface than to a single value. Figure 1.55a on page 32 shows a cross section of a kriging surface. Figure 1.55b shows a similar surface with an erroneous data

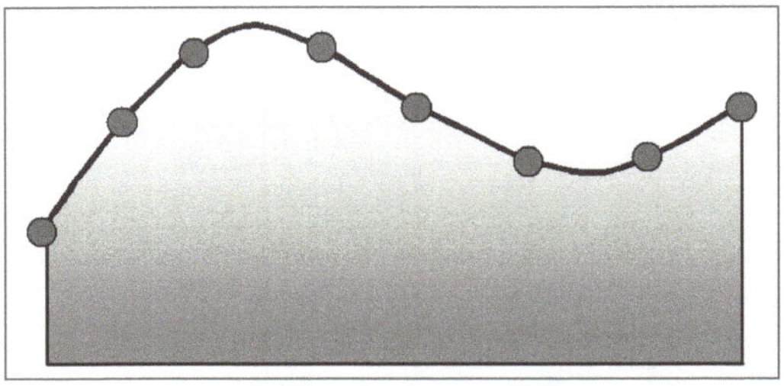

Figure 1.55a. Cross section of a surface from kriging.

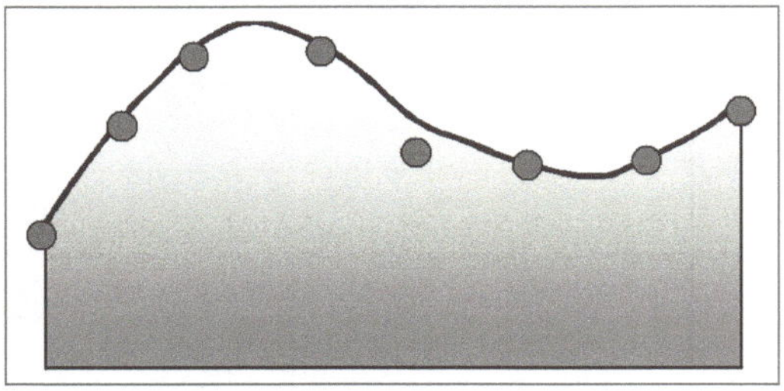

Figure 1.55b. Cross section of the same surface with one of the data points changed, illustrating how the model can create a surface that does not exactly conform to the data points.

point. Notice that the surface maintains its general shape even with the value of one point changed.

Recent developments in kriging techniques avoid some of the difficulties of proper use of kriging by using simulation to fit a spectrum of semivariogram models, which minimizes the effort by the user to fit an optimal model, as well as developed automated techniques to properly address non-Gaussian and non-stationary data (Krivoruchko, 2012). Figure 1.56 shows an example interpolation result from kriging.

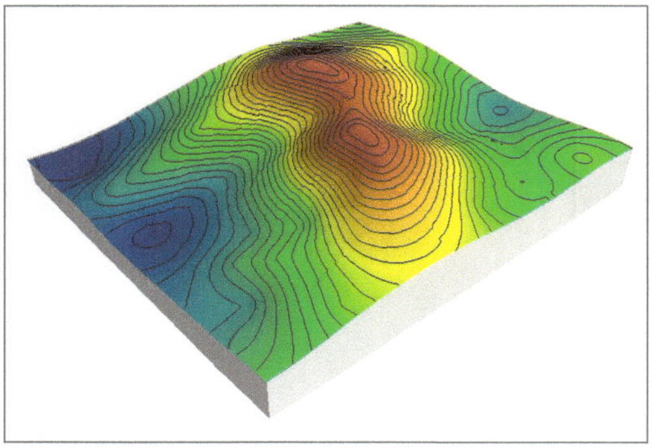

Figure 1.56. Interpolation result from kriging.

Specialized Terrain Interpolation

A variation on the spline technique is an interpolator specially designed for the interpolation of elevation data known as ANUDEM (Hutchinson, 1989, 2011). This interpolator uses a discretized thin plate spline technique, which is optimized for the creation of hydrologically correct terrain models. ANUDEM is unique in both its inputs and technique for building good terrain models.

Most interpolators use only point data as input, and sometimes breaklines. This locally adaptive interpolator uses not only point data, but also breaklines in the form of elevation contours, streams, ridges, and cliffs for drainage, and polygons for lake boundaries. This specialized interpolator works intelligently with contour data by recognizing the critical points in areas of highest contour line curvature, and the lake boundary algorithm (Hutchinson, 2007) automatically determines lake heights that are fully compatible with connecting streams and neighboring elevation values.

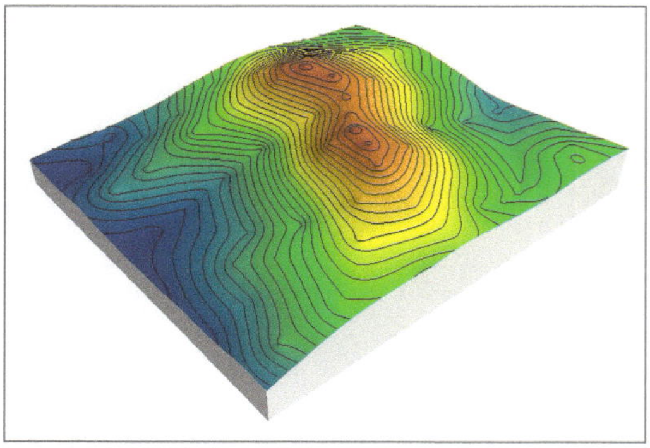

Figure 1.57. Interpolation result from ANUDEM.

ANUDEM also has an option for hydro-enforcement. The hydro-enforcement option attempts to remove all sinks that have not been identified as input sink data. This option will first create a network of valleys and ridges then use a spline interpolator to build a surface while maintaining consistent downslope flow. ANUDEM also produces a series of diagnostic outputs that are useful to refine input parameters and data. Figure 1.57 shows an example of an ANUDEM surface.

General Interpolation Issues

When creating surfaces using any of the previously described techniques, there are some issues that are common to all of them.

The number, distribution, and location of sample points are the key to building a good representative surface model. When building a TIN or using an interpolator such as IDW or linear natural neighbors, where the output surface values are constrained to the range of the input values, it is important that the high and low areas are adequately represented in the sample data. Having more sample points is always better, assuming they are accurate. Generally it is best if these points are well distributed spatially, not all clumped in small areas, but this distribution should vary with surface roughness. The number necessary to create a good, representative surface will vary with the character of the original surface. If the elevation samples are relatively evenly distributed and the surface characteristics do not change across the landscape, a smaller number of points may be needed. If, however, the surface is quite varied such as in a mountainous area, more sample points should be used. If the surface is mixed, mountains and plains, the mountainous area should be more densely sampled than the plain. There are at least two known exceptions to this guideline for civilian and military applications.

For hydrologic and hydraulic (H&H) modeling of flood hazards, the hydrologic modeling of the mountains and hills that drain into the floodplain do not require high-accuracy, high-resolution elevation data to determine the direction in which water will flow into the floodplain; whereas the hydraulic modeling of the relatively flat floodplain requires high-accuracy, high-resolution elevation data to model the subtleties of the terrain to determine where flood waters will encroach. In other words, better elevation data are needed for the relatively flat floodplains than are needed for the surrounding hills that drain into the floodplains.

For military terrain analyses, it is easy to perform intervisibility and cover and concealment modeling, for example, in rugged terrain where elevation errors of several meters usually have little impact on the result; but it is difficult to perform such modeling in relatively flat terrain where differences of a few inches can impact whether or not one can see between points A and B on the terrain. Similarly, for automated terrain correlation guidance for smart weapons, it is easy to correlate the terrain when it is rugged, but it is far more complex with sand dunes and subtle variations in the terrain where there are not clearly defined terrain features on which to base the automated terrain correlation algorithms.

Related to the distribution of sample points, it is important to note that the interpolation at the edge of a study area should include sample data from a reasonable distance outside the study area, not just inside. The same principal holds true for interpolation of surfaces that fit into a map series with finite boundaries. Additional data from neighboring maps should be included so areas along the edge can be accurately portrayed, and if the maps or derived data are stitched together they match along their boundaries.

Contour lines are a special kind of input for interpolation. They are generally not considered to be a good input dataset to an interpolator because of the high density of sample points along the contour and the sparse sampling of data between the contours. Surfaces created from contours are often biased to the contour interval values. In the worst cases, a terraced or stair-step effect may be noticeable In almost all cases, a histogram of the cell values will show spikes of high frequency at the contour interval. This phenomenon is minimized in the ANUDEM interpolator mentioned above because it has been optimized to work intelligently with contour input.

All interpolators, except the TIN-based linear and natural neighbors, have parameters for specifying which nearby sample points will be used in the prediction of a value for a cell. These neighborhoods are often specified as a fixed number of points (e.g., use the 8 nearest sample points to predict a value) or as a radius (e.g., use all points within 50 meters). It is most common to use between 6 and 12 points. Some software also allows the neighborhood to be divided into 4 or 8 sections with a sample taken from each quadrant. This has an effect similar to the input point selection for natural neighbors, and it minimizes the effects of contour biasing.

In addition to the sample points to be used, most interpolators have other parameters to specify. Because the characteristics of a surface change through space, what is a good set of parameters in one area may not be good in another area. In the case of terrain interpolation, changes in geomorphology will often mean changes in parameters. A coastal plain will require different parameters than mountains. If you have two very different landscapes to interpolate it may be better to separate them into two different points sets and interpolate them independently using the best parameters for each, then merge the surfaces back together.

There are many types of errors that can occur in a surface model, some significant, others minor depending upon the application. Contour biasing of a surface, mentioned previously, can be a significant source of error when creating slope or curvature maps, but may not be noticeable when used for visualization. Previous photomechanical methods of creating DEMs sometimes resulted in striping patterns that can still be found in older data.

As noted earlier in this chapter, DEMs that are to be used for water resources applications should be hydrologically enforced, hydrologically conditioned, or both. The ANUDEM interpolator discussed above does this as part of the interpolation process. When creating terrain surfaces, it is important to know the intended purpose of the data, as the end use will bear on how the surface needs to be created. For example, if the DEM will be used for hydraulic modeling, bridges and road fills over culverts must be removed from the surface or they will act as dams, preventing the downstream modeling of the water. If the surface is to be used for transportation modeling or visualization then bridges and road fills would be included.

There are some quick and easy techniques to find errors in DEM data. The first is to look at the minimum and maximum values to see if they are within a realistic data range. A histogram of the data distribution can also be helpful. There should be no values that occur much more frequently than nearby values. In other words, the elevation value of 625 should occur with roughly the same frequency as 623, 624, and 626. An interactive fly-through of the data is also a good technique for finding significant errors. The analysis tools discussed previously in this chapter can also be quite useful, particularly hillshade or shading. Using a low-angle hillshade makes it easy to pick out striping or patterns in the data as well as other errors that may be in error by only a few feet. Creating slope maps can also be useful, by checking for an appropriate range of slope values, and also when analyzed visually.

Because there are so many ways to create a raster surface from point data, it is useful to evaluate how well different methods or parameters for a method perform at creating a good surface. In this case, a good surface is described as one that most closely matches the input data. The techniques used to do this are known as validation and cross validation. The simpler approach is validation, when some percentage of the known input data is withheld from the interpolation, and the value at each withheld sample location is compared to the interpolated surface value and a statistical measure such as a root-mean-square-error (RMSE) is calculated. Cross validation is similar, except that each point is removed individually and the interpolated value at each location recalculated one at a time. In any case, the user should keep in mind that an interpolated DEM is inherently diminished in accuracy and/or detail from the original source data. It is easy to regard a DEM as a true model of the ground when in fact it is merely an approximate representation of the ground, and as has been demonstrated above, no single DEM can be regarded as the most or only correct representative surface.

Terrain Visualization

While digital terrain models provide the benefit of computer based analysis they also offer the equally important ability to provide an intuitively comprehensible visual display of the terrain characteristics. Computers offer a dizzying array of options when it comes to viewing surface models and the data derived from them. Uses include analysis, mapping, simulation, and entertainment. Also see Chapter 14 of this manual for explanations of how terrain visualization is used for DEM quality assessment.

As noted in Sloman (1983), "The human visual system is the most powerful information-processing system known." Through our sight, we can identify patterns and relationships between features and their attributes. Computers enable us to render terrain using both realistic and highly abstract, symbolized, methods. Symbolized representations make it easy for us to focus on a particular aspect of the terrain and not hide the patterns we are searching for behind unrelated noise. On the other hand, certain applications benefit from a high degree of realism.

Figure 1.58 shows an example of terrain visualization on a clear day, whereas Figure 1.59 shows an example of terrain visualization on a cloudy or foggy day. Both images are realistic, but both were produced, not from ground or oblique aerial photography as they appear, but from draping satellite imagery on top of a DTM.

Figure 1.58. RapidEye satellite imagery draped on an airborne IfSAR terrain model showing Mt. Hayes, Alaska. Image courtesy of Dewberry, Fairfax, VA. This is how the landscape would appear on a perfectly clear day.

Figure 1.59. LANDSAT Thematic Mapper satellite imagery draped on a terrain model. Sky haze, and fog are added to provide a sense of realism for a more-typical day with fog and/or clouds.

Similarly, Figure 1.60 (top) shows a RapidEye satellite image draped on an airborne IfSAR digital terrain model (DTM) showing the crest of Denali (previously Mt. McKinley), Alaska, the highest elevation in North America, as compared with an actual fly-by photo (bottom) taken by a pilot looking out the window of his airplane.

This comparison in Figure 1.60 clearly shows how accurate such synthetic views can be in providing realistic terrain visualizations, even when the satellite image has 5-meter pixel resolution and the IfSAR DTM also has 5-meter DTM elevation posts. Such realistic terrain visualizations could be a matter of life or death for a team planning to climb this mountain to its crest.

The DTM could have come from any source, but airborne IfSAR works best in Alaska because so many areas in Alaska have frequent cloud cover, fog or smoke, making it nearly impossible to acquire statewide DTM coverage with any technology other than IfSAR.

Figure 1.60. Denali, Alaska. The top image is a RapidEye satellite image draped over an IfSAR DTM; the bottom image is a fly-by photograph. Image courtesy of E-Terra, LLC, Anchorage, Alaska.

In addition to visualization, terrain data are useful when performing spatial analysis. Most information derived from analysis of terrain data is created as input to site selection models in a GIS. Whether the problem is to find the best location to build a new road tower, or housing development, or identify the best place to find a particular animal or plant species, the analysis of terrain data will almost always be a key ingredient.

Terrain analysis has a substantially different meaning for the military for which there are terrain analyst military occupational specialties that support battlefield intelligence. The U.S. Army and the U.S. Marine Corps combat units have specialized terrain analysts, equipped with computers and GIS software, to perform military terrain analyses of intervisibility, slope, aspect, cover and concealment, and cross-country movement, for example, often based on digital terrain elevation data (DTED) from the National Geospatial-Intelligence Agency (NGA). The intent is to achieve superior use of the battlefield terrain itself in order to effect mission accomplishment and save lives. When lives are in potential jeopardy, military personnel of all services recognize the importance of being able to "see" the battlefield prior to engagement with the enemy.

Virtual Reality

Virtual reality is a term that applies to computer-simulated environments that can simulate physical presence in places in the real world, as well as in imaginary worlds. Most current virtual reality environments are primarily visual experiences displayed either on a computer screen or through special stereoscopic displays, but some simulations include additional sensory information, such as sound through speakers or headphones.

The simulation and video game industries have long been using terrain models as a necessary component of their products. Flight simulators and the like require this information in order to create realistic views of the world. A satellite or orthophoto draped on top of a terrain surface with sky, fog, and haze added for effect goes a long way to ensure realism, as shown in Figure 1.59 on page 35. Computer simulations of the potential battlefield terrain for mission rehearsals and "fly-throughs" are vital for saving lives in actual military combat.

In addition to military flight simulators used for mission rehearsals and training worldwide, flight simulators are also widely used for training civilian pilots. Most flight-training in the U.S. is done with common flight simulators made to simulate Instrument Flight Rule (IFR) conditions and aircraft system failures, rather than Visual Flight Rule (VFR) procedures common in Alaska. Most common flight simulators are made to train the turbine-engine-airline-IFR environment (high above all terrain and weather). Due to many complex factors, Alaska's pilots require different skill sets than pilots elsewhere. In harsh contrast, aircraft simulators for Alaska are made to simulate Alaska's unforgiving mountains and passes in photo-realism and harsh weather conditions that too often lead to Controlled Flight Into Terrain (CFIT) accidents. Simulators, such as the one shown in Figure 1.61, require elevation data draped with imagery. In a sense of "look-before-you-fly," Alaska's pilots learn their way through the mountains in the simulator and gain experience necessary for flight safety. Realistic flight simulators are also vital for training under IFR procedures. With few roads, Alaskans routinely rely on aviation, even when flights must be conducted under poor visibility because of darkness, clouds, fog and/or smoke.

Digital elevation data are also used by the transportation industry in the development of high fuel efficiency and driverless vehicles. When enabled to anticipate upcoming changes in slope, both combustion engine and electric motor vehicles can adjust their output to the power needed more precisely than if operating reactively (as with a human driver). This automatic power adjustment can produce substantial fuel savings and extension of vehicle range.

Figure 1.61. Flight simulator used by E-Terra, LLC, for training pilots in Alaska.

3-D virtual cities are now popular for a large variety of user applications, especially for major cities and urban areas. Figure 1.62 shows the virtual City of Philadelphia, produced by Esri CityEngine®. The finished 3-D city is created by starting with geodatabase/2-D information, to which 3-D information is added for terrain, streets, blocks and parcels. Then 3-D buildings are extruded, roofs are generated and street furniture is added. Next, texturing is applied and building facades are added to finish the 3-D virtual city.

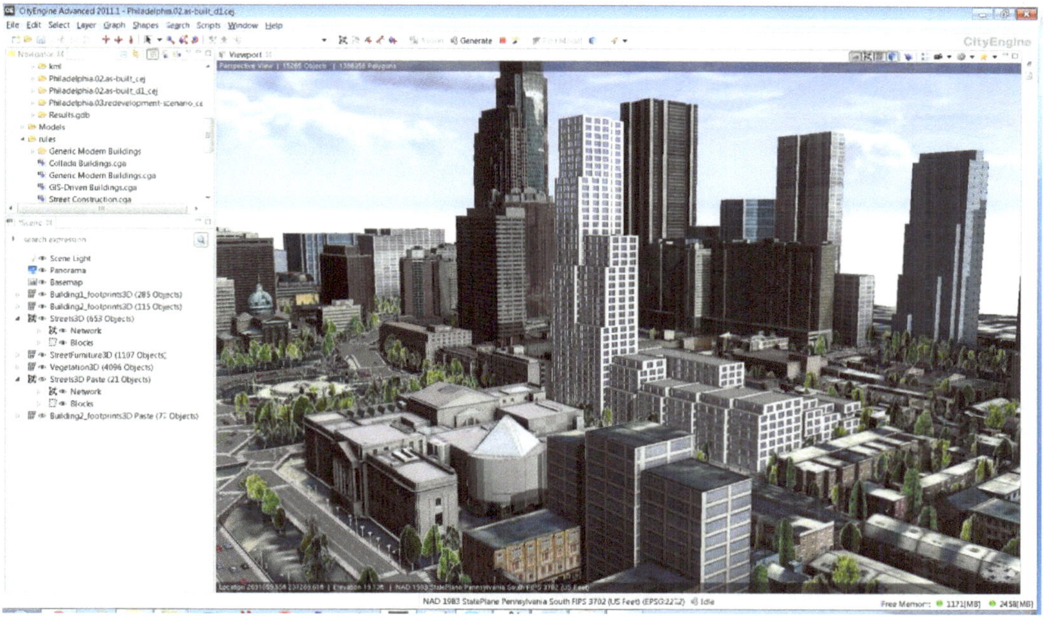

Figure 1.62. Produced by Esri CityEngine, this is a virtual model of the City of Philadelphia. Several existing buildings are modeled applying oblique imagery. Some buildings have been synthesized and added to the city model for the sake of a scenario-based urban planning application.

AUTHOR BIOGRAPHIES

Dr. David F. Maune is a geodesist, photogrammetrist, and Senior Project Manager for Mapping and Remote Sensing at Dewberry in Fairfax, VA. He is a Certified Photogrammetrist, Certified Floodplain Manager, ASPRS Fellow, charter member of the National Geospatial Advisory Committee (NGAC) and a current member of NOAA's Hydrographic Services Review Panel (HSRP). He is a licensed photogrammetric surveyor in three states. He manages Dewberry's geospatial products and services contracts with the Department of Homeland Security (DHS), the U.S. Geological Survey (USGS) and the National Oceanic and Atmospheric Administration (NOAA) that entail production of DEMs and related products from photogrammetry, topographic and bathymetric lidar, IfSAR, and sonar. He is the principal author of the National Enhanced Elevation Assessment (NEEA) published by the USGS, which formed the basis for the USGS 3D Elevation Program (3DEP). He is the editor and principal author of the U.S. Army Corps of Engineers' Engineering Manual 1110-1-1000, Photogrammetric and Lidar Mapping, and he is a principal author of accuracy standards, specifications and guidelines, pertaining to digital elevation data, published by ASPRS, the National Digital Elevation Program (NDEP), and the Federal Emergency Management Agency (FEMA). He was previously an active duty Army Colonel, retiring after 30 years with the U.S. Army Corps of Engineers where he served as Director, U.S. Army Topographic Engineering Center (TEC)—now the Army Geospatial Center (AGC); Director, Defense Mapping School (DMS); and Inspector General, Defense Mapping Agency (DMA)—now the National Geospatial-Intelligence Agency (NGA). He is a remote sensing consultant to Federal, state and local governments, specializing in independent quality assurance/quality control (QA/QC) of lidar data, contours and digital orthophotos. He received his B.S. degree in Mechanical Engineering in 1961 from what is now the Missouri University of Science and Technology, and in 1970 and 1973 he respectively received his M.S. and Ph.D degrees in geodetic science and photogrammetry from The Ohio State University. He was the 1977 recipient of the ASPRS Talbert Abrams Grand Award and the 2016 recipient of the ASPRS Photogrammetric Award (Fairchild). In 2018, he received the International Lidar Mapping Forum (ILMF) inaugural Lidar Leader Award for Outstanding Personal Achievement.

Mr. Hans Karl Heidemann received his B.A. in Geography from the University of North Carolina-Greensboro in 1995 and completed graduate studies in Geographic Information Processing at the University of South Carolina in 1997-98. His professional experience includes extensive GIS work, primarily supporting water resources engineering at Woolpert and Watershed Concepts; development of lidar processing workflows and breakline algorithms at EarthData International of North Carolina; and architecture of commercial lidar software at ProLogic, Inc. Now a Physical Scientist at the U.S. Geological Survey (USGS) Earth Resources Observation and Science (EROS) Center, Mr. Heidemann leads the development of lidar standards and specifications within the USGS and serves as the Bureau's representative on various inter-agency and national and international lidar commissions. He is the author of the USGS-NGP Lidar Base Specification. Mr. Heidemann provides technical expertise on lidar contracting and quality assurance for the National Geospatial Program and 3DEP. He holds GISP certification and is an active member of ASPRS, chairing the Airborne Lidar Committee, and serving on the LAS Committee and other working groups focused on calibration standards and industry best practices. He is a contributing author to the ASPRS Manual of Airborne Topographic Lidar.

Mr. Stephen M. Kopp currently leads the Raster/Modeling Product Development teams at the Environmental Systems Research Institute (Esri), Inc. During his 11 years at Esri, he focused primarily on development of the Arc/INFO GRID and ArcView Spatial Analyst products. He received his M.A. degree in Geography from Indiana State University, Terre Haute, Indiana, in 1990.

Mr. Clayton A. Crawford is currently the Product Lead for the 3-D Development Team at Esri. He received his M.S. degree in Geography from the University of Tennessee, Knoxville, in 1990.

Any use of trade, firm, or product names is for descriptive purposes only and does not imply endorsement by the U.S. Government.

REFERENCES

ASPRS (1990). ASPRS accuracy standards for large-scale maps. Photogrammetric Engineering and Remote ASPRS (1990). ASPRS accuracy standards for large-scale maps. In *Photogrammetric Engineering and Remote Sensing*, 56(7) (pp. 1068-1070). Bethesda, MD: American Society for Photogrammetry and Remote Sensing.

Ayeni, O. O. (1982). Optimum sampling for digital terrain models: A trend towards automation. In *Photogrammetric Engineering and Remote Sensing*, 48(11), (pp. 1687-1694). Bethesda, MD: American Society for Photogrammetry and Remote Sensing.

Bureau of the Budget. (1947). *United States National Map Accuracy Standards*. Washington, D.C.: U.S. Bureau of the Budget.

Carter, J. R. (1988). Digital representations of topographic surfaces. In *Photogrammetric Engineering and Remote Sensing*, 54(11), (pp. 1577-1580). Bethesda, MD: American Society for Photogrammetry and Remote Sensing.

Chen, Z. T. (1988) Break lines on terrain surfaces. In *GIS/LIS '88, Proceedings*, Vol. 1, (pp. 781-790). Bethesda, MD: American Society for Photogrammetry and Remote Sensing.

Franke, R. (1982). Smooth interpolation of scattered data by local thin plate splines. In *Computers & Mathematics with Applications*, 8(4), (pp. 273-281).

Gold, C. (1989). Surface interpolation, spatial adjacency and GIS. In J. Raper (Ed.), In *Three Dimensional Applications in Geographic Information Systems*, (pp. 21-35). London, England: Taylor & Francis Ltd.

Gribov, A., & Krivoruchko, K. (2004). Geostatistical mapping with continuous moving neighborhood. In *Mathematical Geology*, 36(2), (pp. 267-281). New York, NY: Springer.

Gribov, A., & Krivoruchko, K. (2011). Local polynomials for data detrending and interpolation in the presence of barriers. In *Stochastic Environmental Research and Risk Assessment*, 25(8), (pp. 1057-1063).

Heidemann, H. K. (2018). Lidar base specification (ver. 1.3, November 2014). In *U.S. Geological Survey Techniques and Methods*, Book 11(B4).

Hutchinson, M. F. (1989). A new procedure for gridding elevation and stream line data with automatic removal of spurious pits. In *Journal of Hydrology*, 106(3-4), (pp. 211-232).

Hutchinson, M.F. (2011). In *ANUDEM version 5.3 user guide*. Canberra, Australia: The Australian National University, Fenner School of Environment and Society.

Issaks, E. H., & Srivastava, R. M. (1989). In *An introduction to applied geostatistics*.

Jenson, S. K., & Domingue, J. O. (1988). Extracting topographic structure from digital elevation data for geographic information system analysis. In *Photogrammetric Engineering and Remote Sensing*, 54(11), (pp. 1593-1600). Bethesda, MD: American Society for Photogrammetry and Remote Sensing.

Kidner, D. B., Ware, J. M., Sparkes, A. J., Jones, C. B. (2000). Multiscale terrain and topographic modeling with the implicit TIN. In *Transactions in GIS*, 4(4), (pp. 379-408).

Krivoruchko, K. (2012, Fall). Empirical bayesian kriging. In *ArcUser*.

Mitas, L., & Mitasova, H. (1988). General variational approach to the interpolation problem. In *Computers & Mathematics with Applications*, 16(12), (pp. 983-992).

Peng, W., Petrovic, D., Crawford, C. (2004). Handling large terrain data in GIS. In *International Archives of the Photogrammetry*, Remote Sensing, and Spatial Information Sciences, 35(B4), (pp. 281-286).

Peucker, T. K., Fowler, R. J., Little, J. J., Mark, D. M. (1978). The triangulated irregular network. In *Proceedings of the Digital Terrain Models (DTM) Symposium*, (pp. 516-540). Falls Church, VA: American Society for Photogrammetry and Remote Sensing.

Poppenga, S. K., Worstell, B. B., Stoker, J. M., Greenlee, S. K. (2010). Using selective drainage methods to extract continuous surface flow from 1-meter lidar-derived digital elevation data. In *U.S. Geological Survey Scientific Investigations Report 2010-5059*.

Sibson, R. (1981). A brief description of natural neighbor interpolation. In V. Barnett (Ed.), In *Interpolating Multivariate Data* (pp. 21-36). New York, NY: John Wiley & Sons.

Sloman, A. (1983). Image interpretation: The way ahead? In *Physical and Biological Processing of Images*, O. J. Braddick, et. al. (Eds.), (pp. 380-401).

Tarboton, D. G. (1997). A new method for the determination of flow directions and contributing areas in grid digital elevation models. In *Water Resources Research*, 33(2), (pp. 309-319).

USACE (2015). In *Engineer manual 1110-1-1000, Photogrammetric and lidar mapping. U.S. Army Corps of Engineers.*

Watson, D. (1992). *Contouring: a guide to the analysis and display of spatial data*. Oxford, England: Pergamon Press.

Weibel, R., & Heller, M. (1991). Digital terrain modeling. In *Geographic Information Systems: Principles and Applications*, Vol. 1, D. Maguire, et. al. (Eds.), (pp. 269-297). Essex, England: Longman Scientific.

Wysocki, D. A., Schoeneberger, P. J., LaGarry, H. E. (2005). *Soil surveys: A window to the subsurface*. In Geoderma, 126(1-2), (pp. 167-180).

Chapter 2

VERTICAL DATUMS

Dru Smith

INTRODUCTION

The meaning of the word "elevation" in the term Digital Elevation Model (DEM) can be easily misunderstood. Technically, the correct meaning is "height." There are several types of heights, though four are the most commonly encountered in geospatial applications—ellipsoid, orthometric, normal, and dynamic. The definitions of these various types of heights are explained later in this chapter. Furthermore, the vertical datum to which the heights in a DEM refer must also be known. A DEM that does not come with metadata explicitly stating both the type of height and the vertical datum is of significantly limited use.

What is a Vertical Datum?

Strictly speaking a vertical datum is a surface representing zero height However, the term is frequently used to mean a system for the determination of some form of height relative to a zero height surface. It is this latter usage which will be adopted herein. There have been numerous national vertical datums and hundreds of local vertical datums established in the U.S. during the past century, and even more internationally. Users of vertical datums should understand and know the differences between each.

All vertical datums have two components—their definition and their realization. The definition of a vertical datum is the idealized description of the vertical datum. The realization is the reality of the vertical datum in practice. Take for example, the North American Vertical Datum of 1988:

Definition: The surface of equal gravity potential energy to which orthometric heights shall refer in North America, and which is 6.271 meters (along the plumb line) below the geodetic mark at "Father Point/ Rimouski" (NGSIDB PID TY5255).[1]

Realization: Over 500,000 geodetic marks across North America with published Helmert orthometric heights, most of which were originally computed from a minimally constrained adjustment of leveling and gravity data, holding the geopotential value at "Father Point/Rimouski" fixed.

One way to think of it is this—the definition is what one hopes to provide while the realization is what one truly does provide.

It is important that DEM users know both to which datum their data refer and what types of heights are in the DEM. Differences between vertical datums can exceed tens of meters. Differences between different types of heights can also exceed tens of meters. Therefore, combining or mixing vertical data products (either using multiple vertical datums or using multiple types of heights, or both) can result in large errors and cause problems for their users—chiefly modelers and planners. This is especially true in the last 30 years, as new technologies such as the Global Positioning Systems (GPS), and airborne Light Detection and Ranging (lidar) have given users access to ellipsoid heights, a fairly non-traditional type of height, historically speaking.

The next few sections will outline important factors to consider in understanding vertical datums and heights. First, an overview of possible reference surfaces which are used in vertical datums will be given. Then the various heights will be defined. Finally, the actual vertical datums of the U.S. and the world will be discussed. This chapter will then close on how to transform DEM data between the various types of heights, vertical datums or both.

[1] This geodetic mark is also the primary tidal benchmark at the Pointe-au-Pere ("Father Point") tide station in Rimouski, Quebec, Canada, and the numerical value of 6.271 meters chosen as its NAVD 88 height is identical to the LMSL height at this mark, as measured at the tide station.

REFERENCE SURFACES

Sea Level

The term "height above mean sea level" has origins that go back at least a few centuries. The ocean (and any other large body of water) seems to be a very natural reference surface to which humans prefer to measure things. But with tides, currents, waves and other phenomenon affecting the water/atmosphere boundary, the definition of "mean sea level" should be carefully described. To do so requires a careful understanding of tides and non-tidal water level variations.

Ocean Tides

The word "tide" (more specifically "ocean tide") is a generic term used to define the alternating rise and fall of the oceans with respect to the land, produced by differential variations in the gravitational attraction of the moon and sun. Non-astronomical factors such as configuration of the coastline, local depth of the water, ocean-floor topography, and other hydrographic and meteorological influences play an important role in determining the range of tide, delay times of the tide, and the time interval between high and low waters. Although there are many different temporal cycles affecting the tides, the largest tend to produce a cycle of two low tides and two high tides spanning a little over 24 hours (called a "Tidal Day"). See Figure 2.1. In general, the two low tides in one tidal day will not be of an identical magnitude. The lower of the two is called "lower low water" and the higher of the two is called "higher low water." Similarly for the two high tides, which are "higher high water" and "lower high water." Other cycles affect the rise and fall of the sea surface as well. The 1st and 2nd editions of this manual showed characteristic tide curves near port facilities along the U.S. East, West, and Gulf Coasts; those tide curves are not repeated in the 3rd edition in order to contain costs.

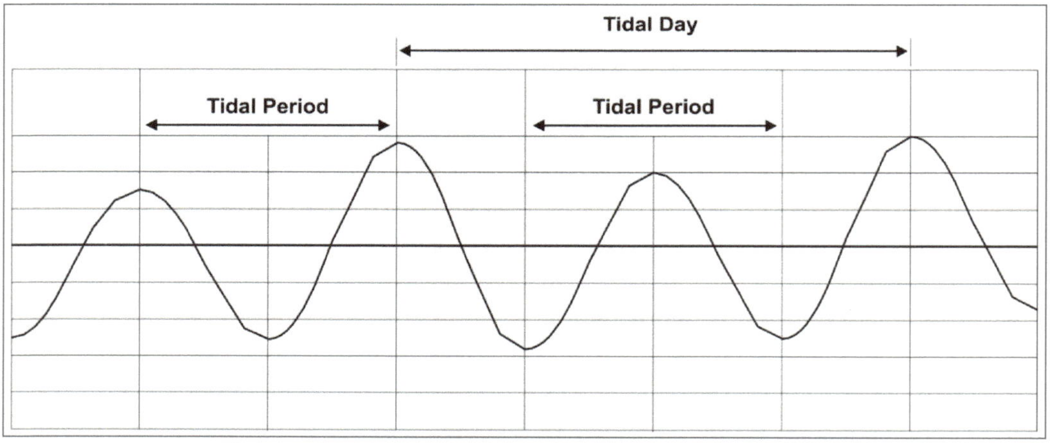

Figure 2.1. Example of a semi-diurnal tide with two high tides and two low tides per tidal day.

Tide gauges are instruments which record these changes in the sea surface relative to some reference mark of the instrument. Over time, averaging up certain parts of the tides will yield a mean value which represents the height of a surface relative to the tide gauge's reference mark. For instance, averaging up all of the "higher high water" values will yield the height of the "mean higher high water" surface relative to the tide gauges reference mark. Averaging up both high water marks (both the higher high water and the lower high water) will yield the "mean high water" location relative to the tide gauge's reference mark. Averaging up all of the sea surface values themselves will simply yield "mean sea level."

However, as tides are very localized, the mean sea level at any given tide gauge is often called "local mean sea level" and reflects only that surface of "mean sea level" surrounding the tide gauge, out to a distance at which the physical attributes of the sea surface no longer behave as they do at the tide gauge.
In this way at a tide gauge, a map or survey may reference heights relative to the reference mark of the tide gauge, and thus to any local "mean" surface generated at that tide gauge. Examples of such mean surfaces (their definitions can be inferred from their names) are (see also Figure 2.2):

- Mean Higher High Water
- Mean Lower High Water
- Mean High Water
- Mean Sea Level
- Mean Low Water
- Mean Higher Low Water
- Mean Lower Low Water

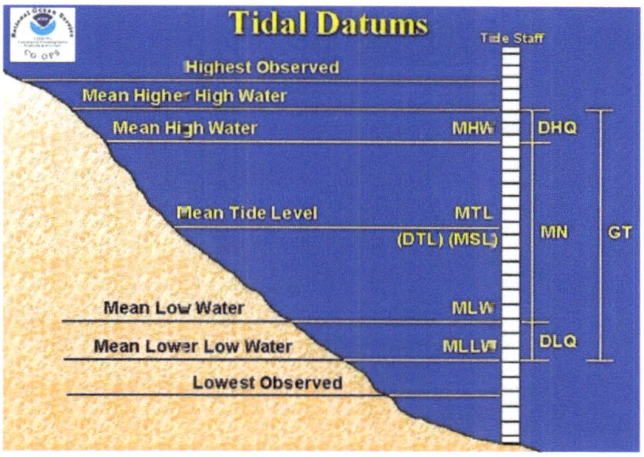

Figure 2.2. The relationship between various tidal surfaces.

As mentioned previously, tides are localized. When one says "Mean Higher High Water," what is more specifically meant is: "That surface surrounding this particular tide gauge where the tides behave identically as at the tide gauge, and which represents the average of all the "higher high water" locations over the time being averaged." Note that two details must still be addressed: Where is the surface valid and over what time is the averaging being done.

First, the "mean surface" is a mean of the ocean surface, and thus only exists over the ocean. Inland, there is no ocean and thus no tides. Mean Sea Level is something that exists only over the oceans. Furthermore, the sort of Local Mean Sea Level taken at a tide gauge is only valid over the oceans and near enough to the tide gauge that the tides behave as they do at the tide gauge. When one gets far enough from a tide gauge that the tides begin to change behavior, this is called a transition zone. It is important to know the location of these transition zones because they limit how far tidal datum computation procedures can be applied successfully. Note therefore that it is, strictly speaking, impossible to truly give a "height above mean sea level" to a point on the land, as there is no "mean sea level" surface directly below any non-oceanic points (This point will be revisited later in the section pertaining to the geoid.). However this has not prevented the colloquial application of the term "height above mean sea level" to land points, despite its effective meaninglessness. It is not, however, incorrect to think of "depths below sea level." In fact, the bathymetry of many nautical charts are perfectly correct in referencing "depths below mean lower low water" for instance.

The question of how long the averaging is done varies by the application. There is no "right" answer to this question. But if the intent is to remove as many of the gravitational fluctuations in the tides as possible (within human limitations), then averages are taken over an 18.6 year time span known as a "tidal epoch." In the U.S., a 19 year period called the "National Tidal Datum Epoch" (NTDE) is used to encapsulate this 18.6 year span. Data from continuously operating water level stations in the National Water Level Observation Network (NWLON) provide the basis for maintaining and updating the NTDE.

The primary point of this primer on tides is this: "Sea Level" has been used colloquially, though incorrectly, as a reference surface for land heights for a very long time. The proper use of "sea level" as a land height reference surface is to actually use the geoid (see later section). However, that issue aside, any reference to "mean sea level" should be clear about what kind of averaging is being done and over how long of a time span.

For additional information, some of the basic references for tides are Hicks, 1985; Hicks, 1989 International Oceanographic Commission, 2000; Marmer, 1951; Mero, 1998; National Ocean Service, 1999; Schultz and Scherer, 1999; Swanson, 1974; and U.S. Department of Commerce, 1970.

Non-Tidal Variations of Sea Level

Tides are not the only factor causing the sea surface height to change. Additional factors include waves and wave setup; ocean and river currents; ocean eddies; temperature and salinity variation of the ocean water; wind; barometric pressure; seiches; and relative sea level change.

In the U.S., the National Ocean Service (NOS), which runs the NWLON network, does not separate non-tidal sea level fluctuations from the tides for computation of tidal datums. As such, mean sea level, for instance, is truly the mean of sea level due to tides and non-tidal variations and so no further discussion about separating tidal from non-tidal issues will be presented.

The Geoid

For reasons mentioned in the previous section, it is incorrect to consider "heights above sea level" on the land, since the surface of the sea does not extend below the land. Sea level thus cannot provide a reference surface from which we would measure the height of a land point. However, it so happens that there is a surface, very closely related to sea level, with a global span extending across the oceans and under the land masses and is thus well-suited for being the height reference surface for both oceanic and land points. This surface is the geoid. To discuss the geoid, it will be worthwhile to discuss gravity and level surfaces.

Level Surfaces

Picture a bathtub being filled half-way up with water. While the faucet is running, the surface of the water is disturbed. But once that external force is no longer acting on the water, and the only thing influencing the water is the pull of gravity, then the boundary between the water and the air will coincide with what is called a "level surface." It is a fact of physics that when water settles down like this, that the gravity potential energy everywhere on a level surface happens to be constant. Such a surface of constant gravity potential energy, or level surface, is also known as an "equipotential surface." Although that surface looks flat, it is not. It is slightly rounded (due to the curvature of the Earth) and slightly undulating (due to the various differences in mass surrounding the tub). At bathtub scale those effects are not visible to the naked eye. The importance of this relationship between the air/water boundary and the equipotential surface with which it coincides is this: that equipotential surface does not stop at the edge of the tub. It is part of Earth's gravity field, and actually extends globally, forming a closed figure. There are an infinite number of such equipotential surfaces, each one fully inside the next larger one, and every time more water is put into (or taken out of) that tub, its air/water boundary will eventually settle down and coincide with a new equipotential surface.

The Geoid as a Level Surface

Now expand that bathtub until it is the size of the world's oceans, and visualize the ocean waters settling down without tides or waves or anything except Earth's gravity field affecting it. The air/water boundary would eventually settle into a level, or equipotential, surface as before. Only this time it settles to a very special surface, called the geoid. The geoid is not, strictly speaking, "global mean sea level" (GMSL) because GMSL contains the average of all tidal and non-tidal effects, and there is no guarantee that such effects average

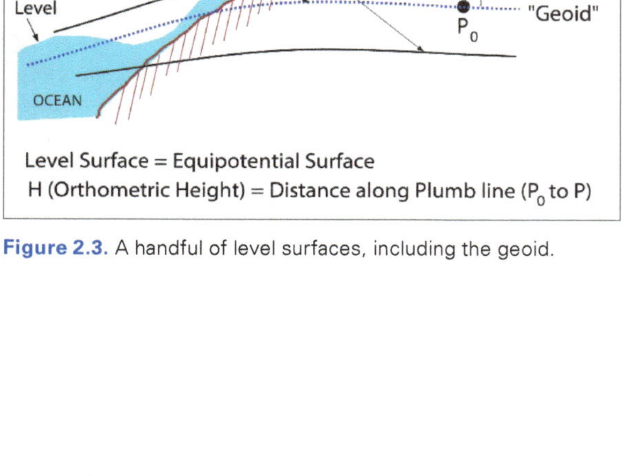

Figure 2.3. A handful of level surfaces, including the geoid.

out to zero. But, GMSL and the geoid are related. Specifically, and by conventional definition, the geoid is that surface of equal gravity potential energy which best fits global mean sea level. See Figure 2.3.

Because the geoid is both closely associated with sea level (which has the intuitive and historic appeal mentioned in the previous section) and because it extends away from the oceans ("cutting" through land masses below the Earth's surface), it is one of the most frequently used surfaces for referencing heights over land.

The Ellipsoid

To a first order approximation, the Earth is a sphere. However, that approximation was challenged in the 1600s and experimental evidence ultimately showed that the Earth is shaped more like an oblate ellipsoid (i.e., "flattened at the poles"). Although not as simple as a sphere, an oblate ellipsoid is a well understood mathematical shape requiring only two values (one for size, one for amount of flattening of the poles) to be fully described. One could fit ever more complicated shapes to the Earth's surface (triaxial ellipsoids, for instance) but the simple elegance of the oblate ellipsoid has had centuries of use and further complexity has not been necessary for most surveying and mapping applications.

Although ellipsoids have been used in surveying and mapping for centuries, they were primarily a tool for understanding latitude and longitude, and did not see significant usage as a height reference surface until the 1980s. This goes back to a historical difference between how latitude and longitude were determined versus how heights were determined. Positioning with GPS changed much of that. With GPS positioning, geocentric Cartesian coordinates (e.g., XYZ) are obtained, and these can quickly and simply be converted to latitude, longitude and ellipsoid height by using a simple oblate ellipsoid. See Figures 2.4 and 2.5.

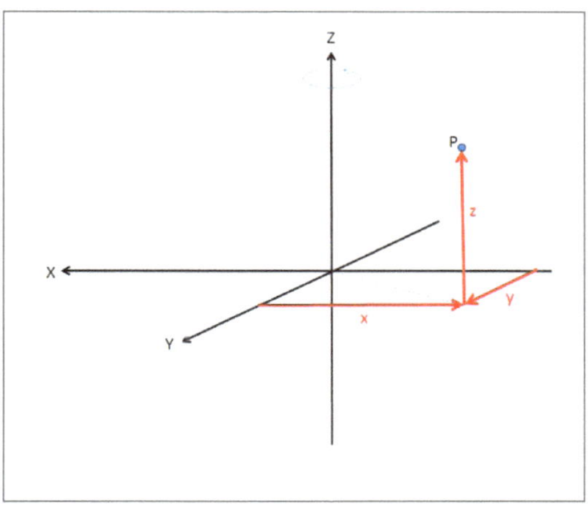

Figure 2.4. Point "P" as described using 3-D Cartesian coordinates.

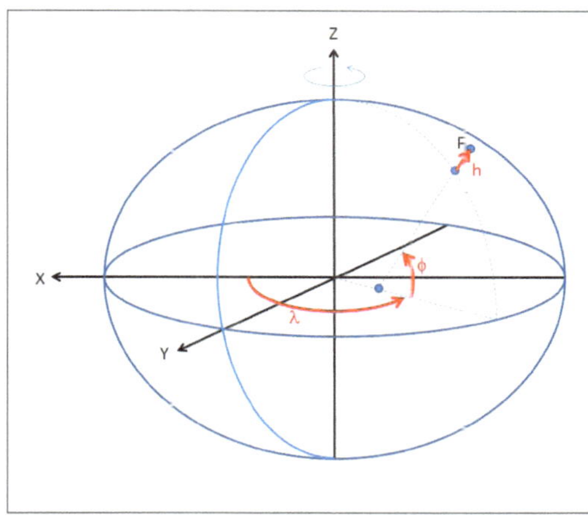

Figure 2.5. The same point "P" as described using ellipsoidal coordinates (ϕ is latitude, λ is longitude and h is ellipsoid height).

HEIGHTS

Height is frequently, but not always, (a) Some length, measured along (b) Some path from (c) Some starting location to (d) Some final location in (e) Some specified "up" direction. This definition may seem a bit complicated, but it is necessary when dealing with all of the types of heights which may be found in DEMs. To break it down to a simple example: People often ask "how tall are you?" This question could be interpreted as "what is your height?" and thus an answer, in accordance with the complicated definition above, might be:

My height is:

a. +6 feet as measured along
b. A straight line from
c. The bottom of my feet to
d. The top of my head
e. Where "up" is the direction from my feet to my head

However, the more usual answer would just be "I'm 6 feet tall" and all of the metadata in the answer would be assumed, but not explicitly stated. Such assumptions should not be made when discussing geospatial data. Asking "how tall is Mt. Everest" requires that every one of these pieces of metadata be clearly defined. Otherwise, the answer can be very easily misunderstood and used in an incorrect way.

As mentioned earlier, there are four primary types of heights encountered in DEMs, though there are many others. Each of the four is outlined below, including their definition, what sort of vertical reference surface is used, and where, geographically, such heights are commonly used.

Orthometric Height

Historically, but incorrectly, called a "height above sea level," the orthometric height of a point of interest is defined (using the full definition from earlier) as:

a. A distance
b. Measured along the plumb line
c. From the geoid
d. To the point of interest
e. Where "up" is the direction along the plumb line pointed away from the center of the Earth

The "plumb line" is a line that always points in the direction of gravity. Alternately, but equally valid, it is the line that always is perpendicular to every level surface through which it passes. The reference surface for an orthometric height is the geoid.

Refer back to Figure 2.3 on page 44. The orthometric height in that figure is the distance along the curved "plumb line" from the geoid (at point P0) to the point of interest (P). Obviously, nobody pulls a tape measure down a hole along a curved plumb line to determine this height. Instead, using various surveying methodologies, the orthometric height is approximated in a variety of ways. The two most commonly encountered are the Helmert orthometric height and the normal orthometric height.

Helmert Orthometric Heights

Helmert orthometric heights are approximated from two types of measurements, taken at the surface of the Earth: geodetic (or "spirit") leveling, and gravimetry. Making use of certain assumptions about the relationship between gravity, density and level surfaces, the Helmert orthometric height is one of the most commonly used approximations to orthometric heights. Helmert orthometric heights are found in the North American Vertical Datum of 1988 (NAVD 88), as described later.

Normal Orthometric Heights

Normal orthometric heights are similar to Helmert orthometric heights, except they do not make use of actual gravity measurements. Rather, they rely on a very simple latitude-based gravity formula. This approximation is frequently found in areas where gravity measurements have not been historically available. Normal orthometric heights are found in many of the U.S.'s island vertical datums (see later) as well as the National Geodetic Vertical Datum of 1929 (NGVD 29), the predecessor to NAVD 88.

Ellipsoid Height

Prior to the rise of GPS, ellipsoid heights were an esoteric type of height known mostly to geodesists but not having much practical application. This has changed significantly in the GPS era. Note first that most GPS positioning tools work in a Cartesian (XYZ) system, centered at the center of the Earth's mass, but this fact is hidden from most end-users. Usually an ellipsoid is mathematically associated with that system (often, but not always, centered at the same origin as the XYZ system). This allows the XYZ coordinates to be transformed exactly into geodetic latitude, geodetic longitude, and an "ellipsoid height" (see Figure 2.4 and Figure 2.5 on page 45). Sticking to the formal height definition template, the ellipsoid height of a point of interest is:

a. A distance
b. Measured along an "ellipsoidal normal"
c. From the ellipsoid
d. To the point of interest
e. Where "up" is the direction along the ellipsoidal normal, pointed away from the center of the Earth

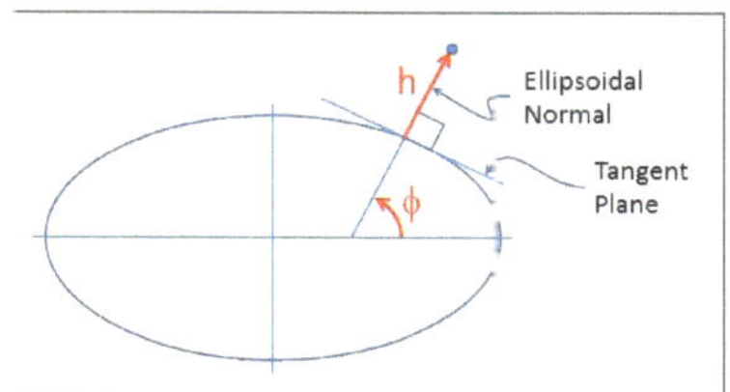

Figure 2.6. The geometry of an ellipsoid height (h).

The "ellipsoidal normal" is simply a straight line that is perpendicular to a plane which is tangent to the ellipsoid. The reference surface of an ellipsoid height is the chosen ellipsoid. "Chosen" ellipsoid is used because, unlike the geoid, which has a unique definition, any sized/shaped ellipsoid, positioned anywhere in the XYZ coordinate frame could be chosen to convert Cartesian coordinates into geodetic latitude, geodetic longitude and ellipsoid height. As such, the magnitude of the ellipsoid height of a point is entirely dependent upon which ellipsoid is chosen. Ellipsoid heights are found in the North American Datum of 1983 (NAD 83), as described later, despite the original intent of NAD 83 being a horizontal datum (i.e., not built to provide heights).

Figure 2.5 shows the ellipsoid height in the 3-D context of a Cartesian coordinate system. Figure 2.6 shows a cross-section of an ellipsoid to further clarify the ellipsoid height (h).

Dynamic Height

Dynamic heights are not lengths. However, dynamic heights are, in fact, reported in length units (meters, for example), but this is only because they have been purposefully scaled that way. Dynamic heights are not generally found in DEMs, but it is possible.

Dynamic heights are scaled versions of gravity potential energy. Recall the relationship between gravity potential energy and any given level surface from earlier: "The gravity potential energy everywhere on a level surface happens to be constant." Because humans seem to more intuitively understand lengths (e.g., "meters") than gravity potential energy units (e.g., "m^2/s^2"), the dynamic height was invented. It is merely the value of gravity potential energy, divided by (i.e., "scaled by") some constant value of gravity (in m/s^2), so that what is left are units of meters.

Now consider the important work of monitoring the level of large bodies of water, such as the Great Lakes. In the absence of wind and other forces working on the lake surfaces, one can state that the value of gravity potential energy everywhere on a lake surface should be constant. Or, in terms actually used in the Great Lakes, one can say with equal validity: "The dynamic height everywhere on a lake surface should be constant." Unlike orthometric heights or ellipsoid heights which have no physical reason to be constant over the surface of a body of water, the dynamic height, by definition, must be constant.

Strictly speaking, there is no requirement for a "reference surface" for dynamic heights, as they are not lengths measured from a surface to a point. However, for practical reasons (not the least of which is providing reasonable magnitudes), dynamic heights are often scaled versions of the difference between gravity potential energy at a point of interest and gravity potential energy of the geoid. This does not make them lengths relative to the geoid! It merely helps put dynamic heights into the same general magnitude as orthometric heights.

Figure 2.7 shows a simplified view of the relationship of ellipsoid, orthometric and dynamic heights for a few points on the surface of a body of water.

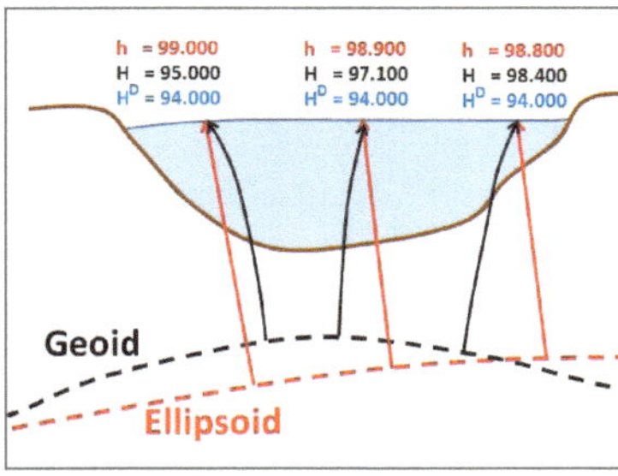

Figure 2.7. Ellipsoid heights (h, red), orthometric heights (H, black) and dynamic heights (H^D, blue, no line shown) for three points across a body of water. Note that only the dynamic heights are constant.

Normal Height

Normal heights, like orthometric heights, are related to the level surfaces of Earth's gravity field. They were originally described in the work of Mikhail Molodenskii. Seen as a way to access heights without all of the approximations of the orthometric height, normal heights became a popular field of study in the 1960s and many European and Asian countries have adopted normal heights for use in their official vertical datums. They are generally not used in any vertical datums in the U.S. (and thus are not found in U.S. DEMs). Normal heights use the "quasi geoid" as a reference surface, which is a projection of the telluroid upon the ellipsoid. Definitions of the quasi geoid and telluroid are beyond the scope of this book.

Transformations between orthometric heights and normal heights are complicated, but they are not large. In general, orthometric and normal heights agree within 1-2 meters. Formal transformations between the two, however, require a deeper geodetic application than this book may adequately present. If working with DEMs in the U.S., one is unlikely to encounter normal heights.

VERTICAL DATUMS

There are a variety of vertical datums in use in the world, and it would be impossible to address them all. Even in the U.S. there are too many datums to properly document herein. This is because many local-use datums have been established in small geographic regions. Therefore this chapter restricts itself to the largest, most common datums first, with some words about the regional datums at the end.

Readers interested in the history of vertical datums or the details about their creation are referred to the 2nd edition of this DEM Users Manual.

NAVD 88

The North American Vertical Datum of 1988 (NAVD 88) was made the official vertical datum of the National Spatial Reference System (NSRS) in 1991. This means that all federal civilian agencies using geospatial data are required to work in NAVD 88 when working with heights. The NAVD 88 contains Helmert orthometric heights (see above). Most DEMs created in the 1990s or later in CONUS or Alaska will be in NAVD 88.

NAVD 88 was created through short line-of-sight leveling surveys spanning across Alaska, Canada, CONUS and Mexico. As such, it is impossible for NAVD 88 to exist outside those four regions. Any DEM purporting to be on NAVD 88 in any region besides these four is absolutely incorrect.

At the time of its adoption, NAVD 88 replaced the then-official vertical datum, NGVD 29 (National Geodetic Vertical Datum of 1929). The new NAVD 88 datum was far superior to NGVD 29 in both accuracy and availability. Specifically, heights were more accurately calculated in NAVD 88 (using better data and less approximations than NGVD 29) and there were hundreds of thousands of more NAVD 88 marks available across the country than NGVD 29 had.

As a special note, Canada never formally adopted NAVD 88, but rather continued with their own Canadian Geodetic Vertical Datum of 1928 (CGVD 28), so it is unlikely (but not impossible) for a DEM to be on NAVD 88 in Canada.

NGVD 29

The National Geodetic Vertical Datum of 1929 (NGVD 29) was replaced in 1991 by NAVD 88. Originally called the "Sea Level Datum of 1929" (because it was created by holding mean sea level as a zero height at 26 tide gauges), the name NGVD 29 was used later as the Coast and Geodetic Survey (C&GS, predecessor to the current National Geodetic Survey or NGS) learned more about the sea level/geoid differences varied from tide gauge to tide gauge. Most DEMs created prior to 1991 in the U.S. will be in NGVD 29.

NGVD 29 was computed for Alaska and CONUS, though leveling could have been expanded into Canada or Mexico after NGVD 29 was originally created. It is impossible for NGVD 29 to exist outside of CONUS, Alaska, Canada or Mexico. Any DEM purporting to be on NGVD 29 in any region besides these four, is absolutely incorrect.

PRVD02, ASVD02, NMVD03, GUVD04, VIVD09

As mentioned in the section about NAVD 88, the historic method by which a vertical datum is defined and accessed is line-of-sight geodetic leveling and it is therefore impossible to use this method to expand a vertical datum across large stretches of water. Therefore, the six primary island-based states or territories of the U.S. require their own vertical datums, distinct from NAVD 88. Of the six (Hawaii being the exception), each has its own datum as follows:

- PRVD02: Puerto Rico Vertical Datum of 2002
- ASVD02: American Samoa Vertical Datum of 2002
- NMVD03: Northern Marianas Vertical Datum of 2003
- GUVD04: Guam Vertical Datum of 2004
- VIVD09: Virgin Islands Vertical Datum of 2009

Hawaii does not have an official statewide vertical datum as of this edition of this book.

Each of these datums provide normal orthometric heights, due to the general lack of available measured gravity values in each territory.

IGLD 85

The International Great Lakes Datum of 1985 (IGLD 85) is the official dynamic height system of the Great Lakes for both the U.S. (as part of the NSRS) and Canada. Because the surveys used for determining Helmert orthometric heights (see NAVD 88) and dynamic heights are identical, the NAVD 88 and IGLD 85 datums were co-defined using the same data in one project. IGLD 85 contains dynamic heights and is only valid in the immediate region of the Great Lakes.

Tidal Datums (Local Mean Sea Level, Mean Lower Low Water, etc.)

At any given tide gauge, and within its immediate area, a tidal datum representing any of the averaged surfaces mentioned earlier (such as "Mean Lower Low Water" or "Mean Sea Level") may be created. While strictly only valid over the ocean areas themselves, such a datum can, theoretically, be extended inland slightly for the purposes of determining flood risks around the tide gauge, for instance. But the most frequent use of such a tidal datum is for representing bathymetric data (i.e., "heights downward from the reference surface") for nautical charts in the oceanic areas.

Other Vertical Datums

Hundreds of vertical datums have been established in the U.S. alone, where individual cities, counties, river valleys or other specific regions have had a need for a localized system. For the most part, such datums are not related to national datums and converting a DEM from one of these local datums into something consistent with a national vertical datum is not easy.

HORIZONTAL DATUMS

Similarly, there are a variety of horizontal datums in use in the world, and it would be impossible to address them all.

NAD 83

The North American Datum of 1983 (NAD 83) was made the official horizontal datum of the NSRS in 1986. At that time, however, its purpose was horizontal positioning, and there were effectively no usable heights as part of the datum. Almost immediately after the release of NAD 83, GPS became a commonly used tool for horizontal positioning. So surveys and maps which used GPS to get into NAD 83 had an almost "side product" of ellipsoid heights that were determined. Years of research were dedicated to improving the height component determinable from GPS. As such, by the mid 1990s, it was common enough to purposefully use GPS to determine ellipsoid heights. Any heights referenced to NAD 83 will be ellipsoid heights.

There have been many "adjustments" to the original NAD 83. At first, these were horizontal corrections as determined from GPS surveys in the early 1990s (known as the High Accuracy Reference Network, or HARN surveys). Later, in the late 1990s, there were attempts to specifically determine ellipsoid heights using GPS (these surveys were known as the Federal Base Network, or FBN, surveys). Both the HARN and FBN surveys modified NAD 83 in patches (usually one state or small group of states at a time). This led to some state-by-state issues which were finally resolved in a 2007 adjustment of the entire network, followed by another adjustment in 2011.

Finally, it should be noted that, despite the words "North American" in its name, there was an expansion of NAD 83 into the Pacific tectonic plate region (for Hawaii and American Samoa) and the Marianas tectonic plate region (for Guam and CNMI).

WGS-84

The original World Geodetic System of 1984 (WGS-84) was a global horizontal datum, and had much in common with the original NAD 83. Over the following decades, WGS-84 changed in two ways. First, it gradually corrected a non-geocentricity which was allowed to remain in NAD 83 (allowing the two ellipsoids to drift away from one another) and secondly, it eventually included an associated geoid model accurate enough to use for height determination. First in 1996, the Earth Gravitational Model EGM96 geoid model was released with WGS-84. Later, in 2008, the EGM2008 geoid model became the official model of the geoid for WGS-84. A DEM purely in WGS-84 should contain ellipsoid heights, while one in WGS-84/EGM96 or WGS-84/EGM2008 should contain orthometric heights. It is currently defined and maintained by the National Geospatial-Intelligence Agency (NGA).

CONVERTING HEIGHTS AND/OR VERTICAL DATUMS

Not every DEM will contain the data that is right for a particular application. In such cases, it is necessary to convert the DEM, either by changing its type of height, its vertical datum, or both. The most commonly encountered transformations are described below.

Ellipsoid/Orthometric: Geoid Undulations

The most frequently encountered transformation involves changing between ellipsoid heights and orthometric heights. In general, this is accomplished using a model of geoid undulations. A geoid undulation is simply the distance of the geoid above the chosen ellipsoid. See Figure 2.8. Although the geoid is shown above the ellipsoid in that figure, the geoid can lie above or below the ellipsoid. In fact, in the U.S., the geoid is almost always below the ellipsoid. This means the geoid undulation (N) would be negative. Applying a negative geoid undulation in the H = h - N formula is perfectly acceptable, provided the algebraic signs are applied correctly. In the case where the geoid undulation is negative, the orthometric height will be larger than the ellipsoid height. Thus:

$$H > h \text{ when } N < 0 \text{ (in the U.S., generally)}$$
$$\text{and } H < h \text{ when } N > 0.$$

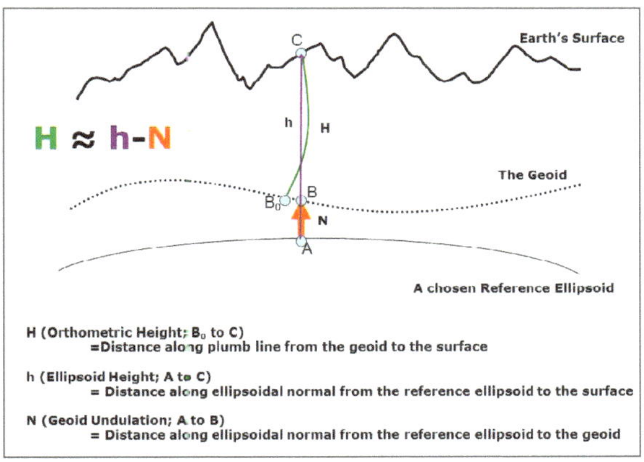

Figure 2.8. The relationship between ellipsoid heights (h, purple), orthometric heights (H, green) and geoid undulations (N, orange).

The geoid undulation forms the linear relationship H ≈ h - N as seen in Figure 2.8 where the curvature in the plumb line segment H is exaggerated. Although the curvature of the plumb line means this relationship is approximate, that approximation is good to 1 mm everywhere in the U.S.

This idealized situation, however, is made difficult by the fact that Helmert orthometric heights in NAVD 88 do not perfectly refer to the geoid, but rather have been found to refer to a surface biased and tilted from the geoid. Nor are the ellipsoid heights of NAD 83 relative to a perfectly geocentric ellipsoid, but one rather off center

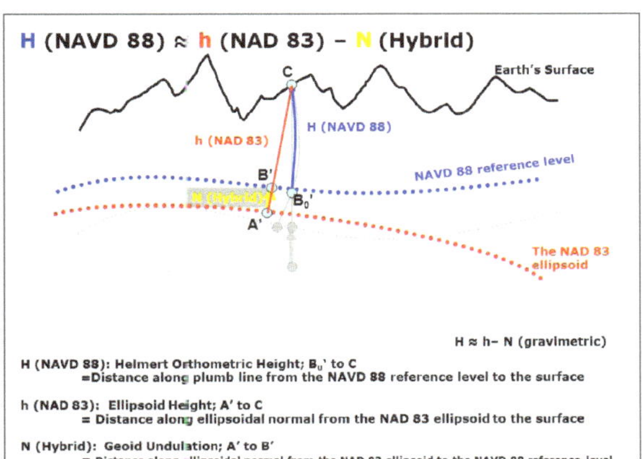

Figure 2.9. The relationship between NAD 83 ellipsoid heights (h, red), NAVD 88 orthometric heights (H, blue) and hybrid geoid undulations (N, yellow). For comparison, these are overlain (gray) against the idealized geometry represented in Figure 2.8.

by over 2 meters. Because of these two differences, the National Geodetic Survey builds hybrid geoid models which specifically transform ellipsoid heights in NAD 83 into orthometric heights in NAVD 88. See Figure 2.9 on page 51 and compare to Figure 2.8.

As of 2016, the current hybrid geoid model is GEOID12B.

Transformation Tools

Most of the tools needed to transform DEMs between height types and/or vertical datums are made available to the public by the National Geodetic Survey (NGS) at their website (*geodesy.noaa.gov*). Use the table below to find the tools applicable to your individual needs. URLs are not provided for individual tools, as web page trees change over time; however, to find these tools, go to *geodesy.noaa.gov* and enter the tool name in the search engine.

Height Type	Datum	Height Type	Datum	Tool
Ellipsoid	NAD 83*	Ellipsoid	NAD 83**	GEOCON v2.0 (or later)
Ellipsoid	NAD 83*	Orthometric	NAVD 88	GEOID12B† (or later)
Ellipsoid	NAD 83*	Tidal	Any	VDATUM
Ellipsoid	NAD 83*	Ellipsoid	WGS-84	HTDP†
Ellipsoid	NAD 83*	Orthometric	NGVD 29	GEOID12B† (or later) and VERTCON 2.0† (or later)
Orthometric	NAVD 88	Orthometric	NGVD 29	VERTCON v2.0† (or later)
Orthometric	NAVD 88	Dynamic	IGLD 85	DYNAMIC_HT†
Orthometric	NAVD 88	Tidal	Any	VDATUM
Orthometric	NAVD 88	Orthometric	WGS-84/EGM96	***
Orthometric	NAVD 88	Orthometric	WGS-84/EGM2008	***
Ellipsoid	NAD 83*	Orthometric	ASVD02 or NMVD03 or GUVD04 or VIVD 09	GEOID12B† (or later)

* = some HARN or FBN year, or NSRS2007 or 2011
** = some HARN or FBN year, or NSRS2007 or 2011 ("*" should not equal "**")
*** = Because WGS-84 and EGM96 and EGM2008 are not controlled by NGS, they do not provide a transformation tool to these datums
† Each of these tools is also found in the NOAA tool "VDATUM"

FUTURE VERTICAL DATUMS

Geodetic leveling, supporting the computation of approximated orthometric heights at passive geodetic control marks was the way that vertical datums have been defined and accessed for centuries. The advent of GPS, especially its speed and accuracy as a surveying tool, as well as the improved accuracy of geoid undulation models has driven an intense scrutiny of this approach toward vertical datums in the last ten years. In some parts of the world this scrutiny has yielded a radical shift in vertical datum approach and soon (2022) the U.S. will follow suit.

Beginning in 2022, all vertical datums of the civilian federal government, as defined by the National Geodetic Survey, will be replaced by one "geopotential datum," which will include a single geoid model which serves as the definitive zero height surface for all areas of the U.S. This means that orthometric heights will be primarily accessed through GPS receivers. For this to work accurately, a geoid undulation model of unprecedented accuracy must be built. For this purpose, NGS has engaged in the GRAV-D (Gravity for the Redefinition of the American Vertical Datum) project. The GRAV-D project is an airborne gravity survey campaign running from 2007 until 2021, as well as the establishment of a geoid monitoring service (to introduce continental scale time dependencies in the gravity field). Once GRAV-D is complete, a new global geopotential model will be built, incorporating all of the GRAV-D surveys and the latest geopotential modeling theory. From that, three gridded geoid undulation models will be computed. The first will encompass North America (including all of Alaska, Greenland, Canada, CONUS, Mexico, Central America, the Caribbean, and Hawaii) and will serve as the new zero elevation surface for that region. A similar approach will be taken for the regions of American Samoa and Guam/CNMI.

The names for the replacements to NAD 83 are:

- North American Terrestrial Reference Frame of 2022 (NATRF2022)
- Pacific Terrestrial Reference Frame of 2022 (PATRF2022)
- Caribbean Terrestrial Reference Frame of 2022 (CATRF2022)
- Mariana Terrestrial Reference Frame of 2022 (MATRF2022)

Each of these four frames will be related to the ITRF through a model which describes the tectonic plate for which each frame is named. As such, they will have different time-varying latitude and longitude (which attempts to yield semi-stable horizontal coordinates through time for the rigid parts of each plate) but will yield identical ellipsoid heights.

All of the vertical datums in the NSRS, including NAVD 88, will be replaced with the North American-Pacific Geopotential Datum of 2022 (NAPGD2022). Then, the primary geodetic formula will be:

$$H_{NAPGD2022} = h_{TRF2022} - N$$

SUMMARY

Proper use of a DEM requires complete understanding of what kind of height it contains, and the vertical datum to which that height refers. Anything short of this minimum set of information makes it impossible to know how the DEM actually relates to the world it purports to model.

There are many types of heights and many datums in use in every country. In the U.S., the official vertical datum is that one defined by the National Geodetic Survey as part of the National Spatial Reference System (NSRS) and approved by the Federal Geodetic Control Subcommittee (FGCS) of the Federal Geographic Data Committee (FGDC). As of 2017, for CONUS and Alaska that vertical datum is NAVD 88 and contains Helmert orthometric heights (which are approximations of true orthometric heights). The island regions of Guam, CNMI, American Samoa, Puerto Rico, and the U.S. Virgin Islands all have their own vertical datums. These will all change in 2022.

Sea Level is an oceanic surface, and one that can be averaged in many ways. But those averages are only good in the local region around a given tide gauge and also are only meaningful over the oceans. Heights (or depths) related to "sea level" should be restricted to the sea.

GPS yields ellipsoid heights, but these differ from traditional orthometric heights. Use a geoid model (or, in the U.S., a hybrid geoid model) to transform them. Transformation tools exist to change the heights and/or datums of most DEMs in the U.S. These tools can be found at the NGS web page *geodesy.noaa.gov*.

AUTHOR BIOGRAPHY

Dr. Dru Smith was Chief Geodesist of NOAA's National Geodetic Survey from 2005 until 2015, and now serves as the NSRS Modernization Manager, responsible for overseeing the replacement of NAD 83 and NAVD 88. He first entered NGS in 1995 after receiving his Ph.D in geodetic science from The Ohio State University. He has published over 40 papers on research topics ranging from geodetic surveying to ionosphere determination to geoid modeling. He is a member of the American Geophysical Union, the International Association of Geodesy and is a Fellow of the American Congress on Surveying and Mapping (now the National Society of Professional Surveyors) and has previously served on the Board of Directors for the American Association for Geodetic Surveying. He has received the Bronze, Silver, and Gold medals from the Department of Commerce for outstanding federal service.

REFERENCES

Hicks, S.D. (1985). Tidal datums and their uses—a summary. In *Shore and Beach*, Volume 53, No. 1 (pp. 27-33). Fort Myers, FL: The American Shore & Beach Preservation Association.

Hicks, S.D. (1989). In *Tide and current glossary*. Silver Spring, MD: NOAA National Ocean Service.

International Oceanographic Commission (2000), In *Manual on sea level measurement and interpretation*, volume 3—reappraisals and recommendations as of the year 2000, Manuals and Guides 14. Paris, France: The United Nations Organization for Education, Science and Culture (UNESCO).

Marmer, H.A. (revised 1951). Tidal datum planes, NOAA National Ocean Service, Special Publication No. 135, *U.S. Coast and Geodetic Survey*. Washington, D.C.: U.S. Government. Printing Office.

Mero, T.N. (1998). NOAA/National Ocean Service application of real-time water levels. In *Proceedings, Volume 2*, Ocean Community Conference, the Marine Technology Society Annual Conference, November 16-19, 1998, (pp. 1036-1039).

National Ocean Service (1999). *NOS hydrographic surveys specifications and deliverables* (pp. 91). Silver Spring, MD: NOAA National Ocean Service.

Schultz, J.R. and W.D. Scherer (1999). *Tidal datums and their applications*, NOAA Technical Report NOS CO-OPS 1. Silver Spring, MD: NOAA Center for Operational Oceanographic Products and Services.

Swanson, R.L. (1974). Variability of tidal datums and accuracy in determining datums from short series of observations, NOAA Tech. Rep. NOS 64 (pp. 41). Silver Spring, MD, NOAA National Ocean Service.

U.S. Department of Commerce (1970). *Demarcating and mapping tidal boundaries*. Silver Spring, MD: NOAA Environmental Science Services Administration, Coast & Geodetic Survey

Chapter 3

STANDARDS, GUIDELINES, AND SPECIFICATIONS

David F. Maune

INTRODUCTION

Standards for geospatial data have broad applications nationally and/or internationally, whereas specifications provide technical requirements and/or acceptance criteria that a geospatial product must conform to in order to be considered acceptable for a specific intended use. Guidelines provide recommendations for acquiring, processing and/or analyzing geospatial data, normally intended to promote consistency and industry best practices. The common goal of each is to promote standardization, usability and interoperability of geospatial data so as to leverage resources to satisfy multiple requirements at minimum cost to taxpayers.

Standards are best when they include accuracy thresholds. The best examples are the legacy National Map Accuracy Standards (NMAS, 1947) and ASPRS Accuracy Standards for Large-Scale Maps (ASPRS 1990)—both designed for hardcopy maps—and the new ASPRS Positional Accuracy Standards for Digital Geospatial Data (ASPRS, 2014).

Guidelines are best when they promote geospatial data consistency and industry best practices. Good examples are the National Digital Elevation Program (NDEP) Guidelines for Digital Elevation Data, (NDEP, 2004), the NOAA Technical Memorandum NOS NGS-58 Guidelines for Establishing GPS-Derived Ellipsoid Heights (Standards: 2 cm and 5 cm) (NOAA, 1997) and the NOAA Technical Memorandum NOS NGS-59 Guidelines for Establishing GPS-Derived Orthometric Heights (NOAA, 2008).

Specifications are best when they clearly document requirements for standard geospatial products made by different producers but going into a common database that serves the public. Good examples are the NOS Hydrographic Surveys Specifications and Deliverables (NOAA, 2015) which documents requirements for hydrographic surveys funded by NOAA; the ASPRS LASer (LAS) 1.4 Format Specification (ASPRS, 2011) which documents requirements for all lidar file formats; and the USGS Lidar Base Specifications Version 1.3 (Heidemann, 2018) which documents all lidar requirements for the nationwide 3D Elevation Program (3DEP) explained in Chapter 5.

Chapter 3 differs from the 2nd edition of this manual in two fundamental ways: (1) It features the ASPRS Positional Accuracy Standards for Digital Geospatial Data (ASPRS, 2014) which supersedes most of the legacy accuracy standards and guidelines published previously, and (2) It also features the USGS Lidar Base Specification V1.3 (Heidemann, 2018), the current nationwide specification in the U.S. for Quality Level Ø (QLØ), Quality Level 1 (QL1) and QL2 lidar data for the 3DEP.

ASPRS strongly endorses the 3DEP which will: (1) Assure nationwide availability and accuracy of "best available" digital topographic data; (2) Enhance data sharing among federal, state and local agencies, the private sector, and academia; (3) Minimize redundant data production; (4) Leverage resources to satisfy multiple requirements; and (5) Enforce accuracy and quality standards common to all users, nationwide.

FEDERAL GEOGRAPHIC DATA COMMITTEE (FGDC) STANDARDS

In 1998, the FGDC published its Content Standard for Digital Geospatial Metadata, FGDC-STD-001-1998 (FGDC, 1998) to provide a common set of terminology and definitions for the documentation of digital geospatial data. The standard established the names and definitions of data elements to document and describe the geospatial data and information about the values that are to be provided for the data elements. The data elements pertaining to data quality, for example, includes horizontal and vertical accuracy, attribute accuracy, and lineage information about the data set including acquisition dates, processing dates, responsible parties and other information about events, parameters and source data.

Geospatial Positioning Accuracy Standards

Between 1998 and 2005, the FGDC published its Geospatial Positioning Accuracy Standards in support of the National Spatial Data Infrastructure (NSDI). These standards have five parts that are relevant to topographic or bathymetric DEMs:

- Part 1, Reporting Methodology, FGDC-STD-007.1-1998 (FGDC, 1998a), provides guidance on how to report the accuracy of horizontal and vertical coordinates at the 95-percent confidence levels. The reporting standard in the horizontal component is the radius of a circle of uncertainty, such that the true or theoretical location of the point falls within that circle 95-percent of the time; this is called $Accuracy_r$ in Part 3 or ACCr in other publications. The reporting standard in the vertical component is a linear uncertainty value, such that the true or theoretical location of the point falls within ± of that linear uncertainty value 95-percent of the time; this is called $Accuracy_z$ in Part 3 or ACC_z in other publications.
- Part 2, Standards for Geodetic Networks, FGDC-STD-007.2-1998 (FGDC, 1998b), provides guidance on how to utilize and supplement the National Spatial Reference System (NSRS) maintained by the National Geodetic Survey (NGS) to establish local accuracy and network accuracy of spatial data. The local accuracy of a control point is a value that represents the uncertainty in the coordinates of the control point relative to the coordinates of other directly connected, adjacent control points at the 95-percent confidence level. The network accuracy of a control point is a value that represents the uncertainty in the coordinates of the control point with respect to the geodetic datum at the 95-percent confidence level. For NSRS network accuracy classification, the datum is considered to be best expressed by the geodetic values at the Continuously Operating Reference Stations (CORS) supported by NGS. By this definition, the local and network accuracy values at CORS sites are considered to be infinitesimal, i.e., to approach zero."
- Part 3, National Standard for Spatial Data Accuracy (NSSDA), FGDC-STD-007.3-1998 (FGDC, 1998c), provides guidance on how to implement a testing and statistical methodology for positional accuracy of digital geospatial data. The NSSDA directed the geospatial community to use RMSEx, RMSEy, RMSEr and RMSEz and different RMSE-multipliers to compute horizontal and vertical accuracies at the 95-percent confidence levels, with the assumption that errors follow a normal distribution. However, the geospatial community subsequently learned that the assumed normal error distribution is not necessarily true for lidar datasets; therefore subsequent documents from ASPRS, USGS, FEMA and others directed the use of the 95th percentile errors for Vegetated Vertical Accuracy, for example, as explained below. The main complaint about the NSSDA is that it did not include accuracy thresholds, leaving users to establish their own accuracy thresholds. In 1990, the ASPRS Accuracy Standards for Large-Scale Maps (ASPRS, 1990) established accuracy thresholds for hardcopy maps, and in 2014, the ASPRS Positional Accuracy Standards for Digital Geospatial Data (ASPRS, 2014) established accuracy thresholds for digital orthoimagery, digital planimetric data, and digital elevation data
- Part 4, Standards for Architecture, Engineering, Construction (A/E/C) and Facility Management, FGDC-STD-007.4-2002 (FGDC, 2002), provides accuracy standards for engineering drawings, maps, and surveys

used to support planning, design, construction, operations, maintenance, and management of facilities, installations, structures, transportation systems, and related projects. The U.S. Army Corps of Engineers (USACE) is responsible for Part 4 and, in 2015, published USACE EM 1110-1-1000, Photogrammetric and LiDAR Mapping (USACE, 2015), with detailed standards, guidelines and specifications including accuracy thresholds for 63 common USACE user applications

- Part 5, Standards for Nautical Charting Hydrographic Surveys, FGDC-STD-007.5-2005 (FGDC, 2005), provides minimum standards for the horizontal and vertical accuracy of features associated with hydrographic surveys that support nautical charting, including water depths, objects on the seafloor, navigational aids, and shoreline. A major advantage of Part 5 is that it establishes accuracy thresholds, as shown in Table 3.1, based upon the International Hydrographic Organization Standard for Hydrographic Surveys as contained in Special Publication No. 44 (SP-44) (IHO, 2008)

Order	Special	1a	1b	2
Description of areas	Areas where under-keel clearance is critical	Areas shallower than 100 metres where under-keel clearance is less critical but features of concern to surface shipping may exist	Areas shallower than 100 metres where under-keel clearance is not considered to be an issue for the type of surface shipping expected to transit the area	Areas generally deeper than 100 metres where a general description of the sea floor is considered adequate
Maximum allowable THU 95% confidence level	2 metres	5 metres + 5% of depth	5 metres + 5% of depth	20 metres + 10% of depth
Maximum allowable TVU 95% confidence level	a = 0.25 metre b = 0.0075	a = 0.5 metre b = 0.013	a = 0.5 metre b = 0.013	a = 1.0 metre b = 0.023
Full sea floor search	Required	Required	Not required	Not required
Feature detection	Cubic features >1 metre	Cubic features >2 metres, in depths up to 40 metres; 10% of depth beyond 40 metres	Not applicable	Not applicable
Recommended maximum line spacing	Not defined as full sea floor search is required	Not defined as full sea floor search is required	3 x average depth or 25 metres, whichever is greater. For bathymetric lidar a spot spacing of 5 x 5 metres.	4 x average depth
Positioning of fixed aids to navigation and topography significant to navigation. (95% Confidence level)	2 metres	2 metres	2 metres	5 metres

Order	Special	1a	1b	2
Positioning of the coastline and topography less significant to navigation (95% Confidence level)	10 metres	20 metres	20 metres	20 metres
Mean position of floating aids to navigation (95% confidence level)	10 metres	10 metres	10 metres	20 metres

Table 3.1. Summary of Minimum Standards for Hydrographic Surveys.

Notes:

1: THU means total horizontal uncertainty, and TVU means total vertical uncertainty. Recognizing that there are both constant and depth dependent uncertainties that affect the uncertainty of the depths, the formula below is to be used to compute, at the 95% confidence level, the maximum allowable TVU. The parameters "*a*" and "*b*" for each Order, as given in the Table, together with the depth "d" have to be introduced into the formula in order to calculate the maximum allowable TVU for a specific depth: $\pm\sqrt{[a^2+(b*d)^2]}$

Where:

a represents that portion of the uncertainty that does not vary with depth
b is a coefficient which represents that portion of the uncertainty that varies with depth
d is the depth
b x d represents that portion of the uncertainty that varies with depth

2: For safety of navigation purposes, the use of an accurately specified mechanical sweep to guarantee a minimum safe clearance depth throughout an area may be considered sufficient for Special Order and Order 1a surveys

3: A cubic feature means a regular cube each side of which has the same length. It should be noted that the IHO Special Order and Order 1a feature detection requirements of 1 metre and 2 metre cubes respectively, are minimum requirements. In certain circumstances it may be deemed necessary by the hydrographic offices/organizations to detect smaller features to minimise the risk of undetected hazards to surface navigation. For Order 1a the relaxing of feature detection criteria at 40 metres reflects the maximum expected draught of vessels

4: The line spacing can be expanded if procedures for ensuring an adequate sounding density are used. "Maximum Line Spacing" is to be interpreted as the:

- Spacing of sounding lines for single beam echo sounders, or the
- Distance between the useable outer limits of swaths for swath systems

5: These only apply where such measurements are required for the survey

Geographic Information Framework Data Content Standards

In 2008, the FGDC also published a series of Geographic Information Framework Data Content Standards, with eight parts:

- Part 0, Base document, FGCD-STD-014.0-2008 (FGDC, 2008)
- Part 1, Cadastral, FGDC-STD-014.1-2008 (FGDC, 2008a)
- Part 2, Digital orthoimagery, FGDC-STD-014.2-2008 (FGDC, 2008b)

- Part 3, Elevation, FGDC-STD-014.3-2008 (FGDC, 2008c) including content standards for elevation points, elevation grids, contours, TINs, and profiles (also called breaklines cross sections or transects)
- Part 4, Geodetic control, FGDC-STD-014.4-2008 (FGDC, 2008d)
- Part 5, Governmental unit and other geographic area boundaries, FGDC-STD-014.5-2008 (FGDC, 2008e)
- Part 6, Hydrography, FGDC-STD-014.6-2008 (FGDC, 2008f)
- Part 7, Transportation, FGDC-STD-014.7-2008 (FGDC, 2008g) with base standards plus standards specific to air, rail, road, transit, and inland waterways

Those portions of Part 3 (Elevation) and Part 4 (Geodetic Control) most relevant to DEM users are included in ASPRS, 2014 and Heidemann, 2018, summarized below.

ASPRS POSITIONAL ACCURACY STANDARDS FOR DIGITAL GEOSPATIAL DATA

The ASPRS Positional Accuracy Standards for Digital Geospatial Data (ASPRS, 2014) replace the ASPRS Accuracy Standards for Large-Scale Maps (ASPRS, 1990) and the ASPRS Guidelines, Vertical Accuracy Reporting for Lidar Data (ASPRS, 2004). This new standard includes positional accuracy standards, with thresholds, for digital orthoimagery, digital planimetric data, and digital elevation data. Accuracy classes, based on RMSE values, were revised and upgraded from ASPRS 1990 to address the higher accuracies achievable with newer technologies. The standard also includes additional accuracy measures, such as orthoimagery seam lines, aerial triangulation accuracy, lidar swath-to-swath relative accuracy, recommended minimum nominal pulse density (NPD), horizontal accuracy of elevation data, delineation of low confidence areas for vertical data, and the recommended number of QA/QC checkpoints based on project area; this standard also draws a clear distinction between lidar Non-vegetated Vertical Accuracy (NVA) and Vegetated Vertical Accuracy (VVA). Other motivations for the new standard include:

- The ASPRS, 1990 standard as well as the prior National Map Accuracy Standard (NMAS, 1947) were both based on printed maps with a fixed scale and contour interval. ASPRS needed to address digital data which are easily manipulated and could give the erroneous impression of higher positional accuracy
- The NSSDA, published by the Federal Geographic Data Committee (FGDC, 1998c), provides no accuracy thresholds and assumes that systematic errors have been eliminated and errors are normally distributed. ASPRS needed to provide relative and absolute accuracy thresholds, place limits on systematic errors, and accommodate lidar in vegetated areas where errors are not normal
- Within recent decades, most standard measures for relating map scale and ground sample distance (GSD) to the final mapping accuracy were inherited from photogrammetric practices using scanned film. New mapping processes and methodologies have become much more sophisticated with advances in technology and our knowledge of mapping processes and mathematical modeling. Photogrammetric map accuracy can no longer be associated with the camera geometry and flying altitude alone
- The accuracy of photogrammetrically-compiled data is influenced by many factors including the quality of camera calibration parameters; quality and size of a charged coupled device (CCD) used in the digital camera CCD array; amount of imagery overlap; quality of parallax determination or photo measurements; quality of the Global Positioning System (GPS) and Inertial Measurement Unit (IMU) data for direct georeferencing; capability of the aerial triangulation (AT) software to use automated image correlation for mass production of tie points and pass points, perform camera self-calibration, and other innovations

- New accuracy standards were needed for elevation products from lidar and IfSAR which differ significantly from image products and photogrammetric processes

The highlights of the new standard are as follows:

- The positional accuracy thresholds for digital orthoimagery and digital elevation data are independent of imagery ground sample distance (GSD), map scale or contour interval
- Primary absolute accuracy thresholds are based on RMSEx, RMSEy and/or RMSEz values for which there are an unlimited number of horizontal and vertical accuracy classes
- Additional accuracy thresholds are provided for aerial triangulation; ground control; orthoimagery seam lines; lidar swath-to-swath relative accuracy; recommended minimum nominal point density (NPD); horizontal accuracy of elevation data; delineation of low confidence areas for vertical data; and required number and spatial distribution of vertical and horizontal QA/QC checkpoints based on project area
- ASPRS no longer recognizes Class 1, Class 2 or Class 3 accuracy levels
- This standard was introduced by a highlight article in the March, 2015 edition of Photogrammetric Engineering & Remote Sensing (PE&RS)

Horizontal Accuracy Standards and Examples

Table 3.2 shows the generic ASPRS, 2014 horizontal accuracy standards for digital geospatial data. This essentially enables an unlimited number of horizontal accuracy classes. Tables 3.3, 3.4 and 3.5 demonstrate how this generic standard is applied to specific user requirements.

Horizontal Accuracy Class	Absolute Accuracy			Orthoimagery Mosaic Seamline Mismatch (cm)
	RMSEx and RMSEy (cm)	RMSEr (cm)	Horizontal Accuracy at 95% Confidence Level (cm)	
X-cm	≤X	≤1.414*X	≤2.448*X	≤2*X

Table 3.2. Generic horizontal accuracy classes.

Table 3.3 presents examples of 24 specific horizontal accuracy classes and associated quality criteria as related to orthoimagery. The full standard also includes a table that lists the most common associations that have been established (based on users interpretation and past technologies) to relate orthoimagery pixel size to map scale and the ASPRS, 1990 legacy standard map accuracy classes.

Horizontal Accuracy Class RMSEx and RMSEy (cm)	RMSEr (cm)	Orthoimage Mosaic Seamline Maximum Mismatch (cm)	Horizontal Accuracy at the 95% Confidence Level (cm)
0.625	0.9	1.25	1.5
1.25	1.8	2.5	3.1
2.50	3.5	5.0	6.1
5.00	7.1	10.0	12.2
7.50	10.6	15.0	18.4
10.0	14.1	20.0	24.5

Horizontal Accuracy Class RMSEx and RMSEy (cm)	RMSEr (cm)	Orthoimage Mosaic Seamline Maximum Mismatch (cm)	Horizontal Accuracy at the 95% Confidence Level (cm)
12.5	17.7	25.0	30.6
15.0	21.2	30.0	36.7
17.5	24.7	35.0	42.8
20.0	28.3	40.0	49.0
22.5	31.8	45.0	55.1
25.0	35.4	50.0	61.2
27.5	38.9	55.0	67.3
30.0	42.4	60.0	73.4
45.0	63.6	90.0	110.1
60.0	84.9	120.0	146.9
75.0	106.1	150.0	183.6
100.0	141.4	200.0	244.8
150.0	212.1	300.0	367.2
200.0	282.8	400.0	489.5
250.0	353.6	500.0	611.9
300.0	424.3	600.0	734.3
500.0	707.1	1,000.0	1,223.9
1,000.0	1,414.2	2,000.0	2,447.7

Table 3.3. Horizontal accuracy/quality examples for digital orthoimages.

Table 3.4 on page 62 provides a general guideline to determine the appropriate orthoimagery accuracy class for three different levels of geospatial accuracy.

Values listed as "Highest accuracy work" specify an RMSEx and RMSEy accuracy class of 1-pixel (or better) and are considered to reflect the highest tier accuracy for the specified resolution given current technologies. This accuracy class is appropriate when geospatial accuracies are of highest importance and higher accuracies are supported by sufficient sensor, ground control and digital terrain model accuracies.

Values listed as "Standard Mapping and GIS work" specify a 2-pixel RMSEx and RMSEy accuracy class; this accuracy is appropriate for a standard level of high quality and high accuracy geospatial mapping applications, equivalent to ASPRS, 1990 Class 1 accuracies of the past—typical of a large majority of existing projects designed to legacy standards.

RMSEx and RMSEy accuracies of 3 or more pixels would be considered appropriate for "Visualization and less accurate work" when higher accuracies are not needed.

Users should be aware that the use of the symbol ≥ in Table 3.4 is intended to infer that users can specify larger threshold values for RMSEx and RMSEy. The symbol ≤ in Table 3.4 indicates that users can specify lower thresholds at such time as they may be supported by current or future technologies. These associations may change in the future as mapping technologies continue to advance and evolve.

It should be noted that in Table 3.4, it is the pixel size of the final digital orthoimagery that is used to associate the horizontal accuracy class, not the GSD of the raw image. When producing digital orthoimagery, the GSD as acquired by the sensor (and as computed at mean average terrain) should not be more than 95% of the final orthoimage pixel size. In extremely steep terrain, additional consideration may need to be given to the variation of the GSD across low lying areas in order to ensure that the variation in GSD across the entire image does not significantly exceed the target pixel size.

Common Orthoimagery Pixel Sizes	Recommended Horizontal Accuracy Class RMSEx and RMSEy (cm)	Orthoimage RMSEx and RMSEy in terms of pixels	Recommended Use
1.25 cm	≤1.25	≤1-pixel	Highest accuracy work
	2.5	2-pixels	Standard Mapping and GIS work
	≥3.75	≥3-pixels	Visualization and less accurate work
2.5 cm	≤2.5	≤1-pixel	Highest accuracy work
	5.0	2-pixels	Standard Mapping and GIS work
	≥7.5	≥3-pixels	Visualization and less accurate work
5 cm	≤5.0	≤1-pixel	Highest accuracy work
	10.0	2-pixels	Standard Mapping and GIS work
	≥15.0	≥3-pixels	Visualization and less accurate work
7.5 cm	≤7.5	≤1-pixel	Highest accuracy work
	15.0	2-pixels	Standard Mapping and GIS work
	≥22.5	≥3-pixels	Visualization and less accurate work
15 cm	≤15.0	≤1-pixel	Highest accuracy work
	30.0	2-pixels	Standard Mapping and GIS work
	≥45.0	≥3-pixels	Visualization and less accurate work
30 cm	≤30.0	≤1-pixel	Highest accuracy work
	60.0	2-pixels	Standard Mapping and GIS work
	≥90.0	≥3-pixels	Visualization and less accurate work

Common Orthoimagery Pixel Sizes	Recommended Horizontal Accuracy Class RMSEx and RMSEy (cm)	Orthoimage RMSEx and RMSEy in terms of pixels	Recommended Use
60 cm	≤60.0	≤1-pixel	Highest accuracy work
	120.0	2-pixels	Standard Mapping and GIS work
	≥180.0	≥3-pixels	Visualization and less accurate work
1 meter	≤100.0	≤1-pixel	Highest accuracy work
	200.0	2-pixels	Standard Mapping and GIS work
	≥300.0	≥3-pixels	Visualization and less accurate work
2 meter	≤200.0	≤1-pixel	Highest accuracy work
	400.0	2-pixels	Standard Mapping and GIS work
	≥600.0	≥3-pixels	Visualization and less accurate work
5 meter	≤500.0	≤1-pixel	Highest accuracy work
	1,000.0	2-pixels	Standard Mapping and GIS work
	≥1,500.0	≥3-pixels	Visualization and less accurate work

Table 3.4. Digital orthoimagery accuracy examples for current large and medium format metric cameras.

Table 3.5 presents 24 common horizontal accuracy classes for digital planimetric data, approximate ground sample distance (GSD) of source imagery for high accuracy planimetric data, and equivalent map scales per legacy National Map Accuracy Standards (NMAS) and ASPRS, 1990 accuracy standards. The values for the approximate GSD of source imagery only apply to imagery derived from common large and medium format metric cameras. The range of the approximate GSD of source imagery is provided as a general recommendation only, based on the current state of sensor technologies and mapping practices. Different ranges may be considered in the future depending on future advances of such technologies and mapping practices.

ASPRS, 2014				Equivalent to map scale in		Equivalent to map scale in NMAS
Horizontal Accuracy Class RMSEx and RMSEy (cm)	RMSEr (cm)	Horizontal Accuracy at the 95% Confidence Level (cm)	Approximate GSD of Source Imagery (cm)	ASPRS 1990 Class 1	ASPRS 1990 Class 2	
0.625	0.9	1.5	0.31 to 0.63	1:25	1:12.5	1:16
1.25	1.8	3.1	0.63 to 1.25	1:50	1:25	1:32
2.5	3.5	6.1	1.25 to 2.5	1:100	1:50	1:63
5.0	7.1	12.2	2.5 to 5.0	1:200	1:100	1:127

ASPRS, 2014				Equivalent to map scale in		Equivalent to map scale in NMAS
Horizontal Accuracy Class RMSEx and RMSEy (cm)	RMSEr (cm)	Horizontal Accuracy at the 95% Confidence Level (cm)	Approximate GSD of Source Imagery (cm)	ASPRS 1990 Class 1	ASPRS 1990 Class 2	
7.5	10.6	18.4	3.8 to 7.5	1:300	1:150	1:190
10.0	14.1	24.5	5.0 to 10.0	1:400	1:200	1:253
12.5	17.7	30.6	6.3 to12.5	1:500	1:250	1:317
15.0	21.2	36.7	7.5 to 15.0	1:600	1:300	1:380
17.5	24.7	42.8	8.8 to 17.5	1:700	1:350	1:444
20.0	28.3	49.0	10.0 to 20.0	1:800	1:400	1:507
22.5	31.8	55.1	11.3 to 22.5	1:900	1:450	1:570
25.0	35.4	61.2	12.5 to 25.0	1:1000	1:500	1:634
27.5	38.9	67.3	13.8 to 27.5	1:1100	1:550	1:697
30.0	42.4	73.4	15.0 to 30.0	1:1200	1:600	1:760
45.0	63.6	110.1	22.5 to 45.0	1:1800	1:900	1:1,141
60.0	84.9	146.9	30.0 to 60.0	1:2400	1:1200	1:1,521
75.0	106.1	183.6	37.5 to 75.0	1:3000	1:1500	1:1,901
100.0	141.4	244.8	50.0 to 100.0	1:4000	1:2000	1:2,535
150.0	212.1	367.2	75.0 to 150.0	1:6000	1:3000	1:3,802
200.0	282.8	489.5	100.0 to 200.0	1:8,000	1:4000	1:5,069
250.0	353.6	611.9	125.0 to 250.0	1:10000	1:5000	1:6,337
300.0	424.3	734.3	150.0 to 300.0	1:12000	1:6000	1:7,604
500.0	707.1	1223.9	250.0 to 500.0	1:20000	1:10000	1:21,122
1,000.0	1,414.2	2447.7	500.0 to 1000.0	1:40000	1:20000	1:42,244

Table 3.5. Horizontal accuracy examples for digital planimetric data, compared with legacy standards.

Vertical Accuracy Standards and Examples

Table 3.6 shows the generic ASPRS, 2014 vertical accuracy standards for digital elevation data, enabling an unlimited number of vertical accuracy classes for Non-vegetated Vertical Accuracy (NVA) and Vegetated Vertical Accuracy (VVA). VVA procedures for computing the 95th percentile are explained in the section, below, on the NDEP Guidelines for Digital Elevation Data (NDEP, 2004) The relative accuracy terms pertain primarily to lidar swaths in non-vegetated open terrain or manmade surfaces such as roads.

Vertical Accuracy Class	Absolute Accuracy			Relative Accuracy (where applicable)		
	RMSEz Non-Vegetated (cm)	NVA at 95% Confidence Level (cm)	VVA at 95th Percentile (cm)	Within-Swath Hard Surface Repeatability (Max Diff) (cm)	Swath-to-Swath Non-Vegetated Terrain (RMSDz) (cm)	Swath-to-Swath Non-Vegetated Terrain (Max Diff) (cm)
X-cm	≤X	≤1.96*X	≤3.00*X	≤0.60*X	≤0.80*X	≤1.60*X

Table 3.6. Generic vertical accuracy classes.

Table 3.7 provides vertical accuracy examples and other quality criteria for ten common vertical accuracy classes. Table 3.8 on page 66 provides the same ten vertical accuracy examples with recommended lidar nominal pulse density (NPD) and nominal pulse spacing (NPS) suitable for each of them. Table 3.9 on page 66 compares the same vertical accuracy examples with contour intervals from legacy ASPRS, 1990 and NMAS, 1947 standards.

Vertical Accuracy Class (cm)	Absolute Accuracy			Relative Accuracy (where applicable)		
	RMSEz Non-Vegetated (cm)	NVA at 95% Confidence Level (cm)	VVA at 95th Percentile (cm)	Within-Swath Hard Surface Repeatability (Max Diff) (cm)	Swath-to-Swath Non-Veg Terrain (RMSDz) (cm)	Swath-to-Swath Non-Veg Terrain (Max Diff) (cm)
1.0	1.0	2.0	3	0.6	0.8	1.6
2.5	2.5	4.9	7.5	1.5	2	4
5.0	5.0	9.8	15	3	4	8
10.0	10.0	19.6	30	6	8	16
15.0	15.0	29.4	45	9	12	24
20.0	20.0	39.2	60	12	16	32
33.3	33.3	65.3	100	20	26.7	53.3
66.7	66.7	130.7	200	40	53.3	106.7
100.0	100.0	196.0	300	60	80	160
333.3	333.3	653.3	1000	200	266.7	533.3

Table 3.7. Vertical accuracy examples for digital elevation data.

Vertical Accuracy Class (cm)	Absolute Accuracy		Recommended Minimum NPD (pts/m2)	Recommended Maximum NPS (m)
	RMSEz Non-Vegetated (cm)	NVA at 95% Confidence Level (cm)		
1.0	1.0	2.0	≥20	≤0.22
2.5	2.5	4.9	16	0.25
5.0	5.0	9.8	8	0.35
10.0	10.0	19.6	2	0.71
15.0	15.0	29.4	1	1.0
20.0	20.0	39.2	0.5	1.4
33.3	33.3	65.3	0.25	2.0
66.7	66.7	130.7	0.1	3.2
100.0	100.0	196.0	0.05	4.5
333.3	333.3	653.3	0.01	10.0

Table 3.8. Vertical accuracy examples and recommended lidar nominal point density and point spacing.

Vertical Accuracy Class (cm)	RMSEz Non-Vegetated (cm)	Equivalent Class 1 Contour Interval Per ASPRS, 1990 (cm)	Equivalent Class 2 Contour Interval Per ASPRS, 1990 (cm)	Equivalent Contour Interval Per NMAS, 1947 (cm)
1.0	1.0	3.0	1.5	3.29
2.5	2.5	7.5	3.8	8.22
5.0	5.0	15.0	7.5	16.45
10.0	10.0	30.0	15.0	32.90
15.0	15.0	45.0	22.5	49.35
20.0	20.0	60.0	30.0	65.80
33.3	33.3	100.0	50.0	109.55
66.7	66.7	200.0	100.1	219.43
100.0	100.0	300.0	150.0	328.98
333.3	333.3	1,000.0	500.0	1096.49

Table 3.9. Vertical accuracy examples for elevation data, compared with legacy standards.

Recommended Number of QA/QC Checkpoints

Table 3.10 lists ASPRS recommendations for the number of QA/QC checkpoints to be used for vertical and horizontal accuracy testing of elevation data sets and for horizontal accuracy testing of digital orthoimagery and planimetric data sets, based on the size of the project area.

Project Area (Square Kilometers)	Horizontal Accuracy Testing of Orthoimagery and Planimetrics	Vertical and Horizontal Accuracy Testing of Elevation Data Sets		
	Total Number of Static 2-D/3-D Checkpoints (clearly-defined points)	Number of Static 3-D Checkpoints in NVA	Number of Static 3-D Checkpoints in VVA	Total Number of Static 3-D Checkpoints
≤500	20	20	5	25
501-750	25	20	10	30
751-1000	30	25	15	40
1001-1250	35	30	20	50
1251-1500	40	35	25	60
1501-1750	45	40	30	70
1751-2000	50	45	35	80
2001-2250	55	50	40	90
2251-2500	60	55	45	100

Table 3.10. Recommended number of QA/QC checkpoints based on area.

For the first 2,500 km² area within the project, ASPRS recommends 60 2-D/3-D checkpoints for horizontal accuracy assessments and 100 3-D checkpoints for vertical accuracy assessments, which provides a statistically defensible number of samples on which to base valid accuracy assessments. Although vertical checkpoints are normally not well-defined, where feasible, the horizontal accuracy of lidar data sets should be tested by surveying approximately half of all NVA checkpoints at the ends of paint stripes or other point features that are visible and can be measured on lidar intensity returns or equivalent.

For horizontal testing of areas >2,500 km², clients should determine the number of additional horizontal checkpoints, if any, based on criteria such as image resolution and urbanization extent.

For vertical testing of areas >2,500 km², add five additional vertical checkpoints for each additional 500 km² area. Each additional set of five vertical checkpoints for 500 km² would include three checkpoints for NVA and two for VVA. The recommended number and distribution of NVA and VVA checkpoints may vary depending on the importance of different land cover categories and client requirements.

Horizontal and Vertical Accuracy Reporting

Horizontal and vertical accuracies shall be reported in terms of compliance with the RMSE thresholds and other quality and accuracy criteria outlined in Tables 3.2, 3.3, 3.5, 3.6, 3.7, and 3.8 previously.

If testing is performed, accuracy statements should specify that the data are "tested to meet" the stated accuracy. If testing is not performed, accuracy statements should specify that the data are "produced to meet" the stated accuracy. Detailed specifications for testing and reporting to meet these requirements are outlined in Annex C and Annex D of the full standard.

The horizontal accuracy of digital orthoimagery, planimetric data, and elevation data sets shall be documented in the metadata in one of the following manners:

- "This data set was tested to meet ASPRS Positional Accuracy Standards for Digital Geospatial Data (ASPRS, 2014) for a ___ (cm) RMSEx/RMSEy Horizontal Accuracy Class. Actual positional accuracy was found to be RMSEx = ___ (cm) and RMSEy = ___ cm which equates to Positional Horizontal Accuracy = +/- ___ at 95% confidence level."
- "This data set was produced to meet ASPRS Positional Accuracy Standards for Digital Geospatial Data (ASPRS, 2014) for a ___ (cm) RMSEx/RMSEy Horizontal Accuracy Class which equates to Positional Horizontal Accuracy = +/- ___ cm at a 95% confidence level."

The vertical accuracy of elevation data sets shall be documented in the metadata in one of the following manners:

- "This data set was tested to meet ASPRS Positional Accuracy Standards for Digital Geospatial Data (ASPRS, 2014) for a___ (cm) RMSEz Vertical Accuracy Class. Actual RMSEz = ___ cm, equating to NVA of +/- ___ cm at 95% confidence level. Actual VVA accuracy was found to be +/- ___ cm at the 95th percentile."
- "This data set was produced to meet ASPRS Positional Accuracy Standards for Digital Geospatial Data (ASPRS, 2014) for a ___ cm RMSEz Vertical Accuracy Class equating to NVA = +/-___cm at 95% confidence level and VVA =+/-___cm at the 95th percentile.

Major Points in Understanding the ASPRS, 2014 Standards

- ASPRS no longer uses Class 1, Class 2 or Class 3 accuracy classes. Horizontal accuracy classes are based on RMSEx and RMSEy values; vertical accuracy classes are based on RMSEz values in open, non-vegetated terrain
- RMSEz is used with a multiplier of 1.9600 to compute the Non-vegetated Vertical Accuracy (NVA) at the 95-percent confidence level. The 95th percentile is used to estimate the Vegetated Vertical Accuracy (VVA) at the 95-percent confidence level. Annex D of the full standard provides accuracy statistics and examples for computing mean errors, sample standard deviations, RMSE errors, and accuracies at the 95-percent confidence levels and 95th percentile
- The full standard also provides guidance on determining or estimating the horizontal accuracy of digital elevation data from photogrammetry and lidar; accuracy requirements for aerial triangulation and INS-based sensor orientation of digital imagery; accuracy requirements for ground control used for aerial triangulation; QA/QC checkpoint accuracy and placement requirements; relative accuracy of lidar and IfSAR using root mean square differences ($RMSD_z$); horizontal and vertical accuracy reporting; designation of Low Confidence Areas for DEMs derived from photogrammetry, IfSAR or lidar where the DTM may not meet the overall data accuracy requirements, including criteria for determining Low Confidence Areas for lidar datasets

ASPRS LAS SPECIFICATION, V1.4

The ASPRS LASer (LAS) file format is a public file format for the interchange of 3-D point cloud data between data users. Although developed primarily for exchange of lidar point cloud data, this format supports the exchange of any 3-D xyz tuplet. This binary file format is an alternative to proprietary systems or a generic ASCII file interchange system used by some companies. The problem with proprietary systems is obvious in that data cannot be easily taken from one system to another. There are two major problems with the ASCII file interchange. The first problem is performance because the reading and interpretation of ASCII elevation data can be very slow and the file size can be extremely large, even for small amounts of data. The second problem is that all information specific to the lidar data is lost. The LAS file format maintains information specific to the lidar nature of the data while not being overly complex.

LAS V1.4 has the following major changes relative to earlier versions:

- Extension of offsets and field sizes to support full 64 bit
- Support for up to 15 returns per outgoing pulse
- Extension of the Point Class field to support 256 classes (see Table 3.11)
- Definition of several new ASPRS standard classes
- Extension of the Scan Angle field to 2 bytes to support finer angle resolution
- Addition of a Sensor Channel bit field to support mobile mapping systems
- Addition of Well Known Text (WKT) definitions for Coordinate Reference Systems
- Addition of an Overlap bit to allow indicating pulses in the overlap region while maintaining the class definition
- Addition of an (optional) Extra Byte Variable Length Record to describe “extra bytes” stored with each point
- Added definitions for “LAS Domain Profile” and “LAS Domain Profile Description.”

Whereas lidar data producers need to fully understand the LAS specification, the typical DEM user primarily needs to understand that each lidar return may be classified in adherence to values in Table 3.11, but they are often simplified to Classes 1, 2, 7, 9, 17, 18, 20, 21, and 22 only to reduce classification costs when the user just needs to separate the bare-earth ground and water from all other classes.

Class Value	Meaning
0	Created, never classified
1	Unclassified, including undefined
2	Ground
3	Low vegetation
4	Medium vegetation
5	High vegetation
6	Building
7	Low point (noise)
8	Reserved
9	Water
10	Rail
11	Road surface
12	Reserved
13	Wire—guard (shield)
14	Wire—conductor (phase)
15	Transmission tower
16	Wire-structure connector (e.g., insulator)
17	Bridge deck
18	High noise
20	Ignored ground (near a breakline)
21	Snow (if present and identifiable)
22	Temporal exclusion (typically non-favored data in intertidal zones)
40	Bathymetric point (e.g., seafloor or riverbed; also known as submerged topography)

Class Value	Meaning
41	Water surface from bathy or topobathy lidar, distinct from Class 9 used for topo lidar
42	Derived water surface (synthetic water surface used in computing refraction at water level)
43	Submerged object, not otherwise specified (e.g., wreck, rock, submerged piling)
44	International Hydrographic Organization (IHO) S-57 object, not otherwise specified
45	No-bottom-found-at (bathy lidar point for which no detectable bottom return was received)
Others to 63	Reserved
64-255	User definable

Table 3.11. ASPRS Standard Lidar Point Classes (LAS Classes).

The lidar mapping community now has the ability to customize the LAS file format to meet their application-specific needs. The mechanism that makes this possible is the LAS Domain Profile which is a derivative of the base LAS v1.4 specification that adds (but does not remove or alter existing) point classes and attributes. For example, the Topo-Bathy Lidar Domain Profile adds point classification values for bathymetric points (e.g., seafloor or riverbed, also known as submerged topography), water surface, derived water surface, submerged object, IHO S-57 object, and bottom-not-found depth. Extra Byte Variable Length Records (EXTRA-BYTES or Extra Byte VLRs) are added for pseudo-reflectance, uncertainty, water column depth, figure of merit, and processing specific flags. It is anticipated that additional domain profiles will be released in the future.

Major Points in Understanding the ASPRS LAS Specifications

- The LAS format is organized into four sections: Public Header Block, Variable Length Records (VLRs), Point Data Records, and Extended Variable Length Records (EVLRs)
- The Public Header Block contains generic data for the entire LAS file, such as number of point records, minimum and maximum xyz values, point data record format (1-10), LAS major and minor version, number of VLRs, number of EVLRs, and other fields
- The VLRs contain variable types of data such as Spatial Reference System (SRS) information and waveform packets. EVLRs have a higher payload than VLRs and have the advantage they can be appended to the end of LAS files without having to rewrite the entire file. The only required VLR/EVLR is for the SRS information
- Point Data Records are the information stored for each individual lidar return, such as the xyz of each return, the return number, the classification, intensity value, scan angle, point source ID, timestamp, and other fields. The same type of information will be stored for every return within a file. The fields that are stored for each return are indicated by the point data record format. Some fields are core fields shared by all point data record formats, but different point data record formats will store slightly different information for each return, such as the ability to store RGB color information for a return
- Most software programs capable of reading and viewing LAS files can display and process the lidar data by point data record information other than classification, such as return number, elevation, and point source ID. Many of these software programs can also parse portions of or the entire Public Header Block, VLRs/EVLRs, and Point Data Records to a text-based file for additional analysis of the data

USGS LIDAR BASE SPECIFICATION, V1.3

Chapter 13 explains the major requirements of the USGS Lidar Base Specification (LBS) v1.3 (Heidemann, 2018) as pertains to metadata, raw point cloud, classified point cloud, bare earth surface (raster DEM), DEM hydro-flattening and breaklines. As the Office of Management and Budget Circular A-16 designated lead agency for topographic elevation data, the USGS, through the National Geospatial Program (NGP), has developed and adopted this specification as the base specification for the National interagency 3D Elevation Program (3DEP), explained in Chapter 5.

Tables 3.12 and 3.13 are amalgamations of five USGS V1.3 tables that specify vertical and relative accuracies, densities and other quality parameters for four Quality Levels (QLs) of lidar data. Whereas QL2 or better is required for lidar data to satisfy 3DEP requirements, the majority of legacy lidar datasets are QL3 or poorer.

Quality Level (QL)	Aggregate Nominal Pulse Spacing (ANPS) (m)	Aggregate Nominal Pulse Density (ANPD) (pts/m²)	RMSEz (non-vegetated) (cm)	NVA at 95% confidence level (cm)	VVA at 95th percentile (cm)
QL0	≤0.35	≥8.0	≤5.0	≤9.8	≤15.0
QL1	≤0.35	≥8.0	≤10.0	≤19.6	≤30.0
QL2	≤0.71	≥2.0	≤10.0	≤19.6	≤30.0
QL3	≤1.41	≥0.5	≤20.0	≤39.2	≤60.0

Table 3.12. Lidar density and absolute vertical accuracy.

Quality Level (QL)	Smooth Surface Repeatability (cm)	Swath Overlap Difference, RMSDz (cm)	Minimum DEM Cell Size (m)	Minimum DEM Cell Size (ft)
QL0	≤3	≥4	0.5	1
QL1	≤6	≥8	0.5	1
QL2	≤6	≥8	1	2
QL3	≤12	≥16	2	5

Table 3.13. Lidar relative vertical accuracy and minimum DEM cell sizes.

To align with the new ASPRS Positional Accuracy Standards for Digital Geospatial Data (ASPRS, 2014), accuracy per Heidemann, 2018 is reported based on Non-vegetated Vertical Accuracy (NVA) and Vegetated Vertical Accuracy (VVA); these two classes replace the previously used Fundamental, Supplemental, and Consolidated Vertical Accuracy (FVA, SVA and CVA) classes (see NDEP, 2004, below). Heidemann 2018 also requires adherence to ASPRS 2014 for the quantity of QA/QC checkpoints required for a project based on its areal extent (see Table 3.10 on page 67).

Another table specifies the minimum point cloud classification scheme for LAS classes 1 (processed, but unclassified), 2 (bare earth), 7 (low noise), 9 (water), 17 (bridge decks), 18 (high noise) and 20 (ignored ground, near a breakline); and another table compares land cover classes and their new reporting group (NVA and VVA) and prior reporting group (FVA and SVA).

The lidar collection specifications include criteria for the collection area, quality level, multiple discrete returns, intensity values, nominal pulse spacing, data voids, spatial distribution and regularity, and collection conditions.

The lidar data processing and handling specifications include criteria for the ASPRS LAS v1.4 file format, full waveform, time of GPS data, datums, coordinate reference system, units of reference, swath identification, point families, swath size and segmentation, scope of collection, positional accuracy validation (checkpoints, relative and absolute vertical accuracy), use of the LAS withheld flag, use of LAS overlap flag, point classification, classification accuracy, classification consistency, and tiles.

The lidar DEM hydro-flattening specifications include criteria for inland ponds and lakes, inland streams and rivers, non-tidal boundary waters, tidal waters, islands, and breaklines.

The lidar deliverables specifications include criteria for metadata, raw point cloud, classified point cloud, bare-earth raster DEM, and breaklines.

Heidemann, 2018 also includes detailed references, glossary, supplemental information and four appendices (common data upgrades, hydro-flattening reference, lidar metadata example, and lidar metadata template).

Major Points in Understanding the USGS Lidar Base Specifications

- Version 1.3 of the USGS Lidar Base Specification provides specifications for QL0, QL1, QL2 and QL3 lidar
- Only QL2 lidar or better (QL1 or QL0) are acceptable for the 3DEP which leverages resources to satisfy multiple requirements, reduces overall costs to taxpayers, and delivers consistent, standardized, authoritative data used by all
- To encourage nationwide production of QL2 lidar or better, USGS has a Broad Agency Announcement (BAA) program in which there are partnership opportunities for joint funding of high-quality lidar data for the 3DEP

NATIONAL OCEAN SERVICE (NOS) SPECIFICATIONS

The National Ocean Service Hydrographic Surveys Specifications and Deliverables (NOAA, 2015) provide technical specifications that detail the requirements for hydrographic surveys to be undertaken either by NOAA field units or by organizations under contract to the Director, Office of Coast Survey (OCS), National Ocean Service (NOS), NOAA, U.S. Department of Commerce. These specifications are based in part on the International Hydrographic Organization's Standards for Hydrographic Surveys, Special Publication 44, Fifth Edition, February 2008, specifically for Order 1a surveys. Hydrographic surveys classified as Order 1a are intended for harbors, harbor approach channels, recommended tracks, inland navigation channels, coastal areas of high commercial traffic density, and are usually in shallower areas less than 100 meters water depth. These specifications can be viewed and downloaded at *http://nauticalcharts.noaa.gov/hsd/specs/specs.htm*. Highlights of major sections are extracted as follows:

Section 2, Datums: "All positions will be referenced to the North American Datum of 1983 (NAD 83). This datum must be used throughout a survey project for everything that has a geographic position or for which a position is to be determined. Those documents used for comparisons, such as charts, junctional surveys, and prior surveys, must be referenced or adjusted to NAD 83. In addition, all software used on a survey must contain the correct datum parameters. All sounding data will be reduced to Mean Lower Low Water (MLLW). Heights of bridges and overhead cables will be referenced to Mean High Water (MHW). Coordinated Universal Time (UTC) will be used for all time records."

Section 3, Hydrographic Position Control: "If any non-USCG differential reference stations are used (e.g., a "fly-away" DGPS beacon established by the hydrographer, the Federal Aviation Administration's Wide Area Augmentation System (FAA WAAS), or commercial satellite-based correction services), the hydrographer shall conduct a certification to ensure that no multipath or other site specific problems exist. The system shall be re-certified annually for each project area in which the non-USCG correctors are utilized. Certification of non-USCG differential correctors shall be accomplished as follows: (1) Identify a point ashore in close proximity to the project area for which the position is known to at least the total horizontal uncertainty (THU) given in Section 3.1.1, (2) Occupy this position with a GPS receiver utilizing the non-USCG correctors for at least 24 hours, logging data at 1-second intervals., (3) Create a plot comparing surveyed positions to the known position. Certification is successful if the surveyed and established positions agree to the requirements of

Section 3.1.1, 4) Include a description of the certification process plots in the Horizontal Control Report for each project (see Section 8.1.5.2)... Real Time Kinematic (RTK) and Post Processed Kinematic (PPK) may be used for positioning during hydrographic surveys. If RTK or PPK techniques are used, the hydrographer must ensure that all positions meet the accuracy requirements of Section 3.1."

Section 4, Tides and Water Levels Requirements: "The present NOAA Nautical Chart Reference Datum for tidal waters is Mean Lower Low Water (MLLW) based on the NOAA National Tidal Datum Epoch (NTDE) of 1983-2001 as defined in the Tide and Current Glossary. All tidal datum computations and water level reductions shall be referenced to this datum. In non-tidal coastal areas soundings will be reduced to Low Water Datum (LWD) which is Mean Sea Level (MSL) - 0.5 ft. In the Great Lakes all soundings will be reduced to Great Lakes Low Water Datum as referenced to the current International Great Lakes Datum (IGLD). In areas which are charted to "Special Datums" such as Columbia River Datum, Hudson River Datum, Mississippi River Datum, etc., the soundings will be reduced to the specific special datum. If a subordinate station has a currently published datum, every effort must be made to set the station datum for the new installation to the historic station datum via re-occupation, so that all newly collected observations are on the same zero reference as the currently accepted datum."

Section 5, Depth Soundings and Accuracy Standards: "Depth values shall be recorded in meters, with a precision of at least centimeters. ...Uncertainty estimates for depth soundings and ancillary measurements shall be recorded with sufficient precision to support Total Propagated Uncertainty (TPU) estimates for depth values at centimeter precision. NOS standards for Total Vertical Uncertainty (TVU) in hydrographic surveys apply to general water depths and least depths over wrecks and obstructions. By extension, they also apply to the elevations of rocks or other features which uncover at low water and to the measurement of overhead clearances. These standards apply regardless of the method of determination; whether by single beam echosounder, multibeam echosounder, lidar, lead line, diver investigation, or other method. The formula below shall be used to compute the maximum allowable TVU (defined at the 95% confidence interval) for all depth estimates included in bathymetric data products or feature attribution after application of correctors for all systematic and system specific errors: $\pm\sqrt{a^2+(b*d)^2}$, Where: a represents that portion of the uncertainty that does not vary with depth, b is a coefficient which represents that portion of the uncertainty that varies with depth, (b x d) represents that portion of the uncertain that does vary with depth, d is the depth. The variables a and b shall be defined as follows: In depths less than 100 meters, a = 0.5 meters and b = 0.013 (IHO Order 1), in depths greater than 100 meters, a=1.0 meters and b=0.023 (IHO Order 2). The maximum allowable uncertainty in depth includes all inaccuracies due to residual systematic and system specific instrument errors; the speed of sound in water; static vessel draft; dynamic vessel draft; heave, roll, and pitch; and any other sources of error in the actual measurement process, including the errors associated with water level (tide) variations (both tidal measurement and zoning errors)."

Subsection 5.2.2, Multibeam Resolution Standards: "In general, there are four classifications of coverage: Object Detection Coverage, Complete Coverage, Set Line Spacing, and Trackline (transit and reconnaissance). Specific requirements of each coverage classification are given ...in Sections 5.2.2.1, 5.2.2.2, 5.2.2.3, and 5.2.2.4.

Subsection 5.2.3, Corrections to Echo Soundings: "To meet the accuracy and resolution standards specified in Section 5.1.3, and to create a Bathymetric Attributed Grid (BAG) that includes an accurate uncertainty layer, the hydrographer must conduct an error analysis of their survey systems. In recognition of the possibility that some discrepancies in soundings may not be detected until the final processing phase of the survey, the determination and application of corrections to echo soundings must be accomplished and documented in a systematic manner. Also, it is preferable that all corrections be applied in such a way that the on line values may be removed and replaced with a revised set of correctors during office processing. Corrections to echo soundings are divided into five categories, and listed below in the sequence in which they are applied:

- "Instrument error corrections account for sources of error related to the sounding equipment itself."
- "Draft corrections shall be added to the observed soundings to account for the depth of the echosounder transducer below the water surface."

- "Dynamic draft corrections shall be applied to soundings to correct the vertical displacement of the transducer, relative to its position at rest, when a vessel is underway."
- "Speed of sound corrections shall be applied to soundings to compensate for the fact that echosounders may only display depths based on an assumed sound speed profile while the true speed may vary in time and space."
- "Attitude corrections shall be applied to multibeam soundings to correct the effect of vessel motion caused by waves and swells (heave, roll, pitch) and the error in the vessel's heading."

Subsection 5.5.3, Crosslines: "The regular system of sounding lines shall be supplemented by a series of crosslines for verifying and evaluating the accuracy and reliability of surveyed soundings and positions. Crosslines shall be acquired and processed to the same accuracy and data quality standards as required for mainscheme lines, and shall be included in the grids that are submitted as the final bathymetric product of the survey. As a quality control tool, the benefits of crosslines are most readily derived when they are acquired at or near the beginning of survey operations...Crosslines shall have good temporal and geographic distribution (across depth ranges, distinct water masses, and vessels) such that maximal nadir-to-nadir comparisons are achieved. Crossline requirements are dependent on bathymetric coverage type assigned and achieved by the field unit, Section 5.2.2"

Section 8, Deliverables, documents requirements for Field Reports (Progress Report, Survey Outline, Danger to Navigation Report, Descriptive Report, and Supplemental Reports), Side Scan Sonar Mosaics and Contact List, and Digital Data (Bathymetric, Side Scan Sonar, Backscatter, and ERS).

Subsection 8.5.3, Shallow-Water Multibeam Data: "The hydrographer's bathymetric data format shall provide complete traceability for all positions, soundings, and correctors including sensor offsets, biases, dynamic attitude, sound speed, position, sensor position, date and time, vertical datum reducers, and sounding data from acquisition through post processing. Data quality and edit flags must be traceable.

- "Full Resolution Echosounding Data. The hydrographer shall submit full resolution echosounding data in a format readable by CARIS HIPS. Full resolution echosounding data shall be delivered fully corrected for tides, sound speed, vessel offsets, draft and dynamic draft. These corrections may be made within CARIS, with data submitted as a complete CARIS project (including HDCS files, sound speed files, Vessel Configuration, CARIS tide files, etc.)... Alternately, non-CARIS HIPS users may submit fully corrected, such that it will be read in CARIS HIPS using a 'zeroed' Vessel Configuration file (.vcf or .hvf) and a 'zero' tide file (.tid), etc.."
- "Full Resolution Lidar Data. The contractor shall submit the full resolution lidar data in CARIS compatible format. The submission will include the appropriate CARIS converter, lidar data before conversion, and all necessary CARIS files so that NOAA can reconvert all files, if desired."
- "CARIS BASE Surface and/or BAG. The final depth information from the survey will be composed of a collection of grids. This collection of grids must reflect the state of the seafloor at the time of the survey, with resolution and attribution as described in Section 5.2, and/or the Hydrographic Survey Project Instructions. The hydrographer must take steps to ensure that all data has been correctly processed and that appropriate designated soundings have been selected (see Section 5.2.1.2, 5.2.2, 5.3.1.2). The collection of grids representing the final reviewed results of the hydrographic survey shall be submitted as CARIS BASE or BAG surfaces...Field units shall not normally submit combined BAGs unless otherwise specified in the Project Instructions."
- "Side Scan Sonar Data. The hydrographer shall submit digital side scan data in a format readable by CARIS SIPS or SonarWiz. Digital side scan sonar shall be geocoded using the towfish position (towfish position corrected)... The hydrographer shall submit a digital image file for each 100 percent coverage. The digital image file shall be in a standard geo-referenced image format (Section 8.3.1)."
- "Backscatter. The hydrographer shall submit raw backscatter data in a format readable by IVS Fledermaus Geocoder Toolbox."

Section 9, Ellipsoidally-Referenced Surveys (ERS). "The ideal technique for NOAA ERS-vessel positioning to obtain centimeter-level [height] accuracy is tightly coupled inertially-aided, post-processed kinematic GPS (IAPPK)... GPS is preferred over other GNSS (e.g., GLONASS); however, if the availability of 5 or more GPS satellites is unacceptably low in a particular survey environment, a hybrid GPS-GNSS solution may be leveraged... Other GPS/GNSS-based highly-accurate positioning techniques may be used, such as real-time kinematic (RTK) or real-time Precise Point Positioning (PPP) on a case-by-case basis."

Subsection 9.1.1.1 GNSS Infrastructure. "...For carrier-phase differential GNSS/GPS (baseline) processing, all International GNSS Service (IGS) products are sufficient: Broadcast, Ultra-Rapid, Rapid and Final—use the best available orbits and clocks at the time of baseline processing. For ERS inertially-aided PPP, Ultra-Rapid (or better) orbit and clock products suffice...do not mix products from different Analysis Centers... a maximum-baseline length of 40 km is used for the pair-wise, vessel-to-single reference station configuration.... A Virtual Reference Station (VRS) system may be used by employing a network of four or more user-installed or continuously operating reference stations "surrounding" the ERS project area. For planning purposes, a maximum inter-station baseline length of 70 km is used... The nominal maximum required data rate for reference station data is equal to that of the vessel (rover) data: 1 Hz. Reference station data rates as low as 30 seconds are permitted when carrier-phase interpolation methods are used in IAPPK software."

Subsection 9.2 ERS Datum Transformation Requirements. "The standard method for the ellipsoid-to-chart datum transformation is to utilize the National VDatum transformation framework and software developed and published by NOAA. However, certain ERS projects may proceed in areas lacking VDatum coverage using one of the alternative ellipsoid-to-chart datum separation (SEP) models..."

Major Points in Understanding the NOS Specifications

- All horizontal positions are referenced to NAD 83
- All sounding data are reduced to Mean Lower Low Water (MLLW)
- Heights of bridges and overhead cables are referenced to Mean High Water (MHW)
- Hydrographic survey data are focused on safety of marine navigation
- Hydrographic surveys with multibeam echosounders have many similarities to topographic surveys with lidar sensors. They both rely upon differential GPS to position sensors relative to known control points. They both collect swaths of elevation data points, either as raw lidar data or raw multibeam data. They both perform extensive post-processing to generate final products either ground (topographic) or navigation surfaces (bathymetric), above and below water levels. They both use crosslines and sun-illuminated DTM images along with uncertainty models for QA/QC purposes.
- Multibeam sonar data depicts dangers to marine navigation (e.g., depths of rocks, shoals, wrecks, and underwater obstructions) much like photogrammetric, IfSAR or lidar digital surface models (DSMs) depict dangers to aviation (heights of buildings, towers, trees, and other elevated obstructions)

NATIONAL DIGITAL ELEVATION PROGRAM (NDEP) GUIDELINES

The National Digital Elevation Program (NDEP) published its "Guidelines for Digital Elevation Data, Version 1.0," on May 10, 2004 (NDEP, 2004). These guidelines were developed by representatives of the U.S. Geological Survey (USGS), National Geospatial-Intelligence Agency (NGA), U.S. Army Corps of Engineers (USACE), National Oceanic and Atmospheric Administration (NOAA), Bureau of Land Management (BLM), National States Geographic Information Council (NSGIC), National Aeronautics and Space Administration (NASA), Natural Resources Conservation Service (NRCS), U.S. Forest Service (USFS), Federal Emergency Management Agency (FEMA), and U.S. Census Bureau. Because FEMA initially had the best base of experience, FEMA requirements were used to draft the initial NDEP guidelines—subsequently expanded to address the needs of the other organizations represented as well as the perceived needs of counties and local communities for digital elevation data. The NDEP guidelines also drew heavily from case studies, user requirements guidance, and accuracy assessment procedures documented in the first edition of this ASPRS manual: Digital Elevation Model Technologies and Applications: The DEM Users Manual (ASPRS, 2001). In turn, the 2nd edition of this manual drew heavily upon the NDEP guidelines.

The NDEP documented the fact that lidar errors in vegetated terrain do not always follow a normal distribution as assumed by the NSSDA (FGDC, 1998c). As such the NDEP introduced three new terms not previously used, i.e., Fundamental Vertical Accuracy (FVA), Consolidated Vertical Accuracy (CVA), and Supplemental Vertical Accuracy. These terms were also adopted in the ASPRS Guidelines: Vertical Accuracy Reporting for Lidar Data (ASPRS, 2004) which were superseded by ASPRS, 2014.

- Fundamental Vertical Accuracy (FVA) uses RMSEz x 1.9600 to define vertical accuracy at the 95-percent confidence level in open, non-vegetated terrain only where errors are expected to follow a normal distribution. Open terrain includes dirt, rock, sand and mowed grass
- Consolidated Vertical Accuracy (CVA) uses the 95th percentile to estimate vertical accuracy at the 95-percent confidence level in all land cover categories combined
- Supplemental Vertical Accuracy (SVA) uses the 95th percentile to estimate vertical accuracy at the 95-percent confidence level in each land cover category, individually. It is common to compute separate SVA values for weeds and crops, shrub and brush lands, and fully forested, for example, where lidar errors may not be normally distributed
- For example, using 20 checkpoint errors in column A of a spreadsheet, the 95th percentile is determined by first computing the absolute value of all errors in column B, and then using a standard spreadsheet command: Percentile (B1:B20, 0.95)

The 95th percentile indicates that 95 percent of the errors in the dataset will have absolute values of equal or lesser value and five percent of the errors will be of larger value.

Major Points in Understanding the NDEP Guidelines

- The NDEP guidelines supplemented the NSSDA standards, specifically with regards to lidar and IfSAR bare-earth elevation errors that typically do not have a normal distribution (bell curve) above and below the true elevations with mean error near to zero
- The NDEP guidelines included other guidance, such as the use of NOAA, 1997 to survey QA/QC checkpoints and the proper and improper locations of such checkpoints.
- Most but not all of the NDEP guidelines are now included in ASPRS, 2014. The main difference between the NDEP guidelines and ASPRS, 2014 is that the NDEP uses FVA, CVA and SVAs whereas ASPRS, 2014 uses NVA and VVA
- ASPRS, 2014 combines the NDEP's FVA (open terrain) + SVA (urban terrain) into a single NVA metric for non-vegetated terrain; and it also combines the NDEP's SVA (weeds and crops) + SVA (brush lands) + SVA (fully forested) into a single VVA metric for all types of vegetated terrain. Users may still specify requirements for FVA, CVA and/or SVA, especially if they need to discriminate vertical accuracies in various forms of vegetated terrain

USACE EM 1110-1-1000, PHOTOGRAMMETRIC AND LIDAR MAPPING

The USACE, in its Engineer Manual 1110-1-1000, Photogrammetric and LiDAR Mapping (USACE, 2015) strongly endorses ASPRS, 2014 while providing specific guidance for 63 common USACE applications.

The generic ASPRS horizontal and vertical accuracy standards (Tables 3.2 and 3.6) leave users to determine their own RMSEx, RMSEy and RMSEz requirements; this is a challenge for some. Whereas Tables 3.3, 3.5 and 3.7 provide dozens of examples for digital orthoimages, planimetric data and elevation data, users still may not know their accuracy requirements for specific applications. For these reasons, USACE established standardized horizontal and vertical accuracy criteria for 63 common USACE applications relevant to military construction, civil works, and real estate activities of the Corps, in both metric and English units. Table 3.14, below, shows the first seven of those 63 applications, including horizontal accuracy criteria (typical target map scale/scale ratio, horizontal RMSE, and horizontal control survey order and class) and vertical accuracy criteria (typical contour interval, vertical RMSE, and vertical control survey order and class).

Project or Activity	Horizontal Accuracy Criteria			Vertical Accuracy Criteria		
MILITARY CONSTRUCTION (MCA, MCAF, OMA, OMAF):	Typical Target (Plot) Map Scale/Scale Ratio 1 in = x ft	Feature Horizontal Accuracy RMSExy cm/ft	Ground Horizontal Control	Typical Contour Interval cm/ft	Feature Vertical Accuracy RMSEz cm/ft	Ground Vertical Control
Design and Construction of New Facilities: Site Plan Data for Direct Input into CADD 2-D/3-D Design Files						
General Construction Site Plan Feature and Topographic Detail	1:400/33 ft.	10 cm/ 0.33 ft	3rd-I	15 cm/0.5 ft	5 cm/ 0.167 ft	3rd
Surface/Subsurface Utility Detail	1:400/ 33 ft.	10 cm/0.33 ft	3rd-I	N/A	5 cm / 0.167 ft	3rd
Building or Structure Design	1:200/ 16.7 ft	2.5 cm / 0.083 ft	3rd-I	7.5 cm/ 0.25 ft	2.5 cm/ 0.083 ft	3rd
Airfield Pavement Design Detail	1:200/ 16.7 ft	2.5 cm/ 0.083 ft	3rd-I	7.5 cm/ 0.25 ft	2.5 cm/ 0.083 ft	2nd
Grading and Excavation Plans (Roads, Drainage, etc.)	1:600/ 50 ft	15 cm/ 0.5 ft	3rd-I/II	30 cm/ 1 ft	10 cm/ 0.33 ft	3rd
Maintenance and Repair (M&R), or Renovation of Existing Structures, Roadways, Utilities, etc., for Design/Construction/ Plans and Specifications (P&S)	1:400/ 33.3 ft	10 cm/ 0.33 ft	3rd-I	30 cm / 1 ft	10 cm/ 0.33 ft	3rd
Recreational Site P&S (Golf Courses, Athletic Fields, etc.)	1:2400/ 200 ft	60 cm/ 2 ft	3rd-I/II	60 cm/ 2 ft	20 cm/ 0.67 ft	3rd

Table 3.14. Surveying and mapping accuracies for selected USACE applications

GEODETIC CONTROL STANDARDS AND GUIDELINES

Appendix B of this manual provides definitions for the absolute accuracy and relative accuracy of any point in a geospatial dataset, as well as definitions for the network accuracy and local accuracy of control points which are typically monumented with some degree of stability and permanency. The prior tables in this chapter specify absolute accuracy when a column refers to RMSE values or accuracies at the 95% confidence levels or 95th percentile. Other columns specify relative accuracy when they compare differences between adjoining images or elevation datasets or differences between overlapping lidar swaths, for example.

Local accuracy is very similar to what has previously been called relative accuracy, and network accuracy is very similar to what has previously been called absolute accuracy, but it is nearly impossible to know the absolute accuracy of any geospatial data because all surveys are relative to something else. Yet, we come as close as we can to determination of absolute accuracy in the U.S. when we use Continuously Operating Reference Stations (CORS) supported by the National Geodetic Survey (NGS). For National Spatial Reference System (NSRS) network accuracy classification in the U.S., the datum is considered to be best expressed by the geodetic values at CORS stations supported by NGS; by this definition, the local and network accuracy values at CORS sites are considered to be infinitesimal, i.e., to approach zero.

Figure 3.1 shows an example NGS Data sheet (PID AJ4599) where the horizontal and vertical network accuracy values are less than one centimeter at 95% confidence levels. If this monument was used as the nearby base station for a local GPS survey, the resultant relative (or local) accuracies would be expected to have excellent absolute (or network) accuracies as well.

The author of this chapter knows of two adjoining lidar datasets in a flat county where both datasets had been independently certified as satisfying 1-foot contour accuracy requirements (RMSEz < 10 cm); yet along the entire seamline between the two lidar datasets, there was an erroneous 2 foot offset between the two surfaces. When hired to independently determine how this could be, we concluded that one lidar dataset had been correctly surveyed relative to an NSRS monument with good network accuracy whereas the second lidar data had not. The second lidar firm, and the survey firm that performed accuracy testing, had both used an inaccurate local monument so that systematic errors canceled out and the dataset appeared to pass.

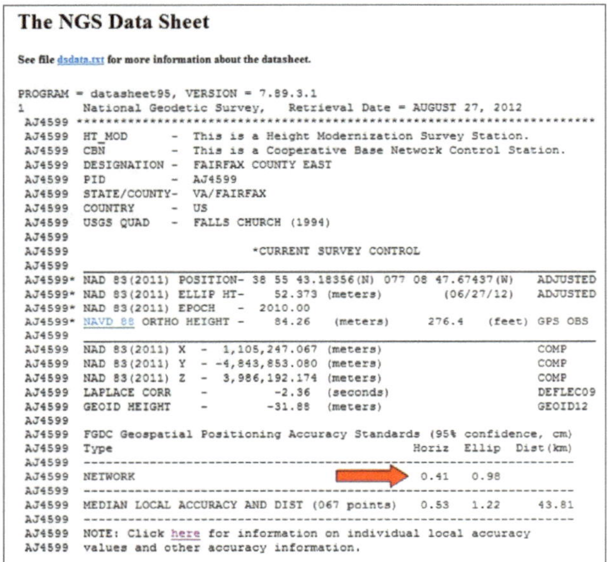

The NGS Data Sheet

See file dsdata.txt for more information about the datasheet.

```
PROGRAM = datasheet95, VERSION = 7.89.3.1
1        National Geodetic Survey,   Retrieval Date = AUGUST 27, 2012
 AJ4599 *********************************************************************
 AJ4599  HT_MOD       -  This is a Height Modernization Survey Station.
 AJ4599  CBN          -  This is a Cooperative Base Network Control Station.
 AJ4599  DESIGNATION -  FAIRFAX COUNTY EAST
 AJ4599  PID          -  AJ4599
 AJ4599  STATE/COUNTY-  VA/FAIRFAX
 AJ4599  COUNTRY      -  US
 AJ4599  USGS QUAD    -  FALLS CHURCH (1994)
 AJ4599
 AJ4599                         *CURRENT SURVEY CONTROL
 AJ4599  ___________________________________________________________________
 AJ4599* NAD 83(2011) POSITION- 38 55 43.18356(N) 077 08 47.67437(W)   ADJUSTED
 AJ4599* NAD 83(2011) ELLIP HT-    52.373 (meters)        (06/27/12)   ADJUSTED
 AJ4599* NAD 83(2011) EPOCH    -  2010.00
 AJ4599* NAVD 88 ORTHO HEIGHT -    84.26   (meters)     276.4   (feet) GPS OBS
 AJ4599  ___________________________________________________________________
 AJ4599  NAD 83(2011) X  -  1,105,247.067 (meters)                     COMP
 AJ4599  NAD 83(2011) Y  - -4,843,853.080 (meters)                     COMP
 AJ4599  NAD 83(2011) Z  -  3,986,192.174 (meters)                     COMP
 AJ4599  LAPLACE CORR    -          -2.36  (seconds)                   DEFLEC09
 AJ4599  GEOID HEIGHT    -         -31.88  (meters)                    GEOID12
 AJ4599
 AJ4599  FGDC Geospatial Positioning Accuracy Standards (95% confidence, cm)
 AJ4599  Type                                        Horiz  Ellip  Dist(km)
 AJ4599  -------------------------------------------------------------------
 AJ4599  NETWORK                                      0.41   0.98
 AJ4599  -------------------------------------------------------------------
 AJ4599  MEDIAN LOCAL ACCURACY AND DIST (067 points)  0.53   1.22    43.81
 AJ4599  -------------------------------------------------------------------
 AJ4599  NOTE: Click here for information on individual local accuracy
 AJ4599  values and other accuracy information.
```

Figure 3.1. Example NGS Data Sheet showing sub-centimeter network accuracies of an NSRS monument relative to horizontal and vertical datums.

In some states, vertical errors of survey monuments and benchmarks in the NSRS are known to be in error by several feet, causing significant errors and inconsistencies in elevation datasets produced relative to these monuments/benchmarks. They can be inaccurate for a number of reasons:

- Some monuments/benchmarks have never been surveyed relative to CORS or other monuments with network accuracy
- Some monuments/benchmarks have never had their coordinates converted from NGVD29 to NAVD88
- Local surveyors often know which benchmarks are high and which are low, but they don't know which are the most accurate

- Some coordinates change over time as a result of subsidence, uplift, crustal movement, construction activities or other local disturbances. Although these changes are mostly gradual, changes can be abrupt as a result of seismic events. Unless monuments are repeatedly surveyed relative to CORS, their changing coordinates are difficult to track
- Even the coordinates of CORS change, per their changing International Terrestrial Reference Frame (ITRF) coordinates. For this reason, NGS has an initiative to establish 12-15 Foundation CORS in the U.S. with state-of-the-art equipment such as GNSS (GPS, GLONASS, Galileo) compatible receivers and antennas, shallow/deep brace monumentation, and improved communications infrastructure. Foundation CORS locations are selected that complement a global effort for an improved geodetic infrastructure; have another geodetic observing system such as VLBI, DORIS, and SLR nearby; and support an operational time frame of 25-50 years so global scientific interests such as sea level change, subsidence can be monitored
- All conventional and GPS surveys are relative, where errors typically depend upon distances and procedures used

Although passive survey monuments are expected to remain in use for some time, they are impossible to maintain at a national level, and there is a movement to rely totally on CORS and Foundation CORS in the future, especially when the new geometric and gravimetric datums are introduced in the 2020s. In the meantime, geospatial firms should ensure that they rely upon CORS stations and/or monuments in the NSRS that have established network accuracies, as demonstrated with the NGS Data Sheet at Figure 3.1 on page 78.

AUTHOR BIOGRAPHY

Dr. Dave Maune, Ph.D of Dewberry worked closely with Dr. Qassim Abdullah of Woolpert, Hans Karl Heidemann of USGS, and Doug Smith of David C. Smith and Associates, Inc. for three years, to develop and coordinate the ASPRS Positional Accuracy Standards for Digital Geospatial Data (ASPRS, 2014) and gain approval for its publication. Hans Karl Heidemann also worked tirelessly for many years to author continued improvements to the USGS Lidar Base Specification (Heidemann, 2018). These two documents are vital for all forms of DEMs and especially lidar data for the 3DEP.

Dr. Maune is the primary author of the NDEP Guidelines for Digital Elevation Data (NDEP, 2004) as well as USACE EM 1110-1-1000, Photogrammetric and LiDAR Mapping (USACE, 2015).

The author also acknowledges the support of Megan Greenaway of NOAA for her contributions to NOS Specifications; the support of Neil Weston of NOAA for his information regarding NGS' Foundation CORS; and the support of Jennifer Novac of Dewberry for her contribution to the section on LAS classification.

REFERENCES

ASPRS (1990). ASPRS accuracy standards for large-scale maps. In *Photogrammetric Engineering and Remote Sensing*, Vol 56, No. 7, (pp. 1068-1070). Bethesda, MD: American Society for Photogrammetry and Remote Sensing.

ASPRS (2004). *ASPRS guidelines, Vertical Accuracy Reporting for Lidar Data*. Bethesda, MD: American Society for Photogrammetry and Remote Sensing, May 24, 2004.

ASPRS (2013). *LAS specification, Version 1.4—R13*. Bethesda, MD: American Society for Photogrammetry and Remote Sensing, 15 July, 2013.

ASPRS (2014). ASPRS positional accuracy standards for digital geospatial data. In *Photogrammetric Engineering and Remote Sensing*, 81(3) (pp. A1-A26), Bethesda, MD: American Society for Photogrammetry and Remote Sensing.

Bureau of the Budget (1947). *United States National Map Accuracy Standards*. Washington, D.C.: U.S. Bureau of the Budget.

FGDC (1998). *Content Standard for Digital Geospatial Metadata* (version 2.0), FGDC-STD-001-1998. Reston, VA: Federal Geographic Data Committee.

FGDC (1998a). *Geospatial Positioning Accuracy Standards*, Part 1: reporting methodology, FGDC-STD-007.1-1998. Reston, VA: Federal Geographic Data Committee.

FGDC (1998b). *Geospatial Positioning Accuracy Standards,* part 2: standards for geodetic networks, FGDC-STD-007.2-1998. Reston, VA: Federal Geographic Data Committee.

FGDC (1998c). *Geospatial Positioning Accuracy Standards*, part 3: national standard for spatial data accuracy, FGDC-STD-007.3-1998. Reston, VA: Federal Geographic Data Committee.

FGDC (2002). *Geospatial Positioning Accuracy Standards*, part 4: standards for architecture, engineering, construction (A/E/C) and facilities management, FGDC-STD-007.4-2002. Reston, VA: Federal Geographic Data Committee.

FGDC (2005). *Geospatial positioning accuracy standards*, part 5: standards for nautical charting hydrographic surveys, FGDC-STD-007.5-2005. Reston, VA: Federal Geographic Data Committee.

FGDC (2008). *Geographic information framework data content standard*, part 0: base document, FGDC-STD-014.0-2008. Reston, VA: Federal Geographic Data Committee.

FGDC (2008a). *Geographic information framework data content standard*, part 1: cadastral, FGDC-STD-014.1-2008. Reston, VA: Federal Geographic Data Committee.

FGDC (2008b). *Geographic information framework data content standard*, part 2: digital orthoimagery, FGDC-STD-014.2-2008. Reston, VA: Federal Geographic Data Committee.

FGDC (2008c). *Geographic information framework data content standard*, part 3: elevation, FGDC-STD-014.3-2008. Reston, VA: Federal Geographic Data Committee.

FGDC (2008d). *Geographic information framework data content standard* part 4: geodetic control, FGDC-STD-014.4-2008. Reston, VA: Federal Geographic Data Committee.

FGDC (2008e). *Geographic information framework data content standard*, part 5: governmental unit and other geographic area boundaries, FGDC-STD-014.5-2008. Reston, VA: Federal Geographic Data Committee.

FGDC (2008f). *Geographic information framework data content standard*, part 6: hydrography, FGDC-STD-014.6 -2008. Reston, VA: Federal Geographic Data Committee.

FGDC (2008g). *Geographic information framework data content standard* part 7: transportation base, FGDC-STD-014.7-2008. Reston, VA: Federal Geographic Data Committee.

Heidemann, Hans Karl (2018). Lidar base specification (ver. 1.3, February 2018). *U.S. Geological Survey Techniques and Methods*, book 11, Chap. B4, 67 p. with appendixes. Reston, VA: U.S. Geological Survey.

IHO (2008). *IHO standards for hydrographic surveys*, special publication No. 44, 5th edition. Monaco International Hydrographic Organization.

NDEP (2004). *NDEP guidelines for digital elevation data*, Version 1.0, May 10, 2004. Reston, VA: National Digital Elevation Program.

NOAA (1997). *NOAA technical memorandum NOS NGS-58*, Guidelines for Establishing GPS-Derived Ellipsoid Heights (Standards: 2 cm and 5 cm), November, 1997. Silver Spring, MD: National Ocean Service.

NOAA (2008). *NOAA technical memorandum NOS NGS-59*, Guidelines for Establishing GPS-Derived Orthometric Heights, 26 March 2008. Silver Spring, MD: National Ocean Service.

NOAA (2015). *NOS Hydrographic Surveys Specifications and Deliverables* April, 2015. Silver Spring, MD: National Ocean Service.

USACE (2015). *Engineer Manual 1110-1-1000*, Photogrammetric and LiDAR Mapping, 30 April 2015. Washington D.C.: U.S. Army Corps of Engineers.

THE NATIONAL ELEVATION DATASET

Dean B. Gesch, Gayla A. Evans, Michael J. Oimoen, and Samantha T. Arundel

INTRODUCTION

The National Elevation Dataset (NED) is a primary elevation data product that has been produced and distributed by the U.S. Geological Survey (USGS). Since its inception, the USGS has compiled and published topographic information in many forms, and the NED is a significant development in this long line of products that describe the land surface. The NED provides seamless raster elevation data of the conterminous United States (CONUS), Alaska, Hawaii, U.S. island territories, Mexico, and Canada. The NED is derived from diverse source datasets that are processed to a specification with consistent resolutions, coordinate system, elevation units, and horizontal and vertical datums. The NED was developed as the logical result of the maturation of the long-standing USGS elevation program, which for many years concentrated on production of quadrangle-based digital elevation models (DEM). The NED contributes to the elevation layer (Gesch et al., 2009) of The National Map (Kelmelis et al., 2003), and it provides basic elevation information for earth science studies and mapping applications in the U.S. and most of North America.

For over 15 years (1999–2014), the NED served as the flagship elevation product of the USGS. In 2015, the 3D Elevation Program (3DEP) was initiated (see Chapter 5 in this Manual). When the 3DEP initiative became operational, the name "National Elevation Dataset" (and the abbreviation "NED") were retired as the USGS elevation activities and data were rebranded under the 3DEP banner. However, elevation data produced and distributed as part of the NED are still widely used (and distributed by other entities), so there is a continuing need for detailed documentation, including how it was produced, its accuracy, and how it is used. This chapter directly addresses that need for detailed information about the NED. The most recent detailed description of the NED appeared in the 2nd edition of the DEM Users Manual (Gesch, 2007a), and because NED production continued through 2014, the details reported herein provide valuable information for data accessed by the user community from 2007 through 2014. The NED has been widely used in operational applications and research studies and is extensively cited in reports on those activities, so it is important for the user community to have access to information about the NED to better judge how its qualities and characteristics might affect results derived from its use as the elevation data source. Additionally, the NED seamless layers serve as one of the input data sources for the current 3DEP elevation production system (details provided below), so, as with any input data source, an understanding of the data characteristics is critical.

Background, Rationale, and History

The history of USGS DEM production is detailed in Osborn et al., (2001). The production of 7.5-minute quadrangle-based DEMs began in the mid-1970s, and once-over coverage of CONUS at a post spacing of 30 meters or 10 meters was completed in 1999. Production of complete coverage of 10-meter resolution DEMs for CONUS was completed in 2012. The geospatial data community has made extensive use of these quadrangle-based DEMs; however, prior to availability of the NED a significant amount of pre-processing was required to assemble multiple DEMs into a seamless dataset suitable for application. Gesch et al. (2002) describe the steps required for production of an elevation layer, including projection transformation and resampling, datum conversions, mosaicking, edge-matching, sliver filling, filtering for artifact removal, and elevation unit conversion. The consistent, seamless structure of the NED removes the requirement for users to handle and individually process multiple input files. Instead, the NED provides application-ready data and allows users to focus on analysis rather than data preparation.

The NED was designed to address the requirement for large-area coverage of the "best available" elevation data. This approach fulfills the framework data concept as promoted by the National Spatial Data Infrastructure (NSDI) (Federal Geographic Data Committee, 2012), in which the most commonly needed geographic data themes are integrated into coherent datasets based upon source data from multiple public and private entities. The USGS has been designated as the lead agency responsible for national elevation data (Office of Management and Budget, 2002, 2010). As such, the USGS assembled the NED as an implementation of the framework data concept. The NED was the first high-resolution seamless digital geospatial dataset assembled by the USGS, and it was the first seamless dataset developed and made available as part of The National Map (Kelmelis, 2003).

The USGS began the development of seamless elevation datasets in the early 1990s, first working on continental and global datasets at a resolution of 1 kilometer (Gesch et al., 1999). The experience gained in assembling multi-source, multi-resolution elevation data into a global seamless dataset proved to be valuable, as the methods were adapted, refined, and further developed for use in producing regional U.S. elevation framework prototypes in 1996. The first complete seamless coverage of CONUS was finished in 1997 and was based on 10-meter, 30-meter, 2-arc-second, and 3-arc-second resolution DEM source data. In 1999, for the first time, the NED was assembled completely (for CONUS) from 7.5-minute DEM source data (10-meter and 30-meter data). The first "non-standard" (non-USGS produced) source data were integrated into the NED in 2003 with the addition of lidar-derived DEMs over the Puget Sound region. The bimonthly update cycle for the NED began in 2000 and continued through February 2014. Beginning in April 2014 some features of the NED production system changed as a result of a modernization effort (Arundel et al., 2015). One result of the system modernization is a continuous update process to include new source data instead of a bimonthly update cycle. Also, in the modernized system the February 2014 version of the NED serves as the baseline source data layers so that older source data do not have to be continuously reprocessed.

SPECIFICATIONS AND PRODUCTION

The NED is a seamless multi-resolution raster dataset that was regularly updated to integrate newly available, improved elevation source data. To maintain seamlessness in its national coverage, the NED model is cast in a geographic coordinate system (horizontal locations are referenced in decimal degrees of latitude and longitude). NED data are available at grid spacings of 1 arc-second (approximately 30 meters) and ⅓ arc-second (approximately 10 meters) for CONUS, Hawaii, portions of Alaska, and the U.S. island territories. Canada and Mexico are also covered by 1-arc-second data, which were one-time additions to the NED, and no updates are done to the NED for these areas. Some parts of Alaska are only available at a 2-arc-second grid spacing where lower resolution source data are used, although the state is undergoing a significant upgrade in base map data and NED has been upgraded accordingly. NED data at a grid spacing of ⅑ arc-second (approximately 3 meters) derived from high-resolution source data are available for about 23% of CONUS (in the final bimonthly release of February 2014), and this portion continues to grow as higher resolution source data are added to the 3DEP elevation products.

The NED production approach ensured that georeferencing of the layers resulted in properly nested and coincident data across the three resolutions used for CONUS. In the context of the raster data model used for the NED, the area represented by one elevation post in the 1-arc-second layer is represented by nine elevation posts in the ⅓-arc-second layer, and by 81 elevation posts in the ⅑-arc-second layer (Figure 4.1). In the original NED production system, where all three resolution layers could be produced each layer was constructed independently from the same high-resolution source data using an aggregation method appropriate to the grid spacing being produced. In the modernized system, the same procedure is used in all but two cases. In the first case (for production of ⅓-arc-second data), the values are taken directly from the baseline February 2014 NED ⅓-arc-second layer where no higher resolution data exist, or the values are aggregated where higher resolution data cover the entire cell. In locations where higher resolution data occupy but do not fill the entire cell they are aggregated and blended with values from the baseline February 2014 NED ⅓-arc-second layer. In the second case (for production of ⅑-arc-second data), the values are calculated in a similar fashion to the ⅓-arc-second layer but the February 2014 NED ⅑-arc-second layer serves as the baseline source data.

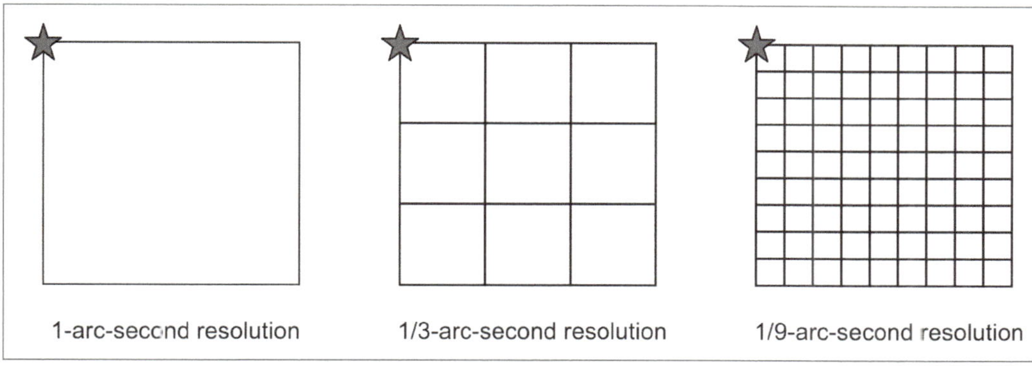

Figure 4.1. NED nested multi-resolution raster elevation layers. The area represented by one elevation post (or cell) in the 1-arc-second layer is represented by nine elevation posts in the ⅓-arc-second layer and by 81 elevation posts in the ⅑-arc-second layer.

In all three NED resolution layers, the elevation units are standardized to decimal meters. The horizontal datum for the NED is the North American Datum of 1983 (NAD 83), and the vertical datum is the North American Vertical Datum of 1988 (NAVD 88), except for some islands where a local sea level datum is used. Federal Geographic Data Committee (FGDC) compliant metadata (Federal Geographic Data Committee, 1998) are provided for efficient documentation. Additionally, spatially referenced metadata are supplied as an attributed geospatial data layer that captures all metadata from the source DEMs and NED processing steps on a polygonal footprint basis (Figure 4.2).

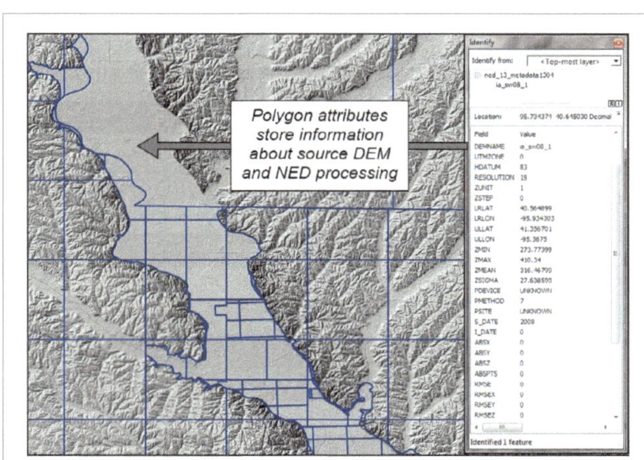

Figure 4.2. NED spatially referenced metadata layer.

The NED is a logically seamless dataset. To address practical concerns of data processing and storage, the NED was assembled, processed, and stored on a tiled basis, and the tiles collectively form the national mosaic. In the original NED production system, a 1- by 1-degree tile (with integer degree boundaries) was used as the unit for assembly and processing of the 1-arc-second and ⅓-arc-second layers. Adjacent tiles were produced with a small amount of overlap to ensure the seamless nature of the dataset. In its entirety, the NED comprises more than 3,800 1- by 1-degree tiles. Additional tiles were added as required to accommodate new areas of coverage (Table 4.1 on page 86).

Release Date	Number of Tiles	Note
June 2000	1,367	CONUS: 925 tiles; Alaska: 428 tiles; Hawaii: 14 tiles
April 2001	1,375	8 tiles added: Puerto Rico and U.S. Virgin Islands
June 2001	1,387	12 tiles added: Pacific islands
August 2001	1,392	5 tiles added: Pacific islands
October 2008	1,651	259 tiles added: Mexico
April 2012	2,073	422 tiles added: Canada data adjoining CONUS
June 2012	2,159	86 tiles added: Canada data adjoining Alaska
October 2012	2,188	29 tiles added: Canada data over Gulf of Maine
August 2013	3,836	1,648 tiles added: completed Canada data coverage

Table 4.1. Number of NED 1- by 1-degree tiles and changes by release date.

The 1/9-arc-second layer is processed and stored on a source data project area basis, but it is packaged into 15- by 15-minute tiles for distribution. Currently in the modernized production system, the NED is maintained as a single integrated mosaic dataset, and seam lines are processed (blended) for new data on-the-fly as they are added to the NED. Sorting and manipulation of the raster source data occur in this storage mechanism, and then tiles are produced as required to reflect updated areas.

Source Data

When a NED tile was assembled, the best available source data were selected according to the following criteria (ordered from highest priority to lowest priority):

- High-resolution data most often derived from lidar, but they may also be derived from digital photogrammetry or interferometric synthetic aperture radar (IfSAR) data. These data often have water bodies flattened. If collected at a ground sample distance of better than 5 meters, which was most often the case, then the source data were used to update all the resolution layers of the NED
- 10-meter quadrangle map-based DEMs derived from 1:24,000-scale cartographic contours and mapped hydrography. In most cases, these data have been produced by the USGS as a standard elevation product. In 2012 coverage of 10-meter DEMs was completed for CONUS, and these data are the source used for most of the NED where there is no lidar or other high-resolution source available.
- 2-arc-second (approximately 60-meter) DEMs derived from 1:63,360-scale cartographic contours. These DEMs are a product of the USGS and are used as the source data for much of Alaska
- 1-arc-second Shuttle Radar Topography Mission (SRTM) data (Farr et al., 2007) are used for a portion of the Aleutian Islands in Alaska where there are no other better quality source data available
- 0.75-arc-second DEMs derived from cartographic and/or remote sensing sources for Canada. These data are known as Canadian Digital Elevation Data (CDED) (see *http://www.geobase.ca/geobase/en/data/cded/*)
- 1-arc-second DEMs derived from cartographic sources for Mexico
- USGS 30-meter (Level 1 and Level 2) DEMs (Osborn et al., 2001) and 3-arc-second DEMs are only used as a source of fill values over large water bodies

The continuing goal has been to upgrade the NED, emphasizing high-resolution lidar data as the primary source. As specified in the NED spatially referenced metadata, in the final bimonthly release of February 2014 more than 20% of the 1/3-arc-second and 1-arc-second NED layers for CONUS were derived from lidar or other high-resolution source data, with the remaining 80% derived from standard 10-meter DEMs based on cartographic contours. The resolution and quality of the source data determined the highest resolution NED layer (1/9, 1/3, or 1-arc-second) that was produced for any given area (Figure 4.3).

NED was updated regularly to incorporate all new DEM production, thus retaining the "best available data" framework concept. With the original NED production system, a new version of the NED was released every two months and contained updated areas for which new source DEMs became available since the previous release. On average, approximately 5 to 10% of the total number of 1- by 1-degree tiles were updated at each release. An update was required for a tile if even one new DEM file was available. The spatially referenced metadata provided with each NED bimonthly release indicated the specific areas where new source DEMs were incorporated since the previous release. In the modernized production system, the addition of a new source DEM triggers an automatic update of all required layers and tiles, sending the output products directly to The National Map store.

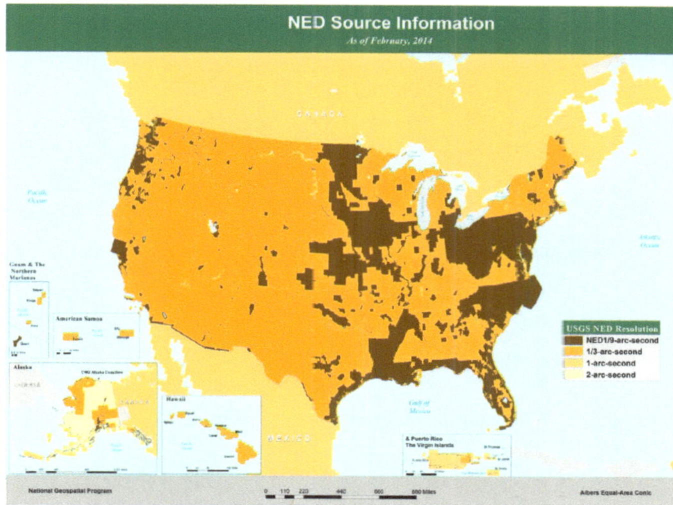

Figure 4.3. The highest resolution NED layer produced is based on the resolution of the NED source data.

Given the variability in the quality of available high-resolution source data, a set of decision rules is applied to determine the best source data for each of the three NED resolution layers. Newer collections of high-resolution data, especially lidar data, have improved quality because the data are collected to a USGS specification (Heidemann, 2012) that is designed to result in data suitable for direct integration into the NED. Ideally, the available high-resolution source data would have been used to update all three NED layers, but in some cases lower data quality does not allow for upgraded NED data in all layers. The newly available source data may have improved resolution over the current source data, but artifacts in the new data outweigh the advantage of the improved resolution. In these cases, the data were used to produce NED 1/9-arc-second data so that the user community has ready access to the data in the format of the NED data model, but the new source data were not migrated into the 1/3- and 1-arc-second layers. The release notes previously issued with each NED bimonthly update cycle showed the status of the migration of high-resolution source data into the 1/3- and 1-arc-second layers. The modernized production system provides this information via downloadable status graphics that are revised with each new update. The following criteria were used to determine the "best available" data for use as NED source data:

- Reject for all NED layers:
    - Water bodies higher than adjacent ground
    - Large data voids
    - Severe mismatches along tile edges
    - Poor bare earth processing (excessive residual vegetation and/or buildings)

- Use only in ⅑-arc-second layer:
    - Water bodies not flattened
    - Excessive and/or large triangulated irregular network (TIN) artifacts in areas of sparse lidar point density
    - Some residual vegetation and/or buildings in bare earth grids
- Use in ⅑-arc-second layer, and migrate to ⅓-arc-second and 1-arc-second layers:
    - Water bodies flattened
    - Very minimal TIN artifacts
    - Very minimal residual vegetation and/or buildings in bare earth grids
    - Replaces very poor current source data
- Use only in ⅓-arc-second and 1-arc-second layers:
    - Nominal resolution is coarser than 5 meters

These rules are a general guideline and are applied on a case-by-case basis to determine the path for each candidate source data collection. Other factors than those listed above may weigh heavily in some cases, such as the size of the area covered by a collection (for example, an entire state) or agreements made with partners to include their contributed data in the NED. In some cases, the USGS performed post-processing on source data to make them acceptable for inclusion in the NED. Prior to April 2014, when the quality of high-resolution source data allowed for integration into all three NED layers, the finished NED ⅑-arc-second data served as the source for updating the corresponding areas in the ⅓-arc-second and 1-arc-second layers. In areas updated since April 2014, all output layers are derived directly from the highest resolution source data (except as previously noted).

Production Processing

In the original NED production process, NED assembly and updating were accomplished with a highly automated production process (Stoker et al., 2009) that was developed specifically for the seamless elevation layers. The main steps in the process are outlined on the top of Figure 4.4. The selection of source data was accomplished by an automated query of a database of USGS and partner-supplied DEMs. The "best available" data criteria outlined above were applied to select specific file-based DEMs from the database. The headers of the selected DEMs were read to create an index, which controlled production and triggered subsequent processing steps. Industry standard tools from the National Geodetic Survey (NGS), NADCON and VERTCON, were used to perform horizontal and vertical datum transformations, respectively.

The source DEMs were projected from their native coordinate system to a geographic coordinate system (decimal degrees of latitude and longitude) using standard cartographic transformation software. Resampling of the original elevation values was done with a windowed sinc interpolation function similar to cubic convolution, which was optimized for elevation data by maintaining the integrity of shorelines and water bodies. Additional low pass or anti-alias filtering was selectively applied to avoid the introduction of false striping patterns in the output tile. Note that the artifact removal step in Figure 4.4 (top) was an important part of the NED production process for early versions of the NED that used older USGS 30-meter Level 1 DEMs that were produced with photogrammetric methods. These DEMs often exhibited high frequency linear artifacts, and the "mean profile filter" (Oimoen, 2000), which was specifically designed to process these DEMs, was quite effective at improving these source data. In the more recent and current versions of the NED, the 30-meter Level 1 DEMs are no longer used for any land areas because they have been replaced by better source data.

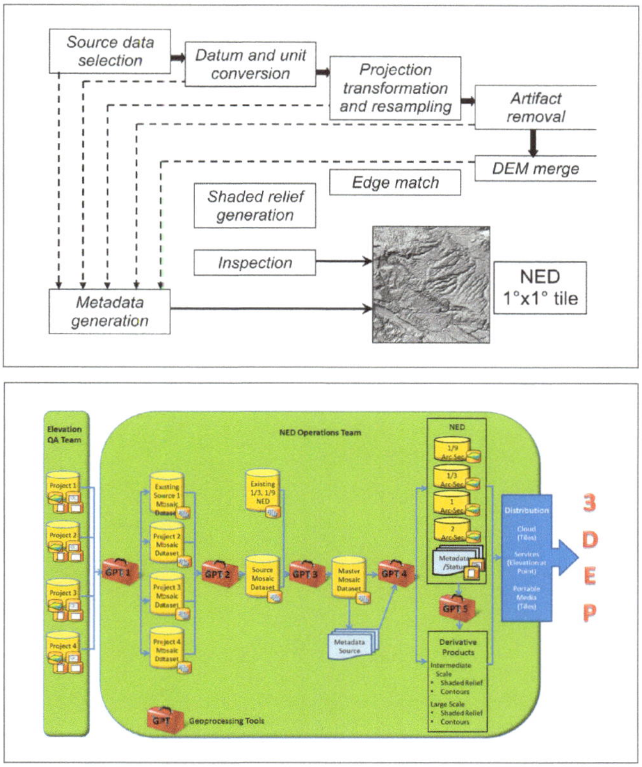

Figure 4.4. The original NED production process (top), and the current modernized process (bottom).

The final steps in the production process included paneling the DEMs to fill the 1- by 1-degree tile, filling slivers of missing data along DEM boundaries as necessary by interpolating adjacent values, and edge matching the seams along the DEM boundaries. The edge match algorithm used a distributed residual feathering approach that maintained local slope continuity across the seam (Stoker et al., 2009). Finally, a shaded-relief image of the tile was generated for inspection by an analyst to verify successful processing. In some cases, erroneous values occurred in the DEM data or the source files contained incorrect header information, and were not acceptable for inclusion into the NED. These cases were detected in the visual inspection step, the DEM was edited, if possible, or rejected entirely, and the tile was submitted for reprocessing. Tiles passing inspection were transferred to the next release version, and the spatially referenced metadata were updated accordingly.

In the modernized production system (Figure 4.4, bottom), the February 2014 NED tiles are managed through a source baseline mosaic dataset, which is in turn managed by the master NED mosaic dataset. A candidate source dataset is added temporarily to the master mosaic dataset and compared against existing source data. After the new source data are accepted for inclusion in the NED and ordered among the overlapping DEMs (according to the "best available" data criteria outlined above), the required tiles are exported with the mosaic dataset properties so that any required projection transformation, resampling (with bilinear interpolation), and pixel alignment are accomplished simultaneously in creating the output products. An inspection cycle verifies the validity of the final products before they are released for distribution. Data distribution is accomplished through The National Map (see *http://nationalmap.gov/viewer.html#download_ned*).

The National Map interactive Viewer (Figure 4.5 on page 90) allows a user to display the NED data source index, which indicates the date of the most recent update, the resolution of the source data, the production method of the source data, and the currency (date) of the source data for specific areas. The user can also query the spatially referenced metadata to examine additional information about each file-based DEM used to assemble the NED. The NED website (*ned.usgs.gov*) contains documentation on the NED assembly process, accuracy, standards, metadata, data distribution, release notes, research, publications, and answers to frequently asked questions.

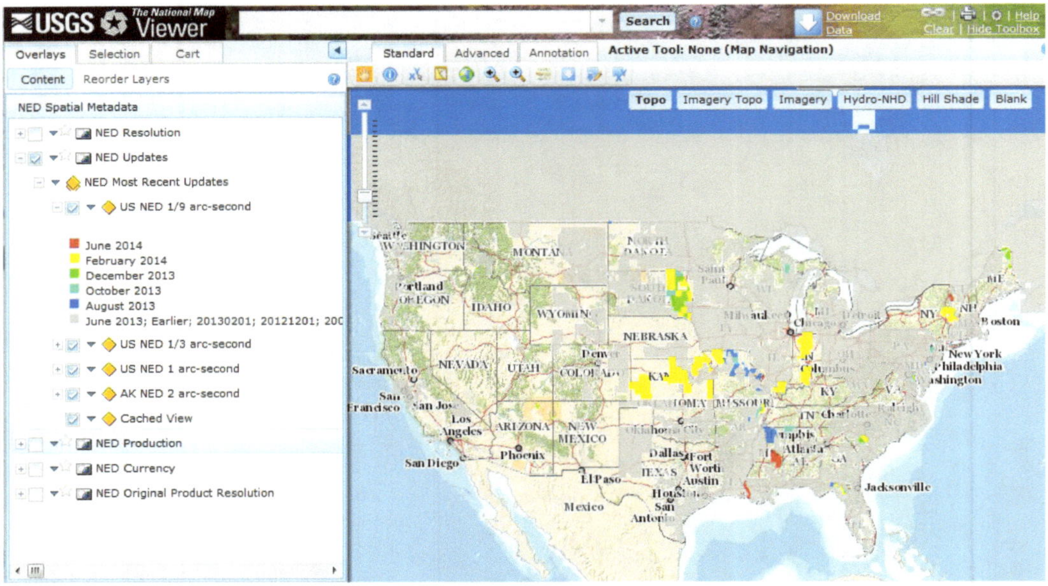

Figure 4.5. NED data source index on The National Map Viewer.

Spatially Referenced Metadata

The NED spatially referenced metadata, an attributed geospatial data layer that captures all metadata from the source DEMs and NED processing steps on a polygonal footprint basis (Figure 4.2 on page 85), are a unique resource for information that can be very beneficial to the NED user. The spatially referenced metadata (often referred to as NED "spatial metadata") are repackaged with each update of the NED. The spatial metadata allow every elevation value in each of the NED layers to be traced back to its source. The attributes of the polygons fully describe the source data used to produce the NED. A full description of the attribute fields and valid values is given in the NED Data Dictionary (see *http://staging-ned.usgs.gov/documents/NED_DataDictionary.pdf*). The most commonly used attributes are source data resolution, production method, source date, and date of most recent NED update. The information displayed in the NED data source index (Figure 4.5) is pulled directly from the NED spatial metadata. Attributes also include other fields that describe the source data, including horizontal and vertical datums, elevation units, accuracy (if provided by the data producer), elevation statistics, and production site and date. In the case of USGS 7.5-minute quadrangle-based DEMs, many of the attribute fields of the NED spatial metadata are filled with values read directly from the header record of the native format DEM file. The spatial metadata also include attribute fields that record parameters used to transform the source data during NED production processing, including horizontal and/or vertical datum adjustment values.

Because the NED 1/9-arc-second data are derived from high-resolution source data that do not share many of the descriptors found in the file header records of USGS quadrangle-based cartographic source DEMs, the spatial metadata for the 1/9-arc-second layer include additional fields. An important field in this case is the link to the online version of the full FGDC-compliant metadata for the source data provided by the vendor.

An example of the usefulness of the information contained within the NED spatial metadata is seen in the National Biomass and Carbon Dataset for the year 2000 (NBCD2000) developed by the Woods Hole Research Center (*http://www.whrc.org/mapping/nbcd/*). The NBCD2000 provides a high-resolution (30-meter) baseline estimate of forest canopy height, aboveground biomass, and standing carbon stock for CONUS. A key component in the development of the NBCD2000 was the differencing of SRTM data and the NED to derive raw estimates of canopy heights (Kellndorfer et al., 2004). The NED spatial metadata provided important information on the spatially variable data quality of the NED (primarily due to source DEM production method) that helped in the analysis of observed anomalous features in the SRTM-NED difference images (Walker et al., 2007). The 1-arc-second NED layer provided the required bare earth elevation model for comparison with SRTM to derive canopy height/biomass/carbon data for CONUS, and the NED spatial metadata provided critical information to help interpret the raw elevation difference datasets.

In addition to the spatial metadata, as part of the modernization effort the NED collection-level metadata have been through an extensive review and improvement process, resulting in output that not only meets FGDC standards but also directly feeds Data.gov, which allows the data to be discovered by any public metadata search. New metadata returns include a direct link to the FTP data download, eliminating the need to repeat the search in The National Map Viewer.

ACCURACY AND DATA QUALITY

Users of the NED are often interested in its accuracy, as that characteristic is a primary factor in how suitable the data are for a specific application and in the quality of derivative products. For example, the vertical accuracy of the NED directly controls the water level increment that can be effectively modeled in sea-level rise vulnerability assessments (Gilmer and Ferdaña, 2012; Gesch, 2013; Thatcher et al., 2013.) The accuracy of the NED varies spatially because of the variable quality of the source data. As such, the NED inherits the accuracy of the source DEMs. In an effort to provide detailed information to users on the vertical accuracy of the NED, the April 2013 version of the dataset was tested by comparing it with an independent reference source of very high accuracy (Gesch et al., 2014.) The reference data are the geodetic control points that NGS uses for development of its hybrid geoid models (Smith and Roman, 2001), in this case model GEOID12A. The distribution of this set of more than 25,000 survey points, referred to by NGS as "GPS on bench marks" (GPSBM) (National Geodetic Survey, 2012), across North America is shown in Figure 4.6.

Absolute Vertical Accuracy

To complete the accuracy assessment, the NED elevation value at each NGS control point location was derived through bilinear interpolation, and error statistics were calculated. The assessment was conducted by area and NED resolution for CONUS (⅓-arc-second data) (Table 4.2 on page 92), Alaska (2-arc-second data), and Canada and Mexico (1-arc-second data) (Table 4.3 on page 92). The overall absolute vertical accuracy for each area is expressed as the root mean square error (RMSE), a widely used error metric for documenting elevation data accuracy (Maune et al., 2007). Table 4.2 also contains the accuracy expressed

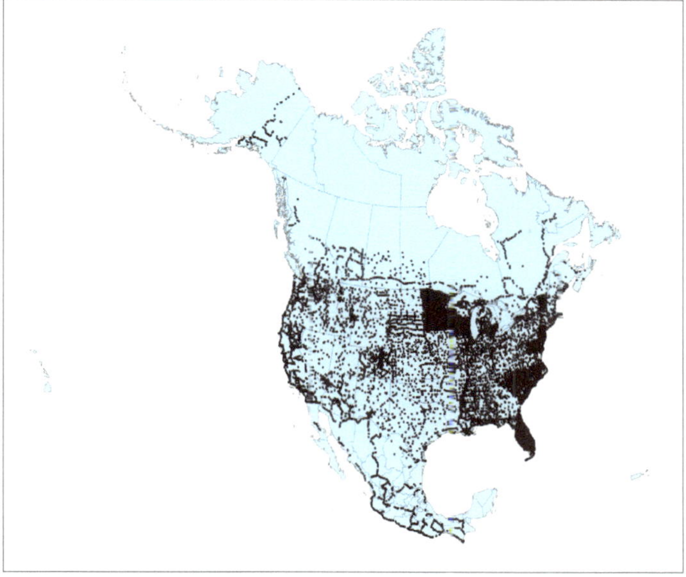

Figure 4.6. The reference control point dataset used for accuracy assessment of the NED includes more than 25,000 points from the "GPS on bench marks" associated with NGS's GEOID12A model.

in terms of the National Map Accuracy Standards (NMAS), which use a 90% confidence level, and in terms of the FGDC National Standard for Spatial Data Accuracy (NSSDA), which uses a 95% confidence level. The methods described in Maune et al. (2007) were used to convert the measured vertical RMSE to equivalent NMAS and NSSDA expressions. Although the RMSE is a widely used error metric, it has been noted that in many cases elevation errors do not follow a normal distribution, so a key assumption in the NSSDA approach to computing the 95% confidence level based on the RMSE may be violated (Liu et al., 2012). To address this condition, the National Digital Elevation Program (NDEP) has recommended use of the 95th percentile method of expressing accuracy (National Digital Elevation Program, 2004; Maune et al., 2007). For completeness, Table 4.2 also includes the CONUS NED accuracy expressed as the 95th percentile error, which is interpreted as stating that 95% of the errors have absolute values less than or equal to the reported value.

NED Version	Number of Reference Points	Minimum	Maximum	Mean	Standard Deviation	RMSE	NMAS (90%)	NSSDA (95%)	NDEP (95th percentile)
April 2013	25,310	-24.64	15.57	-0.29	1.52	1.55	2.55	3.04	3.02
June 2003	13,305	-42.64	18.74	-0.32	2.42	2.44	3.99	4.75	5.59

Table 4.2. NED absolute vertical accuracy for CONUS as measured vs. reference geodetic control points. Error statistics are in meters. Accuracy assessment for the April 2013 ⅓-arc-second version of the NED was done with bench marks associated with GEOID12A from the National Geodetic Survey (NGS). For comparison purposes, also reported are the numbers from an assessment of the June 2003 version of the 1-arc-second NED (Gesch, 2007a) using bench marks associated with GEOID03 from NGS.

NED Area and Resolution	Number of Reference Points	Minimum	Maximum	Mean	Standard Deviation	RMSE
Alaska 2-arc-second	106	-19.64	16.17	-1.44	4.66	4.85
Canada 1-arc-second	578	-18.94	12.29	-0.63	3.53	3.64
Mexico 1-arc-second	675	-20.68	36.66	0.98	6.68	6.74

Table 4.3. NED absolute vertical accuracy as measured vs. reference geodetic control points. Error statistics are in meters. Accuracy assessment for the April 2013 version of the NED was done with bench marks associated with GEOID12A from NGS.

For CONUS, there appears to be no correlation of NED error and GPS bench mark elevation value (Figure 4.7), and there is no preference for negative or positive errors. The scatter plot in Figure 4.7 shows the data points uniformly distributed about the zero error axis, so users can expect a consistent range of errors regardless of elevation magnitude for a given area.

The RMSE of 1.55 meters for CONUS is a marked improvement over the previously published accuracy value of 2.44 meters for an earlier version of the NED (Gesch, 2007a.) The time interval between the two versions of the NED is nearly 10 years, and the source data for much of CONUS has changed significantly, which has resulted in the observed improvement in overall accuracy (Table 4.2). Nearly 61% of CONUS was updated between the June 2003 and April 2013 releases of the NED. About two-thirds of the updated area is due to 10-meter 7.5-minute quadrangle-based DEMs derived from 1:24,000-scale hypsography (contours and spot heights) replacing older 30-meter versions of the source DEMs. The remaining one-third of the updated area is where lidar or other high-resolution source DEMs replaced 30-meter or 10-meter quadrangle-based DEMs.

Examination of the NED error statistics in just the updated area indicates the effects of integrating improved source data between the June 2003 and April 2013 versions of the NED. In June 2003 the area exhibited an RMSE of 2.43 meters, and in April 2013 the same area has an accuracy of 1.29 meters RMSE. Looking in more detail at the new source data types for the updated area exhibits an RMSE of 1.62 meters for the two-thirds of the area where 10-meter DEMs replaced 30-meter DEMs, and an RMSE of 0.87 meter for the one-third of the area where lidar and digital photogrammetry source data replaced USGS quadrangle-based DEMs (10- or 30-meter).

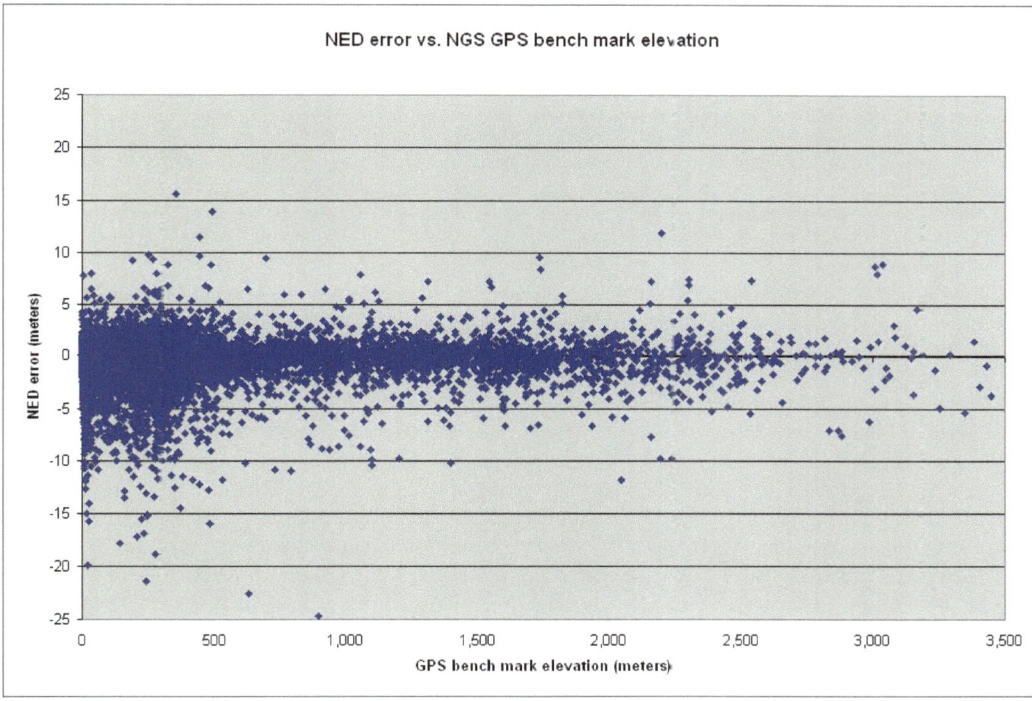

Figure 4.7. NED errors (in meters) plotted against elevation.

Use of the NED spatially referenced metadata allows for calculation of accuracy statistics segmented by source DEM characteristics. Because the NED is derived from source DEMs that were produced with several different methods, it may be important for a user to know what levels of accuracy can be expected for areas based on DEMs produced with the various methods. Table 4.4 on page 94 shows the error statistics for the areas of the NED derived from each of the four primary production methods. The advantages of newer, high-resolution source data are recognized in the better accuracy compared to the accuracy of the 1:24,000-scale quadrangle-based data. In each case, the RMSE for the second, third, and fourth sources in Table 4.4 is smaller than that for the first source.

For the areas in the NED derived from USGS 7.5-minute quadrangle-based DEMs (produced from 1:24,000-scale maps), the 1.89-meter RMSE (Table 4.4) as an accuracy statement is an actual measured quantity, so it has advantages over the quoted RMSE of 7 meters for USGS DEMs. The 7-meter RMSE, often cited as the vertical accuracy of USGS 7.5-minute DEMs, is a production goal described in the USGS Data Users Guide 5—Digital Elevation Models, last published in 1993 and traditionally known by many DEM users as the "blue book" (see also U.S. Geological Survey, 1997). The 7-meter RMSE is not, nor has it ever been, a measured accuracy assessment of the NED.

Source Data Production Method	Number of Reference Points	Minimum	Maximum	Mean	Standard Deviation	RMSE
LT4X (1:24,000 hypsography)	14,813	-24.64	15.57	-0.36	1.85	1.89
Lidar	10,281	-22.53	4.66	-0.19	0.85	0.87
Photogrammetric mass points and breaklines	146	-6.81	1.69	-0.28	1.12	1.15
Airborne digital stereo image correlation	70	-2.15	1.13	0.18	0.45	0.48

Table 4.4. NED absolute vertical accuracy for CONUS as measured vs. reference geodetic control points. The results have been segmented by the four primary types of source data. Error statistics are in meters. Accuracy assessment for the April 2013 ⅓-arc-second version of the NED was done with bench marks associated with GEOID12A from NGS.

The vertical accuracy of the NED also varies by land cover. The National Land Cover Database (NLCD) (Fry et al., 2011) provides land cover information for each reference control point location. Table 4.5 shows how the accuracy for the NED for different land cover types compares to the overall NED absolute vertical accuracy of 1.55 meters RMSE.

For Alaska, segmenting the accuracy assessment results by source data type (based on the spatially referenced metadata) shows the advantages of integrating new, improved source DEMs into the NED. The source data for a majority of Alaska are DEMs derived from 1:63,360-scale topographic maps, but there is an ongoing program to collect improved elevation data for the state, in this case 5-meter spatial resolution DEMs derived from airborne IfSAR. Where available, these new data have been integrated into the NED, resulting in a significant improvement over the older cartographic source data, which have been shown to be lacking in the required accuracy and quality for many applications (Maune, 2008). The overall accuracy for Alaska is 4.85 meters RMSE (Table 4.3 on page 92), but for the areas derived from the IfSAR source DEMs the RMSE is 1.63 meters, which is much better than the RMSE of 6.32 meters for areas derived from the cartographic source DEMs.

One caveat to note about the accuracy assessment presented here is that even though the reference control point dataset is large, the number of source DEMs on which the points are located is relatively small. For CONUS, approximately 12% of the source DEMs have at least one point located within. Thus, if users need very specific accuracy information for the NED for a local area, a separate assessment should be done with suitable reference data just for that area. In addition, even though the reference control points are located broadly across CONUS, the distribution of elevations and terrain conditions within the dataset is not completely representative of the nation's topography. This stands to reason, as surveyed bench marks are generally located in open, accessible areas. Thus, high elevation, steep slope locations are under-represented in the reference dataset. Also, for CONUS about 40% of the reference points are located on 20% of the area covered by high-resolution source data, and 80% of the area covered by cartographic-based source data contains about 60% of the reference points, so proportions of source data and reference points do not match. Such a mismatch in proportions may affect the overall vertical accuracy number because more data points are located in areas of higher accuracy source DEMs. Despite these limitations with the reference data, the overall vertical accuracy reported here is useful for applications that need to factor in the quality of the NED over large areas. As the NED is continually upgraded based on new acquisitions of high-resolution data, the overall vertical accuracy will improve. In many cases, the source datasets will have comprehensive error reports supplied with them, and these statistics will be captured and preserved and linked to the NED metadata.

Land Cover Class	Number of Reference Points	Minimum	Maximum	Mean	Standard Deviation	RMSE
Developed, Open Space	8,247	-24.64	15.57	-0.15	1.31	1.32
Developed, Low Intensity	6,680	-14.91	13.92	-0.27	1.48	1.50
Developed, Medium Intensity	3,081	-15.12	7.88	-0.54	1.80	1.88
Developed, High Intensity	695	-21.43	6.62	-0.79	2.32	2.45
Barren Land	246	-19.88	7.79	-0.86	2.42	2.57
Deciduous Forest	356	-11.80	3.57	-0.33	1.48	1.52
Evergreen Forest	197	-13.09	5.17	-0.62	2.22	2.30
Mixed Forest	83	-9.82	3.66	-0.88	2.20	2.36
Shrub/Scrub	878	-17.84	8.87	-0.51	2.17	2.22
Grassland/ Herbaceous	856	-11.82	7.34	-0.33	1.55	1.58
Pasture/Hay	1,295	-8.82	5.77	-0.18	1.18	1.19
Cultivated Crops	1,889	-11.48	5.77	-0.14	1.01	1.02
Woody Wetlands	414	-15.68	8.91	-0.42	1.59	1.64
Emergent Herbaceous Wetlands	393	-10.10	3.17	-0.51	1.50	1.59
All	25,310	-24.64	15.57	-0.29	1.52	1.55

Table 4.5. NED absolute vertical accuracy for CONUS as measured vs. reference geodetic control points. The results have been segmented by land cover classes from the National Land Cover Database (NLCD) 2006. Error statistics are in meters. Accuracy assessment for the April 2013 ⅓-arc-second version of the NED was done with bench marks associated with GEOID12A from NGS.

Relative Vertical Accuracy

For some applications of elevation data, the relative, or point-to-point, vertical accuracy is more important than the absolute vertical accuracy. Whereas absolute accuracy accounts for the combined effects of systematic and random errors, relative accuracy is a measure of just random errors. The relative vertical accuracy of a dataset is especially important for derivative products that use the local differences among adjacent elevation values, such as slope and aspect calculations. To characterize the relative vertical accuracy of the NED, the same set of reference geodetic control points used in the assessment of absolute vertical accuracy was processed and analyzed. For CONUS, each point in the reference control point dataset was processed to identify its closest neighboring point, and this resulted in 15,509 unique point pairs for which the NED elevation at each point location and the distance between the points were recorded. The relative vertical accuracy, RV, was calculated for each point pair with the following formula (National Digital Elevation Program, 2004):

$$RV = |\Delta ref - \Delta NED|$$

$$\text{where } \Delta ref = |\text{reference elevation difference}|$$
$$\Delta NED = |\text{NED elevation difference}|$$

Because assessing relative accuracy across very long distances can have the effect of averaging random errors (thereby reducing the overall error), a subset of the point pairs with closely spaced points was used to characterize the relative vertical accuracy of the NED (Table 4.6). Averaged over the 1,068 point pairs (with distances of less than 500 meters), the relative vertical accuracy is 0.81 meter. Expressed as the 95th percentile, the relative vertical accuracy is 2.93 meters, meaning that 95% of the point pairs exhibit a relative difference of 2.93 meters or less. Table 4.6 also shows the slope accuracy (expressed as the mean and 95th percentile) as derived from the relative vertical accuracy fit within a 3-by-3 window of raster elevation cells (Gesch, 2007a.) As seen previously in the absolute vertical accuracy reports, integration of high-resolution source data into the NED between June 2003 and April 2013 has improved NED quality, in this case better relative vertical accuracy and the corresponding slope accuracy.

NED Version	Number of Unique Point Pairs	Minimum (meters)	Maximum (meters)	Mean (meters)	Standard Deviation (meters)	95th Percentile (meters)	1-Arc-Second Slope Accuracy (mean)	1-Arc-Second Slope Accuracy (95th percentile)
April 2013	1,068	0.00	10.71	0.81	1.19	2.93	0.77° (1.35% slope)	2.79° (4.88% slope)
June 2003	700	0.00	21.53	1.30	1.79	4.87	1.24° (2.17% slope)	4.64° (8.12% slope)

Table 4.6. NED relative (point-to-point) vertical accuracy for CONUS measured using closely spaced (< 500 meters) NGS geodetic control points as the reference information. Assessment of the April 2013 version of the ⅓-arc-second NED was done with bench marks associated with GEOID12A, and assessment of the June 2003 version of the 1-arc-second NED was done using bench marks associated with GEOID03. Slope accuracy was calculated based on the relative vertical accuracy as described in Gesch (2007a).

Comparison with Other Large-Area Elevation Datasets

A common question from users of the NED is how it compares with other similar seamless elevation datasets that cover broad areas. In particular, two other datasets of interest are SRTM (Farr et al., 2007) and the ASTER Global Digital Elevation Model (GDEM) (Abrams et al., 2010). Both SRTM and GDEM are 1-arc-second datasets that have near global coverage and represent significant advances in freely available higher resolution global elevation data (Gesch, 2012). SRTM data became available within a few years after the space shuttle mission flew in 2000. The ASTER GDEM was first released in 2009, and an improved version was released in 2011. After the second version of GDEM was produced, both SRTM and GDEM were assessed against NGS GPS bench mark control points, in this case the points associated with the GEOID09 model. For comparison purposes, the 1-arc-second layer of the NED was also assessed with the same reference control points. Prior to comparison against the reference points, SRTM and GDEM were adjusted to be in the same vertical datum (NAVD 88) as the NGS bench marks, as the native vertical datum for SRTM and GDEM is the Earth Gravitational Model 1996 (EGM96) geoid. The accuracy assessment results (Gesch et al., 2012) show both SRTM and GDEM to be less accurate than the NED, with SRTM exhibiting an RMSE of about 4 meters and GDEM exhibiting an RMSE of nearly 9 meters, compared to an RMSE of better than 2 meters for the NED. SRTM and GDEM also both show a positive elevation bias in built-up and forested areas, as they are both derived from systems that collect "first return" data that do not measure ground elevations in the presence of buildings and vegetation canopies. The NED does not exhibit a positive vertical bias in these areas because by definition it is a bare earth elevation model

Further analysis comparing SRTM data with the NED shows distinctions that may be important for users to consider, especially for derivative products. Slope and aspect comparisons at the locations of reference control points reveal that values derived from the 1-arc-second NED layer and SRTM data can differ significantly. In this case, the reference data were the NGS GPS bench marks associated with GEOID03, numbering more than 13,000 points in CONUS. The NED used in the analysis was an earlier version (June 2003) constructed only from USGS 7.5-minute quadrangle-based DEMs. Corresponding NED and SRTM slope values at the reference control point locations show significant variability (Figure 4.8). SRTM shows overall higher slopes compared to the NED, especially in lower slope categories (1-5 degrees) (Figure 4.9), which is likely due in part to the spaceborne IfSAR source data for SRTM that tend to have higher noise content (local variability) than other systems that collect elevation data. The slope analysis further reveals that both the NED and SRTM errors are strongly correlated with slope (Figure 4.10 on page 98), although the NED exhibits lower overall error than SRTM across the range of slope categories. Corresponding aspect values derived from the NED and SRTM can also vary significantly between the two datasets (Figure 4.11 on page 98). In this comparison, aspect values have been categorized into 45-degree bins centered on the cardinal and diagonal compass directions, which is a common approach to aspect mapping. More than 40% of the corresponding aspect values differ by 90 degrees or more between the NED and SRTM.

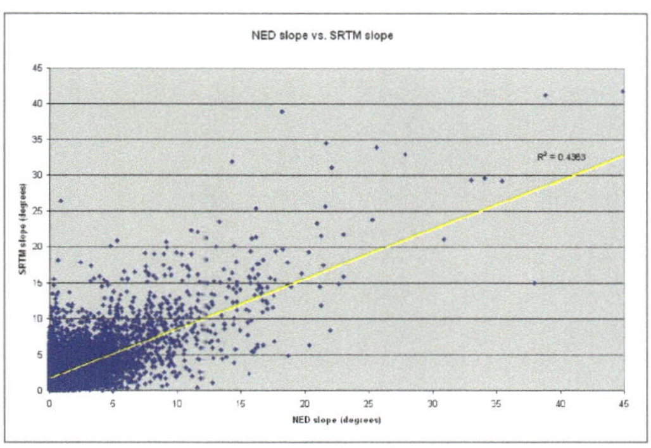

Figure 4.8. Comparison between corresponding NED and SRTM slope values at reference control point locations.

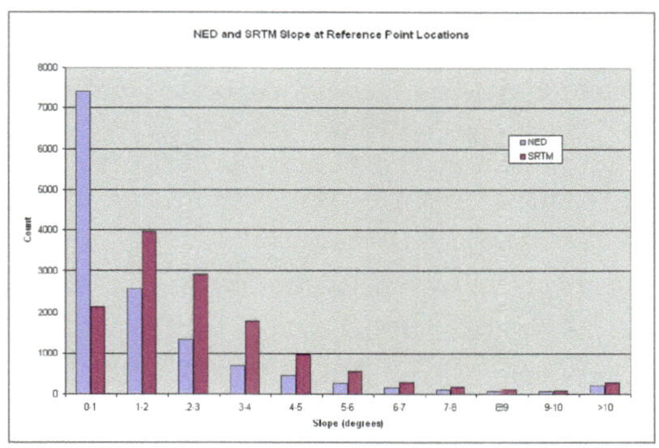

Figure 4.9. A comparison of corresponding NED and SRTM slope values shows that SRTM exhibits overall higher slopes compared to the NED, especially in lower slope categories.

It is clear that users should not expect consistent data characteristics and error patterns between the NED and SRTM, so each dataset would not be a direct replacement or substitute for the other. However, the datasets can be viewed as being complementary, and some applications have exploited this unique pairing of seamless, large-area coverage elevation datasets to derive information products. Several investigators have taken advantage of the bare earth nature of the NED and the first return nature of the SRTM to calculate vegetation canopy heights (Kellndorfer et al., 2004; Walker et al., 2007; Yu et al., 2010; Ni et al., 2013.) The temporal differences between the acquisition dates for the NED and SRTM source data have been exploited to map areas of significant topographic changes in CONUS due to surface mining, road construction, urban development, landfills, and dam construction and reservoir filling (Gesch, 2005; Gesch, 2006; Gesch, 2014; see also: *http://topochange.cr.usgs.gov/*.)

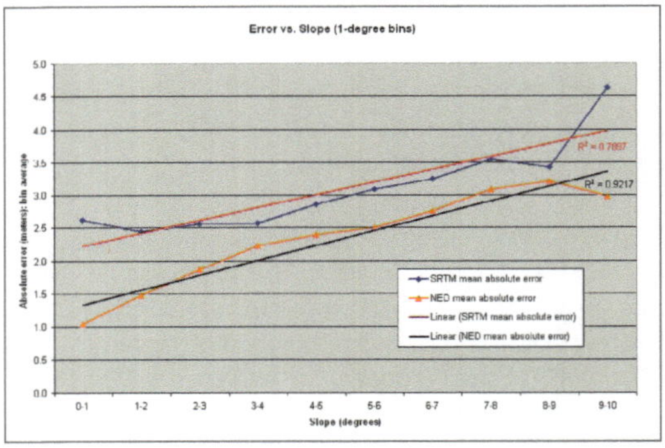

Figure 4.10. NED and SRTM errors are strongly correlated with slope (increasing error with increasing slope), although the NED exhibits lower overall error than SRTM across the range of slope categories.

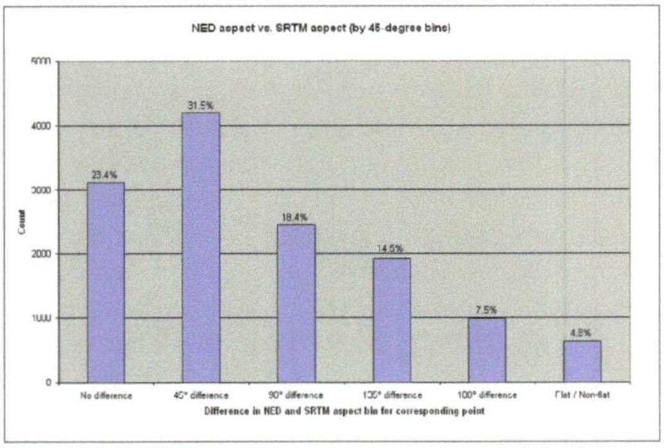

Figure 4.11. Corresponding aspect values derived from the NED and SRTM can vary significantly between the two datasets.

APPLICATIONS

Because topographic information is a fundamental requirement for so many earth science studies and operational geospatial applications, the NED has gained wide acceptance and attained broad usage by the user community. Among the characteristics of the NED that support its broad applicability are its national coverage, multi-resolution design, regular update cycle (with integration of new high-resolution, high-accuracy source data), thorough documentation (including accuracy assessment results), multiple product formats, and data distribution and use policy (no-cost data and no redistribution restrictions). The NED is a highly cited dataset, and in addition to those already mentioned above the following applications are examples that have made effective use of the NED:

- Large-area land cover mapping (Homer et al., 2004; Fry et al., 2011)
- Regional soil erosion vulnerability modeling (Das et al., 2004)
- Climate data analysis (Gallo, 2005; Jefferson, 2012; Wickham et al., 2012)
- Wildland fire mapping products (Rollins, 2009)
- Elevation contours for The National Map Viewer (Tyler and Greenlee, 2012) and for US Topo quadrangle maps (Arundel-Murin, 2011; see also: *http://nationalmap.gov/ustopo/index.html*)
- Hurricane storm surge inundation mapping (Gesch, 2007b; Stoker et al., 2009; Gesch, 2009b)
- Hydrologic derivative products from elevation data (Berelson et al., 2004 ; Franken, 2004)
- Extraction of surface drainage features (Poppenga et al., 2009)
- Hydrography change detection (Poppenga et al., 2013)
- Sea-level rise vulnerability assessment (Gesch, 2009a; Lam et al., 2009; Neumann et al., 2010; Emrich and Cutter, 2011 ; Weiss et al., 2011; Gilmer and Ferdaña, 2012; Strauss et al., 2012; Thatcher et al., 2013; Lentz et al., 2015)
- Mapping forest biomass (Blackard et al., 2008; Lamsal et al., 2012)
- Mapping of landslides (Scheingross et al., 2013)
- Glacier monitoring (Le Bris and Paul, 2013)
- Species distribution modeling (Moisen et al., 2006; McGarvey, 2011; Wilson et al., 2013)

- Landform and geomorphological feature mapping (Arbogast et al., 2009; Fornaciai et al., 2012; Howell et al., 2012; Julian et al., 2012; Theobald et al., 2015)
- Terrestrial ecosystem mapping (Sayre et al., 2009)
- Wildlife habitat mapping (Telesco et al., 2007 ; Suzuki et al., 2008; Tirpak and Giuliano, 2010)
- Estimation of forest canopy height (Kenyi et al., 2009; Huang et al., 2010)
- Geologic structure and processes (Oakley and Boothroyd, 2012; Miller et al., 2013; Rovere et al. 2015)

CURRENT DEVELOPMENT AND FUTURE DIRECTIONS

From the beginning of development of the NED and continuing to the present, the goal has always been to provide the best publicly available elevation data for the nation. Many sources of elevation data are available that could potentially be used for the NED; however, the focus has been, and will continue to be, on integrating the highest resolution, highest accuracy data that are free from licensing or other redistribution restrictions. The NED is a public domain resource, and it must remain so to fulfill the requirements of being a national framework geospatial data theme. To make sure that its seamless elevation products fully meet the intent of the NSDI framework approach, the USGS elevation program is taking the direction that the primary source for updating its elevation holdings is recent collections of high-resolution data acquired through partnerships with other organizations.

In contrast to the situation several decades ago, there are many elevation data producers in addition to the USGS. Numerous federal, state, local, and tribal agencies either produce or contract for elevation data. Much of the data are actually generated by private sector mapping and engineering companies. For the NED to truly reflect the best publicly available data, elevation data from all these entities must be considered as potential information sources. The USGS has a cadre of liaisons to state and federal agencies that are dedicated to developing and implementing partnerships that benefit both parties by improving USGS geospatial data production, integration, and distribution. Much of the newer, high-resolution elevation data that have been integrated into the NED are the result of partnering arrangements fostered by the liaison staff.

Although the NED, as the flagship elevation product of the USGS, has been quite successful in providing quality controlled, well documented seamless elevation data for many applications, it has been clearly recognized that much of its source data are dated and do not provide the resolution and accuracy needed for some applications (National Research Council, 2007, 2009). It has also been widely recognized that newer elevation technologies, especially lidar, can provide elevation data with significantly improved levels of resolution and accuracy that can facilitate new and enhanced applications of terrain information. More than a decade ago the USGS recognized the growing advantages of lidar (Crane et al., 2004), and it began a concerted effort to utilize the technology (Stoker et al., 2006), which eventually led to its prominence as the primary source for update of the NED.

Despite broad acceptance of lidar as the technology of choice for high-resolution, high-accuracy elevation information, to date only about one-third of CONUS is covered by lidar collections through diverse projects conducted by all levels of government, private entities, and universities. In recognition of the fact that a more coherent national lidar strategy would be advantageous in upgrading national elevation data holdings and be beneficial to many organizations, the USGS hosted a series of workshops in 2007 (Stoker et al., 2007) and 2008 (Stoker et al., 2008) to initially explore the concept of a national lidar collection, especially the requirements and possible implementation approaches. At the same time, the Committee on Floodplain Mapping Technologies of the National Research Council (2007) recommended a new national elevation data collection termed "Elevation for the Nation" that should employ lidar technology as the primary method of data acquisition. In 2010, the USGS and other members of the NDEP, a consortium of federal agencies working collaboratively to meet common elevation requirements (Osborn et al., 2001), followed up on this momentum toward improved national elevation information. The group sponsored a study to document national requirements for improved elevation data, estimate the costs and benefits of meeting those requirements, and explore multiple implementation scenarios (Snyder, 2012). The study participants included 34 federal agencies, all 50 states, some tribal and local government offices, and private organizations. The National Enhanced Elevation Assessment (NEEA) was completed in 2011, and its primary conclusions indicate that significant benefits could be realized by systematically upgrading national elevation data, and many improved elevation applications would benefit all levels of government and numerous industries (Dewberry, 2011). In response to the NEEA results, the USGS has developed the 3D Elevation Program (3DEP) initiative, and its primary objective is to systematically collect enhanced elevation data on an 8-year cycle, specifically lidar data for CONUS, Hawaii, and the U.S. island territories, and IfSAR data for Alaska (Snyder, 2013). Unlike the NED, these data include 3-D measurements of aboveground features (such as vegetation canopies and built structures) in addition to bare earth surfaces. Such data will expand the usefulness of elevation information and are becoming increasingly critical for broad-area landscape mapping and monitoring (Stoker, 2013). USGS elevation and related mapping activities will benefit by integrating data collected under the auspices of 3DEP, and the USGS has already begun investigating and prototyping how its seamless elevation products will support the enhanced elevation information afforded by these new data (Stoker et al., 2013). For further information on the NEEA and 3DEP, see Chapter 5 in this Manual.

As a primary USGS elevation data product, the NED has been at the center of much of the topographic research and development conducted by the USGS. An important focus is on the development of seamless merged topographic-bathymetric datasets at the coastal land/water interface. Many applications of geospatial data in coastal environments require knowledge of the near-shore topography and bathymetry. However, because existing topographic and bathymetric data have been collected independently for different purposes, it has been difficult to use them together. A tool called VDatum (Parker et al., 2003; Myers, 2005), developed by the National Oceanic and Atmospheric Administration (NOAA), is used to transform disparate datasets into a common vertical datum, which then facilitates generation of hybrid topographic-bathymetric elevation models. Shortly after the NED became an operational product, USGS and NOAA began investigating the feasibility of merging coastal elevation datasets, with the NED supplying the topographic component and NOAA's hydrographic survey database supplying the bathymetric component for a successful prototype product of the Tampa Bay, Florida, region (Gesch and Wilson, 2002.) Hurricane storm surge modeling (Weisberg and Zheng, 2006) and sea-level rise vulnerability assessment (Gesch, 2013) are examples of applications that have benefitted from merged topographic-bathymetric elevation models. The National Research Council (2004) recommended collaborative efforts among federal agencies to produce such products for all U.S. coastal regions, and the National Ocean Council (2013) in its National Ocean Policy Implementation Plan recently renewed the call for a seamless representation of coastal elevations to improve coastal change analysis products. To address the requirements for improved coastal elevation information, the USGS is developing a new line of science products known as the Coastal National Elevation Database (CoNED). The CoNED builds upon and enhances the NED by integrating recent very high resolution coastal lidar data (both topographic and bathymetric) and by including a temporal component that is captured in coastal lidar collections from multiple dates. As regional CoNED products are completed, they are added to the USGS seamless elevation data holdings and are distributed as part of The National Map, with the spatial metadata updated accordingly. Figure 4.12 shows an example of the CoNED over the Mobile Bay, Alabama, region (Danielson et al., 2013.)

Another area of development is Web services based upon the NED. The National Map Viewer includes spot elevation (at interactively selected locations), contour, and shaded relief services that are derived from seamless elevation layers. Other federal agencies are successfully using NED-based Web services internally to meet some of their specific business requirements. The Natural Resources Conservation Service (NRCS) serves the NED and derivative products (including shaded relief, slope, and aspect) via Web services to its field offices to support applications such as engineering field tools and conservation planning and design. The Federal Aviation Administration (FAA) uses a NED-based Web service that returns the x, y, z coordinates of the highest elevation point within a user submitted polygon. The elevation information is used as one input for FAA's airport approach planning model (Poppenga et al., 2010.) Realizing that Web services that fulfill operational business needs rely upon well maintained data resources, USGS is working toward other enhancements including consideration of more frequent accuracy assessments.

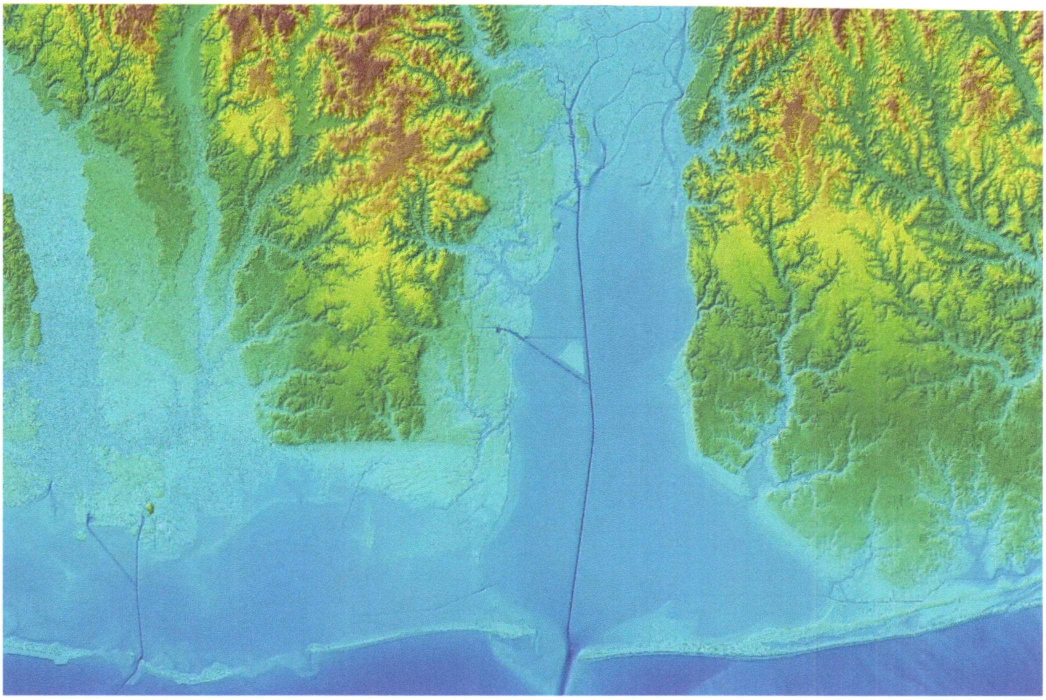

Figure 4.12. The Coastal National Elevation Database (CoNED) coverage of Mobile Bay, Alabama. CoNED is a merged topographic-bathymetric dataset that facilitates applications at the coastal land/water interface and is distributed as part of The National Map elevation products.

The NED is a longstanding and successful USGS product line that has evolved over its lifetime and its legacy continues in 3DEP elevation products, which will continue to advance based on new and emerging methods and techniques in elevation data acquisition, processing, and provisioning. The usefulness of the NED approach has been recognized not just in the U.S. but also in Australia where Geoscience Australia (2013) has developed and implemented the National Elevation Data Framework (NEDF) to provide access to the "best available" elevation data for Australia just as the NED does for the U.S. Staff members from Geoscience Australia have consulted with USGS scientists and developers on many aspects of the NED model for a comprehensive elevation data program including lidar data acquisition specifications, data processing, data management and distribution, data services, and documentation. These consultations have benefitted the development of the NEDF for Australia. The advantages of a seamless, multi-source, multi-resolution, collaborative approach to elevation data are many, and the continuing goal for USGS elevation products is to provide quality controlled, well documented elevation data that are useful for many geospatial applications.

AUTHOR BIOGRAPHIES

Dr. Dean Gesch is a research physical scientist with the U.S. Geological Survey. Since 1992 he has worked at the Earth Resources Observation and Science Center (EROS) in Sioux Falls, South Dakota. At EROS he is a senior member of the Topographic Science team where he has worked on a variety of geographic science topics including leading the design, development, and application of USGS baseline elevation datasets at national and global scales. His research topics include large-area topographic change analysis and monitoring, hurricane storm surge mapping, elevation data accuracy assessment, and sea-level rise vulnerability assessment. His academic background includes a bachelor's degree in geography received in 1982 from Carroll College, Waukesha, Wisconsin; a master's degree in geosciences received in 1984 from Murray State University, Murray, Kentucky; and a Ph.D. in Geospatial Science and Engineering received in 2006 from South Dakota State University in Brookings, South Dakota.

Ms. Gayla Evans is a geographer working with the U.S. Geological Survey EROS Center in Sioux Falls, South Dakota. Since 1985, she has worked with digital geospatial data developing expertise in lidar quality assessment and elevation modeling, vector and raster image processing, research, and project management. She has served as task lead for the NED with responsibility for daily operations, status, and future innovations of this multi-resolution elevation dataset. Her academic background includes a bachelor's degree in geography received in 1995 from South Dakota State University in Brookings, South Dakota.

Mr. Michael Oimoen has worked in the GIS industry since 1992, and has worked on the NED from its inception, performing much of the algorithm development, including methods of artifact removal and edge matching, as well as establishing procedures for regular updates and developing tools to assist in visual quality inspection. He received his BS in Meteorology and MS in Civil and Environmental Engineering from the University of Wisconsin, Madison, and is currently a senior scientist at Stinger Ghaffarian Technologies (contractor to the U.S. Geological Survey).

Dr. Samantha Arundel is a research geographer who has worked at the U.S. Geological Survey National Geospatial Technical Operations Center in Rolla, Missouri since 2009. She is the elevation theme development lead. Her previous academic research career as assistant and associate professor at Northern Arizona University prepared her well for her current challenges. She has a B.S. in geography from Northern Arizona University, and an M.A. and Ph.D. in geography from Arizona State University where she specialized in raster modeling of plant species distribution related to climate change.

ACKNOWLEDGEMENTS

Many individuals across the USGS have contributed to the success of the NED. This includes the efforts of surveyors, field technicians, and cartographers throughout the history of the USGS mapping program that resulted in the production of the base map series from which quadrangle-based DEMs were derived. In the recent past as high-resolution, high-accuracy data became the primary source for NED updates, the USGS state geospatial liaisons provided valuable contributions in working with partners to acquire new data. The staff of the USGS National Geospatial Technical Operations Center (NGTOC) provided important assistance in preparing NED source data. The dedicated and conscientious efforts of the Topographic Science team at the USGS Earth Resources Observation and Science (EROS) Center in developing, producing, and maintaining the NED are gratefully acknowledged. Special thanks go to Sue Greenlee, formerly of the USGS EROS Center, and to Jim Mauck, formerly of the USGS National Geospatial Program, for their unwavering support and promotion of the NED as the flagship elevation product of the USGS.

REFERENCES

Abrams, M., B. Bailey, H. Tsu, and M. Hato (2010). The ASTER global DEM. In *Photogrammetric Engineering & Remote Sensing*, 76(4), (pp. 344–348). Bethesda, MD: American Society for Photogrammetry and Remote Sensing.

Arbogast, A., A. Shortridge, and M. Bigsby (2009). Volumetric estimates of coastal sand dunes in lower Michigan: explaining the geography of dune fields. In *Physical Geography*, 30(6), (pp. 479–500). Taylor & Francis Online.

Arundel, S. T., L. A. Phillips, A. J. Lowe, J. Bobinmyer, K. S. Mantey, C. A. Dunn, E. W. Constance, and E. L. Usery (2015). Preparing The National Map for the 3D Elevation Program—products, process and research. In *Cartography and Geographic Information Science*, 42(S1), (pp. 40–53) Taylor & Francis Online.

Arundel-Murin, S.T. (2011). Challenges in integrating The National Map themes of hydrography and elevation. In *Proceedings of the U.S. Geological Survey Eighth Biennial Geographic Information Science Workshop and first The National Map Users Conference, Denver, Colorado, May 10-13 2011* (J.B. Sieverling and J. Dietterle, editors), U.S. Geological Survey Scientific Investigations Report 2011-5053 (p. 80).

Berelson, W.L., P.A. Caffrey, and J.D. Hamerlinck (2004). Mapping hydrologic units for the national watershed boundary dataset, *Journal of the American Water Resources Association*, 40(5), (pp. 1231–124).

Blackard, J.A., M.V. Finco, E.H. Helmer, G.R. Holden, M.L. Hoppus, D.M. Jacobs, A.J. Lister, G.G. Moisen, M.D. Nelson, R. Riemann, B. Ruefenacht, D. Salajanu, D.L. Weyermann, K.C. Winterberger, T.J. Brandeis, R.L. Czaplewski, R.E. McRoberts, P.L. Patterson, and R.P. Tymcio (2008). Mapping U.S. forest biomass using nationwide forest inventory data and moderate resolution information. In *Remote Sensing of Environment*, 112(4), (pp. 1658–1677).

Crane, M.P., T. Clayton, E.S. Raabe, J.M. Stoker, L. Handley, G. Bawden, K. Morgan, and V. Queija (2004). *Report of the U.S. Geological Survey lidar workshop sponsored by the Land Remote Sensing Program and held in St. Petersburg, FL, November 2002*, U.S. Geological Survey Open-File Report 2004-1456, 72 p.

Danielson, J.J., J.C. Brock, D.M. Howard, D.B. Gesch, J.M. Bonisteel-Cormier, and L.J. Travers (2013). *Topobathymetric model of Mobile Bay, Alabama*. U.S. Geological Survey Data Series 769.

Das, C., W.J. Capehart, H.V. Mott, P.R. Zimmerman, and T.E. Schumacher (2004). Assessing regional impacts of conservation reserve program-type grass buffer strips on sediment load reduction from cultivated lands. In *Journal of Soil and Water Conservation*, 59(4), (pp. 134–142).

Dewberry (2011). *Final Report of the National Enhanced Elevation Assessment*, Fairfax, Virginia, 84 p. plus appendices (revised 2012).

Emrich, C.T., and S.L. Cutter (2011). Social vulnerability to climate-sensitive hazards in the southern United States. In *Weather, Climate, and Society*, 3(3), (pp. 193–208).

Farr, T.G., P.A. Rosen, E. Caro, R. Crippen, R. Duren, S. Hensley, M. Kobrick, M. Paller, E. Rodriguez, L. Roth, D. Seal, S. Shaffer, J. Shimada, J. Umland, M. Werner, M. Oskin, D. Burbank, and D.E. Alsdorf (2007). The shuttle radar topography mission. In *Reviews of Geophysics*, 45, RG2004, doi:10.1029/2005RG000183.

Federal Geographic Data Committee (1998). *Content standard for digital geospatial metadata.*

Federal Geographic Data Committee (2012). *NSDI framework.*

Fornaciai, A., M. Favalli, D. Karátson, S. Tarquini, and E. Boschi (2012). Morphometry of scoria cones, and their relation to geodynamic setting: A DEM-based analysis. In *Journal of Volcanology and Geothermal Research*, 217-218, (pp. 56–72).

Franken, S., (2004). USGS EROS Data Center produces seamless hydrologic derivatives with GIS. In *ArcNews*, Fall 2004, Environmental Systems Research Institute, Inc.

Fry, J.A., G. Xian, S. Jin, J.A. Dewitz, C.G. Homer, L. Yang, C.A. Barnes, N.D. Herold, and J.D. Wickham (2011). Completion of the 2006 National Land Cover Database for the conterminous United States. In *Photogrammetric Engineering & Remote Sensing*, 77(9), (pp. 858–864).

Gallo, K.P. (2005). Evaluation of temperature differences for paired stations of the U.S. Climate Reference Network. In *Journal of Climate*, 18(10), (pp. 1629–1636).

Geoscience Australia (2013). *National Elevation Data Framework.*

Gesch, D.B. (2005). Analysis of multi-temporal geospatial data sets to assess the landscape effects of surface mining. In *Proceedings, 22nd Annual National Conference of the American Society of Mining and Reclamation*, Breckenridge, Colorado, June 19-23, 2005, (pp. 415–432) (R.I. Barnhisel, editor).

Gesch, D.B. (2006). *An inventory and assessment of significant topographic changes in the United States: Ph.D. dissertation*, Geospatial Science & Engineering Program, South Dakota State University, Brookings, South Dakota, 217 p.

Gesch, D.B. (2007a). The National Elevation Dataset, in *Digital Elevation Model Technologies and Applications: The DEM Users Manual, 2nd Edition* (D.F. Maune, editor), (pp.99–118). Bethesda, MD: American Society for Photogrammetry and Remote Sensing.

Gesch, D. (2007b). Topography-based analysis of Hurricane Katrina inundation of New Orleans. In *Science and the Storms: the USGS Response to the Hurricanes of 2005* (G.S. Farris, G.J. Smith, M.P. Crane, C.R. Demas, L.L. Robbins, and D.L. Lavoie, editors), U.S. Geological Survey Circular 1306.

Gesch, D.B. (2009a). Analysis of lidar elevation data for improved identification and delineation of lands vulnerable to sea-level rise. In *Journal of Coastal Research*, SI53, (pp. 49–58).

Gesch, D.B. (2009b). *Mapping and visualization of storm-surge dynamics for Hurricane Katrina and Hurricane Rita.* U.S. Geological Survey Scientific Investigations Report 2009-5230, 18 p.,

Gesch, D.B. (2012). Global digital elevation model development from satellite remote-sensing data. In *Advances in Mapping from Remote Sensor Imagery—Techniques and Applications* (X. Yang and J. Li, editors), (pp. 91–117).

Gesch, D.B. (2013). Consideration of vertical uncertainty in elevation-based sea-level rise assessments: Mobile Bay, Alabama case study. In *Journal of Coastal Research*, SI63, (pp. 197–210).

Gesch, D. B. (2014). An inventory of topographic surface changes: the value of multi-temporal elevation data for change analysis and monitoring, *ISPRS—International Archives of the Photogrammetry, Remote Sensing and Spatial Information Sciences*, XL-4, (pp. 59–63).

Gesch, D., G. Evans, J. Mauck, J. Hutchinson, W.J. Carswell Jr. (2009). The National Map—*Elevation, U.S. Geological Survey Fact Sheet 2009-3053*, 4 p.

Gesch, D.B., M.J. Oimoen, and G.A. Evans (2014). Accuracy assessment of the U.S. Geological Survey National Elevation Dataset, and comparison with other large-area elevation datasets—*SRTM and ASTER. U.S. Geological Survey Open-File Report 2014–1008*, 10 p.

Gesch, D., M. Oimoen, S. Greenlee, C. Nelson, M. Steuck, and D. Tyler (2002). The National Elevation Dataset, *Photogrammetric Engineering & Remote Sensing*, 68(1), (pp. 5–11). Bethesda, MD: American Society for Photogrammetry and Remote Sensing.

Gesch, D.B., M.J. Oimoen, Z. Zhang, D.J. Meyer, and J.J. Danielson (2012). Validation of the ASTER Global Digital Elevation Model Version 2 over the conterminous United States. In Imaging a Sustainable Future, Congress, 22nd, Melbourne, Australia, August 25-September 1, 2012 (M. Shortis and N. El-Sheimy, editors), *The International Archives of the Photogrammetry, Remote Sensing, and Spatial Information Sciences*, XXXIX, pt. B4, (pp. 281–286). Lemmer, Netherlands, International Society for Photogrammetry and Remote Sensing.

Gesch, D.B., K.L. Verdin, and S.K. Greenlee (1999). New land surface digital elevation model covers the earth. In *Eos, Transactions, American Geophysical Union*, 80(6), (pp. 69–70).

Gesch, D., and R. Wilson (2002). Development of a seamless multisource topographic/bathymetric elevation model of Tampa Bay, *Marine Technology Society Journal*, 35(4), (pp. 58–64).

Gilmer, B., and Z. Ferdaña (2012). Developing a framework for assessing coastal vulnerability to sea level rise in southern New England, USA, in Cities and Adaptation to Climate Change—Proceedings of the Global Forum 2011 (K. Otto-Zimmermann, editor), *Resilient Cities 2, Local Sustainability Volume 2*, (pp. 25–36).

Heidemann, Hans Karl (2018). Lidar base specification (ver. 1.3, February 2018). In *U.S. Geological Survey Techniques and Methods*, Book 11(B4), 101 p.

Homer, C., C. Huang, L. Yang, B. Wylie, and M. Coan (2004). Development of a 2001 national land-cover database for the United States, *Photogrammetric Engineering & Remote Sensing*, 70(7), (pp. 829–840). Bethesda, MD: American Society for Photogrammetry and Remote Sensing.

Howell, J.K., S.M. White, and D.R. Bohnenstiehl (2012). A modified basal outlining algorithm for identifying topographic highs in gridded elevation data, part 2: Application to Springerville volcanic field, In *Computers and Geosciences*, 49, (pp. 315–322).

Huang, S., R.L. Crabtree, A.K. Swanson, K.Q. Halligan, and J.A. Harmsen (2010). Error analysis and correction for extracting the forest height from airborne C-band interferometric SAR and national elevation datasets. In *International Journal of Remote Sensing*, 31(10), (pp. 2493–2512).

Jefferson, A.J. (2012). Seasonal versus transient snow and the elevation dependence of climate sensitivity in maritime mountainous regions, *Geophysical Research Letters*, 38(16): article no. L16402.

Julian, J.P., A.J. Elmore, and S.M. Guinn (2012). Channel head locations in forested watersheds across the mid-Atlantic United States: A physiographic analysis. In *Geomorphology*, 177-178, (pp. 194–203).

Kellndorfer, J., W. Walker, L. Pierce, C. Dobson, J. Fites, C. Hunsaker, J. Vona, and M. Clutter (2004). Vegetation height estimation from Shuttle Radar Topography Mission and National Elevation Datasets. In *Remote Sensing of the Environment*, 93(3), (pp. 339–358).

Kelmelis, J. (2003). To The National Map and beyond. In *Cartography and Geographic Information Science*, 30(2), (pp. 185–198).

Kelmelis, J.A., M.L. DeMulder, C.E. Ogrosky, N.J. Van Driel, and B.J. Ryan (2003). *The National Map*—from geography to mapping and back again. In *Photogrammetric Engineering & Remote Sensing*, 69(10), (pp. 1109–1118).

Kenyi, L.W., R. Dubayah, M. Hofton, and M. Schardt (2009). Comparative analysis of SRTM-NED vegetation canopy height to lidar-derived vegetation canopy metrics. In *International Journal of Remote Sensing*, 30(11), (pp. 2797–2811).

Lam, N.S.-N., H. Arenas, Z. Li, and K.-B. Liu (2009). An estimate of population impacted by climate change along the U.S. coast. In *Journal of Coastal Research*, SI56, (pp. 1522–1526).

Lamsal, S., D.M. Rizzo, and R.K. Meentemeyer (2012). Spatial variation and prediction of forest biomass in a heterogeneous landscape. In *Journal of Forestry Research*, 23(1), (pp. 13–22).

Le Bris, R., and F. Paul (2013). An automatic method to create flow lines for determination of glacier length: A pilot study with Alaskan glaciers. In *Computers and Geosciences*, 52, (pp. 234–245).

Lentz, E.E., S.R. Stippa, E.R. Thieler, N.G. Plant, D.B. Gesch, and R.M. Horton (2015). *Evaluating coastal landscape response to sea-level rise in the northeastern United States—Approach and methods*. U.S. Geological Survey Open-File Report 2014–1252, 26 p.

Liu, X., P. Hu, H. Hu., and J. Sherba (2012). Approximation theory applied to DEM vertical accuracy assessment. In *Transactions in GIS*, 16(3), (pp. 397–410).

Maune, D.F. (2008). *Digital elevation model (DEM) data for the Alaska Statewide Digital Mapping Initiative (SDMI)*, 161 p.

Maune, D.F., J.B. Maitra, and E.J. McKay (2007). Accuracy standards and guidelines, in *Digital Elevation Model Technologies and Applications: The DEM Users Manual, 2nd Edition* (D.F. Maune, editor), (pp. 65–97). Bethesda, MD: American Society for Photogrammetry and Remote Sensing.

McGarvey, D.J. (2011). Quantifying ichthyofaunal zonation and species richness along a 2800-km reach of the Rio Chama and Rio Grande (USA). In *Ecology of Freshwater Fish*, 20(2), (pp. 231–242).

Miller, S.R., P.B. Sak, E. Kirby, and P.R. Bierman (2013). Neogene rejuvenation of central Appalachian topography: Evidence for differential rock uplift from stream profiles and erosion rates. In *Earth and Planetary Science Letters*, 369-370, (pp. 1–12).

Moisen, G.G., E.A. Freeman, J.A. Blackard, T.S. Frescino, N.E. Zimmermann, and T.C. Edwards Jr. (2006). Predicting tree species presence and basal area in Utah: A comparison of stochastic gradient boosting, generalized additive models, and tree-based methods. In *Ecological Modelling*, 199(2), (pp. 176–187).

Myers, E.P. (2005). Review of progress on VDatum, a vertical datum transformation tool In *Proceedings of Oceans 2005 MTS/IEEE Conference*, September 17–23, 2005, Washington, DC, (Vol. 2) (pp. 974–980).

National Digital Elevation Program (2004). *Guidelines for digital elevation data*—Version 1.0, 93 p.

National Geodetic Survey (2012). *GPS on Bench Marks* (GPSBM) GEOID12A.

National Ocean Council (2013). *National Ocean Policy Implementation Plan Appendix*,

National Research Council (2004). *A geospatial framework for the coastal zone*. Washington D.C., The National Academies Press, 149 p.

National Research Council (2007). *Elevation data for floodplain mapping*. Washington D.C.: The National Academies Press, 151 p.

National Research Council (2009). *Mapping the zone: Improving flood map accuracy*. Washington D C.: The National Academies Press, 122 p.

Neumann, J.E., D.E. Hudgens, J. Herter, and J. Martinich (2010). Assessing sea-level rise impacts: A GIS-based framework and application to coastal New Jersey. In *Coastal Management*, 38(4), (pp. 433–455).

Ni, W., G. Sun, Z. Zhang, Z. Guo, and Y. He (2013). Co-registration of two DEMs: impacts on forest height estimation from SRTM and NED at mountainous areas. In *IEEE Geosciences and Remote Sensing Letters*, DOI: 10.1109/LGRS.2013.2255580.

Oakley, B.A., and J.C. Boothroyd (2012). Reconstructed topography of Southern New England prior to isostatic rebound with implications of total isostatic depression and relative sea level. In *Quaternary Research (United States)*, 78(1), (pp. 110–118).

Office of Management and Budget (2002). *Circular No. A-16 Revised*—Coordination of geographic information and related spatial data activities.

Office of Management and Budget (2010). *Circular A-16 Supplemental guidance* —Geospatial line of business.

Oimoen, M.J. (2000). An effective filter for removal of production artifacts in U.S. Geological Survey 7.5-minute digital elevation models. In *Proceedings of the Fourteenth International Conference on Applied Geologic Remote Sensing*, November 6–8, 2000, Las Vegas, Nevada: Veridian ERIM International, Ann Arbor, Michigan (pp. 311–319).

Osborn, K., J. List, D. Gesch, J. Crowe, G. Merrill, E. Constance, J. Mauck, C. Lund, V. Caruso, and J. Kosovich (2001). National digital elevation program (NDEP). In *Digital Elevation Model Technologies and Applications: The DEM User's Manual* (D.F. Maune, editor) (pp. 83–120). Bethesda, MD: American Society for Photogrammetry and Remote Sensing.

Parker, B., K. Hess, D. Milbert, and S. Gill (2003). A national vertical datum transformation tool. In *Sea Technology*, 44(9), (pp. 10–15). Arlington, VA: Compass Publications.

Poppenga, S., G. Evans, D. Gesch, J. Stoker, V. Queija, B. Worstell, D. Tyler, J. Danielson, N. Bliss, and S. Greenlee (2010). *Topographic Science, U.S. Geological Survey Fact Sheet 2010-3049*, 4 p.

Poppenga, S.K., D.B. Gesch, and B.B. Worstell (2013). Hydrography change detection—the usefulness of surface channels derived from LiDAR DEMs for updating mapped hydrography. In *Journal of the American Water Resources Association*, 49(2), (pp. 371–389).

Poppenga, S.K., B.B. Worstell, J.M. Stoker, and S.K. Greenlee (2009). *Comparison of surface flow features from lidar-derived digital elevation models with historical elevation and hydrography data for Minnehaha County, South Dakota*. U.S. Geological Survey Scientific Investigations Report 2009-5065, 24 p.

Rollins, M.G. (2009). LANDFIRE: a nationally consistent vegetation, wildland fire, and fuel assessment. In *International Journal of Wildland Fire*, 18, (pp. 235–249).

Rovere, A., P. J. Hearty, J. Austermann, J. X. Mitrovica, J. Gale, R. Moucha, A. M. Forte, and M. E. Raymo (2015). Mid-Pliocene shorelines of the US Atlantic coastal plain—an improved elevation database with comparison to Earth model predictions. In *Earth-Science Reviews*, 145, (pp. 117–131).

Sayre, R., P. Comer, H. Warner, and J. Cress (2009). A new map of standardized terrestrial ecosystems of the conterminous United States. *U.S. Geological Survey Professional Paper 1768*, 22 p.

Scheingross, J.S., B.M. Minchew, B.H. Mackey, M. Simons, M.P. Lamb, and S. Hensley (2013). Fault-zone controls on the spatial distribution of slow-moving landslides. In *Bulletin of the Geological Society of America*, 125 (3-4), (pp. 473–489).

Smith, D.A., and D.R. Roman (2001). GEOID99 and G99SSS: 1-arc-minute geoid models for the United States. In *Journal of Geodesy*, 75(9–10), (pp. 469–490).

Snyder, G.I. (2012). *National enhanced elevation assessment at a glance*. U.S. Geological Survey Fact Sheet 2012-3088, 2 p.

Snyder, G.I. (2013). The benefits of improved national elevation data. In *Photogrammetric Engineering & Remote Sensing*, 79(2), (pp. 105–110).

Stoker, J.M. (2013). Special Issue Foreword—Are we moving past the pixel? The third dimension in national landscape mapping. In *Photogrammetric Engineering & Remote Sensing*, 79(2), (pp. 133–134).

Stoker, J.M., S.K. Greenlee, D.B. Gesch, and J.C. Menig, (2006). The new USGS center for lidar information coordination and knowledge. In *Photogrammetric Engineering & Remote Sensing*, 72(6), (pp. 613–616).

Stoker, J.M., D. Harding, and J. Parrish (2008). The need for a national lidar dataset. In *Photogrammetric Engineering & Remote Sensing*, 74(9), (pp. 1066–1068).

Stoker, J.M., H.K. Heidemann, G.A. Evans, and S.K. Greenlee (2013). *A conceptual prototype for the next-generation national elevation dataset*. U.S. Geological Survey Open-File Report 2013-1023, 52 p.

Stoker, J.M., J. Parrish, D. Gisclair, D. Harding, R. Haugerud, M. Flood, H.-E. Andersen, K. Schuckman, D. Maune, P. Rooney, K. Waters, A. Habib, E. Wiggins, B. Ellingson, B. Jones, S. Nechero, A. Nayegandhi, T. Saultz, and G. Lee (2007). *Report of the first national lidar initiative meeting, February 14-16, 2007, Reston, Va.*, U.S. Geological Survey Open-File Report, 2007-1189, 69 p.

Stoker, J.M., D.J. Tyler, D.P. Turnipseed, K. Van Wilson Jr., and M.J. Oimoen (2009). Integrating disparate lidar datasets for a regional storm tide inundation analysis of Hurricane Katrina. In *Journal of Coastal Research*, SI53, (pp. 66–72).

Strauss, B.H., R. Ziemlinski, J.L. Weiss, J.T. Overpeck (2012). Tidally adjusted estimates of topographic vulnerability to sea level rise and flooding for the contiguous United States. In *Environmental Research Letters*, 7(1): article no. 014033.

Suzuki, N., D.H. Olson, and E.C. Reilly (2008). Developing landscape habitat models for rare amphibians with small geographic ranges: A case study of Siskiyou Mountains salamanders in the western USA. In *Biodiversity and Conservation*, 17(9), (pp. 2197–2218).

Telesco, R.L., F.T. Van Manen, J.D. Clark, and M.E. Cartwright (2007). Identifying sites for elk restoration in Arkansas. In *Journal of Wildlife Management*, 71(5), (pp. 1393–1403).

Thatcher, C.A., J.C. Brock, E.A. Pendleton (2013). Economic vulnerability to sea-level rise along the northern U.S. Gulf coast. In *Journal of Coastal Research*, SI63, (pp. 234–243).

Theobald, D. M., D. Harrison-Atlas, W. B. Monahan, and C. M. Albano (2015). Ecologically-relevant maps of landforms and physiographic diversity for climate adaptation planning. In *PLoS One*, 10(12): e0143619.

Tirpak, J.M. and W.M. Giuliano (2010). Using multitemporal satellite imagery to characterize forest wildlife habitat: The case of ruffed grouse. In *Forest Ecology and Management*, 260(9), (pp. 1539–1547).

Tyler, D.J., and S.K. Greenlee (2012). *Creation of digital contours that approach the characteristics of cartographic contours*. U.S. Geological Survey Scientific Investigations Report 2012–5167, 31 p. with appendices.

U.S. Geological Survey (1997). *Part 1: General—Standards for digital elevation models*, 11 p.

Walker, W.S., J.M. Kellndorfer, and L.E. Pierce (2007). Quality assessment of SRTM C- and X-band interferometric data: implications for the retrieval of vegetation canopy height. In *Remote Sensing of Environment*, 106(4), (pp. 428–448).

Weiss, J.L., J.T. Overpeck, and B. Strauss (2011). Implications of recent sea level rise science for low-elevation areas in coastal cities of the conterminous U.S.A. In *Climatic Change*, 105(3-4), (pp. 635–645).

Weisberg, R.H., and L. Zheng (2006). Hurricane storm surge simulations for Tampa Bay. In *Estuaries and Coasts*, 29(6 A), (pp. 899–913).

Wickham, J.D., T.G. Wade, and K.H. Riitters (2012). Comparison of cropland and forest surface temperatures across the conterminous United States. In *Agricultural and Forest Meteorology*, 166-167, (pp. 137–143).

Wilson, J.W., J.O. Sexton, R.T. Jobe, and N.M. Haddad (2013). The relative contribution of terrain, land cover, and vegetation structure indices to species distribution models. In *Biological Conservation*, 164, (pp. 170–176).

Yu, Y., S. Saatchi, L.S. Heath, E. Lapoint, R. Myneni, and Y. Knyazikhin (2010). Regional distribution of forest height and biomass from multisensor data fusion. In *Journal of Geophysical Research G: Biogeosciences*, 115(3): article no. G00E12.

Chapter 5

THE 3D ELEVATION PROGRAM

Jason M. Stoker, Vicki Lukas, Allyson L. Jason,
Diane F. Eldridge, and Larry J. Sugarbaker

INTRODUCTION

For over 15 years the National Elevation Dataset (NED) was the standard of elevation data in the U.S. Geological Survey (see Chapter 4 in this Manual). The NED was a multi-source, multi-resolution bare earth seamless digital elevation model (DEM) for the entire U.S. The 3D Elevation Program (3DEP) takes the concept of the NED one step further—to provide not just bare earth DEMs, but a full suite of 3-D elevation products that are standardized and consistent across the U.S. and its territories. 3DEP was initiated in 2012 with the preparation of standards for collection of three-dimensional (3-D) elevation data and the modernization of data management and delivery systems. For the program to be successful, broad support is essential, and the effort included a call for action (Sugarbaker et al., 2014) to accelerate the collection of remotely sensed high-resolution data of two types: (1) High-quality light detection and ranging (lidar) data for the conterminous United States (CONUS), Hawaii, and the U.S. territories and (2) Interferometric synthetic aperture radar (IfSAR) data for Alaska. Lidar is one of two primary technologies used in the United States to map elevation and other Earth-surface characteristics currently. Lidar is an essential remote sensing technology needed to support high-value applications, including flood-risk management, agriculture and precision farming, infrastructure and construction management, natural resources management and conservation, geologic resource assessment, and hazard mitigation. IfSAR, while lower in absolute accuracy and resolution, has the ability to penetrate cloud cover and can also be obtained for a lower cost because the sensors can be flown by aircraft flying at higher altitudes and ground speeds than those needed for lidar. For these reasons, IfSAR has been the chosen technology for Alaska, where remoteness and persistent cloud cover often preclude the use of lidar. The primary goal of 3DEP is to provide nationwide coverage of new products and services to augment and improve the bare earth seamless layers formerly called the NED.

The lidar and IfSAR data being collected for 3DEP are useful for many business applications that require more robust data than a basic bare-earth digital elevation model (DEM). Lidar data are collected from aircraft by using sensors that detect the reflections of a pulsed laser beam. The reflections are recorded as millions of individual points, collectively called a "point cloud," that represent the 3-D positions of objects on the surface of the Earth (Figure 5.1 on page 112). The objects can be buildings, vegetation, the ground, or any other feature. The points are usually classified by the type of feature from which they were reflected, allowing various surface models to be created, such as a digital surface model (DSM; a first-return surface) or a DEM or a digital terrain model (DTM; a bare-earth surface). In addition, the intensity of the reflected light is measured and stored as an attribute of the point, allowing a type of image to be created from the lidar data. The lidar point cloud data can be further processed to model the shape and volume of buildings, the size of trees or density of vegetation, land cover categories, line of sight, and many other physical phenomena to support a wide variety of applications.

IfSAR, a type of imaging radar, uses longer wavelength radio wave pulses from the sensor. The IfSAR system produces lower accuracy, less detailed data than lidar. It does not penetrate vegetation as thoroughly as lidar, and it is limited to producing DEMs, DSMs, and radar backscatter images, without ancillary filtering for DTM generation. However, IfSAR has the advantage of being able to penetrate cloud cover to operate in less than ideal weather conditions. IfSAR systems also operate from higher altitudes and at higher flight speeds, allowing them to collect data from large areas quickly, making IfSAR an attractive alternative to lidar for large collections in remote and challenging environments when high precision is difficult to obtain.

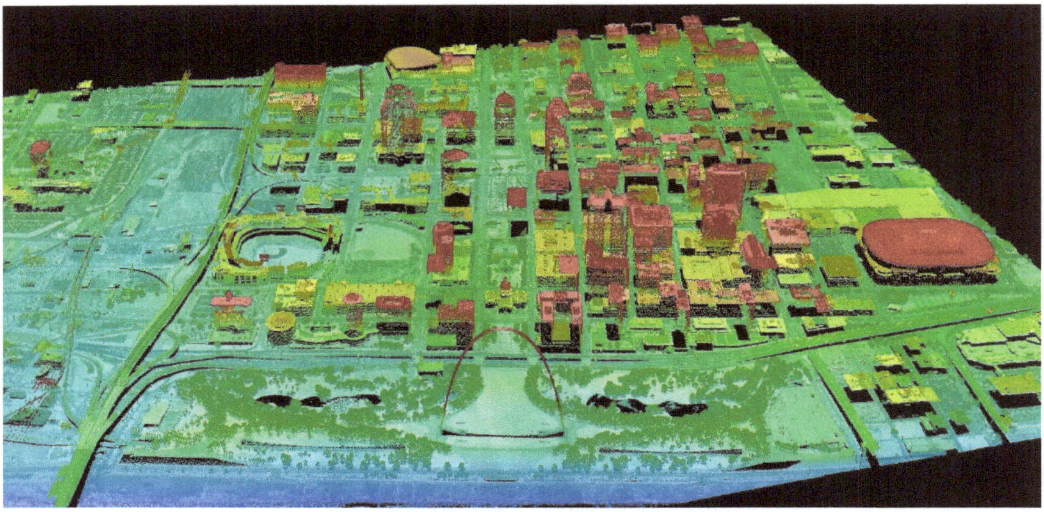

Figure 5.1. A typical lidar point cloud being collected for the 3D Elevation Program (3DEP). St. Louis, MO. Source: USGS.

Background, Rationale, and History

The 3DEP is based on the results of the National Enhanced Elevation Assessment (NEEA) that was funded by National Digital Elevation Program (NDEP) agencies and published in 2012 (Dewberry, 2012; Snyder et al, 2014). The study, led by the USGS, identified more than 600 applications that have a need for enhanced elevation data to address mission-critical information requirements of 34 Federal agencies, all 50 States, and a sample of private sector companies and Tribal and local governments. Many requirements were identified where high-quality 3-D elevation products would never be affordable if the data were acquired to solely meet a specific need. For example, the wind power industry requires volumes of very high quality surface data for large areas to plan wind farms and to determine ideal placement of wind turbines. It would be costly for this industry to fund the acquisition of high-quality lidar data solely to support that purpose. Yet, when all of the overlapping needs from multiple users are considered, a national program is more than justified. The NEEA study identified many examples like the wind farm application where a single use could not justify the necessary investment in lidar or IfSAR data. In contrast, funding for data collection today comes from government agencies where mission-specific needs, with justifiable lidar investments, are driving data acquisitions. Although the mission-driven data collection strategy has benefited individual projects and government programs, it will not result in the national dataset needed to achieve the benefits, as documented in the NEEA report. Also, if individual organizations were collecting data only to benefit them, a nationally consistent, standardized product would be extremely difficult and costly to create.

Both increased funding and improved program efficiencies and advancements in technology are necessary to increase the rate of collection of elevation. The 3DEP is designed to achieve a 25 percent cost-efficiency gain by collecting data in larger projects. To achieve national coverage in 8 years, however, the overall rate of data collection will need to increase threefold compared to 2015 rates. In order to increase the level of participation from cooperating agencies, and to ensure that mission needs can be addressed, 3DEP includes the following proposals:

- Increase overall investments through budget and other initiatives in order to provide greater incentive for increased partner engagement
- Ensure that the mission needs of partner agencies are addressed by soliciting multi-year acquisition priorities and considering those needs in the 3DEP acquisition process
- Encourage Federal, State, and Tribal government participation through advantageous partnerships that adhere to accepted quality standards while recognizing the need for flexibility among partners
- Realize efficiencies and lower costs through economies of scale achieved by larger project sizes
- Take advantage of ongoing, improved industry capabilities for collecting and managing elevation data
- Provide improved elevation-based products and services and make them freely available for everyone

The first set of 3DEP products and services became available to partners and the public in 2015. The strategy is to pool funding from partners and from an increase in contributions from Federal agencies. If fully funded, it is estimated that 3DEP can conservatively provide new benefits of $690 million per year and has the potential to generate $13 billion per year in new benefits through applications that span the economy (Dewberry, 2012). Current investments primarily by Federal, State, and local agencies combined in publicly available lidar and IfSAR data are approximately $50 million annually.

As the lead Federal agency for terrestrial elevation data, the U.S. Geological Survey managed the NED and coordinated its activities through the NDEP for more than 15 years. The NED, managed as part of The National Map (*http://nationalmap.gov/*), represented the standard of quality for elevation data in the United States during this time. Today, new elevation data are acquired using lidar and IfSAR technologies to replace elevation data that are, on average, more than 30 years old.

While the NDEP efficiently coordinated among Federal agencies and other partners, the rate of data collection and the typical project specifications in the past were insufficient to address many data requirements of government, the private sector, and others. Technologies for collecting and managing elevation data have changed significantly. Lidar and IfSAR data have made it feasible to collect large quantities of data that map the Earth's surface, including infrastructure and vegetation. These technology changes create an opportunity to address a wider range of needs and a challenge to modernize the way the USGS and its partners define and carries out their mapping activities. The potential to address requirements for natural hazards assessment and mitigation, natural resource conservation, infrastructure development, agricultural production, national security and environmental protection, and other applications requires a fresh perspective.

USGS Commitment to Manage National Elevation Data Assets

The OMB Circular A-16, "Coordination of Geographic Information and Related Spatial Data Activities" (*https://www.whitehouse.gov/omb/circulars_a016_rev*), provides for improvements in the coordination and use of spatial data and describes effective and economical use and management of spatial data assets in the digital environment for the benefit of the Federal Government and the Nation. Circular A-16 designates the U.S. Geological Survey, part of the Department of the Interior (DOI), as the lead agency for the coordination and management of the Nation's terrestrial elevation data and designates the National Oceanic and Atmospheric Administration (NOAA), part of the Department of Commerce as the lead agency for bathymetric data. The Elevation Theme agencies are leading on improving the national mapping foundation by coordinating the topographic and bathymetric mapping data contributions of the agencies and partners in the 3DEP, the Interagency Working Group on Ocean and Coastal Mapping (IWG-OCM), and relevant elevation data activities of other Federal, state, academic and private sector groups. The 3DEP continues and improves upon the USGS commitment to coordinate the collection of terrestrial elevation data and to manage these national elevation data assets.

The supplemental guidance issued in 2010 for OMB Circular A-16 was used throughout the planning and program development process of 3DEP. Though terminology may vary, the steps, intermediate products, and expected outcomes of 3DEP are intended to fully implement and be compliant with the guidance as it relates to the management of National Geospatial Data Assets (NGDAs). The USGS has a long history of managing elevation data, and 3DEP is a natural extension to address a wide range of new and emerging needs. The 3DEP embodies the model of portfolio management and cooperative programs that will improve efficiency across all levels of government. In fulfillment of its responsibilities to manage NGDAs and to track program implementation progress, the USGS plans to publish annual reports on the state of the elevation data.

The 3DEP includes a program of data collection, data management, and delivery of products and services that encompasses one complete geospatial data lifecycle as described by the OMB A–16 Supplemental Guidance. The definition of the first lifecycle should not be interpreted to mean that 3DEP will end after 2022. The plan intentionally addresses a defined period of time and an accomplishment that can be realized. USGS managers expect that technology will change and that data sensors will become more prolific and may include space platforms and may be cheaper to deploy in the future. Ten years from now, 3DEP will look quite different, and it will be time again to assess and evaluate where this important national program needs to go. It is also expected that billions of dollars in benefits to the Nation will have been realized and that ongoing and new needs will shape the future of the next-generation 3DEP.

Assessment of Requirements

The NEEA, completed in December 2011 (Dewberry, 2012), was conducted to (1) Document national-level requirements for improved elevation data, (2) Estimate the benefits and costs of meeting those requirements, and (3) Evaluate multiple national-level program-implementation scenarios. The assessment was sponsored and funded by the NDEP's member agencies. The study participants included 34 Federal agencies, agencies from all 50 States, selected local government and Tribal offices, and private and not-for-profit organizations. In all, 602 mission-critical activities were summarized by major business uses that need significantly more accurate data than are currently available. The results of the assessment indicate that these improved enhanced elevation data have the potential to generate $13 billion in new benefits annually.

All organizations identified key functional activities, described in their own words, with mission-critical requirements for enhanced elevation data, along with their elevation data requirements by quality level (Table 5.1 and on page 71), update frequency, and geographic area. Each functional activity was linked to the business use that was most similar to the described activity. These data were collected by an online questionnaire, interviews, and workshops; they were finalized during a validation process that resulted in the formal documentation of each organization's requirements and anticipated benefits. All of these data were entered into a master geodatabase. The full NEEA report (Dewberry, 2012) includes detailed documentation for 104 functional activities from 34 Federal agencies, 329 functional activities from 50 States and U.S. territories, and 144 functional activities from local and Tribal governments. Twelve private companies and one not-for-profit organization documented 25 more functional activities.

Benefit and Cost Analysis and Recommended Data-Acquisition Scenario

The geographic extents of mission-critical data needs were mapped for each functional activity, and a geodatabase was created to support the analysis. The data requirements were aggregated into cells measuring one degree of latitude by one degree of longitude, and the probable benefits and costs of a national program were computed for every cell for each of 25 implementation options. The 25 options were based on various combinations of data quality levels (QLs) and refresh cycles. Where data were required for a portion of a cell at a particular QL, the benefits were prorated such that benefits would not be overstated for any cell. The estimated costs for each option included those for data collection and life-cycle management. The benefit analyses used the conservative benefit estimates. In addition, environmental and ecosystem services were generally not quantified and, therefore, did not contribute to the benefit analysis.

On the basis of the 25 options, a set of national program implementation scenarios was developed that would optimize every cell for QL and replacement cycle (or some combination of the two) for a Federal-only program and national strategies that considered all needs. Other scenarios were developed that would hold QLs and replacement cycles constant. The final analysis yielded 10 leading scenarios. Each scenario would implement a national data-collection strategy to achieve cost efficiencies and meet the requirements of multiple organizations. While it was possible to lower costs slightly, none of the optimized scenarios offered the benefits of national consistency. All of the scenarios assumed IfSAR data coverage in Alaska, where cloud cover and remoteness preclude consideration of lidar data acquisition for much of the State. A detailed overview of each of the leading scenarios can be found in the full NEEA report (Dewberry, 2012).

A midrange scenario offers a high benefit/cost ratio of 4.7/1, uniform QL2 data, and an 8-year acquisition cycle. For Alaska, IfSAR QL5 data is being collected over 10 years as part of the Alaska mapping initiative. Nearly all cells have a positive benefit/cost ratio in this scenario. The exceptions include arid regions of the western United States that are predominantly in Federal ownership. If this scenario is viewed from the perspective of return on investment, a positive return on investment is realized in less than 14 years in all cells. There are many business uses in the U.S. arid regions where NEEA study participants were unable to assign dollar benefits. If quantified, these benefits would increase the benefit/cost ratios in those cells. Many of the NEEA study participants indicated that QL consistency was important for their applications. Analysis of the return-on-investment periods indicates that extending the acquisition cycle is better than collecting lower quality data in these areas. The added benefits of national consistency led to the recommendation of the QL2, 8-year-cycle scenario.

According to the benefits identified in the NEEA study, the Federal agencies poised to realize the highest benefits to their mission from enhanced elevation data include the Natural Resources Conservation Service, the U.S. Army Corps of Engineers, the Defense Installation Spatial Data Infrastructure, the U.S. Geological Survey, the National Oceanic and Atmospheric Administration, the Federal Emergency Management Agency, the Environmental Protection Agency, the U.S. Forest Service, the Federal Aviation Administration, and the National Geospatial-Intelligence Agency. States and other partners will benefit from and be able to fully participate in 3DEP.

3D ELEVATION PROGRAM (3DEP) DEVELOPMENT

Following the analysis and vetting of the NEEA study, the QL2, 8-year-cycle scenario for collecting elevation data was recommended by the NDEP committee and its 12 Federal member agencies. The 3D Elevation Program was endorsed by or received letters of budget support from the American Society for Photogrammetry and Remote Sensing (ASPRS), the Association of American State Geologists (AASG), the Association of State Floodplain Managers (ASFPM), the Coalition of Geospatial Organizations (COGO), the Management Association for Private Photogrammetric Surveyors (MAPPS), the National Geospatial Advisory Committee (NGAC), the National Society of Professional Surveyors (NSPS), and the National States Geographic Information Council (NSGIC).

The primary activities from 2012 until 2015 were program growth, new data collection, product and service development, and technology modernization. Data collection rates have been increasing as a growing number of state and Federal agencies participate in cooperative data acquisition projects. In 2015 the final year of program preparation, the focus was mainly on the expansion of data collection activities and initiation of product delivery services. FY2016 saw a significant ramp-up in data acquisition and delivery, as a result of continued year-over-year increases in funding for 3DEP from USGS and partners. Total funding for 3DEP from all partners increased from $39M in FY15 to $49M in FY16, an increase of 26%. The number of square miles of lidar contracted for acquisition in the lower 49 states and territories increased from 149,884 square miles to 203,188 square miles in FY16, an increase of 36%. Looking forward, it is clear that the combined efforts of a broad partnership community are achieving the envisioned 3DEP goal. In particular, partners are pooling funds to collect lidar data over larger areas and at lower costs than previously achieved.

DATA ACQUISITION PROCEDURES FOR 3DEP

There are several steps that participants in 3DEP take as relates to lidar acquisitions that meet the goals of systematic, consistent data across the conterminous United States. There are 3 main processes that participants follow when acquiring data for 3DEP:

1. Determining if data has been acquired already over a proposed area for 3DEP. This involves querying the U.S. Interagency Elevation Inventory (USIEI) to see if publicly available lidar already exists
2. Agreeing to collect or provide data to meet requirements detailed in the USGS Lidar Base Specification (Heidemann, 2018). This ensures that data collected is consistent and standardized across the country, so processes run in one part of the country can be applied to others
3. Procuring data via the Broad Agency Announcement (BAA), via a collaborative partnership with a 3DEP partner, or contributing data for inclusion in to the 3DEP holdings

U.S. Interagency Elevation Inventory

The USGS and the National Oceanic and Atmospheric Administration (NOAA) maintain the U.S. Interagency Elevation Inventory (*https://coast.noaa.gov/inventory/*) of all publicly available lidar and IfSAR data to guide new data acquisition and avoid duplication. The fiscal year 2015 (FY15) update of the U.S. Interagency Elevation Inventory indicated that 113,896 square miles (mi^2) of new quality level 2 (QL2) or better lidar data were acquired in FY15 by Federal, state, and other organizations covering 3.8 percent of 49 states (not including Alaska) and the U.S. territories. The locations where lidar were or are being collected that meet the QL2 requirements of 3DEP in 2018 are shown in Figure 5.2. The USIEI includes public sources of lidar in all states that meet 3DEP specifications (FY08-FY18). Information is provided by users during the annual U.S. Interagency Elevation Inventory (USIEI) survey and was updated in September, 2018. The USIEI also includes lidar point cloud data that are in production and unpublished as of the report release date. Visit the USIEI at *https://coast.noaa.gov/inventory/* to find availability from state and Federal agencies. Visit *https://www.usgs.gov/core-science-systems/ngp/3dep/3dep-status-maps* to see the latest status reports and maps for all USGS published lidar point cloud data.

Data collection rates by FY (2007-17) are shown in Figure 5.3. The collection of QL2 or better lidar data continues to increase substantially, whereas the collection of QL3 and lower quality lidar continues to decline. In FY15, the inventory reported that 20,517 mi^2 of QL3 and lower quality lidar were collected, a decrease from 54,414 mi^2 in FY14. The trend toward QL2 or better data is expected to continue as users realize greater utility in the higher quality data and acquisition costs decline (see the "3D Elevation Program Data Collection Contracted in Fiscal Year 2015" section). To accomplish the program goal in eight years, 12.8 percent would need to be collected each year.

Because of weather and remote conditions, IfSAR is the technology best suited for statewide data collection in Alaska. IfSAR data were collected in Alaska for 74,634 mi^2 covering 11 percent of the State in FY15, and 80,574 mi^2 covering 12 percent of the State in FY16. The IfSAR collection activity in Alaska is shown in Figure 5.4 on page 118. As of 2018, 98% of Alaska has been collected.

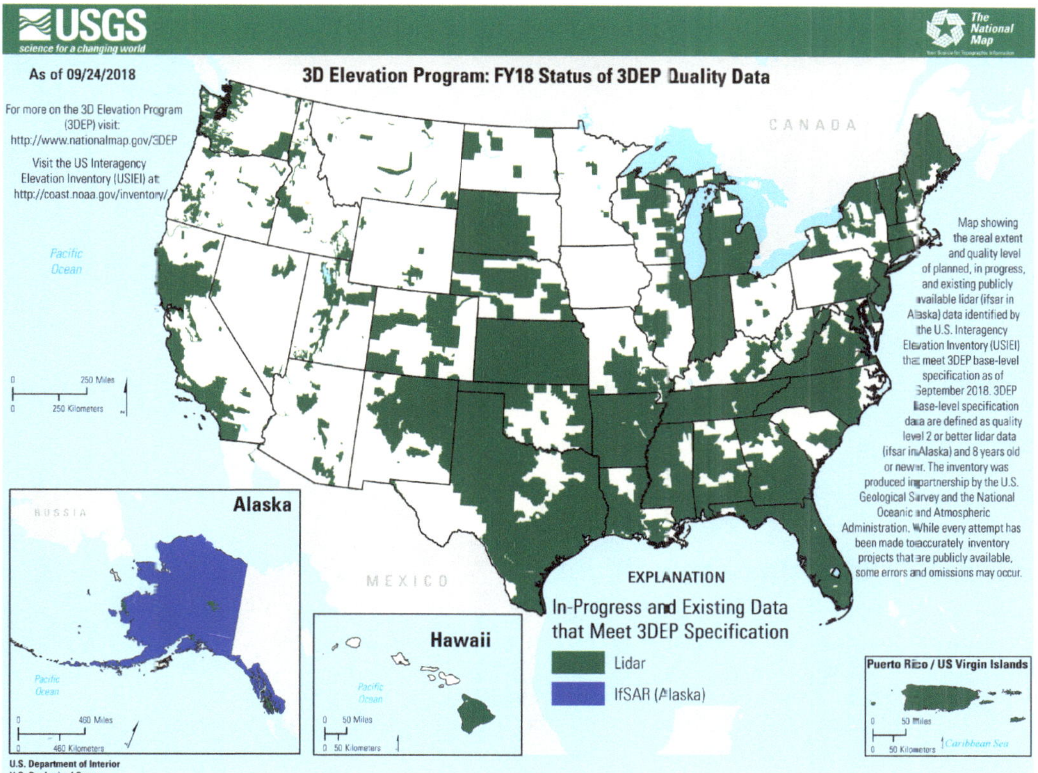

Figure 5.2. Lidar Point Cloud Data Availability. Map depicts public sources of lidar in all states that meet 3DEP specifications (FY08-FY18). Information is as provided by users during the annual U.S. Interagency Elevation Inventory (USIEI) survey and was last updated in September, 2018. Note that some lidar point cloud data may be in production and unpublished as of this Manual's release date. Visit the USIEI at *https://coast.noaa.gov/inventory/* to find availability from state and Federal agencies. Visit *https://www.usgs.gov/core-science-systems/ngp/3dep/3dep-status-maps* to see the latest status reports and maps for all USGS published lidar point cloud data.

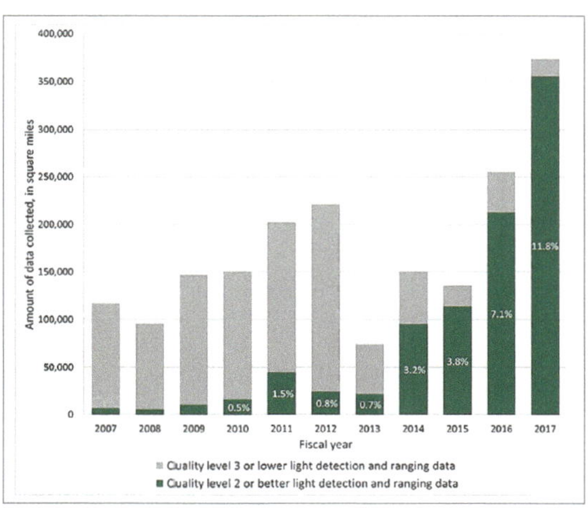

Figure 5.3. Acquisition of lidar data by year. Total square miles of lidar by Quality Level acquired since FY07. Percent indicates the portion of the 49 states and U.S. Territories (not including Alaska) that were acquired (dark green) each year for QL2 and better lidar meeting 3DEP requirements. Percentage labels are not shown in years 2007-2009 where the area collected was less than 0.5 percent of the country. To accomplish the program goal in eight years, 12.8 percent would need to be collected each year. Source: 2018 U.S. Interagency Elevation Inventory (USIEI).

Alaska IfSAR Elevation Data—92% Available or In-Work

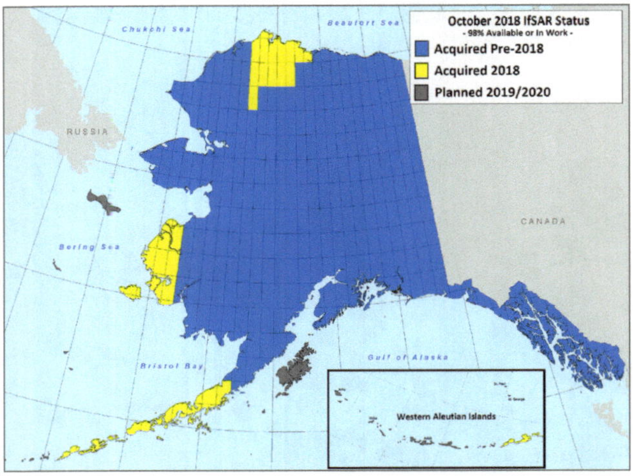

Figure 5.4. Public sources of interferometric synthetic aperture radar data collected in Alaska before and in fiscal year 2018 that meet 3D Elevation Program base-level specifications (quality level 5).

The USGS Lidar Base Specification

To ensure that data that is ingested by the 3D Elevation Program is standardized and consistent, the 3DEP requires that any data contracted (directly or indirectly), or contributed for inclusion must meet the requirements documented in the USGS Lidar Base Specification, currently version 1.3 (Heidemann, 2018). Unlike most other lidar data procurement specifications, which largely focus on the products derived from lidar point cloud data such as the bare-earth digital elevation model (DEM), this specification places particular emphasis on the handling of the source lidar point cloud data. These specifications are intended to ensure that the complete source dataset remains intact and viable to support the wide variety of DEM and non-DEM science and mapping applications that can benefit from lidar technology. The source dataset includes the data, metadata, descriptive documentation, quality information, and ancillary data, all collected in accordance with the minimum parameters described within this specification.

By requiring data meet this specification, and by applying rigorous quality assurance methods to check for compliance, users of data with the 3DEP brand can be fairly certain what information can and cannot be extracted. This specification will undergo periodic revisions to update content in the specification, incorporate new data types and technologies, and provide open-source methods for ensuring that anyone can check compliance to the specification prior to delivering to USGS for processing and dissemination.

The 3DEP Data Acquisition Process

The USGS issues a Broad Agency Announcement (BAA) that provides detailed information on how to partner with the USGS and other Federal agencies to acquire high-quality 3-D elevation data. The BAA process was established to make Federal partnerships with non-Federal partners more competitive, fair, and transparent. These efforts ultimately lead to cost efficiencies and a more consistent product. Information and contacts for the BAA are now available at Fed Biz Opps and Grants.gov web pages. Applicants may contribute funds toward a USGS lidar data acquisition activity via the Geospatial Products and Services Contracts, or they may request 3DEP funds toward a lidar data acquisition activity where the requesting partner is the acquiring authority. Federal agencies, state and local governments, tribes, academic institutions and the private sector are eligible to submit proposals.

In July of 2014 the USGS issued the first Broad Agency Announcement to solicit funding partnerships for 3DEP. There was tremendous interest in the program. USGS received 72 pre-proposals and ultimately funded 29 projects. The federal offices (USGS, FEMA and NRCS) that provide funds to support BAA awards collectively contributed $10.8M. BAA partners (regional federal agencies, state and local offices) contributed $13.1M to increase the value of the BAA projects to just under $24M.

In 2015 project awards ranged from $61K to $1.3M, and the average award was approximately $355K. The average BAA award covered 36% of the cost of the project, ranging from a low of 14% to the maximum allowable contribution of 50%. The average project size was 3,100 square miles, with a range from 550-11,900 square miles (Figure 5.5).

In FY16, BAA numbers were very similar. USGS supported 29 projects. The USGS, FEMA and NRCS direct contributions to BAA projects have been consistent, however the overall federal contribution to 3DEP rose significantly. 3DEP BAA projects resulted in $26.8M of project work which will result in 123,000 square miles of new lidar data As of July 2016, total FY16 contributions to 3DEP projects by 3DEP stakeholders is up around $58M, supporting a total of 245,000 square miles of project work. These are preliminary figures that will be modified. The final contracts for some of these projects will likely not be completed in FY16 but likely will be contracted before the end of the calendar year in preparation for the 2017 spring acquisition window.

The growth in federal contributions outpaced the growth in BAA submissions. Many federal acquisitions moved forward without state and local partners. For FY17, Federal agencies have worked hard to identify their high priority areas in sufficient time to include their areas of interest in the BAA to advertise these partnership

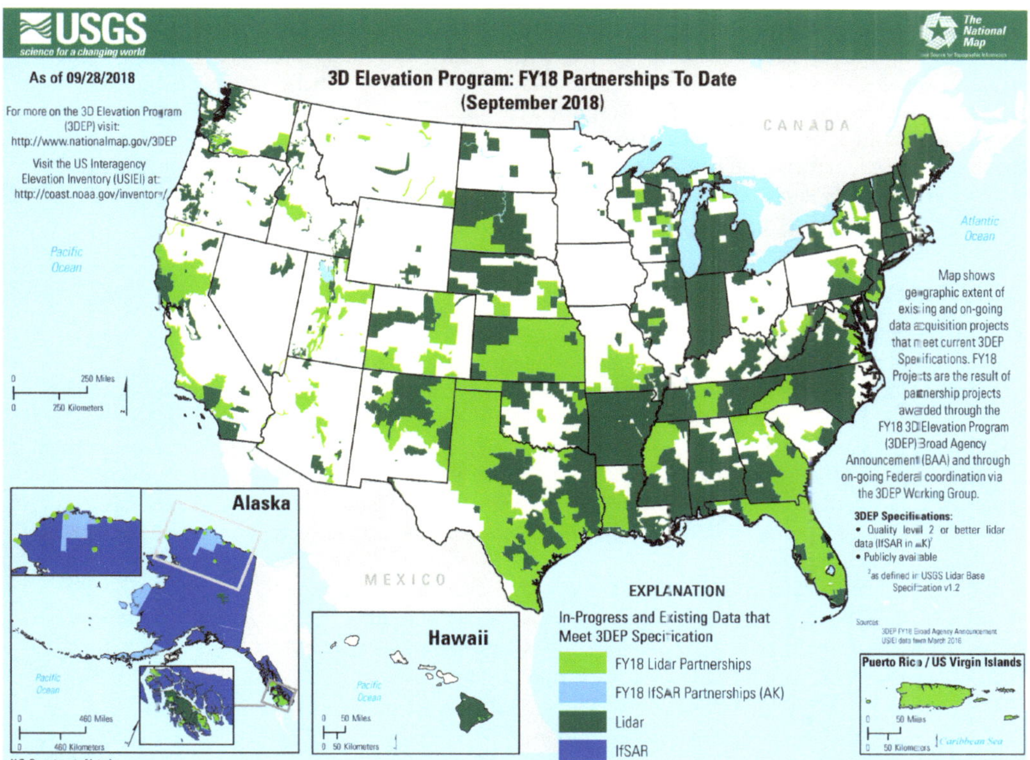

Figure 5.5. 3D Elevation Program acquisition awards for new light detection and ranging and interferometric synthetic aperture radar data acquisition projects planned for fiscal year 2018 and that meet 3D Elevation Program base-specifications in the United States and U.S. territories.

opportunities. The average project size increased to 4,220 square miles, an increase of over 1,000 square miles over the FY15 average project size. The cost efficiencies gained with larger acquisitions reduce the cost per square mile of the project.

The 3D Elevation Program and the Interagency Working Group on Ocean and Coastal Mapping have established a system to share information about areas of interest and proposed and planned elevation projects. Federal, State and Local requirements are available at the NOAA sponsored SeaSketch site: U.S. Federal Mapping Coordination, A Demonstration Site for Federal Mapping Data Acquisition (*http://seasket.ch/hwpR3E-MxO*). This mapping site is comprehensive, showing areas of interest, proposed, planned, in-work, and completed elevation and related projects. Federal Agencies annually update their High Priority Areas of Interest. Each Federal Agency is responsible for updating their requirements. State and Local agencies may enter their own requirements or may utilize the 3DEP Lidar Priorities Collector Tool to enter their Areas of Interest. Areas entered into the project collector tool are collated into a service with a direct feed to SeaSketch (State and Local 3DEP Interest (1-3 years), 3DEP Mapping Areas of Interest.

The program offers applicants two different contract mechanisms. Applicants may choose to make use of the USGS Geospatial Products and Services Contracts (GPSC) or they may choose to apply for financial assistance in the form of a cooperative agreement. For the GPSC, the USGS manages the contract, while the applicant provides their portion of the project funding to the USGS through a Joint Funding Agreement (JFA) or an Interagency Agreement (IA). While each JFA is unique to the funding partner, there is a standard template that can be used by all. The GPSC is a multiple award acquisition vehicle that is designed to utilize the teams of firms on the contract for services needed to accomplish 3DEP data acquisition. Firms on the GPSC have been selected based on their qualifications and performance in providing professional services needed for 3DEP. The contracts include acquisition, processing, and quality assurance of lidar and other source geographic data. The second option is where the USGS provides financial assistance to the applicant by means of a cooperative agreement. The applicant is responsible for securing a vendor to acquire the data. The data must adhere to the data specifications and deliverables as defined in the USGS Lidar Base Specification, and partners must agree to allow the data be put in to the public domain. The grants.gov website must be used by those requesting a cooperative agreement. For more information on the BAA process, please go to *http://nationalmap.gov/3DEP/FY17BAA.html*

Acquisition Trends

There are several notable trends that are happening among 3DEP GPSC and partner contracts. The total investment in data collection among all partners continues to increase. In FY13, USGS and partner investments in 3DEP totaled about $24 million, and in FY17 they increased substantially to about $86 million (Table 5.1). Although the increased investments mean greater lidar coverage, the overall quality level of data collected has shifted as well. In FY15 no QL3 contracts were awarded through the BAA process, a trend which is being mirrored throughout the industry. In FY13, 49,000 mi^2 of QL1-QL2 lidar data collection contracts were awarded. In FY15, that number increased to 150,000 mi^2 of QL1-QL2 lidar data. Investments in IfSAR data, part of the Alaska Mapping Initiative, have remained stable at about $7 million per year with an increase in FY17 (Table 5.1). The funding mix for USGS and other partners include both annual appropriations and other funds. For example, USGS received a onetime appropriation for lidar to respond to Hurricane Sandy recovery efforts in FY13 and FY14. Partner funds represent the value as identified in cooperative agreements. Actual partner investments, which would include administrative and QA/QC costs are likely higher. In addition, some data were contributed to USGS and are being made available for download. These contributed datasets are not included in the summary information.

	USGS	FEMA	NRCS	Other partners	Total	Total, Lidar and IfSAR
FY13						$20.2M
Lidar	$4.0M	*	*	$9.3M	$13.4M	
IfSAR	$3.5M	*	*	$3.2M	$6.8M	
FY14						$30.0M
Lidar	$7.3M	*	*	$15.8M	$23.1M	
IfSAR	$2.8M	*	*	$4.1M	$6.9M	
FY15						$46.3M
Lidar	$7.2M	$11.2M	$7.1M	$13.5M	$39.0M	
IfSAR	$3.6M	$0.0M	$0.1M	$3.7M	$7.4M	
FY16						$59.2M
Lidar	$9.2M	$13.6M	$12.4M	$16.2M	$51.4M	
IfSAR	4.8	0.0	0.3	2.8	7.8	
FY17						$86.1M
Lidar	$11.6M	$24.6M	$23.6M	$16.7M	$76.5M	
IfSAR	$7.2M	$0.0M	$0.7M	$1.7M	$9.6M	

Table 5.1. Total acquisition contracts awarded by fiscal year (2013-17) including U.S. Geological Survey and partner organization contracts for light detection and ranging (all quality levels) in the U.S. and U.S. territories and interferometric synthetic aperture radar data in Alaska. * FEMA and NRCS funds are included in the "Other partners" category in FY13 and FY14

The USGS monitors data collection costs and evaluates the average cost per square mile for representative projects each year. In the past, this has been reported as the total acquisition cost consisting of collection contracts, quality assurance/quality control (QA/QC), and program administration (Lukas et al, 2015). Beginning in FY15, the USGS tracked and reported the actual contract collection cost. The QA/QC and other program administration costs are part of the base program budget and were monitored separately. The change normalized the cost and resulted in reporting that is consistent with the way Federal and state partners report data collection cost. The FY15 QL2 data collection cost was based on an evaluation of 47 USGS and partner data collection contracts. Lidar projects with unusual costs or missing cost components were not included in the analysis nor were projects in Hawaii, Puerto Rico, and the U.S. Trust Territories which are highly variable due to high costs of aircraft mobilization, weather, and other factors related to operating in remote locations. The costs identified can be used as a baseline for developing project estimates; however, care should be exercised when using the average cost figures for data-collection cost estimates in remote locations or for areas having difficult terrain.

The lidar data collection costs continue to decline. The average cost per square mile for QL2 lidar was $289 in FY13 and $203 in FY18. The average cost per square mile in FY15 was $226, and the average project size was 2,836 mi². Increasing project size is a specific goal of 3DEP and is one of the criteria used when making awards through the BAA process. Larger projects produce consistent data across large areas and lower program management costs. The IfSAR data collection costs have increased in Alaska primarily because of the remoteness and rugged terrain associated with the remaining collection cells and because many of the flight lines in later years were over water in coastal areas.

Average cost per square mile (number of projects evaluated, average size of project in square miles)[2]						
Quality level	2012	2013	2014	2015	2016	2017
QL1 lidar	$698 (1, 2021)	$520 (6,934)	$680 (7,1009)	$478 (6,1429)	$496 (5,1984)	$406 (10, 2231)
QL2 lidar	$378 (3,648)	$289 (20,1665)	$284 (23, 2504)	$226 (47, 2836)	$203 (41,4299)	$203 (44, 6491)
QL3 lidar	$228 (28, 2183)	$142 (9, 2632)	$220 (4, 6225)	N/A	N/A	N/A
QL5 IfSAR	$95	$100	$110	$127	$97	$105

Table 5.2. Comparison of average acquisition costs of light detection and ranging data and interferometric synthetic aperture radar data by 3D Elevation Program quality level, fiscal years 2012–17.

QUALITY ASSURANCE OF DATA FOR 3DEP

Once data are acquired, either via GPSC or partnerships, or contributed, the resulting data go through a rigorous QA process to ensure data meet the USGS Lidar Base Specification, among other checks. Once delivered, QA staff make sure that all deliverables were sent. These include:

- Lidar Point Cloud data
    - Calibrated Swath point cloud in LAS format
    - Tiled point cloud LAS format (usually classified)
- Breaklines
    - As referenced in the current standards and specification
- Digital Elevation Models (DEMs)
- Metadata
- Any other deliverables required by the Task Order or Statement of Work

Metadata are verified, including project and product level metadata. These are checked to ensure that all metadata are compliant with the FGDC Content Standard for Digital Geospatial Metadata (*https://www.fgdc.gov/metadata/csdgm/*). Spatial metadata, either in shapefile or geodatabase format are checked and verified that they meet the Task Order or Statement of Work.

From here, automated checks of the point cloud data are conducted. These include verifying that the Non-Vegetated and Vegetated Vertical Accuracies meet the requirements defined in the USGS Lidar Base Specification. Tests are run to verify that LAS files are compliant. Currently data must be compliant with LAS version 1.4. Examples of items that are checked include conformity to the classification scheme, scan angles populated correctly, intensity values populated correctly, and File Source and Point Source IDs are populated correctly.

The next QA steps involve visual reviews of the data. Staff check the lidar swath data for calibration issues, and to ensure Overlap and Withheld flags are properly assigned. Checks on the classified tile data include visualizing classifications and assuring the tiles conform to the tiling scheme that was defined. Breaklines are loaded in with the classified tiles to ensure proper conflation, and to verify that water features are acceptable voids in the point clouds.

Derivative products, such as bare earth DEMs, are visually inspected to look for seamline and interpolation issues, and artifacts caused by inadequate breaklines, such as artifacts across water bodies that are supposed to be flattened, and bridges that are supposed to be removed. Improper classification of bare earth points are much easier to detect while inspecting the derived DEMs than the point clouds.

Chapter 15 of this manual explains 3DEP quality assurance steps in great detail.

PRODUCTS AND SERVICES FROM 3DEP DATA

The information technology infrastructure necessary to quality control, manage, and deliver 3DEP products and services was the primary systems development focus from 2013 to 2015 (Sugarbaker, 2014). The data management and delivery system was built to support lidar point cloud (LPC) projects, IfSAR orthorectified radar images (ORIs) and digital surface models, and standardized digital elevation models (DEMs) at multiple levels of resolution. Additionally, online services to support elevation point query and viewing of elevation surfaces were created. The 3DEP implementation phase also included the transition of all legacy data previously managed under the umbrella of the National Elevation Dataset. Most of this work was completed by early FY15, with the remainder of the year spent accelerating production and improving lidar and IfSAR processing efficiency. For example, once LPC data have been through the QA/QC process, a single workflow catalogs, stores, derives products, and releases data for download to the general public. The expectation is that the time from data acceptance to publication will become very short. Although LPC and IfSAR production processing began in FY15, the infrastructure to support higher production rates in FY16 and later years was also being established (Figure 5.6). The LPCs and derived products were published for over 60,000 mi^2 in FY15. The projected production in FY16 was 132,500 mi^2, while the actual production rate was closer to 190,000 mi^2. In FY18, production rates are projected to exceed 225,000 mi^2 based on the volume of new data collection awards issued in FY15 and FY16. Visit *https://www.usgs.gov/core-science-systems/ngp/3dep/3dep-status-mapsall* to see the latest status reports and maps for all USGS published lidar point cloud data.

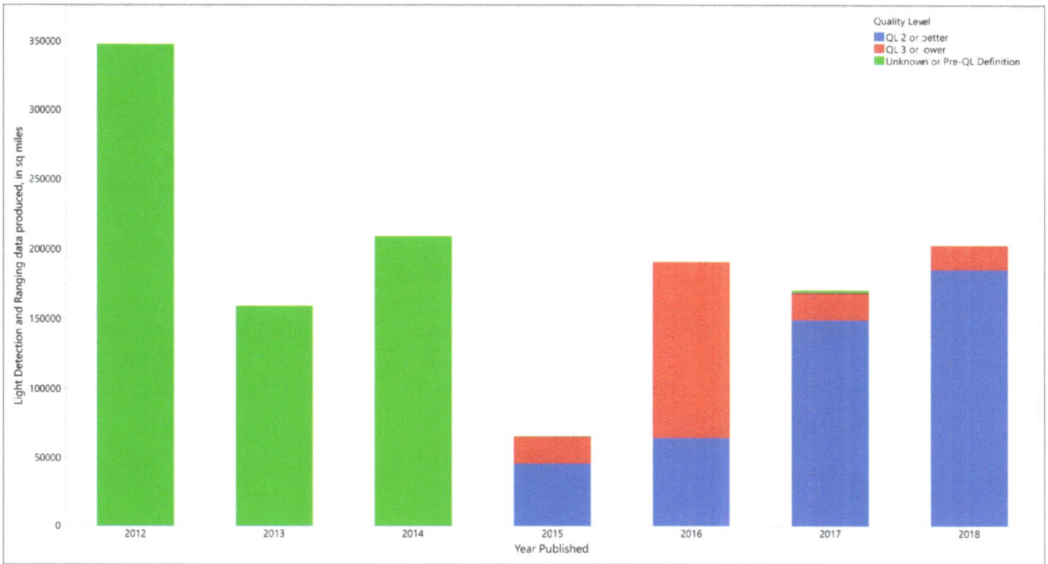

Figure 5.6. New light detection and ranging point cloud data available (2012–18) for download by fiscal year and quality level.

Source Data

Source data products for 3DEP are defined as the source input data that was produced by the contractor, provided by a partner, or contributed for inclusion in to 3DEP. Source data products include:

- Lidar point cloud (LPC): These data are the foundational data for 3DEP in the conterminous U.S., and contain the original three-dimensional information from which the DEM products are derived. Most of the data collected in 2014 and later meet 3DEP specifications for quality level 2 nominal pulse spacing and vertical accuracy, and data collected prior to 2014 often do not meet the quality level 2 specification. Distinctions in nominal pulse spacing are provided in the lidar point cloud status graphics and in the download platform; however, other qualities such as vertical accuracy must be examined to determine if the data meet particular 3DEP quality level specifications
- IfSAR digital surface model (DSM): These 5 meter rasters, available only in Alaska, are the initial IfSAR product, before bare-earth filtering is done to create the DEMs. Man-made structures and vegetation are modeled in the DSM
- IfSAR orthorectified radar image (ORI): These rasters (resolutions vary), available only in Alaska, are radar reflectance intensity recordings detected by the IfSAR sensor
- Source resolution DEMs: These data are the original bare earth DEMs derived from lidar point cloud source. Source DEM's processed by the USGS after January 2015 are provided where the original DEM horizontal resolution or projection differ from the 3DEP standard DEM datasets

Standard Products

Standard products are created from the input source DEMs that are contracted, provided via a partnership, or contributed. The standard DEMs represent the topographic surface of the earth and contain flattened water surfaces. Each DEM dataset is identified by its horizontal resolution and is produced to a consistent set of specifications. Standard DEMs are characterized either as project-based or seamless. Project-based DEMs are available for the full areal extents of projects when produced from light detection and ranging (lidar), or as one-degree blocks with overedge when produced from IfSAR. Seamless DEMs are produced by blending only the highest quality project data into a continuous terrain surface for the U.S. These data are distributed in tiles that can be merged to support analysis across large geographic areas. There are several types of standard DEMs provided by the 3DEP. Project-based data are data that have been produced on a project-by-project basis. They are assumed to be seamless within projects, but not necessarily between projects. Due to the high resolution of these data if you merge disparate adjacent projects together you may notice discrepancies along the boundaries due to different sensors, processing and temporal issues. The seamless datasets are created from elevation data that are sourced from multiple technologies, positional accuracies and collection dates. Although these datasets are being continuously updated as new elevation source is acquired, much of the national coverage is derived from topographic map contours that may be decades old. The spatial metadata for each tile in a seamless DEM layer must be examined to fully determine the source datasets used to create each tile. The following are the categories of standard product DEMs:

Project-based

- 1-meter: This dataset was introduced in 2015 with limited coverage of the U.S., but will be expanding as new DEMs from 3DEP quality level 2 or better lidar data are acquired. Horizontal coordinates are referenced to the Universal Transverse Mercator projection.
- ⅑ arc-second: This dataset covers about 25 percent of the conterminous U.S. and is produced from 3-meter or higher resolution DEMs acquired by the USGS prior to January 2015. Horizontal coordinates

are referenced to geographic coordinates (longitude, latitude). The ⅑ arc-second dataset will no longer be updated with newly acquired DEMs; however, it will continue to be distributed

- 5 meter: This dataset is comprised of 5-meter IfSAR-derived DEMs (3DEP quality level 5) over Alaska only. Horizontal coordinates are referenced to Alaska Albers Equal Area Conical projection

Seamless

- ⅓ arc-second: This is the highest resolution seamless DEM dataset for the U.S. with full coverage of the 48 conterminous states, Hawaii, and U.S. territories. Alaska coverage is partially available now and is being expanded to statewide coverage as part of the Alaska Mapping Initiative. Ground spacing is approximately 10 meters north/south, but variable east/west due to convergence of meridians with latitude
- 1 arc-second: This is a lower resolution seamless dataset providing complete coverage over the conterminous U.S. and partial coverage of Alaska. Most of Canada and Mexico are also covered by the 1 arc-second dataset. Ground spacing is approximately 30 meters north/south, but variable east/west depending on latitude
- 2 arc-second: This seamless dataset is the lowest resolution seamless dataset available and covers only Alaska. Although ground spacing is approximately 60 meters north-south, east-west spacing can vary from 35 meters in southern Alaska to 20 meters on the North Shore

DISSEMINATION OF DATA FOR 3DEP

All data and derived products collected for 3DEP are made available, for free, via The National Map (*http://nationalmap.gov/3DEP/*). All source data and standard products are available for querying and downloading via The National Map viewer. Also, all data and metadata are available for direct access via FTP (*ftp://rockyftp.cr.usgs.gov/vdelivery/Datasets/Staged/Elevation/*), as well as via Amazon's S3 cloud service (*http://prd-tnm.s3.amazonaws.com/index.html?prefix=StagedProducts/Elevation/*) The elevation index of data is also available via services, such as REST, WMS, and via ArcGIS.com.

Metadata about the individual source datasets used to assemble the standard DEM layers are presented in a spatially-referenced polygonal footprint. The attributes of each source dataset, such as original spatial resolution, production method, and date entered into the standard DEM, are linked to this footprint. A list and description of these attributes for the 3DEP DEMs (previously known as the National Elevation Dataset) and source datasets can be found in the 3DEP Data Dictionary. Please note that the data dictionary has not been updated to new 3DEP product terminology; however, National Elevation Dataset (NED) references are directly applicable to the newly branded 3DEP seamless DEM layers. Spatial metadata in the form of an Esri shapefile is supplied for 3DEP elevation products, and can be found at *http://nationalmap.gov/3DEP/3dep_prodmetadata.html*.

CURRENT DEVELOPMENT AND FUTURE DIRECTIONS

The 3DEP is constantly looking for ways to improve efficiencies by investigating and adopting new technologies and methodologies that have the potential to reduce costs while still producing data that satisfies 3DEP requirements. Currently (2018), two new lidar technologies—Geiger-mode (Aull et al, 2004) and Single-photon counting lidar (Degnan et al., 2007) are being evaluated. These instruments have the potential to collect data at much higher altitudes than traditional lidar instruments and, as a result, could provide data at a much lower cost. The 3DEP is testing the accuracies and noise of these sensors because it has not been conclusively determined if they can meet QL2 or better specifications routinely met by linear-mode lidar (Stoker et al., 2016).

Other technologies are beginning to be evaluated as well; for example, topographic-bathymetric lidar has the potential to collect inland bathymetric data under certain conditions related to water clarity, water depth, and other factors. This technology can provide new elevation data for submerged topography where 3DEP has traditionally required hydro-flattened water surfaces in DEMs. In addition, structure-from-motion and other image-based photogrammetric processes for deriving 3-D information have the potential to collect data that meet 3DEP accuracy requirements. The use of unmanned aircraft systems as platforms for data acquisition could further affect the utility of these technologies.

The USGS is currently porting all 3DEP point cloud data and derived products to Amazon S3 cloud, to enable users the ability to tap in to 3DEP data in the cloud and extract custom information from the products without necessarily having to download all data to a local computer to create informational products used in decision making. The 3DEP hopes to transition from a product and download-first strategy to a service and information derivative strategy in the near future. Using Amazon S3, in cooperation with some form of cloud computing will allow third parties to tap in to the source 3DEP data and produce value added products or information used in decision making much easier than the current download-first paradigm. Services, such as streaming DEM and point cloud derivative generations and visualizations are currently in development, and the goal is to have these services operational by the end of FY17.

AUTHOR BIOGRAPHIES

Jason Stoker is the 3DEP Product and Service Lead for the USGS National Geospatial Program.
Vicki Lukas is the Chief of Topographic Data Services for the USGS National Geospatial Program.
Allyson Jason is the 3DEP Data Analyst for the USGS National Geospatial Program.
Diane Eldridge is the 3DEP Data Acquisition Coordinator for the USGS National Geospatial Program.
Larry Sugarbaker is the Senior Strategic Advisor for the USGS National Geospatial Program.

ACKNOWLEDGEMENTS

The 3D Elevation Program (3DEP) initiative is a partnership among many individuals within the U.S. Geological Survey (USGS) and across our partner Federal, State, and professional organizations. The authors wish to acknowledge the 3DEP Executive Forum and Working Group members for their guidance, technical expertise, and enthusiastic participation in implementing 3DEP. We thank Kevin Gallagher, USGS Associate Director for Core Science Systems, for his ongoing commitment and leadership to make 3DEP a success.

Note: Any use of trade, product, or firm names is for descriptive purposes only and does not imply endorsement by the U.S. Government.

REFERENCES

ASPRS (2012). ASPRS LAS specification, version 1.4–R12. Bethesda, Md., American Society for Photogrammetry and Remote Sensing.

ASPRS (2014). *ASPRS positional accuracy standards for digital geospatial data*. In Photogrammetric Engineering and Remote Sensing, 81(3) (pp. A1-A26), Bethesda, MD American Society for Photogrammetry and Remote Sensing.

Aull, B.F., Loomis, A.H., Young, D.J., Stern, Alvin, Felton, B.J., Daniels, P.J., Landers, D.J., Retherford, Larry, Rathman, D.D., Heinrichs, R.M., Marino, R.M., Fouche, D.G., Albota, M.A., Hatch, R.E., Rowe, G.S., Kocher, D.G., Mooney, J.G., O'Brien, M.E., Player, B.E., Willard, B.C., Zong-Long Liau, and Zayhowski, J.J. (2004). *Three-dimensional imaging with arrays of geiger-mode avalanche photodiodes*. In Semiconductor Photodetectors, Linden, K.J. and Dereniak, E.L, eds. (pp. 105-116). San Jose, CA: International Society of Optical Engineering, SPIE Proceedings, v. 5353.

Degnan, J.J.; Wells, David; Machan, Roman; and Leventhal, Edward (2007). *Second generation airborne 3D imaging lidars based on photon counting*. In Advanced photon counting techniques II, Becker, Wolfgang (ed). Boston, MA: International Society of Optical Engineering, SPIE Proceedings, v. 6771.

Dewberry (2012). *Final report of the national enhanced elevation assessment*. Fairfax, VA., Dewberry, 84 p. plus appendixes A-J.

Heidemann, Hans Karl (2018). Lidar base specification (ver. 1.3, February 2018). In *U.S. Geological Survey Techniques and Methods*, Book 11(B4), 101 p.

Lukas, Vicki, Eldridge, D.F., Jason, A.L., Saghy, D.L., Steigerwald, P.R., Stoker, J.M., Sugarbaker, L.J., and Thunen, D.R. (2015). *Status report for the 3D Elevation Program*, 2013–2014. U.S. Geological Survey Open-File Report 2015–1161, p.17

Snyder, G.I., Sugarbaker, L.J., Jason, A.L., and Maune, D.F. (2014). *National requirements for improved elevation data*. U.S. Geological Survey Open-File Report 2013–1237, p.371.

Stoker, J.M.; Abdullah, Q.A.; Nayegandhi, A.; Winehouse, J. (2016). *Evaluation of single photon and geiger mode lidar for the 3D elevation program*. Remote Sensing, 8, 767.

Sugarbaker, L.J., Constance, E.W., Heidemann, H.K., Jason, A.L., Lukas, Vicki, Saghy, D.L., and Stoker, J.M. (2014). *The 3D elevation program initiative—a call for action: U.S. Geological Survey Circular 1399*, p.35.

U.S. Geological Survey (2014a). *Budget justifications and performance information—Fiscal year 2015: U.S. Department of the Interior.*

U.S. Geological Survey (2014b). *National enhanced elevation assessment*, U.S. Geological Survey.

U.S. Geological Survey (2015a). *3D Elevation Program (3DEP): U.S. Geological Survey*, The National Map Web page.

U.S. Geological Survey (2015b). *Budget justifications and performance information—Fiscal year 2015: U.S. Department of the Interior.*

U.S. Geological Survey (2016a). *Budget justifications and performance information—Fiscal year 2016: U.S. Department of the Interior.*

U.S. Geological Survey (2016b). *Budget justifications and performance information—Fiscal year 2017: U.S. Department of the Interior.*

Chapter 6

PHOTOGRAMMETRY

J. Chris McGlone and Scott Arko

TECHNOLOGY OVERVIEW

The photogrammetric production of DEMs has a long history, dating from the earliest days of photogrammetry, and involves a chain of technologies under continuous development. It has been transformed in recent years by the adoption of digital image acquisition, storage, and processing technologies, as well as Airborne Global Positioning System (ABGPS) and Inertial Measurement Unit (IMU) developments. The photogrammetric process includes:

1. Project planning: Selecting data acquisition parameters to ensure a satisfactory product
2. Data acquisition: Obtaining the imagery and any ground control or other auxiliary information required
3. Image georeferencing: Determining the position and orientation of all images, to allow their use for mapping
4. Data collection: Extracting the required products from the imagery
5. Post-processing and quality control: Preparation of the products for delivery

This chapter covers the process and considerations in its use for the generation of digital elevation data.

SENSOR TYPES

Photogrammetric elevation mapping is currently done using digital sensors on airborne and satellite platforms. While terrestrial platforms (Mobile Mapping Systems) are used for corridor mapping applications, no elevation mapping is done using such systems.

Both large- and medium-format airborne sensors are currently in use, built using a number of imaging configurations. While satellite sensors were originally suitable only for small-scale mapping, their increasing resolution has made their use feasible for larger scale applications. The increasing availability and consequent decreased cost of high-resolution satellite imagery has further encouraged wider applications.

While the aerial film camera was once the workhorse of photogrammetric mapping there are currently few in service and none in production, due to their economic and logistical disadvantages in fully-digital workflows.

Airborne Digital Systems

An all-digital workflow has a number of advantages, both economic and practical. The elimination of film handling, processing, duplication and scanning greatly reduces production timelines and material expenses. Digital sensors can capture a broader dynamic range of scene intensities, allowing data collection from scenes collected with less than ideal lighting, shadows or highlights. In addition, many current sensors have the ability to collect both color-infrared (CIR) and near-infrared (NIR), increasing the potential applications of the imagery. The main obstacle to adoption of airborne digital sensors up to now has been the inability to build an affordable sensor with enough pixels to give sufficient coverage and resolution for mapping applications. Advances in chip technology have now made large-format digital sensors possible.

Current chip technology is well established for area arrays of about 4Kx4K pixels, with some up to 9Kx9K. In many applications this is not a sufficient number of pixels for mapping purposes. A number of camera geometric designs have been developed to provide large area, high-resolution coverage using current chip technology, including:

- Mosaicked area (frame) arrays
- Linear pushbroom sensors
- Stepped frame arrays
- Scanning linear arrays (panoramic)

In many applications, medium-format cameras based on commercial cameras have been modified for mapping use or new cameras have been designed around commercial imaging chips. These medium-format cameras are suitable for low- and medium-altitude mapping mounted on UAVs or used alongside lidar. An increasingly common configuration is to use multiple medium-format cameras mounted such that one camera images at nadir while up to four other cameras obtain oblique views.

The mosaicked area array approach uses a number of area imagers with overlapping fields of view, arranged so that the combined image is equivalent to a single frame image. A current example is the Vexcel Imaging UltraCam Condor (Figure 6.1) which includes two panchromatic sensors, five multispectral sensors, and two NIR sensors with electronic motion compensation and real-time tie point generation to ensure registration between the images. The final RGB image is 37,800 by 5,200 pixels and the mosaicked panchromatic image is 13,400 by 9,200 pixels (Figure 6.2). Once the final image has been formed (panchromatic, color, false-color, or pan-sharpened) the frame geometry allows the use of standard photogrammetric software.

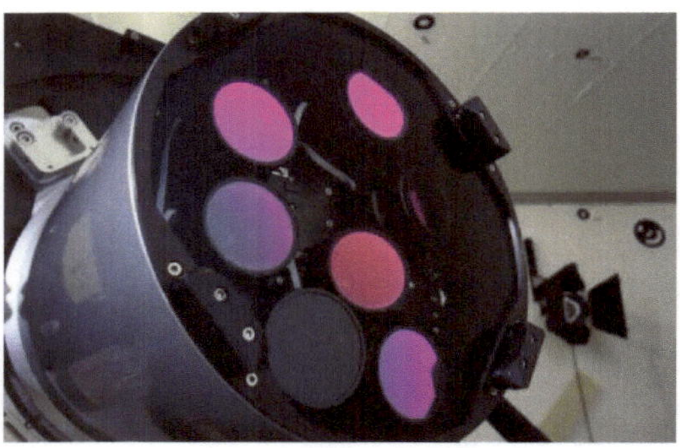

Figure 6.1. Vexcel UltraCam Condor large-format digital camera.

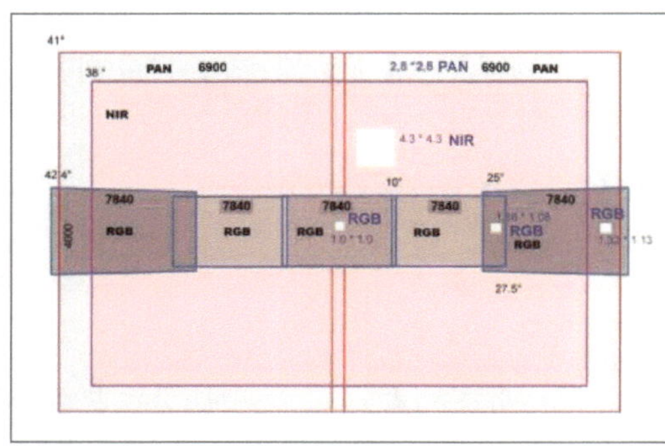

Figure 6.2. Arrangement of RGB, panchromatic, and NIR sensors in Vexcel UltraCam Condor.

The Leica ADS100 is a current example of a pushbroom aerial sensor. Multiple linear arrays, facing forward, down and aft, enable along-track stereo coverage as well as color and multispectral imaging (Figure 6.3). Three separate stereo viewing combinations are available, due to the three panchromatic arrays. Ground sample distances (GSDs) of 10cm are obtained from a flying height of 1,250m across a 20,000 pixel swath. The exploitation of pushbroom imagery requires special sensor models and continuous GPS/IMU information.

The VisionMap A3 Edge (Figure 6.4) uses the stepped frame acquisition geometry to acquire high-resolution imagery over an angular field of view up to 106°. The camera contains two separate optical trains and image sensors which simultaneously scan perpendicular to the flight direction and generate frames which overlap in both the scan direction and the flight direction. The frames from the two scans are merged into a 78,000 by 9,600 pixel "super large frame" (Figure 6.5) covering the entire field of view, which is then used for photogrammetric stereo processing and DEM generation.

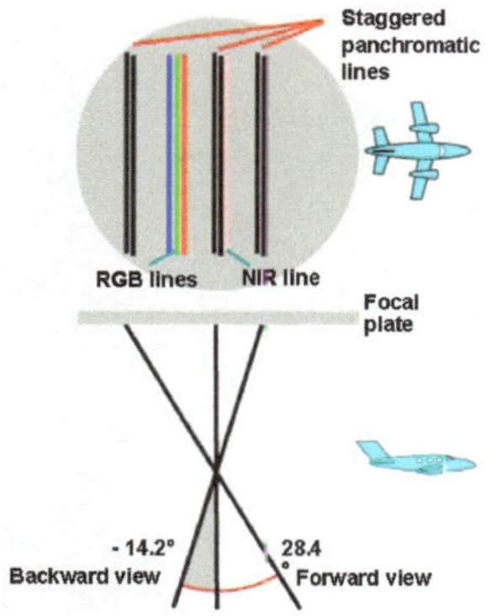

Figure 6.3. Focal plane of the Leica ADS40 (predecessor of the ADS100) showing the layout of the linear arrays.

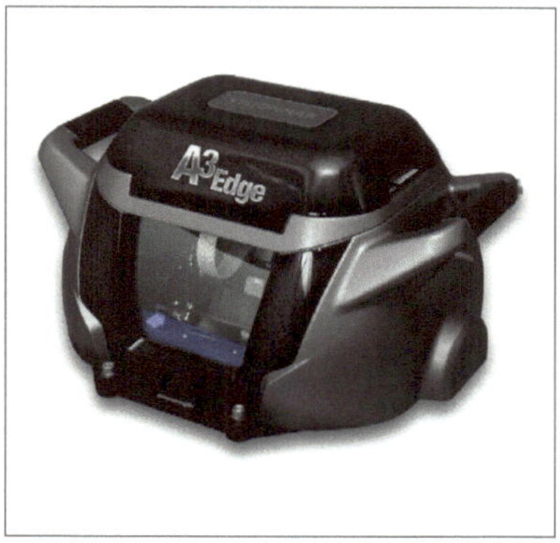

Figure 6.4. The VisionMap A3 Edge camera.

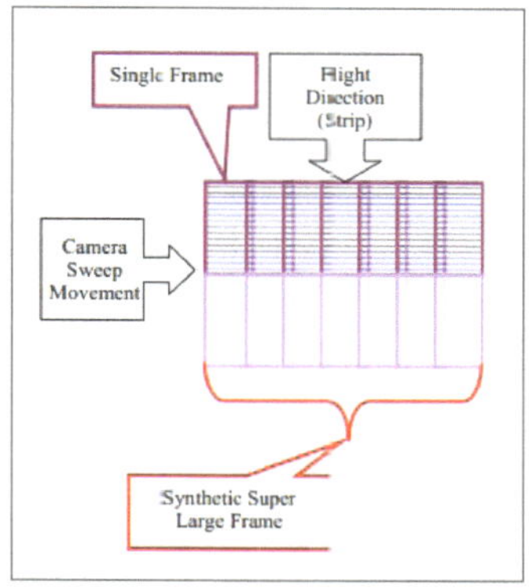

Figure 6.5. VisionMap A3 Edge image formation, showing the combination of frames from the two simultaneous scans into a super large frame.

Satellite Imagery

High-resolution satellite imagery, 1m or less, is now available from a number of U.S. and foreign commercial providers (Table 6.1). The highest ground sample distance (GSD) currently available commercially is 0.3 meters/pixel. The maximum resolution of imagery commercially available from U.S. satellite companies, currently 0.25m, is regulated by U.S. Government licensing. Satellite imagery is typically used for smaller-scale mapping, with 1m imagery usually specified for 1:24,000 scale and smaller.

Satellite	Company (Original/Current)	Launch Date	Panchromatic GSD @ nadir (m)
Ikonos*	SpaceImaging(GeoEye/DigitalGlobe)	9/1999	0.82
Quickbird-2*	DigitalGlobe	10/2001	0.65
Orbview 3*	Orbimage (GeoEye/DigitalGlobe)	6/2003	1.0
Worldview-1	DigitalGlobe	9/2007	0.5
GeoEye-1	GeoEye (DigitalGlobe)	9/2008	0.46
WorldView-2	DigitalGlobe	10/2009	0.46
WorldView-3	DigitalGlobe	8/2014	0.31
GeoEye-2/ WorldView-4	GeoEye (DigitalGlobe)	11/2016	0.31
SkySat-1/2	SkyBox Imaging (Google Terra Bella)	11/2013–7/2014	0.90
KOMPSAT-3A	Korean Aerospace Research Institute	3/2015	0.55
Pleiades-1A/B	Airbus Space and Defense	12/2011–12/2012	0.5

Table 6.1. Examples of commercial high-resolution satellites (*no longer functioning).

Most current high-resolution commercial satellites employ a linear pushbroom imaging geometry in which the single-line image sensor is effectively a straight line perpendicular to the satellite track and scans a linear swath along the orbital track. The pointing direction of the sensor is steerable to either side of the orbital track and also fore and aft along the orbit. This capability increases the number of opportunities to cover any point on the ground, because a given area may be imaged from more than one orbit, and also allows stereo imaging. Satellite images have 11-12 bits of dynamic range and can also collect multispectral imagery, albeit at a four times lower GSD than the panchromatic images.

Exploitation of satellite sensors is more complex than for frame cameras because of the dynamic acquisition geometry. Due to the complexity of the full physical model and the proprietary design information it reveals, the physical sensor models are usually not distributed or implemented; instead, a replacement or generic sensor model which is simpler to implement and use is derived from the original physical model (McGlone, 2013). The generic sensor model takes the form of a set of rational polynomials which express the image coordinates in terms of latitude, longitude and elevation.

Stereo satellite imagery can be collected in two modes: same-orbit or along-track, and different-orbit or cross-track. For same-orbit stereo the sensor must look forward and then pivot to look backwards at the same area on the ground. This provides the best stereo viewing, because the two images are nearly simultaneous, but the area which can be covered is limited by the maximum forward and backward look angles and the amount of time required to change the pointing angle. Different-orbit stereo provides the most flexibility since new imagery can be acquired to form stereo coverage with archived imagery, but the differing acquisition times may result in changes in the scene or atmospheric conditions.

Satellite imagery may be supplied as uncorrected imagery, as geo-rectified images corrected to a uniform elevation, orthorectified images corrected to an elevation model, or as stereo pairs ready to exploit. Some commercial triangulation packages support the block adjustment of multiple satellite images, possibly in conjunction with images of other sensors.

As with aerial imagery, the amount of cloud cover must be taken into account when acquiring satellite imagery. However, because satellites have less flexibility in timing overflights, it can be harder to obtain cloud-free stereo scenes. The ability to image from multiple orbits mitigates this somewhat, but it still can be an issue.

PROJECT PLANNING CONSIDERATIONS

When 6-inch focal length film cameras and analog stereoplotters were in universal usage the expected accuracy of contour extraction was directly tied to photographic scale and the C-factor of the stereoplotter. However, with the variety of current camera system configurations, the integration of GPS/IMU systems into aerotriangulation, and the range of products required the determination of project parameters which will meet product specifications has become much more complex.

Planning for photogrammetric data acquisition begins with product accuracy requirements and includes considerations such as:

- Image resolution (GSD)
- Leaf-on/leaf-off acquisition
- The number and distribution of ground control points (GCPs) and/or checkpoints
- Airborne GPS/INS usage
- Atmospheric conditions and sun angle

Project planning has been made easier by in-flight control systems using GPS navigation which can now maintain very precise flight lines and trigger each exposure at pre-planned points to optimize coverage and overlap. This reduces the necessity of adding extra frames or overlap to allow for possible navigation errors.

Image Resolution and GSD

For satellites or digital airborne cameras, the concept of photographic scale has been replaced with the concept of ground sample distance (GSD) or image resolution. The GSD of an image is the size of an imaging sensor element (pixel) projected to the ground and is reported as linear units/pixel, e.g., 1 meter/pixel, or as a

GSD of 1.0 meter. The GSD is a function of the focal length, altitude (or range from sensor to ground), and the size of a pixel.

The GSD should not be confused with interpretive resolution. While it is true the GSD represents the ground footprint of a pixel, interpretation or identification of linear features is often better than the GSD. Conversely, poor imaging conditions or low-contrast targets may mean that the effective resolution is less than the GSD.

The accuracy of photogrammetric elevation extraction is determined by the ability of the operator or an automated system to locate corresponding features or surface patches in the overlapping imagery. The resolution of the imagery determines how accurately this can be done. Well-defined point targets or an extended feature such as a curb or building edge can be located more accurately than the GSD of the image, due to the averaging effect along the feature, but in general the GSD is the limiting bound on attainable accuracy.

Leaf-On/Leaf-Off Acquisition

Nearly all image acquisition must be performed during leaf-off periods to allow collection of bare-earth elevations. At smaller map scales, average tree crown heights may be used to reduce data to bare earth. If the elevation data are used only for digital orthoimage production, the leaf-off requirement may be dropped.

Ground Control Points

The number and distribution of ground control points (GCPs) is primarily based on the presence of water bodies within the project and whether ABGPS/IMU data acquisition is employed. If ABGPS and IMU data are available, aerotriangulation can even be forgone for some projects and the control points will serve as quality control checkpoints.

ABGPS data dramatically diminishes the number of ground control points required for all types of projects because it essentially serves as additional control points during the aerotriangulation. Although the cost of aerotriangulation is generally less than 10% of the total cost of a project, obtaining ground control points can be a cost and logistics driver.

It is often difficult to measure and panel control points, especially in dense urban areas. Paneling must be synchronized with flight missions, complicating project logistics, and can be time-consuming and costly. In some cases, photo-identifiable points can be used as control provided they can be easily identified and accurately measured at all required photographic scales for the project.

ABGPS Considerations

Flight planning with ABGPS in mind will alter the overall imagery coverage plan. Traditional cross-strips which maintain the rigor of the block are not generally required with ABGPS, only for irregularly shaped areas and/or included bodies of water. Time of day must be considered to ensure an adequate geometric configuration of GPS satellites. Current flight planning packages contain models of GPS satellite orbits and availability to ensure an adequate constellation. Periods of solar flare activity, which interfere with GPS signals, must also be avoided.

Image Geometry and Overlap

Traditional photogrammetric practice utilized stereo image overlap between adjacent frames in a flight line, usually with 60% forward overlap and 10-30% sidelap. With the adoption of digital cameras and automated elevation extraction software, some projects are flown with highly-overlapping imagery, up to 80-90% forward overlap and sidelap. The additional redundant viewpoints provide more reliable automated image matching due to the additional viewpoints for each ground point and the reduced impact of occlusion. The cost of obtaining and processing the additional digital imagery is minimal compared to the cost savings due to improved automated matching.

Atmospheric Conditions and Sun Angle

While atmospheric conditions cannot be accurately predicted, they must be monitored as the data acquisition proceeds. Cloud-free days are required and, preferably, low haze conditions. For DEM generation, acquisition of imagery below the clouds can sometimes be considered provided illumination is still adequate to measure and interpret in low-contrast areas. Cloud coverage must be no more than 10% for most satellite applications. As noted earlier, this is perhaps the most limiting factor for satellite imagery in many areas of the world.

The sun should be at least 30° above the horizon to avoid long shadows within the scene which affect interpretability and image appearance, if orthoimagery is to be produced. The sun angle varies with the latitude of the project and time of year and can be calculated using flight planning software.

GEOREFERENCING AND AEROTRIANGULATION

Georeferenced images are defined as those in which 3-D ground coordinates can be mathematically projected into the 2-D image space, and vice-versa. No mapping of any type can begin until all acquired imagery is accurately tied to the ground reference frame. In most cases this requires aerotriangulation, but can sometimes be accomplished with the collection of ABGPS and IMU data.

Georeferenced images require a geometric sensor model that accurately models a number of components:

- Interior orientation
- Exterior orientation
- Atmospheric affects
- Earth curvature or coordinate system compensation

Interior orientation establishes the internal geometry of the camera. For a digital frame camera geometry this includes the focal length, the location of the principal point in the image, and lens distortion. The principal point is defined as the intersection of the optical axis of the lens with the image and serves as the origin of the image coordinate system. Lens distortion includes both radial and tangential components, although current lens designs reduce distortions to nearly-negligible levels. For digital imaging systems, both airborne and space-borne, the image coordinate system is defined by the lines and samples of the digital image itself. For a dynamic camera, such as a scanning linear pushbroom or stepped frame, the interior orientation will also include the scanning velocity, any image motion compensation applied, or other camera-specific parameters.

The exterior orientation of an image refers to the position of the exposure station (camera lens center, or nodal point) and the angular attitude of the image coordinate system with respect to the ground coordinate frame. For a frame sensor, the exterior orientation is valid for the whole image, while in linear pushbroom or other dynamic acquisition geometries the exposure station position and orientation change with each image line. The exterior orientation must therefore be modeled as a function of time, based on inputs from GPS and IMU measurements. Because GPS measurements occur at finite intervals, splines, polynomials, or other interpolation methods are used to model the behavior of the exterior orientation parameters between the discrete readings. When triangulating dynamic imagery, the parameters of the interpolation model are solved instead of the parameters themselves.

Atmospheric refraction alters the image coordinates of a ground object because the light ray bends as it moves through the atmosphere. Thus, image points need corrections to determine their actual (non-refracted) position. Standard atmospheric models and correction formulae are available (McGlone, 2013) and are applied by the triangulation program.

There are a number of choices for the project coordinate system, including latitude-longitude or various projected coordinate systems, such as UTM or state plane. If the coordinate system is a projection and the project covers a large area, earth curvature effects will be present. These are caused by representing the Z coordinate relative to a plane tangent to the earth's curved surface, instead of modeling Z relative to the ellipsoid or geoid.

Some triangulation packages do their computations in geocentric or local tangent coordinates to avoid this problem; otherwise, an earth curvature correction must be applied.

ABGPS and IMU data can be combined to define the exterior orientation of images as long as the camera boresighting and antenna offsets are well calibrated and the ground coordinate and datum transformations are well-understood. If the ABGPS and IMU data are not sufficiently accurate to georeference the images involved in the project, then the aerotriangulation process will be required.

Aerotriangulation determines consistent orientation parameters for a block of imagery, minimizing errors between images which would lead to unacceptable parallax in the stereomodels and accurately relating the entire block to the world coordinate system using the minimum amount of control information possible. When ABGPS/IMU data are available, aerotriangulation may still be used to refine the relative orientation of the imagery, as a check on the ABGPS/IMU data, and to relate the navigation information to the world coordinate system datum.

Imagery to be triangulated is acquired in strips along flight lines and the strips aggregated into a block of imagery covering the project area. Imagery is overlapped along each strip, with at least 60% overlap required for stereo viewing. The strips, in turn, are flown parallel to one another and generally overlap adjoining strips by about 30% of the image width. With digital sensors, where acquiring additional images does not incur additional costs, many blocks are flown with 80–90% overlap in both directions. The additional redundant images enable automated matching methods to work better and allow a choice of images from different viewpoints for viewing particular points on the ground and for orthoimage production.

As input to the triangulation solution, control and tie points are measured on the overlapping imagery. Control points have known horizontal, vertical, or both horizontal and vertical positions determined by ground survey, while tie points are common points between overlapping images in the same or different strips.

Manual tie point measurement procedures measure a 3 by 3 array of points on each image, which places points in the overlap regions with images in the same strip and in adjacent strips. However, due to the labor required for manual measurement, current triangulation packages incorporate automated tie point measurement techniques. The software uses image operators to identify well-defined tie point candidates in the overlap areas on each image, then matches the candidate points between images using the approximate image orientations and image correlation. Large numbers of tie points can be produced, yielding a very accurate triangulation solution due to the solid geometry between images while allowing bad tie point matches to be removed.

Control points may be well-defined photo-identifiable points, such as sidewalk corners, or may be paneled prior to flying. Wood or plastic panels are used on ground surfaces, while painted marks are used if the point is on a street or parking lot. Patterns used to signalize ground control points include three- or four-armed crosses, as well as discs, and must be sized for accurate measurement based on the image scale. Control points must be placed in locally flat areas which are free of obstructions such as overhanging trees or nearby structures.

Aerotriangulation packages simultaneously determine the ground coordinates of all measured points and exterior orientation elements using the method of least squares adjustment and the sensor's geometric model. The solution yields a best fit of all parameters—the most probable values given the assumed statistical properties of the input estimates and measurements. The solution also estimates the precision of the parameters which can be used to evaluate the quality of the solution. Large parameter standard deviation values may indicate a weakness in block geometry, such as missing tie points or insufficient overlap, or may point to errors in the input data. The input parameters typically include:

- Ground control point coordinates
- Tie point coordinates
- Estimates of exposure station position, such as from ABGPS
- Estimates of image attitude, such as from IMU data
- Modeling of ABGPS drift parameters and their estimated precision
- Camera interior orientation parameters (if self-calibration is being exercised)
- INS-camera boresight parameters

The use of GPS for exterior orientation significantly strengthens the aerotriangulation results, particularly for smaller scale projects. Because the GPS positions effectively act as additional control points, blocks using GPS require fewer ground control points. Incorporation of IMU information, which measures the attitude of the sensor, also results in a stronger adjustment. Ground control point distribution throughout the block of images is generally focused on the number of horizontal/vertical control points along the edges of the block and a lesser number of points scattered in the center of the block. It is not unusual for only 10–12 control points to control a block of hundreds of images when ABGPS is employed.

Aerotriangulation software packages also support self-calibration, in which some of the sensor interior orientation parameters, the sensor boresight parameters, and the ABGPS/IMU drift parameters are determined as part of the solution. This models any changes between the calibrated parameters and the actual in-flight conditions.

With the ever-increasing reliance upon ABGPS/IMU instrumentation, automated triangulation provides cost-efficient quality control for the navigation data by checking the results prior to production. At a minimum, this capability can be used to assess the relative orientation of images to one another before detailed operator inspection of suspect areas is required. At its fullest, it stands to decrease triangulation time frames and cost if they are required.

PHOTOGRAMMETRIC DATA COLLECTION METHODS

Both manual and automated methods are used for photogrammetric collection of digital elevation models. In practice, collection tends to be a combination of both, with the exact proportions determined by the project scale, the characteristics of the scene, and the available production facilities.

In the past, elevation extraction was primarily focused on producing topographic contours. However, the increasing emphasis on digital orthoimage production over the past decade, as well as GIS and other applications which utilize digital terrain data, have made elevation rasters the most common product.

Orthoimage DEMs require significantly less time and cost to produce as compared to DEMs for contour generation. Because many projects involve feature (planimetric) data production in addition to general elevation data and digital orthoimages, it is difficult to determine the cost of the DEM generation itself. Many production techniques will use planimetric collection to initialize the DEM generation by extracting elevations from the planimetric data, which is generally collected in 3-D with feature coding useful in determination of usable "bare earth" points.

Automated elevation extraction techniques have been integrated into photogrammetric workstations for several years and are increasingly being used as the algorithms improve and new approaches are developed and as operators gain experience in selecting scene and processing parameters to guide the algorithms.

Softcopy Stereoplotters

Softcopy systems were first introduced into the commercial production environment in the early 1990's. Rapid improvements in the storage, processing, and display capabilities of standard computer workstations have led to increases in functionality and productivity while simultaneously reducing prices. Working in a purely-digital environment with digital imagery, feature, and elevation data enables great flexibility in data collection modes.

For instance, collected data or reference datasets, either vector or raster, can be superimposed on the monoscopic or stereoscopic scene for reference, review, and editing. For elevation extraction, elevation points or contours can be superimposed for review against the scene viewed in stereo. Most workstations are now integrated with GIS or CAD packages and allow the data to be superimposed on the scene. Collecting and editing data directly within the final application package has the advantage of avoiding errors and effort in data translation.

A fully-digital environment enables flexibility in data collection modes. Elevation data can be extracted as contours, profiles, rasters, or mass points for TINs, while features can be collected in stereo or from an orthoimage with a DEM to assign elevation. Automated extraction aids or tools can also be integrated. For instance, some workstations will automatically keep the cursor on the ground while the operator is tracing planimetry by performing image matching in the background. Many current workstations have automated linear feature extraction or 3-D building model generation, usually requiring the operator to provide a starting point or "cue."

Workstation costs have decreased as functionality has become more specialized. Early soft-copy workstations often included all triangulation, collection, and editing functions; the current approach is to provide lower-cost workstations optimized only for collection, supported by a smaller number of general purpose workstations or by specialized triangulation, editing, or management stations.

The all-digital workflow has enabled new efficiencies, but managing the massive volume of image and product data has led to its own set of problems. The requirements of distributing imagery to the various workstations, then collecting, merging, and editing the extracted data products, while tracking the progress of the project through the system, has motivated the development of digital data and project management systems.

Data management systems include large amounts of data storage, often with terabyte capacities, and a database system which maintains the metadata. This metadata includes the acquisition parameters, sensor and GPS data, as well as versioning and processing history for the imagery itself. Due to the critical nature of the data and the impact on production should the server have problems, the systems are usually configured with redundant and distributed disk storage to cope with the failure of any one component. The availability of cheap storage, backup, and processing in the "cloud" has motivated the migration of photogrammetric operations to cloud services such as those from Amazon or Google. Project management software allows the manager to assign specific images or areas to each operator, to monitor progress, and to generate cost and time estimates for completion of the project.

Manual Elevation Compilation

There are several standard approaches to collecting elevation data:

- Capturing spot elevations and breaklines
- A digital elevation matrix is collected, augmented with breaklines and spot elevations
- If planimetric compilation is also being performed, the elevation data may be initialized from the planimetric data, with spot elevations and breaklines added as needed to finalize the DEM

If planimetric data are not part of the overall project, or if it is scheduled at a later date, most mapping firms will use skilled operators to focus on the collection of breaklines and spot elevations. A breakline is defined as a polyline that defines a slope change along its extent and may include both naturally-occurring and man-made features. Common examples of breaklines include ridges, valleys, edges of embankments, edges of rivers, curbs/gutters, etc. Breaklines are augmented by spot elevations in areas devoid of natural or man-made breaks in the terrain surface. A number of spot elevations are collected at identifiable points for later checking and are also collected at local maxima and minima that are not represented by a contour line. The advantage of this approach is that the minimum amount of points required to describe the terrain is collected. The fewer points collected, the more quickly editing can be performed and finishing completed. If a digital workstation is used, it is common to periodically produce the contours in the stereoscopic view to ensure that sufficient points and breaklines have been collected.

Some mapping firms use a grid or profile approach to the collection of elevation data, augmenting as needed with breaklines and spot elevations. This is most common in very smooth areas where it is difficult to ensure that subtle changes in the terrain elevation are captured. It is also the most common approach and final format for smaller map scales.

If planimetric data are part of the overall project, in addition to a DEM and/or contours, then it is common to first collect the attributed feature data and then derive the elevation points by filtering feature codes for appropriate data. Because most planimetric features are collected at the ground level, their codes can be used to extract elevations and even breaklines (such as along a river edge or road edge). Generally only a small number of additional breaklines are collected. However, a relatively large amount of spot elevations may be added, sometimes using a rough grid to ensure the terrain form is captured.

Automated Elevation Collection

Automatic collection of DEMs by means of digital image correlation has been a topic of research and implementation for over 40 years. As more powerful desktop computers have become available, more robust and intensive algorithms have been implemented. Automated collection has mostly been applied to smaller-scale projects, with wider applicability as users gain experience.

Many issues dealing with the management and editing of automatically-collected elevation data are similar to those with the processing of data derived from lidar systems. Both approaches yield a large number of elevation points, including erroneous values and noise, which may tax standard terrain elevation modeling packages tailored to traditional compilation approaches. Advances in thinning large numbers of points and reducing elevations on man-made structures to "bare earth" in both disciplines will inevitably pay off for both.

Automatically deriving elevations by digital correlation amounts to directly comparing pixel patches on conjugate images or indirectly comparing features derived from the digital images. Direct comparison techniques were the first to be researched, with increasing work done in comparing derived information, such as edge features in the images. Edge/feature collection techniques tend to be superior in large scale projects, with area-based techniques more common for smaller scale projects. Newer techniques employ hybrid approaches that attempt to take advantage of the best of each.

The recently-developed technique of semi-global matching (SGM) (Hirschmüller, 2011) is finding increasing application. In this technique, a global cost function based on disparity is optimized at each pixel along eight separate paths, thereby striking a balance between surface continuity and precise disparity extraction.

Successfully employing automated approaches in day-to-day production requires an understanding of the strengths and weaknesses of the technology as a whole. General strengths include:

- Ability to generate a very dense array of elevation points in short periods of time in an unattended mode
- Generally good capture of the overall surface of a ground scene
- Generally high quality results at high accuracies
- Very good results at smaller map scales (lower resolutions)

General statements on weaknesses are:

- The need to edit large amounts of data that does not necessarily reflect what would be captured interactively by a human stereocompiler. Skilled stereocompilers collect mass points and breaklines that model the terrain with essentially a minimal set of data. Digitally correlated data produces a sampling of elevations without regard to breaklines, etc.
- Even small percentages of erroneous data can require editing times in excess of the time required to interactively capture the data
- Production systems require the operator to select parameters or a scene type to ensure best performance for any particular area, which may change across a large project

Strengths and weaknesses can be parlayed into a number of practical usages in a production system:

- Generation of elevation models for large areas at smaller product scales (generally 1:20,000 and smaller) where less effort is required to reduce elevations to a "bare earth" model. This also applies to extremely flat or smooth areas

- Capture of digital surface models (DSMs), as opposed to bare earth digital terrain models (DTMs), for visualization database generation
- Generation of DSMs adequate for fast production of interim digital orthoimages at most scales
- Automatic change detection in surfaces over periods of time—such as for production planning efforts or elevation model updates

Digital correlation is often used to quickly develop a surface over large areas, leaving final editing to soft-copy workstations. This approach supports interim or expedited delivery of digital orthoimages, with the soft-copy compilation used to develop breakline information required for topographic contours. For smaller map scales, this approach can be very effective. At larger map scales, the approach becomes marginally effective because extensive editing of the digital correlation data may exceed the time to perform soft-copy compilation from the beginning.

A significant amount of work in computer vision has been expended on structure from motion (SfM): determining the structure of a moving object in a sequence of images from a stationary camera, or, equivalently, determining the structure of a stationary object using images taken from a moving camera. Most approaches to SfM are based on the simultaneous matching of multiple images instead of pairwise image matching, using scale- and rotation-invariant feature extraction operators such as scale-invariant feature transform (SIFT) (Lowe, 2004). The photogrammetric community has applied multi-image matching to highly-overlapping acquisitions from calibrated cameras (Leberl, et al., 2010), producing denser 3-D point clouds and more accurate and reliable results. Computer vision research has concentrated on images taken from arbitrary viewpoints using uncalibrated and unknown cameras (Hartley and Zisserman, 2003), which has found photogrammetric applications in mapping from unmanned aerial systems (UAS), as shown in Figure 6.6, and for architectural documentation, among others.

Figure 6.6. DJI Matrice 600, typical of UAS platform used for Structure from Motion (SfM) photogrammetric techniques.

Structure from Motion (SfM)

Structure from Motion (SfM) encompasses a set of computer vision algorithms and techniques that have come into wide usage and gained wide acceptance in surveying, archaeology, and the geosciences during the past decade. SfM, combined with the proliferation of high-quality, non-metric cameras, small GPS/GNSS chips, and aerial platforms, including UAS, has allowed the generation of high-resolution DEMs and orthoimagery at a fraction the cost of traditional techniques (Westoby, et al., 2012). Furthermore, given the relative computational ease of SfM processing can be accomplished on commodity systems or in the cloud.

Sometimes referred to as close-range photogrammetry, SfM has its roots in many of the same principles as traditional photogrammetry. However, it differs in one fundamental way. There is no requirement in SfM for any a priori knowledge of the camera's interior or exterior orientation or the scene geometry. These quantities are all recovered empirically through a redundant network of matched feature points during the bundle block adjustment process. The ability to derive these quantities sets SfM apart from traditional techniques and allows for the use of lower cost cameras for which the calibration is unknown. In fact, SfM has been shown to work well in the presence of many cameras of varying quality, calibration, and type (Agarwal., et al., 2009).

Background

While the mathematical problem of SfM with known correspondences shares many of the same principles as traditional photogrammetry, SfM has a unique history. In the 1950s Wallach and O'Connell (Wallach, et al., 1953) attempted to develop a complete theory of 3-D vision in humans. A fundamental component was how human 3-D vision was derived from primarily a series of orthographic 2-D images projected to the retina. The problem of parallax-derived distance from stereoscopic vision was already solved, but a remaining issue, and special case, persisted. How does human 3-D vision function in the monocular case (i.e., the case of 3-D vision with only one eye). To solve this problem, Wallach and O'Connell presented the theory of the kinetic depth effect. The principle of kinetic depth was that if points on a body in motion could be tracked over time then 3-D structure could be derived.

In 1979 S. Ullman of the Massachusetts Institute of Technology formalized the mathematical theory of the kinetic depth effect and termed this Structure from Motion (Ullman, 1981). In presenting his theory, Ullman demonstrated that given a rigid configuration, four non-coplanar points in three views were sufficient to extract a complete understanding of the structure and motion of the views. This formalization, along with the rigidity assumption, continues to be fundamental in SfM implementations today. Ullman also showed that the complexity of calculation only grew linearly with the number of points. In modern SfM techniques and software, where millions of points are easily generated, this linear growth allows for processing large numbers of scenes on standard, commodity hardware.

The final required precursor to modern SfM systems came in 1999 when David Lowe published and demonstrated the utility of the scale-invariant feature transform (SIFT) (Lowe, 1999). SIFT provides the ability to generate a large number of matched key points in image sequences even in the presence of rotations, scale variations, and to lesser extent obscurations. When combined with outlier rejection techniques like random sample consensus (RANSAC) (Fischler, et al., 1987) modern SfM systems can generate millions of high-quality image key points with relative ease.

SfM Processing Workflow

In most applications today SfM is used to extract 3-D surface information (i.e. a DEM) from a series of images collected by a moving platform. This is the case of aerial photography, whether from a manned or unmanned system, and is the case to be discussed in this text. However, the inverse problem can also be solved, that being the 3-D structure of a moving target collected by a stationary camera. This is the situation of Wallach and O'Connell's original experiments. A key component of SfM is that only the target or the camera should be moving, never both.

Preliminary to any processing workflow is the collection of the image data themselves. SfM can make use of data collected in either nadir-view or oblique-view geometries, with a variety of cameras. If possible, both geometry types should be collected during fieldwork in order to have the greatest range of viewing angles for the various targets. Image overlap is key in SfM, as the matching is not performed in a pair-wise sense, but rather features are expected to be available in a large number of images with variable viewing angles. SfM quality improves with more images and higher levels of overlap and sidelap. It is not uncommon for SfM projects to make use of image sets with 80–90% along-track overlap and 60–70% cross-track sidelap.

The first SfM processing step is the generation and matching of image features, or key points in the computer vision vernacular. There are many modern techniques for feature detection and matching, both proprietary and open-source. These techniques are generally referred to collectively as SIFT or SIFT-type features, though the SIFT algorithm itself is a patented algorithm requiring a license for commercial use. In the SIFT approach key points are detected and ranked based on their stability through a multi-scale Gaussian image reduction process. This provides key points with descriptors that are consistent across a variety of image scales, hence scale invariance.

Key point density and the complexity of the matching process vary with image resolution and the scene complexity and texture. Many SfM processing systems will allow the user to down sample images prior to initial matching or use embedded GPS information in order to improve the overall efficiency of the process. Initial point matching generates a sparse point cloud and using the point correspondences initial estimates of the camera positions, orientations, and calibrations, along with the scene geometry are recovered.

Considering the point cloud and geometric model are determined without knowledge of a real-world coordinate system, the models derived from SfM are inherently in their own local coordinates. The translation from local coordinates to real-world coordinates is accomplished via inclusion of 3-D ground control points. Control points are tied to image-identifiable features and, presumably, have a higher accuracy than the camera positions, if known. Following insertion of control points a 7 degree of freedom (3 translation, 3 rotation, and 1 scale) similarity transformation is typically applied to the local model to transform it to real-world coordinates. The optimization, or local minimization problem, attempts to minimize the overall error across the known camera positions and ground control. This last point makes clear the need for careful collection of ground control and an understanding of the relative accuracy of all GPS information in your project. Optimization is usually performed as a local minimization problem using a general technique, such as Levenberg-Marquardt (Oliensis, 2000). If optimization does not recover the true global error minimum due to misunderstood control or camera positions the resulting DEM can be grossly in error. As with any technique proper check points and validation of your result are critical.

The sparse point cloud is typically insufficient for most applications given that the density may be low and highly spatially dependent. Once the camera positions and orientations are known, full resolution matches between images can be established and a point cloud with significantly higher densities generated. Established triangulation procedures, such as Delaunay triangulation, can be used to reconstruct the scene geometry and generate a high-resolution geometric mesh or DEM. In many cases the point cloud, and resulting DEM, may be decimated to reduce the resolution and corresponding storage requirements.

A powerful aspect of SfM is that many of these operations can be automated and parallelized for computational efficiency. Control point inclusion typically remains a manual process, however, many software packages help to make this efficient as well. The SfM-derived point cloud and DEM can be used in any application requiring elevation data.

The multitude of uses of SfM for mapping and scientific applications is outside the scope of this text. However, the reader is encouraged to search the latest publications in their field to see how this exciting technology can be used in their specific application.

Figures 6.7 and 6.8 show examples of orthoimagery and DEM produced using SfM from a UAS for a small construction site in Alaska. Whereas the collection was imperfect, the results were useful as engineers could clearly see heavy machinery tracks and the slope in the culvert area.

Figure 6.7. Orthophoto of a small construction site produced by a UAS using SfM procedures.

Figure 6.8. DEM of the same construction site produced by a UAS using SfM procedures.

POST-PROCESSING AND QUALITY CONTROL

Post-processing of photogrammetrically-derived DEMs may be required for several reasons:

- Editing of elevation data generated by digital correlation
- Generation of contours or TINs from elevation data
- Conversion between elevation data file formats

The amount of interactive editing required in the case of digital correlation becomes the limiting factor on this automated approach. Typical problem areas include:

- Reduction of elevations to bare earth in urban or heavily vegetated areas. Automated bare earth extraction has become much more capable, due to the development work expended for the same problem in lidar processing
- Editing of error-prone areas, such as those containing minimal contrast and extensive localized variations in elevation that produce "noisy" elevations
- Failure to capture sufficient detail around natural and/or man-made breaklines
- Failure to provide a sufficiently detailed boundary between land and bodies of water

Most digital correlation applications have correlation parameters which can be tuned to optimize the processing for particular scene types, or even areas within projects. For instance, elevation capture in urban areas with many vertical areas uses different slope tolerances than for areas with rolling terrain. In addition, familiarization with the characteristics of each digital correlation algorithm and package yields editing procedures that focus in on known problems. Bulk editing functions such as correcting water body boundaries and reducing elevations from the treetops in a forest to bare earth are often developed and tailored to the digital correlation approach.

Generation of contours from elevation data is mostly automated but still involves some manual editing to produce contours which are aesthetically pleasing (smoothed) and conform to cartographic contour representation principles, as demonstrated at Figure 1.23 in Chapter 1 on page 17.

Depending upon the end-user requirements, native DEM formats may require conversion or translation. For example, mapping at larger scales nearly always requires breaklines and spot elevations to ensure proper contour generation, which generally involves some type of TIN format. Some deliverables, however, require a gridded elevation data set which must be extracted from the TIN data.

Quality control of photogrammetrically-produced DEMs requires both statistical and visual procedures, the statistical sampling of elevations measured at checkpoints and visual inspection of the DEM, TIN, or contours overlaid on stereo imagery.

In most cases, checkpoints are provided by the mapping service customer and are checked by the mapping firm and/or the customer. If aerotriangulation is utilized, most mapping firms also select photo-identifiable tie and pass points to be carried along in the production process as quality control points. These points are then used to check compilation measurements, as elevation check values against those interpolated from the DEM model, and as horizontal check points on digital orthoimages.

Visual inspection of the DEM, TIN, or contours overlaid on stereo imagery provides a qualitative check on the quality of the DEM itself. A number of software packages also provide the capability to visualize the TIN surface (triangle sides) to determine conformance to the ground surface.

DATA DELIVERABLES

Data deliverables are a function of the end-user's requirements. A DEM may itself be a deliverable or may be delivered as supporting data, e.g., as part of an orthoimage project. Standard DEM formats include

- Gridded (raster) elevation data, including elevation profiles
- TIN (triangulated irregular network) data
- Mass point and breakline data

DEMs are defined as either "bare earth," in which case they are usually referred to as digital terrain models (DTMs), or "surface," referred to as digital surface models (DSMs) and include treetops and rooftops Topographic contouring requires bare earth elevations, whereas DEMs for digital orthoimages do not necessarily require this in all areas. The term "surface model" is often used for DEMs in which the tops of all surface features are captured as opposed to bare earth; this is common for raw lidar and digitally correlated data sets.

For digital orthoimage generation, it is entirely acceptable to use elevations on the top of forest canopies at smaller scales (as opposed to lowering elevations to the underlying bare earth) if the required planimetric (horizontal) accuracy of the resulting digital orthoimage can still be achieved. If large man-made features such as buildings and bridges are not included in the DEM they will appear distorted in the orthoimage. For examples, bridges often have an erroneous "hourglass" shape when draped over a bare earth DTM, requiring breaklines at the edges of the bridge deck to correct the distortion, as shown in Figures 6.9 and 6.10.

Figure 6.9. Digital orthophoto showing bridge distorted by use of bare earth DTM only.

Figure 6.10. Digital orthophoto showing bridge correctly by use of DTM with breaklines.

With increased use of orthoimagery in urban areas, the "true" orthoimage has become more popular (Figure 6.11 on page 144). In a true orthoimage, 3-D building models are included in the DEM and areas which had been occluded by building lean are filled in from adjacent images.

Gridded models are often used for smaller map scale projects, particularly state, province and national mapping programs at scales of 1:20,000 or smaller. These are easy to use (simple indexing) and to convert from one file format to another. However, they are not always appropriate for contour generation, which requires breaklines of some type. If contours are required, or if breaklines are required around bridges and large structures, the gridded data is augmented with at least localized TIN information.

Mass points are generally intermediate products and are relatively easy to convert from one specific data file format to another. However, they are more difficult to deal with in terms of indexing. They are often imported into or exported from TIN packages for use and are not commonly delivered unless accompanied by breaklines.

TIN models are the best way to represent terrain for contouring and are also most suitable for digital orthoimages at larger map scales. A TIN represents the terrain surface as a set of triangles; the stored data consists

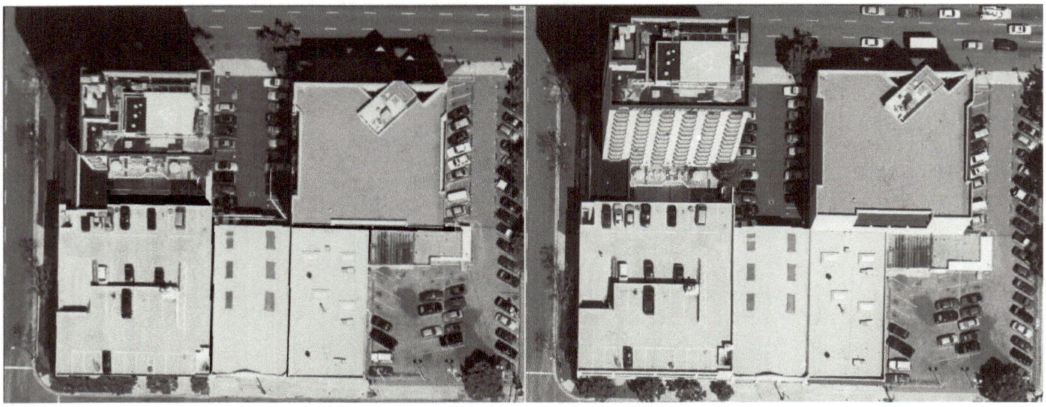

Figure 6.11. True orthoimage (left) and standard orthoimage (right). In the left image the tall building is in its true X,Y position and data from another image has been mosaicked into the occluded area. In the right image, windows and sides of buildings are clearly visible.

of the triangle vertices and the links between them. TINs are a very efficient terrain representation, but TIN data file formats are more difficult to convert to/from because they tend to be software package-specific.

ENABLING TECHNOLOGIES

The most important enabling technology is ABGPS/IMU instrumentation, which has greatly reduced ground survey and aerial data acquisition and processing costs.

ABGPS/IMU Instrumentation

Photogrammetrists first began using ABGPS to augment aerotriangulation roughly a decade ago to record accurate positions for all exposure stations. With ABGPS, fewer ground control points are necessary because exposure stations can effectively function as control points in the aerotriangulation process. Though this reduced the amount of ground control necessary, lack of known camera attitude still prevented attainment of the final goal: elimination of the aerotriangulation step. Advanced IMU systems can now provide attitudes accurate to 0.2-0.4 arc-minutes. Combining this with the better than 0.5 foot GPS accuracy, aerotriangulation can be eliminated for larger map scales for some applications. For example, for 1"=660' photographic scale, IMU measurements will provide an angular pointing and positioning accuracy of roughly 0.4' on the ground, satisfying requirements for 2' contour generation.

The 1990 *ASPRS Accuracy Standards for Large-Scale Maps* have been superseded by the *ASPRS Positional Accuracy Standards for Digital Geospatial Data* (ASPRS, 2014) which specifies RMSEx and RMSEy values for Standard Mapping and GIS work (previously ASPRS Class 1). At a minimum, integrated ABGPS/IMU can provide sufficient accuracy, without aerotriangulation, to meet nearly all standard mapping and GIS accuracy requirements for some products, including DEMs. In addition, aerotriangulation based on ABGPS/IMU instrumentation would require only minimal control points to attain standard mapping and GIS accuracy requirements for all projects (see section 7.7 of ASPRS, 2014). This technology is also extremely critical to the proper execution of lidar and digital airborne imaging systems. When using ABGPS/IMU systems, care must be taken to achieve reliable results. Atmospheric disturbances, poor GPS satellite constellation geometry during data collection, or other system discrepancies can lead to poor results.

CALIBRATION PROCEDURES

Cameras and any instrumentation used to obtain the supporting data must be calibrated to ensure the accuracy and validity of the data obtained. Calibration may be performed on the camera and individual components, or on the mapping system as a whole including the navigation sensors and the installation in the platform.

Digital Airborne Systems

Frame digital systems, in which all pixels are captured in a manner analogous to a film camera exposure, can use calibration procedures analogous to those of film cameras (Kröpfl, et al., 2004).

However, current digital frame camera systems with their varying imaging formats and architectures are not easily accommodated in a single calibration facility as film cameras were in the USGS Optical Science Laboratory. An additional complication is the fact that dynamic digital cameras such as linear pushbroom or scanning sensors require relative motion between the scene and camera to form an image, which cannot be easily duplicated in a laboratory.

If ABGPS and IMU instrumentation is to be used with a camera system, as is the case with nearly all current mapping systems, the spatial relationships between the GPS antenna, the IMU, and the camera must also be determined. The offsets between the GPS antenna position, the IMU device, and the front nodal point of the camera in the aircraft reference frame are typically determined by land survey techniques, yielding accuracies on the order of 0.1 foot or less. However, determining the angular offsets between the IMU and camera coordinate systems, known as boresighting, requires an *in situ* system calibration procedure.

Due to these considerations *in situ* calibration has become the most common method for system calibration. ASPRS guidelines (ASPRS, 2015) outline a calibration procedure using a right-angle crossroads as a test field with targets placed along the intersecting roads (Figure 6.12). Flights are flown with six headings to ensure that control points are imaged across the entire format of the camera, as shown in Figure 6.13. The calibration solution is a rigorous triangulation adjustment which includes the boresight angles, the offsets, and camera interior orientation parameters.

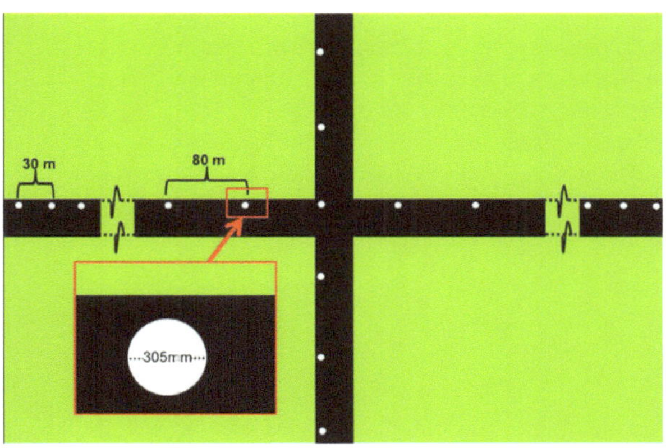

Figure 6.12. Schematic of targets and crossroads calibration range for in situ calibration.

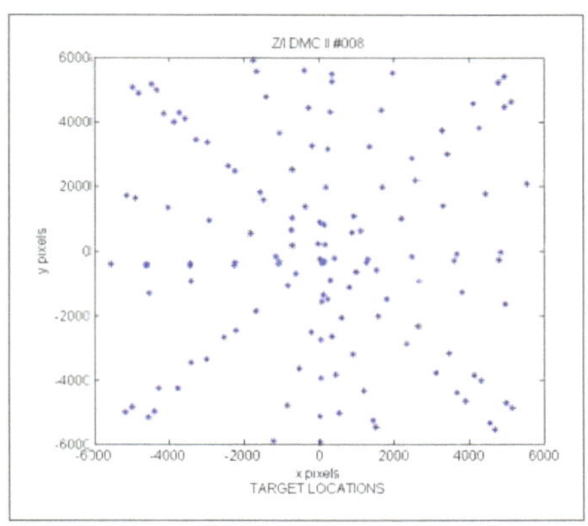

Figure 6.13. Plot of ground control image point distribution. Courtesy of Midwest Aerial.

Depending on the mounting of the IMU device, boresighting may have to be carried out both at a local range and also prior to, or following, missions over project areas. If the IMU device is mounted in a position and manner that precludes inadvertent contact, it will be stable for long periods of time and generally is only boresighted every few months. If, on the other hand, it is mounted in an open area susceptible to being bumped or interfered with, it may require boresighting for every project or flight.

Satellite Systems

Satellite imaging systems are calibrated on-orbit by imaging test ranges at periodic intervals (Mulawa, 2004). Satellite system calibration includes orbital parameters as well as geometric and radiometric parameters, allowing long-term monitoring and characterization of sensor performance. Due to the proprietary nature of the sensor-specific physical parameters and the complex nature of the sensor models, this calibration is performed only by the imagery provider. End-users must use statistical sampling of checkpoints to ensure overall adherence to accuracy specifications, or perform pilot projects to prove system accuracy.

CAPABILITIES AND LIMITATIONS

If bare earth elevations are required, DEM generation by photogrammetric means is generally only hampered or cost-constrained by dense vegetative coverage. Leaf-off conditions are required if deciduous trees are present. Even with leaf-off conditions, however, obtaining sufficient measurements through the trees will be time-consuming and costly.

The primary advantages of the photogrammetric approach are:

- The imagery can be re-used for other purposes, such as planimetric data extraction
- The imagery can always be used to correct errors of commission and omission at a later time
- Steps such as editing and finishing can always rely upon a stereomodel to resolve discrepancies and correct errors to the highest accuracy the system provide
- Photogrammetry is a proven and well-understood approach

COMPARISONS WITH COMPETING/COMPLEMENTARY TECHNOLOGIES

Ground surveying is generally less cost-effective than photogrammetric mapping and has largely been abandoned for all but the smallest projects. The main alternative to photogrammetric mapping is therefore lidar data capture or some combination of the two (Baltsavias, 1999; Leberl, 2010).

Lidar instrumentation and processing has moved from infancy to maturity over the last few years. This technology is another tool for use by mapping companies and has advantages and disadvantages for virtually every project relative to the photogrammetric approach. As time and experience with lidar collection has progressed, it is common to see both it and photogrammetric means utilized in the same project to optimize schedule and cost.

Advantages of lidar include:

- Fast capture of large amounts of elevation data
- Capture of high amounts of detail
- Smaller field of view yielding less problems with obscuration (from buildings, rugged terrain, etc.)
- Direct digital capture of data
- Ability to "see through" the trees substantially better than photogrammetric approaches
- Adequate accuracy for all but the very largest scales
- More leeway on flight conditions (under cloud cover, day/night, etc.)

Lidar's disadvantages, relative to the photogrammetric approach are:

- Processing large amounts of data—generally hundreds of times more information than collected with photogrammetric means. This means more editing time and more specialized software. Editing of data can become extensive enough that it exceeds that of photogrammetric capture using skilled operators
- Most elevations are at the surface of man-made features (such as tops of buildings) and must be reduced to "bare earth" for use in topographic mapping and for DEM generation in general. Though some automated techniques have been developed for filtering data down to bare earth, this increases editing times in areas that can be quickly collected by skilled operators—vegetation-free areas and urban areas specifically
- The narrow field of view and lower operating altitudes require more flight time. This is countered, however, by the day/night and somewhat all-weather collection capability of lidar. Furthermore, Geiger mode lidar is acquired from much higher altitudes than linear mode lidar, having both disadvantages and advantages
- Accuracy is acceptable for nearly all products, but high accuracies at large scales are still problematical. This is partly due to smoothing of the data to an accurate, but representative surface, and also the fact that lidar elevations are an average over a finite spot size on the ground. If the spot size is too large relative to the vertical accuracy requirement, especially in rugged areas, the product specification may not be met
- Lidar requires high-quality ABGPS and IMU data. Balky equipment, atmospheric conditions, or poor ABGPS results can invalidate a flight and require re-collection. Photogrammetric approaches often use IMU data as an enhancement to processing, whereas lidar requires it to project range data to ground coordinates. Re-collection may not be required by photogrammetric approaches provided sufficient ground control is available for aerotriangulation
- Without imagery, the checking of data and resolving of elevation ambiguities becomes problematical. Photogrammetric mapping provides a fall-back to re-collecting elevation data at a later time. For this reason, many lidar systems utilize companion digital cameras

Lidar's ability to see through trees is one of the primary reasons it may be selected over photogrammetric methods. The ground surface beneath trees can often be automatically determined by inspecting the multiple returns associated with each point. Dense broad-leaf trees in leaf-off condition will likely be amenable to lidar whereas a leaf-on condition may preclude the sensor seeing the ground. Densely populated areas of evergreens generally hamper the use of both lidar and photogrammetry.

DEM USER APPLICATIONS

Uses of photogrammetrically-derived DEMs generally fall into several classes:

- Topographic mapping
- GIS applications
- Generation of digital orthoimages
- Engineering requirements

Elevation data is delivered in digital form as raster DEMs, TINs, or as polylines and spot elevations. Contours are typically generated as a secondary product. Other than smaller map scales, contours are nearly always derived from TIN data models.

GIS applications of photogrammetrically-derived DEMs have become widespread for uses such as slope and runoff analysis and viewshed delineation. Digital orthoimages have become popular for use as base layers in GIS systems because they provide visual context for the other data layers and allow additional planimetric detail to be derived as required. Accurate digital orthoimage databases can be used to derive reasonably accurate planimetric data in an expedient manner by monoscopic plotting.

Some projects call for the generation of "true orthoimages" where building lean is removed and the building sides are not visible (Figure 6.11 on page 144). To produce a true orthoimage, building models are added to the terrain surface. This allows the displacement of buildings due to their height to be corrected, because the roof is modeled at its correct elevation. Occluded and shadowed areas are filled in using imagery from other viewpoints.

Engineering efforts at larger scales nearly always require the delivery of a TIN-like dataset, minimally as a set of breaklines and spot elevations. At smaller scales, generally for planning, gridded elevation data sets are used. Some engineering projects also require the delivery of elevation profiles, which may be derived from TIN data. In some cases, such as highway projects requiring profiles along planned centerlines or perpendicular to the centerlines, data is sometimes directly compiled as profiles. In this case, elevations are measured along programmed azimuths in stereocompilation equipment at each break in the terrain.

COST CONSIDERATIONS

Costs for photogrammetrically-derived elevation models are mainly determined by scale requirements and the area covered, with additional expense added by the amount of dense urban areas and vegetative cover within a project. Dense urban areas may present challenges to automated DEM generation although recent developments with highly-overlapping imagery and well-tuned correlation parameters may allow some success. However, skilled compilers are often required to make urban collection cost-effective. If planimetric compilation is part of the overall project, feature codes are used to determine elevations and breakline information from previously collected planimetric data. In this case, the incremental cost to producing DEMs supporting contouring is relatively low. Dense vegetative cover or trees will always increase the cost of elevation model extraction but leaf-off imagery exploited by experienced compilers can reduce the additional cost.

TECHNOLOGICAL ADVANCEMENTS

Completely-digital workflows are now standard, making the incorporation of new data sources or algorithms much easier and faster. Integration of cloud processing and storage capabilities will reduce production costs while improving throughput and enabling more processor-intensive algorithms.

Sensor developments can be expected to improve the acquisition speed and incoming data quality. Combinations of sensors such as multiple oblique cameras or synergistic combinations of visible/lidar/multispectral sensors will be optimized to support automated feature extraction as well as new types of products.

Attempts to improve automated elevation extraction will continue and will be successful. Ongoing research can be expected to further improve algorithms, while the availability of more processing power will make higher-performing but previously too expensive algorithms feasible. Sensors and data acquisition procedures will be optimized to deliver more usable inputs for automated processes, e.g., highly redundant imagery.

Innovations in photogrammetric platforms have the potential of revolutionizing many types of photogrammetric mapping. Unmanned aerial systems (UAS) are already practical for many small engineering and mapping projects. Applications can be expected to expand rapidly as more experience is gained in their operation and their datasets are characterized in terms of accuracy. Smallsats have similar potential for revolutionizing satellite mapping, especially at smaller scales.

There have already been dramatic changes in the types of standard elevation products required, from strictly contours to digital elevation rasters and realistically-rendered 3-D scenes. Future products will require the ability to integrate many types of data; while elevation products will still be judged on vertical accuracy and horizontal resolution, capabilities such as realism in depicting urban scenes will be similarly important for some users. The customer base for elevation data will continue to expand from the engineering and mapping world to other scientific fields and consumer interests.

As more of the Earth's surface is mapped digitally, the emphasis in photogrammetric production will shift toward database maintenance and update as opposed to starting from a blank sheet. This will call for shifts in thinking about production flows, database maintenance, and even contracting.

The increasing availability of off-the-shelf elevation data for large areas will also affect the market for photogrammetrically-derived elevation data. While current datasets have post-spacings of 5 to 15 meters, it can be expected that these will improve over time.

Crowdsourcing has been successfully applied in photogrammetric feature extraction, as in OpenStreetMap, but has yet to be applied to elevation extraction and processing due to the training and equipment infrastructure required for stereo viewing. The introduction of virtual reality headsets may provide a hardware platform for stereo extraction, assuming a use case can be identified.

AUTHOR BIOGRAPHIES

Dr. J. Chris McGlone received a Ph.D. in Photogrammetry from Purdue University in 1980 and since then has focused his career on the integration of photogrammetry and computer vision techniques. His employers have included H. Dell Foster Associates, Loral Fairchild, Carnegie Mellon University, and SAIC/Leidos. He was a co-founder of TerraSim, Inc., a producer of visual simulation database construction software. Dr. McGlone was Editor-in-Chief of the Fifth and Sixth Editions of the *Manual of Photogrammetry*, published by the American Society for Photogrammetry and Remote Sensing, and a co-author of the textbook *Introduction to Modern Photogrammetry*. He was the 2004 recipient of the ASPRS Photogrammetric Award (Fairchild).

Mr. Scott Arko received his B.A. and M.S. in Physics from Lawrence University and Colorado State University, respectively, and has over 20 years of experience in airborne and spaceborne remote sensing. Having worked for government, industry, and academia, he has gained a unique view into the needs of a variety of user communities. He is currently the Deputy Director of the Alaska Satellite Facility at the University of Alaska Fairbanks, where his current research interests include disaster/hazard monitoring with spaceborne radars and Structure from Motion using historic and modern aerial photography for change detection applications.

REFERENCES

Agarwal, S, N. Snavely, I. Simon, S. Seitz, and R. Szeliski (2009). *Building Rome in a day*, Proceedings of the International Conference on Computer Vision.

ASPRS (2000). *Camera calibration panel report*. Bethesda, MD: American Society for Photogrammetry and Remote Sensing.

ASPRS (2014). ASPRS positional accuracy standards for digital geospatial data. In *Photogrammetric Engineering and Remote Sensing*, 81(3) (pp. A1–A26). Bethesda, MD: American Society for Photogrammetry and Remote Sensing.

ASPRS (2015). *Guidelines for the in situ geometric calibration of the aerial camera system*. Bethesda, MD: American Society for Photogrammetry and Remote Sensing.

Baltsavias, E.M. (1999). A comparison between photogrammetry and laser scanning. In *ISPRS Journal of Photogrammetry and Remote Sensing*, 54 (pp. 83–94).

Belward, A.S. and Skøien, J.O. (2015). Who launched what, when and why; trends in global land-cover observation capacity from civilian earth observation satellites. In *ISPRS Journal of Photogrammetry and Remote Sensing*, 103, (pp.115–128).

Carrivick, J.L., M. W. Smith, and D.J. Quincey (2016). *Structure from motion in the geosciences*. Wiley-Blackwell. 208 pages. ISBN 978-1-118-89584-9.

Dallaert, F., S. Seitz, C. Thorpe, S. Thrun (1999). *Structure from Motion without Correspondence*.

Fischler, M., R. Bolles, (1987). Random sample consensus: a paradigm for model fitting with applications to image analysis and automated cartography. In *Readings in Computer Vision: Issues, Problems, Principles and Paradigms*, (pp. 726–740).

Hartley, R. and A. Zisserman (2003). *Multiple View Geometry in Computer Vision*. Cambridge University Press. ISBN 0-521-54051-8.

Hirschmüller, H. (2011). Semi-global matching—motivation, developments and applications. In Dieter Fritsch (Ed), *Photogrammetric Week '11*, Wichmann/VDE Verlag, Belin & Offenbach.

Kröpfl, M., Kruck, E., and Gruber, E. (2004). Geometric calibration of the digital large format aerial camera Ultracam. In *International Archives of Photogrammetry and Remote Sensing* XXXV-B1, (pp. 42–44).

Leberl, F., Irschara, A., Pock, T., Meixner, P., Gruber, M., Scholz, S., and Wiechert, A. (2010). Point clouds: lidar versus 3-D vision. In *Photogrammetric Engineering & Remote Sensing*, 76(10), (pp.1123–1134).

Lowe, D. G., (1999). Object recognition from local scale-invariant features. In *Proceedings of the International Conference on Computer Vision*, 1999.

Lowe, D.G., (2004). Distinctive image features from scale-invariant keypoints. In *International Journal of Computer Vision*. 60, (pp. 91–110). doi:10.1023/b:visi.0000029664.99615.94.

McGlone, J.C. (Ed) (2013). *Manual of photogrammetry*, Sixth Edition. Bethesda, MD: American Society for Photogrammetry and Remote Sensing.

Mulawa, D. (2004). On-orbit geometric calibration of the Orbview-3 high resolution imaging satellite. In *International Archives of Photogrammetry and Remote Sensing* XXXV-B1, (pp. 1–6).

Oliensis, J. (2000). A critique of structure-from-motion algorithms. In *Computer Vision and Image Understanding*, 80, (pp. 172–214).

Pacey, R., Scheidt, M., Walker, A. S. (1999). Calibration of analog and digital airborne sensors at LH Systems. Proceedings of the 1999 ASPRS Annual Conference, Portland, Oregon, May 17–21. Bethesda, MD: American Society for Photogrammetry and Remote Sensing.

Ullman, S. (1981). The interpretation of structure from motion. In *Proceedings of the Royal Society of London, Series B, Biological Sciences*, Vol. 203, No 1153, (pp. 405–426).

Wallach, H. and D. N. O'Connell (1953). The kinetic depth effect. In *Journal of Experimental Psychology*, Vol. 45, No 4, (pp. 205–217).

Westoby, M., Brasington, J., Glasser, N, Hambrey, M., and Reynolds, J. (2012). Structure from motion photogrammetry: A low-cost, effective tool for geoscience applications. In *Geomorphology*, 179, (pp. 300–314).

Chapter 7

INTERFEROMETRIC SYNTHETIC APERTURE RADAR (IFSAR)

Scott Hensley and Lorraine Tighe

TECHNOLOGY OVERVIEW

Remotely sensed methods are the most efficient methods for generating digital elevation models, especially over large geographic areas. While the success of traditional optical sensors has been well documented, in regions of persistent cloud cover, they are not able to obtain full coverage data. Radar sensors on the other hand, can penetrate clouds and operate at night.

Combining conventional synthetic aperture radar (SAR) and interferometry creates what is commonly known as IfSAR (also written as InSAR), providing a 24/7 all-weather three-dimensional mapping technique that has been commercially available since the mid-1990s. Since then, the IfSAR technique has been used to map the entire landmass of the Earth (from pole-to-pole) providing a homogeneous high resolution digital elevation model with an unrivalled accuracy and quality that surpasses that of any global optical-based DEM available today.

The main objectives of this chapter are to provide DEM users with an overview of the IfSAR topographic mapping technique and the operating principles as they relate to commercially available systems to include the STAR systems operated by Intermap Technologies Inc., the GeoSAR system operated by Fugro EarthData Inc., and various satellite SAR systems. The advantages and disadvantages of such systems are reviewed. Practical information regarding the collection, processing, products, and quality assessment of IfSAR DEM data are discussed. The chapter closes with examples of applications for use with IfSAR derived DEMs, to include Dewberry' ongoing aerial IfSAR mapping of Alaska with data acquired by Intermap and Fugro.

This chapter has been modified considerably from the first and second editions of the DEM Users Manuals. Readers interested in more of the physics and history of IfSAR techniques, hardware, system engineering or signal processing are recommended to consult (Richards 2007, Rosen 2000) and the previous editions of the DEM User's Manual.

SAR Explained

RADAR, which stands for "radio detection and ranging," subsequently referred to as radar, is an active remote sensing instrument that operates in the microwave portion of the electromagnetic spectrum. It provides its own source of illumination in the form of microwave pulses that are transmitted to the terrain being imaged. Therefore, radar sensors operate independently of the sun, often collecting data during night.

To obtain high resolution data from radar sensors, SAR configuration is implemented. SAR is a form of radar where the radar antennae, onboard a platform (air or space), transmit radio waves in the form of a wide beam of high-power pulses ("echoes") of microwave energy to the area being imaged (Elachi, 1988). Radar pulses are transmitted toward the terrain of interest in a vertically oblique direction that intersects the terrain with an oval footprint elongated in the range direction (e.g., across track; Figure 7.1 on page 154). Consequently, two different scanning mechanisms are employed: One in the range direction, the second in the azimuth direction (direction of flight). In the range direction each transmitted pulse sweeps across the swath at the velocity of light. In the azimuth direction, the swath is scanned in the along-track direction at the speed of the antenna footprint (or forward movement of the platform). These two mechanisms differ in timescales from each other by several orders of magnitude, which allows us to treat them as mutually independent (Bamler and Hartl, 1998).

SAR sensors record the interference (Doppler) pattern from echo signals over several hundred to several thousand meters along the flight path. The received echoes form the raw data matrix or complex image containing amplitude and phase. The coordinates of the 2-D raw signal image are range for the distance of the scatterer from the SAR (or equivalently echo delay time) and azimuth for the position of the scatterer along the sensor path (Figure 7.1). The target of interest is imaged (illuminated) several hundred times from different locations along the flight path generating numerous echoes that are recorded coherently (i.e., amplitude and phase as a function of time).

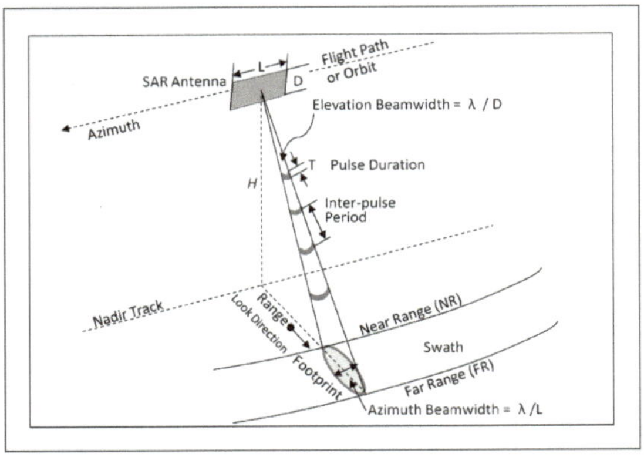

Figure 7.1. Scanning configuration for a right-hand looking rectangular SAR antenna, modified after Olmsted (1993). L is the antenna length, D is the antenna width, T is the pulse duration, and H is the Nadir Range. The recorded signal data (amplitude and phase) can produce a two-dimensional radar image. The distance measured between the antenna and the target is known as slant range.

The accepted parameter for discussion of a radar system's "resolution" performance is impulse response. The microwave pulses are combined to synthesize a linear array with angular resolution equivalent to an antenna several hundred meters in length, resulting in what is called the azimuth resolution, in the direction parallel to the sensor flight path and the range resolution, perpendicular to the flight path direction. Azimuth impulse response is a function of antenna beam-width and the number of azimuth looks processed, whereas range resolution is a function of the range bandwidth. Azimuth resolution is equal to approximately one-half the actual (real) antenna length (L/2) and does not depend on platform altitude (distance). Range resolution, on the other hand, is determined by the transmitted pulse width (T in Figure 7.1); e.g., narrow pulses yield fine range resolution (Elachi, 1988)). Range is determined by precisely measuring the time from transmission of a pulse to receiving the echo from a target.

Microwaves are transversal waves, for example, in the direction of propagation the electric and magnetic fields are mutually orthogonal. By convention, polarized electromagnetic radiation is described by specifying the orientation of the wave's electric field at a point in space over one period of oscillation. Most earth observation SARs systems use Linear (or planar) polarizations where the vibration of the electric field vector is in a parallel direction to the propagating wave. The radar signals can be set up to vibrate in either a horizontal (H) or vertical (V) plane or in a circular fashion. Antennae can transmit in one plane and receive in another. Possible variations are HH, VV, HV, and VH for: Horizontal transmit, horizontal receive; vertical transmit, vertical receive; and so on. All four polarizations are typically denoted as "Quad" polarization.

Targets on the Earth's surface scatter microwave radiation differently depending on the polarization of the wave transmitted. Volume scattering (illustrated in Figure 7.10 on page 163) is the main interaction in vegetated canopies. It causes the incoming signal to be depolarized, for example, the signal is scattered in other directions, at different relative vibrational planes. A H-polarized wave might hit a tree, bounce back and forth among leaves/twigs, and be backscattered to the SAR sensor in a V-polarized state. Using an HV setup, the antenna could receive these V-polarized returns. Though the overall radar backscatter received would be less than, say, a VV setup, the resulting image would have increased variation between regions of volume scattering (higher potential to return V-polarized backscatter—usually vegetation) and regions of surface scattering.

SAR systems are not impeded by darkness, atmospheric conditions such as haze, clouds, light rain, or smoke (Bamler and Hartl, 1998). The resulting microwave energy can produce two-dimensional cloud free radar images. These have significant advantages over optical systems providing a suitable sensor technology for nearly all weather and day or night data acquisitions, across the globe. SAR systems require a complex integrated array of onboard navigational and control systems with location accuracy provided by both Doppler and inertial navigation equipment. The outputs produced are typically two-dimensional SAR imagery.

Basic Concept and Operating Principles

As with many topographic mapping techniques the location of a point on the ground is determined in three dimensions by solving for an unknown component of a triangle associated to the observation geometry. In the case of IfSAR, the triangle is created by the physical separation (in space or in time) of two SAR apertures' (SAR Antenna A1 and SAR Antenna A2) in the across-track plane (perpendicular to the flight direction) and the distance from the apertures to the point on the ground (P) being measured (illustrated by the red triangle in Figure 7.2). Radar pulses are transmitted from one or both SAR antenna(s), radar echoes are received by both the SAR antennas. If the received signals from the two antennas are combined coherently for each imaged point to measure the phase difference, then the system forms an interferometric SAR. More detail on IfSAR systems and processing can be found in (Rosen et al., 2000), (Madsen and Zebker, 1999), and (Bamler and Hartl, 1998).

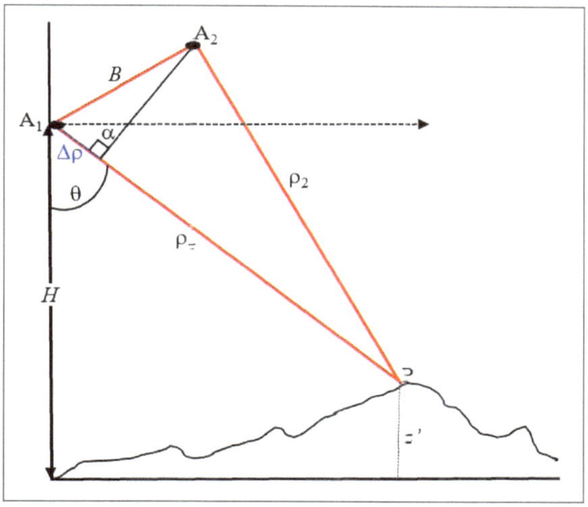

Figure 7.2. Simplified broadside looking (zero Doppler) radar interferometry geometry. The difference in range from the two observing antennas (A1, A2) to the target (P) is approximately equal to the projection of the baseline vector onto the line-of-sight vector shown in blue ($\Delta\rho$). This range difference can be related to a phase measurement using equation 7.1 and forms the primary interferometric observable used to derive topographic height measurements. Parameters in this figure are defined following discussion of Equation 7.1.

The interferometric phase difference is essentially related to the geometric path length ($\Delta\rho$) difference to the image point, which depends on the topography. With knowledge of the interferometer geometry, the phase difference can be converted into an altitude for each image point. By having a third measurement, the interferometric phase, in addition to the standard cross-track location of an image point obtained with the SARs (A_1 and A_2), it is possible to determine the three-dimensional location of a point (P) shown in Figure 7.2.

By measuring the time, it takes a radar pulse to propagate from the SAR antenna to a target and return, the distance or range measurements are related to phase measurements by converting the distance to units of wavelength and recalling each wavelength corresponds to 2π radians or 360° of phase (Figure 7.3 on page 156). Phase measurements in interferometric systems can be made with degree level accuracy, and with typical wavelengths in the range of 2-20 cm, corresponds to measurements having millimeters to centimeter accuracy.

The path length difference, $\Delta\rho$, of the signals received by the two antennas is approximately given by:

$$\Delta\rho = |\rho_2| - |\rho_1| \approx B\sin(\theta-\alpha) \qquad (7.1)$$

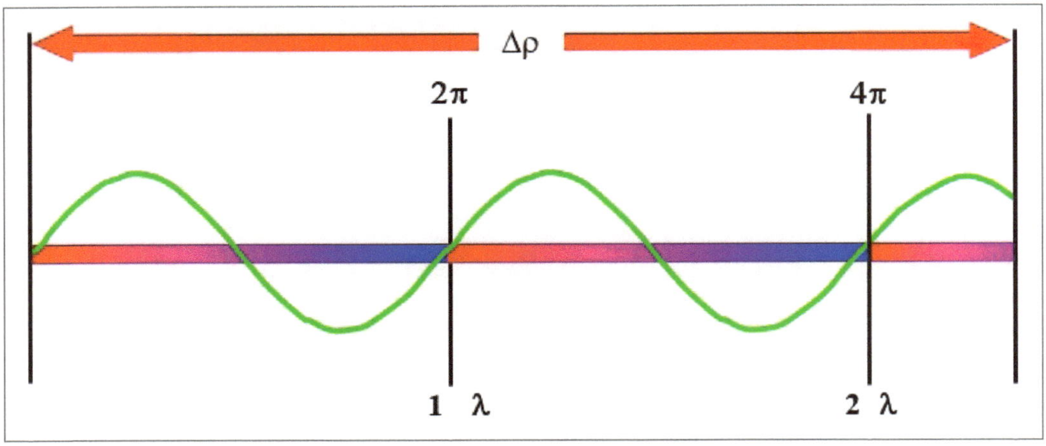

Figure 7.3. Illustration of the relationship between phase, distance, and wavelength.

where ρ_1 indicates the vector from antenna 1 to the target (P), B is the length of the baseline vector which is the vector pointing from antenna 1 to antenna 2, θ is the desired elevation (or look) angle and the baseline orientation angle, α is the angle the baseline vector makes with respect to the horizontal. Observe that the range difference to a good approximation for most systems is simply the length of the projection of the baseline vector onto the line-of-sight. The range difference, Δρ, may be obtained by measuring ϕ, the phase between the two interferometer signals, using the relation:

$$\phi = -\frac{2\pi m \Delta\rho}{\lambda} \quad m = 1,2 \qquad (7.2)$$

where λ is the radar wavelength and m equals 1 when using a single transmit antenna, or 2 when transmitting from two antennas, the so-called Ping-Pong mode, or for repeat pass systems described below. Using the simplified geometry of Figure 7.2 the height of a target, z', is given by:

$$z' = H - \rho\cos(\theta) \qquad (7.3)$$

where H is the altitude of the radar antenna and ρ is the slant range from the antenna to the target. Since the signal phase is sensitive to displacements between images of a fraction of a wavelength, the interferometric technique provides a very accurate means of determining topographic heights. Using Equations 7.1 and 7.2 the elevation angle can be determined to be:

$$\theta = \sin^{-1}\left(\frac{\lambda\phi}{2\pi m B}\right) + \alpha \qquad (7.4)$$

It is immediate from Equations 7.3 and 7.4 that determining the height of a target requires knowledge of the platform position, the range, the interferometric baseline length, the baseline angle and the interferometric phase. Generation of accurate topographic maps using radar interferometry places stringent requirements on the knowledge of the platform and baseline vectors. Figure 7.4 provides an example of interferometric phase measurements (interferogram) and elevation data.

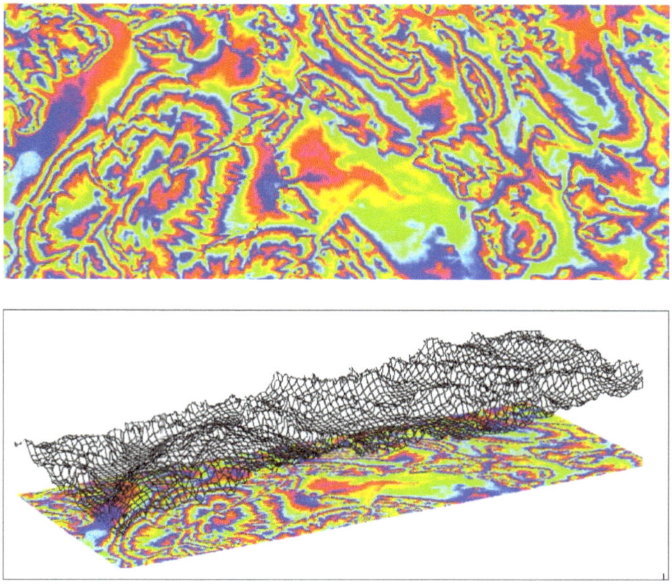

Figure 7.4. Interferometric phase and associated elevation data generated from single pass IfSAR observations. Top image: IfSAR interferogram where one set of colors (pink, red, yellow, green, and blue) represents one 0-2π interferometric fringe. Bottom Image: The interferogram has been processed to unwrap the phase data to derive a DSM represented by the wire mesh and Lower image) generated from single pass IfSAR observations (modified after Gens, 2002).

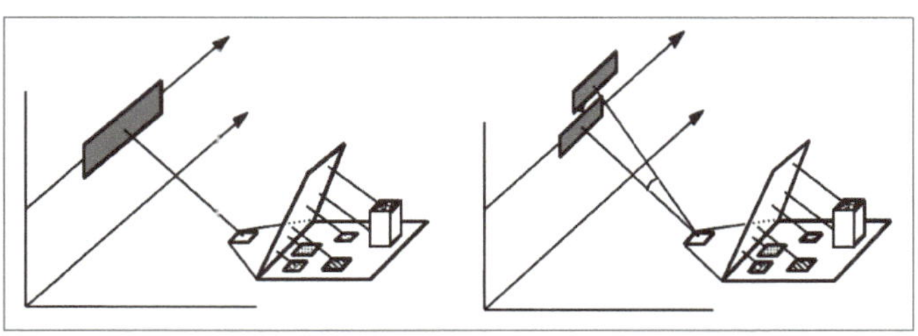

Figure 7.5. The left portion of the figure shows a notional arrangement of elemental scatterers within an imaging cell. Each elemental scatterer may have different surface roughness and dielectric properties as indicated by the different shading patterns. The right portion of the figure shows that if the imaging geometry is nearly the same the relative distance between scatterers is preserved and cancels out in the interferometric phase measurement.

Understanding conditions when interferometric phase measurements useful for topographic mapping are possible requires us to examine more closely what happens to radar signals within a resolution element. Consider a resolution cell with elemental scatterers arranged throughout as shown in Figure 7.5. Each elemental scatterer will contribute a portion of the backscatter that is added coherently with the other elemental scatterers to produce the return from the cell. Since the return from the elemental scatterers adds coherently, the relative phase or distance between the scatterers affects the magnitude and phase of the total signal. Conceptually, the phase can be decomposed into a systematic and random component by selecting the center of the cell as reference. The systematic component is the phase from the antenna to the center of the cell and is the portion of the signal needed for interferometry. The random component is the coherent sum of the signals from the randomly arranged elemental scatterers within the cell to the center of the cell. This component, although random from resolution element to resolution element, remains the same (or nearly the same) if the viewing geometry is nearly identical and if the relative position of the elemental scatterers within a cell remain the same (or nearly the same) as shown in Figure 7.5.

For interferometric applications, where the viewing geometry of the interferometric pair is nearly the same, the random component arising from the elemental scatterer arrangement cancels and leaves only the difference between systematic components. This phase difference is the interferometric phase measurement and equals the phase in Equation 7.2 modulo 2π. Another random component, which does not cancel, is thermal noise.

Thermal noise is different for each receiving antenna, and depending on its magnitude relative to the desired signal degrades interferometric phase measurement. The above discussion is summarized below:

$$\text{phase} = \underbrace{\begin{array}{c}\text{range from antenna}\\ \text{to center of cell}\end{array}}_{\begin{array}{c}\text{Systematic component}\\ \text{desired by interferometric}\\ \text{measurement}\end{array}} + \underbrace{\begin{array}{c}\text{coherent sum of elemental scatterers}\\ \text{arranged randomly in cell}\end{array}}_{\begin{array}{c}\text{Random component that if look direction is}\\ \text{nearly the same and scatters within cell do not}\\ \text{move relative to each other this component}\\ \text{cancels in the interferogram formation process}\end{array}} + \underbrace{\text{thermal noise}}_{\begin{array}{c}\text{Random component}\\ \text{that does not cancel and}\\ \text{results in interferometric}\\ \text{phase noise}\end{array}}$$

It is important to appreciate the consequences of the fact the interferometric phase measurement is made modulo 2π. The total range difference between the two observations that the phase represents in general can be many multiples of the radar wavelength, or expressed in terms of phase, many multiples of 2π. It is this value that is required to make height measurements. The standard approach for determining the unique phase that is directly proportional to the range difference is to first determine the relative phase between pixels via the so-called "phase unwrapping" process. Unwrapping of IfSAR imagery is a non-trivial process for which several algorithms have been developed. Complications arise in avoiding unwrapping errors in regions of shadow, layover and low signal return. The connected phase field after unwrapping may still need to be adjusted by an overall constant of 2π. The step that determines the overall constant of 2π is referred to as absolute phase determination.

Interferometric correlation, a measure of the similarity of the signal received at the two antennas, can be estimated directly from the image data of the two interferometric channels (Zebker and Villasenor, 1992). Correlation measurements have values between 0 and 1, with 1 designating perfect correlation between the channels. Sometimes it is more convenient to refer to the amount of interferometric decorrelation, which is defined as one minus the correlation. The amount of decorrelation due to the slightly different viewing geometry is called geometric decorrelation. Thermal noise induced signal decorrelation is called noise decorrelation. Shadowed regions suffer from noise decorrelation and areas on steep slopes exhibit geometric decorrelation that increases phase noise and can preclude useful phase measurements altogether. Another form of decorrelation occurs when there is a vertical distribution of scattering elements within a resolution element as shown in Figure 7.6. Not only is the signal decorrelated, the point within the resolution cell corresponding to the interferometric phase measurement depends on the wavelength and the scatterer distribution in the cell. This form of decorrelation is called volumetric decorrelation and can be used to infer information about the vertical structure of the volume. Recently, there has been a great deal of activity using volumetric correlation to estimate tree and canopy structure within the interferometric SAR community.

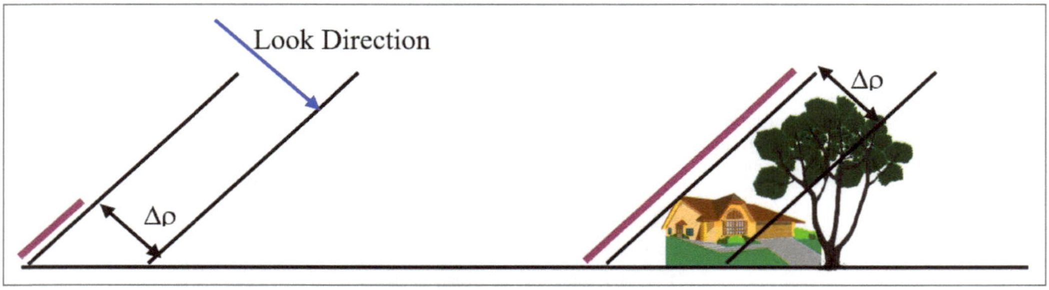

Figure 7.6. Vertical structure of scatterers within a resolution element due to vegetation or other structures present in the cell cause increased decorrelation. This form of decorrelation can be used to infer information about vertical structure within a resolution element. The increased decorrelation results from the increased size of the range cell projected back toward the direction of the radar (shown in magenta) when compared to a flat surface.

There is an upper value on the useful baseline length known as the critical baseline (Rosen et al., 2000). The critical baseline is reached when the amount of phase change per resolution element exceeds 2π radians. This limitation is a result of the fact that the interferometric phase measurement is made only modulo 2π. As the baseline approaches the critical baseline, the phase values from the two antennas become completely decorrelated. However, as the baseline increases, the sensitivity of phase to height increases improving the accuracy of interferometric SAR systems. Practical mapping systems must select baselines with a balance between adequate phase to height sensitivity to meet mapping requirements and excessive decorrelation with corresponding processing difficulties.

After the multi-looked interferogram has been generated the phase for each complex sample is computed. To generate a continuous height map, the two-dimensional phase field must be unwrapped. After the unwrapping process an overall multiple of 2π is estimated and added to the unwrapped phase (the estimated value may be 0).

Combining the SAR image formation process, interferometric phase measurement, unwrapping and height determination into an automated processing algorithm has a process flow that is shown in Figure 7.7 (Madsen et al., 1993). An understanding of the steps involved is useful for understanding IfSAR calibration and for identifying potential error sources in IfSAR generated DEMs. Raw data are collected and stored onboard for batch processing. The first processing step is decoding the byte data, followed by range compression for each of the two interferometric channels. Using the platform motion information obtained from Inertial Navigation System (INS) and Global Positioning System (GPS) measurements, as well as any other baseline metrology devices, the data are compensated for perturbations in aircraft motion from a reference path[1] and then azimuth compressed. This generates two single-look complex images.

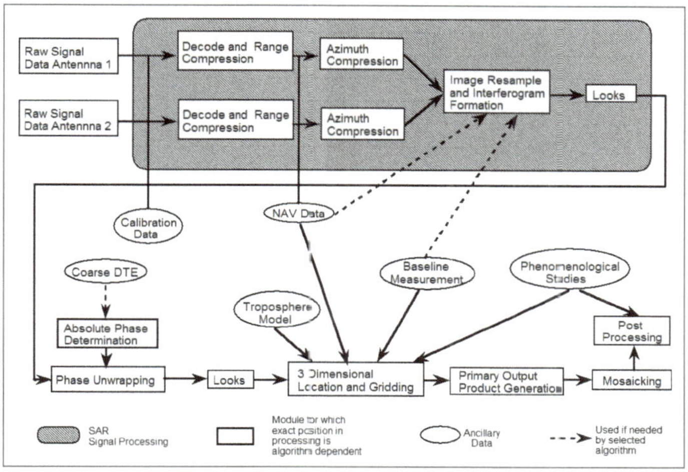

Figure 7.7. Interferometric processing block diagram.

One image of the single-look complex image pair is resampled to overlay the other. This registration must be done to a small fraction of a pixel (typically less than .1 of a pixel) to avoid phase decorrelation. Multiplying the complex pixel value in one image by the complex conjugate of the corresponding pixel value in the second image forms an interferogram. The resulting interferogram is multi-looked, by spatially averaging the complex pixels in a box about a given pixel to reduce the amount of phase noise.

After determining the absolute phase for each pixel in the interferogram and possibly taking additional looks, the 3-dimensional target position can be determined. Phase corrections are applied to the interferometric phase to account for tropospheric effects, and the range is corrected to account for the speed of light changes in the atmosphere. Using accurate baseline and platform position information, the phase and range information for the target position is computed. A relief map is generated by gridding the unevenly sampled 3-dimensional target locations in a natural coordinate system aligned with the flight path. The gridded products include the target heights, the SAR image which has been orthorectified, a correlation map,

[1] The process of correction for motion away from an ideal path is referred to as motion compensation. Motion compensation corrects for motion on the order of fractions of a wavelength to generate well focused SAR imagery. The requirements for motion compensation are even more demanding for IfSAR systems and place stringent requirements on the platform and baseline metrology systems.

and a height error map described below. These four products will be referred to as primary mapping or strip map products. The resulting radar relief map may be measuring the heights above the ground, within the vegetation canopy or beneath the surface, in arid regions. To convert this map into a true ground surface DEM, corrections based on phenomenological studies, e.g., using scattering or semi-empirical curves to correct elevation measurements based on the amount of decorrelation in the canopy, must be incorporated into either the 3-dimensional location algorithms or into a post processing step.

One of the unique aspects of interferometric SAR systems is the ability to determine the statistical height precision, that is the degree of height noise from pixel to pixel, estimated from knowledge of the correlation, λ (Hensley and Webb, 1994). The amount of phase noise between the two channels is simply and directly related to the correlation and number of looks used to reduce phase noise[2]. The Cramer-Rao bound relating the phase variance, σ_ϕ, to the correlation coefficient, λ, is given by:

$$\sigma_\phi = \frac{1}{\sqrt{2N_L}} \frac{\sqrt{1-\lambda^2}}{\lambda} \tag{7.5}$$

where N_L is the number of looks. From Equations 7.3 and 7.4 the height error, σ_h, as a function of the phase noise is found to be:

$$\sigma_h = \frac{\lambda \rho \sin(\theta)}{2\pi m B \cos(\theta-\alpha)} \sigma_\phi \tag{7.6}$$

Equations 7.5 and 7.6 allow the generation of an error map showing the local height accuracy for each post in an interferometrically derived DEM.

Typically, the post spacing of the IfSAR topographic data are comparable to the fine spatial resolution of SAR imagery while the altitude measurement accuracy generally exceeds stereoscopic accuracy at comparable resolutions. The registration of the two SAR images for the interferometric measurement, the retrieval of the interferometric phase difference and subsequent conversion of the results into digital elevation models of the terrain can be highly automated, representing an intrinsic advantage of the IfSAR approach. The performance of IfSAR systems is largely understood both theoretically and experimentally enabling these systems to be designed and built to meet specific mapping objectives. These developments have led to airborne and spaceborne IfSAR systems for routine topographic mapping.

For the remainder of this chapter IfSAR is defined as an airborne or spaceborne interferometric radar system, flown aboard rotary or fixed wing aircraft, or any space based platform, that is used to acquire 3-dimensional coordinates (these coordinates must be convertible to a specified geographic datum) of terrain and terrain features that are both manmade and naturally occurring. IfSAR systems consist of a platform, GPS and attendant GPS base station(s) if needed, INU and interferometric radar system including commanding and data acquisition systems. The system may also include other ancillary equipment such as baseline metrology systems as necessary for accurate map generation. These systems form synthetic aperture images of terrain surfaces from two spatially separated antennas over an imaged swath that may be located to the left, right or both sides of the imaging platform.

SAR Wavelength

Most geoscientists refer to SAR systems in terms of their wavelength, λ, denoted by a letter code (assigned in World War II for security reasons). The most common wavelengths for SAR remote sensors are labeled K, X, C, L, S, and P, listed in order of increasing wavelength size (Table 7.1). To appreciate why particular radar frequencies (wavelengths) are selected for a given application, it is necessary to have a cursory

[2] The Cramer-Rao bound used to relate the phase noise to correlation and number of looks is only valid when the number of looks exceeds 4 or 5. The number of looks in most interferometric systems used to generate topographic maps usually is much larger than 4. A notable exception is the SRTM system (described later in the chapter) where the number of looks varied between 1 and 4.

SAR Band Identification	Wavelength Range (cm)	Frequency Band (mHz)
Ka	1.13-.75	26,500-40,000
K	1.66-1.13	18,000-26,500
Ku	2.4-1.66	12,500-18,000
X	3.75-2.4	8,000-12,500
C	7.5-3.75	4,000-8,000
S	15-7.5	2,000-4,000
L	30-15	1,000-2,000
P or UHF	100-30	300-900
VHF	1,000-100	30-300
HF	10,000-1,000	3-30

Table 7.1. SAR Band, Wavelength and Frequency Relationship.

understanding of how radar signals interact with terrain (Elachi, 1988), (Raney, 1999). Each pixel in a SAR image is a complex number having a magnitude and phase determined by the terrain surface properties and the image geometry. A radar signal impinging on a resolution element (area of the surface contained within a single range and azimuth bin) will in general scatter energy in all directions. The signal reflected toward the radar is referred to as the backscatter. Backscatter strength is a function of the composition of the surface and its structure. Wavelength selection is often a trade among scientific considerations, technology readiness, platform constraints, availability and cost.

Elevations derived from IfSAR methods are determined from the scattering phase centre height (denoted as h_{spc}). The h_{spc} is a type of amplitude weighted average of all vertically distributed scattering elements within a resolution cell. The SAR wavelength, baseline and vertical structure of the scattering medium determines the h_{spc}. The location of h_{spc} depends on the penetration depth of microwaves into the canopy (Figure 7.8), which depends primarily on the SAR wavelength in addition to other parameters such as the size and density distribution of the scattering elements, the geometric arrangement of the scatterers, canopy moisture condition, surface roughness, and moisture content of the ground layer. For example, at X-band wavelength the h_{spc} is at or very near bare ground in barren areas whereas, in forest it is closer to the canopy top (Figure 7.8).

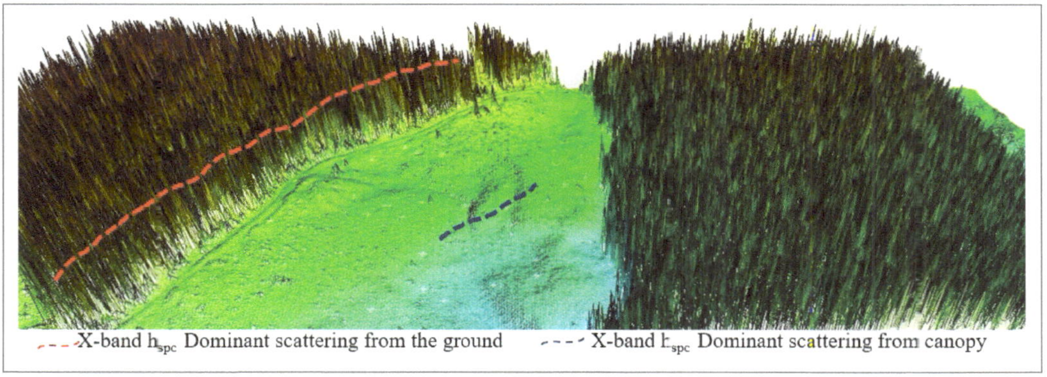

Figure 7.8. Relative position of X-Band IfSAR h_{spc} for bare ground (blue-dashed line) and over dense forest (red-dashed line). The data presented in this illustration represents the DSM lidar data over a section of Edson, Alberta Canada (Tighe, 2012).

Generally, the main scattering elements tend to be of comparable size to the wavelength of the radiation (X-band (λ = 2.4 to 3.75 cm) and C-band (λ = 3.75-7.5 cm) energy, while tree trunks and the ground surface will act as strong scatterers at L-band (λ = 15-30 cm) and P-band (λ = 30 cm to 120 cm), all else being equal (e.g., density, cover type, incidence angle, polarization, etc.). Moreover, the penetration depth of the microwave signal into porous medium increases from K-band (λ = 1.13 cm to 1.66 cm) through P-band or with increasing wavelength (Figure 7.9). For example, the DSM IfSAR height for X-/C-band short wavelength IfSAR may be in the upper part of a forest canopy and can be used directly as a rough estimate of canopy height (Andersen et al., 2008). Conversely, longer wavelengths (such as L- and P-band) penetrate deeper, interacting with the ground, tree trunks, and other lower canopy elements, depending on the characteristics of the canopy, and can provide rough estimates of bare ground surface elevation.

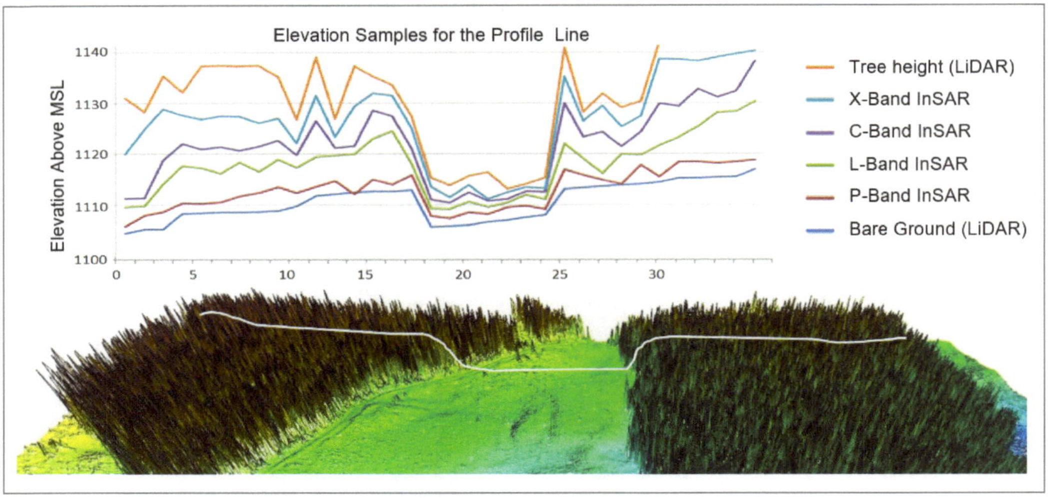

Figure 7.9. IfSAR signal penetration over dense vegetation and bare ground at different wavelengths for a profile line in grey, superimposed on a lidar DSM (Tighe, 2012).

Ideally, the choice of wavelength would be tailored to electromagnetic properties of the surface of interest. However, the varied nature of the Earth's terrain precludes any single frequency from satisfying all possible application requirements. For example, SAR signal penetration depth within vegetation varies with wavelength as follows: 1-7 m at X-band, 3-10 m at C-band, 3-15 m at L-band, and 7-35 m at P-band (Lewis and Henderson, 1999). In sparse vegetation canopies, all wavelengths will produce signal returns from the ground and lower canopy elements, resulting in a lower overall scattering phase center height, that which is closer to bare ground elevations One estimate or the other may be more desirable depending on the application. For repeat-pass interferometry topographic systems, shorter wavelength systems are usually preferred because temporal decorrelation is less than for higher frequency systems, particularly in vegetated regions.

Electrical composition of a surface is characterized by its dielectric constant. The dielectric constant of a material determines how much energy is absorbed or reflected from the surface and depends on the radar frequency. Surface structure is usually characterized by roughness, a measure of how much the surface varies in a resolution element. Roughness is measured in terms of the incident radiation's wavelength, so surfaces that are smooth at one wavelength may appear rough at another wavelength. As a rule of thumb, the rougher the surface the greater the backscatter.

Surface variations on the order magnitude of the radar wavelength will scatter radar as specified by the Rayleigh criterion. The Rayleigh criterion can be used to differentiate between, for example, surface topography in non-vegetated areas or extent of vegetation in others. The Rayleigh criterion defines a surface as smooth if h (in equation 7.7) is less than 1/8th the radar wavelength (Balmer and Hartl, 1998):

$$h = \frac{\lambda}{8 \sin \theta} \tag{7.7}$$

Where:

- h: Surface roughness, defined as the root-mean-square (RMS) height relative to a perfectly smooth surface
- λ: Wavelength of the SAR
- θ: Incidence angle

Horizontal smooth surfaces such as calm water reflect nearly all incident energy away from the radar antenna (called specular reflection) at the same angle as the incidence angle per Snell's law (Figure 7.10—specular reflection) returning little to no signal back to the SAR receiver and resulting in a dark image tone or decorrelation in phase. In contrast, microwaves incident upon a rough surface are scattered in many directions (called diffuse scattering/reflectance).

Reflections can bounce again off other objects and be redirected back toward the SAR receiver, resulting in a stronger return signal (Figure 7. 10—volume scattering and dihedral reflection). In the case of volume scattering, which occurs in vegetation where (high water content) leaves reflect radar signals onto other leaves and branches, multiple scattering occurs until the signal exits the vegetation. A percentage of the incoming radar is therefore volumetrically scattered back to the SAR sensor, giving the vegetation a brighter signature than a smooth lake.

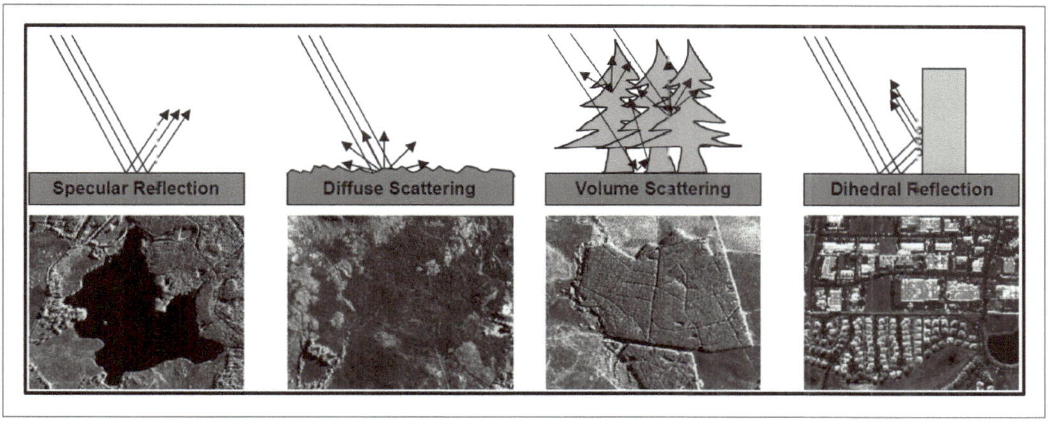

Figure 7.10. Surface roughness effects on radar backscatter for four land cover types. L-R: A body of calm water (specular reflection), grass (diffuse scattering), forest (volume scattering), and buildings.

IfSAR Sensor Viewing Geometry

IfSAR configurations use oblique side looking viewing geometry that can lead to distortions in the SAR Imagery and derived elevations. Unlike some optical sensors that look directly below the imaging platform, radar and IfSAR sensors "view" the ground per a perspective beam that looks out to the side of the platform (Figure 7.11 on page 164).

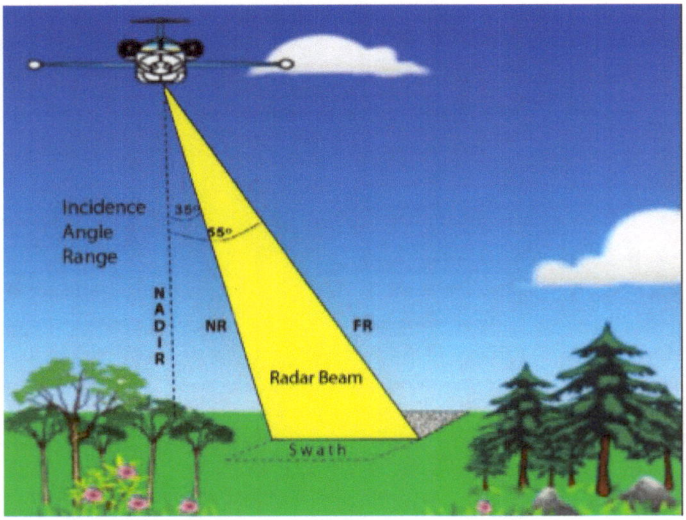

Figure 7.11. Figure showing the side looking viewing geometry of an IfSAR system in which the radar beam (in yellow) has an incidence angle range from 35° to 55°. The location of the near range (NR) and far range (FR) and swath width are also indicated.

The beam shown in yellow in Figure 7.11 corresponds to the line between the radar antenna and the target on the ground. The radar sensor points the radar beam out to one side of the aircraft, defining an incidence angle range. This configuration is an optimized viewing geometry for an IfSAR topographic mapping system. As the aircraft (or spaceborne system) flies over the terrain, an image strip or swath is collected. This beam, called the slant range, is the distance as measured by the radar directly, in effect along each line perpendicular to the flight vector and directly with the radar and each scatter. The slant range is further defined by two terms: near range (NR) and far range (FR). A target located in the NR is closer to the antenna than a target positioned in the FR location. This viewing geometry creates distortions radiating out from the NR to the FR, rather than radiating out from nadir, as with aerial photography data collections.

SAR Image Distortion

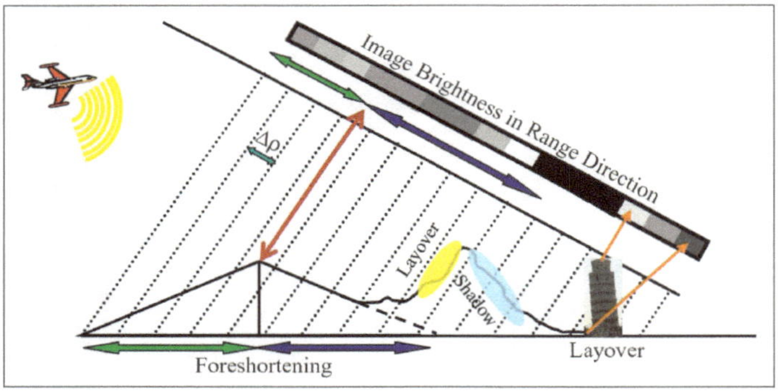

Figure 7.12. The three-dimensional world is collapsed to two dimensions in conventional SAR imaging. After image formation, the radar return is resolved into an image in range-azimuth coordinates. This figure shows a profile of the terrain at constant azimuth, with the radar flight track into the page.

Because the range direction is not parallel to ground coordinates, SAR images are distorted relative to a planimetric view. Three common distortion phenomena observed in SAR imagery that bear mention are foreshortening, layover and shadow, shown in Figure 7.12.

Foreshortening in radar imagery results from the fact that relief displacement is towards the direction of the radar. Because the range increases more slowly than ground coordinates on slopes facing toward the radar (higher elevations contend with increasing ground distance, slowing the range increase) they tend to appear bunched relative to a planimetric view. The opposite occurs on slopes facing away from the radar (lower elevations coupled with increasing ground distance speeds the range increase) where they tend to expand out when compared to a planimetric view Both situations are illustrated in Figure 7.12. Note that foreshortening in radar images is opposite to that of optical imagery where relief displacement is away from the direction of the camera.

Layover is a limiting case of foreshortening where points arranged with increasing ground coordinates appear reversed in the radar imagery. Layover occurs because the range to objects with larger ground coordinates is less than the range to other objects with smaller ground coordinates. Geometrically this happens when the slope of the terrain is greater than the angle the incident radiation makes with respect to vertical. More importantly for our purposes is to note for interferometric radar systems layover causes a loss of useful signal and therefore precludes the determination of elevation in layover regions.

Shadow occurs when the radar beam cannot reach a portion of the terrain being imaged because it is occulted by other parts of the terrain or other objects in the scene. Where the terrain is shadowed, the radar image will appear dark and the signal to these range cells is only due to thermal noise. As with layover regions, shadowed regions have no useful interferometric signal and consequently no elevation values can be determined.

IfSAR systems, being able to resolve the three-dimensional coordinates of points in SAR imagery, can produce imagery having correct planimetric placement in regions where there are good interferometric phase measurements.

IfSAR DEM Distortion

The side looking IfSAR sensor configuration maps a look or incidence angle range (θ—Figure 7.2 on page 155) away from nadir. Incidence angle range is arbitrarily denoted as near-range (NR; closest to nadir; or steep—Figure 7.13)), mid-range (MR, or moderated—Figure 7.13), and far-range (FR; furthest from nadir; shallow—Figure 7.13) across a flight line strip of data or an orbital pass. Incidence angle variations across the swath are usually more pronounced in airborne data collections and result in variations in varying relative signal contributions from ground and surface features as illustrated in Figure 7.13. The resulting elevation measurements as described earlier reflects this relative signal strength of the various scattering elements and therefore the vertical location within the medium for which the elevation corresponds.

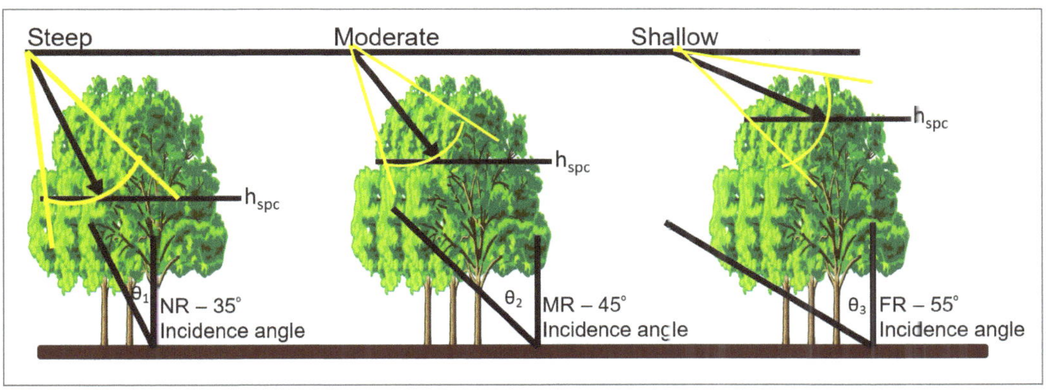

Figure 7.13. Incidence angle effect on h_{spc} retrieval of surface elevation data for single-pass IfSAR.

In airborne collections, for example, at steep incidence angles (e.g. NR—θ ~ 35°; Figure 7.13, left-hand tree) more exposure of the lower portion of a vegetation canopy and ground occurs because of greater signal penetration into the canopy, resulting in a h_{spc} or estimated elevation height that is closer to the ground elevation (Hensley et al., 2001, Andersen et al., 2005; 2008, Garestier et al., 2008, Hajnsek et al., 2008, Tighe et al., 2012). The opposite effect occurs for the same tree located, in the FR (e.g., NR—θ ~ 55°; Figure 7.13, right-hand tree) such that the h_{spc} or estimated elevation height that is closer to the top of the tree canopy elevation. Interferometric mapping systems can reduce the impact of incidence angle sensitivity by combining multiple flight lines strips of data.

Because radar primarily interacts with structures that have lengths comparable to the wavelength or larger, longer wavelength radars tend to penetrate deeper into the vegetation canopy or ground surface. The amount of penetration in a vegetation canopy, bare ground or snow or ice layers depends on the structure of the scattering medium as described earlier. Radar wavelengths less than roughly 10 cm mostly sense the upper portions of canopies while wavelengths longer than 20 cm sense deeper into a canopy. Longer wavelength radars have been known to penetrate several meters or more in dry sandy soil and even deeper into certain types of ice (10 s of meters). Commercial IfSAR providers attempt to minimize this effect by processing and averaging several data collections (Figure 7.14) or using systems designed specifically to compensate or minimize this effect.

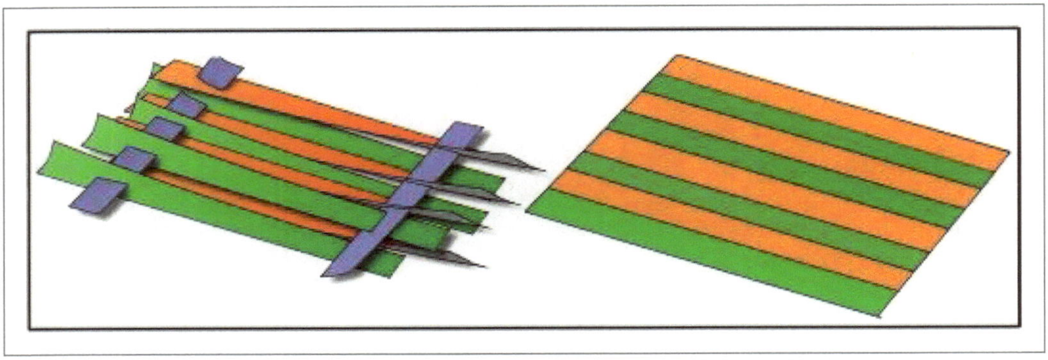

Figure 7.14. Example of airborne IfSAR strip DSM data collection merging of primary (green and orange single take data strips) and tie (blue single take strips) flight line strips that are merged to derive the final DSM given in the right mosaic.

DEVELOPMENTAL HISTORY

The IfSAR mapping for Earth based topography measurement had its genesis with an airborne SAR system by Graham in the early 1970s (Graham, 1974). Later IfSAR techniques for non-military applications was undertaken by various groups including the U.S. Jet Propulsion Laboratory (JPL), the Canada Centre for Remote Sensing (CCRS), the Environmental Research Institute of Michigan (ERIM), European Space Agency (ESA), German Aerospace Centre (DLR), and others. These investigations utilized airborne platforms including the CCRS CV-580, the JPL DC-8, and space platforms, such as SEASAT, European Remote Sensing-1/2 (ERS), and the space shuttle Shuttle Imaging Radar-A to -C (SIR-A-C) radars. In 1996, Intermap Inc. fielded the world's first commercial implementation of a high-performance airborne interferometer, called STAR-3i whose development was funded by DARPA and used algorithms developed by the Jet Propulsion Laboratory. There have been many spaceborne SAR instruments flown in recent decades: ERS-1/-2 and ENVISAT satellites operated by the European Space Agency, JERS-1 and ALOS PALSAR operated by the National Space Development Agency of Japan, RADARSAT-1 and -2 operated by the Canadian Space Agency, SIR-C/X-SAR operated by the U.S., German, and Italian space agencies, TerraSAR-X and Tandem-X operated by Airbus and the Italian Cosmo/SkyMed operated by the Italian space agency. SAR constellations such at Copernicus (C-band) and urthecast along with spaceborne SAR instruments like the NASA/ISRO NISAR (L and S-bands), European BIOMASS (P-band), German/Brazilian MAPSAR (L-Band), and Japanese ALOS II (L-band) and will continue to spawn new IfSAR technology and applications. Commercially available IfSAR systems are detailed in the IfSAR operation sensors section found in this chapter.

IFSAR DEM MAPPING CONSIDERATIONS

When considering the acquisition of IfSAR generated DEMs or IfSAR data to generate DEMs, the following key IfSAR mapping considerations should be considered. Constructing accurate DEMs using radar interferometry requires precise knowledge of the IfSAR sensor platform position, attitude, and interferometric baseline as well as knowledge of the radar operating parameters. Phase stability and tracking of any phase changes not a result of topographic variations are also key considerations for IfSAR mapping systems. Here we outline elements common to all interferometric mapping systems and describe some of the tradeoffs between the various configurations. This treatment is geared for DEM users who need to understand how best to select a system that accommodates their mapping requirements.

IfSAR Sensor Configuration

There are two standard approaches to obtain the two SAR images necessary for IfSAR topographic elevation mapping: (1) Single-pass interferometry (SPI) and (2) Repeat-pass interferometry (RPI) illustrated in Figure 7.15.

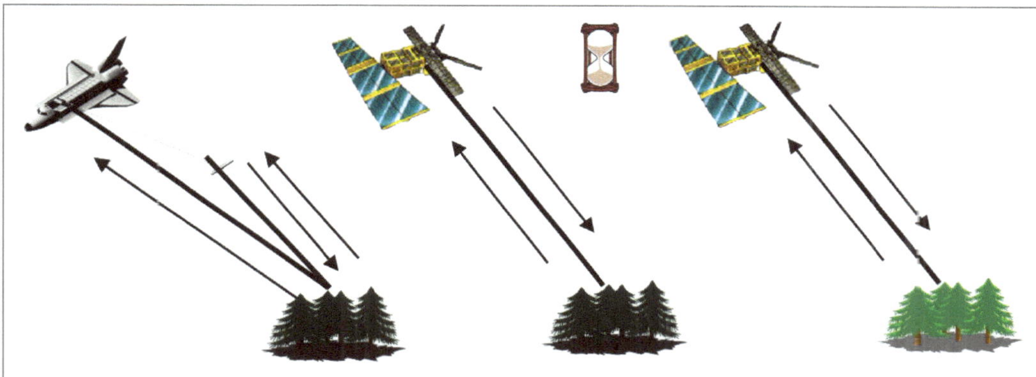

Figure 7.15. Interferometric radar topographic measurements are made with single-pass (SPI) and repeat-pass (RPI) systems. SPI configuration consists of two spatially separated antennas (usually on the same platform) arranged to collect echoes from the terrain simultaneously (left image—the SRTM mapping IfSAR with two antennas separated by a 60 m retractable boom (interferometric baseline). In RPI (middle and right images), the scene is imaged at least twice separated by an interval of time that may be seconds to years.

SPI configurations occur when interferometric observations are made with both antennas on the same platform[3]. In the case of RPI multiple observations are separated in time. RPI observations may be separated by as little as a fraction of a second or may be many years apart.

RPI is possible when the flight tracks are separated by less than the critical baseline length and when the surface has not changed enough to cause decorrelation. The time lapse between the two acquisitions creates a temporal interferometric baseline which typically ranges from minutes (for airborne systems) to many days (even years) representing the repeat orbit period of the satellite (e.g., 24 days for RADARSAT). Atmospheric changes between observations, particularly those attributed to tropospheric water vapor, can dramatically alter interferometric phase measurements (Goldstein, 1995), (Zebker et al., 1997). Spatial scales for atmospheric phase distortion effects are typically on the order of kilometers and scale in direct proportion to wavelength. The effect on interferometric height measurements can be a meter to hundreds of meters depending on the amount of distortion and the baseline length. Spaceborne SARs flying above the ionosphere (orbits above 300 km) also experience phase distortions due to changes in the ionosphere between repeat observations, however

[3] In principle, the antennas could be on separate platforms flying in formation (Zebker et al., 1994).

these changes typically have larger spatial scales of 10-100 kilometers and have a non-linear wavelength dependency. The non-linear wavelength dependence offers the possibility of removing ionospheric distortion by flying a multi-frequency system, like the way GPS corrects for the ionosphere using two frequencies.

IfSAR systems grouped by the platform type, airborne or spaceborne, and method of data collection, single-pass or repeat-pass, yields four major implementations (Table 7.2) with various relative strengths and weaknesses.

	Single Pass	Repeat Pass
Airborne	• Regional Coverage • Affordable Fine Resolution • Necessary Motion Compensation • Benign Atmospheric Effects • No Temporal Decorrelation • No Ionospheric Effects • Possible Need for Active Baseline Metrology	• Regional Coverage • Affordable Fine Resolution • Necessary Motion Compensation • Problematic Atmospheric Effects • Temporal Decorrelation Effects • No Ionospheric Effects • Difficult Track Repeatability • Difficult Baseline Estimation • Change Detection Possible • Increased Baseline Flexibility
Spaceborne	• Global Coverage • Costly Fine Resolution • Benign Motion Compensation • Benign Atmospheric Effects • No Temporal Decorrelation • Problematic Ionospheric Effects • Baseline Metrology Required	• Regional Coverage • Costly Fine Resolution • Benign Motion Compensation • Problematic Atmospheric Effects • Temporal Decorrelation Effects • Problematic Ionospheric Effects • Difficult Track Repeatability • Benign Baseline Estimation • Change Detection Possible

Table 7.2. Summary of Types of IfSAR Sensors.

Airborne Single Pass Interferometry (SPI)

Single pass aircraft systems (e.g., STAR and GeoSAR) are well suited for generating fine resolution regional scale DEMs. Single pass systems are best suited for generating high quality topographic maps to a specified absolute accuracy since they do not suffer from temporal decorrelation or from atmospheric phase distortion problems. Aircraft systems have a great deal of flexibility in scheduling data acquisitions, orientation of flight lines, and modes of operation.

Spaceborne Single Pass Interferometry (SPI)

Spaceborne platforms have the advantage of global and rapid coverage and accessibility. Increased coverage for spaceborne systems comes about from the combination of the faster velocity by a factor of 30 and the larger swath widths ranging from 50-500 km. Spaceborne systems also avoid airspace restrictions that make aircraft operations difficult in certain parts of the world. Typical baselines for spaceborne IfSAR system making topography measurements range from 10-1,000 meters. This poses a difficult metrology problem regardless of whether the antennas are connected to the same platform or are on separate platforms flying in formation.

Tracking phase instability of the radar hardware and antennas, which may go through a hundred degrees Celsius or more of temperature change in an orbital period, requires special hardware. SRTM and Airbus's Tandem-X satellites are the only spaceborne single-pass IfSAR systems flown to date.

Spaceborne-Repeat Pass Interferometry (RPI)

Repeat pass interferometric observations have their greatest utility in measuring surface deformation over wide areas for geophysical applications such as earthquake monitoring, volcano inflation and deflation, and glacier motion rather than topographic elevations. Nonetheless, repeat pass IfSAR (e.g., RADARSAT-1/2, TerraSAR-X and COSMO SkyMed) has been used to make topographic maps in many parts of the world, often exceeding the accuracy of the best topographic maps currently available in those regions. Besides tropospheric or ionospheric propagation effects to the interferometric phase (explained in detail in previous DEM User Manuals), the other major limiting factor to repeat-pass IfSAR topographic map generation is temporal decorrelation.

Polarimetric-Interferometric SAR (PolInSAR)

Another IfSAR sensor configuration called PolInSAR combines polarimetry (defined below) and IfSAR to derive bare ground elevations and canopy height measurements. Polarimetry deals with the full vector nature of polarized (vector) electromagnetic waves throughout the frequency spectrum from Ultra-Low-Frequencies (ULF) to above the Far-Ultra-Violet (FUV) (Boerner et al., 1998). A complete description of radar polarimetry is beyond the scope of this chapter; however, we mention a few relevant aspects about radar polarization are needed for our heuristic development polarimetric interferometry. The electric and magnetic fields comprising an electromagnetic wave are vector quantities meaning they have both a magnitude and direction (Figure 7.16). The vector nature of electromagnetic radiation gives rise to sensitivity to the orientation of an object from which it scatters and hence provides additional information about the scatterer.

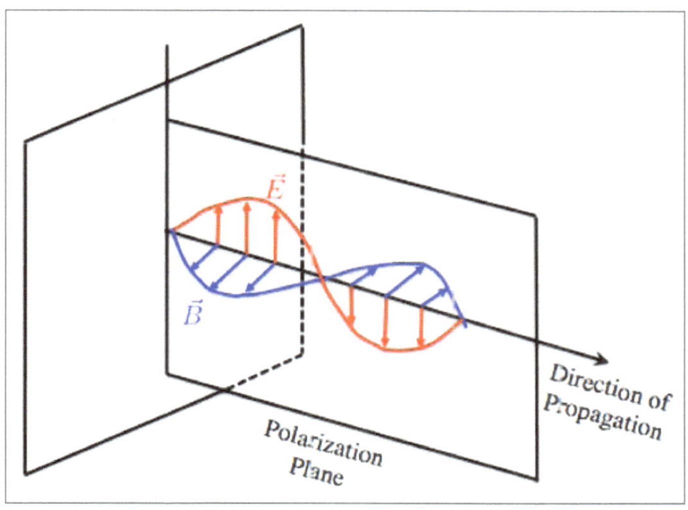

Figure 7.16. The direction of the electric field (and the magnetic field) lies in the plane perpendicular to the direction of propagation and defines the polarization of the wave.

Microwave polarization is characterized by the shape of the path that the electric field vector traces out in the plane perpendicular to the direction of propagation. The most general path is an ellipse that can be characterized by two parameters. Tilt measures the orientation of the ellipse with respect to the horizontal direction and ellipticity is a measure of the amount of deviation from a circular shape. Now although there are an infinite number of polarization states, it turns out that by transmitting and receiving data on two suitably chosen polarizations (a sufficient requirement is that they be orthogonal) it is possible to synthesize all other polarizations from a combination of these representative polarizations. The most common representative set used by most radars are horizontal polarization and vertical polarization denoted by H and V respectively. Polarimetric radars operate by transmitting H and V polarization on alternate pulses and receiving both H and V polarization on every pulse. Radars can also operate in restricted polarization combinations, say transmitting only H and receiving H and V or transmitting H and receiving H and transmitting V and receiving V. When the sensor is operated in a restricted polarization combination it is not possible to synthesize all possible polarizations any longer. The amount of backscatter depends on polarization because it is sensitive to the orientation of scatterers within a resolution element. Polarization diversity can also be used to identify dominant scattering mechanisms within a resolution element as well as determine quantitative information about the orientation of objects.

Collecting polarimetric data from two spatially separated SAR antennas generates scattering diversity that can be exploited to solve for information about the volume. Figure 7.17 illustrates polarimetric interferometric returns from two polarizations for the same resolution element. Note, the backscatter magnitude variation as a function of height in the volume and is different for the two polarizations and hence the volumetric correlation is different for different polarizations. Using models such as the random volume over ground (RVoG) or polarization coherence tomography (PCT) information about vegetation height, ground surface topography and canopy shape can be ascertained (Cloude, 2010).

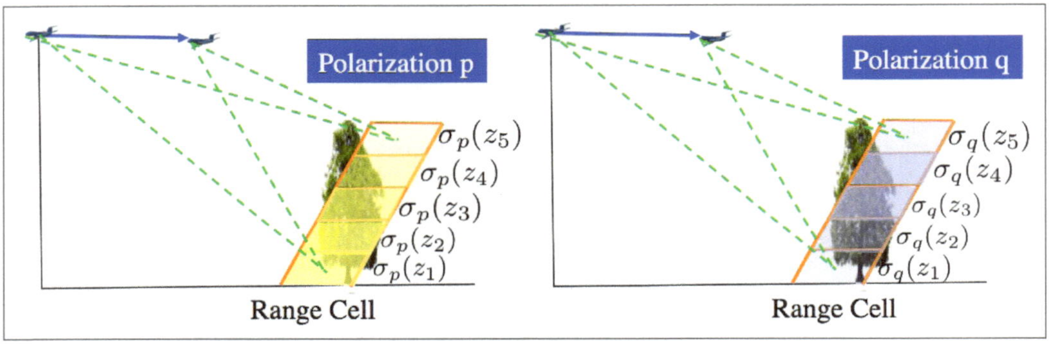

Figure 7.17. Different polarizations backscatter differently as a function of height within a canopy and hence additional information on the structure of the volume can be obtained using polarimetric interferometry.

The power of the PolInSAR technique is best illustrated with a tree example, Figure 7.18. A tree contained within a single resolution element, highlighted in shades of yellow in Figure 7.18, is imaged from two vantages separated by an interferometric baseline shown in blue. The various shades of yellow show the amount of backscattered energy that varies depending on height within the canopy, e.g., thicker portions of the canopy in the intermediate heights may scatter more than spare vegetation at the top of the canopy. Recall, that the interferometric phase is proportional to the differential path length from the two antennas which is illustrated in red, D_b, for the bottom of the canopy and in orange, D_t, for the top of the canopy.

Because the differential paths lengths are varying as a function of height within the canopy so is the interferometric phase. Note, if the baseline length goes to zero then the path lengths are identical regardless of location within the canopy and there is no interferometric sensitivity to the volume. The interferometric signal from a resolution element is a backscatter weighted combination with phase varying as a function of height to the canopy elements contained in the cell. Thus, radar interferometry has sensitivity to the volume but with only one observation per resolution element has limited visibility to the internal structure within a range cell.

We see that interferometric measurements from a resolution element essentially consists of two components: First, a scattering component that depends on the strength of the reflected signal that varies as a function of height within a resolution element, and second a geometric component that depends on the interferometric baseline and determines how quickly the interferometric phase varies as a function of height within a resolution element. Thus, to obtain additional information about the vertical structure within a

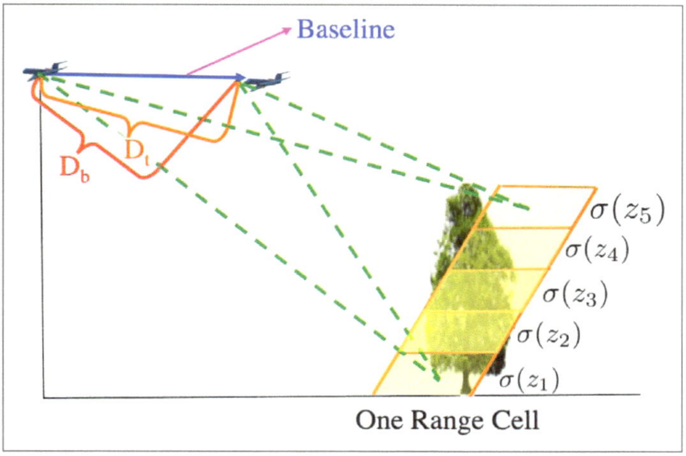

Figure 7.18. A tree contained within a single radar resolution cell is imaged from two vantages separated by a baseline shown in blue. The amount of backscattered signal and phase are a function of height within the tree

resolution element we need to either increase scattering diversity or the geometric diversity. Increasing geometric diversity is accomplished by imaging the scene with multiple baselines. If enough multi-baseline observations are made it is then possible to use tomographic imaging techniques to get fine vertical resolution information of the backscatter. Alternatively, one can obtain backscatter diversity by varying the polarization of the electromagnetic radiation.

We present an example using data from the NASA/JPL L-band UAVSAR radar system collected in La Amistad National Park, Costa Rica in February of 2010 (Hensley et al., 2014). UAVSAR is a NASA/JPL L-band fully polarimetric synthetic aperture radar employing an electronically scanned array whose primary design goal was to enable robust repeat pass radar interferometric measurements of deforming surfaces either from natural or anthropogenic causes. The radar is housed in a pod mounted to the fuselage of a Gulfstream III jet. Nominally, the aircraft flies at an altitude of 12.5 km and maps a 22 km swath with incidence angles ranging from 25° to 60°. The 80 mHz range bandwidth results in single look complex (SLC) imagery with range and azimuth resolutions of 1.66 m and 1 m respectively. Electronic steering of the antenna is tied to the inertial navigation unit so that consistent pointing is achieved regardless of the platform yaw. The platform was modified to include a platform precision autopilot (PPA) that allows the aircraft to fly a specified trajectory within a 5 m tube. This enables UAVSAR platform to fly a series of flight lines with well prescribed interferometric baselines.

On February 8, 2010 UAVSAR collected a series of repeat pass lines with a variety of physical baselines. Data were collected from coast to coast to cover a wide range of biomes and terrains. Figure 7.19 shows polarimetric radar imagery and a multi-color correlation image for a site in La Amistad National Park. Flat open areas have high correlations as shown in the right image as there is no volumetric correlation whereas forested areas exhibit lower lowers depending on the amount of volumetric correlation at that polarization. In addition to the volumetric correlation because the data were collected in a repeat pass mode there is also

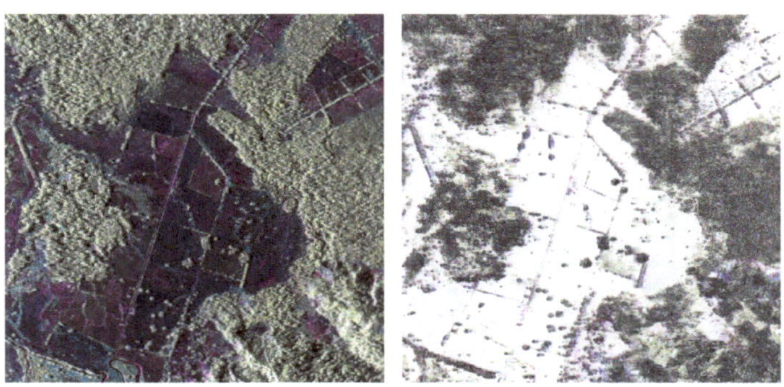

Figure 7.19. Polarimetric imagery and interferometric correlations for a selected portion of the La Amistad National Park. Blue denotes hh+vv, red hh-vv and green is hv collected on February 8, 2010.

temporal decorrelation. The smaller the physical baseline the bigger the impact temporal decorrelation has on tree height estimate. Temporal decorrelation biases tree height estimates high. Several methods have been proposed to mitigate the impacts of temporal decorrelation, however their efficacy is situation dependent.

Figure 7.20 shows tree height estimates for 6 physical/temporal baseline combinations. Physical baselines varied from 60-100 m and the temporal baselines varied from 0.7-1.9 hrs. The tree height estimates are very similar, except for Baseline 5 (60 m, 0.7 hr) which is biased high compared to the other estimations. This is because Baseline 5 has the smallest baseline, making it more sensitive to temporal correlation (that degraded the tree height estimates).

Figure 7.21 shows histograms of the tree estimates. The histogram of Baseline 5 is the yellow curve that has an extended tail for larger tree heights that is a result of the increased impact of temporal decorrelation.

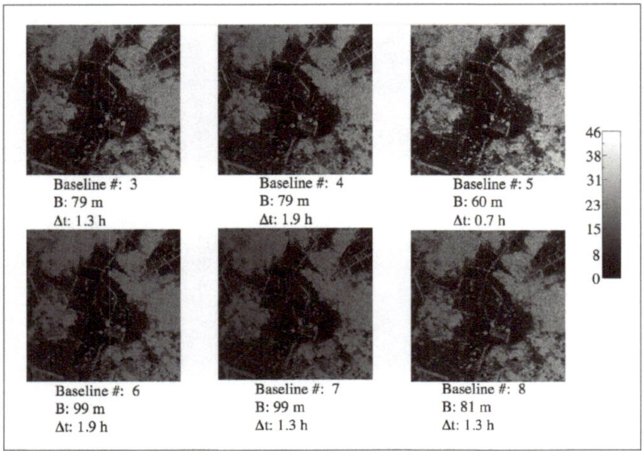

Figure 7.20. Tree height estimates for six spatial/temporal baseline combinations. Tree height estimates are very similar except for Baseline 5 where increased temporal decorrelation results in increased tree heights.

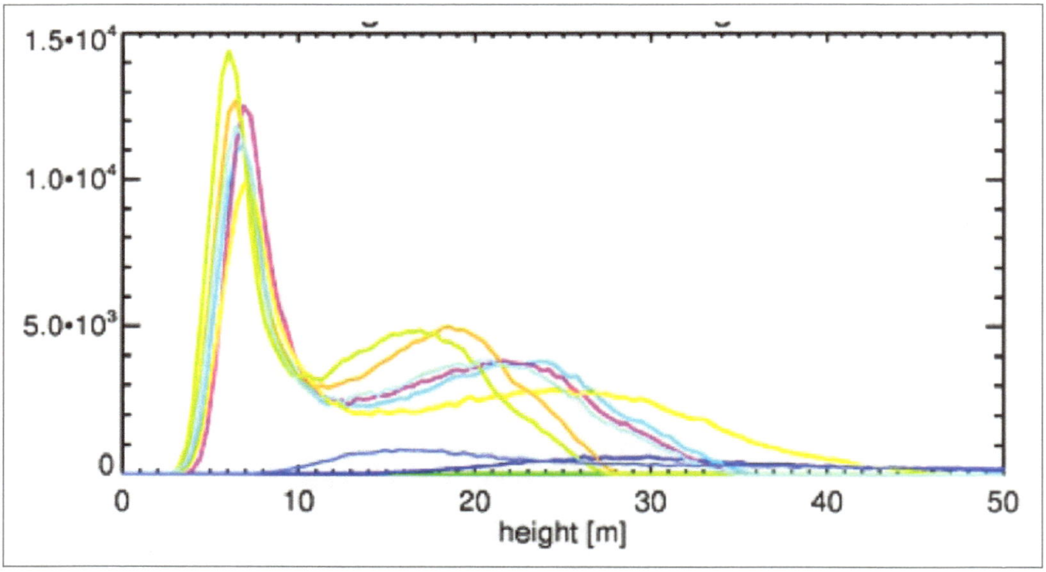

Figure 7.21. Histogram of the tree height estimate for the six spatial/temporal baselines. The yellow line corresponds to the histogram of the baseline which has an extended in tail for larger tree heights resulting from the increased effect of temporal correlation.

Platform Position

While it is straightforward to measure distances (e.g., phase information needed to derive topographic height measurements) using IfSAR techniques, precise knowledge of the location of the SAR antennas is required, consequently requiring the need of two additional technologies GPS (or Global Navigation Satellite System—GNSS) and inertial navigation system (INS), explained in Chapter 12. Position measurement with the advent of GPS and differential GPS (DGPS) systems are very good; INSs enables very accurate motion measurement on a broad range of time scales. Kinematic GPS used with either airborne or spaceborne platforms achieves decimeter absolute position accuracy with a 2 Hz sampling rate. Although 2 Hz sampling of the motion works for satellites that have relatively smooth trajectories, faster sampling of 20 Hz or greater is required for airborne IfSAR applications. Inertial measurement unit (IMU) systems provide the faster motion and attitude update rates needed for motion compensation. Blending of kinematic DGPS and IMU data is an increasingly common method of optimizing position and attitude data to have a high effective sampling rate and excellent absolute position accuracy necessary for airborne and spaceborne applications.

For interferometric applications, it is knowledge of the antenna locations that is essential and required for baseline determination and motion compensation. Rarely are the motion metrology systems mounted to the center of the antenna; therefore, platform attitude measurements coupled with measurements of the antenna location relative to the motion sensors is needed. Absolute angle determination with accuracy of approximately a few thousandths of a degree is off-the-shelf technology today with tightly coupled IMU/GPS systems, significantly improving the critical determination of the baseline orientation angle. For spaceborne platforms, star trackers provide absolute attitude measurements with 1-10 arc second accuracy. If faster update rates are needed then star tracker measurements are coupled with IMU measurements.

It is not always possible to design IfSAR systems such that the antennas remain fixed relative to the motion measurement system. When this is the case additional metrology devices are required to track antenna motion with respect to the platform. Very few systems thus far have been fitted with active baseline metrology systems but those that have used a combination of optical ranging and target tracking devices. Update rates for these systems are matched to the expected motion of the antenna relative to the platform to insure proper baseline determination.

Spaceborne radars, until recently, have relied upon Doppler tracking for orbit determination Doppler tracking can determine satellite positions with accuracy from 10 cm to 100 m depending on the orbit and the amount of tracking data available. The highest accuracy Doppler derived orbit position data may fall short of the accuracy requirements needed for repeat pass interferometry (RPI) topographic mapping. Almost all modern spaceborne radar systems are equipped with GPS receivers, vastly improving the orbit determination. Orbital positions with a 3-10 cm accuracy are routine. For those baselines that must be known to a centimeter or even sub-centimeter accuracy cannot be determined strictly using Doppler tracking data. Ground truth in the form of existing DEMs or radar identifiable ground control points are used to determine the baseline for RPI topographic applications when adequate metrology is not available.

Platform Calibration

IfSAR data collection systems are operated in a similar manner to other survey sensors. Sensor error models are created through a calibration process and used within a production process. Further, data acquisition includes the collection of redundant data that are used to validate and if necessary refine the calibration parameters.

IfSAR calibration is a process with a goal that is not unlike the concept of calibrating a photogrammetry camera. As with these metric cameras and their interior orientation parameters (focal length, lens model, and perspective center offset), IfSAR sensors require an understanding of certain physical behaviors—primarily timing and impulse response of the transmitted SAR signals. As is the case in photogrammetry, calibrating cameras utilizing the co-linearity equations, IfSAR systems utilize a set of performance equations based on equations that determine a scattetter's location from the interferometric observables. With a calibrated IfSAR system, a similar collection strategy to photogrammetry is typically utilized. Thus, one of the keys to acquiring accurate and reliable terrain elevation data from IfSAR platforms lie in the knowledge of the radar antenna locations essential for interferometric baseline determination and motion compensation.

The concept of tie point and ground control point used in block adjustments are employed as "parallel profiles", tie lines, and ground control points. These are used to establish the equivalent of the exterior orientation parameters for each flight. The ultimate accuracy budget for the IfSAR is then based on a fully calibrated sensor with a loosely coupled navigation solution combining GPS and INS streams, all measured and controlled by profiles and control points. The IfSAR product vertical accuracy has been tested and independently verified many times and consistently achieves 3 m vertical RMSE for the DEMs in low slope unobstructed areas, and 3 m CE95 for the ORIs without the use of in-scene ground control. Note, the height precision for IfSAR mapping systems is often considerably better than the height accuracy.

IfSAR DATA POST PROCESSING

Processing and post processing of IfSAR data to derive commercial DEMs (DSM and DTM) affect the accuracy and quality of final map products. The division between processing and post processing is somewhat arbitrary, however for the purposes of this chapter post processing will be defined to begin after the interferometric phase is transformed to a position and elevation measurement. Commercial companies utilize proprietary software to generate the digital elevation models from interferometric data. Many aspects of IfSAR post processing are like standard photogrammetric or lidar post processing. The type and amount of post processing is application dependent and is tailored to meet specific user requirements. The most common IfSAR post processing steps are described.

IfSAR DEM Void Fill

Data gaps resulting from too much decorrelation, e.g., low backscatter from calm water, layover or shadow regions, may still be present after mosaicking is complete. Most commercial vendors fill these data gaps to provide fully populated DEMs. There are three basic methods used to fill gaps in the final DEM. Data may be specially acquired over the gap regions, data from alternate sources may be used, or analytical methods for filling gaps may be employed. Data from previous data collections or other data sources (e.g., USGS DEMs) may also be used to fill in gaps in the data. These data in general may not meet the same accuracy or resolution requirements of the IfSAR data, however alternate source data often proves adequate for small gaps. By incorporating these alternate sources during the mosaicking process, a seamless final product is achieved, provided there has not been extensive change to the topography between the time of the IfSAR data collection and the time the alternate source data were acquired. Specially acquired data to fill residual gaps is the most expensive means of filling gaps in the DEM and may require a delay in product delivery to allow for data collection and processing. This option has the benefit of maintaining uniform quality data throughout the DEM but is usually only warranted when there are excessively large data voids in critical portions of the DEM. A myriad of algorithms is used to analytically fill gaps in topographic data. Surface fitting, kriging methods, and polynomial interpolators are among the most commonly employed algorithms for hole filling. The choice of algorithm depends on the size of holes to be filled and the intended application for the DEM.

IfSAR Data Editing

Data editing is used to correct errors in the DEM detected during the quality control process or to manipulate height values so that they conform to a user prescribed mapping standard. Unwrapping errors occur when an incorrect multiple of 2π is added to the interferometric phase measurement. This results in the IfSAR elevation measurements being too high or too low by a multiple of the ambiguity height, a quantity that is determined from the interferometric system parameters and mapping geometry. Unwrapping errors are detected and edited from the DEM by searching for height discontinuities that are multiples of the ambiguity height. Spikes and wells are isolated points in the DEM whose elevation differs from surrounding heights by an unphysical amount. These points are edited from the final DEM and marked as data voids or filled in using a combination of the surrounding elevation values. Some map products (e.g., DTED products) require that water bodies have single elevation value. IfSAR DEMs over water are usually noisy and have intrinsic height variation that depends on the amount of thermal noise. Water body editing consists of identification of the water body and setting the elevation to the desired value. Water body identification using IfSAR data is a difficult problem and is a sizable portion of the editing process.

IfSAR Vegetation Removal

Applications that require bare surface DEMs (e.g., DTM) need to have IfSAR reflective surface elevation measurements corrected to bare surface elevations. Correction of reflective surface elevation measurements is called vegetation removal. Vegetation removal involves identifying vegetated regions and then correcting the elevation measurements to the bare surface. Identification of vegetated region uses combination of imagery, elevation measurements and correlation data. Correction to bare surface elevations may employ algorithms like lidar and photogrammetric sensors where elevation measurements that penetrate to the bare surface are used in combination with surface fitting algorithms to make elevation adjustments. More sophisticated algorithms that use the image brightness, correlation and elevation measurements along with a model of the vegetation are now being employed by some IfSAR sensors for vegetation removal. Additionally, polarimetric-interferometric or multiple baseline interferometry as described above can directly measure the vegetation height and used to correct the data for bare surface elevations. Correction of IfSAR elevations to bare surface elevations in dense urban environments can be difficult.

IFSAR QUALITY CONTROL

Quality control is a task or series of tasks that scrutinizes all, or a sample, of the IfSAR products issued during or at the end of the IfSAR map generation process to ensure that the final product meets or exceeds requirements. This scrutiny involves a combination of review, inspection and quantitative measurements, against well-defined criteria that are outlined in references. Additional quality controls determined by the data provider are used for other map products such as SAR imagery and other IfSAR specific derived products. Many data providers certify their QA/QC processes to the ISO standard. An overview of some of the standard qualitative and quantitative quality control procedures is presented in the following sections.

Visual Accuracy Checks

IfSAR visual quality control begins by looking for gross processing errors associated with incomplete phase unwrapping, large spikes and wells, large tilts on water bodies and features that seem out of place. The maximum and minimum elevations in the IfSAR DEM are compared to the maximum and minimum values represented by contours or spot elevations available on the most recent available map. A check for completeness in the project size and for continuity along mosaic seams and data gap boundaries is also made. Overlay the IfSAR map products on available map data to check if geo-referencing is correct Spot check selected pixel values such as corner and center pixel values against heights on published maps. Use a DEM viewing workstation with the appropriate software tools to aid in the identifications of blunders such as spikes and holes. Blunders are generally identified through a combination of color banding of elevation contours, stereoscopic viewing using anaglyphic filters, shaded relief enhancements and use of histograms. Artifacts identified during the visual accuracy checks are documented and quantified noting the location and source, for example: terrain masking, radar shadow, DEM sub-patch boundary, land/water boundary, vegetated regions, wind motion, or other factors.

Ground Truth

Quantitative assessment using ground truth data is an important component of the quality control procedure. A selected set of ground control points, typically greater than 20 for any region or sub-region to be tested, is compared with the corresponding IfSAR generated height measurements. For a rectangular area that is believed to have uniform positional accuracy, check points may be distributed so that points are spread at intervals of at least 10% of the diagonal distance. At least 20% of the points are to be in each quadrant. The independent source of higher accuracy shall be of the highest accuracy feasible and practical to evaluate the accuracy of the IfSAR data. To make a rigorous accuracy assessment usually requires truth data that is three times more accurate than the product tolerance. Although vertical checkpoints are often not well-defined, each horizontal checkpoint must be well defined (see Chapter 3 of this manual, and section 7.9 of ASPRS, 2014). A well-defined point represents a feature for which the horizontal position is known to a high degree of accuracy and for which the absolute position with respect to the map product geodetic datum is known. Kinematic GPS measurements taken along major highways and trunk roads provide excellent data sets for quality control and accuracy assessments. Kinematic GPS transects should be collected away from urban areas where multi-path in the radar and kinematic GPS data is often problematic.

Control points are usually derived from either kinematic GPS measurements or a high accuracy photogrammetric or lidar DEM with relative height accuracy at least three times better than the expected IfSAR mapping accuracy.

IFSAR DATA DELIVERABLES

Data deliverables can be divided into three categories, pre-project deliverables, post-project deliverables, and map products including DEMs, imagery and other derived products. Depending on whether data is specifically collected to support a project or purchased from archived sources, not every deliverable category is applicable. The main goal of this chapter is to provide the DEM user with a list of the type of deliverables that may be available throughout the course of a project. Selection of those deliverables most useful for a project will depend on the type of data needed and the application.

COST CONSIDERATIONS

As with most mapping techniques, the cost of the DEM products depends on many factors. In the case of IfSAR, some of the factors are accuracy, project size, geographic location, ground post spacing, terrain type and density, and type of the vegetation cover. Current reported prices for IfSAR generated DEMs are grouped into two categories: (1) Project specific, where data is collected by the data provider to meet a specific customer mapping requirement, and (2) Archival data where data is purchased from previously collected (and possibly processed) IfSAR mission data.

For project specific IfSAR, DEM prices range from $35/km^2 to $80/km^2 depending upon the area size, site location, terrain ruggedness, foliage density and platform. Archival data prices are lower than project specific prices and depend on whether the data must be reprocessed or can be used as previously processed. Archival data purchases are more common today with the advent of cloud storage and SaaS (Software as a Service) options. Data archive prices range from $1/km^2 to $10/km^2 for DEMs and typically require a minimum area purchase.

IFSAR PRESENT OPERATING STATUS

DEM generation from the IfSAR mapping technique is offered by a small number of commercial companies in comparison to the plethora of aerial photography and lidar companies available today. Commercial companies offer DEM users with options to purchase data from an archive (e.g., Intermap airborne IfSAR and Airbus spaceborne world DEM) as well as offer project specific data collections (e.g., Furgo or Intermap). Commercially available IfSAR providers are listed in Tables 7.3 and 7.4.

Sensor	Country	Λ	POL	DEM	GSD	RMSE	IP	Data
Star	Canada	X	HH	DSM/DTM	5 m	0.5-6 m	S	P/A
GeoSAR	US	X	VV	DSM/DTM	3-5 m	0.5-5 m	S	P
	US	P	HH/VV	DSM/DTM	5 m	2-6 m	S	P
BRADAR	Brazil	X	Quad	DSM	5 m	10 m	S	P
	Brazil	P	Quad	DSM	5 m	10 m	S	P
Glisten	US	Ka	HH	DSM	3 m	0.15-2 m	S	P

Table 7.3. Summary of the key parameters of airborne IfSAR systems. Table abbreviations: wavelength (λ); polarization (POL); ground sample distance (GSD); elevation vertical accuracy (RMSE); IfSAR Pass is IP; Single is S; Repeat is R, data collection type project (P) and archived (A).

Sensor	Country	Λ	POL	DEM	GSD	RMSE	IP	Data
Tandem-X	Germany	X	Quad	DSM/DTM	12 m	10 m	R	P/A
COSMO-SkyMed	Italy	X	Quad	DSM	5-30 m	10 m	R	P
SRTM	US	C	VV	DSM	30-90 m	8 m	S	A

Table 7.4. Summary of the key parameters of select spaceborne SAR systems suitable for deriving elevation data via IfSAR methods. Table abbreviations: wavelength (λ); polarization (POL); ground sample distance (GSD); elevation vertical accuracy (RMSE); IfSAR Pass is IP; Single is S; Repeat is R, data collection type project (P) and archived (A).

Many commercial companies now have software that allows users themselves to generate products from raw data. The quality of these products depends on the type of data collection (single or repeat-pass), the number of interferometric pairs available over the region of interest, the amount of decorrelation and atmospheric distortion, baseline length for the interferometric pairs, and the wavelength of the operating sensor(s). The principal companies marketing these technologies and products are Intermap, Fugro, Bradar, Airbus, e-Geos, and, MDA. In addition to commercial companies that market software there are several freeware packages (e.g., ROIPAC or ISCE) that can be used to generate DEMs from spaceborne IfSAR data for non-commercial purposes. The most accurate and reliable sources of IfSAR DEM data are generated using single-pass systems.

Airborne IfSAR Systems

STAR—Intermap Technologies Inc.

Intermap is a long-time provider of airborne radar mapping products. The foundation of the company's success lies in its innovations in 3-D digital modeling, geometric imaging, data acquisition, and data aggregation. Today, Intermap is a global geospatial solutions leader focused on improving the ways people, businesses, and governments use location-based information. Intermap does this through its products and services, which include the NEXTMap® five-meter elevation datasets, NEXTMap World 30™, AdPro®, LinkPro®, RiskPro™, Geospatial Services, and Data Fusion. Foundational to these products and services is the NEXTMap 3-D digital models, created using IfSAR technology and integrated third-party data.
Intermap's IfSAR systems (Figure 7.22. STAR-3 and STAR-6) configuration consists of a synthetic aperture radar sensor with two X-band radar antennas, a global positioning system (GPS) and inertial navigation system (INS) components. The radar pulse signals are collected from two radar antennas and compressed by digital match filtering to extract the interferometric phase information that is then processed to derive elevation data and radar imagery. The GPS/INS system provides highly accurate position and orientation information without the need for in-scene ground control. Positioning technologies, such as GPS and INS coupled with differential GPS (DGPS) post processing software have reduced ground control survey costs, while simultaneously reducing the need for ground control surveys by making direct sensor orientation possible. Combining the SAR interferometric information and the GPS/INS navigation results allow for the creation of a highly accurate ortho-rectified radar image and digital elevation model, without the need for the deployment of in-scene ground control. Table 7.5 lists the system parameters of the STAR-3/6 radars.

Figure 7.22. The STAR-3 (left image) and STAR-6 (right image) radars are flown on a Learjet 36 with the antennas mounted to an invar frame contained in the large dome on the underside of the aircraft. (Images courtesy of Intermap Technologies Inc.).

Parameter	STAR-3	STAR-6
Peak Transmit Power	8 kW	8 kW
Bandwidth	67/135 mHz	67.5/135 mHz
Center Frequency	9.55 mHz	9.55 gHz
Baseline Length	0.92 m	1 m
Baseline Tilt Angle	0°	0°
Platform Altitude	10,000 m	10,000 m

Table 7.5. STAR-3 and -6 System Parameters.

Antennas are mounted to a solid invar frame with a ~1 m physical separation that provides a temperature and mechanically stable interferometric baseline. The frame is attached to the bottom of the aircraft on an azimuth steerable pedestal thus allowing for mapping on either side of the aircraft. Motion measurement data is obtained using a combination of DGPS data collected using an Ashtech Z-12 receiver and a Honeywell 770 IRU. The IMU is mounted to the same invar frame as the radar antennae to minimize lever arm errors between the IMU and antenna phase centers.

Flight plans are generated using a customized flight planning tool and uploaded to the radar prior to flight. The radar operates in the Ping-Pong mode with the interferometric channels sharing a single receiver. Radar signal data and motion data are recorded directly to an onboard disc system. Data are processed on an array of 3-D workstations, quality control checked prior to mosaic and post processing into final products delivered to the customer. Post processing steps include interferometric processing using proprietary software that included averaging of multiple IfSAR flight lines (from overlapping flight lines and tie lines) where possible, and filtering of the interferogram to reduce phase noise using a Gaussian filter. Reference ground elevation points are used as seed points to assist with phase unwrapping, to provide elevations that link ground elevations to the interferogram. This process creates IfSAR data that requires merging and editing to produce the final NEXTMap DSM and DTM products with a 5m ground sampling distance (GSD) in grid format using horizontal datum NAD83, the latest geoid model, and georeferenced to latitude and longitude in decimal degrees and seconds. Table 7.5 provides a summary of the main system parameters.

The DEM data editing is employed to produce the commercial NEXTMap products. Editing included the following steps. All voids in the data were populated with tie line and/or overlapping flight-line data where possible, and recalculated using ancillary elevation data (data are tied and warped to the DSM bare ground), or interpolated where no ancillary data were available. All water bodies were smoothed to the elevation of the surrounding shoreline. For example, lakes of 400 m or more in length were flattened and set to a constant elevation. Single line streams up to 20 m in width were delineated using 3-D vectors, whereas double line streams greater than 20 m wide were collected as 3-D polygons. Both types of streams were stepped down in height to match the surrounding water levels to maintain monotonicity (e.g., water flowing downstream). DSM and DTM examples are provided in Figure 7.23.

Additional processing was applied to create the DTM where elevations of features such as vegetation and buildings were removed, based on a set of edit rules described in the Intermap Handbook (Intermap, 2011). Intermap offers a variety of IfSAR derived DEMs and SAR image products around the globe, under the NEXTMap product name. The NEXTMap products vary in accuracy and specification worldwide as outlined in Table 7.6 on page 180.

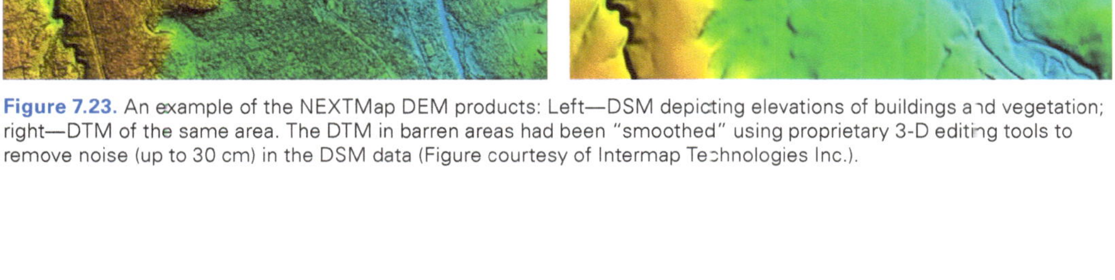

Figure 7.23. An example of the NEXTMap DEM products: Left—DSM depicting elevations of buildings and vegetation; right—DTM of the same area. The DTM in barren areas had been "smoothed" using proprietary 3-D editing tools to remove noise (up to 30 cm) in the DSM data (Figure courtesy of Intermap Technologies Inc.).

Data Type	North America			Puerto Rico, Jamaica, Central/South America		
	Resolution	Accuracy	Coverage Area*	Resolution	Accuracy	Coverage Area*
DSM/DTM	5 m	<1 m LE90% (vertical)	40%	5 m	<3 m LE90% (vertical)	40%
		1-3 m LE90% (vertical)	40%		3-5 m LE90% (vertical)	40%
		>3 m LE90% (vertical)	20%		>5 m LE90% (vertical)	20%
ORI	0.625 m	3 m CE90% (horizontal)	4%	—	—	—
	1.25 m	4 m CE90% (horizontal)	94%	1.25 m	4 m CE90% (horizontal)	90%
	2.50 m	5 m CE90% (horizontal)	2%	2.50 m	5 m CE90% (horizontal)	10%
Data Type	Western Europe			Democratic Republic of Congo		
	Resolution	Accuracy	Coverage Area*	Resolution	Accuracy	Coverage Area*
DSM/DTM	5 m	<1 m LE90% (vertical)	40%	5 m	<3 m LE90% (vertical)	40%
		1-3 m LE90% (vertical)	40%		3-5 m LE90% (vertical)	40%
		>3 m LE90% (vertical)	20%		>5 m LE90% (vertical)	20%
ORI	0.625 m	3 m CE90% (horizontal)	4%	—	—	—
	1.25 m	4 m CE90% (horizontal)	94%	1.25 m	4 m CE90% (horizontal)	90%
	2.50 m	5 m CE90% (horizontal)	2%	2.50 m	5 m CE90% (horizontal)	10%
Data Type	Malaysia, Parts of Indonesia, and Australia			Philippines, Solomon Islands, Vanuatu, and Parts of Indonesia		
	Resolution	Accuracy	Coverage Area*	Resolution	Accuracy	Coverage Area*
DSM/DTM	5 m	<1 m LE90% (vertical)	40%	5 m	<3 m LE90% (vertical)	40%
		1-3 m LE90% (vertical)	40%		3-5 m LE90% (vertical)	40%
		>3 m LE90% (vertical)	20%		>5 m LE90% (vertical)	20%
ORI	0.625 m	3 m CE90% (horizontal)	4%	—	—	—
	1.25 m	4 m CE90% (horizontal)	94%	1.25 m	4 m CE90% (horizontal)	90%
	2.50 m	5 m CE90% (horizontal)	2%	2.50 m	5 m CE90% (horizontal)	10%

Table 7.6. STAR-Derived NEXTMap Mapping Products Specifications.

GeoSAR—Fugro EarthData Inc.

Fugro EarthData is a global remote sensing, mapping, and GIS services organization that provides customized products and solutions to support a wide range of land-use and natural resource management activities. GeoSAR, their primary radar remote sensing mapping system, came about from a program initiated to develop a dual frequency airborne radar interferometric mapping instrument designed to meet the mapping needs of a variety of users in government and private industry. Program participants consisted of the Jet Propulsion Laboratory (JPL), EarthData International (previously Calgis, Inc.), and the California Department of Conservation with funding provided initially by DARPA and subsequently by the National Geospatial-Intelligence Agency (NGA). The program was begun to address the critical mapping needs of the California Department of Conservation to map seismic and landslide hazards throughout the state. GeoSAR operates X-band and P-band radars designed to measure the terrain elevation at the top and bottom of the vegetation canopy. The GeoSAR radar flies onboard a Gulfstream-II aircraft and is a dual-frequency (P- and X-band) interferometric Synthetic Aperture Radar (SAR), with HH and HV (or VV and VH or full quad) polarization at P-band and VV polarization at X-band (Hensley et al., 2001) that collects data on both sides of the aircraft simultaneously. The radar hardware onboard the Gulfstream-II aircraft is supplemented with a Laser-Baseline Measurement System (LBMS) which provides real-time measurements of the antenna baselines in a platform based coordinate system that is tied to the onboard Embedded GPS Inertial (EGI) units. The onboard data collection via the Automatic Radar Controller also records navigation data from the aircraft's GPS/INU system, the laser-based antenna-baseline measurement system, and raw signal data from X- and P-band radars.

The P-band antenna system is mounted in the port and starboard wingtip pods providing a long antenna-baseline of about 20 meters. X-band antennas are mounted in pairs under the wings with an antenna-baseline of 2.5 meters. The system can collect data in continuous strips greater than 500 km in length. Also, upgraded at the same time were the analog-to-digital converters to provide higher fidelity radiometric and height mapping products. GeoSAR maps a 20 km swath by collecting two 10-km swaths on the right and left sides of the plane as shown in Figure 7.24.

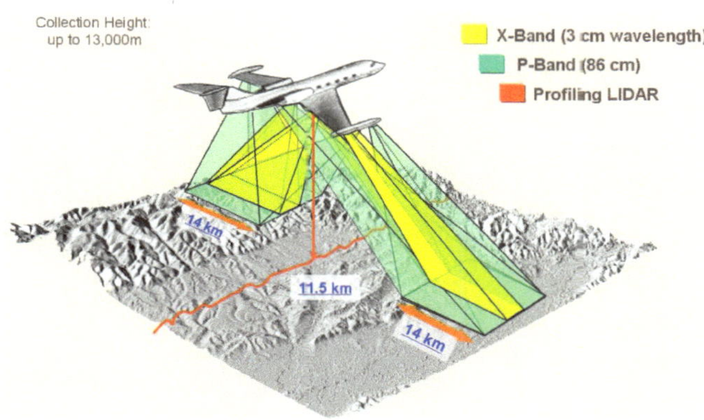

Figure 7.24. The GeoSAR radar flies aboard a Gulfstream-II modified jet at high altitude and high-speed. This configuration makes it efficient for wide-area mapping applications acquiring data on both sides of the plane simultaneously in X-band and P-band at a rate of approximately 300 km^2 per minute, per band (~400 MB per second).

Radar operations are controlled by a command disk generated preflight by the Mission Planning Software. Real-time data collection is controlled in-flight via an Automatic Radar Controller (ARC) that sets data collection windows, performs Built-In Tests (BITs) before and after each data take, and automatically turns the radar on and off during data acquisition. Raw radar data is recorded on high-density digital tape recorders for subsequent, post-flight processing. The processor incorporates several algorithms used to remove radio frequency interference (RFI) at P-band, focus the P-band data with its large synthetic aperture, and for regridding and mosaicking the data. The expected map accuracy of the X-band system is sub-meter in bare surface regions and 1-4 m in vegetated areas through a combination of X-band and P-band data. The system has undergone several upgrades since it began operational service. First the system incorporated a smaller and more accurate second generation LBMS to track the interferometric baselines, greatly increasing the mapping swath at X-band. Secondly, the system was augmented with a lidar mapping system that collects nadir pointing elevation data with 15-20 cm mapping accuracy. Lidar data are used to provide very accurate control for large area topographic mapping projects. Upgrades to the digital system replaced the high-speed tape based storage system with new disk based storage devices, thereby increasing the amount of data that could be collected in a single flight line. Table 7.7 provides a summary of the main system parameters.

Parameter	P-Band Value	X-Band Value
Peak Transmit Power	4 kW	8 kW
Bandwidth	80/160 mHz	80/160 mHz
Center Frequency	350 mHz	9.755 gHz
Baseline Length	20 m and 40 m	2.6 m or 5.2 m
Baseline Tilt Angle	0°	0°
Platform Altitude	5,000 m to 13,000 m	5,000 m to 13,000 m

Table 7.7. GeoSAR System Parameters.

Figure 7.25 provides an example of the DEMs generated using the X-band and P-band IfSAR over Monarch Grove, CA. The eucalyptus tree stand highlighted in the imagery shows up extremely well at X-band but is nearly undetectable in the P-band data because of the increased penetration into the canopy.

Figure 7.25. GeoSAR X-band DSM (left) and P-band (right) DEM data.

BRADAR

Bradar Industrial S.A. is a technology based company, specialized in remote sensing and radar aerial and ground surveillance. Bradar is engaged in the development, construction, service, and sale of radars and other sensors for use in defense, and security and surveillance sectors. It also offers remote sensing solutions for mapping and monitoring the earth surface. The company was founded in 1984 and is based in São José dos Campos, Brazil. Bradar Industrial S.A. operates as a subsidiary of Embraer S.A. The OrbiSAR-1 was designed for this purpose. This system generates high-precision altimetry and planimetric maps through the X and P radar bands (Figure 7.26). Table 7.8 gives a summary of the main system parameters.

Figure 7.26. Illustrates the X and P bands signals. The X-band waves are reflected and scattered through the treetops, while the P-band penetrate the foliage and is reflected by the ground and the dense trunks, allowing the soil mapping under the vegetation.

Parameter	P-Band Value	X-Band Value
Peak Transmit Power	4 kW	8 kW
Bandwidth	80/160 mHz	80/160 mHz
Center Frequency	350 mHz	9.755 gHz
Baseline Length	20 m and 40 m	2.6 m or 5.2 m
Baseline Tilt Angle	0°	0°
Platform Altitude	5,000 m to 10,000 m	5,000 m to 10,000 m

Table 7.8. BRADAR System Parameters.

GLISTIN Ka-Band Ice Topography Mapper

As interferometric mapping technology evolves, systems are emerging to meet specific operational or scientific needs. An example of this is the development of the Ka-band (8 mm wavelength) airborne Glacier and Land Ice Surface Topography Interferometer (GLISTIN) mapping radar developed by NASA/JPL as part of the International Polar Year (IPY) to map glacier and ice topography (Moller, et al., 2011, Hensley, 2016).

The GLISTIN instrument was developed as an adaptation of the NASA/JPL UAVSAR L-band radar that was designed for repeat pass radar observations of deforming surfaces (Hensley et al., 2008). This radar is flown in a pod mounted underneath the fuselage of a Gulfstream III aircraft as shown in Figure 7.27 on page 184. The lower insert shows the mounting of Ka-band antenna that have a baseline length of 25 cm and are oriented 45° from the horizontal. Because the along track dimension of the antennas is 50 cm detailed along track SAR imagery of about 30 cm is generated. The cross-track ground resolution (GSD) of the SAR is 3 m. Ka-band was chosen to minimize the amount of penetration into snow and ice. Measurements of interferometric penetration into dry snow, i.e., the amount the interferometric elevation measurements were below the air/snow interface, were made at Summit Camp in Greenland and found to be on the order of 25 cm (Hensley, et al., 2016).

Figure 7.27. The NASA Gulfstream III with pod configured for Ka-band interferometry. Lower insert shows close-up with details of the IfSAR antennae.

This system was designed to observe ice topographic measurements. For ice topography applications measurements with 100 m spatial resolution are needed and thus spatial averaging to this resolution allows the system to meet its 15-20 cm height precision requirement. A Ka IfSAR derived DEM collected on a May 1, 2009 near the Greenland coast having about 800 m of topographic height variation is presented in Figure 7.28, along with the corresponding IfSAR correlation and IfSAR DEM height precision map derived from the correlation data. Elevation precision decreases across the stripmap mode swath (from left, near range (NR) to right, far range (FR)) due to lower signal to noise. This reduction in elevation precision from NR to FR is typical of IfSAR mapping systems. Operational IfSAR mapping systems collect multiple overlapping flight lines of data to minimize the across-track height error. For example, a final DEM maintains a 10 cm RMSE elevation accuracy almost everywhere in the output DEM.

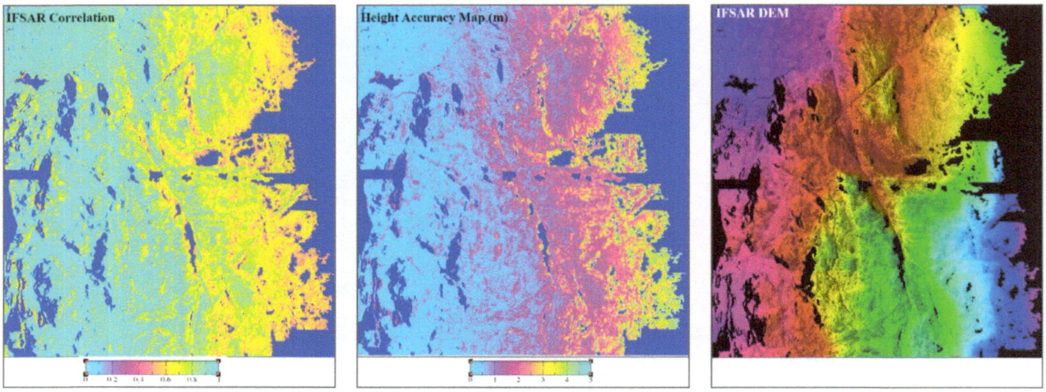

Figure 7.28. Interferometric correlation data (left image), IfSAR derived height error map (middle image) generated from the correlation data. Note the height accuracy varies from 30 cm in the NR (left handed side of the middle image, light blue in color) to about 3 m in the FR (right handed side of the middle image, yellow in color). The far-right image is the Ka-band IfSAR derived DEM where the elevations are color coded with a wrap of 800 m (from pink to red to green to pink again) to represent an elevation change of 800 m.

Change in ice sheet topography is an important metric in quantifying and understanding the impacts of climate change. Most measurement of ice sheets to monitor topographic change has used either spaceborne or airborne lidar systems. Satellite radar altimetry is most accurate over flat areas, but performs poorly over the steep coastal regions where substantial amounts of change are located. Airborne laser altimetry is better suited to these steep regions but is limited in spatial coverage and swath width (500 m), thus making it impractical for use at the continental scale over Antarctica. GLISTIN was designed to provide a wide swath ice topography measurement with high precision to augment lidar ice topography measurements. Key to the design of the GLISTIN instrument was the desire for the radar signal to have minimal penetration into the ice so that measurements could be easily cross compared with lidar measurements. This was accomplished using a very high frequency of 35 gHz (0.008 m wavelength) SAR. Table 7.9 gives a summary of the main system parameters.

A Ka IfSAR derived DEM collected on a May 1, 2009 near the Greenland coast having about 800 m of topographic height variation is presented in Figure 7.28, along with the corresponding IfSAR correlation and IfSAR DEM height precision map derived from the correlation data. Reduction in elevation precision from NR to FR is depicted in Figure 7.28. This phenomenon is typical of IfSAR mapping systems and mitigated by collecting multiple overlapping flight lines so that final DEM products maintain a consistent RMSE elevation accuracy almost everywhere in the output DEM.

Parameter	P-Band Value
Peak Transmit Power	75 W
Bandwidth	80 mHz
Center Frequency	35.66 gHz
Baseline Length	0.0084 m
Baseline Tilt Angle	45°
Platform Altitude	8,000 m to 12,000 m
Boresight Look Angle	35°
Polarization	Horizontal

Table 7.9. GLISTIN System Parameters.

Spaceborne IfSAR Systems

TanDEM-X Sensor—Airbus

TanDEM-X ("TerraSAR-X add-on for Digital Elevation Measurement") is a SAR Earth observation satellite system that consists of a pair of almost identical TerraSAR-X radar satellites that form a high-precision IfSAR system. It marks the first bistatic IfSAR mission, formed by adding a second, almost identical spacecraft, to TerraSAR-X and flying the two satellites in a closely controlled formation with typical distances between 250 and 500 m. Primary mission objective is the generation of a consistent global digital elevation model with an unprecedented accuracy. This system generated the WorldDEM product which was made possible because of a public private partnership between Airbus and the German Aerospace Centre (DLR). The approximately 150 million square kilometers of land surface were scanned from space by the TanDEM-X IfSAR mission. The global coverage achieved by TanDEM-X is also unparalleled—all land surfaces were scanned multiple times and the data was then processed to create elevation models. The WorldDEM accuracy surpasses that of any global satellite-based elevation models available today and defines a new industry standard. The WorldDEM is a complete pole-to-pole coverage DEM with unrivalled horizontal and vertical accuracy and quality. An example of the World DEM is presented in Figure 7.29. Table 7.10 on page 186 gives a summary of the main system parameters.

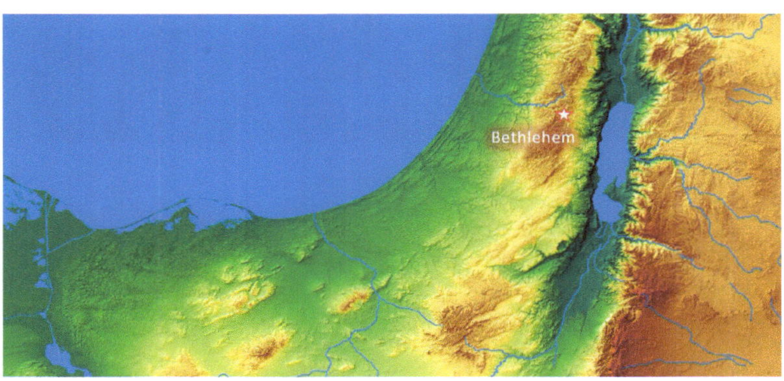

Figure 7.29. Bethlehem WorldDEM example, completed December 23, 2016.

Airbus holds the exclusive commercial marketing rights for the data and refines the elevation models per the needs of commercial users worldwide. Airbus provides an expanded IfSAR derived product

Parameter	Value
Peak Transmit Power	10-15 W
Bandwidth	100 mHz
Center Frequency	9.65 gHz
Baseline Length	200-500 m
Platform Altitude	513 km
Incidence Angle Range	20°-55°
Polarization	Horizontal

Table 7.10. Tandem-X System Parameters.

portfolio based on the WorldDEM data, including DEM, DTM, Global Ocean Shoreline, Waterbody Map, and Global Airport/Harbor Map.

COSMO-Sky-Med—e-Geos

COSMO (Constellation of small Satellites for Mediterranean basin Observation) SkyMed is a constellation of SAR satellites that can be utilized in RPI. COSMO SkyMed was commissioned and funded by the Italian Space Agency (ASI) and the Italian Ministry of Defense (MoD) as an end-to-end earth observation system civilian and defense system. The SAR constellation is comprised of four low earth orbit mid-sized satellites each equipped with multi-mode high resolution X-band SARs (Figure 7.30). Table 7.11 gives a summary of the main system parameters.

e-GEOS is the worldwide exclusive data distributor for COSMO-SkyMed data and products. e-GEOS, an ASI (20%)/Telespazio (80%) company, is a leading international player in the earth observation and geospatial information business. e-GEOS offers a unique portfolio of application services thanks to the superior monitoring capabilities of the COSMO-SkyMed constellation and provides environmental monitoring, rush mapping in support to natural disaster management, specialized products for defense and intelligence, maritime surveillance, interferometric products for landslide and ground subsidence analysis, and thematic mapping for agriculture and forestry. e-GEOS operates the Matera Space Centre for acquisition, archiving and processing of multi-mission satellite data.

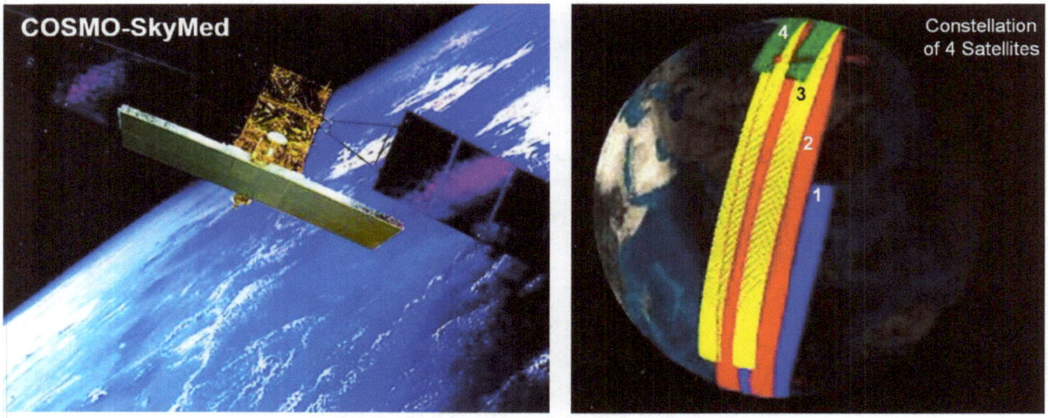

Figure 7.30. COSMO-SkyMed X-band constellation.

The DEM products are derived by mean interferometric processing of the SAR Level 1A co-registered products, in any acquisition mode except polarimetric in an automatic way. The DEM products consist of the ellipsoidal height map and the associated height error map. The attributes defining the DEM products are derived from the SAR image couple, with some substantial changes (e.g., due to the change of the image projection). The DEM product is presented in UTM cartographic coordinate system with respect to the WGS84 ellipsoid, different from the input geometry (slant-range). In the case of a DEM product originated from ScanSAR interferometric couple, output is presented in a single layer having elementary beams mosaicked in the range direction. The same constraints already shown for interferometric product also exist in the generation of the

Parameter	Value
Peak Transmit Power	17 W
Bandwidth	400 mHz
Center Frequency	9.6 gHz
Baseline Length	151 km
Platform Altitude	619 km
Incidence Angle Range	25°-50°
Polarization	Horizontal

Table 7.11. COSMO-Sky-Med System Parameters.

DEM products. The DEM and error map are represented in the same geometry with the same pixel spacing and have the same size. The accuracies are strongly dependent on the coherence value and the geometric configuration of the acquisition and scene, as well as the quality of the input ground control points used during the geometric calibration. There are a four IfSAR products for the end user to select from to create a DEM using COTS software. They are called L1A, L1B, L1C, and L1D. Each product type has a distinct set of processing applied to them.

The L1A product processing is aimed at generating Single-look Complex Slant (SCS) products. The SCS product, obtained after the L1A processing, contains focused data in complex format, in slant range and zero Doppler projection. Gain receiver compensation, internal calibration, data focusing, statistics estimation of the output data and data formatting into output are applied to this product.

The L1B product processing is aimed at generating Multilook Detected Ground (MDG) products, starting from input (L1A) data. An MDG product, obtained after L1B processing, contains focused data, detected, radiometrically equalized and in ground range/azimuth projection, and multi-looking for speckle reduction. Image detection (amplitude), ellipsoid ground projection, statistics evaluation, and data formatting are applied to this product.

The L1C product processing is aimed at generating Geocoded Ellipsoid Corrected (GEC) products. A GEC product, obtained after L1C processing, contains focused data, detected geo-located on the reference ellipsoid and represented in a uniform pre-selected cartographic presentation.

The L1D processing is aimed at generating Geocoded Terrain Corrected (GTC) products. A GTC product, obtained after L1D processing, contains focused data projected onto a reference elevation surface in a regular grid obtained from a cartographic reference system. The processing performed on L1B input data is GEC processing with the use of the DEM for map projection.

SRTM—NASA-NGA

The Shuttle Radar Topography Mission (SRTM) C-HH IfSAR sensor was flown on board the Space Shuttle Endeavour during mission STS-99. It was a single IfSAR mission that operated from February 11-22, 2000, making it the world's first IfSAR to acquire spatially-continuous elevation information over 80% of the Earth's land mass (Farr et al., 2007). SRTM utilized two SAR antennae with a fixed-baseline of 60 m at an altitude of 233 km and an orbit inclination angle of 57°, collecting data between 60° N and 57° S latitude (Farr et al., 2000). The C-band antennas were operated in a two-beam ScanSAR mode such that four narrow but overlapping sub-swaths were generated sequentially with incidence angles ranging between 30° and 60° (Farr et al., 2000). Of the Earth's land mass covered by SRTM, 99.97% was mapped with at least one Shuttle overpass, 94.59% with at least two passes, 49.25% with at least three passes, and 24.10% with at least four passes.

The C-band IfSAR data were interferometrically processed by the Jet Propulsion Laboratory (JPL, Pasadena CA) and made available to the public via the National Map Seamless Data Distribution System administered by the USGS (*http://seamless.usgs.gov*). The interferometric processing included averaging of multiple data acquisitions where possible, and filtering of the interferogram to reduce phase noise and improve the phase unwrapping process (Hensley et al., 2000; Smith and Sandwell, 2003, Rodriguez,).

NASADEM is a modernization of Shuttle Radar Topography Mission elevation data designed to improve DEM quality and provide additional data layers. The project is reprocessing raw radar signal data using improved unwrapping algorithms and mosaicking the strip data using spaceborne lidar data from ICESat data for improved ground control that was unavailable at time of original processing. The most

Parameter	Value
Peak Transmit Power	1.7 W
Bandwidth	160 mHz
Center Frequency	5.3 gHz
Baseline Length	60 km
Platform Altitude	233 km
Incidence Angle Range	57°
Polarization	Horizontal

Table 7.12. SRTM System Parameters.

significant improvements involve void reduction through the improved phase unwrapping and using ICESat data for control. This allowed NASADEM to make a so-called height ripple error correction that reduced elevation artifacts in the strip elevation data resulting from under sampled motion metrology of the antenna boom. Remaining voids are primarily filled with PRISM and ASTER DEM data. A new post-processing module creates DEM-derived layers from void-filled and merged elevation data. New output layers include imagery, height precision, incidence angle, slope and curvature and vegetation layers. Data were released to the LPDAAC in 2017. Table 7.12 gives a summary of the main system parameters.

Similar DSM finishing techniques performed on the X-HH IfSAR data provided by Intermap were performed on the SRTM data. The SRTM data were segmented into 1-degree cells (equivalent to 64 of the Intermap 7.5' tiles) comprised of elevations generated by averaging all data from the multiple passes that fell within that cell. The National Geospatial-Intelligence Agency (NGA) performed the quality assurance checks and carried out several additional editing steps (Slater et al., 2006). First, spikes and wells were removed if they exceeded 100 m compared to surrounding elevations. Second, small voids (16 contiguous posts or less) were filled by interpolation of surrounding elevations. Large voids were left in the data. Third, the ocean elevation was set to 0 m, lakes of 600 m or more in length were flattened and set to a constant height, rivers over 183 m wide were delineated and monotonically stepped down in height, and islands were depicted if they had a major axis exceeding 300 m or their relief was greater than 15 m. The C-band DSM has a 30 m and 90 m ground sample distance (GSD) both with specifications of 16 m absolute vertical height accuracy, a 10 m relative vertical height accuracy (at a local scale of ca. 200 km), and 20 m absolute horizontal accuracy, all at the 90% confidence level (Rabus et al., 2003).

COMPARISON WITH OTHER TECHNOLOGIES

Commercially available IfSAR systems can acquire data in the absence of the sun (day or night operation), in most weather and atmospheric conditions, through clouds, at high collection rates and over large and remote geographic areas. These characteristics make these systems a viable alternative to conventional aerial photography and lidar technologies for large area topographic and land cover mapping.

Each type of sensor is sensitive to different aspects of the surface under observation and therefore measures height differently depending on surface type and ground cover. All the above sensors make height measurements that depart from an "idealized height sensor" that gives the height measurement at precisely one point. Only for simple surfaces or after appropriate post processing of the data can the height measurements of the three types of sensors be directly compared with point measurements like those obtained by DGPS surveys.

The desired elevation measurement is application dependent. Flood plain applications require DEMs to have all vegetation and buildings removed, and water constriction features such as bridges, fences and power poles edited from the data. However, for flight obstruction or forest mapping it is desired to leave some or all the elevations unaltered. Each of the different sensor technologies has strengths and weaknesses depending on the desired height measurement for an application. In fact, a synergistic combination of measurements from two or more of the above sensors can produce the best possible product.

As the variety of IfSAR, lidar, and photogrammetric sensors is quite numerous, and performance parameters continuously improving for all three types of sensors, only very general comparisons of the sensor characteristics are presented.

IfSAR systems can acquire data in the absence of the sun (day or night operation), in most weather and atmospheric conditions, through clouds, and at high collection rates. These characteristics make these systems a viable alternative to conventional aerial photography and lidar technologies for large area topographic and land cover mapping.

Photogrammetry

Fully explained in Chapter 6, photogrammetric sensors, like IfSAR systems, generate both imagery and height data and have been operated on both airborne and spaceborne platforms. Unlike aerial or spaceborne photogrammetry, IfSAR missions can be flown without regard to sun angle. Flights may take place at night or in conditions of inclement weather provided the conditions are such that the image formation process is not degraded. Airborne optical cameras continue to generate extremely fine resolution (often sub-meter) imagery without the troublesome layover and shadow problems of radar. However, radar interferometers are proving to be a cost-effective method for wide area, rapid mapping applications, and do not require extensive hand-editing and tie pointing. Additionally, because IfSAR systems often fly at greater altitudes, they can operate in congested air-traffic corridors that are often difficult to image photogrammetrically. Mapping in tropical regions that are often cloud-covered can be done more reliably with IfSAR systems that penetrate clouds. Urban mapping is a challenging venue for mapping by IfSAR systems due to the extremely complex scattering environment. Although some high-resolution systems have shown promise for urban mapping, photogrammetry still has inherent advantages for this application.

Densely vegetated surfaces can be problematic for both sensors if bare surface elevations are desired. Photogrammetric true ground surface heights can only be obtained if sufficiently large gaps are present in the canopy. These points, usually determined manually, are then extrapolated to other portions of the canopied area to produce bare earth height maps. Heights measured by IfSAR systems are reflective surface heights and can lie anywhere within the canopy. Longer wavelength systems penetrate deeper into the canopy but the exact location within the canopy corresponding to the height measurement is not easily determined. IfSAR correlation has information about the vertical structure of the canopy and has the potential of providing corrections to measure bare surface elevations (Hokeman and Varekamp, 2001).

Unlike photogrammetric and lidar systems, IfSAR systems can generate height error maps on a pixel-by-pixel basis (Hensley and Webb, 1994). These images provide an estimated relative statistical height error as described earlier. These products are extremely useful for ascertaining whether an IfSAR derived DEM is suitable for a mapping application and for locating problematic regions within the DEM.

Lidar

Airborne topographic lidar is fully explained in Chapter 8. Lidar like IfSAR is an active sensor providing its own illumination and similarly records the time delay between transmit and receipt of reflected signals from the surface Employing a very narrow beam so that the projected footprint on the ground is typically 75 meters or less from space (several feet from airborne sensors), lidar systems obtain one or more height measurements per pulse. The number of height measurements is dependent upon the vertical structure of objects within the beam and the type of lidar system. Some lidar systems are equipped to only record a single time delay per pulse whereas most lidar systems record time delays for multiple samples exceeding a signal level threshold. By scanning cross track to either side of the nadir point of the aircraft and rapidly pulsing the laser, reasonable mapping swaths are obtained. Operating at optical instead of microwave frequencies, lidar systems do not penetrate clouds and other atmospheric obscurants.

Unlike IfSAR systems lidars are not imaging sensors, though they do provide intensity returns or reflectance images. Applications, where contextual information that is not easily derived from height data alone, favor IfSAR and photogrammetric mapping sensors. Classification studies and resource inventory surveys are examples where imagery plays a vital role in separating and identifying vegetation and crop types, geologic structures, and various anthropogenic uses. Water body mapping and delineation is often possible with IfSAR systems depending on the frequency of operation and the roughness of the water body surface, whereas topographic lidar systems do not obtain good height measurements over water. Chapter 10 explains the advantages of bathymetric and topobathymetric lidar sensors for mapping water elevations and submerged topography.

Lidar systems have become the sensor of choice for mapping vegetated regions when elevation measurements beneath the canopy are needed. True ground surface elevation measurements are obtained after a post processing step to insure the reflective surface does not arise from within the vegetation canopy. As previously noted, heights measured by IfSAR systems are reflective surface heights and can lie anywhere within the canopy, but with proper frequency selection and use of correlation data has the potential of providing corrections to measure bare surface elevations. Larger swaths possible with IfSAR mapping systems can substantially reduce the time to collect data over large areas.

USER APPLICATIONS

Fine resolution topographic measurements have applications throughout the commercial, civilian, and military sectors. Applications include, for example, land slope stability and land-slide characterization, land-use classification and change monitoring for agricultural and military purposes, flood plain and hydrologic modeling, littoral zone mapping, and archeological and geological applications. The accuracy and resolution depend on application, and several technologies are usually available that can meet any application requirements. The ability to generate accurate IfSAR DEMs at regional to global scales quickly and at a reasonable cost has seen the rapid infusion of this technology into a variety of applications.

Topographic Mapping

Topographic maps similar or identical in nature to aerial photo derived maps are being successfully generated from IfSAR data (Tighe, 2000; Tighe and Chamberlain, 2009). IfSAR technologies have advanced to the point where centimeter-level accuracies are possible. These high-resolution systems are making their way into aerial photography niches, especially in areas inaccessible by photogrammetric systems due to cloud cover or congested airspace.

Vegetation Mapping and Land Use Classification Maps

The use of interferometry for land use classification and vegetation parameter determination is a rapidly expanding area of research (Engdahl, M., 2013). The use of multi-frequency IfSAR systems that exploit the relative penetration into the canopy, and the use of interferometric correlation which is sensitive to the vertical structure of the canopy, have shown great promise for extracting canopy parameter elevations (Rodriguez, Martin and Michel, 1999). Land use maps that have classification accuracy in the 90% level have been demonstrated using data from airborne IfSAR systems (Rodriguez, Martin and Michel, 1999) and (Tighe, 2012). Although significant ambiguities were observed under certain conditions. Specifically, problems arose due to the sensitivity to the absolute calibration of the radar backscatter and from changes in backscatter as a function of incidence for the same ground cover type. Using multi-frequency, multi-temporal or other optical data sources can significantly reduce classification error.

Urban Mapping

Urban mapping varies from the relatively low density and simple structures of a suburb to the extremely complex and high density environment of a modern major city. Multiple scattering, shadow and layover make urban mapping a challenging application environment for IfSAR systems. High resolution airborne IfSAR systems have shown some utility for this application particularly in medium to low density urban areas (Mercer and Gill, 1998). System resolution and the degree of algorithm optimization for urban environments greatly affect the achievable mapping accuracy (Grey et al., 2003).

Geological Mapping

Topographic maps have traditionally played a key role in geological applications. IfSAR system height and image data can simultaneously provide topographic information at two scales. Using the topographic data directly, topography at the DEM posting provides information about geologic structures such as faults, volcanic structures, and alluvial fan size and extent. The associated SAR imagery, which is sensitive to the surface roughness on the scale of the radar wavelength, provides information about the micro-topography of the surface. The combination of the two scales can be used to infer information about the surface geology such as the relative age of lava flows and lithology (Tighe et al., 2006).

Ice Mapping

Change in ice sheet topography is an important metric in quantifying and understanding the impacts of climate change. Most measurement of ice sheets to monitor topographic change has used either spaceborne or airborne lidar systems. Satellite radar altimetry is most accurate over flat areas, but performs poorly over the steep coastal regions where substantial amounts of change are located. Airborne laser altimetry is better suited to these steep regions but is limited in spatial coverage and swath width (500 m), thus making it impractical for use at the continental scale over Antarctica. Airborne IfSAR systems (e.g., Glisten) have been designed to provide a wide swath ice topography measurement with high precision to augment lidar ice topography measurements (Hensley et al., 2009, 2010, 2016).

Littoral Zone Mapping

Littoral zone mapping is an area where IfSAR mapping is playing an increasingly key role because it maps from the near shore regions out onto the water surface [4]. Photogrammetric and topographic lidar systems have greater difficulty mapping these regions because neither type of system makes height measurements over water bodies. By mapping at low tides IfSAR systems should provide some of the most complete and detailed maps of the littoral zone.

Hydrology

The insurance industry as well as local and national governments have become increasingly concerned about proper risk assessment associated with flooding. Risk assessment is needed for flood warning, flood plain management and financial liability studies. Flood plain mapping and risk assessment uses a combination of topography and surface cover type along with propagation models to determine depth of flooding. Airborne derived IfSAR DEMs with the combination of accuracy and resolution and cost have proved ideal for regional flood risk assessment as has been shown in several studies both in the United Kingdom (Galy and Sanders, 2000) and in the U.S.

[4] Mapping water body surfaces is usually only possible for higher frequency systems (C-band and above) and when there is sufficient wind or current to make the water appear rough at the imaging wavelength.

Seismic Hazards

Slope and along-slope and cross-slope curvature estimates are needed for slope hazard analysis. Special care must be taken in computing slope and surface curvature from interferometric DEMs because point-to-point height noise can be comparable to a significant fraction of the post spacing. Studies have shown that when this is considered, IfSAR derived DEMs improve classification of areas of landslide induced seismic risk (Real et al., 1997).

Environmental Disasters

Data and products supplied by the COSMO-SkyMed system represent a valid and important instrument, to carry on studies about causes and phenomena preceding environmental disasters as well as to improve the monitoring and evaluation of damages in case, for example, of landslips, flood, earthquakes and volcanic eruptions. Observation within a specific area can be made during day or night and even in case of clouds, thus allowing evaluations of superficial alterations of the territory and providing institutions in charge of managing the risk with a new and useful instrument of prevention and control.

Archeology

Understanding where, how and when ancient civilizations modified and controlled their physical environment is an aspect of archeological research where SAR and IfSAR systems have made important contributions. Multiple frequency observations that penetrate dense vegetation coupled with accurate topographic information is providing archeologists with unique regional scale observations of ancient sites such as Angkor Wat in Cambodia and the Great Wall in China. Because of the unique perspective, SAR and IfSAR systems can provide, the use of these data in future investigations is expected to increase.

MAPPING ALASKA—AMERICA'S LAST FRONTIER

This section details a large area mapping project using the IfSAR technology. Here the reader will gain an insight on how this technology provided the only available DEM generating technique solution to map America's last frontier.

The Alaska DEM Whitepaper

At the 2008 Alaska Surveying and Mapping Conference in Alaska, surveying and mapping firms in attendance were asked to provide suggestions on the best way to map Alaska. This information was invited because the state of Alaska had never been mapped at any scale meeting the National Map Accuracy Standards, thus, leaving it the only U.S. state without statewide digital orthophotos or elevation data comparable to accuracies held by Continental United States (CONUS). Immediately it became apparent the importance of collecting an accurate elevation dataset, prior to producing orthophotos (which benefit from accurate and homogenous elevation data). Dewberry was hired, based on their knowledge of varied DEM technologies from many commercial companies and their quality assurance procedures that are highly regarded in the mapping industry, to prepare what is commonly known as the Alaska DEM Whitepaper (Dewberry, 2008).

In researching the Alaska DEM Whitepaper, Dewberry learned that the state and various federal agencies had been trying for decades to determine the best solution for mapping the state. Dewberry learned that Alaska had America's worst geodetic and geospatial infrastructure; had the largest area to be mapped (covering 10 UTM zones); had the longest distances from suitable airfields where aircraft could be based; had the tallest mountains and longest shorelines in the U.S.; had the highest incidents of Controlled Flight

Into Terrain (CFIT) aircraft accidents; had the fewest roads and the most reliance on airfields (but with small aircraft not powerful enough to fly over mountains, but through mountain passes that must be mapped accurately); had the most impact from climate and sea level changes and the highest impact from post-glacial rebound that changed elevations and gravity. This whitepaper recommended statewide aerial IfSAR as the best solution for acquiring DTMs and DSMs, prior to production of digital orthoimagery and updates to the National Hydrography Dataset (NHD), transportation and other geospatial layers needed in Alaska. The use of IfSAR technology due to its ability to operate day and night, in such a remote large area (limited road access), varied terrain, and with little in-scene ground control placement, and short data collection timeframes (a couple of months in the summer due to weather) was the only mapping technology that could collect accurate elevation data in a short mapping season.

The Geographic Information Network of Alaska (GINA) provided examples of what happened when attempting to drape satellite imagery over the existing DEM from the National Elevation Dataset (NED). Some mountains were said to be horizontally displaced by over a mile in the NED, making it erroneously appear as though streams flowed over mountains and hills rather than through the valleys. In addition to extended periods of darkness in winter months, Alaska's valleys are often obscured by clouds and/or fog, but flights must continue under Instrument Flight Rules (IFR) to get patients to health clinics or otherwise conduct daily operations routinely executed under adverse weather conditions. IFR procedures rely on accurate elevation data. Imagine flying in the dark or through clouds, intending to be flying up a valley, but because of DEM errors in the cockpit display, the plane is unknowingly flying into an unseen mountainside. If it were a clear day, this would not be a problem whey flying under Visual Flight Rules (VFR), but when flying through clouds or in the dark under IFR, having mountains mapped in the wrong place can be a fatal error. In addition to aviation safety, there was an obvious and immediate need to collect accurate statewide elevation data for production of statewide digital orthophotos as the accuracy of the DEM generated from the NED was insufficient for orthorectification.

In analyzing DEM user requirements, although some cities and high value corridors had needs for higher-accuracy, higher-cost lidar data, aerial IfSAR was Dewberry's recommendation for statewide mapping because it was the only technology that could map through clouds, because it could produce DEMs with RMSEz ≤2 meters and RMSEr ≤8 meters, and because it was the most affordable. Clouds had long been Alaska's nemesis for statewide mapping, but with IfSAR this was a non-issue.

The Alaska Statewide Digital Mapping Initiative (SDMI)

Officials of the Alaska Statewide Digital Mapping Initiative (SDMI) then asked USGS to use its Geospatial Products and Services Contract to contract with Dewberry to develop an overall plan for statewide IfSAR mapping, to manage IfSAR contractors and to quality control their data. With Dewberry, as prime contractor, Dewberry, Fugro EarthData and Intermap Technologies developed a joint plan in which Intermap would map the easiest (least rugged and least forested) 75% of the state with X-band IfSAR and Fugro would map the hardest 25% of the state with X-band and P-band IfSAR. Because terrain slope impacts the accuracy of IfSAR data (see Figure 7.12 on page 164), both IfSAR firms felt they could satisfy the 2 m RMSEz requirement only for slopes up to 20 degrees, assuming larger vertical errors in steeper terrain.

The IfSAR mapping started in 2010 and will continue until approximately 2020, depending on available funding. As of the end of 2017, IfSAR data has been acquired for approximately 90% of the state, with the southwestern part of the state, including the Aleutian Islands, remaining to be mapped during the next 2-3 years. The principal IfSAR deliverables include a hydro-enforced DTM with 5-meter posts, a DSM with 5 meter posts, ortho-rectified radar imagery (ORI) with 5 meter pixels or better (being delivered down to 62.5 cm pixels), in multiple file formats and projections required by the state, USGS and NGA. FGDC-compliant metadata files, a certified ISO 9001 data-quality report for each 15 minute tile, void masks and slope masks are also provided. Blind QA/QC checkpoints have been surveyed statewide by JOA Surveys.

Figure 7.31 shows an example hydro-enforced DEM from Intermap and Figure 7.32 shows the comparable DEM previously available from the NED. Both images are courtesy of Intermap. Because hydrologic modeling, and updating the obsolete NHD datasets, are among the most critical user requirements in Alaska, the new IfSAR data are vital for management of Alaska's vast natural resources, including water. Water boundaries are "loud and clear" with IfSAR data whereas they are often questionable with optical imagery or lidar.

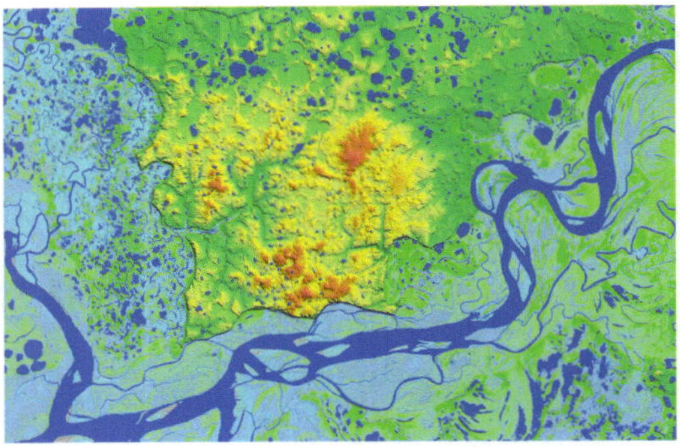

Figure 7.31. IfSAR data showing complex hydrographic features very important in Alaska where water resources are of critical importance.

Figure 7.32. NED data covering the same geographic extent as the IfSAR in Figure 7.31 but with minimal hydrographic features shown.

DIFFERENTIAL INSAR METHOD

This section provides an example of using IfSAR technology in an advanced configuration called differential InSAR or DInSAR for short. Radar targets can be Permanent Scatter (PS) (e.g., buildings, rock outcrops) or Distributed Scatter (DS) (e.g., fallow fields, roadways). The processing of a long series of radar images using the DInSAR technique allows the motion history of a radar target to be observed (including non-linear motion), and increases measurement precision to millimeter level (Ferretti, et al., 2000; Ferretti, et al., 2011). DInSAR technique is typically used to monitor subsidence/uplift or lateral deformation.

The DInSAR technique involves using multi-interferometric outputs created from using many radar images (at least 15-20) acquired over the same area with the same acquisition mode and geometry. The resulting DInSAR interferogram series is then, a comprehensive history of changes in ground movement or displacement (Ferretti, et al., 2000).

Surface movement is determined by the spatial variation of phase across the image over several pixels. When using the DInSAR technique to identify surface movements of millimeter to centimeter scale with high spatial resolution it is necessary to remove topographic and atmospheric effects. To ensure that the interferogram measurements have topography removed from the DInSAR data, the topographic phase is often simulated using a reference DEM (or if possible an IFSAR derived DEM is used) and then removed from the interferogram.

Atmospheric artifacts show a strong spatial correlation within every single SAR acquisition, but are uncorrelated in time. Conversely, target motion is usually strongly correlated in time and can exhibit different degrees of spatial correlation depending on the phenomenon at hand. When long time-series of SAR images are combined, atmospheric effects can be estimated, modelled and removed based on these different correlation patterns. To exploit all the available images, and improve the accuracy of estimation, only coherent scatters (targets) are selected.

DInSAR Example—Coastal Subsidence Monitoring over Hampton Roads

Coastal risk management is a complex equation that must take into consideration changing land and maritime conditions. Within this equation, subsidence can strongly impact sensitive and at-risk areas, particularly when coupled with sea level rise.

Hampton Roads is located within the southeastern corner of the state of Virginia, U.S. This coastal inlet is predominantly urban and has the largest concentration of federal bodies and facilities of any metropolitan area in the world. This area is experiencing some of the highest rates of sea level rise, in part due to land subsidence related to compaction of aquifers within.

A DInSAR analysis was carried out over Hampton Roads to assess and quantify subsidence rates. This was accomplished using 130 Cosmo-SkyMed SAR images covering a six-year period between July 13, 2011 and August 28, 2017. A considerable number of localized movement features and wider-area trends were detected, such as the localized subsidence observed over the East Ocean View neighborhood as shown in Figure 7.33.

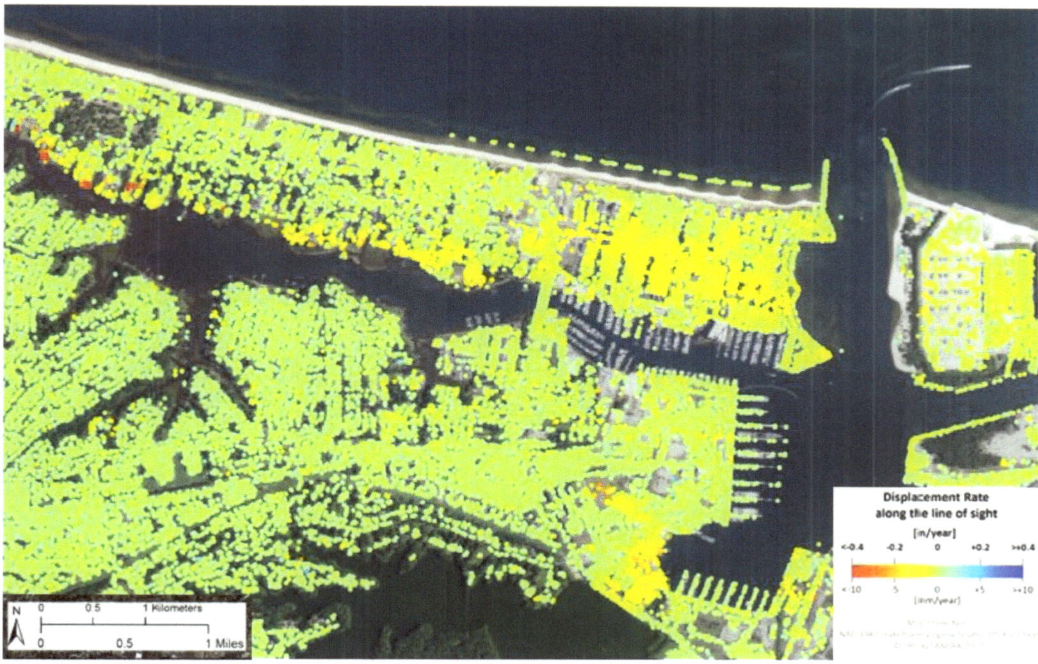

Figure 7.33. A total of 4,150,248 measurement points was identified over the processed area, exhibiting an average displacement rate of -0.04 inches/year (-1.0 mm/year) with an associated confidence limit of ±0.01 inches/year (± 0.35 mm/year).

Examples of time series are shown in Figures 7.34 and 7.35 on page 196.

A cross section over the George Washington overpass was selected to illustrate the DInSAR technique used to observe subsidence (Figure 7.36 on page 197). DInSAR results viewed as surface profile cross-sections (Figure 7.36) illustrate subsidence dynamics over time where each individual profile corresponds to ground displacement compared to the first acquired SAR image. While the George Washington overpass is stable, both ramps are showing over -4.5 inches (-115.0 mm) of subsidence over the period of this analysis (Figure 7.37 on page 198).

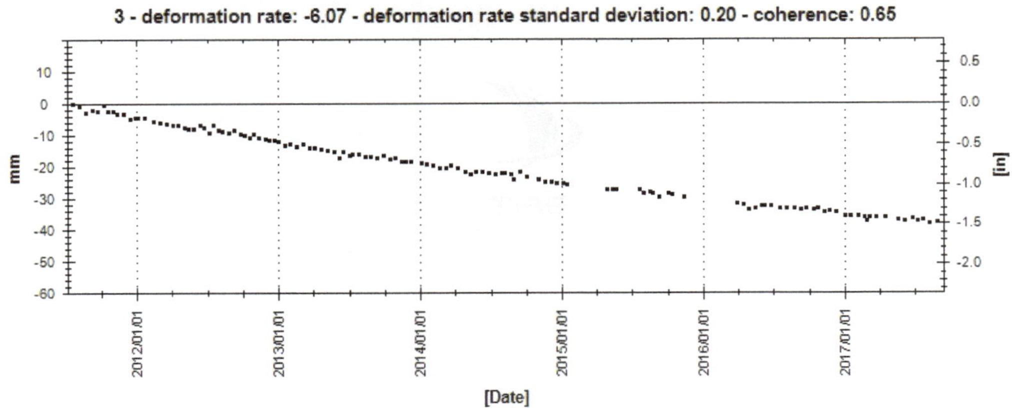

Figure 7.34. Example of time series identified over Tanner Point, within the Virginia Port area.

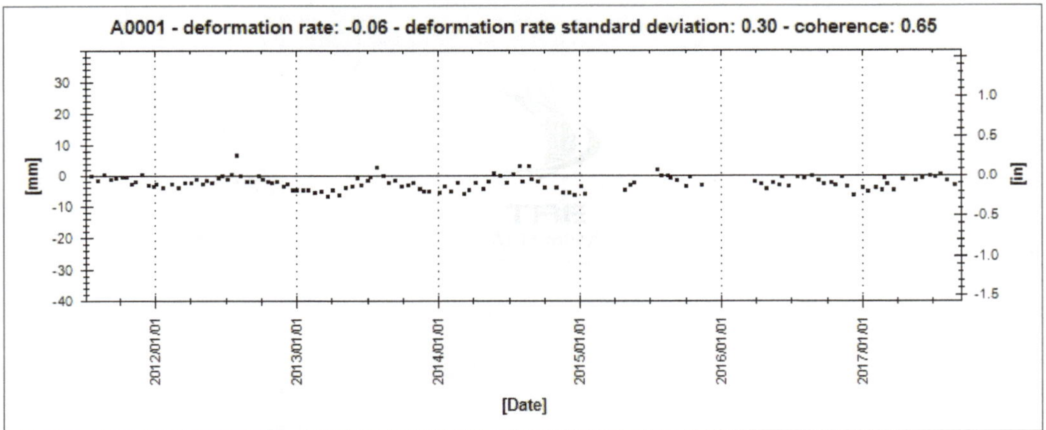

Figure 7.35. Seasonal variation captured from measurement points identified over part of the W Norfolk Bridge.

Figure 7.36. A cross-section over the George Washington overpass illustrating differential subsidence patterns in this area.

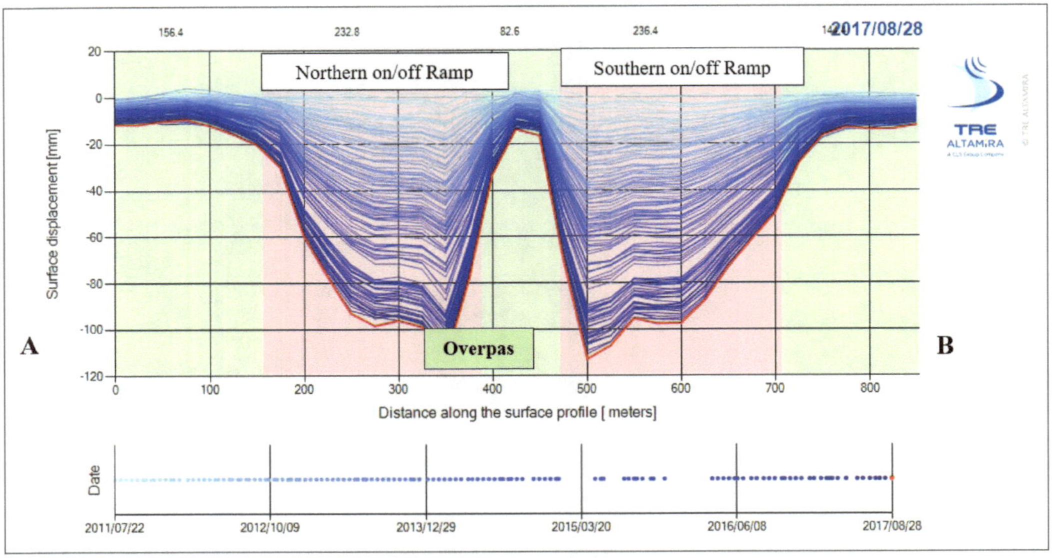

Figure 7.37. Surface profile cross-section corresponding to the trace illustrated in Figure 7.35.

TECHNOLOGICAL ADVANCEMENTS

The success of the IfSAR/InSAR technology over the past two decades has provided a plethora of DEM generation platforms. Continued advances in IfSAR/InSAR technology have led the way for the deployment of SAR satellite constellations such as Europe's Copernicus program (10 Satellites: 5 tandem pairs, 2 orbital planes) and Urthecast OptiSAR™ Constellation (16 Satellites: 8 tandem pairs, 2 orbital planes) which will continuously observe the earth every day, every night, rain or shine, revolutionizing the world's ability to monitor our planet regardless of obstacles like tree cover and weather phenomena. Such constellations provide high reliability, improved revisit time, geographical coverage, and rapid SAR and IfSAR/InSAR data collection and dissemination to support operational applications in the priority areas of marine monitoring, land monitoring, and emergency services. To promote wider use of IfSAR data products, to increase research on interferometric techniques and to make it easier for people to produce accurate IfSAR measurements, IfSAR providers will develop web-based tools which, through a user-friendly interface, which enable users to generate IfSAR products from the spaceborne data repositories, in a matter of minutes.

What's Next with SRTM? Engineers and scientists at JPL are currently working on a complete reprocessing of the original SRTM radar data to produce an improved near-global digital elevation model to be called NASADEM. As with SRTM Plus, this work is funded under NASA's "Making Earth System Data Records for Use in Research Environments" (MEaSUREs) Program. In brief, the expected improvements include (1) Fine vertical adjustments within and among individual shuttle data takes via reference to precise ICESat (Ice, Cloud, and land Elevation Satellite) laser profiles, (2) Void reduction via improved radar interferometric processing, (3) Use of better fill data in the remaining voids, especially ASTER GDEM3 when available, and (4) Improved quality assessments and adjustments. This project was completed in 2017.

With the success of the Tandem-X mission, DLR has demonstrated its pioneering role and satisfied the prerequisites for the next major development step in satellite-based Earth observation—the Tandem-L radar mission. The Tandem-L successor mission could provide a current elevation image of Earth's entire landmass every eight days and thereby capture dynamic processes in a timely manner. This would also make it possible to contribute to the review of international climate and environmental agreements. New radar methods and innovative missions such as Tandem-L are set to contribute to gaining a better understanding of dynamic processes to protect and preserve Earth. Completion of the TanDEM-X global elevation model has now paved the way for the next dimension of radar remote sensing.

ACKNOWLEDGEMENTS

The principal authors for this chapter are Dr. Scott Hensley, Radar Science and Engineering Section, Jet Propulsion Laboratory and Dr. Lorraine Tighe, Esri. Contributing authors include Dr. David Maune of Dewberry who contributed the Mapping Alaska section; Jessica Morgan of TRE Altamira Inc who contributed the section on the Differential InSAR (DInSAR) method; and Dr. Riadh Munjy and Dr. Paul Rosen who co-authored some of the materials re-used from the 1st and 2nd editions of this manual.

Scott Hensley (SM'10-F'14) received his B.S. degrees in Mathematics and Physics from the University of California at Irvine and the Ph.D in Mathematics from Stony Brook University where he specialized in the study of differential geometry. In 1992, Dr. Hensley joined the staff of the Jet Propulsion Laboratory where he is currently a Senior Research Scientist studying advanced radar techniques for geophysical applications. He has worked on most of the SAR systems developed at JPL over the past two decades including the Magellan and Cassini radars. He was the GeoSAR Chief Scientist, a simultaneous X-band and P-band airborne radar interferometer for mapping above and beneath the canopy that is now commercially operated by EarthData International. He led the SRTM Interferometric Processor Development Team for a shuttle based interferometric radar used to map the Earth's topography between ±60° latitude. He also employed the Earth-based Goldstone Solar System Radar to generate topographic maps of the lunar surface. He was Principal Investigator and is currently the Project Scientist for the NASA UAVSAR program which is an electronically scanned active array L-band fully polarimetric radar designed for repeat pass applications.

Lorraine Tighe has a Ph.D degree in Earth Sciences, a graduate diploma in Remote Sensing, and a B.Sc. degree in Physics and Geology. Dr. Tighe is currently the Senior Remote Sensing Product Strategy and Marketing lead at Esri. She has been involved with IfSAR/InSAR research for topographic and geological mapping and forestry applications for the past 23 years. Dr. Tighe's research activities include studying the amount of X-band Interferometric Synthetic Aperture Radar (IfSAR/InSAR) and fully polarimetric C- and L-band polarimetric InSAR (PolInSAR) penetration into open barren terrain and vegetated canopy to classify vegetation type and estimate canopy and terrain height over diverse environments and large geographic extents common in North America. She has delivered lectures, workshops and training sessions on IFSAR mapping in Canada, U.S., Jamaica, Brazil, Equator, Honduras, England, Scotland, Poland, Germany, Indonesia, Philippines, Malaysia, and Australia.

REFERENCES

Andersen, H.-E., Reutebuch, S.E., and McGaughey, RJ. (2005). Accuracy of an IfSAR derived digital terrain model under a conifer forest canopy. In *Canadian Journal of Remote Sensing*, 3, I, (pp.1-6).

Andersen, H. -E., R. J. McGaughey, and S. E. Reutebuch (2008). Assessing the influence of flight parameters, interferometric processing, slope and canopy density on the accuracy of X-band IfSAR-derived forest canopy height models. In *International Journal of Remote Sensing*, 29(5), (pp.1495-1510).

ASPRS (2014). ASPRS positional accuracy standards for digital geospatial data. In *Photogrammetric Engineering and Remote Sensing*, 81(3) (pp. A1-A26), Bethesda, MD: American Society for Photogrammetry and Remote Sensing.

Bamler, R. and Hartl, P. (1998). Synthetic aperture radar interferometry. In *Inverse Problems*, Vol.14, No. 4: R1-54.

Boerner, W-M, H. Mott, E. Lüneburg, C. Livingston, B. Brisco, R. J. Brown and J. S. Paterson with contributions by S.R. Cloude, E. Krogager, J. S. Lee, D. L. Schuler, J. J. van Zyl, D. Randall P. Budkewitsch and E. Pottier (1998). Polarimetry in radar remote sensing: basic and applied concepts, Chapter 5 in F.M. Henderson, and A.J. Lewis, (eds.), Principles and Applications of Imaging Radar, Vol. 2 of *Manual of Remote Sensing*, 3rd Ed., New York, NY., John Willey & Sons.

Cloude, S. R., and K. P. Papathanassiou (1998). Polarimetric SAR interferometry. In *IEEE Transactions on Geoscience and Remote Sensing*, 36(5), (pp.1551-1565).

Dewberry (2008). *Digital elevation model (DEM) data for the Alaska Statewide Digital Mapping Initiative (SDMI)*, September 18, 2008.

Elachi, C. (1988). Spaceborne Radar Remote Sensing: Applications and Techniques. New York: IEEE Press, 288p.

Engdahl, M.(2013). *Multitemporal insar in land-cover and vegetation mapping.*

Farr, T. G., Paul A. Rosen, Edward Caro, Robert Crippen, Riley Duren, Scott Hensley Michael Kobrick, Mimi Paller, Ernesto Rodriguez, Ladislav Roth, David Seal, Scott Shaffer, Joanne Shimada, Jeffrey Umland, Marian Werner, Michael Oskin, Douglas Burbank, and Douglas Alsdorf (2007). The shuttle radar topography mission. In *Reviews of Geophysics.*, 45, RG2004, doi:10.1029/2005RG000183.

Feretti, A., Fumagalli, A., Novali, F., Prati, C., Rocca, F., and Rucci, A. (2011). A new algorithm for processing interfermetric data-stacks: SqueeSAR. In IEEE *Transactions on Geoscience and Remote Sensing*, (pp. 3460-3470).

Ferretti, A., Prati, C., & Rocca, F. (2000). Non- linear subsidence rate estimation using the permanent scatters in differential SAR interferometry. In *IEEE Transactions on Geoscience and Remote Sensing*, (pp. 8-20).

Garestier, F., Dubois-Fernandez, P., Papathanassiou, K. (2008). *Pine forest height inversion using single-pass X-band PolInSAR data.*

Gens, R. (2002). *Introduction to SAR interferometry. Presentation given at the University of Fairbanks.*

Graham, L. C. (1974). Synthetic interferometric radar for topographic mapping. In *Proc. IEEE,* 62(6), (pp. 763-768)

Grey, W.M.F., Luckman, A. J., D Holland (2003). Mapping urban change in the UK using satellite radar interferometry. In *Remote Sensing of Environment,* Volume 87, Issue 1, 15 September, (pp. 16–22).

Hajnsek, I., Kugler, F., Lee, S. K., and Papathanassiou, K. P. (2008). Tropical forest parameter estimation by mean of Pol-InSAR: the INDREX-II campaign. In *IEEE Transactions on Geoscience and Remote Sensing,* Vol. 47, No. 2 (pp. 481-493).

Hensley, S. and Webb, F. H. (1994). Comparison of long valley TOPSAR data with kinematic GPS measurements. In *IGARSS Proceedings,* Pasadena, CA.

Hensley, S., Chapin, E., Freedman, A., Le, C., Madsen, S., Michel, T., Rodriguez, E., Siqueira, P. and Wheeler, K. (2001). First P-band results using the GeoSAR mapping system. In *Proceedings of IGARSS 2001,* Sydney, Australia.

Hensley, Scott, Eric Gurrola, Leif Harcke, Martin Slade, Kevin Quirk, Meera Srinivasan, Clement Lee, Sang-Ho Yun, Joseph Jao, Barbara Wilson, Eric De Jong, and Nick Marechal (2010). Lunar topographic mapping using a new high-resolution mode for the GSSR radar. In *Radar Conference, 2010 IEEE.*

Hensley, Scott, Eric Gurrola, Paul Rosen, Martin Slade, Joseph Jao, Mike Kobrick, Barbara Wilson, Curtis Chen and Raymond Jurgens (2009). *An improved map of the lunar south pole with earth based radar interferometry.* Radar conference, 2008).

Hensley, Scott, Thierry Michel, Maxim Neumann, Marco Lavalle, Razi Ahmed, Ron Muellerschoen and Bruce Chapman (2014). A comparison of multi-baseline polarimetric interferometry at La Amistad and La Selva, Costa Rica with a modified PolSARProSim scattering tool. In *Proceedings of EUSAR 2014.*

Hensley, S., K. Wheeler, G. Sadowy, C. Jones, S. Shaffer, H. Zebker, T. Miller, B. Heavey, E. Chuang, R. Chao, K. Vines, K. Nishimoto, J. Prater, B. Carrico, N. Chamberlain, J. Shimada, M. Simard, B. Chapman, R. Muellerschoen, C. Le, T. Michel, G. Hamilton, D. Robison, G. Neumann, R. Meyer, P. Smith, J. Granger, P. Rosen, D. Flower, R. Smith (2008). The UAVSAR instrument: description and first results. In *IEEE Radar Conference,* 6 pp.

Hensley S, Moller D, Oveisgharan S, Michel T & Wu X (2016). Ka-band mapping and measurements of interferometric penetration of the Greenland ice sheets by the GLISTIN radar. In *IEEE Journal of Selected Topics in Applied Earth Observations and Remote Sensing,* 9(6), DOI: 10.1109/JSTARS2560626.

Intermap (2011) *Intermap product handbook.*

Kobayashi, Y., K. Sarabandi, L. Pierce, and M. C. Dobson (2000). An evaluation of the JPL TOPSAR for extracting tree heights. In *IEEE Transactions on Geoscience and Remote Sensing,* 38, (pp. 2446-2453).

Lewis, A. J. and Henderson, F. M. (1999). Radar fundamentals: the geoscience perspective. In *Principles and Application of Imaging Radar: Manual of Remote Sensing,* 3rd edition, Robert A. Ryerson (Ed.), New York, NY, Wiley.

Li, F. and Goldstein, R.M. (1990). Studies of multibaseline spaceborne interferometric synthetic aperture radar. In *IEEE Trans. Geosci. Remote Sensing,* 28(1), (pp. 88-97).

Madsen, S. N., Zebker, H. A. and Martin, J. (1993). Topographic mapping using radar interferometry: processing techniques. In *IEEE Trans. Geosci. and Rem. Sens.,* 31(1), (pp. 246-256).

Mercer, J. B. and Gill, M. (1998). Radar-derived DEMs for urban areas. In *Proceedings of the ISPRS Commission IV Symposium,* Stuttgart, Germany.

Moller, D., Hensley, S., Sadowy, G.A., Fisher, C.D., Michel, T., Zawadzki, M., Rignot, E. (2011). The Glacier and Land Ice Surface Topography Interferometer: An Airborne Proof-of-Concept Demonstration of High-Precision Ka-Band Single-Pass Elevation Mapping. In I*EEE Transactions on Geoscience and Remote Sensing,* IEEE Transactions on, vol.49, no.2, (pp.827-842), Feb. 2011, doi: 10.1109/TGRS.2010.2057254.

Positional Accuracy Handbook (1999). Minnesota Planning Land Management Information Center.

Schneider, R. Z., K. P. Papathanassiou, I. Hajnsek, and A. Moreira (2006). Polarimetric and interferometric characterization of coherence scatters in urban areas. In *IEEE Transactions on Geoscience and Remote Sensing,* 44(4), (pp. 971-984).

Raney, K. (1999). Radar fundamentals: technical perspective. In *Manual of Remote Sensing,* 3rd edition, Ch. 2, Boston, MA: Artech House.

Real, C., Wilson, R. I. and McCrink, T. P. (1997). Suitability of airborne-radar topographic data for evaluating earthquake-induced ground failure hazards. In *12th Int. Conf. on Appl. Geol. Rem. Sens.,* Denver, CO.

Rabus, B, M. Eineder, A. Roth, and R. Bamler (2003). The shuttle radar topography mission—a new class of digital elevation models acquired by spaceborne radar. In *ISPRS Journal Photogrammetry and Remote Sensing,* 57(4), (pp. 241–262).

Richards, M. A. (2007). A beginner's guide to interferometric SAR concepts and signal processing. In *IEEE Aerospace and Electronic Systems Magazine,* 22(9), (pp. 5-29).

Slater, J. A., G. Garvey, C. Johnston, J. Haase, B. Heady. G. Kroenung, and J. Little (2006). The SRTM data "finishing" process and products. In *Photogrammetric Engineering and Remote Sensing,* 72(3), (pp. 237-247).

Smith, B., and D. Sandwell (2003). Accuracy and resolution of shuttle radar topography mission data. In *Geophysical Research Letters,* 30, (pp. 1467-1470).

Tighe, M. L. (2012). *Empirical assessment of multi-wavelength, synthetic aperture radar for land cover and canopy height estimation. Ph.D Thesis,* Carleton University, 251p.

Tighe, M.L., D. King, H. Balzter, A. Bannari, and H. McNairn (2012). Airborne X-HH incidence angle impact on canopy height retrieval: implications for spaceborne X-HH Tandem-X global canopy height models. In *XII Congress of the International Society for Photogrammetry and Remote Sensing,* Melbourne, Australia.

Tighe, M. L. and D. Chamberlain (2009). *Accuracy comparison of the SRTM, ASTER, NED and NEXTMap USA digital terrain model over several USA study sites.* ASPRS/MAPPS Fall Conference, San Antonio, Texas, 12 p.

Tighe, M. L., T. Said, and I. Ipranta (2006). Three-dimensional visualization of orthorectified radar imagery and digital elevation models used to generate a 1:50,000-scale geology map for Amurang, Sulawesi, Indonesia. In *Map Asia Conference,* 9 p.

Tighe M. L. (2000). Topographic line maps production using high resolution airborne interferometric SAR. In *International Archives of Photogrammetry and Remote Sensing.* Vol. XXXIII, Part B4. Amsterdam, 2000, (pp. 602-608).

Zebker, H. A. and Goldstein, R. M. (1986). Topographic mapping from interferometric SAR observations. In *J. Geophys. Res.,* 91, (pp. 4993-4999).

Zebker, H. A., Rosen, P. A. and Hensley, S. (1997). Atmospheric effects in interferometric synthetic aperture radar surface deformation and topographic maps. In *J. Geophys. Res.,* 102, (pp. 7547-7563).

Zebker, H.A. and Villasenor, J. (1992). Decorrelation in interferometric radar echoes. In *IEEE Trans. Geosci. Rem. Sens.,* vol. 30, (pp. 950-959).

Chapter 8

AIRBORNE TOPOGRAPHIC LIDAR

Amar Nayegandhi and Joshua Nimetz

INTRODUCTION

Airborne light detection and ranging (lidar) system, sometimes referred to as Airborne Laser Scanning (ALS), is a remote sensing technique used to measure the distance to an object by determining the time of flight for an emitted laser beam. A scanning mechanism (such as an oscillating mirror) is employed to steer a series of laser pulses (typically over 100 kHz) over a wide area from an airborne platform. All airborne lidar systems use enabling technologies such as Global Positioning System (GPS) and Inertial Measurement Units (IMUs) to determine the location and orientation of the remote sensor located on the airborne platform. Airborne topographic lidar systems typically use a near infrared laser pulse to measure topography of the land surface, features and objects on land, such as building structures and vegetation, as well as bare earth topography under vegetation. See Figure 8.1.

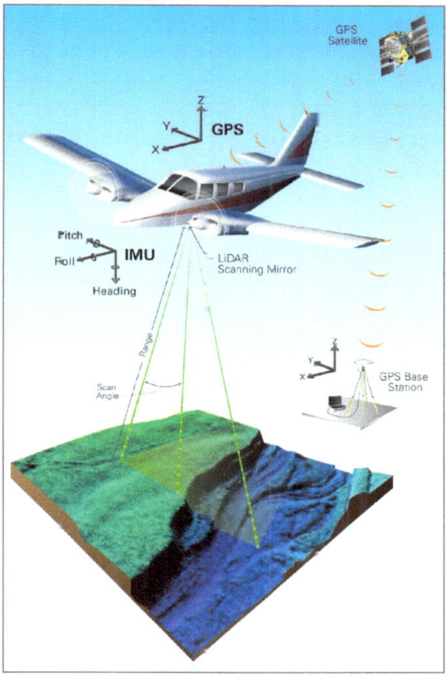

Figure 8.1. Airborne lidar technology is used to measure topography using laser beams directed towards the ground with GPS and IMU systems providing the location and orientation of the airborne platform.

DEVELOPMENTAL HISTORY

When the first working laser was reported in 1960, it was described as "a solution looking for a problem." But before long the laser's distinctive qualities—its ability to generate an intense, very narrow beam of light of a single wavelength—were being harnessed for science, technology, and medicine (Townes, 2010). Today, lasers are everywhere—from research laboratories at the cutting edge of quantum physics to medical clinics, supermarket checkouts, and of course, as a precise ranging device for remote sensing.

The earliest versions of laser remote sensing systems date back to the mid-1960s; in 1965, the first airborne laser profiler was introduced for use in commercial topographic mapping operations. The rangefinder part of the system was based on a helium-neon gas laser operating as a continuous wave (CW) device at the wavelength of 632.8 nanometers. The system was operated in the range of 1,000-5,000 ft (300-1,500 m). Pulsed lasers, in those days, generated a very low pulse repetition rate—only a few pulses per second. As a result, the CW rangefinder was preferred. By the mid-1970s, pulse laser rangefinders based both on neodymium-doped yttrium aluminum garnet (Nd:YAG) solid-state lasers and gallium arsenide (GaAs) semiconductor lasers had been developed that could be used to form the basis of airborne laser profiling systems (Shan and Toth, 2008).

In 1975, Avco Everett Research Laboratory prepared a report to summarize the results under contract with NASA Wallops Flight Facility, the purpose of which was to prepare the specifications and preliminary design for an Airborne Oceanographic Lidar (AOL) System (NASA, 1975). The AOL was subsequently developed by NASA to conduct airborne bathymetric surveys for "coastal water charting, fish school detection, and tracking." The sensor also included a flourosensing capability for oil detection and identification; mineral resource survey; crop growth and disease identification; air sea rescue; ocean temperature, salinity and current determination; identification of hazardous materials and pollutants; algae detection and measurement. The AOL project continued through the 1980s and 1990s as new generations of the AOL were designed and developed at NASA (Wright et al., 2001). NASA also developed the Airborne Topographic Mapper (ATM) in the early 1990s. The Airborne Topographic Mapper (ATM) is a scanning lidar developed and used by NASA for observing the Earth's topography for several scientific applications, foremost of which is the measurement of changing arctic and Antarctic icecaps and glaciers. In 1993, NASA conducted its first ATM mission to measure the Greenland ice sheets, and has been conducting surveys in Greenland every year since. The ATM sensor has been used in scientific coastal studies through a collaboration between NASA, NOAA, and USGS (Brock et al., 2001). NASA also developed laser-based remote sensing systems for vegetation applications and prototypes for developing sensors that would provide high-resolution data from space. SLICER (Scanning Lidar Imager of Canopies by Echo Recover), RASCAL (Raster Scanning Airborne Laser Altimeter), and SLA (Shuttle Laser Altimeter) are examples of the many sensors developed by NASA (Baltsavias, 1999).

As laser ranging devices improved over time, the ability to use scanning mechanisms enabled the mapping of large contiguous areas. In the mid-1990s airborne terrain profilers were transformed into airborne laser scanners and provided terrain elevation values over large swaths of terrain. Furthermore, the practical implementation of the airborne laser scanning technique was also made possible through the successful concurrent development of integrated GPS and IMU technology. GPS/IMU systems determined the position and orientation of the aircraft, which enabled the precise location of the object being mapped by the laser scanning/ranging system. After their introduction in the mid-1990s, airborne laser scanners were at first adopted gradually and rather cautiously (Shin and Toth, 2008). However, since 2002, as the technology has improved and experience gained, they have been used operationally all over the world, and the technology has developed exponentially to provide high resolution ranging devices capable of tracking multiple laser pulses in the air, scanners and other electro-optical devices that are able to transmit and detect individual photons, and create topographic maps at very high spatial resolution and accuracies that can be compared with ground-based survey methods.

Over the past 15 years, the Massachusetts Institute of Technology, Lincoln Laboratory (MIT/LL), Defense Advanced Research Projects Agency (DARPA), and private industry have been developing airborne lidar systems based on arrays of Geiger-mode Avalanche Photodiode (GmAPD) detectors capable of detecting a single photon (Aull et al., 2004; Clifton et al., 2015). The extreme sensitivity of GmAPD detectors allow them to fly from high altitudes and potentially increase acquisition efficiency rates. Up until now the primary emphasis of this technology has been limited to defense applications, despite the potential benefits of applying this technology to non-military uses such as mapping, monitoring critical infrastructure and disaster relief. The first commercial Geiger-mode lidar system, by Harris Corporation, began advertising its capability in early 2015.

Photon-counting lidar systems were in development shortly after some of the first commercially available linear-mode systems began operating (Priedhorsky et al., 1996). The first successful photon-counting airborne laser altimeter was demonstrated in 2001 under NASA's Instrument Incubator Program (IIP) (Degnan et al., 2007). This instrument was flown from altitudes up to 6,700 m above ground level (AGL), operated at a wavelength of 532 nm, and imaged terrestrial and shallow water bathymetry to depths of a few meters over the Atlantic Ocean and Assawoman Bay off the Virginia coast. This instrument was designed primarily to produce, over a mission life of three years, a globally contiguous map of the Martian surface, with 5 m horizontal resolution and decimeter vertical accuracy, from an altitude of 300 km. This type of photon counting instrument is planned to replace the Geoscience Laser Altimeter System (GLAS) on the ICESAT-II mission, slated for launch in June 2018 (Moussavi et al., 2014; Awadallah et al., 2013; Abdalait et al., 2010). The first commercially available single photon lidar (SPL) system was the High-Resolution Quantum Lidar System (HRQLS) by Sigma Space Corporation, which is now known as the SPL-100 system offered by Leica Geosystems.

BASIC CONCEPTS

Light and the Electromagnetic Spectrum

In theory, any light source can be used to create a lidar instrument, but in practice all modern lidar instruments use a definable laser as the source. Laser is an acronym for light amplification by stimulated emission of radiation. When activated by an external energy source, a laser produces and emits a beam or pulse of monochromatic radiation in which all the waves are coherent and in phase. The emitted radiation is also highly collimated and directional, and is emitted as a narrow beam or pulse (constituting a multitude of photons) traveling in a specific direction.

When laser light strikes a surface, some of the light is transmitted, some of the light is absorbed, and the rest is reflected. The amount of each is a function of the type of target material along with the wavelength of light. Lasers are usually classified according to the type of material that is being used as the radiation source. The most common types are gas lasers, solid-state lasers, fiber lasers, and semiconductor lasers. For topographic mapping using airborne laser scanning, where high energy pulses are required to perform distance measurements over long ranges, only certain types of solid-state, semiconductor, and fiber lasers have the specific characteristics—ability to produced high intensity collimated beams—that are necessary to carry out these operations. Many airborne topographic lidar systems that use solid-state crystalline material such a Nd:YAG lasers operate in the near-infrared wavelength range (typically 1064 nm). In the early days of laser ranging, semiconductor material such as GaAs was also used to create lasers for airborne laser ranging. There are other instruments using Er-doped fiber lasers (sometimes referred to as glass lasers) that operate in the 1540 nm range. The laser ranging unit in airborne laser scanning will include the actual laser; the transmitting and receiving optics; and the receiver with its detector, time counter, and digitizing unit.

Absorption and Reflectance

Once light enters the medium, the amount of absorption is a function of the distance traveled within the medium (Renslow, 2012). Hence absorption is expressed as a ratio per unit distance (m-1) and is a function of the thickness of the target object. Absorption also varies by wavelength. For example, the absorption of light by water varies significantly based on wavelength. At the green wavelength of 520-560 nm, the absorption is 0.14-0.04 m-1. At 1064 nm wavelength, the absorption is ~60 m-1. At 1,540 nm the absorption is over 5,000 m-1. Using a laser wavelength that is strongly absorbed by water has disadvantages and advantages. The disadvantage is the inability to penetrate water and potentially detect the bottom of a stream or lake. An advantage is that the laser can be eye safe if it is easily absorbed by water at higher wavelengths. The retinal fluid in the human eye absorbs most of the radiation at higher wavelengths and therefore protects the retina from radiation damage. Fiber lasers operating at 1,540 nm are considered much more eye safe than solid-state lasers operating at 1,064 nm. The green lasers are the least eye safe and special precautions are taken when operating green lasers for airborne topography and bathymetry (see Chapter 10 for more information).

Reflectance is another important property that affects lidar performance. The amount of energy that arrives back to the lidar receiver is directly proportional to the percentage of energy that reflects off the object, or in other words the object's reflectance. The reflectance of the object is wavelength-dependent, and because lidar systems are monochromatic, the reflectance at that particular wavelength determines how detectable an object is given the laser power. Figure 8.2 on page 208 shows the relative spectral reflectance of various common landscape materials. Vegetation is most reflective in the near infrared wavelength range making the 1,064 nm wavelenth solid-state laser most suitable for mapping vegetation. Although the reflectance is low in the 532 nm (green wavelength) range, results from green-wavelength lidar sensors depict good performance in vegetated areas. This is primarily because the green laser power is usually higher than the infrared laser power, and green lasers are typically flown at lower altitudes (for better water penetration) than the near-infrared lasers.

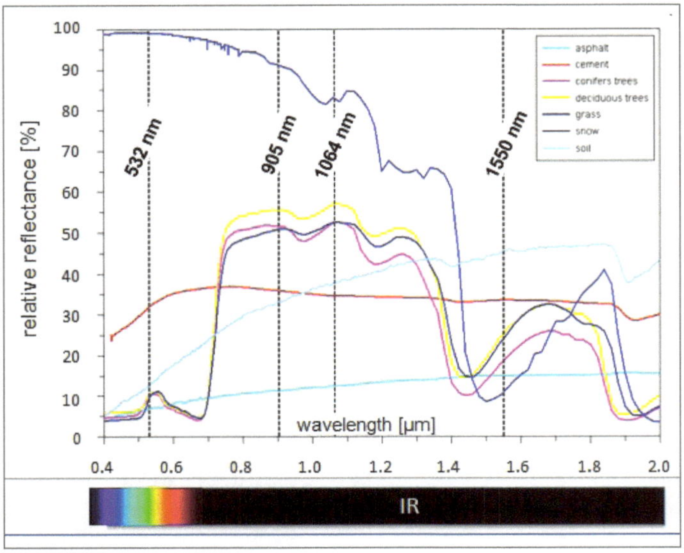

Figure 8.2. Spectral reflectance of various common landscape materials. The common wavelengths used by lidar sensors is also shown. Image courtesy of Riegl, Inc.

Components of an Airborne Laser Scanning System

There are several disparate components in an airborne laser scanning system, that when integrated together help collect the desired 3-D information. The primary components of this system include: (1) The laser and the detector, which provide the ranging information from the source to the object(s); (2) The scanning system, which provide the ability to direct laser pulses in many directions from the aircraft; (3) The inertial navigation system (INS) or inertial measurement unit (IMU), which provides the angular orientation of the laser pulse; and, (4) The global navigation satellite system (GNSS), which provides precise positioning of the aircraft in 3-D space.

Laser Ranging

The laser ranger (or rangefinder) is the device or instrument that is used to measure the slant range or distance from the ranging device to the reflective object. There are two characteristics of laser ranging—the pulse and the detection of the returning pulse. There are also two measurement techniques that are used in topographic applications—(1) The time of flight (TOF) of a pulse or groups of photons within a pulse that bounce off an object and return to the receiver, and (2) The multiple frequency phase comparison or phase shift method for Continuous Wave operation.

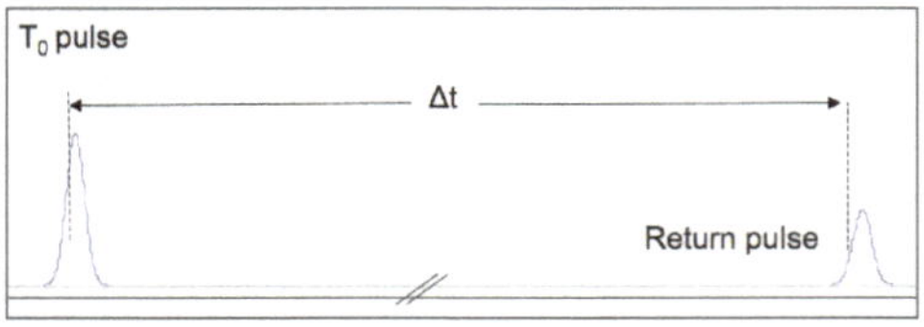

Figure 8.3. Determining time of flight using a pulsed ranging method.

1. TOF: Time of Flight method is the most common method used in airborne topographic lidar to measure distances. The distance (or range) to the target is determined by the one-way time of flight of the laser pulse multiplied by the speed of light (Figure 8.3). The slant range from the measuring instrument to an object requires the very accurate measurement of the time required by a short pulse of laser radiation to travel to the target and back. The timer is a high-speed counter that is controlled by a very stable oscillator. To achieve a 1 cm ranging resolution, the time should be able to measure a 66 pecosecond (ps) interval, which would require a clock rate of about 15 gHz (Shan and Toth, 2008)

2. Phase Comparison Method: The phase comparison method is used with those rangers where the laser radiation is emitted as a continuous wave—often referred to as a CW laser—instead of discrete pulses. In practice, due to the limited power of the CW laser, there are very few laser rangers, profilers, or scanners of this type in operation from the airborne platform as compared with those rangers using the TOF method. The phase comparison technique is often used in short-range terrestrial or ground-based scanners where the distances to be measured are typically less than 100 meters. The transmitted beam in the CW approach comprises a basic carrier signal on which a modulation signal has been superimposed for the purpose of measurement (Shan and Toth, 2008). This modulation signal or measuring wave is derived from and held at a constant value using a stable frequency oscillator. The measuring wave is used to control the amplitude of the carrier wave—a process known as amplitude modulation. The reflection of the signal for a CW laser is similar to that for a pulsed laser, except it is continuous. The transmitted beam strikes the object being measured; a small part of that signal is reflected back along the transmit path for that beam, where it is detected and amplified. The signal is then demodulated—which is the separation of the measuring and carrier waves. The reflected signal is then compared with the signal of the original transmitted beam. The travel time is directly proportional to the phase difference between the received and transmitted signal. The difference in phase between the two signals is measured and represents the fractional part of the total slant range. The integer number of wavelengths between the rangefinder and target is calculated by resolving the ambiguity using a multi-frequency phase measuring system, where the frequency of the emitted beam is changed rapidly within the system to count the wavelengths

Transmit Laser Pulse Characteristics

The pulse energy, measured in micro Joules (µJ), is simply the total energy of the laser pulse. Pulse duration or pulse width, measured in nanoseconds (ns), is typically defined as the time during which the laser output pulse power remains continuously above half its maximum value. Time characteristics of the laser pulse, such as rise time, fall time, and pulse width, are very important for ranging systems. Small pulse width and short rise time enables precise range measurements. The full width at half maximum (FWHM) is used to define the duration or width of the laser pulse.

Beam divergence refers to the angular increase in beam diameter that occurs as the distance between the laser instrument and a plane that intersects the beam axis increases. Even a highly coherent laser beam is diverging with the distance from the emitting source, where its diameter is minimal and known as laser beam waist (Renslow, 2012). The beam divergence and the diameter of the beam waist characterize the quality of a laser beam. Beam divergence is typically defined as the half-angle beam divergence and expressed in milliradians. A wider beam divergence will create a bigger "spot size" on the target and thereby reduce precision and accuracy, especially when the target is sloping or changing rapidly.

The pulse energy of topographic lidar systems are typically low (10-100 µJ) to allow for a tightly focused beam with low beam divergence that is also eye safe. Bathymetric lidar systems have pulse energies up to 7 mJ, which are typically much higher than the near-infrared lasers used in topographic applications. The higher power is needed for the laser pulse to penetrate through the water column to map the bottom. The bathymetric sensors with very high laser pulse power also have a large footprint so that the energy is spread across a larger area for eye-safety reasons. Refer to Chapter 10 for more information on airborne lidar bathymetry.

The pulse width determines the range resolution of the pulse in multiple return systems (explained below), or the minimum distance between consecutive returns from a pulse. Pulse width is usually measured full width at half maximum (FWHM), as shown in Figure 8.4 on page 210. Traditionally, pulse widths for topographic systems have been in range of 10 ns. Light travels a distance of approximately 30 cm in 1 ns, thereby measuring a two-way distance of ~15 cm after it bounces off a target in 1 ns. If the pulse width is 10 ns, the laser pulse itself is ~3 m long, which will affect its ability to discriminate between targets that are close to each other (such as vegetation structure). Newer laser technology has enabled the use of much shorter pulse widths (1-2 ns) for topographic and topobathymetric applications. For topobathymetric applications, a short pulse width laser enables the separation of a return from the water surface and bottom in very shallow water

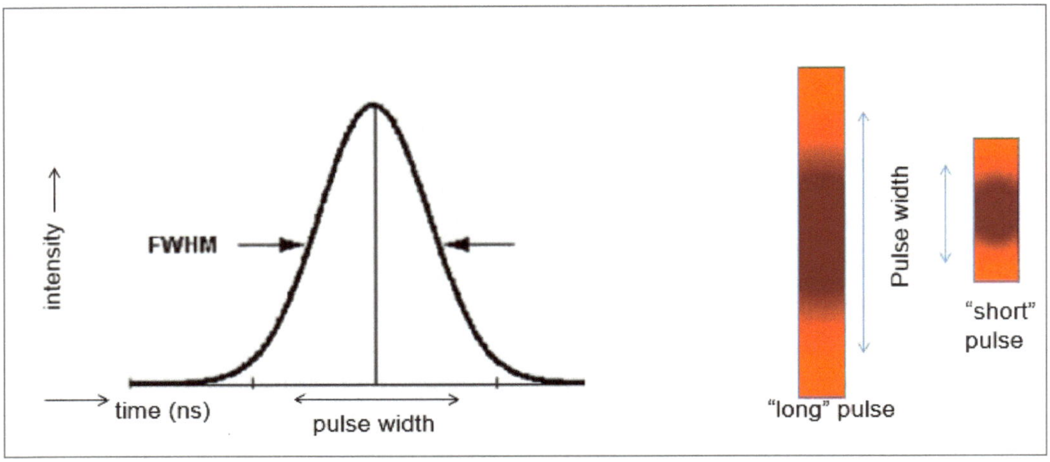

Figure 8.4. Characteristics of a transmitted laser pulse estimated as a Gaussian. The total energy in a pulse is not uniform, but is maximum at the center of the pulse.

depths. If the pulse energy is kept low, the pulse width can be reduced significantly. Photon-detecting sensors typically have very low laser power and hence can have pulse widths in the 600-800 picosecond range.

Pulse Repetition Frequency (PRF): The PRF, measured in kilohertz (kHz), is the number of pulses emitted by the laser instrument in 1 second. Older instruments emitted a few thousand pulses per second. Modern systems can support frequencies of 500 kHz and newer technologies are now enabling two lasers channels to be used in conjunction with the same scanning mirror, thereby producing effective PRF of 1 mHz or higher. Many systems allow different settings for the PRF. This is usually done to allow the systems to fly at different flight altitudes and speed. The PRF is directly related to the point density on the target. For example, a system operating at 167 kHz from the same flying altitude will have higher number of returns than when operating at 100 kHz. Equivalently, a high PRF system can generate desired return densities by operating on an aircraft that flies higher and faster than an aircraft carrying a lower PRF system, thereby reducing flying time and acquisition costs. PRF may be limited by a sensor's ability to track multiple pulses in air, often referred to as MPiA or Multiple Time Around (MTA). Most new sensors have the ability to track multiple pulses in air by closely tracking the zones within which to expect each pulse to return to the sensor. Knowledge of the flight altitude plays a key role in determining the MTA zones. The sensor manufacturers use proprietary techniques during data collection and in post-processing to resolve the MPiA/MTA issue.

Quantitative relations between pulse width, pulse energy, and pulse peak power as well as relations between these parameters and range measurement precision and maximum measurement range are well documented (Baltsavias, 1999). Systems that have higher PRF typically have lower pulse energy. The peak power is inversely proportional to the pulse width, hence narrow pulses (which improve resolution and accuracy) can increase the peak power of the pulse. Modern lasers operate at high PRF and use very narrow pulses (1-2 ns) so as to increase the peak power to maximize operational range.

Laser Scanning Techniques

An optical scanning mechanism is attached rigidly in front of the output end of the laser ranging unit. As the name suggests, it uses an optical element such as a rotating plane or polygon mirror or a fiber-optic linear array to send a stream of laser pulses at known angles and at high speed in the cross-track direction relative to the airborne platform's flight path. The combination of the cross-track scanning pattern along with the forward motion of the aircraft allows the elevations of the ground surface and its objects to be determined for a wide swath of the terrain. Several scanning techniques have been used in airborne lidar systems. In theory there are no special reasons why one scanning technique is preferable to another. However, the primary goal of the scanning technique is to create a wide swath with consistent along- and across-track point spacing, and reliable and accurate elevations for the entire swath. The most common scanning techniques are the oscillating mirror and rotating mirror.

In systems using an oscillating mirror, the mirror rotates back and forth between limited extents, producing a zigzag (i.e., sinusoidal pattern) line on the surface of the target area (Figure 8.5). The mirror is always pointing downwards towards the ground so data collection can be continuous and theoretically all pulses of the laser can be used. The field of view and scan rate can be set by the operator prior to acquisition. Changing the field of view provides additional flexibility as it allows laser pulses to be collected over a shorter span (denser data) or a wider span (further apart). Although the oscillating mirror is the most widely used scanning mechanism for airborne lidar systems, there are inherent disadvantages of using of the oscillating mirror principle. The changing velocity and acceleration of the mirror as it oscillates from one end to the other causes unequal spacing of the laser pulses on the target. The point density increases along the edges of the scan where the mirror slows down, and decreases along the center in the along-scan direction. The forward motion of the aircraft causes the zigzag pattern with varying point spacing along the edges of the scan in the cross-scan direction. Manufacturers have solved these problems by essentially ignoring the outlier points on a scan and modeling the distortions caused by changing speed using a computer algorithm.

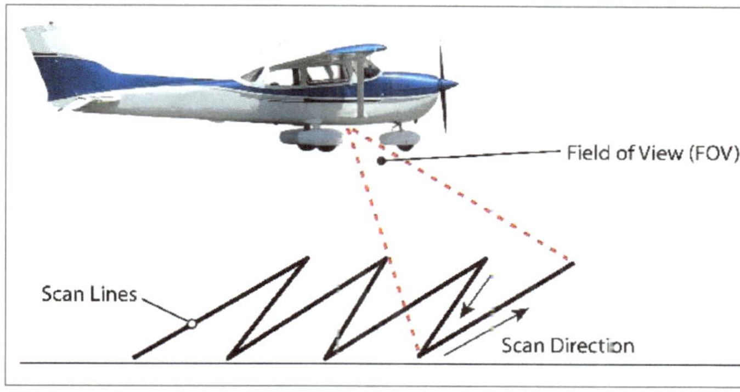

Figure 8.5. Sample scanning pattern produced by an oscillating mirror.

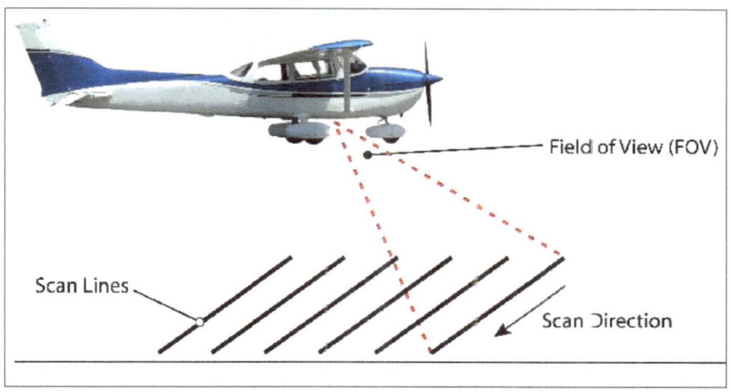

Figure 8.6. Sample scanning pattern produced by a rotating mirror.

The rotating mirror is another commonly used scanning mechanism for airborne lidar systems. In this approach, the mirror is rotated continuously at a constant velocity in one direction producing a parallel line scan (Figure 8.6). The constant velocity ensures that there are no acceleration type errors encountered in the oscillating mirror scanner. The point spacing is also more regular both along and across the scan. However, the biggest disadvantage is that observations cannot be taken during a significant time during each mirror rotation when the mirror is pointing away from the target. Typically, 30-40% of the emitted laser pulses are not aimed at the target area and are essentially lost to the scanning mechanism.

Other scanning mechanisms less commonly used include the push broom (fiber scanning) pattern where the laser pulsed energy is transmitted into one of the fibers arranged in a circle producing a nutating scan pattern (Figure 8.7) and the Palmer scanner that produces an elliptical scanning pattern with redundant data that can be used for calibration or to get a forward and aft view of the same target (Figure 8.8).

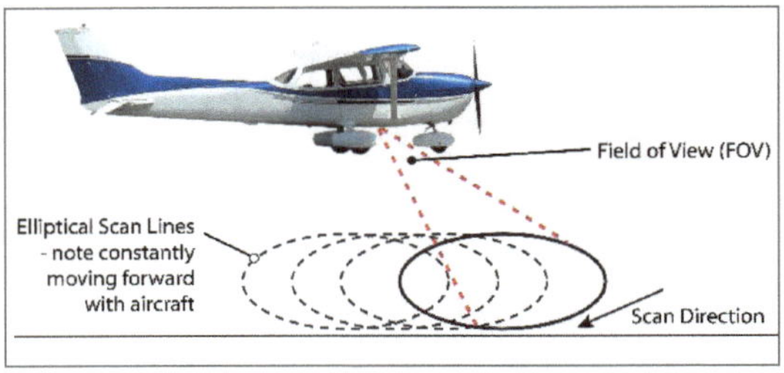

Figure 8.7. Sample scanning pattern produced by a push broom fiber scanner.

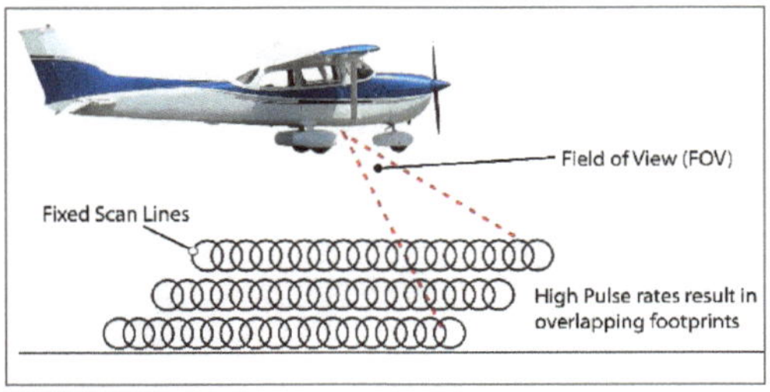

Figure 8.8. Sample scanning pattern produced by a Palmer or elliptical scanning pattern.

Detection Methodology

After the laser pulse is sent and directed via some kind of scanning technology, the return pulse is detected and timing information is recorded. This detection can be performed in several different ways, and methods usually depend on the applications being used.

Detector Types

The type of detector used in an airborne laser ranging system is a very important consideration during the design of the system. There are several factors that determine the type of detector that would be used for a lidar system. The optical detector used in linear or photon-based lidar is typically either an Avalanche Photodiode Detector (APD) or a Photo-Multiplier Tube (PMT). The primary function of the optical detector is to convert the optical signal into an electrical current in the most efficient manner. Detectors can be triggered at different amounts of returned laser energy—from being triggered after detection of a single photon, to recording the entire signal returned per pulse (e.g., waveforms). Most linear mode systems operating in the near-infrared wavelength range currently use the APD detector to receive the returned laser pulse.

Detection methodologies of airborne lidar systems are generally categorized as follows:

1. The detection of the entire laser pulse after it reflects off an object and returns to the receiver, often referred to as "linear" or "linear mode" lidar
2. The detection of a photon or groups of photons from the returned laser pulse, generally referred to as "photon based" or "photon counting" lidar. Photon based lidars are typically used in conjunction with a focal plane array detector, referred to as 3-D flash lidar technology, and can be further categorized into "single photon" and "Geiger-mode" lidar

Linear vs. Photon-Sensitive and Geiger-Mode Lidar

In an airborne lidar system, a laser pulse is transmitted to illuminate the target. The laser pulse consists of thousands of photons transmitted at a specific wavelength. A subset of these photons bounce off the target and are received by the sensor in the aircraft. The received signal is composed of reflected laser energy and noise sources—internal detector noise (also called "dark counts"), solar backscatter noise from the target and the intervening atmosphere within the FOV, and laser backscatter from the atmosphere within the selected range gate. If the optical detector on the receiver detects the entire laser pulse as a single entity, by receiving thousands of reflected photons per pulse in order to exceed the detector noise floor and have sufficient signal to accurately determine the range to the illuminated target, then the system is referred to as a linear mode (or linear or conventional) lidar sensor. In this case, the optical receiver is sensitive to and will trigger when several hundred or thousands photons are detected. The detection could be based on a first return, last return, multiple return, or as a waveform. The term linear mode, as used in this chapter and in the airborne lidar industry in general, really came about to differentiate itself from the photon-sensitive lidar sensors. The original reference to linear systems related to an older analog signal processing version of lidar, and not the digital and waveform detection capabilities available today. However, linear lidar is more broadly used for any lidar sensor that detects the received pulse as a single entity. Linear lidar generally has a high signal-to-noise ratio and therefore does not record single photon solar or dark count events.

As the name suggests, photon-sensitive lidar sensors detect individual or groups of photons that bounce off the target and are received by the sensor. Only a few photons are required to trigger the receiver to detect a return. Photon counting detectors record the arrival of single photons. The detector is highly sensitive to individual photons and can be easily triggered by stray photons from the same laser source or photons from another light source, such as sunlight. Photon counting detectors have very low dark counts to reduce the rate of the detector noise compared to the rate of received signal photons. The single photon is a sample from the full distribution of surfaces illuminated by the laser pulse, and accumulation of many single photon ranges can recreate the height of the target. During daytime operations, solar illumination reflected from the Earth's surface or from clouds is a significant source of noise for photon counting systems, more so than for linear systems in which the detector is the dominant noise source (Shan and Toth, 2008). The amount of solar background noise is typically controlled by a combination of narrow bandpass filtering to block detection of energy at all wavelengths other than that of the laser, and using a small receiver field-of-view, thereby restricting collection of light to the location illuminated by the laser. Photon-sensitive lidars generally have a very low signal-to-noise ratio.

The term Geiger-mode comes from the use of the Geiger-mode Avalanche Photo-Diode (APD) detector for laser ranging applications. It is also known as Single Photon Avalanche Diode (SPAD). It is a solid-state photodetector in which a photon-generated carrier can trigger an avalanche current by exceeding the breakdown voltage of the p-n junction between two types of semiconductors. The device is able to detect low-intensity signals (down to the single photon) and measure the arrival time of the photons with a jitter of a few tens of picoseconds. The biggest advantage of using a Geiger-mode detector is the ability to detect range using very few photons. This allows ranging to be done using a low-power pulse with short pulse-width from very high altitudes by using an array of Geiger-mode avalanche photo diodes. Each of the APDs in the detector array is biased above the breakdown voltage so that a single photon may trigger the APD with a certain detection probability. Geiger-mode APD arrays are highly sensitive to dark count rate and also photons due to solar background radiation. The Geiger mode sensor offered by Harris Corporation uses the Geiger-mode APD detector and is explained in the section, below, entitled: Harris Geiger-Mode Lidar. The single photon lidar system SPL100 offered by Leica Geosystems also uses an APD detector and is explained in the section, below, entitled: Leica Geosystems. Both these photon-sensitive sensors use a variant of focal-plane array detector technology, also known as 3-D imaging, as described below and in the section on Current Sensor Technology.

Photon-sensitive 3-D imaging lidars have several advantages over linear lidars. Each range measurement requires only a single detected photon as opposed to hundreds or even thousands in linear-mode pulse time of flight or waveform lidars. The high photon sensitivity enables higher flying heights and spatial resolution, larger swaths, and greater areal coverage. The output laser power for photon-sensitive lidars is much lower than linear lidars, which also allows for much faster system response time and thereby capable of recording returns from objects that are very close to each other (e.g., within a canopy structure). However, there are several disadvantages as well. Photo-sensitive sensors are much more sensitive to noise originating from dark counts, solar backscatter, as well as laser backscatter from the atmosphere. Data editing for photon-sensitive lidar sensors is also a lot more challenging. Linear lidars operate at high detection thresholds, thereby nullifying solar noise. However, photon-sensitive systems requires a substantial amount of noise editing. Furthermore, a lot more data are collected and stored by the photon-sensitive sensors due to the presence of noise in the data.

3-D Flash Lidar Technology

3-D flash lidar systems are analogous to a digital camera with a flash illumination, but with the flash being provided instead by a pulsed laser and by use of a flash-focal-plane array (FFPA) detector with clocks in each readout integrated circuit pixel cell to determine the time needed for the flash to depart, reflect from a target, and return (Renslow, 2012). The TOF of the reflected laser pulse gives a range measurement along with intensity for each pixel in the image. Analogous to scanning lidars, this information set over the range of all detector pixels is referred to as a 3-D lidar image or point cloud because the scene includes both spatial and range information. In the context of imaging lidar, the term image and point cloud are synonymous.

The entire scene within the sensor's field of view (FOV) is imaged with a single flash of the laser on the order of a nanosecond (Renslow, 2012). This produces an image that correlates each pixel in time, which makes the sensor insensitive to any relative motion between the platform and target and eliminates the need for complex processing algorithms to account for relative motion or laser scan time over the FOV. There are several advantages to 3-D flash array technology. The only limitation on the data rate of a flash lidar is the size of the focal-plane array. Consequently, there is no fundamental upper limit to the resolution as there is with raster scanning lidars. One only needs to look at the remarkable progression in pixel density in the digital 2-D camera world to see what is possible. The sensor frame rate and laser pulse rate can be synchronized to produce full motion 3-D video imagery. Each individual point in the point cloud can be fused (colored) with red-green-blue (RGB) information from a visible wavelength camera or with intensity information from a camera sensitive to other wavelengths. Perhaps the most significant advantage of a 3-D flash lidar is that each pixel of the resulting 3-D imagery array is correlated in time. This, and the fact that all mechanical motion is slow compared to the pulse width, makes the sensor insensitive to relative motion between the sensor and the target. This eliminates the need for complex processing algorithms that account for relative motion or laser scan time of the FOV. This is in sharp contrast to a spot scanning lidar where a laser beam is raster-scanned over the sensor FOV and each pixel intensity and range must be separately correlated with and filtered against platform motion. The flash lidar system has an advantage in that the platform relative motion can be applied to the entire array of pixels, greatly reducing processing requirements and enabling real time processing of the data into actionable intelligence products (Renslow, 2012). Flash lidar technology is also inherently efficient with mass, power, and volume. This is because the system requires only one set of receiver optics, one detector, and no mechanical scanner. Because the laser is only used for broad area illumination, there is no complex mechanism needed to control laser pointing nor is there a need for precision alignment between the laser and receiver optics. Use of an imaging focal plane array yields a self-aligning system, thus significantly simplifying assembly, integration and testing.

Flash imaging lidar cameras have been incorporated into a wide variety of remote sensing instruments by Ball Aerospace and Technologies Corp (BATC), and with their video co-processing architecture, form the basis of Ball's real-time technology (Renslow, 2012). 3-D flash lidar cameras are available from Advanced Scientific Concepts, Inc. (ASC 3-D) and Raytheon Vision Systems (RVS). The technology is being used for Space Exploration, unmanned air and ground vehicles, and surveillance.

Note that 3-D flash lidar is essentially a type of detector and is also used in other systems and sensors, including single photon and Geiger-mode sensor technologies. In these sensor technologies, flash lidar is used in conjunction with a scanning mirror. The combination of a focal plane image array, the pulse repetition frequency of the laser, and the scanning rate can significantly increase the effective data collection rate (and hence data density) for the scanner. For example, a 128 x 128 image array combined with a 30 kHz laser pulse rate used in conjunction with a scanner operating at 30 Hz, will produce over 14 million first return elevation measurements in 1 second.

Discrete Return vs. Waveform-Resolving Lidar

As the laser beam, although usually collimated to a divergence of less than 1 mrad, may hit not just a single target object, it is beneficial from a user's point of view to get all of the ranges to the targets that the laser pulse has interacted with and the respective echo signals exceed the detection threshold of the receiver (Ullrich and

Pfennigbauer, 2011). Providing more than just one target range per laser pulse is usually addressed as multi-target capability. Laser range finders based on the pulsed time-of-flight principle are capable of providing multiple targets per laser pulse.

However, there are fundamental limits to the multi-target capability: the width of the laser pulse and the receiver's bandwidth, both having an impact on the so-called system response of a lidar system. System response is usually defined as the response of the lidar to a single target, either a flat target perpendicular to the laser beam axis or to a "point target," where the area of the target's extension is small compared to the laser footprint across the beam axis and small with respect to the laser pulse length along the beam axis. This system response is mainly determined by the laser pulse width and the signal bandwidth of the receiver. In every well-designed lidar system the signal bandwidth will nearly match the signal bandwidth of the laser pulse to optimize signal-to-noise ratio and thus maximum range. The system response limits the power to resolve echo pulses from nearby targets, as the finite pulse width of the system response will lead to merging of the target echoes if the temporal difference is less than the pulse width. The waveform of the interaction of the laser pulse with complex target constellations can be seen as the convolution of the lidar system response with the backscatter profile of the complex targets. Even if the backscatter profile has very high "frequency components," the resulting echo waveform bandwidth is limited to the bandwidth of the system response due to convolution.

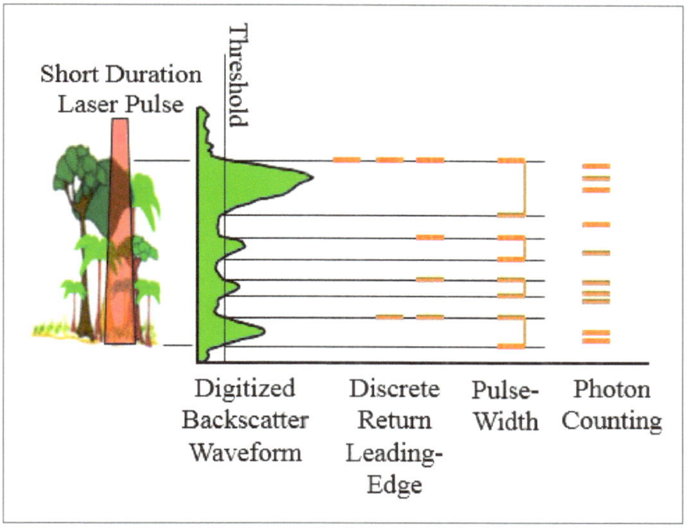

Figure 8.9. Schematic showing various detection techniques used by airborne topographic lidar systems. Image courtesy of D. Harding, NASA GSFC.

Early versions of airborne lidar systems were capable of recording first-return-only pulse at low pulse repetition rates. However, more advanced lidar systems can record simultaneously multiple returns for each transmitted pulse, and the reflected intensity at each return (Figure 8.9). Multiple return lidar systems are very useful in forestry applications or even to derive bare Earth topography under vegetation. When the laser beam from a multi-return system interacts with a tree canopy, then the first return is usually assumed to arrive from the top of the tree (or where the transmitted laser beam first interacts with the tree canopy). The last return may interact with the ground underneath the tree, although the ability to map the ground is largely dependent on the density of the vegetated canopy. Intermediate returns, perhaps 2nd, 3rd, and 4th, are expected to be caused by tree branches and understory vegetation between the top of the canopy and the ground.

A case of multiple returns from a single pulse can also happen when a laser beam reaches the roof of the building. If the beam is broad enough, it can partly hit the roof edge of a building and the rest of the beam can continue along its slant range to reflect off the ground. The receiver on the aircraft can detect two returns from that same laser pulse, where the "first return" can be used to compute the elevation of the rooftop, and the "last return" can be used to compute the elevation of the ground.

A number of factors limits the capability of a discrete-return lidar sensor in deriving forest measurements (Nayegandhi and Brock, 2008): (1) The inability of a narrow-beam laser pulse to strike the tops of trees consistently, especially in conifer stands; (2) The narrow-beam pulse may penetrate through gaps all the way through to the ground without intercepting the canopy, thereby indicating the absence of a canopy; (3) Conversely, the pulse may be entirely blocked by dense canopy materials (leaves and branches), thereby not reaching the ground; (4) Most discrete-return systems require a minimum vertical object separation to register consecutive returns from the pulse separately, thereby being blind to canopy material within the "dead zone." This dead zone, which typically ranges from 0.5 m to 4 m in vertical height is usually caused by the pulse width and other hardware limitations in the sensor design (Nayegandhi et al., 2006). Figure 8.10 is a schematic illustrating the effect of the dead zone between successive discrete returns D1 and D2, wherein reflected laser backscatter from canopy materials between these reflections are not distinguishable by the sensor unless the backscatter is digitized to represent a waveform.

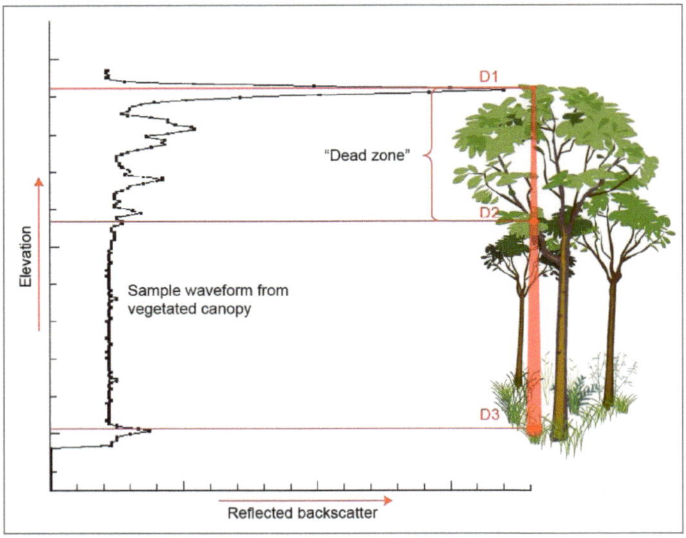

Figure 8.10. Schematic showing the dead zone effect in discrete return lidar systems. "D1," "D2," and "D3" are successive reflections from a discrete return laser pulse. The laser pulse is effectively blind to canopy material between "D1" and "D2." The vertical structure information is, however, captured in the sample waveform from a vegetated canopy acquired by a waveform-resolving system (from Nayegandhi and Brock, 2008).

A full-temporal waveform-digitizing lidar system senses and records lidar backscatter returned to the sensor in a series of equal time intervals, usually in 1 nanosecond increments (Nayegandhi and Brock, 2008). The high sensitivity of the receiver optics and electronics results in the characterization of small volume changes in vegetation and a detailed description of vertical canopy volume distribution, which are not always separated by discrete-return lidar systems (Figure 8.10). Digitized return waveforms can give detailed insights into the vertical structure of surface objects, surface slope, roughness, and reflectivity. In coastal forestry applications, vertical information from waveform-resolving lidar gives details on canopy and sub-canopy structures and the potential to derive accurate ground information under short herbaceous vegetation (Nayegandhi et al., 2006).

Full waveform digitization of a laser pulse creates a pulse-by-pulse reflection that is highly sensitive to even minor changes in vegetation structures, especially when using a short-temporal-pulse-width laser pulse (Figure 8.10). Small variations in the vertical structure can be detected by processing the resulting waveforms using a variety of pulse detection methods. Wagner et al. (2004) report a number of detection methods that can be applied on the backscatter waveform: threshold, center of gravity, maximum, zero crossing of the second derivative, and constant fraction. Determining the range in post-processing software has the advantage of selecting one or more pulse detection algorithms based on the application, analyzing the intermediate results, and considering neighborhood relations of pulses. For example, canopy-height estimation may be approved by detecting small inflections or "sub peaks" in the waveform. Figure 8.11 illustrates the result of using different pulse detection methods to derive the range to the first and last return from short vegetation. Reflections from short shrubs, grass, or other herbaceous vegetation cause the return pulse to widen. The early inflection in the waveform at height A is detected by determining the zero

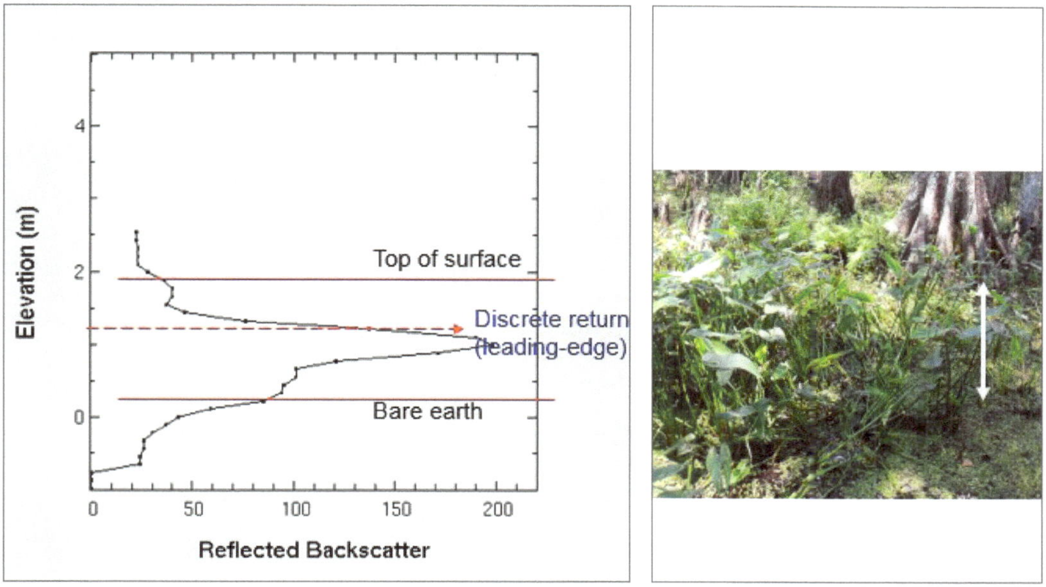

Figure 8.11. Sample waveform illustrating the widening of the return pulse from low herbaceous vegetation. The red lines indicate the height of the top of the vegetated surface and the bare earth derived using post-processing of the waveform. The dotted red arrow indicates the range being resolved to only 1 discrete return.

crossing of the second derivative at a low backscatter threshold. Similarly, the range to the last return can be determined by analyzing the waveform starting from the trailing edge, thereby selecting the small inflection at height B from the convolved waveform (Figure 8.11). The first return represents the height of the vegetated surface, and the last return represents ground topography. If this area were mapped by a traditional discrete return lidar sensor using real-time leading-edge threshold detection to determine the range, it would have resolved the range to only one significant return (dashed line in Figure 8.11).

One of the main disadvantages of full waveform data is simply the sheer volume of the data that is recorded. This impacts storage and the time it takes to convert these data to information. Also, currently there is not an agreed upon standard file format across platforms; however the PulseWaves specification seems to have the highest amount of adoption by the community *(github.com/PulseWaves/Specification/blob/master/specification.rst)*. Use of waveforms directly are still relatively rare. Most users tend to convert waveform information into discrete returns for actionable use.

Georeferencing

Unlike traditional photogrammetry, which can use ground control and the collinearity condition to georeference aerial imagery to a known coordinate system, airborne lidar requires direct sensor-assisted georeferencing from the GNSS-inertial subsystem. The GNSS-inertial subsystem is primarily comprised of the GNSS antenna (mounted on the airframe) and receiver, IMU, and an onboard processor. The GNSS data provides a tie to real-world coordinates and the IMU captures the attitude of the lidar sensor.

The GNSS antenna-lidar sensor reference frame is tied together by way of the external lever arm values X, Y, and Z. The external lever arm values define the vector between the antenna phase center and the external lidar sensor reference point. A laboratory-calibrated lever arm between the external sensor reference point and the internal sensor reference frame, which may be the optical center of the scanning mirror, is also defined. If these lever-arm vectors are improperly measured or signed, the projected lidar points will be linearly displaced proportional to the measurement error. Also, an improperly measured external lever arm will compromise the processing algorithms in the GNSS-inertial processing software. The GNSS-lidar sensor reference frame vectors define the exterior orientation of the lidar system. The interior orientation is defined by the vector and mounting angles between the IMU reference point (perhaps the theoretical center) and the lidar sensor reference frame. Accounting for the above lever arms and IMU mounting angles are necessary for the lidar ranges to be georeferenced.

Creation of the point cloud data begins with processing of the airborne GNSS-inertial data. There are a variety of processing methodologies associated with the GNSS-inertial subsystem. Nearly all involve differential correction, which is also known as relative carrier phase positioning. One particular methodology, precise point positioning (PPP) does not rely on the use of static reference stations on the ground. PPP relies on use of the precise ephemerides (published by the International GNSS Service) and a wide-area augmentation system (WAAS). However PPP is seldom used in airborne lidar operations because the absolute accuracy of the processed positions are often not sufficient for airborne lidar data requirements.

At the present time, there are two market-share leaders in the airborne lidar GNSS-inertial subsystem arena. These are Applanix, a Trimble company headquartered in Richmond Hill, Ontario, Canada and NovAtel, a Hexagon company headquartered in Calgary, Alberta, Canada. Optech and Riegl employ the Applanix position and orientation system, airborne vehicle (POS-AVTM) while Leica employs the NovAtel synchronized position attitude navigation system (SPANTM). Both of these GNSS-inertial subsystems are very similar: They rely on modern GNSS receiver and antenna technology and accurate IMUs. There are different models of POS-AVTM and SPANTM GNSS-inertial subsystems used by the airborne lidar industry. The primary difference between these models is the quality of the IMUs. A recent trend in the industry is to use the most accurate IMUs available and therefore the highest grade GNSS-inertial subsystems.

Applanix and NovAtel have proprietary software to process the GNSS-inertial data. At present time, this software is Applanix POSPacTM MMS 8 and NovAtel Inertial Explorer®. The software integrates the airborne GNSS observable data with the IMU data. Generally speaking, a majority of the user interface within these software packages involves the differential correction of the airborne antenna position using the ground reference station(s). An important but less involved function is the aligning of the IMU data to the airborne GNSS data by way of precise timestamps.

Applanix and NovAtel hardware and software enable an integrated processing routine known as tight-coupling. Tight-coupling involves using real-time and post-processed data from the GNSS and IMU to overcome limitations occurring from each technology. During the lidar flight, real-time GNSS positions and velocity are used to constrain IMU drift, which is the propagation of error within the IMU sensor sets. In tightly-coupled post-processing, the IMU data are used to resolve satellite integer ambiguity (the number of carrier wave cycles between satellite and airborne receiver) during periods of signal outage, which can occur during banking or pitching of the aircraft.

Tightly-coupled post processing integrates the GNSS, typically recorded at 1-2 Hz, and the IMU, typically 200 Hz for IMUs used in airborne lidar sensors and digital cameras, in a single 1 Hz Kalman filter. This Kalman filter process "smoothes" the data and results in interpolated GNSS-derived geographic positions at 0.005 second intervals (200 Hz) aligned with the IMU data. This process is typically reliable and results in the most accurate trajectory file. However, sometimes the Kalman filtering can smooth the data incorrectly and result in excessive trajectory error. When this happens, it may become necessary to reprocess the trajectory data in a loosely-coupled fashion. Loosely-coupled processing involves processing the airborne GNSS independently of the IMU data. Tightly-coupled processing has been commercially available since late 2007, early 2008 when Applanix came out with the POSPac MMS software.

As mentioned above, the majority of the user interface with the GNSS-inertial software involves the differential correction of the antenna position. Differential correction is the approach behind all GNSS surveying applications requiring the highest degree of positioning accuracy. The underlying principle with differential correction is bias correlation using both the code signal and (primarily) carrier waves transmitted by the satellites. It's important to note that not all biases can be correlated. For example, multipathing of the satellite signals prior to landing on the antenna are not correlated because this particular error source is antenna-dependent.

Biases in aircraft position are correlated to one or more static reference stations on the ground. Bias correlation makes it possible to correct positional error in the airborne GNSS data by applying the same correction vectors used to move the reference station back to its known coordinates. Error sources affecting the positional accuracy of the GNSS data are caused by satellite clock bias and drift, satellite orbits, and other sources such as tropospheric delay. In order for bias correlation to work, the airborne GNSS antenna should be in close proximity to the ground GNSS reference receiver both at the beginning and the end of the lidar flight. Additionally, the airborne GNSS receiver should never be too distant from the ground reference receiver. Exactly how distant is too distant can vary significantly depending on several factors. These factors include maintenance of carrier phase integer lock or the ability to quickly resolve loss of integer lock within the airborne receiver and the severity of error sources acting upon the receivers (e.g., clock, orbital).

The distance between GNSS receivers is known as a baseline. According to Van Sickle (2008), if the carrier phase observable is used in relative positioning, baseline measurement accuracies of ± (1 cm + 2 ppm) are achievable. It is possible for GPS measurements of baselines to be as accurate as 1 ppm or even 0.1 ppm. For example, a 64 km (~40 mile) baseline may have measurement accuracies for the relatively corrected antenna position on the order or 2-13 cm; quite a large swing! It's worth noting that the above calculation by Van Sickle pertains to relative positioning with GPS (not GNSS) in a static environment, under ideal conditions. Modern lidar systems employ GNSS-enabled receivers and operate in a kinematic environment. Conversations with Van Sickle indicate that accuracies have not necessarily improved with the modernization of the GPS constellation nor the integration of other global and regional augmentation constellations including the Russian GLONASS, European Galileo, Chinese Beidou, Japanese QZSS or American satellite-based augmentation system (SBAS)—collectively, part of the GNSS. What has improved with the inclusion of these newer constellations are efficiencies in carrier phase positioning as there are a far greater number of satellites from which to observe signals.

In practice, relatively corrected positions for GNSS receivers connected to modern lidar systems are found to be closer to 5-7 cm absolute Z accuracy on average (half that for X,Y accuracy) for baselines on the order of 40 miles or less. This is evident in the post-processed statistics and plots provided by the GNSS processing software. An inference can also be made here to correlate the accuracy of the relative positioning of the lidar's GNSS-inertial subsystem to the typical Gaussian distribution and small standard deviations found in the absolute accuracy assessment of properly processed lidar point clouds to ground surveyed points of higher accuracy.

Another benefit of differential correction is the static reference station(s) establishes the processing reference frame for the remote airborne receiver during post processing. For example, holding NAD83 (NA2011 2010.0) coordinates for the static reference station make it possible to process the airborne GNSS data on the same NAD83 (NA2011 2010.0) reference frame, even if the native reference frame of the GNSS data is referenced to the International GNSS Service's (IGS) International Terrestrial Reference Frame (ITRF). This is very relevant to airborne lidar mapping projects done in the U.S. because data deliverables are often required on the most current realization of NAD83 (2011 2,010.0 at the time of this writing).

When collected and processed correctly, the airborne GNSS will comprise a relatively small and consistent portion of the lidar point cloud error budget. Also, the error associated with the Assisted GNSS (AGNSS) is primarily in the vertical (typically, 2-3 times the horizontal). A caveat here: It is possible that an incorrectly fixed cycle integer occurring with one or more of the satellites can adversely affect the positional accuracy of the AGNSS positions. Likewise, poor Kalman filtering can introduce positional error in the final airborne trajectory data. Both of these scenarios are related to the AGNSS and can result in positional error of the lidar point cloud of up to several decimeters.

Generally speaking, the IMU portion of the error budget can be more significant and is directly correlated to the slant ranges of the airborne lidar sensor. As slant ranges increase, so too does the error budget stemming from the IMU. IMU error is known as pointing error and is angular in nature. As such, the error vectors affecting the lidar point cloud tend to be greater in the horizontal component.

Modern large-format airborne lidar sensor IMUs employ optical gyroscope technology. Smaller form-factor lidar sensors such as those found on sUAS or drones typically employ micro electro mechanical system (MEMS) IMUs, which are far less accurate than their optical counterparts. Older generation airborne lidar sensors employed IMUs with mechanical gyroscopic technology. Optical gyroscopic technology includes fiber optic gyroscopes and ring laser gyroscopes (FOG and RLG). Both of these optical IMUs operate by sensing the difference in propagation time between beams of light traveling in clockwise and counter-clockwise directions about some closed optical path. A rotationally induced variance in path length produces a phase difference between the light beams propagating in opposing directions. This difference is generally known as the Sagnac effect and forms the basic operating principle of all optical gyroscopes *(asee.org/documents/sections/middle-atlantic/fall-2009/01-Evaluation-Of-Ring-Laser-And-Fiber-Optic-Gyroscope-Technology.pdf)*.

The angular rate measured by the gyros is used by the navigation processor to maintain the GNSS-inertial system's attitude solution. The accelerometers, however, measure specific force, which is the acceleration due to all forces except for gravity. Thus, the measurements produced by stationary accelerometers comprise the reaction to gravity. In a strapdown INS (i.e., typical airborne lidar configuration), the accelerometers are aligned with the navigating body (i.e., aircraft), so the attitude solution is used by the navigation processor. A gravity model is then used to obtain the acceleration from the specific force using the position solution (Groves, P. D. (2013)).

The previous paragraph loosely describes how the local modeled gravity vector is used during the processing of the IMU data. If the gravity input is not modeled correctly, the subsequent attitude solution may be compromised. There are other factors that affect the accuracy of the IMU inputs in the final GNSS-inertial solution. These include (from NovAtel (2014)):

- Scale factor: Scale factor is the relation between the input and the output of the different IMU sensor sets. It is typically a consistent error and proportional to the input. Scale factor effects are most apparent in times of high acceleration and rotation
- Random walk (sensor noise): A random noise is always present in the measurement. This error source is mitigated using statistical techniques, however may still present a problem in the final solution. Random walk has a direct effect on the performance of the GNSS-inertial during periods of GNSS outage, when the sensor noise causes the solution error to grow unbounded
- IMU repeatability bias: Consistency of bias from power-on to power-on, each time IMU is powered up. This bias is due to changes in physical properties of the IMU (i.e. changes to inertial reference frame orthogonality) and initial conditions of signal processing
- IMU stability bias: The change in the initial bias over time, for each start-up. This bias may be related to temperature and/or mechanical stress on the system. An INS filter constantly estimates the bias by making use of the external correction data such as that coming from the GNSS
- Sensor non-orthogonality within the IMU: This can be specific to each of the sensor sets (i.e., accelerometer or gyroscopes) or between the sensor sets and the IMU enclosure. Care in manufacturing and factory calibration can help minimize non-orthogonality, however it cannot be entire removed
- Sensor non-orthogonality between IMU and sensor: Can be largely resolved through a sensor boresight

Even in an optimized scenario, the IMU is rated to be only so accurate. Large format lidar sensors typically employ navigation grade or (the less accurate) tactical grade IMUs. Accuracy specifications for IMUs characterize the rate of drift and angular tolerance.

At the time of this writing, the most accurate IMUs commercially available have performance specifications of 0.0025 deg (RMS) for Roll and Pitch and 0.0050 deg (RMS) for Heading. In a hypothetical scenario of purely Roll error (no Pitch, Heading, GNSS, or other error), at 2,000 m height above mean terrain, 0.0025 deg translates to approximately 9 cm of combined horizontal and vertical error in the point cloud. Discussions with lidar sensor manufacturers and data providers indicate these performance specifications are conservative. However, it is certainly possible that poorer results are possible under normal operating conditions.

Boresight and Adjustment

This section is intended to address the production of the point cloud data with good geometry (i.e., positional accuracy). That said, it is not feasible to cover all of the permutations affecting the positional accuracy of the point cloud data within the scope of this text. There are a multitude of sensor calibration issues affecting the various commercial sensors available today. These include errors associated with the different sensor scanning mechanisms (i.e., oscillating, constant velocity, and palmer or nutating scanners), receiving optics, signal quantization, electronic latency, and scanning mirror curvature.

Several modern lidar production software include capabilities for adjusting the point cloud geometry. There are several different philosophies and technical approaches to performing this adjustment. In most applications, comparisons are made of conjugate features within regions of overlap between two or more swaths. These conjugate features are defined by planes or lines (typically, rooflines). The software determines the magnitude and direction of point to plane, plane to plane, or line to line offsets between the conjugate features defined by each swath. The magnitude and direction of these offsets form the basis of the geometric adjustment.

The discussion should begin on a stable platform wherein systematic sensor calibration errors have been identified and resolved (Figure 8.12). Likewise, an assumption should be made that post processing of the point cloud data was done using proper transforms and geodetic principles. Specifically, production of the point cloud data should be made on the proper projected grid coordinate system including proper horizontal and vertical datums, geoid height model and units.

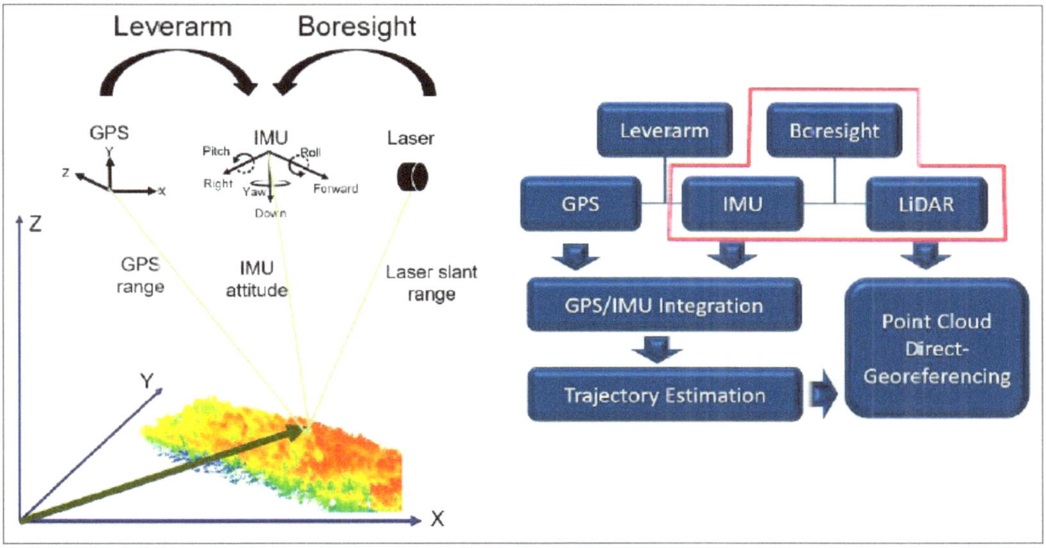

Figure 8.12. Schematic explaining the boresight calibration to produce a point cloud.

Improperly calibrated lidar sensor components can affect geometric, temporal, and radiometric characteristics of the point cloud data. Fortunately, errors associated with improperly calibrated sensor components can often be identified through careful analysis of the point cloud data. Once identified, laboratory-based or *in situ* calibration of sensor components can be done to resolve the problem. Also fortunate is the fact that large-format lidar sensors are designed and built well so calibrated components are typically stable. Put another way, once identified and calibrated, these sensor-components should not be factors in the geometric characteristics of the point cloud data.

It is always necessary to boresight the lidar sensor in order to generate an accurate point cloud. The term boresight is a carryover from the aerial photography industry, which in turn borrowed the term from the process of sighting in a gun's scope. In essence, boresighting a lidar sensor is done to resolve misalignment angles between the IMU and the lidar reference frame, although it can also include the resolution of other variables specific to the lidar sensor model. Once properly boresighted, a lidar system should be free of systematic error.

Modern point processing software packages have the ability to boresight the sensor by performing line to line, point to plane, or plane to plane analyses in regions of overlap between flight line swaths. Modern boresight procedure can be done *in situ* on a project. The need to fly separate calibration ranges is no longer required; however it may still be necessary in certain situations. All that is required for the modern boresight procedure are 3-4 overlapping parallel lines with 1 or more cross lines over a built-up area. The 'built-up' part is critical as man-made above-ground features have sloped planar surfaces, which are necessary for the algorithms to be optimized. It is possible to boresight with data collected over an area that is not built-up, however results may not be optimal.

In the boresight process, vectors of direction and magnitude between corresponding (i.e., conjugate) planar features occurring in two or more overlapping flight lines are recorded. Typically, a least-squares bundle adjustment algorithm then determines which boresight-specific variables to modify to best align the planar features derived in the overlapping areas. These variables are applied globally as they are sensor-specific. Put another way, one set of correction variables are applied to multiple swaths within a dataset.

The boresight algorithms primarily address the theoretical angular offsets of the orthogonal axes of the IMU. The reason these angular offsets need to be periodically modified has to do with the initialized state of the IMU upon start up and the aforementioned sensor non-orthogonality, bias, and scalar errors acting upon the IMU sensor sets. Error in the local gravity model at the time of IMU initialization, whether on the ground or in flight, may also adversely affect the boresight misalignment angles.

In general, this boresight process should greatly improve the quality of the data. However, it is possible that the algorithms may incorrectly solve for the wrong variables. This is especially the case if boresighting of the IMU misalignment angles is done with data that contains systematic lidar sensor errors. If lidar sensor-related errors are present in the data, these must be resolved before the boresight process.

In order for the boresight process to be optimized, it is important that (1) The GNSS-inertial trajectory data has minimal error and (2) There is a good distribution of built-up features (preferred) or non-vegetated terrain relief for the extraction of well-defined planar surfaces with enough slope.

Once the sensor model has been sufficiently boresighted, a routine similar to the boresight process can be done on a line-by-line basis to best fit the alignment between overlapping flight lines. This process works by averaging the misalignment between the overlapping swaths over the entire area of overlap. Then a set of correction variables are determined to best fit the overlapping swaths. These variables are not sensor-specific but rather line-specific which means they will be different for each line. This process applies only one set of correction variables to each line. Therefore some level of residual misalignment between the overlapping lines will remain.

The line-by-line adjustment should be controlled so that the adjustment block does not warp. Warping of the block can significantly affect the positional accuracy of the point cloud. When this warping occurs, relative fit between correlating features between overlapping swaths may improve, however absolute accuracy is compromised. A common process for controlling the adjustment to mitigate warping involves the use of a cross swath. The cross swath is flown perpendicular to, or near perpendicular to the azimuth of the parallel block of lines requiring adjustment and should overlap the entire block. Once defined as a cross swath within the adjustment software, variables that may otherwise be adjusted, such as roll, pitch, heading, scale, are typically held fixed.

The boresight is sensor-specific and involves resolving IMU misalignment angles. These angles may be thought of as omega, phi, and kappa (Ω φ K), which are terms carried over from traditional photogrammetry and relate directly to the IMU—lidar sensor reference frame. The line-by-line adjustment is data-specific and typically involves resolving GNSS-inertial related errors such as roll, pitch, heading, and elevation drift.

There are some lidar adjustment programs that are capable of performing dynamic "in-time" adjustments to the point cloud. This approach may use a Bayesian or other probabilistic statistical approach to effectively rubber-sheet data from overlapping swaths to align features. It's important to note that these dynamic adjustments are not likely to be traceable. Put another way, once the data are adjusted, it is difficult to impossible to define and quantify the variables modified to effect the adjustment. As with the boresight and line-by-line adjustment, care must be taken during the dynamic adjustment to ensure absolute accuracy has not been significantly compromised to improve relative fit between overlapping features.

PLANNING AN AIRBORNE LIDAR PROJECT

There are numerous requirements to assess when planning an airborne lidar project. Project planning always starts with the basic questions:

- Why is this dataset needed?
- What are the specific deliverables that are needed?
- When are the deliverables needed?

From a data provider's perspective, planning is performed after careful review of the project specifications and answering a series of questions:

- What is the vertical accuracy needed for this dataset?
- What horizontal and vertical datums are required for this project?
- What is the coordinate reference system to be used—State Plane Coordinate System or Universal Transverse Mercator (UTM) coordinates, or other?
- Should the units be feet or meters? If feet, should U.S. Survey Feet or International Feet be used?
- What should be the nominal point density?
- What classifications should be included i.e., ground, water, buildings, vegetation, etc.?
- Do planimetric features such as roads or buildings need to be extracted from the lidar data?
- Are there limits on environmental factors such as shadows, clouds, topography, climate, snow cover, standing water, tidal, and river levels?
- Will DEMs, DSMs, contours, or other derivative products be produced?
- What are the metadata requirements? How are accuracies to be reported in the metadata?
- Are waveform data needed?

Lidar Density

Lidar data can be collected with a wide variety of point densities depending on the needs of the project. The selection of point density is a big driver of the overall cost of a lidar project and should be selected with consideration to the end uses for the lidar. Modern lidar sensors are capable of acquiring lidar data with a higher density than previously available and can do so at higher altitudes and with less overlap. A lidar product with 2-4 points per square meter (ppsm) is sufficient for many applications such as flood mapping and generation of data for the 3D Elevation Program (3DEP). Higher point densities (8 ppsm) allow for greater utilization of the data for mapping planimetric features such as roads and structures as well as for vegetation analysis such as biomass and canopy studies. Additionally, specialized lidar at very high densities >20 points per square meter are often used for mapping infrastructure in greater detail such as powerlines, pipelines, and DOT significant features such as mile posts and signs. The ground conditions should be considered when selecting a point density as well. If the area is covered with dense vegetation such as a coniferous forest, a higher density and more overlap would be required to penetrate to the ground than an area where leaf-off conditions exist.

Swath Overlap

Planning for swath overlap should also be included in the overall planning of the point density. A higher percentage of overlap may be beneficial in an area with dense vegetation as there will be more look angles from the sensor to the ground at any given point. The result would mean that there could be a lower point density requirement in an individual swath with the overlap accounting for an overall higher aggregate point density on the project. Depending on the scanning pattern, data from the extreme edges of the swath may be unusable due to geometric nature of the scan pattern. Typically, 10% overlap between swaths is the minimum requirement for an airborne lidar collect. However, most lidar flights are conducted with an average of 25% overlap to account for variations in the terrain and aircraft drift during flights, and those that require higher pulse density in vegetated areas are often flown with 55% overlap.

Flight Planning

Flight planning is always the responsibility of the acquisition provider. Flight planning for lidar will vary greatly depending on the sensor utilized for the acquisition. Parameters such as flying height, ground speed, scan rate, scan angle, etc. will be different for each sensor and focus should be put on ensuring that the results meeting the project requirements for nominal point density. Table 8.1 shows the operational parameters for a sample lidar project performed by Dewberry and Axis Geospatial using a Riegl LMS-Q1560 system in Palm Beach County, Florida.

Operational Parameters					
Project Number	215047	Ground Speed	125 kts	MTA Zones	4
Altitude	3,501 ft	Flying Height AMSL	3,501'	County	West Palm
Lateral Overlap	30%	PRR	0.6 m	State	Florida
Forward Overlap	60%	Point Density AVE	800 kHz	Project Area	2,027 sq miles
Total Length	3,992.1nm	Laser Power	5.78 ppms	Control Points	30
Total Models	22,533	Scan Rate	258 lps		
Total Photos	22,728	FOV	58.5 deg		

Table 8.1. Example of relevant operation parameters for lidar data collection.

The lidar flight planning process is mostly automated after entering basic information such as point density, overlap requirements, and scan angles. Trajectories are planned for each flight line. Furthermore, modern Flight Management Systems (FMS) enable the pilot to fly these trajectories with close tolerance. Lidar sensors are actively acquiring data when the pilot is "on-line." Sensor operators are often able to view the acquisition in real time and assess areas where voids or sensor anomalies may be present during the flight. While lidar sensors also have some forms of stabilization, the roll, pitch and yaw of the aircraft still depends upon wind conditions. Regardless of sensor to be used, flight planning also includes the assessment of military and other controlled air space where special permits may be required. Aviation Sectional Charts are often used to determine flight restrictions and controlled airspace when planning flight lines.

Survey Control Planning

Standards and guidelines, addressed in Chapter 3 of this book, are relevant to the three forms of mapping control required for airborne lidar mapping, i.e., airborne GPS control, ground control points for calibration, and ground checkpoints for independent quality control. It is important to note that control points used for calibration must never be used as checkpoints for independent data verification and quality assessment.

In the U.S., Continuously Operating Reference Station (CORS) can be used for airborne GPS control. Base-stations are also often set up on well-documented permanent control monuments or benchmarks from the National Geodetic Survey's National Spatial Reference System (NSRS). If a control network of horizontal control monuments and vertical control benchmarks do not exist, a control network will first need to be established per references cited above. Figure 8.13 shows an example NGS Data Sheet with the red arrow pointing at the horizontal and vertical network accuracy at the 95% confidence level. In addition to data shown here, Data Sheets typically also include additional information such as: State Plane and UTM coordinates; U.S. National Grid spatial address; explanations of how horizontal coordinates, ellipsoid heights and orthometric heights were determined; station description and instructions for finding the monument; station recovery history, etc.

The NGS Data Sheet

See file dsdata.txt for more information about the datasheet.

```
PROGRAM = datasheet95, VERSION = 7.39.3.1
1        National Geodetic Survey,   Retrieval Date = AUGUST 27, 2012
 AJ4599 ************************************************************
 AJ4599  HT_MOD        -  This is a Height Modernization Survey Station.
 AJ4599  CBN           -  This is a Cooperative Base Network Control Station.
 AJ4599  DESIGNATION -  FAIRFAX COUNTY EAST
 AJ4599  PID           -  AJ4599
 AJ4599  STATE/COUNTY-  VA/FAIRFAX
 AJ4599  COUNTRY       -  US
 AJ4599  USGS QUAD     -  FALLS CHURCH (1994)
 AJ4599
 AJ4599                          *CURRENT SURVEY CONTROL
 AJ4599
 AJ4599* NAD 83(2011) POSITION- 38 55 43.18356(N) 077 08 47.67437(W)   ADJUSTED
 AJ4599* NAD 83(2011) ELLIP HT-    52.373 (meters)         (06/27/12)   ADJUSTED
 AJ4599* NAD 83(2011) EPOCH    -  2010.00
 AJ4599* NAVD 88 ORTHO HEIGHT -    84.26   (meters)     276.4    (feet) GPS OBS
 AJ4599
 AJ4599  NAD 83(2011) X  -  1,105,247.067 (meters)                     COMP
 AJ4599  NAD 83(2011) Y  - -4,843,853.080 (meters)                     COMP
 AJ4599  NAD 83(2011) Z  -  3,986,192.174 (meters)                     COMP
 AJ4599  LAPLACE CORR    -          -2.36  (seconds)                   DEFLEC09
 AJ4599  GEOID HEIGHT    -         -31.88  (meters)                    GEOID12
 AJ4599
 AJ4599  FGDC Geospatial Positioning Accuracy Standards (95% confidence, cm)
 AJ4599  Type                                          Horiz  Ellip  Dist(km)
 AJ4599  ------------------------------------------------------------------
 AJ4599  NETWORK                                        0.41   0.93
 AJ4599  ------------------------------------------------------------------
 AJ4599  MEDIAN LOCAL ACCURACY AND DIST (067 points)    0.53   1.22     43.81
 AJ4599  ------------------------------------------------------------------
 AJ4599  NOTE: Click here for information on individual local accuracy
 AJ4599  values and other accuracy information.
```

Figure 8.13. Example NGS Data Sheet showing sub-centimeter network accuracies of an NSRS monument relative to horizontal and vertical datums.

Airborne lidar is acquired with the use of ABGPS for recording the 3-D (X/Y/Z) coordinates of each pulse, plus an IMU for recording the roll, pitch and yaw of the sensor, when each pulse is transmitted and received. When six exterior orientation parameters of each pulse (X/Y/Z and roll/pitch/yaw) are known and can be accurately measured at the flying height, requirements for surveyed ground control are greatly reduced. Planning for ABGPS control includes consideration of the project area (size and site access for base stations); flight times and maximum allowable GPS baseline distance between GPS base stations and aircraft; satellite availability during the collection period; location of suitable GPS base stations; location and number of ground GPS receivers required; data collection rate for the receivers; aircraft and ground crew logistics (base stations and aircraft receivers must use the same satellite configuration and limitations); calibration of antenna/sensor "lever arm" offset; ABGPS system cost; and experience of pilot, operator and post-processing personnel. ABGPS receivers must be capable of tracking both coarse acquisition (C/A) and pseudorange (P-code) data. They must provide dual frequency (L1 and L2) and multi-channel capability with on-the-fly ambiguity resolution and be able to log GPS data at 1-second epochs or better. GLONASS receivers capable of receiving satellite information from GPS and GLONASS constellation are preferred over GPS-only receivers.

Accuracy

The various components of an airborne laser scanning system contribute to the final accuracy of the elevation data (Vosselman & Mass, 2010). The main error sources that contribute to the total error budget include:

- Erroneous calibration of the GPS, IMU, and scanner assembly
- Limited accuracy in determining the aircraft trajectory from GPS and IMU measurements
- Complexity of the target; note that a sloping terrain can cause a greater vertical error due to an error in locating the position of the ground
- Ranging errors caused by multipath reflections and timing/digitizing electronics in the detector
- Coordinate and datum transformations

Accuracy is often the most important aspect of a lidar project. Lidar system calibration, including creation of precise GPS trajectory solutions that minimize any swath to swath misalignments, and absolute accuracy based on comparisons with or adjusting to ground control points are essential for a successful project. In support of this, a "report of survey" usually accompanies the data collection describing the system calibration, GPS log, any ground survey points referenced, the design accuracy, validation process, and statistical summary of the results.

Although accuracy reports that compare lidar data to independent ground checkpoints are necessary and must be done, the accuracy of a specific data set is more than simply the accuracy of the point at a single location. Comparisons with survey data are essential, but the ability to map an area with several points per square meter enables a better representation of a surface, even if each point is less accurate, compared to a handful of points acquired at greater accuracy by ground survey methods.

CURRENT SENSOR TECHNOLOGY

At the present time, there are three main commercial suppliers of Airborne Lidar Systems—(a) Teledyne Optech International, headquartered in Toronto, Canada with offices in New York and Mississippi; (b) Leica Geosystems, a Hexagon Company, headquartered in Switzerland with over 20 offices in the U.S. and worldwide; and (c) Riegl Gmbh, headquartered in Austria with offices in Florida and worldwide. Recently, Harris Corporation has entered the commercial airborne lidar industry, offering Geiger-mode lidar technology through its Geospatial Solutions suite of products and services.

In this section, we provide an overview of current sensor technology offered by the system manufacturers with a focus on the newer sensors developed by each manufacturer. Airborne lidar technology is evolving rapidly with new sensors entering the market on a yearly basis. Our goal is to provide the reader with information on current available technology for a variety of applications, while keeping an eye on next generation technology. All information provided in this section is based on published literature or information available on websites from the manufacturer. The authors have made a significant attempt to remove any marketing campaigns that suggest a particular sensor is "better" than another sensor. Instead, some of the highlights of the sensors are presented, with distinctions on the sensor technology based on its operations and system characteristics. The "extraction" software that converts the raw data acquired from the aircraft to a point cloud is also discussed.

Teledyne Optech

Established in 1974 under the name Optech, Teledyne Optech has been designing, developing, and manufacturing advanced lidar instruments for over 40 years. Teledyne Optech works closely with commercial, government, military, and space-based organizations to meet their specialized application requirements. They offer standalone and fully integrated lidar and camera solutions in airborne mapping, airborne bathymetry, mobile mapping, terrestrial laser scanning, and mine cavity monitoring, as well as space-proven sensors. Optech systems come with software that can be used for survey planning, operation and post-processing, allowing clients to collect, manage and deliver survey data to their customers.

Teledyne Optech offers four airborne lidar topographic mapper (ALTM) systems and two coastal and marine topographic/bathymetric systems. They also offer various camera systems. The four ALTMs are the Galaxy, Eclipse, Pegasus and Orion; the coastal and marine systems are the Titan and CZMIL Nova. The CZMIL and Titan systems are described in Chapter 10. The systems differ in size, weight, and operational characteristics and are intended to support a variety of applications such as wide-area mapping, engineering-grade surveys, corridor mapping and hydrographic mapping. The ALTMs and the Titan are based on discrete-return technology; Teledyne Optech offers an optional Intelligent Waveform Recorder (IWR-3) as a small strap-on peripheral that provides waveform capture capability to these systems.

Teledyne Optech develops software that enables airborne survey planning, airborne navigation and control, real-time 3-D visualization and quality control, automated pre-processing, and a complete bathymetric workflow from planning to data deliverables. The ALTM and Titan systems use the Optech Flight Management Suite (FMS) and Lidar Mapping Suite (LMS). FMS is an integrated workflow tool that handles flight planning, multiple-sensor system control, real-time in-air monitoring, and navigation. LMS offers high-volume production processing (including parallel processing), lidar calibration, geodetic conversions and accuracy reporting. The LMS Pro suite offers additional calibration tools based on extracted features in overlap areas (such as building rooftops).

Galaxy

The ALTM Galaxy (Figure 8.14) is the latest of Teledyne Optech's airborne solutions, packing more power and accuracy into a tiny footprint than any of their other sensors. It is designed for a wide range of applications including wide-area mapping and corridor surveys. Optech introduced PulseTRAK™ and SwathTRAK™ technologies (patents pending) with the Galaxy. PulseTRAK technology replaces the conventional multi-pulse technologies for high PRFs at high altitude. SwathTRAK™ maintains a fixed-width data swath on the ground in complex terrain by varying the scan FOV dynamically in-flight, thereby maintaining more consistent point density and XY point distribution across the entire data set. The Galaxy has a 1-MHz effective PRF using a single 1,064-nm Class IV laser, and its scan FOV can vary between 10°-60°. The operational envelope for the Galaxy T1000 model is 150-4,700 m AGL on 20% reflective targets. The newly introduced Galaxy PRIME model can achieve a further 25-30% improvement in range performance over the T1000 at the higher sampling rates. The sensor and power distribution unit weigh 33.5 kg, making it suitable for installing on small airborne platforms.

Figure 8.14. ALTM Galaxy.

Eclipse

The Optech Eclipse (Figure 8.15) is an autonomous airborne system that focuses on operator-less data collection from tightly integrated passive and active imaging sensors on low-cost platforms. Designed specifically for efficient data collection in smaller project areas and corridor applications, the Eclipse can be used with only a pilot for navigation. The Eclipse combines a

Figure 8.15. Optech Eclipse.

class 1 eye-safe, 1.5-micron laser with an automatic imaging system for collection of high-resolution lidar data and imagery via its sensor motion detection system. A small dash-mounted status bar feeds sensor status information to the pilot with LEDs and incorporates a simple switch to start recording data. Lidar and image data are simultaneously captured on a single removable SSD cartridge for fast and efficient data transfer to Optech LMS. Eclipse has an operational envelope of 50-1,000 m AGL. At the maximum operating altitude of 1,000 m (3,300 ft.) and flying at 100 knots (115 mph), the pulse density is 4 points/m^2. A rotating polygon scanner is used to produce parallel scan lines with near-even point distribution over a 60° scan field of view. Applications include small-area surveying, powerline mapping, pipeline monitoring, construction site monitoring, mine monitoring, stockpile surveying, and archaeology.

Orion

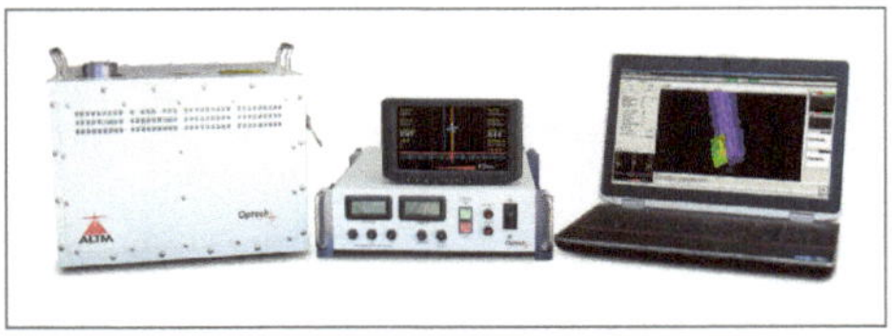

Figure 8.16. Optech Orion ALTM.

The Optech Orion ALTM (Figure 8.16) is a compact total mapping solution that offers modular components to allow scalable and flexible configurations based on the application. Three models are offered: Orion C, which is a low-altitude system for corridor and engineering mapping applications; Orion M, which is a low- to mid-altitude system with a wider operational envelope for applications that include urban modeling, engineering and transportation, and small topographic surveys; and Orion H, which is a high-performance sensor that is also suitable for wide-area mapping. The Orion M and H use a 1064-nm laser whereas the Orion C uses a 1.5-micron laser. The Orion C can be used as low as 50 m AGL whereas the Orion H can be used at a flying height of 4000 m AGL. The sensor and power distribution unit weigh 31.5 kg combined and are compact in size.

Pegasus

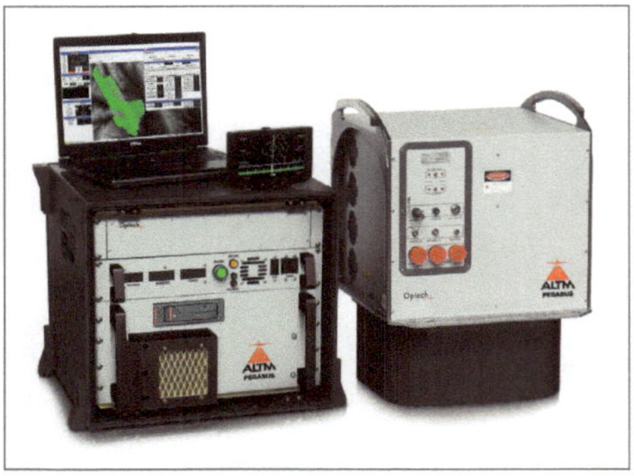

Figure 8.17. Optech Pegasus.

The Optech Pegasus (Figure 8.17) is a high-performance lidar mapping solution that is designed for maximum collection efficiency using a dual-laser sensor approach. The HA500 sensor uses a dual output laser system at 1,064 nm wavelength with a maximum laser repetition rate of 500 kHz. Its operational envelope is 150-5,000 m AGL with a scan width programmable from 0° to 75°. The HD500 sensor is ideally suited for applications that require high density and operates up to 2,500 m AGL. The Pegasus was developed primarily for wide-area mapping applications.

Leica Geosystems

Leica Geosystems is part of Hexagon, a global provider of information technology solutions across geospatial and industrial landscapes. Leica Geosystems develops geospatial solutions for a diverse mix of industries, such as surveying and engineering, building and heavy construction, safety and security. Leica Geosystems products and solutions include EDM (Electronic Distance Measurement), GPS/GNSS technology, 3-D laser scanning, tilt and angle measurement, and point cloud generation and analysis tools.

Leica Geosystems offers the following airborne topographic lidar systems: SPL100—a single-photon topographic lidar for large-area acquisitions; TerrainMapper, an oblique-scanning waveform-based linear-mode lidar for flexible high-pulse-rate acquisitions at a local or regional level, and ALS80—a linear-mode discrete-return topographic lidar for high-flying-height and corridor mapping applications. Related to Terrain Mapper is the CityMapper, an oblique-scanning waveform-based linear-mode lidar with nadir and oblique cameras for urban mapping. In addition, Leica also offers two bathymetric lidar mapping systems—the Chiroptera II topo-bathymetric lidar for shallow water and coastal mapping; and HawkEye III for deep water bathymetric mapping. The Chiroptera and Hawkeye sensors are described in Chapter 10.

Leica HxMap is a high-performance, multi-sensor processing workflow developed by Leica that enables basic data management, calibration, QC of raw data and ability to process and integrate data from various Leica lidar and camera systems, including SPL100, TerrainMapper and CityMapper. Although Leica ALS-series systems currently use Leica CloudPro and Chiroptera/Hawkeye bathymetric systems currently use Lidar Survey Studio (LSS) for integrating raw scan data with GPS/IMU data, post-processing for these systems will ultimately be incorporated into HxMap. Leica MissionPro and FlightPro are the companion software packages developed for mission planning and mission execution, respectively, with FlightPro offering both flight navigation and sensor control functions.

SPL100

Figure 8.18. Leica SPL100.

The Leica SPL100 (Figure 8.18) is a single-photon airborne lidar sensor ideal for large area terrain mapping projects. This system, which is accompanied by an 80 MP RGBN camera, captures 6 million measurements per second using 100 output beams and can be used during day or night, leaf-on or leaf-off conditions and in dense vegetation. The SPL100 enables point densities of 12-30 points/m^2 (depending on flying height and aircraft speed) and can penetrate semi-porous obscurations, such as vegetation, ground fog, and thin clouds.

The information collected with the SPL100 single-photon lidar is especially useful for applications that require high point density over large areas, such as country and state wide mapping, disaster risk planning, forest inventory and flood mapping. Leica SPL100 is the wide-area sensor component employed in the Leica RealTerrain solution package, combining SPL sensors technology with the multisensor processing workflow HxMap. Combining the SPL100 and HxMap, Leica RealTerrain offers the technical foundation for large lidar mapping projects.

SPL100 uses a green wavelength laser at 532 nm wavelength and a PRF of 60 kHz, with the laser output split into 100 beamlets, arranged in a 10 x 10 array. Each beamlet is then imaged onto a pixel in a matching 10 x 10 array detector which, in turn, is input to a timing channel able to record multiple stop events for each pixel with a few picosecond accuracy (Degnan, 2016). The single photon sensitivity, combined with a 400 picosecond laser pulse width and a 1.6 nanosecond receiver recovery time, is therefore capable of recording returns from objects differing by only 24 cm in range (Degnan, 2016). The system uses an oblique scan pattern with a scan speed programmable to 25 Hz. An RCD30 camera is included in the system configuration, allowing RGB or CIR encoding of the developed point clouds. The operating altitude is 2000-4,500 m AGL resulting in typically 20 points/m^2 at 4,000 m AGL and 200 knots speed over ground. The sensor is mounted on the PAV100 heavy-load gyro-stabilized mount. The weight of the scanner system along with the lidar and camera controllers and gyro stabilized mount is ~143 kg and it draws an average of 600 W (max. 1,000 W) at 28 VDC. Mission planning is conducted using Leica MissionPro and flight navigation and sensor operation uses FlightPro. All post-processing is done in HxMap.

TerrainMapper

Leica TerrainMapper is the general-purpose linear-mode counterpart to the SPL100 in the Leica RealTerrain solution package. Primary applications for the TerrainMapper include all high-point-density acquisitions at a corridor, local or regional level. The Leica TerrainMapper (Figure 8.19 on page 230) utilizes the same oblique scanning concept as the related CityMapper sensor, allowing high definition on vertical surfaces. It was designed to survey complex urban landscapes as well as utility corridors containing objects that are obscured by vegetation. This high-performance lidar sensor features a 2.0 MHz maximum pulse rate that can be attained at flying heights to 2,000 m AGL. The TerrainMapper also introduces a gateless MPiA function that allows for scanning over extreme terrain variations where numerous (up to 35) MPiA zones can be crossed in a single scan. This feature is particularly productive, as it allows high pulse rates (normally implying narrow range gates) to be achieved whether or not the terrain covered is flat.

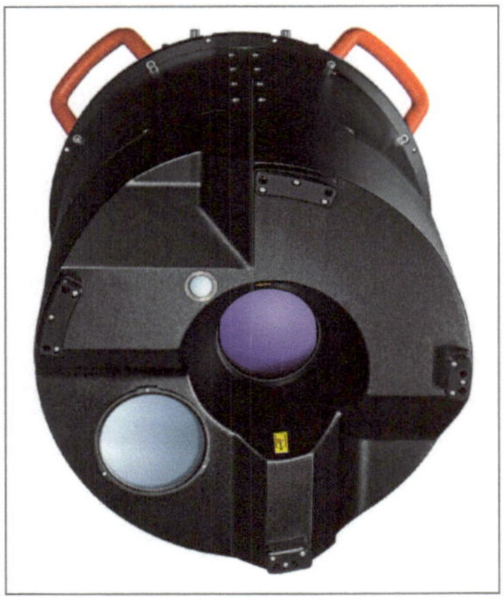

Figure 8.19. Leica TerrainMapper.

The TerrainMapper detects targets using full waveform data capture, with real-time analysis and conversion of the waveform to range, intensity, and a number of additional waveform characteristics, resulting in robust information without the data bulk of full-waveform recording. In TerrainMapper, full-waveform recording is also possible, at reduced pulse rates. With on-the-fly waveform-to-range conversion, up to 15 discrete returns with vertical separations as small as 0.4 m can extracted. Maximum flight altitude is 5,000 m AGL.

The system is normally equipped with the Leica RCD30 digital camera, allowing downstream colorization of the point cloud with RGB or CIR information in addition to the native intensity information. The PAV100 stabilized platform improves productivity by ensuring straight swaths along the flight ling and consistent along-track point density, along with reducing motion-induced blur in the camera images.

As with it's wide-area mapping counterpart, Leica SPL100, data from Leica TerrainMapper is processed using the Leica HxMap workflow, and is offered as the regional-level RealTerrain solution package. For lidar sensors like TerrainMapper, HxMap provides project management, point cloud generation, calibration, line-to-line registration, data QC, colorization with co-collected RGB or CIR camera data and ultimate generation of output data files in virtually any projection.

CityMapper

The Leica CityMapper (Figure 8.20) is the lidar/imaging sensor component in the Leica RealCity solution package. Consisting of a 700 kHz oblique-scanning lidar module, 4-band/80 MP nadir camera and four 80 MP RGB oblique cameras, CityMapper is designed for single-flight airborne data collection in urban environments. The lidar module features a high-pulse-rate laser and waveform-based range data acquisition. As in TerrainMapper, CityMapper detects targets using full waveform data capture, with real-time analysis and conversion of the waveform to range and intensity data without the data bulk of full-waveform recording. In TerrainMapper, full-waveform recording is also possible, at reduced pulse rates. With on-the-fly waveform-to-range conversion, up to 15 discrete returns with vertical separations as small as 0.4 m can extracted.

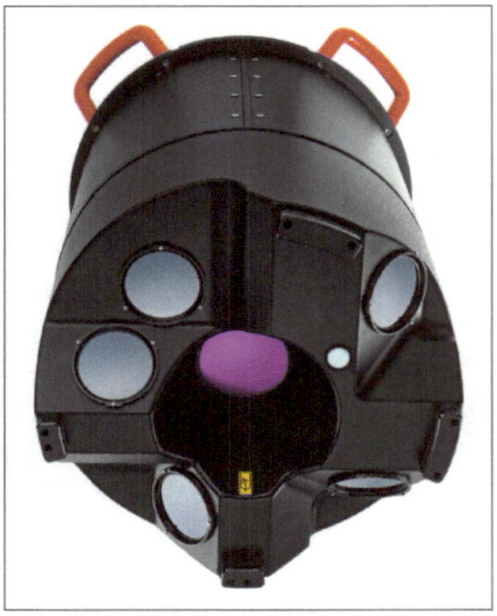

Figure 8.20. Leica CityMapper.

CityMapper is designed for simultaneous acquisition of oblique lidar data and high-resolution oblique image data at industry-standard 45-degree incidence angles. The lidar data optimizes data collection on vertical surfaces and vertical-to-horizontal surface transitions that are often obscured by street-side vegetation, while the image data provides high spatial and spectral resolution. The sensor head is normally installed in the PAV100 stabilized platform, reducing image blur and simplifying downstream processing.

As with SPL100 and TerrainMapper, data from Leica CityMapper is processed using the Leica HxMap workflow, and is offered as the regional-level RealCity solution package mentioned above. HxMap provides project management, point cloud generation, calibration, line-to-line registration, data QC, colorization with the co-collected RGB or CIR camera data and ultimate generation of output data files.

ALS80

Figure 8.21. Leica ALS80.

The Leica ALS80 (Figure 8.21) is a single family of linear-mode discrete-return airborne lidar sensors that offers 1.0 MHz pulse rates to produce high density point clouds. The ALS series is a widely used topographic sensor with Leica reported to have sold over 260 units worldwide. It is used in applications such as county and state-wide mapping, damage assessment, forestry and extreme terrain mapping in all environments. The ALS80s offers three planar scan patterns (sine, triangle, and raster) to allow for even point density or maximum density, as desired, across the swath. The field of view is also user-settable in flight, up to 72°. Leica offers three variants of the ALS80:

- Leica ALS80-CM: For smaller and/or high-density and corridor mapping applications from lower flying heights up to 1,600 m AGL
- Leica ALS80-HP: Supports general purpose mapping at the most widely used flying heights up to 3,500 m AGL
- Leica ALS80-UP: High-altitude variant for low-density wide-area mapping or mapping where high-flying heights (up to 5,000 m AGL) are otherwise dictated

The modular design of the Leica ALS80 offers the flexibility to convert an ALS80-CM to an ALS80-HP at the user's site or be factory-reconfigured into a Leica ALS80-UP. Similar to the Optech systems, the ALS80 is developed with discrete-return technology; an optional Waveform Recorder can be added for forestry and other applications. The system weighs about 80 kg for the scanner and control electronics and draws ~950W at a nominal 28 VDC input. Mission planning, navigation, and processing to point cloud is done using Leica's MissionPro, FlightPro and CloudPro software, respectively.

RIEGL

With 40 years' experience in the research, development and production of laser rangefinders, distance meters and scanners, RIEGL offers state-of-the-art hardware for terrestrial, industrial, mobile, airborne, bathymetric and UAV-based laser scanning with appropriate RIEGL software packages for data acquisition and processing for multiple fields of application. RIEGL's headquarters are located in Austria, with worldwide offices that deliver sales, training, support, and services offices in the USA, in Japan, and in China, and by a worldwide network of representatives covering Europe, North and South America, Asia, Australia, and Africa.

RIEGL offers a wide range of airborne laser scanners and laser scanning systems All RIEGL systems digitize the echo signal online during data acquisition. The LMS series of laser scanners require subsequent waveform analyses. The VQ line of laser scanners offer multiple target capability by online waveform analysis based on echo digitization. No digital data recorder and off-line waveform processing is necessary.

RIEGL provides proprietary companion software for RIEGL laser scanners. Separate programs designed for the different applications of terrestrial and airborne/mobile laser scanning enable the use of the scanners' capabilities in the best possible way, from optimizing the acquisition workflow in the field via providing tools for data control and data processing. RiACQUIRE is the system integration and data acquisition software; RiANALYZE performs full waveform analysis; RiWORLD offers coordinate system transformations;

RiPROCESS is the primary software that is used for data management, processing, and visualization; RiHYDRO is the processing software for hydrographic and bathymetric surveying; RiMTA ALS is the software for automated resolution of range ambiguities; and RiVLIB is the offline/online library for V-line scanners.

RIEGL currently offers the following airborne lidar scanners and systems, as well as others:

LMS-Q680i

The RIEGL LMS-Q680i airborne laser scanner (Figure 8.22) uses a 1,064 nm Class 3R laser source, up to 266 KHz effective measurement rate with multiple time around (MTA) processing, and digital full waveform analysis. This combination allows the operation at varying flight altitudes up to 1,600 m AGL and is suited for aerial survey of complex terrain. The LMS-Q680i gives access to detailed target parameters by digitizing the echo signal online during data acquisition, and subsequent off-line waveform analysis. The operational parameters of the LMS-Q680i can be configured to cover a wide field of applications. Fast opto-mechanical beam scanning with a maximum scan angle range of 60° provides linear, unidirectional and parallel scan lines. The instrument is rugged, compact, lightweight (18 kg), and draws only 7A at 24 V DC, therefore ideally suited for the installation in small twin- or single-engine planes, helicopters or even UAVs.

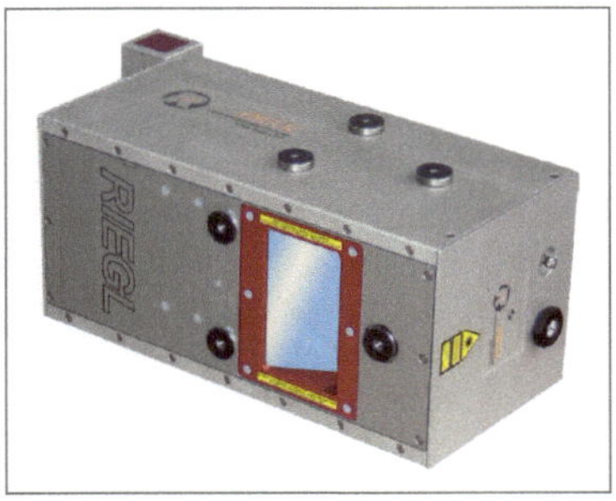

Figure 8.22. RIEGL LMS-Q680i.

VQ-780i

The RIEGL VQ-780i (Figure 8.23) is a high performance, rugged, lightweight, and compact airborne mapping sensor. This system is designed for high efficient data acquisition at low, mid, and high altitudes, covering a variety of different airborne laser scanning applications from high density to wide area mapping. A high speed rotating mirror design ensures reliability and uniform point distribution across its entire wide field of view and at all flying altitudes. A pulse repetition rate of up to 1 mHz enables an effective measurement rate of 666 kHz with 60° maximum scan angle. The system uses online waveform processing as well as offers full-waveform recording. It can perform MTA processing of up to 25 pulses simultaneously in the air. The RIEGL VQ-780i scanning mechanism—based on a continuously rotating polygon mirror wheel—delivers straight parallel scan lines, resulting in a regular point pattern on the ground. With equal spatial sampling frequency along and across track, object extents are well defined and even small objects may be detected. The operating altitude can be varied based on pulse repetition rate or the laser power level. At 100% power level and PRR of 1MHz, the maximum operating flight altitude is 2,700 m AGL; however, the measuring range to detect targets with reflectivity ≥ 20% is 2,050 m AGL. The system consumes 150 W at 18-32 V DC and is approximately 20 kg.

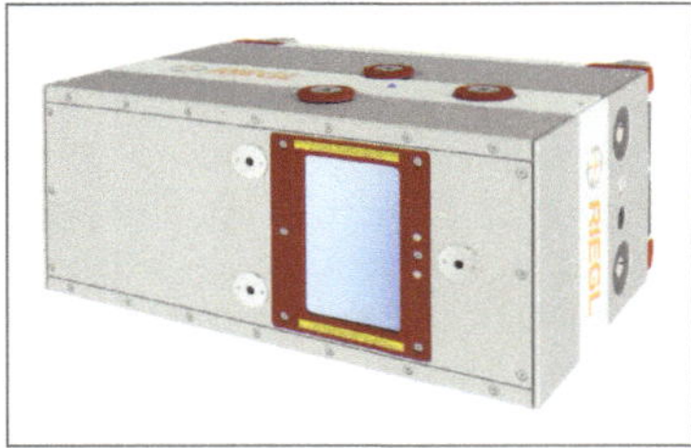

Figure 8.23. RIEGL VQ-780i.

VUX-1LR

The RIEGL VUX-1LR (Figure 8.24) is a very lightweight, compact and rugged laser scanner that was especially designed for airborne surveying missions from helicopter, gyrocopter, and other small aircraft. The sensor provides a maximum measurement range of 1,350 m, an effective measurement rate of up to 750,000 measurements/sec., and an operating flight altitude of up to 530 m (1,740 ft) AGL. In conjunction with an appropriate IMU/GNSS unit with antenna, a control unit, and a digital camera, the sensor is integrated into the VP-1 helicopter pod. This user-friendly turnkey solution can be easily mounted on standard hard points and typical camera mounts of manned helicopters for applications like corridor mapping, power line, rail track and pipeline inspection. The system is very compact and lightweight (3.75 kg) and consumes only 65 W at 11-34 V DC.

Figure 8.24. RIEGL VUX-1LR.

VQ-1560i and VQ-1560i DW

The dual channel airborne mapping system, the RIEGL VQ-1560i, (Figure 8.25) offers online waveform processing as well as full or smart waveform recording, using two laser sources, each emanating pulses 1 mHz, which results in more than 1.3 million measurements per second on the ground and ability to operate at an altitude of up to 15,500 ft. The system is ideally suited for aerial survey of ultra-wide areas as well as of complex urban environments. The VQ-1560i can also be configured to offer two lidar channels of different wavelengths, green and infrared (IR). These wavelengths are well chosen to allow the acquisition of scan data of complementary information content, thus delivering two independent reflectance distribution maps, one per laser wavelength. The dual wavelength system is called the VQ-1560i DW.

The RIEGL VQ-1560i comes with a unique and innovative forward/backward scan angle. This enables capturing data from multiple angles more effectively and more accurately at high point density. With its large field of view of 58° and its variable scan parameters, the system can maintain a consistent point density on the ground both across and along the scan direction. The MTA processing enables 20 pulses simultaneously in the air.

Figure 8.25. RIEGL VQ-1560i.

The system is equipped with a seamlessly integrated high performance IMU/GNSS system and an optional 100 megapixels RGB. A secondary camera, e.g., a thermal camera or a 100 megapixel near-infrared camera can also be integrated into the system. All individual components are integrated into a compact design and equipped with a mounting flange for simple interfacing with typical hatches or gyro-stabilized leveling mounts.

Harris Geiger-Mode Lidar

The Harris Geiger-mode lidar sensor (Figure 8.26) is the first commercial airborne lidar system that takes advantage of the single photon capabilities of the Geiger-mode avalanche photodiode (Clifton et al., 2015). Harris' system uses an array of 32 × 128 detectors with an instantaneous field of view (IFOV) of 35 μrads. The average laser power is 20 W with a pulse width of 550 picoseconds. A Class IV Nd:YAG diode pumped solid state laser of wavelength 1,064 nm is used at a pulse repetition frequency of 50 kHz to produce approximately 205 million elevation measurements per second. Altitudes of up to 11,000 m AGL is possible, although typical flights are conducted at 4,000-5,000 m AGL. A high efficiency narrow bandpass filter reduces solar background noise; however, most flights are typically conducted at night to reduce the effect of solar noise.

Figure 8.26. Harris Geiger-mode lidar sensor.

The Harris Geiger-mode lidar sensor uses a conical Palmer scan pattern produced by a direct drive, hub driven HOE (holographic optical element) scanner. HOE scanners have been used in the past (Stoker et al, 2016) but this particular implementation has several advantages over previous designs. When flown with 50 percent overlap, the scan pattern provides four looks from four different directions. With a scan half angle of 15°, the sides as well as roofs of structures are sampled, which significantly increases the potential for interpretability of the point-cloud data.

Lidar processing to create point clouds is conducted exclusively by Harris using their proprietary software which converts the raw photon data to a calibrated point cloud based on the required point density.

SUMMARY

This chapter provides an overview of the background and the state of the art of current airborne topographic lidar systems. Airborne topographic lidar is really a catch-all term for any lidar system mounted on a manned airborne platform. While the components may vary, the fundamental building blocks in general are the same and include: The laser and detector, scanning mirror (for scanning systems), GPS and IMU. Understanding the laser energy going out from the aircraft, where that laser is pointed, what energy is returned, and the position of the aircraft when that laser was fired, are key to producing accurate point clouds from these systems. Calibration and adjustments of these instruments are critical to producing good data, and will be discussed in detail in following chapters. There are many variations to these components, for example different types of lasers, detectors, digitizers, GPS units and IMUs, but in essence these various components are integrated in to a system that communicates across and between these pieces to eventually produce the 3-D information from aircraft that are invaluable for an incredible array of applications.

AUTHOR BIOGRAPHIES

Amar Nayegandhi is Vice President and Director of Remote Sensing at Dewberry. He oversees the geospatial and technology services line for Dewberry's contracts with federal, state, and commercial clients. With over 18 years of experience, he is a recognized expert in topographic and bathymetric lidar data acquisition and processing. He has authored 15 refereed manuscripts in various peer-reviewed journals and over 65 reports for the U.S. Geological Survey. Amar has a Bachelor's degree in Electrical Engineering from University of Mumbai and a Master's degree in Computer Science from University of South Florida. He is the Director of the ASPRS Lidar Division and is an ASPRS Certified Photogrammetrist and Certified Mapping Scientist-Remote Sensing.

Joshua Nimetz is the Elevation Program Area Lead for the USGS National Geospatial Technical Operations Center, the operational branch of the National Geospatial Program. Joshua provides technical and logistical guidance and support to center staff and external stakeholders in matters related to the 3D Elevation Program. Prior to joining the USGS, Joshua worked in the private sector and has nearly 15 years of professional experience in airborne lidar mapping operations. He received his Bachelor's degree in environmental science from California State University, Monterey Bay and a Master's degree in GIS/Geomatics from University of Colorado, Denver. Joshua is an ASPRS Certified Mapping Scientist—Remote Sensing.

REFERENCES

Abdalati, W.; Zwally, H.J.; Bindschadler, R.; Csatho, B.; Farrell, S.L.; Fricker, H.A.; Harding, D.; Kwok,R.; Lefsky, M.; Markus, T.; et al., (2010). The ICESat-2 laser altimetry mission. *Proc. IEEE* 2010, 98, 735-751.

Aull, B.F.; Loomis, A.H.; Young, D.J.; Stern, A.; Felton, B.J.; Daniels, P.J.; Landers, D.J.; Retherford, L.; Rathman, D.D.; Heinrichs, R.M.; et al., (2004). Three-dimensional imaging with arrays of geiger-mode avalanche photodiodes. *Proc. SPIE* 2004, 5353, 105-116.

Awadallah, M.; Abbott, A.L.; Thomas, V.; Wynne, R.H.; Nelson, R. (2013). Estimating forest canopy height and biophysical parameters using photon-counting laser altimetry. *Proceedings of the 13th International Conference on Lidar Application for Assessing Forest Ecosystems* (SilviLaser 2013), Beijing, China, 9–11 October 2013; pp. 137-144.

Baltsavias, E.P. (1999). Airborne laser scanning: existing systems and firms and other resources. *ISRPS Journal of Photogrammetry and Remote Sensing* 54, pp. 164-198.

Brock, J.C., Wright, C.W., Sallenger, A.H., Krabill, W.B., Swift, R.N. (2002). Basis and methods of NASA airborne topographic mapper lidar surveys for coastal studies. *Journal of Coastal Research*, Vol. 18, No. 1, pp. 1-13.

Clifton, W.E.; Steele, B.; Nelson, G.; Truscott, A.; Itzler, M.; Entwistle, M. (2015). Medium altitude airborne geiger-mode mapping lidar system. *Proc. SPIE* 2015, 9465, doi:10.1117/12.2193827.

Degnan, J. Wells, D.; Machan, R.; Leventhal, E. (2007). Second generation airborne 3-D imaging lidars based on photon counting. *Proc. SPIE 2007, 6771,* doi:10.1117/12.732086.

Degnan, J.J. (2002). Photon-counting multikilohertz microlaser altimeters for airborne and spaceborne topographic measurements. *J. Geodyn.* 2002, 34, 503-549.

Degnan, J. (2016). Scanning, multibeam, single photon lidars for rapid, large scale, high resolution, topographic and bathymetric mapping. *Remote Sensing,* vol. 8, issue 11, pp. 958.

Groves, P. D. (2013). *Principles of GNSS, Inertial, and Multisensor Integrated Navigation Systems, Second Edition.* Boston, London: Artech House.

Juang, Jeng-Nan; Radharamanan, R. (2009). Evaluation of ring laser and fiber optic gyroscope technology. *Proceedings of the American Society for Engineering Educations.*

Moussavi, M.S.; Abdalati, W.; Scambos, T.; Neuenschwander, A. (2014). Applicability of an automatic surface detection approach to micro-pulse photon-counting lidar altimetry data: Implications for canopy height retrieval from future ICESat-2 data. *Int. J. Remote Sens.* 2014, 35, 5263-5279.

NASA, (1975). *Airborne oceanographic lidar system final report,* Contract No: NAS6-2653, October 1975. Prepared for NASA by Avco Everett Research Laboratory, Inc.

Nayegandhi, A., Brock, J.C. (2008). Assessment of coastal vegetation habitats using lidar. In: Yang X. (ed.) *Lecture Notes in Geoinformation and Cartography - Remote Sensing and Geospatial Technologies for Coastal Ecosystem Assessment and Management:* Springer Publication Ch. 16 pp. 365-389.

Nayegandhi, A., Brock, J.C., Wright, C.W., O'Connell, M.O. (2006). Evaluating a small-footprint, waveform-resolving lidar over coastal vegetation communities. *Photogrammetric Engineering and Remote Sensing* 2006-12:1408-1417.

NovAtel (2014), *IMU errors and their effects.*

Priedhorsky, W.C.; Smith, R.C.; Ho, C. (1996). Laser ranging and mapping with a photon-counting detector. *Appl. Opt.* 1996, 35, 441-452.

Renslow, M., (ed.) (2012). *Manual of airborne topographic lidar,* American Society for Photogrammetry and Remote Sensing.

Shan J., Toth, C.K. (2008). *Topographic laser ranging and scanning: principles and processing,* CRC Press.

Stoker, J.M.; Abdullah, Q.A.; Nayegandhi, A.; Winehouse, J. (2016). *Evaluation of single photon and geiger mode lidar for the 3D Elevation Program.* Remote Sens. 2016, 8, 767.

Townes, C.H. (2010). *The first laser. From: A Century of Nature: Twenty-one discoveries that changed science and the world* (ed. Laura Garwin and Tim Lincoln).

Ullrich, A., Pfennigbauer, M. (2011). *Echo digitization and waveform analysis in airborne and terrestrial laser scanning.* Photogrammetric Week, 2011, Dieter Fritsch (ed.).

Van Sickle, J. (2008). *GPS for Land Surveyors, Third Edition.* Boca Raton: Taylor & Francis Group.

Vosselman, G. and Mass, H.G. (2010). *Airborne and Terrestrial Laser Scanning,* First Edition.

Wagner, W., Ullrich, A., Melzer T., Briese, C., and Kraus, K. (2004). From single-pulse to full-waveform airborne laser scanners: potential and practical challenges. In *International Archives of Photogrammetry and Remote Sensing,* v. 35, Part B3, pp. 201-206.

Wright, C.W., Hoge, F.E., Swift R.N., Yungel, J.K., Schirtzinger, C.R. (2001). Next-generation NASA airborne oceanographic lidar system. In *Applied Optics,* 40 (3), pp. 336-42.

Chapter 9

LIDAR DATA PROCESSING

Joshua M. Novac

INTRODUCTION

The intent of this chapter is to provide users with an overview of lidar processing concepts and techniques. This will include filtering the lidar point cloud to determine which points represent ground, noise, structures, vegetation, and other features. Additionally, concepts related to the collection and use of breaklines and the development of digital elevation models will be covered. This chapter makes an assumption that the lidar data has already been calibrated to ensure the points are geometrically accurate when compared to each other and when compared to ground control. It also assumes that the lidar being processed is represented by discreet returns. Waveform processing will not be covered in this chapter nor will the specifics of bathymetric lidar processing be covered. The methods used to calibrate lidar and verify its accuracy are covered in Chapters 8 and 10 respectively. Additionally, the discussion of bathymetric lidar and other types of lidar is not covered in this chapter. The concepts and methods described in this chapter should be useful for lidar acquired using traditional sensors as well as Geiger mode or single photon sensors.

As lidar data has become adopted as a primary source for elevation data a significant amount of emphasis has been placed on improving the methods for processing the lidar point clouds. These improvements have focused on the accuracy of the data, the number and types of features that can be automatically extracted, and the speed at which the data can be processed to final products. One of the main drivers for lidar data in the U.S. is the USGS 3DEP program which is covered in detail in Chapter 5. Lidar data processed to the USGS Lidar Base Specifications (Heidemann, 2018) required for 3DEP will meet the majority of the specifications for airborne lidar where deriving a bare earth elevation model is the end goal. The processing steps in this chapter will cover those base requirements as well as common add-ons typically found in lidar projects such as building extraction, contours, and various derivative products.

Whereas Chapter 8 focuses on the core concepts of how lidar sensors operate, this chapter will focus on the practical processing of lidar. As a result it will focus on the use of commercial off-the-shelf software to process lidar and derivative products thereby providing a practical approach for all users to begin processing the data. The concepts behind how these steps work will be provided to give a background in how the process works.

LIDAR PROCESSING

Prior to any lidar processing it is important to understand the end use of the lidar and derivative products. For example, will the lidar data be used for hydrologic and hydraulic modeling, line of site analysis, vertical obstruction surveys, or any of the other products described elsewhere in this book? Additionally, it is critical to understand the collection parameters used for the lidar acquisition in order to properly set the parameters for processing. Table 9.1 on page 239 below provides an overview of some of the standard lidar information that users will need prior to the final processing. If this information is not available some of it can be gained by performing the steps outlined in chapter 15 related to the quality assurance of the lidar data. Having a good understanding of how the lidar was acquired will allow users to more quickly process the data and achieve better automated results.

Parameter	Necessary Information	Why this Information is Important
Sensors	What sensor was used during the acquisition? Does this sensor use an oscillating mirror or one that rotates?	Understanding what sensor was used during the acquisition will provide users with insight into how the lidar point cloud will be represented. For example, some sensors use a different configuration which allows the sensor to perform at a higher scan angle than others. Sensor such as those built by Riegl use a rotating mirror and as such typically have a functional scan angle closer to +/-30° whereas sensors that using an oscillating mirror have scan angles closer to +/-20°
Nominal Pulse Spacing	What is the nominal pulse spacing for the lidar data?	This information is critical for automated processing of the lidar data. It also will inform users as to what types of features and level of detail will be captured in the lidar data. For example, a product with a nominal pulse spacing (NPS) of 0.71 m (2 points per square meter) will provide less detail than an NPS of 0.22 m (20 points per square meter). As a result the definition of features will be better in the higher density data. See Figures 9.1 and 9.2 on page 240 for examples of the different densities
Signal Returns	How many returns per pulse is a sensor capable of and were these returns recorded?	Most modern sensors are capable of collecting multiple echoes (returns) per pulse with the latest generation recording 4+ returns in areas of dense canopy. There are some sensors such as the current generation of the Geiger mode technology that only collect one return per pulse as a function of how that system operates. It is important to understand this because some software that classifies lidar data will recommend using the last return only as it most closely represents the bare earth. While this is true under some circumstances it should be noted that there will often be last returns within tree canopy and certainly on open roof tops. See Figure 9.3 on page 241 for an example of the lidar return numbers
GPS Times	Is the time set to adjusted or weekly GPS time? What is the precision of the recorded time?	Understanding the time is important to determine if there are unique times for each record. This can be used later on to determine which data is the latest in areas of temporal change. Additionally, for coastal projects understanding the exact time when the lidar was collected can enable the end user to determine the tidal impact or temporal change in the tidal areas

Parameter	Necessary Information	Why this Information is Important
Signal Strength	Has the intensity been recorded?	Most sensors collect the intensity values of each pulse and return (though some only collect unique values for the first few returns). The intensity is useful for providing a visual snapshot of the ground at the time of collection. It can also be used when determining noise or other types of points that have similar signatures. It should be noted though that with an active sensor the radiometric properties of the intensity will vary depending on the time of day and local conditions meaning differences between swaths will exist if the data has not been adjusted to provide a more consistent histogram
Overlap	How much overlap was used when flying the project? Is the overlap necessary to meet the project density requirements?	The amount of overlap will impact how the overlap bit is set and may determine how the final products are generated. If a very consistent surface model is required users may not wish to use the overlapping points if the percent of overlap is less than 50%. The reason for this is that there will be inconsistencies in the point density which may be visible in the final DEM and potentially other derivative products such as contours
Ground Conditions	What were the ground conditions at the time of collection? Was there any flooding or significant rain events that would cause large areas of temporal change in the data?	Understanding the ground conditions at the time of collection will enable the user to determine how to appropriately use the data. For example if there was a flood event that was collected by the lidar the water levels may be higher than normal and not represent the actual banks of a hydrographic feature
Vegetation Requirements	Was the data collected during a time of year when the leaves were at a minimum? Was the lidar collected to be used for vegetation analysis?	These questions will help users determine if there are additional uses for the lidar such as biomass calculations or other vegetation information. Often, lidar is collected to determine ground points and so leaves should be at a minimum during that collection. Much of the data collected under the USGS 3DEP program will fit into this category

Table 9.1. Lidar Acquisition Parameters.

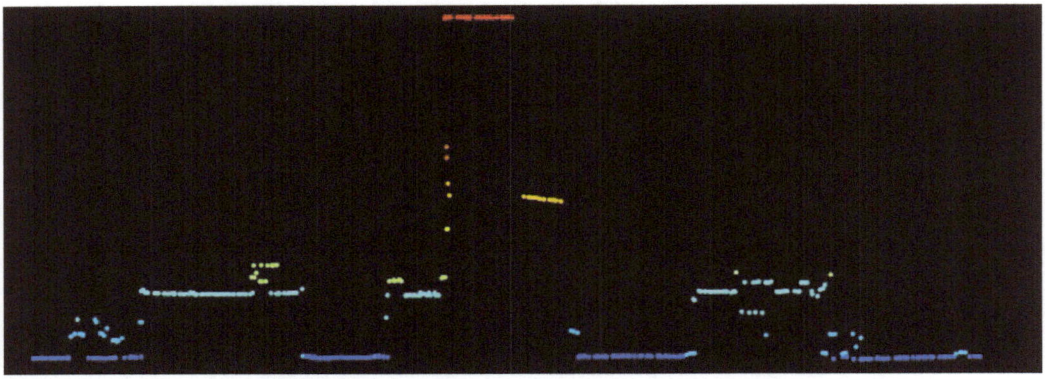

Figure 9.1. One meter cross section of industrial building from lidar with an NPS of 0.7 meters (2 points per square meter).

Figure 9.2. One meter cross section of a water control structure from lidar with an NPS of .22 meters (20 points per square meter). Note that you are able to see a significant amount of detail on the sides of the feature as well as railing and other smaller structures. The ability to detect points on the sides of features will be largely dependent on the type of sensor being used in the collection. On this project a Riegl 1560i was used which collects the data at a forward and reverse angle instead of directly at nadir.

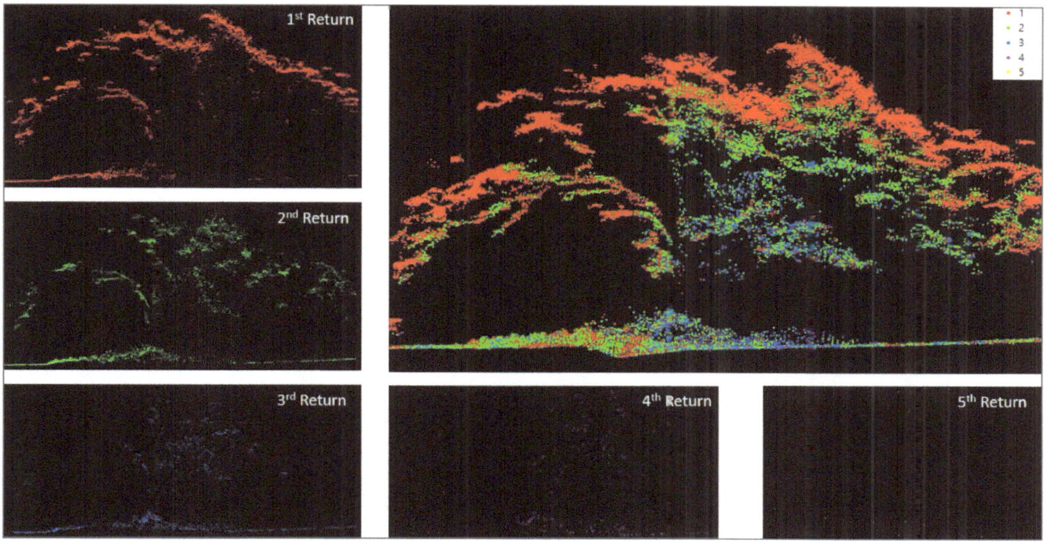

Figure 9.3. Example of vegetation with multiple returns; the breakdown of each return shows how some pulses are capable of having enough energy to register as a return and provide foliage penetration.

Concepts and Approaches to Lidar Processing

This portion of the chapter will provide an overview of the concepts related to processing the lidar data to gain information related to the ground elevation, structures, vegetation and other features. There are many commercial software products available to aid in the processing of lidar; this chapter will not cover how to use any software. Instead this section is intended to provide guidance on what to look for and how to better understand what each type of classification is actually doing in the software. Much of this section focuses on the automated classification steps. It is important to note that at the time of this writing there is no automated process that will get you 100% of the way to a finished product. As outlined previously, the key to successfully processing lidar data is to understand how the data will be used and then understanding what level of error in the classification is acceptable. Many projects will deal with millions or billions of individual points so an expectation of 100% accuracy is typically not achievable. Most automated filtering is followed by some level of manual clean-up and editing, depending on the needs of an individual project.

Automated Filtering of Point Clouds

The first step in processing lidar data is to apply some amount of automated filtering, depending on project needs. This section will cover the concepts related to processing a point cloud to some of the more typical requirements including: Ground, noise, buildings/structures, vegetation, and other above-ground features. The standard topographic lidar point classes as defined by ASPRS in the LAS Specification Version 1.4—R13 (ASPRS, 2013) will be referenced in this section. Table 9.2 on page 243 provides an overview of the standard point classes from the ASPRS specifications.

Class	Description	Details
1	Unclassified	Points that have not been placed in any of the other classifications. Often lidar is classified to only a few classes and much of the data (especially above-ground features) are in this class
2	Ground	Points that have been classified by automated and/or manual methods that represent the bare earth topography
3	Low Vegetation	Points representing vegetation that fall within a low height above ground threshold. The points are usually classified within a height above-ground window or bin, e.g., points that are .05 meters-3 meters in height
4	Medium Vegetation	Points representing vegetation that fall within a medium height above ground threshold. The points are usually classified within a height above ground window or bin, e.g., points that are 3 meters-9 meters in height
5	High Vegetation	Points representing vegetation that fall within a high height above ground threshold. The points are usually classified within a height above-ground window or bin, e.g., points that are greater than 9 meters in height
6	Building	Points representing the tops of buildings. May also include sides of buildings depending on project requirements or needs
7	Low Point (Noise)	Noise points that are below the average ground surface
8	Reserved	This class is often used for model key points generated from the lidar data if a flag or bit is not used instead
9	Water	Points that represent water surface elevations. These are usually classified using breaklines
10	Rail	Points representing the surface of rail lines
11	Road Surface	Points representing the surface of roads. Usually classified through semi-automated means or breaklines
12	Reserved	N/A
13	Wire—Guard (Shield)	Points that often represent portions of the electrical transmission and distribution system. Often requires higher point density lidar to support this classification

Class	Description	Details
14	Wire—Conductor (Phase)	Points that often represent portions of the electrical transmission and distribution system. Often requires higher point density lidar to support this classification
15	Transmission Tower	Points representing the transmission tower on an electrical transmission or distribution system
16	Wire-structure Connector	Points that often represent portions of the electrical transmission and distribution system. Often requires higher point density lidar to support this classification
17	Bridge Deck	Points representing the bridge deck. These are only the points on the deck and guard rails; vegetation or other above-ground features should be removed from this class
18	High Noise	Noise points that are above the average ground surface and can be related to atmospheric noise, sensor noise, or other objects whose source causes unusually high returns
19-63	Reserved	Other non-specific points. If working with bathymetric data water surface, column, and submerged topography are in this window
20	Ignored Ground (USGS)	This classification is used on USGS 3DEP project for points that are near the hydrographic breaklines to aid in DEM generation. Point are classified as ignored ground if they are within two times the nominal point spacing
21	Snow (USGS)	Used for USGS 3DEP projects where snow is present and required to be classified
22	Temporal Exclusion (USGS)	Used on USGS 3DEP projects in areas with temporal change. The points reflecting the surface not desired are classified as temporal exclusion. These points are usually in tidal areas or areas that undergo significant change (6 cm or greater) during the course of the acquisition
64-255	User Definable	Can be used to classify other types of data during processing or as a specific requirement for a project. These can be defined as needed and are considered non-standard

Table 9.2. ASPRS Point Classification Schema.

Ground/Non-Ground

One of the primary uses of lidar is to determine the elevation of the bare earth or ground. Many common geospatial software include options to process lidar to ground (class 2) and non-ground (class 1). There are several key considerations for processing the ground using automated means; these are: General terrain changes (slope, curvature, etc.), point density, vegetation coverage, and urban coverage. These are important because software used to classify lidar usually rely on an approach that looks to find the lowest point within a window and then compare that point to its neighbors. Using that approach the automated process classifies the points as ground that fall within the set parameters.

As the points are being compared to neighbors, users will need to know first how large of a window the process should be looking at which will be based on how far apart the points are. Higher densities can use smaller windows to look for deviations or changes in the surface. Additionally, some software allows users to set a maximum deviation from the ground to be considered a ground point. This could be in the form of a height parameter or slope depending on the software. This is the reason it is important to understand what types of features in the project area represent ground. For example, if working in a very flat environment, a lower threshold can be set compared to mountainous terrain or areas with steep embankments where larger deviations from neighboring points will be necessary to accurately classify the ground.

Additionally, most software allows users to set parameters for buildings and vegetation so that they can be excluded from the ground (see Figures 9.4 and 9.5). One frequent error in this automated process is that large buildings with flat roof tops may be classified as ground. This is a result of the filter identifying those points and surrounding ones as relatively flat and large enough to be mistaken for the top of a ground feature. In some cases these remaining features will require either additional iterations of the ground to be run in order to minimize them or they can be classified manually to ensure they are not remaining in the ground.

Figure 9.4. Correctly classified ground points (brown) and unclassified vegetation (gray).

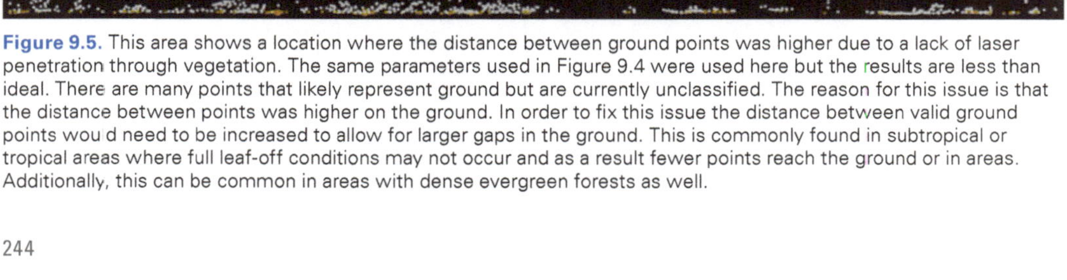

Figure 9.5. This area shows a location where the distance between ground points was higher due to a lack of laser penetration through vegetation. The same parameters used in Figure 9.4 were used here but the results are less than ideal. There are many points that likely represent ground but are currently unclassified. The reason for this issue is that the distance between points was higher on the ground. In order to fix this issue the distance between valid ground points would need to be increased to allow for larger gaps in the ground. This is commonly found in subtropical or tropical areas where full leaf-off conditions may not occur and as a result fewer points reach the ground or in areas. Additionally, this can be common in areas with dense evergreen forests as well.

The automated filtering process often struggles in some very specific areas but users can create additional steps to identify and mitigate these issues early in the process. One area where automated classification often struggles is on steep embankments. Because these can have high slopes it is common for the lidar data to be classified in such a way that the tops of these features are missing. This is also true on manmade embankments and retaining walls. In these areas users can either manually edit the data to more accurately reflect the ground or try additional steps to add these features back to the ground. Often this is done by looking for small voids at the tops of features with steep slopes and then creating polygons around them and re-running the ground within those areas using settings that allow for steeper angles between points. Figure 9.6 shows an example where a steep embankment contains errors in the classification.

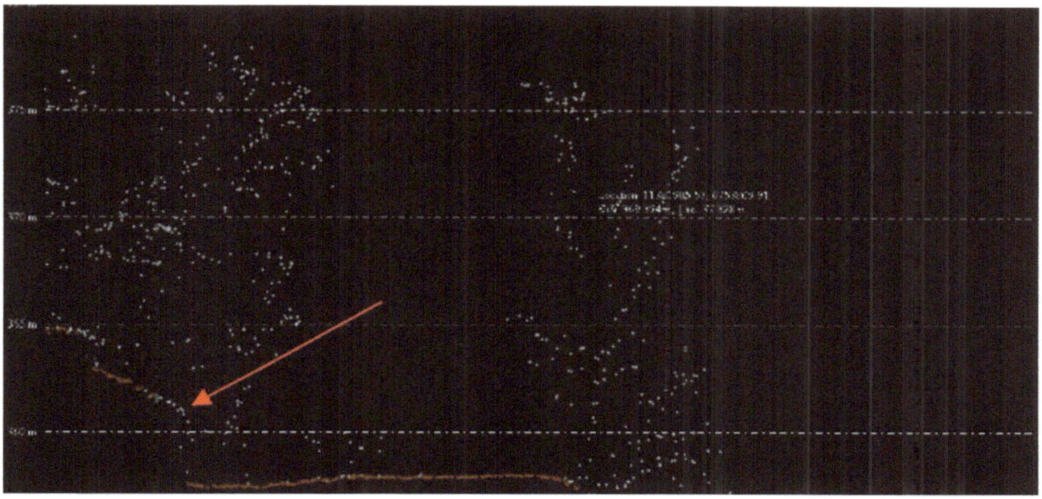

Figure 9.6. Example of lidar data where the ground (brown) was processed using an automated method. In this example a steep embankment was not classified as ground. This embankment can either be outlined and reprocessed based on the slope of the surrounding terrain or manually edited to improve the classification.

One aspect to lidar that can be leveraged for ground processing is the use of last returns (or single returns) to help filter out some of the above-ground features prior to starting ground classification. This is by no means a clean ground as it will contain points on the tops of solid features (buildings, cars, etc.) as well as single return points within vegetation. However, this can help to remove a fair amount of data that would otherwise introduce additional complexities to the classification.

Ground processing can also be improved through the use of iterative ground classification steps where multiple ground filtering steps are run on a dataset. Each ground filtering step can have different parameters set to best identify ground points.

Key Considerations for Ground Processing:

- What is the average point spacing for the project?
- What type of slopes are generally in the project area and should the project be divided up into separate areas for automated classification?
- How dense is the vegetation in the project area and how well did the lidar penetrate through the foliage to reach the ground?
- Are there last return points that can be used as a starting point for the ground classification?

Noise

Noise within a lidar point cloud can come from a number of sources including atmospheric noise, objects between the aircraft and ground (birds, clouds, etc.), sensor anomalies, and other features that cause the returns to be unlikely to reflect the actual ground conditions. The classification of the noise is typically divided between low and high noise, usually relative to the ground, i.e., low noise is below the ground surface and high noise is above the ground surface.

The identification of noise points in the lidar point cloud can be somewhat subjective but is usually focused on points that are isolated or appear to be in error and not reflective of the actual surface. For high noise, objects like birds or floating balloons, will cause spikes in a first return surface (top of objects) and should be classified as noise to make the automated processing of the data perform better and provide better digital surface models using the point cloud. Many software provide a way to identify isolated or groups of isolated points that can be automatically classified as noise. In some cases the grouping of points, or the way they interact with the surface, may require additional manual corrections. Figure 9.7 shows an example of noise along the side of a parked vehicle.

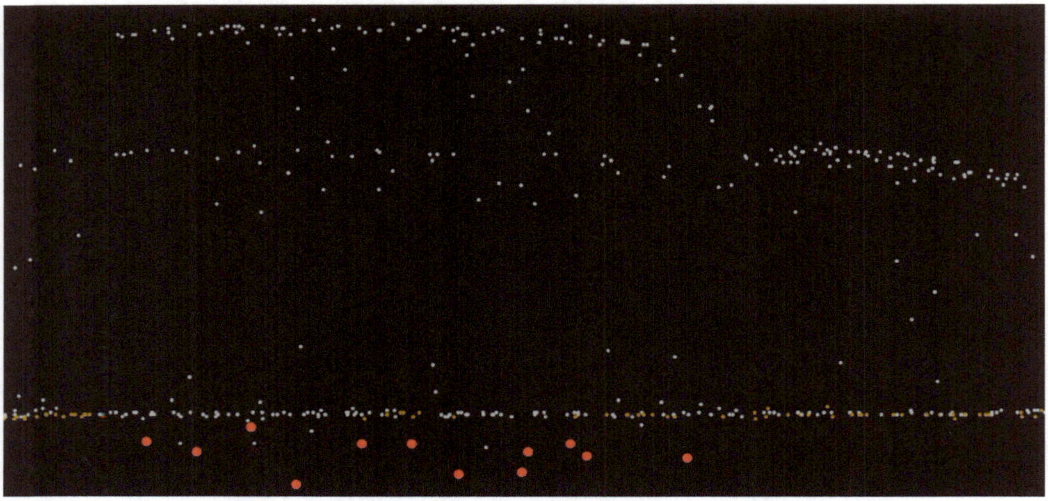

Figure 9.7. Example of low noise (red points) along the side of a parked vehicle. This data was collected at a higher density of about 22 points per square meter and the reflection from the lidar off the vehicle to the ground and back up to the sensor causes some points to appear lower than the actual ground surface. In this case these points were classified as noise using manual revisions.

Noise is often filtered before any other classification steps, including ground classification, to remove these outliers from the group of points available for the remaining classification steps. Additional noise filtering steps may also be used during the ground classification steps to help identify additional noise in relation to the initial ground.

Key Considerations for Noise Processing:

- What constitutes an isolated point? How many neighbors and how far apart are they?
- How much higher than the average reflective (first return) surface would a point need to be in order to be classified as high noise?
- Is there anything specifically known about the sensor used to acquire the lidar data that would lead to additional noise being present?

Structures

The classification of structures in lidar data has become much more common in recent years as lidar data over urban areas has become more pervasive. Classifying buildings can provide a significant enhancement to existing or new lidar data collections and provide end users with information related to the number and size of structures as well as roof types (gables, flat, or complex) and heights. See Figure 9.8. Being able to classify structures can also provide information for impervious surface models and flood risk assessments. In order to automatically classify buildings there are a number of important factors to consider. The first is the size of the buildings to be captured and the expectation of what the lidar can support. Most modern lidar datasets are collected with a minimum point spacing of 0.7 meters or 2 points per square meter as measured from the first return.

Figure 9.8. Example of buildings classified using automated methods. Red points represent the building rooftops, orange points are ground, and gray points are unclassified. For this example a minimum buildings size of 100 square feet was used to define a structure.

While it is possible to collect very small structures using this density of lidar, it may be impractical to automatically derive small features consistently. Most solutions that provide automated processing for building classification do so using algorithms that look for features that have steep slopes and then relatively planar (flat or consistent slope) surfaces on top. If the building size threshold is set too small (10 square feet) users will end up with a significant number of commission errors where points that do not represent true structures have been classified as such. These will typically be found within vegetation where the points are relatively flat or on bridges or other elevated features. Most software has a minimum height parameter which will help to minimize the errors found on generally flat surfaces like roads or parking lots.

It is important to consider the point spacing because users will not get consistent points at that rate along the edges of structures. Features may also be partially obscured by vegetation and have fewer points representing the top of the structure. As a result, when classifying structures in lidar, it is not typical to see results with planimetric quality, i.e., with the edges exactly matching the edge of the feature. Higher density lidar collects will provide much greater fidelity to the building classification and ultimately to building outlines or 3-D buildings.

One aspect of structure classification to keep in mind when performing automated processing is that very irregular features are usually omitted from the classification because they do not fit within the parameters of the automated classification. This is important to consider if working in areas with large industrial structures or other non-regular features.

When performing the classificaiton of buildings starting with a point cloud of the last returns (or filtering this within the software) can provide a better starting point for the building classification. Because these are solid surfaces, the points that reflect them will likely be single returns or the the last return of multiple returns. This may not always be the case at the very edge of the feature but that small number of points will not significantly impact the classification. Additional exports from the classified structure points will be covered in the breakline section and will discuss some of the considerations when converting these points into vector data for building oultines or 3-D features.

Key Considerations for Structure Processing:

- What is the minimum size structure wanted for collection?
- How tall are the structures in the project area (what is the range of building heights)?
- Is there anything specifically known about the sensor used to acquire the lidar data that would lead to additional noise being present?
- Will building outlines or 3-D features also be processed from this dataset?

Vegetation

The classification of vegetation in lidar data is usually performed using a rather simple approach. After the ground and structures have been classified, the remainder of the above-ground data is divided into bins based on the height above ground. These bins are then used to classify low (class 3), medium (class 4), and high vegetation (class 5). The actual heights used are based on the individual needs of the project and the vegetation that falls within the area. Low vegetation generally consists of shrubs, tall grasses, and some crops with a height above ground usually up to 3 meters; medium vegetation is for smaller trees usually up to 10 meters; and high vegetation captures the remainder of features above 10 meters in height.

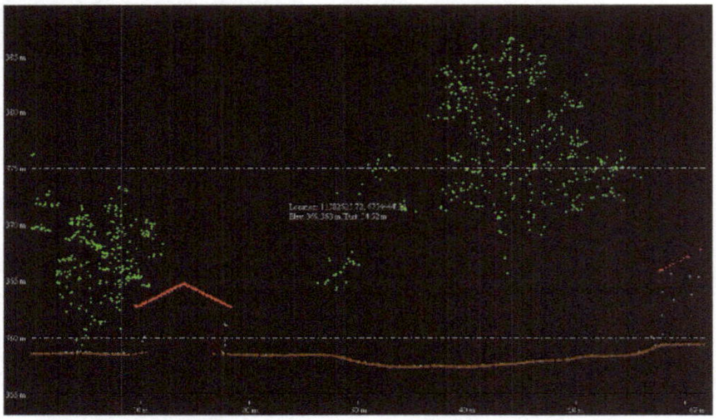

Figure 9.9. Example of lidar data that has been classified to include vegetation (green points). In this example all above-ground features with the exception of buildings have been classified in the vegetation classes.

One consideration when classifying vegetation is that in general it captures all above-ground feature that are currently not classified. See Figure 9.9. This usually includes vehicles, light poles, fences, and any other above-ground feature. This method therefore is likely not ideal when processing vegetation classes for projects that have a focus on forestry. Additionally these bins will divide up an individual tall tree into its low, medium, and high components and will not classify the entire vegetative structure as a single entity. If the accuracy of the vegetation classification is critical it may be necessary to leverage other inputs to aid in the classification and classify accordingly.

Key Considerations for Vegetation Processing:

- What height bins should be used for the vegetation classification?
- How accurate does the vegetation classification need to be and can it include other above-ground features that are not truly vegetation?
- Do individual trees or vegetative structures need to be classified consistently, e.g., should a 20 meter tall tree be classified as high vegetation from the top to base?

Other Above-Ground Features

There are quite a few features collected in lidar datasets that could be considered for additional classification should it be necessary. One item to consider when looking at additional classification is the consistency for which it will be identifiable in the lidar. Items that are consistent and discernable from the surrounding points may be better candidates for automated classification. For example, if roads and/or parking lots are classified, a height bin above those classes can be set to also classify points that represent vehicles as they would likely be within 1-10 feet above the road surface. Other features can be more difficult to classify in that manner as they are identified more inconsistently or their locations make it difficult to automate an extraction. Power poles, fences, and guard rails are consistently difficult to automatically extract in lidar data and generally require more manual effort.

As shown in Figure 9.10, power transmission and distribution lines are features that are often extracted from lidar but usually require a higher density collection to ensure a consistent feature extraction.

Figure 9.10. Example of power lines (blue) classified using automated methods in Global Mapper 19.0. The resutls are relatively good in open areas where there is little vegetation above or below the distribution line. For transmission lines where large buffers from vegetation are required this process can work fairly well.

Key Considerations for Above-Ground Feature Processing:

- Is there something about the feature that makes it stand out from other points? Is it isolated like a power line with little above or below it for some defined distance? Does it fall within a certain height above ground? Are there adjacent features that can be used such as roads or buildings to help narrow down where this type of feature is located?
- Is there sufficient density in the lidar point cloud to consistently map the feature?
- How accurately does the classification need to be?

Manual Editing of Lidar

In the previous section, information about the different types of automated classification that are typical of a lidar dataset, including key considerations for each one, was covered. The reality of current automated processing is that it will only take the data so far. There have been great improvements over the years which have significantly reduced the amount of time required to accurately process the data, and the processing tools themselves have become more common in software that is easy to use and available to large parts of the geospatial community. However, depending on the individual needs of a project, automated processing is likely not enough to ensure a high quality dataset. In this case additional manual editing is required to clean up artifacts or other inconsistencies to ensure the product meets project requirements. A good example of required manual clean-up is when performing ground classification users are likely to still find a few buildings or portions of vegetation that were classified as ground. When dealing with millions or billions of points the error margin may only be a few points incorrectly classified but they have a big impact on the final DEM or DSM. In those cases manual editing should be used to clean up those issues. Chapter 15, covering QA/QC, goes into detail on how to identify these issues. As with automated filtering, many off-the-shelf software programs provide tools for manual clean-up and editing.

BREAKLINE PROCESSING

One common addition to lidar projects is the collection and processing of breakline data. Breaklines are used in conjunction with the lidar points to enforce edges of features. In the case of lidar collected for DEM or DSM generation, breaklines are commonly collected to enforce water elevations to create a more cartographically pleasing surface model or for hydro-enforcement for flow modeling. This would include the collection of stream centerlines and culverts used for enforcement and the footprint of structures which would be used to condition a DEM so that the surface model under buildings is consistent. Additionally, it is common to collect breaklines under bridges or elevated features to ensure that the DEM does not create a saddle of higher elevations (see Figure 15.79 on page 528 in Chapter 15) due to interpolation where lidar points are not present. This chapter will focus on breaklines that are typically collected for wide-area DEM generation.

Concepts and Approaches

Breaklines add information to the lidar point cloud for the express purpose of enforcing elevations in the data that the lidar may not be able to support due to density, vegetation overhang, or because lidar does not accurately portray water surfaces due to absorption of the NIR wavelength used by many topographic lidar sensors. One of the most common examples of this is breaklines that are collected for the hydro-flattening or hydro-enforcement of elevation models.

There are a number of ways to approach the collection of the breaklines for DEM generation. Prior to starting breakline collection however it is important to consider the final application of the data. The applications for which the data will be used will impact the scale at which the data is both collected and reviewed and will impact how the analyst will go about collecting these types of features. The applications that require the most detailed breaklines are likely those generated from very high densities of lidar such as those for electrical transmission and distribution or road and bridge work. These are very specific cases and will require a much more manual approach to the collection of the breaklines.

This section will focus on a workflow which follows that of a standard collection process for hydro-flattened or hydro-enforced DEMs. In this case hydrographic breaklines are collected from the lidar and elevations are assigned to them to enforce down-hill flow throughout the hydrographic network. This is completed in most cases through a semi-automated process where the features are extracted in 2-D using the point cloud, intensity imagery, surface models, and other ancillary data. The 2-D features are then put through additional processing to add elevation information derived directly from the lidar data and enforced to ensure that they have a downhill flow. The process for features that do not have flow (ponds, lakes, and some reservoirs) is similar but these features are set to a constant elevation.

One of the first steps in collecting breaklines for a project is to set the general collection specifications and rules. Table 9.3 is an example of the collection criteria and rules for a project that requires both area features as well as stream centerlines and culverts for hydrographic enforcement. The important thing to capture with these types of rules is the minimum mapping requirement (size) for each of the features where applicable. This will ensure consistency throughout a project and enable all parties involved to work off of the same rules thus minimizing confusion during the process. Table 9.3 uses the USGS 3DEP standards for many of these features as they are the most common national standards available, but information from other standards is also included where applicable. Each project should be evaluated based on its individual needs and this table is just to provide a general overview.

Feature	Capture Rules
Water Bodies	Water bodies represent features that are ponds, lakes, and other features where in general the elevation will be consistent. These types of features can have varying minimum requirement depending on the project. The USGS 3DEP program typically requires these features to be collected if they are 2 acres or greater. The National Hydrography Dataset (NHD) program has smaller requirements with features 0.25 acres or greater being captured. Only features containing water are captured as the resulting polygon will be used to classify the water points (class 9) and flatten the water surface in the DEM Breaklines for water body features should be captured at or just below the elevations of the immediately surrounding terrain. Under no circumstances should a feature be elevated above the surrounding lidar points which would create an artificial plateau
Area Rivers or Streams	These features are rivers or streams that are wide enough to show open water and generally have flow. Only features carrying water are captured as the resulting polygon will be used to classify the water points (class 9) and flatten the water surface in the DEM Features are captured for USGS 3DEP projects if they are on average greater than 100 feet in width. It is very common to collect features that are less than this requirement for NHD and other uses. For example NHD projects will generally collect streams 10-12 feet in width and some state programs require features as narrow as 8 feet to be collected in this manner. The important consideration is to ensure that lidar density will support the identification of the feature Rivers are collected as a dual line (one on each side of the feature) or closed polygon. Each vertex placed should maintain vertical integrity. Generally both banks shall be collected to show consistent downhill flow. There are exceptions to this rule where a small branch or offshoot of the stream or river is present The banks of the stream must be captured or enforced at the same elevation to ensure flatness of the water feature. Breaklines must be captured at or just below the elevations of the immediately surrounding terrain. Under no circumstances should a feature be elevated above the surrounding lidar points Dual line features shall break at road crossings (culverts). In areas where a bridge is present the dual line feature shall continue through the bridge
Tidal Waters	The collection of tidal waters can be a greater challenge for lidar projects. The reason is that the water elevations are constantly changing between flight lines and really represent elevations at that time of collection only. For each project analysts should determine what the end use will be. Below are a few options to consider when collecting tidal water 1. Collect breaklines that represent the elevation of the water at the time of collection. In this case, analysts will have polygons representing the tidal water for each flight line. The final elevations will be set to the actual elevation at the land/water interface. For streams that are tidally impacted they should be enforced to flow downhill to the coast and snap to the open water elevation. Variations in elevation may exist from flight line to flight line 2. For smaller projects it may be sufficient to collect the tidal area at the land water interface and assign a single elevation to the entire feature based on the lowest points on the landward side of the breakline 3. For larger coastal collections it may be applicable to collect the lidar data when the water is below a certain elevation or within a specific tide stage. In this case the exposed ground is used to create a contour at a specified elevation and all points that would be on the water side are classified as water and simply flattened to that elevation in the DEM
Islands	Islands are collected on area features following a general rule that the island should be collected if it is ¼ to ½ the size of the requirement of the water body features. This can vary greatly and is dependent on the needs of the project. Some areas of rocky coast may require an additional collection requirement that defines islands that exceed a certain height would be collected

Feature	Capture Rules
Single Line Streams	Some lidar projects will require the additional collection of stream centerlines or flow lines that represent areas where water is either currently collecting or is likely to collect during precipitation. This is most common for projects related to hydrographic modeling or for programs such as the NHD. There are many good resources that cover how to collect these types of features and this chapter will just be providing an overview and not specific details on this process For this example features that represent drainage (wet or dry) will be collected as centerlines for features that exceed 1,200 feet in length or drain an area of 0.5 square miles. Streams are to be enforced monotonically and connect with other features at a vertex of the same position and elevation
Culvert	Culverts should be collected where streams cut through roads or other features. Depending on the project, culverts can be collected only where other hydrographic features are located or can be collected throughout a project area. Culverts should be collected with only two vertices (start/end) and must snap to adjacent features
Under-Bridge Saddle	In areas where the DEM produces a saddle under the bridge, additional breaklines are collected to support the generation of the DEM. These breaklines are placed perpendicular to the bridge at the bottom of the embankment or as needed to help model the slope. Vertices are to be placed in areas where the density of the ground points is sufficient to model the surface which is usually on either side of the feature and not directly under. One vertex should be placed on either side of the bridge and the elevations derived directly from the lidar surface. The key to remember is that the ends of the breaklines need to be placed where lidar points are present and not within the interpolation area

Table 9.3. Breakline Overview and Collection Rules.

Area Hydrographic Features

The identification of open water using lidar data can be challenging for a number of different reasons. The first is that because lidar is an active sensor (for this discussion we are focusing on lidar using a NIR laser) and the laser interacts inconsistently with water surfaces there is no consistent starting point that holds true for every swath or area. Lidar sensors between manufactures themselves operate slightly differently as well which adds further complication to the extraction of water from lidar data. A second challenge is that to determine the land/water interface one needs to typically have good foliage penetration from the lidar as the vegetation tends to grow along these water edges.

Because lidar can be relatively inconsistent on water surfaces, and automated extraction of the features may require more specialized software and testing, many users look to available data sources that they can leverage. Unfortunately, in general, existing data was either not collected to the level of detail required to accurately outline the hydrographic features or the feature has changed in a significant way from when it was previously collected. Figure 9.11 is an example of an area with breaklines collected using the information from the lidar and the existing NHD data for that same area. In this case the NHD data is temporally very inconsistent with the current water levels and extents of the feature. In many cases, if the NHD was utilized, areas where valid ground points have been collected by the lidar would be flattened as water and additional detail from the ground would be lost. Keep in mind that this does not mean that the NHD data is incorrect for the time of its collection only that it temporally does not match with the current lidar dataset. Additionally, ancillary layers like NHD data can be leveraged to help identify areas where hydrographic features are likely to exist.

Good foliage penetration is especially necessary for automation of river and stream delineation. Identifying bank edges where vegetation is not present tends to be much simpler than identifying bank edges covered by vegetation because the density of the ground points will be higher in non-vegetated areas. Figure 9.12 shows an example of a stream running through a moderately forested area with some open areas along the bank. Figure 9.13 on page 254 is a similar example over an area with a much higher degree of overhanging vegetation. Figure 9.13 shows areas where automation is much more difficult to achieve because the intensity and DSM both show a significant amount of vegetation obscuring the bank edges.

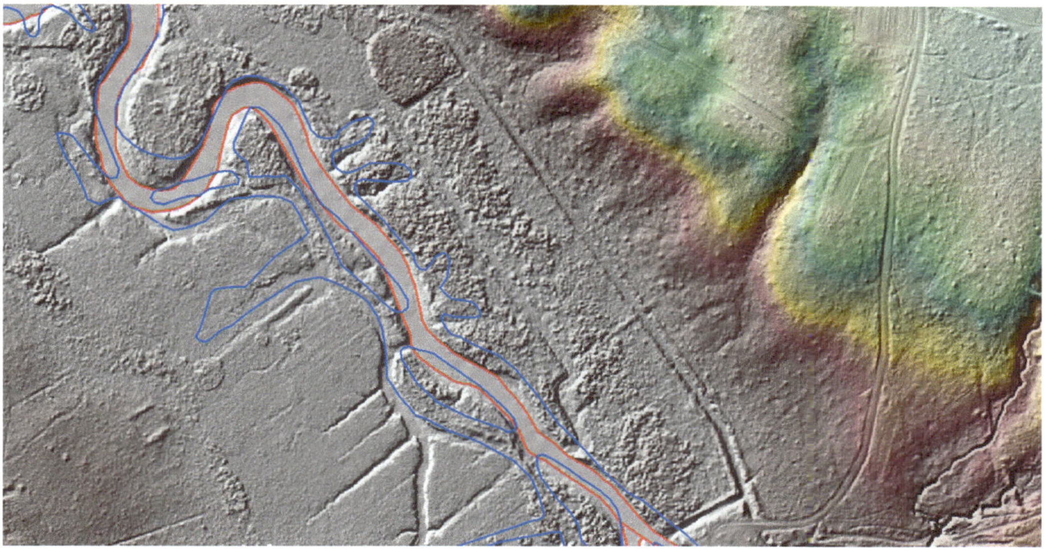

Figure 9.11. Hydrographic breaklines collected using the lidar information are shown in red and represent the actual extent of the water surface at the time of collection. The existing NHD data is shown in blue and temporally does not match with this surface model.

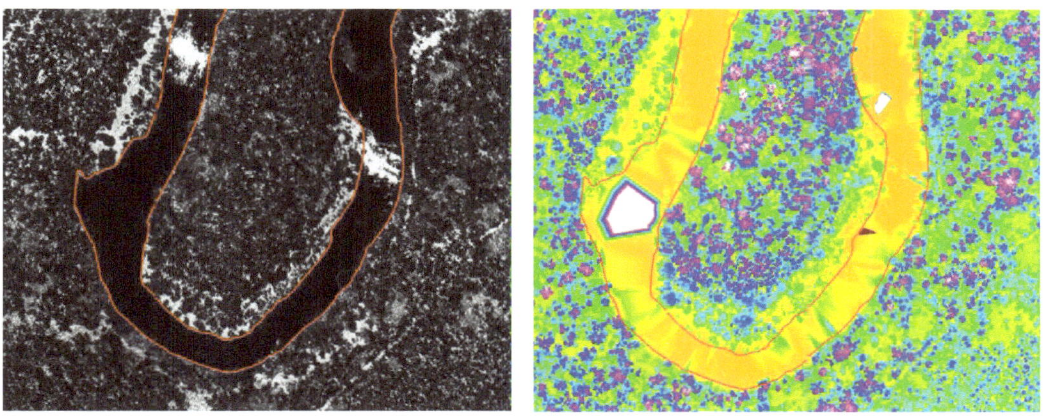

Figure 9.12. Example of stream running through a moderately vegetated area. The image on the left shows the stream (orange) overlaid on the intensity imagery. The image on the right is the same location overlaid on the DSM. There are areas of both vegetation overhanging the bank of the feature and areas of non-obscured banks.

In some cases because of the dense vegetation the user will need to utilize additional information from the grounded surface (bare earth DEM) in order to inform where the edge of the feature is actually located. After the vegetation and above-ground features have been separated from points representing ground it is possible to use the density to aid in defining where water is present. The reason for this is that in general NIR lidar is absorbed by the water resulting in a lower point density on those surfaces. Recall that this is not always true and is highly dependent on the sensor, flying height, scan angle, turbidity of the water, and other general spectral reflectance. However, it holds true enough that in many instances this information can be used. Figure 9.14 on page 254 is an example of the ground lidar points (brown points) that show the lower density

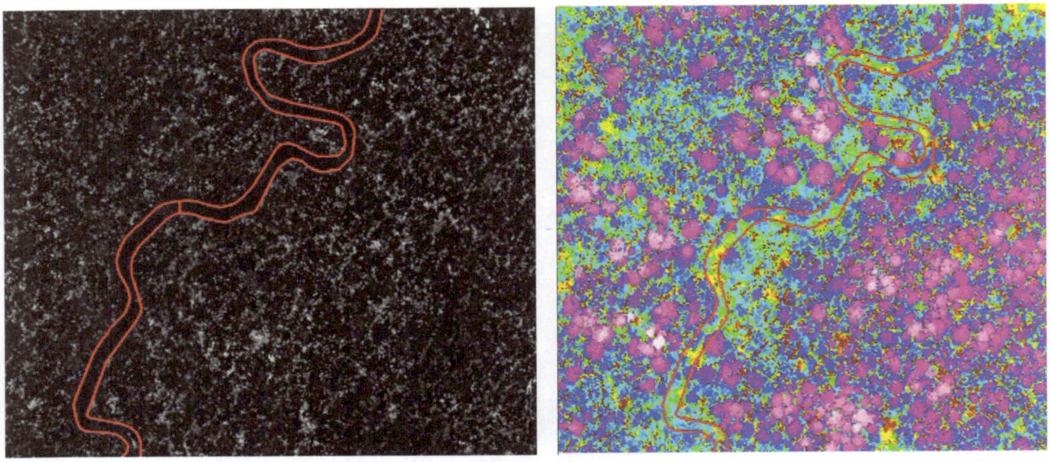

Figure 9.13. Example of stream running through very dense vegetation with significant portions of the feature obscured in the intensity and DSM.

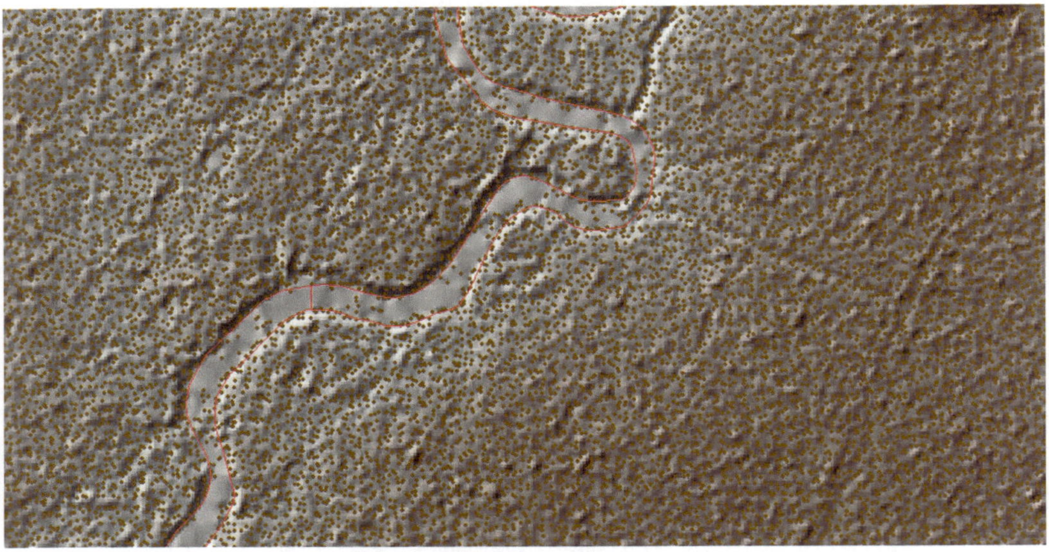

Figure 9.14. Stream (red) overlaid on a bare earth DEM and class 2 (ground) lidar points. This is the same feature from Figure 9.13 that was difficult to identify in the DSM and first return intensity imagery. In this example the stream is only about 30' in width and covered by dense vegetation making it difficult to automatically detect using intensity imagery alone.

of the lidar data on the stream surface. Note that there are points on the surface and it's not a complete void, but the sparse nature of the points helps to identify that feature.

While both intensity and density can be useful for identifying water surfaces it is important to reiterate at this point that because lidar is an active sensor and inconsistently interacts with water there is not a method that works consistently across all features. Figures 9.15 and 9.16 are in the same project area as the previous examples and represent a very large river about 900' in width. The intensity information shows a feature that is relative dark (low intensity) on the northern section and brighter to the south (higher intensity) due to spectral reflectance off of the water surface. As a result of the spectral reflectance, likely due to small waves or ripples on the water surface, the density on that southern section of the stream segment does not look all that different from the surrounding ground.

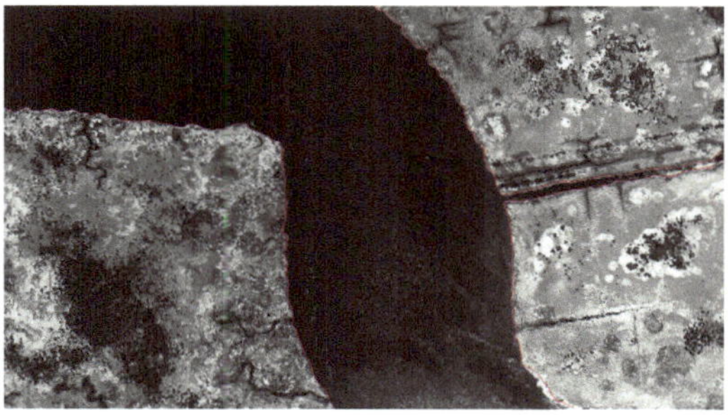

Figure 9.15. Example of large river (red outline) overlaid on the intensity image. This example shows some of the slight variability in the intensity values on water which can complicate the automated extraction of these types of features.

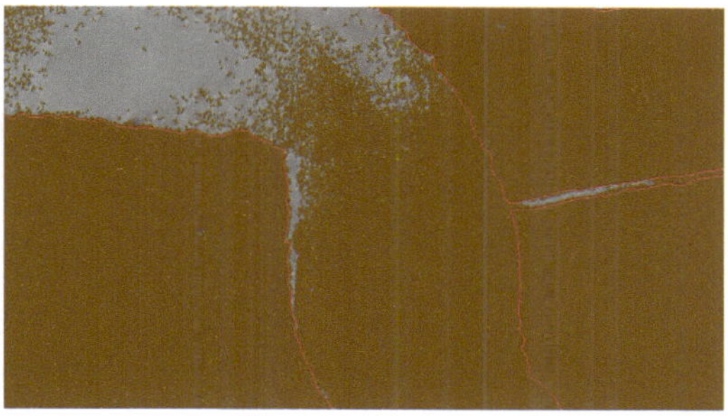

Figure 9.16. Example of large river (red outline) overlaid on the bare earth DEM with ground lidar points shown. This is intended to provide an overview of how the density on water surfaces can be incredibly variable based on how the light has interacted with that water surface.

Understanding that the lidar data is inconsistent in how it interacts with water is a critical component of automated processing because it enables the user to make more informed decisions about how water surfaces can be identified. This may change for each project so take the time to review the lidar data including the bare earth DEM, intensity imagery, density information, height, and other ancillary data to determine the best course of action for delineating these features.

With an understanding of the individual project the user can now build a strategy to identify these types of features. The following paragraphs provide information related to some of the layers that can be useful when attempting to identify area hydrographic features in lidar data. This information can be used and incorporated into the software for automated extraction of features. The software used will determine exactly how this information is used and will vary greatly depending on the complexity of that interface. Some software allows the user to utilize multiple layers simultaneously and weight those layers to build a network for feature identification. Other software may use simpler strategies for categorizing the pixels and require only raster inputs.

A model of the bare earth surface in the form of a DEM, TIN, or Terrain can be very useful to aid in the identification of hydrographic features. This feature is also used as an input in creating some of the other useful models. A primary use of this elevation model is in the visual review of where features should be located and extract the edge of that feature. Typically, hydrographic features can be identified by locating the toe of bank or where the slope of the bank ends and within two banks it is either relatively flat from points located on the water surface or is interpolated due to lack of lidar returns. Figure 9.17 is an example of the hydrographic feature placed using the bare earth surface model. There are a few considerations to be mindful of when using the bare earth surface for extracting hydrographic features:

- Are the lidar points classified accurately along the feature edge?
- Low lying areas do not necessarily represent water and may be mudflats, sandbars, or other features that should be maintained as ground
- Leverage other sources to help identify the edge when it is not clear using only the bare earth model

- Depending on the software the user will want to display the bare earth surface model in addition to a hillshade which will better define the edges. The actual display parameters for the hillshade will vary by software but very steep sun angles should be avoided as they will create areas that cast longer shadows. In general a good sun angle is in the 60 to 70 degree range for this purpose

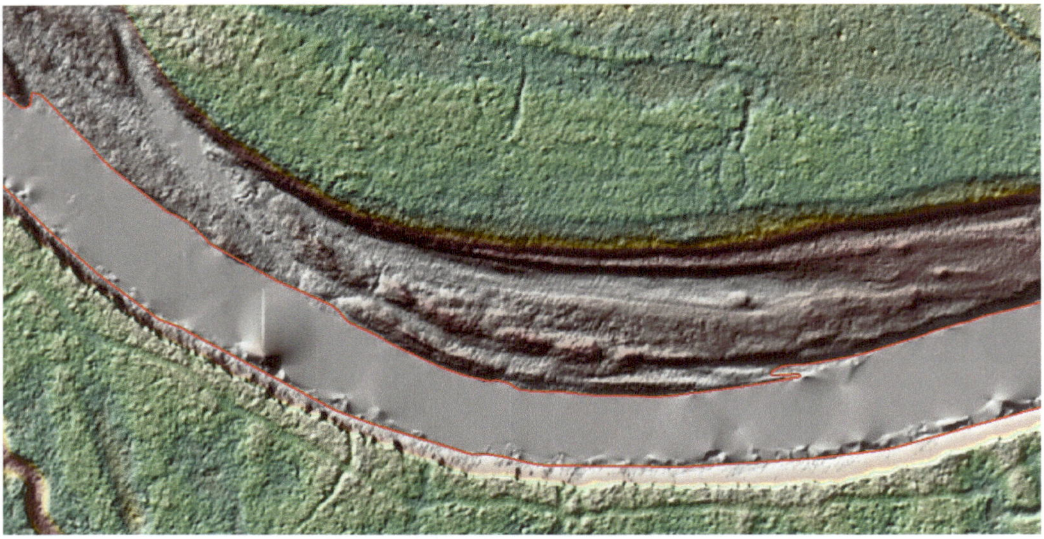

Figure 9.17. Hydrographic feature (red line) overlaid on a bare earth DEM that has been colorized and enhanced using a hillshade to better show the edge of feature. The very edge of the feature is well defined by the lack of points within the feature and the slope along edge. The breakline was placed at the very bottom of the slope.

In addition to the bare earth surface model there are a number of inputs that are helpful to use while attempting to automate the feature exaction. Table 9.4 provides a brief overview of some of these models and how they can be used in the automated process.

Input	Description of Use
First Return Surface Model	This model is used to identify areas where vegetation is present and other layers will be required. It can help with identifying the edge of some features such as mudflats, sandbars, and islands. Additionally, this is used as an input for the height model. This can also be useful for identifying areas of open water but caution should be used in vegetated areas as the vegetation often overhangs the water surface
Height Model	Generated by subtracting the bare earth surface model from the first return surface model. This is usually used to identify elevated structures. For the purposes of hydrographic feature detection the user can leverage this model to identify relatively flat surfaces. In this case water surfaces will usually be relatively flat in nature and when used with other models can aid in the detection of water surfaces
Intensity	As shown in many of the examples in this chapter, the intensity values of the lidar data can be used to help identify where water is located as it usually has a lower value than the surrounding data. Usually the intensity information is rasterized for ease of use. If this is the method chosen, use a cell size that is applicable to the collection of the feature. The larger the cell size, the more jagged the edge of the resulting feature will be. Most software is able to interpolate some information between the points when creating a raster

Input	Description of Use
Density Model	The use of density was described earlier and illustrated in Figures 9.14 and 9.15. While somewhat inconsistent, a model of the ground density can provide information related to where water has a higher probability of being located based on lower densities. Because the density is generated from the ground, areas of very dense vegetation which have few ground points and other elevated features which are not part of the ground surface, such as buildings, are also voids in the ground density layer. To mitigate low density areas caused by structures, users can create a density layer from both ground and the building class (class 6), which will fill many of the ground voids caused by structures
Slope and Curvature	In many cases hydrographic features are surrounded by a sloped bank. By creating models showing the slope and curvature of the surface it is possible to identify the bottom edge of the slope and define the edge of water at that location. This model may not be as useful in very flat terrain or when trying to distinguish mudflats or sandbars but can be very useful in areas with higher variability in the terrain

Table 9.4. Descriptions and uses of models for automated area hydrographic feature identification.

Through the utilization of these input it is possible to automate the extraction of some of the hydrographic features within a project area. Typically the automation works best on features that have well defined edges and are not obscured by vegetation. The intent of this section was to illustrate that the user will need to leverage multiple inputs when attempting to automate the hydrographic feature extraction and that a single layer will not yield good results on all types of features.

Manual Review and Editing

After the automated process has been completed it is important to review the results and either make changes to inputs or perform manual editing to correct areas where the automation failed. The important factor when making final edits is to perform them at the correct scale. The final review and edit scale is dependent on the density and the accuracy to which you require the breaklines to be collected. There may be some testing at different scales to determine which one best fits the needs of the project. Be aware that using larger review scales or being "zoomed in" results in longer edit time and some of the minor changes may not be visible in the final generation of the surface model.

Key Considerations for Area Hydrographic Feature Processing

- What types of features are being extracted and what are the minimum mapping units for those features (area for water bodies and width for rivers)?
- Does the lidar support the features that are desired for extraction? Is the density and foliage penetration sufficient?
- To what level do the features need to accurately represent the surface at the time of collection?
- What input layers appear to best show the area hydrographic features visually? Does available software allow the analyst to weight inputs to give higher priority to those that better show the water?
- How much time can be dedicated to manually editing the features to ensure consistency and correctness?

Uses for Area Hydrographic Features

The resulting hydrographic features of this process will be used for two primary processes. The first is for the classification of the water (class 9) in the point cloud. The area hydrographic features can be used to classify ground points within these polygons as water surface points. The second primary use for the area hydrographic features is for hydro-flattening during DEM processing. The concepts for assigning elevations is discussed a little later in this chapter as is the final incorporation of the breaklines with the lidar data to create a DEM.

Linear Hydrographic Features

The collection of linear hydrographic features is one of the common additions to lidar projects as it enables the user to perform hydro-enforcement of the surface for further uses. The focus for this section is related to collecting centerlines for narrow streams along with culverts that are used to enforce flow through raised features such as roads.

As with the area hydrographic features, reviewing existing data from sources like the NHD can be very useful to help determine where streams are located and determine if any of that information spatially or temporally matches with the lidar data. It is often the case that with new lidar collections the data will not line up with existing features. For linear features it is often an issue with the spatial location of the existing data not lining up well with the higher resolution lidar. Figure 9.18 is an example comparing streams that were collected from the lidar and those that currently existed in the NHD.

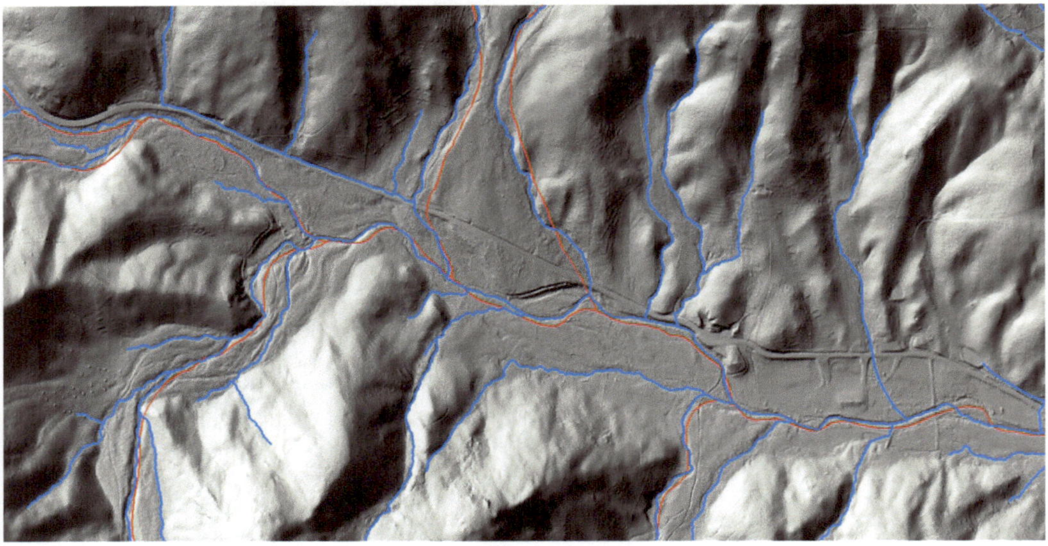

Figure 9.18. Automated streams generated from the lidar data are shown in blue and existing NHD data is shown in red. In this example the features represent flow lines and not necessarily streams that are currently carrying water. Culverts have not been extracted in this example. The streams from the NHD were likely collected using a lower resolution surface or imagery and often cut through banks and valid ground in this example.

The extraction of single line features (either manually or through automation) relies much more heavily on the bare earth surface model. The reason for this is that these features tend to be obscured by overhanging vegetation and not visible in the intensity imagery. If an automated strategy is going to be employed, following a process like the ArcHydro method is usually a good start and has extensive documentation. The majority of methods for automating the extraction of these features look at the surface and attempt to identify where water is going to pool and flow. As a result, modification of the surface model is necessary as lidar data provides a much greater level of detail regarding small fluctuations in the surface. Often this modification is in the form of identifying and filling sinks (low points) in the model so that all cells in the raster flow into other cells that are downhill.

It is important to keep in mind that what these automated models will actually provide are flowlines representing the potential path for water. The features themselves may not actually be carrying water at the time of the collection of the lidar. For each project it is therefore necessary to define what features are going to be kept as actual streams. This will vary for each project and be dependent on the end use.

If an automated approach to extracting these features has been used, the data will require additional review and editing to ensure it is consistent and represents the drainage network accurately. An area where errors are commonly introduced is where culverts cut through a road or other elevated feature and the lidar remains in those locations as ground. Often what will be observed is the automated stream will meander around that area to wherever the surface shows an outflow. In these cases the feature will need to be cut and a culvert inserted to maintain the flow correctly. One approach to improving the results of the automated extraction is to identify and collect the culverts prior to running the automated extraction. The culvert endpoints can be assigned elevations from the lidar data (lower elevations on either side of the elevated feature) and enforced in the surface model prior to automating flowlines in order to provide a location for the water to flow at those locations.

There are a few additional rules to follow for the extraction of linear hydrographic features:

- Collect the feature in the direction of flow. The starting point of the feature should be the highest elevation on that feature. Doing this during feature extraction will aid in the assignment of elevation values and ultimately in the enforcement of monotonicity and connectivity of the network
- Make certain that the features that are collected snap to one another at a vertex. This is also applicable when these features intersect area hydrographic features
- Assess the network while collecting these features. Isolated features or networks are usually not desirable if the outflows cannot be identified
- Be cautious of over-collecting ditches and other features that may not actually have flow. These features may be desired in some instances and their collection requirements should be determined prior to the automated processing or manual editing

Figure 9.19 is an example of a stream that has undergone the extraction process to ensure that it fit the surface model. For narrow channels the intent is to collect the feature in the center of the channel and ensure it does not cross the banks and cut into valid ground around bends. Additionally, a culvert has been cut through the road to maintain flow through that feature.

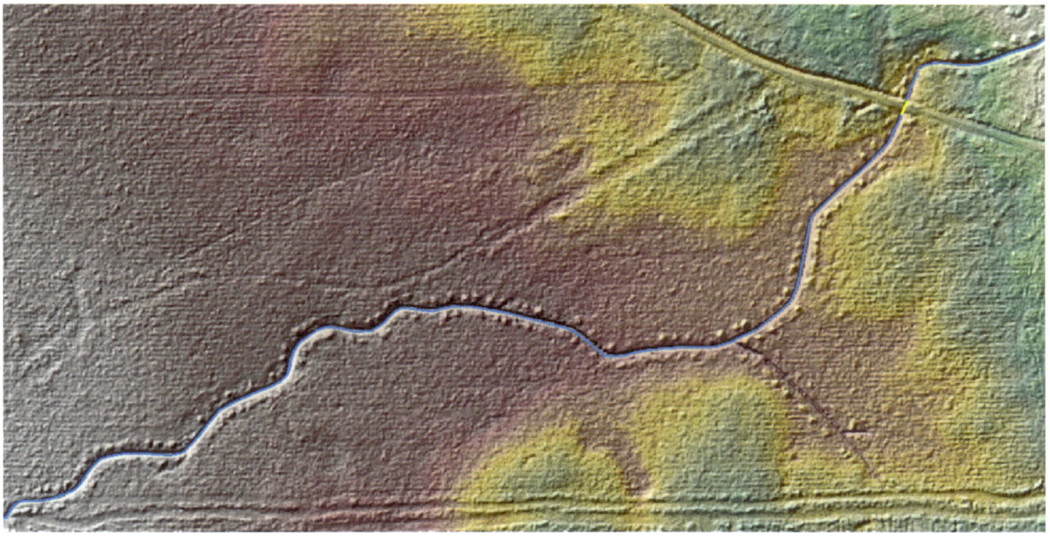

Figure 9.19. Stream centerline (blue) correctly collected in the center of the stream channel. A culvert (yellow) has been extracted where the feature crosses a road.

Stream networks can get extremely complex and by following some of the simple collection rules outlined above it is easier to evaluate the final results and maintain flow throughout a complicated network. Figure 9.20 shows an example of a very complex network where small field drainage was included as it was important to that specific client.

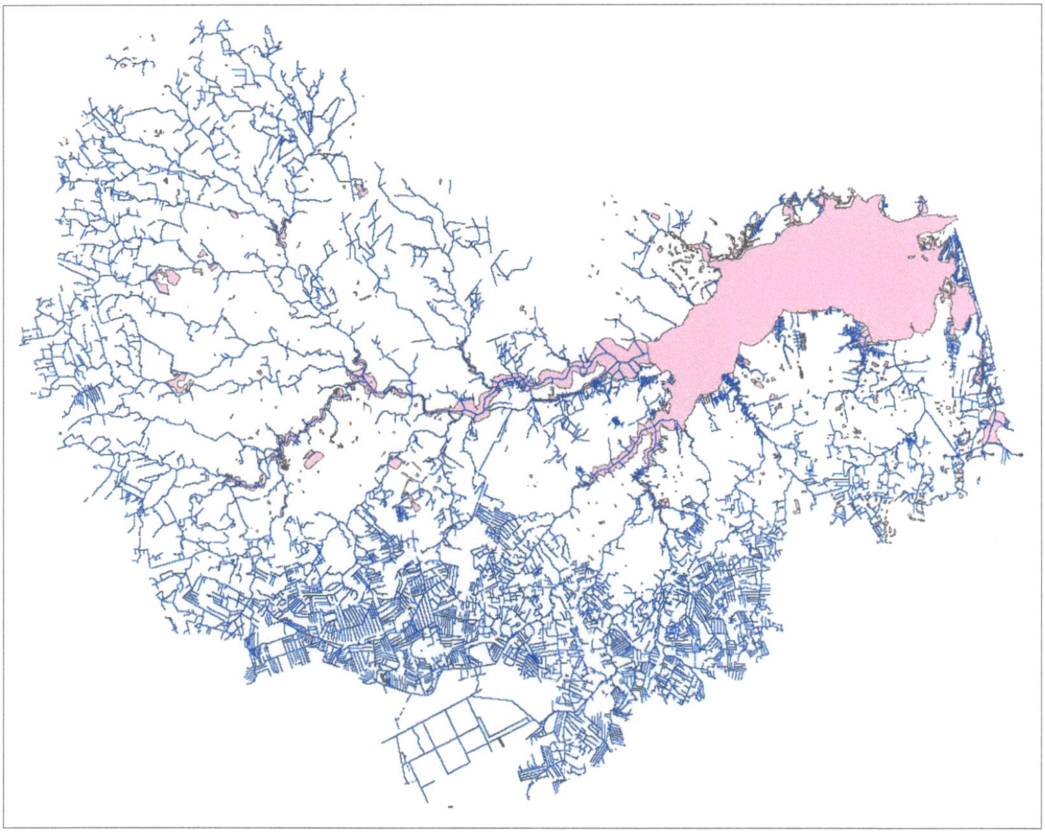

Figure 9.20. Stream network that was extracted from lidar data using a combination of automated and manual processes. The area features for this project were provided by the USGS 3DEP program and the linear drainage network was collected to meet the NHD program requirements. Complex drainage patterns in the fields present specific challenges for automated extraction because of limited flow in these features.

Key Considerations for Linear Hydrographic Feature Processing

- What are the minimum lengths of features that should be collected?
- Are both wet and dry features being collected?
- Do culverts require collection only along conveyances that are collected or will all culverts in the project area be collected?
- Should only the natural drainage be collected or will ditches be included in the collection?
- Does all of the drainage need to flow to a single outflow location or are there multiple networks within the project area

Structures

Previously in this chapter an overview of building classification was provided. Often the classification is helpful but not the final product for users of the lidar data. Instead these classified points are used to create building outlines that can then be used in other processes such as flood modeling or hydro-conditioning of the DEM. The process used to extract buildings can be easy or very complex depending on the level of detail and accuracy required of the buildings. As discussed, the lidar data provides information about the edges of features only to the extent that a point was located at or near the edge of that feature.

For the most simplistic building polygons, the classified points can simply be converted to a vector file that follows that outside edge of the points. In effect this aggregates the points into a polygon feature whose extents are based on the point file. Depending on the density of the lidar data and accuracy of the classification this can produce results that are acceptable for many uses. Figure 9.21 is an example of buildings that were classified from a higher density (8 points per square meter) lidar project.

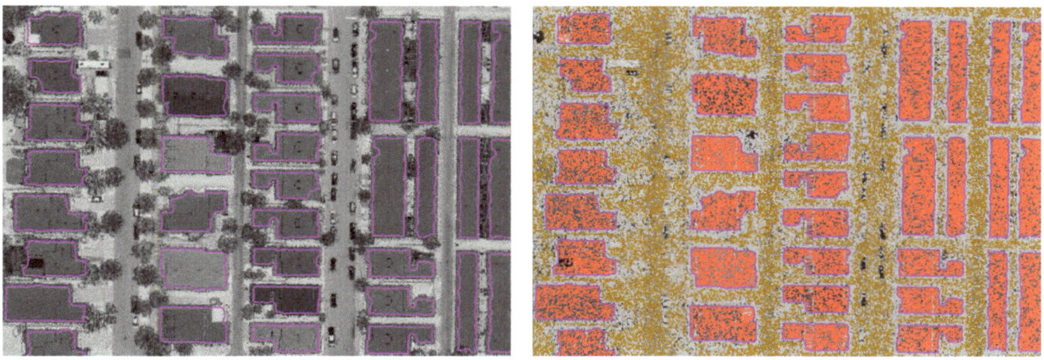

Figure 9.21. Building extraction completed using the outline of the class 6 points from the lidar data. The buildings (purple) are shown overlaid on the intensity imagery on the left and on the point cloud on the right. In this case the direct conversion of the points to a polygon was an acceptable product for this process.

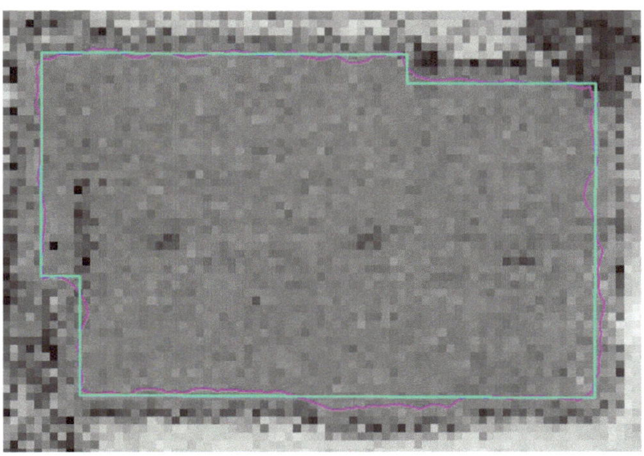

Figure 9.22. Building outline showing initial results directly from the points in purple and the final geometry after additional automated processing to clean and square the feature in blue.

It is possible to perform additional post processing of the resulting polygons to produce improved feature geometry that is close in visual representation to that of a planimetric feature. This process involves simplifying the edges of the feature so that they are straight and represented by fewer vertices as well as squaring corners to 90 degrees where applicable. There are some software packages available that have various settings to drastically improve the quality of the structures. Figure 9.22 is an example of one building that was additionally processed to create a more regular feature geometry.

Key Considerations for Structure Processing

- What is the minimum size of the features to be extracted?
- What level of detail is required for the building outlines? Do they require additional squaring to create a more visually correct feature?
- Are there any other attributes that should be collected for the structure such as height, area, or roof type?

Other

While this section on breakline development primarily focused on hydrography and structure outlines, as they are the most common forms of breaklines used in DEM development, there are many other types of breaklines and features that can be extracted from lidar data. Some common features include:

- Vegetation extraction: This includes both the extent of the crown as well as center points estimating the trunk of the tree. These can be used for a variety of forestry analytics as well as simple counts of features
- Roads: The extent of the paved surface on a road is a relatively frequent update request for lidar projects and is done using primarily the intensity imagery. However, there are automated methods similar to those employed on the hydrographic feature detection that can be used for road surface extraction
- Electrical transmission lines/towers: Higher density lidar is frequently collected for this purpose and some of this information can be extracted from standard lower density lidar as well though the results may be somewhat inconsistent based on the collection

Elevation Assignment to Breakline Features

The final step in creating breaklines for DEM development is to assign elevations and ensure that the features flow monotonically, are flat from bank to bank, and are at or slightly below the surrounding terrain. There are a number of ways to perform this task. One way is to collect the features in a stereoscopic environment so that the location of each vertex is assigned during collection with these requirements in mind. This method requires specialized software and equipment and is also relatively time consuming. The majority of lidar projects approach the assignment of elevations by first extracting the features in 2-D as described in this chapter and assigning elevation values based on the lidar data after the horizontal placement is complete. To aid with this process there are multiple software vendors that provide tools to complete these tasks. Because the way that each software approaches this process is slightly different this section will just provide a general overview of the process and some considerations for performing it.

When assigning elevations to breakline data from the lidar it is important to keep in mind that there is some variability in the point cloud elevations themselves. As a result, a feature that is draped on the lidar ground surface will not have consistent downhill flow. This is mostly due to interpolation where no points are located and the variation of the terrain within the spot size of the point. To counter this issue and to ensure points are not floating above the lidar surface it is usually best to assign the elevation of each vertex based on the lowest adjacent point. This can be done on the point cloud by using a search radius or on a TIN looking for the lowest node on the triangle. This approach will typically ensure that the feature is at or below the surrounding terrain consistently. From this point the process changes slightly depending on the feature that is being converted to 3-D.

Water Body Elevation Assignment

Water body features such as ponds or lakes require the least amount of effort to ensure that they are flat and located at or just below the surrounding terrain. Once the elevations have been assigned to the vertices of the feature, simply select the lowest value and assign that to all vertices on the feature. Many software packages have the ability to perform this step in an automated way so assigning the final elevation to the feature is simple and quick. The features can then be validated to ensure that they are flat and just below the surrounding terrain using the information in the QAQC chapter. Figure 9.23 shows a pond that has been flattened using this method.

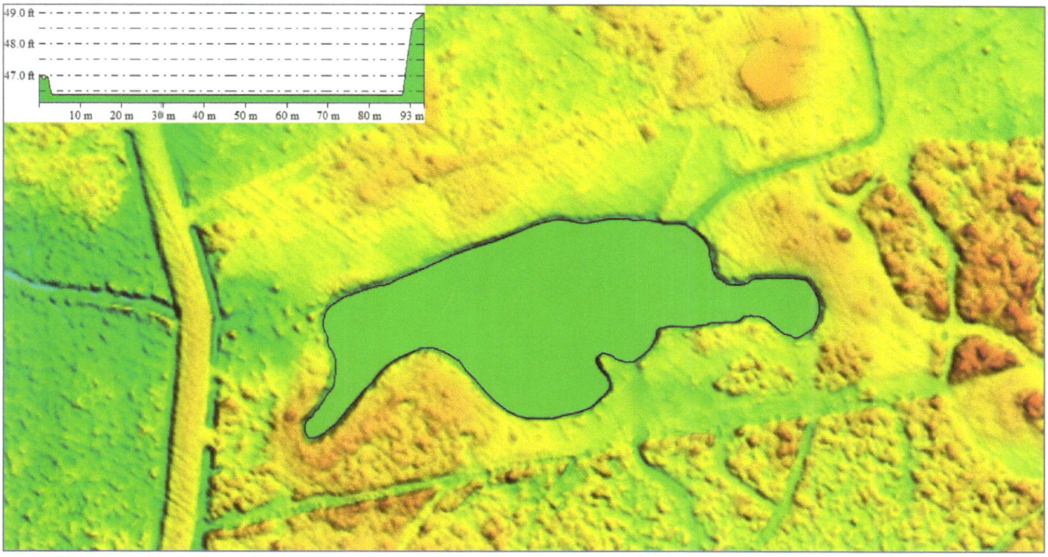

Figure 9.23. Example of a pond that has been flattened by first assigning the minimum elevation for each vertex based on the surrounding lidar points and then re-assigning the lowest of those elevations to the entire feature. As shown in the cross section the feature is approximately 0.5 feet below the surrounding terrain on one side and 1.5 feet below the terrain on the opposite side.

Linear Hydrographic Feature Elevation Assignment

When assigning the elevations to single line streams the challenge is usually with ensuring that the streams flow in a consistent downhill manner (monotonically) and maintain connectivity between features so that the horizontal and vertical coordinates are consistent at that intersection. There are a few software packages available to aid in this process. Regardless of the software that is being implemented there are several considerations that can be helpful when enforcing monotonicity and connectivity on these types of networks. The enforcement of monotonicity in its most simplistic form will look at each vertex along a feature from the starting point to the end and make sure that the elevations are always either equal to or lower than the previous vertex as the tool moves from start to end, which is why the previous section recommended collecting single line streams in downhill manner. This ensures that as tools enforce monotonicity from start to end, the tool is correctly enforcing downhill flow.

One approach to simplifying the additional processing is to use a staged approach for networks. This would entail dividing the network up into main streams and tributaries. In this approach, as users move up the network from the final outflow point up to the source, each branch would be ordered with a higher value. The main stream would be value one, the first branch two, and each subsequent branch would be given the next value. Figure 9.24 on page 264 is an example of what that ordering looks like on a single branch. The reason that this ordering can help with the process is that it will enable the user to enforce each segment independently after the main downstream feature has been processed to enforce monotonicity. Because the monotonicity enforcement changes the actual elevations of the feature, vertices are often slightly lower, post monotonicity enforcement, compared to prior to enforcement. This approach will allow a little more control when using software that cannot intelligently analyze the network and do this automatically.

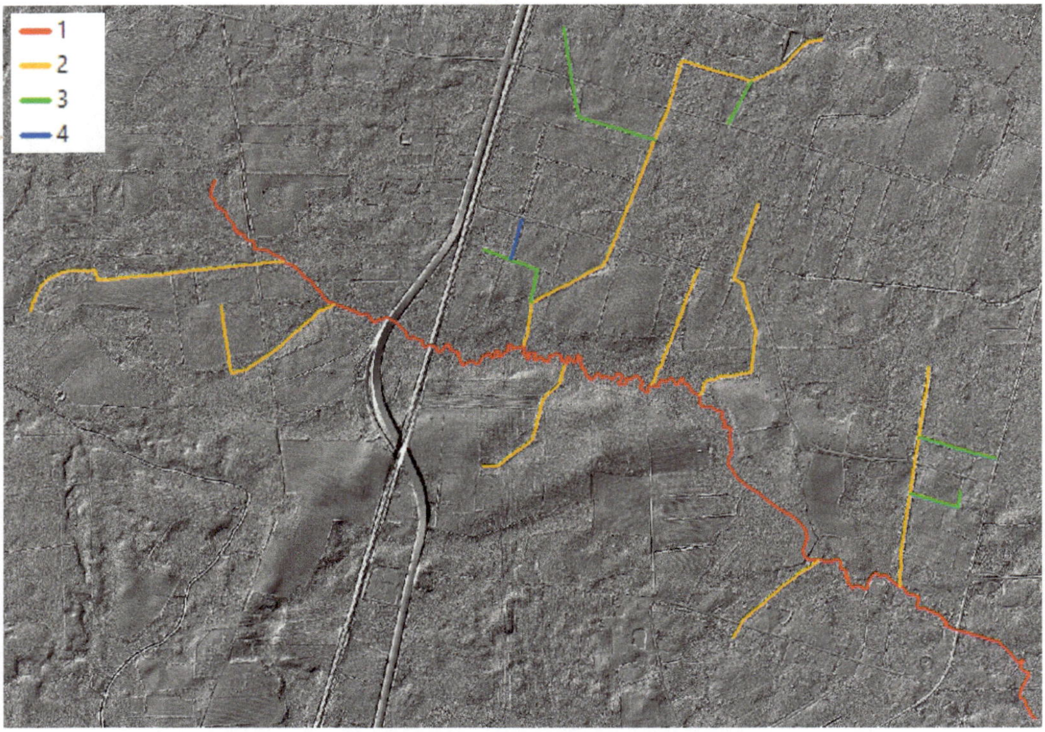

Figure 9.24. Example of stream ordering for monotonicity enforcement. In this process streams with lowest values would be enforced first and then the next value in the order would be enforced, and so on. This enables the surface model to be enforced at each step minimizing the chances of issues with connectivity at intersections.

Area Hydrographic Feature Elevation Assignment

Processing rivers is similar to the approach of processing single line streams. In this case however there is the added complexity that the banks need to be enforced consistently and flat across the channel. In order to perform this process, the majority of available software uses a centerline which is enforced using the same approach as the single line drains. Once that centerline is complete those elevations are assigned back to the bank lines to ensure that the feature flows correctly and is flat from bank to bank.

Using a similar approach as described above, to divide features into their components and enforce the main channel followed by the tributaries, may be necessary if the software that is being used cannot analyze the entire network. This approach will ensure connectivity and flow throughout the network. Figure 9.25 is an example showing a river that has been flattened using this approach.

Assignment of Elevations for Other Features

Using the processes described for the water body breaklines, area river breaklines, and linear features it will be possible to assign elevations to all other types of features which may be collected for a project including tidal features. Depending on the desired results one of these approaches will provide accurate 3-D breaklines that can be used to flatten or enforce the elevation models.

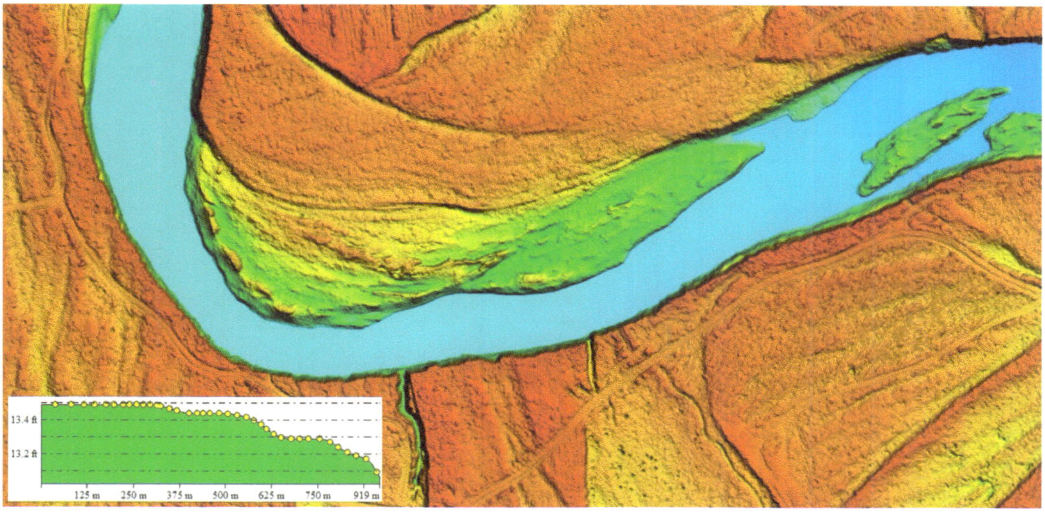

Figure 9.25. Example of a flattened river and profile of the downstream flow along the center of the feature. This river was flattened using the process described above where a centerline was collected and enforced for monotonicity and those elevations were applied to the banks.

For features that do not require enforcement for monotonicity the elevations derived from the lidar as described earlier may be sufficient for that purpose. This is often the case when 3-D attributes are desired for features like roads. Additionally, if a desired result of the DEM processing is to have building footprints remain consistent, the elevation for those footprints can be gathered in a similar way as the water body features—only in this case it will be more beneficial to use the average elevation of the feature or potentially the maximum elevation of the vertices to assign elevations.

DEM PROCESSING

The overview of the concepts related to the processing of lidar and breakline data in this chapter has been leading to the creation of digital elevation models using those inputs. DEMs are one of the most common ways users will ultimately work with elevation data and can be used as inputs on many other derivative products that will be discussed in this section. Prior to starting DEM processing it is generally a good approach to perform a detailed quality evaluation of the lidar and breaklines. The concepts related to the QA/QC of this type of data is provided in Chapter 15.

Concepts and Approaches

DEM creation is essentially the process of taking vector data in the form of points from the lidar and breaklines and creating a raster model. This makes the data easier to transfer between users and provides a useful model for additional processes and analytics. The goal of this section is to introduce the concepts of creating various forms of the elevation models using the lidar and breakline information. There are multiple software vendors that provide tools to perform these processes. The specific way each software works will vary greatly between vendors but each performs very similar tasks in deriving raster pixel elevations based on the source lidar points within those cells and the available breakline information. Some software may require the breakline data be converted to points to enforce the surface at those locations. In those cases it is likely necessary to densify the breakline vertices so that the distance between points is less than or equal to the lidar NPS.

Regardless of the software that is used to generate the DEMs the process itself is relatively similar. There are a few aspects of the process that should be considered prior to starting which will help to create a consistent final product. The first consideration should be the cell size used for the final raster. The cell size should be greater than or equal to the NPS of the lidar. While it is possible to create a raster with a cell size smaller than the NPS it should be noted that this will not provide additional detail as many of the cells will not have a lidar point within them. What users will typically find from projects like 3DEP would be lidar collected with an NPS of 0.7 meters and a final DEM cell size of 1 meter for QL2 projects. Project with higher densities such as those collected for QL1 projects where the point spacing is 0.35 meters will often have final DEM products with a cell size of 0.5 meters.

The cell size is also an important consideration when looking at how the DEMs will be organized. For some projects a single area wide DEM can be created and used without significant issues. For larger projects it may become necessary to tile the DEM so that it is more usable. In this case it is often the best practice to have a tile grid and DEM cell size that are compatible to ensure that the raster files will create without gaps or overlap between them. To ensure this the cell size should be evenly divisible into the final tile size. For example a 1,000 meter square tile will evenly incorporate 1,000 x 1,000 1-m pixels into it. Even if the final data will be tiled, single raster files covering large areas are often generated and then split into tiles to avoid seam issues along tile edges. In general most software will allow the user to generate a single raster file covering very large areas with the only limitations typically being computer resources and output type limitations. Some file format types can more easily handle very large files while others may struggle. Final outputs that can be used by multiple software should also be considered an advantage depending on the needs of the end user.

The next consideration is to determine how the elevations from the lidar and breaklines will be converted into raster values, or how each pixel elevation will be assigned. This is where some of the differences in software will really become more evident as some may offer more flexibility in how the surface is enforced and how the lidar points are used. One of the ways that software may be different is in how the raster cell is assigned an elevation, or the type of interpolation used to transfer the elevations from vector data to a raster. Most software will offer some options on how the lidar and breaklines are interpolated to create the raster. This is typically in the form of allowing the user to set the type of interpolation method used to convert the vector data into raster data. Two common interpolation methods which most software offer are linear and natural neighbors. These methods will result in slightly different results for the final DEM elevations with natural neighbors tending to provide a little more smoothing to the surface because adjacent cells are considered for the final elevation assignment. An example of the difference between interpolation methods can be seen in figure 9.26.

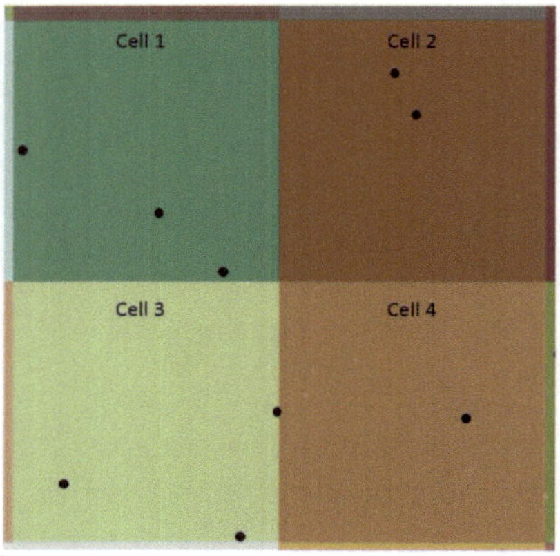

Cell	1	2	3	4
Point 1 Elevation	8.92	8.99	8.87	9.04
Point 2 Elevation	8.92	9.03	8.89	N/A
Point 3 Elevation	8.96	N/A	8.98	N/A
Average Elevation	8.93	9.01	8.91	9.04
Linear Cell—Elevation	8.95412	9.028034	8.914589	9.022964
Natural Neighbors Cell—Elevation	8.949254	9.016303	8.910993	9.019096

Figure 9.26. Example of raster cells containing between 1 and 3 ground elevation points from the lidar. The table shows that there are slight differences in the way that the linear and natural neighbor interpolations calculate the final elevation for an individual cell.

Once the inputs have been determined along with the cell size and interpolation method, the actual processing of the DEMs can begin. The following sections will provide an overview of the various approaches to creating the DEMs for various types of products including the standard bare earth DEMs without breakline enforcement, DEMs that have been enforced with a variety of breaklines, and digital surface models to show above-ground features.

Key Considerations prior to DEM Processing

- What is the cell size desired for the DEM and can the lidar support that cell size?
- Will this DEM be tiled or a single raster?
- What lidar points will be used in the processing (ground, first return, vegetation, etc.)?
- What breaklines are going to be used in the processing?
- What interpolation method will yield the desired results?

Processing Techniques

The processing of lidar data to a DEM can be undertaken in a few different ways depending on what the inputs are and the desired results. In general once the user has answered the questions related to pixel size and inputs the process itself is mostly dependent on the software. It is likely that most users will encounter two general ways to produce DEMs from the input lidar and breaklines. The most common approach is for the inputs to be the lidar points in LAS format (or potentially a proprietary format for processing) and the breaklines in a format that the software can read. In this case the lidar data is usually filtered within the software so that only the necessary points are being used for the processing. In the case of a bare earth DEM only points classified as ground (class 2) would be used in the generation of the DEM. Figure 9.27 is an example of a bare earth DEM that does not incorporate breaklines into the process.

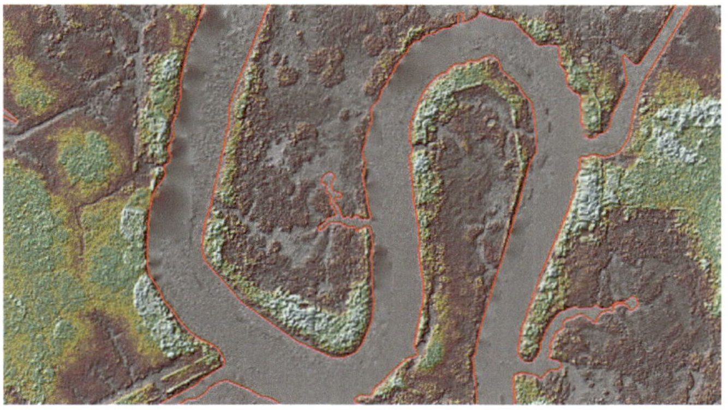

Figure 9.27. Bare earth DEM processed using only the lidar points that had been classified as ground. The rivers are shown in red for reference to where water is located within the example. Figure 9.28 shows the same area but with the hydrographic feature flattened using breaklines.

As with other topics in this chapter there are alternate methods that can be used to generate DEMs other than directly from the lidar. When processing the lidar to a DEM, be sure to review the software guidance and understand how it is actually producing the DEM so that the results match the expectation. For example, even when generating the DEM from the lidar point cloud, different software may have options for binning the points into the cell size and calculating the values or it could run an internal TIN and convert that to a raster. Understanding how the surface is created will better enable the user to determine why errors may exist in the resulting DEM and how to address those issues.

The other common way that DEMs are generated are through the use of an intermediary TIN or Terrain if using ArcGIS software. This process follows the same general principles of converting the lidar to a DEM using the point cloud but exposes that internal step where the data is triangulated to create a continuous surface. The benefit of this approach is that there is no loss of detail as each node on the TIN represents an actual ground point and it is very simple to evaluate how the breaklines (when incorporated) interact with the surface.

Incorporating Breaklines

The enforcement of breaklines is usually done to enhance the usability of the DEM or to create a more cartographically pleasing product. Common uses for breaklines include hydro-flattening and hydro-enforcement of the DEM. Figures 1.33 thru 1.38 on page 23 in Chapter 1 provide a very good overview of how breaklines have been incorporated into DEMs including the use of building footprints to enforce the surface model in Figure 1.36 on page 23. The incorporation of breaklines into DEMs will vary depending on the software used, but most generating software will allow the user to input features as points, lines, or polygons with some flexibility in how those features are used. The most common is that the elevations of the breakline are used to replace the elevations of the lidar in that specific location. In order to ensure that the DEM generates properly, lidar points in the proximity of the breakline can be classified to ignored ground (class 20). The reason that this is necessary is that it avoids the possibility of lidar ground points and breaklines directly overlapping which can cause issues in the processing because there would be two different elevations at the same location and the software would need to determine which one should take priority. The example in Figure 9.28 shows a stream that has been enforced in the DEM using a breakline and is the same location shown in Figure 9.27.

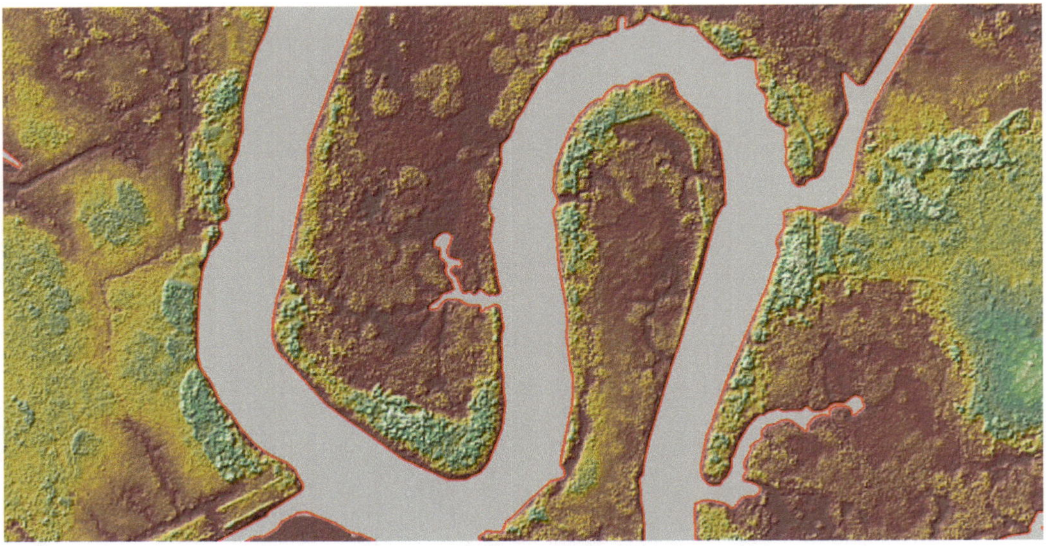

Figure 9.28. Hydro-flattened DEM processed using the lidar points that had been classified as ground and area hydrographic feature breaklines, shown in red. The rivers breaklines have been incorporated into this DEM and are now hydro-flattened compared to Figure 9.27 where the river is not hydro-flattened.

Performing hydro-enforcement is another primary use of the breakline data. In most cases breaklines are collected to enforce lower elevations through features such as culverts which would otherwise create an impedance to water flow in the DEM. One approach to enforcing these types of narrow breaklines is to create a TIN or Terrain and use these features as breaklines to effectively force the surface to follow the line. The TIN or Terrain can then be converted to DEM with those features enforced. One common issue users find with this process is that depending on the cell size of the DEM some features may interpolate improperly. The way to counter this is either with a smaller cell size or by converting the breaklines to a separate raster using only those elevations and mosaicking that on top of the hydro-flatted DEM. If that approach is used, be sure to snap the cell extents of the raster representing only the streams with that of the base DEM so that the final datasets align appropriately. Figure 9.29 is an example of a stream and culvert that have been enforced into the DEM.

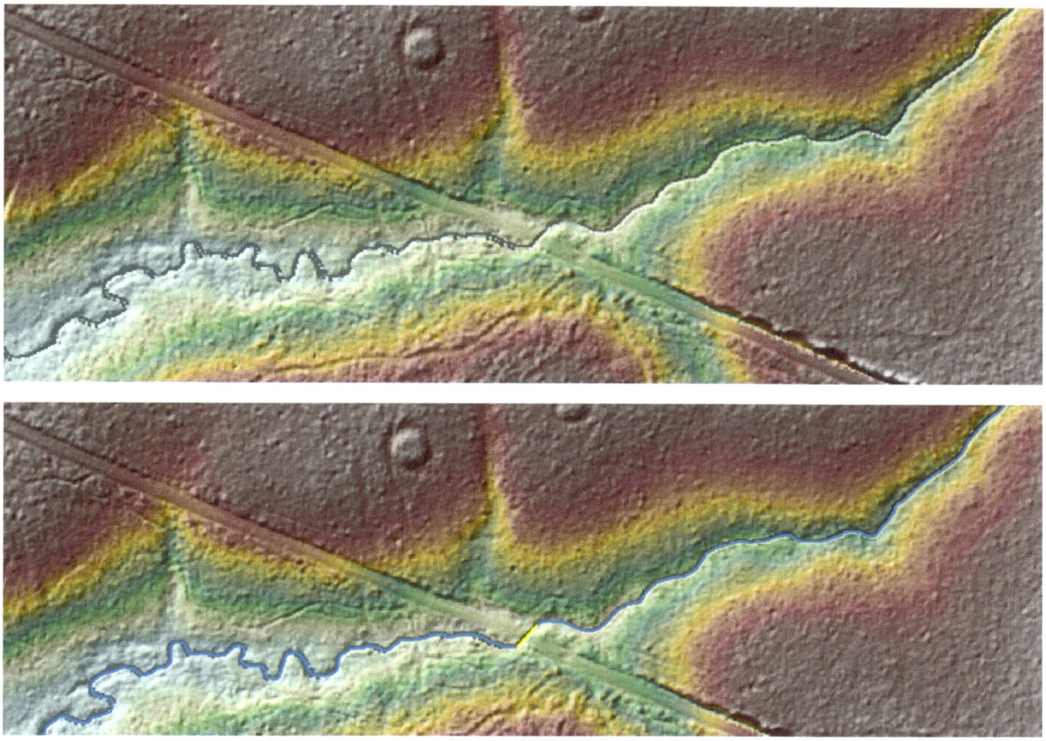

Figure 9.29. Example of a stream and culvert that have been enforced into the DEM. The top image shows the resulting DEM without the breaklines overlaid and the bottom image shows the breakline placement (linear hydrographic breaklines are blue and culvert breaklines are yellow).

In addition to the hydro-flattened and hydro-enforced DEMs, some users require additional modification to the DEM to support products for modeling or other efforts. In these cases the most frequent additions to the DEM processing are for incorporating building footprints so that DEM is consistent underneath these features (where voids and interpolation are typically present) and conditioning of the DEM. The hydro-conditioning process is used to fill sinks in the DEM that prevent water from flowing properly. Individual projects will have unique criteria for what is considered a sink. Figure 9.30 on page 270 is an example where buildings have been enforced into the DEM surface so that these features are slightly above the surrounding terrain.

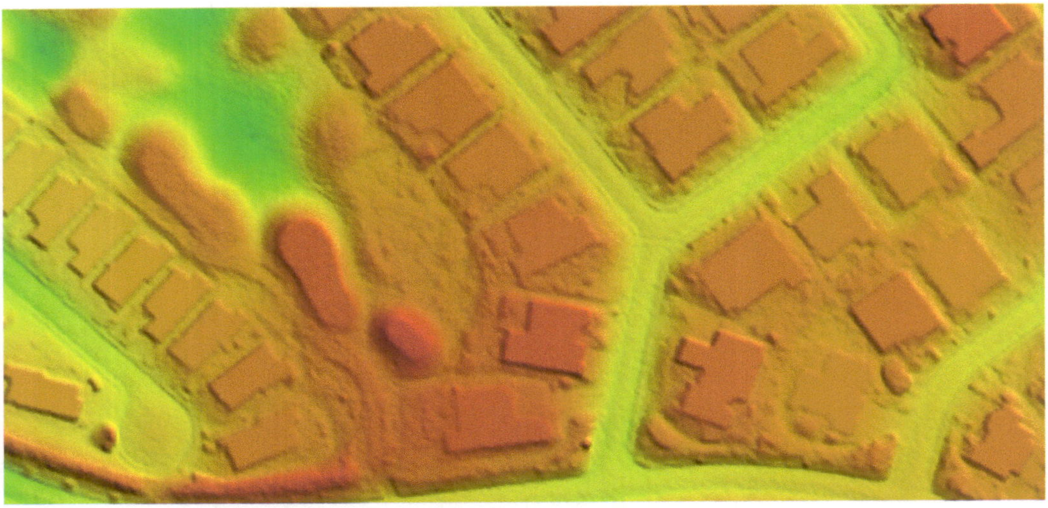

Figure 9.30. DEM with building footprints enforced into the surface at an elevation just above the surrounding terrain. This was completed by draping the building outlines on the lidar ground and assigning an elevation to the feature based on the highest value of the vertices.

DSM Processing

While DEMs representing the ground are a primary product from lidar data, equally as common are requirements for creating DSMs representing the first return surface. These models are created in a very similar way as the bare earth DEM only in this case, instead of selecting lidar points that represent ground, the first returns from all classes excluding noise classes would be selected. The first return of the lidar point cloud usually represents the reflective surface which is comprised of the tops of vegetation, buildings, and open terrain. DSMs can also be generated using any combination of above-ground features that are classified. For example some users may not want any ground to be present in the DSM even if that is the first return of the lidar. In this case a DSM of the building points (class 6) and vegetation points (classes 3, 4, and 5) can be created. The user will still want to only utilize the first returns of these features as vegetation will consist of multiple returns. Figure 9.31 is an example of a DSM generated using the first returns of the lidar, with the exception of noise, over a neighborhood.

Another example of a DSM is one that is generated to help identify power lines or other above-ground structures. That may be more difficult to detect using just the first return surface. In Figure 9.32 a DSM was generated by looking for the highest point within a specific cell size. This highlights above-ground features such as utilities while minimizing the effects of averaging that can be present in some DSMs.

Many times the DSM will be used for additional processing such as the generation of height models, inputs into feature detection, and general feature extraction. In these cases the DSM should be generated so that the pixel size is equal to the bare earth DEM and shares the same processing extents so that the pixels line up between layers.

Key Considerations for DSM Processing:

- What types of features are to be shown in the surface model? Do I only want to show buildings or vegetation?
- Has high noise been properly classified in the lidar data so that the DSM will reflect only actual returns?
- Will this DSM be used for other derivative products such as height models?

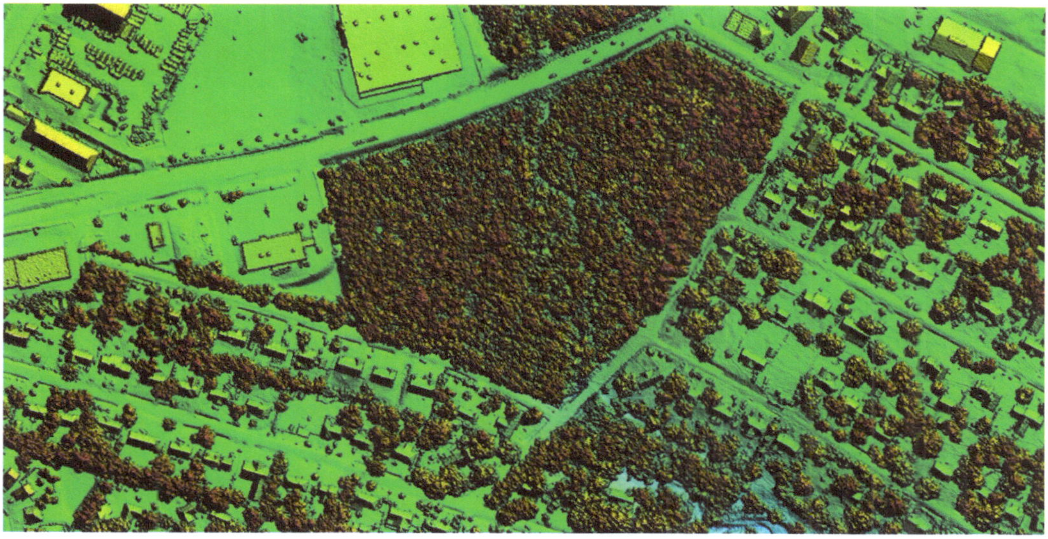

Figure 9.31. Example of a DSM generated using the first returns of the lidar.

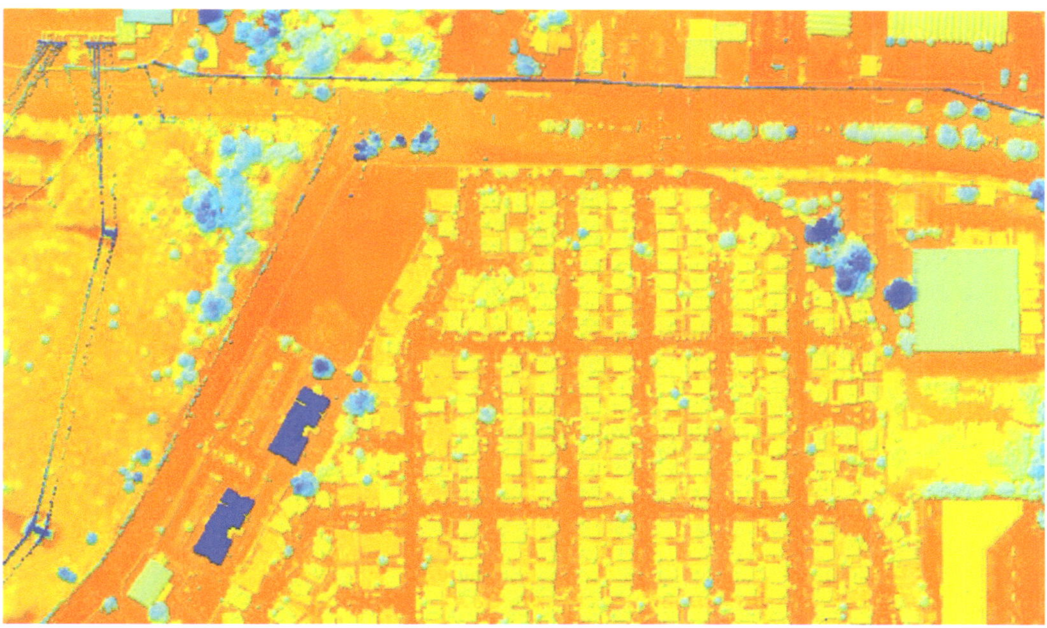

Figure 9.32. DSM generated with the purpose of helping to identify electrical transmission and distributions lines. The DSM was created using a 0.5 meter grid and selecting the highest lidar point (other than noise) within the cell.

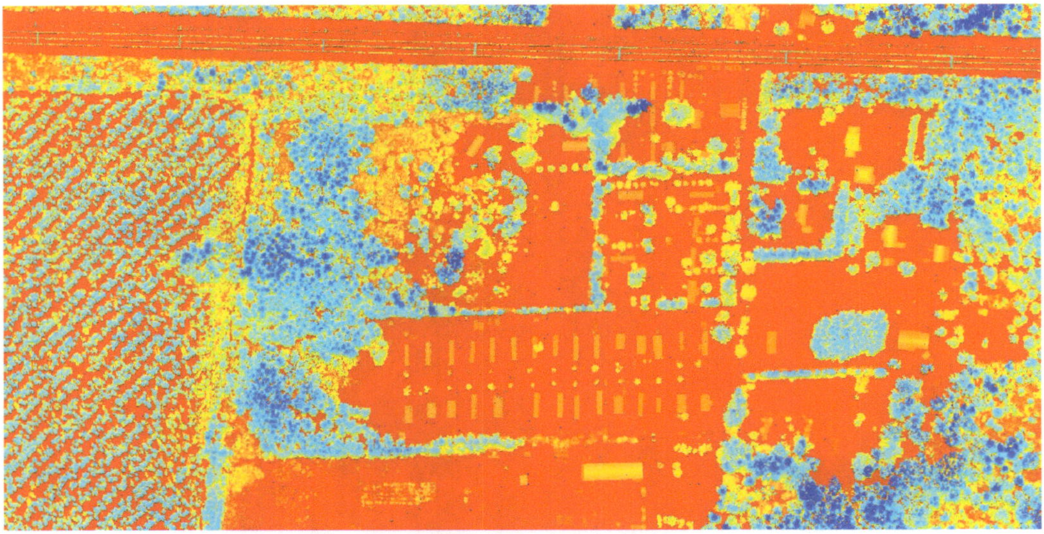

Figure 9.33. Height model based on the difference between the DSM and DEM. Features that are very visible include vegetation, structures, and electrical infrastructure.

OTHER DERIVATIVE PRODUCTS

There are many derivative products that can be generated from the lidar data other than DEMs and DSMs. This section is going to cover the two most common derivative products which are height models and contours. Height models are generated by subtracting the bare earth DEM surface from the first return DSM surface. The result is a raster that provides the user with information related to the heights of objects such as vegetation, buildings, infrastructure, and other above ground features. These models can be useful additions to other processes that help in the automated identification of breaklines and features in the dataset. As shown in Figure 9.33 many specific features begin to stand out from the general terrain when reviewing the height model.

The second derivative product that is very commonly requested is contours. Contours representing the bare earth elevations are still required by many end users as they are good additions to many mapping applications. The generation of contours will vary depending on the software used and this section is only intended to provide an overview of some of the items to consider when producing contours. The first item to consider is the contour interval. Contours should only be produced at the interval for which the lidar is capable of supporting. This will be based on the accuracy of the lidar which should be known prior to generating the contour products. For reference on what contour interval is appropriate based on the lidar accuracy the ASPRS Positional Accuracy Standards for Digital Geospatial Data (ASPRS, 2014) cover this information.

Another consideration is to determine how cartographically pleasing the contours need to be. Because lidar data samples the ground at a much higher rate than traditionally photogrammetry, the contours that are derived from lidar surfaces tend to be relatively angular. For some applications, especially those where the contours are going to be printed on maps, the base contours provided by the lidar data may not be suitable. Figures 9.34 thru 9.36 provide examples of initial contours produced by the lidar data and ones that have undergone additional smoothing to create a more cartographically pleasing product. The key with any additional processing of contours is to not introduce additional error and ensure the accuracy of the resulting contour.

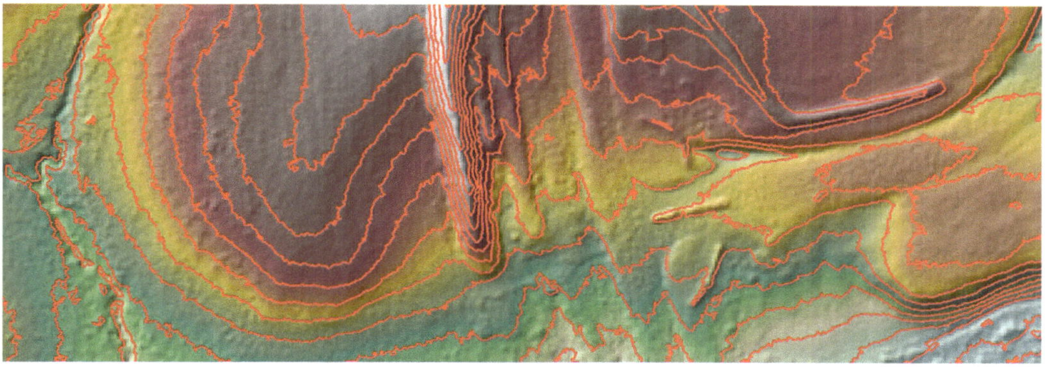

Figure 9.34. Example of contours derived using the DEM from the lidar. The lines have a significant amount of small detail showing the more subtle variations in the surface. While very accurate these small variations along the line may not be desired for a cartographic product.

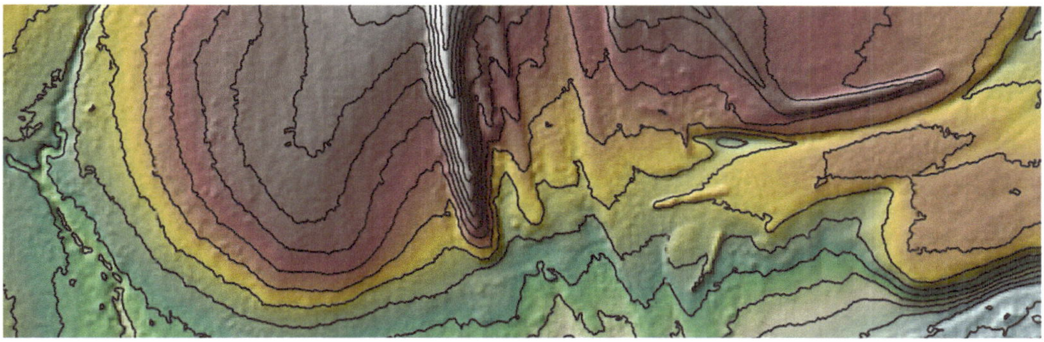

Figure 9.35. The same contours that have undergone additional smoothing to generalize the lines slightly while still maintaining the correct placement. The changes are subtle and are primarily located along the main path of the line where some of the small variations have been smoothed out.

Figure 9.36. This is a close up example of the contours from figure 9.34 and 9.35. The red contours are the original versions from the DEM and the black ones represent the smoothed features. The actual placement of the lines has not changed in any significant way but the feature is now more suitable to cartographic processes.

AUTHOR BIOGRAPHY

Joshua Novac is a Technical Manager at Dewberry for the remote sensing department and provides oversight for geospatial products including topographic lidar, bathymetric lidar, orthoimagery, and feature data extraction. He received his B.A. in anthropology and certificate in Geographic Information Systems from Indiana University-Purdue University in Indianapolis, Indiana in 2006.

ACKNOWLEDGEMENTS

The author acknowledges the help and dedication of the entire Dewberry geospatial team for producing the products that were used in the examples in this chapter.

REFERENCES

ASPRS (2013). *LAS specification, Version 1.4—R13.* Bethesda, MD: American Society for Photogrammetry and Remote Sensing, 15 July, 2013.

ASPRS (2014). ASPRS positional accuracy standards for digital geospatial data. In *Photogrammetric Engineering and Remote Sensing, 81(3)* (pp. A1-A26), Bethesda, MD: American Society for Photogrammetry and Remote Sensing.

Heidemann, Hans Karl (2018). Lidar base specification (ver. 1.3): *U.S. Geological Survey Techniques and Methods,* book 11, chap. B4, 101 p.

Chapter 10

AIRBORNE LIDAR BATHYMETRY

Jennifer Wozencraft and Amar Nayegandhi

INTRODUCTION

Airborne laser (or lidar) bathymetry (ALB) is a technique for measuring the depths of moderately clear, near-shore coastal waters, lakes, and rivers from a low-altitude aircraft using a scanning, pulsed laser beam. The round-trip time-of-flight of each laser pulse to the water surface and sea floor is measured by receivers in the aircraft. With this information, and the speed of light in air and water, accurate water depth can be calculated (Irish and Lillycrop, 1999). This technique is also known as airborne lidar hydrography (ALH) when used primarily for nautical charting. The more generic term, ALB, will be used predominantly throughout this chapter. The term "lidar" is an acronym which stands for LIght Detection and Ranging, but, as with "sonar" and "radar," it is in such common usage that it has become a word in its own right and is no longer capitalized. Why is this technology important? It is well known that hydrographic charts for many of the world's coastal areas are either out of date or nonexistent (Nordstrom, 2000). Almost 50% of the maritime states have no national hydrographic capability; developing coastal nations often lack adequate data and charts The overall status of hydrographic surveying and nautical charting world-wide is rated in the range from "poor" to "fair" (UN, 1989). There is reason to believe that hydrography, in the U.S. and around the world, is in a state of crisis regarding the ability of professional hydrographic organizations to provide the needed and desired products within their budgets and in a timely manner (Sutherland, 2013). Conversely, the use of coastal areas by commercial and recreational concerns is growing at a rapid pace. Coastal zone engineers and managers need coastal bathymetric data for a wide variety of engineering and environmental applications. A technology that can deliver faster and cheaper shallow-water surveying for both hydrographic and bathymetric purposes is critically needed. ALH/ALB is that technology.

Based on three decades of operations, ALB has proven to be an accurate, cost-effective, rapid, safe, and flexible method for surveying in shallow water and on coastlines where sonar systems are less efficient and can even be dangerous to operate (LaRocque and West, 1999; Wellington, 2001; Skogvik and Axelson, 2001). In addition to traditional nautical charting, applications for bathymetric data, such as monitoring engineering structures and the movement of sand, environmental protection, and resource management and exploitation, are expanding rapidly (Wozencraft and Millar, 2005; Reif et al. 2012). The growth in the recognition, utilization, and demand for ALH and ALB surveys has exploded around the world.

The costs of operations for all current ALB systems are reported most often as 15-30% of the standard acoustic survey cost, depending on location, depth, and survey density. Soundings are densely spaced, typically on a 0.5-5 meter grid, within a wide swath under the aircraft, whose width is roughly greater than half of the aircraft altitude. The major limitation is water clarity. For areas with very clear water, the advantage of surveying a wide swath at aircraft speeds can be obtained for depths as great as 50 meters or more. The fact that airborne lidar can also measure land topography and survey simultaneously on both sides of the land/water boundary (Wozencraft and Lillycrop, 2003, Nayegandhi et al., 2006) is highly beneficial and attractive to coastal engineers. Figure 10.1 on page 276 presents a graphic comparison of lidar and sonar operations in shallow water.

Figure 10.1. Depiction of lidar and multi-beam sonar operation in shallow water to emphasize lidar capabilities and efficiency.

The essential qualities required for a successful ALB system are accuracy, capability, and cost effectiveness. Over the past thirty-five years, developments in lasers, optics, electronics, and computers have made it somewhat easier to construct viable airborne lidar systems with varying purposes, and an increasing number are being constructed. Fewer than thirty airborne lidar bathymeters exist in the world today, however, because of the complexity, the large initial monetary investment which must be amortized, relatively limited (but rapidly growing) demand, and because it remains a challenge to meet vertical accuracy standards. It is relatively easy today to build a lidar system that can detect the sea bottom, and it is not hard to get approximate depths from a system. It takes a great deal of understanding and effort, to obtain depths (and elevations) that will meet international vertical accuracy standards and the operational requirements of the typical customer (Guenther, 2007). A great deal of care, time, and money has been put into the design, construction, testing, calibration, and operation of the present ALB systems to ensure that they meet the accuracy standards of the International Hydrographic Organization (IHO 2008) and for federal programs such as the US Army Corps of Engineers (USACE), National Oceanic and Atmospheric Administration (NOAA), and U.S. Geological Survey (USGS).

There are six variants of bathymetric lidars in operation today. They are each unique and defy categorization, but the bathymetric lidar community groups them based on laser beam width. You will learn more about bathymetric lidar sensor characteristics in this chapter, but one important characteristic is beam divergence, which controls the size of the laser footprint on the water surface. Systems with larger beam divergence (broad-beam) and higher laser power have greater depth penetration capability, up to 60 m water depths in clear water, and use sophisticated waveform analysis-based algorithms to determine water depth. Systems with smaller beam divergence (narrow-beam) and lower laser power have less penetration capability, up to 20 m water depths in clear water, and rely on a simple refraction correction technique to determine water depth. This type of system is relatively new to the market, with most coming online in the past 5-10 years, and in some cases offers a lower purchase and operational price tag than the large beam-divergence systems. These smaller beam divergence systems are generally referred to as topobathymetric lidar or topobathy systems. This is a bit of a misnomer, since most large beam divergence systems also have a topographic capability, or are flown with a near-infrared (NIR) topographic lidar in conjunction with the green lidar. Newer systems are often paired with complementary passive airborne sensors and more powerful lidar processing algorithms that characterize more aspects of the marine environment than just depths or elevations (Tuell et al., 2005a).

In this chapter, the history of the technology, how it works, in terms of requirements, theory, hardware design strategies, performance, and limitations, is described. Basic physical concepts of the technology are presented along with a summary of the important water column properties that can affect system performance. The features and capabilities of systems currently in use are examined. Operational considerations and factors affecting costs are analyzed, and comparisons with overlapping systems and technologies are discussed for bathymetry and shoreline mapping. The chapter ends with an overview of overlapping technologies and a summary of this inviting new approach to shallow-water surveying.

The primary considerations in the design, construction, and operation of an airborne bathymeter must be data quality, depth measurement accuracy, and automation. Both the physical environment and system hardware components contribute error sources that must be overcome. This requires thoughtful hardware and software system design and construction, as well as the prediction, modeling, and application of appropriate correctors. Operational procedures for quality control, calibration, and maintenance must be established and followed. Although initial data processing must be largely automated, as dictated by data collection rates as high as ½ billion soundings per hour, manual interaction during data processing continues to be a critical requirement.

The accuracy, capabilities, and cost-effectiveness of these novel bathymeters are now widely recognized and respected by the user community, and demand for contract surveys and new systems is increasing rapidly. As of August 2018, over 25 systems are engaged in ALB operations worldwide. Many of the current-generation bathymetric systems have an operational legacy of over 20 years for diverse applications. New systems have recently become operational, and prospects for expanded services and new products are very positive.

DEVELOPMENTAL HISTORY

The concept of ALB grew out of efforts in the mid 1960's to use the newly invented laser to find submarines (Sorenson et al., 1966). The seminal paper confirming the ability to perform near-shore bathymetry was written by Hickman and Hogg (1969) based on work done at the Syracuse University Research Center.

In the early 1970s a number of first-generation airborne lidar systems were successfully tested by the U.S. Navy, by the National Aeronautics and Space Administration (NASA), in Canada, and in Australia (Cunningham et al., 1988). Several symposia, co-sponsored by NOAA and NASA, were convened to establish user requirements and design goals for the use of the second-generation NASA Airborne Oceanographic Lidar (AOL) for hydrography (Goodman, 1976). Successful bathymetric field testing of the AOL was conducted in 1977 (Hoge et al., 1980). As a result, the existence of environmentally-induced biases in both surface and bottom returns were discovered (Guenther and Thomas, 1981). Other second-generation systems were built and tested in Canada, in Australia, and in the Soviet Union and the attractiveness of this technology was documented as it promised faster and cheaper accurate shallow-water surveys (Enabnit, 1981).

The 1980s began with the 4th Laser Hydrography Symposium in Australia; a great many breakthroughs were reported (Penny and Phillips, 1981). Thereafter, the LARSEN-500 (Banic et al., 1986) was developed in Canada by Optech Incorporated for the Canadian Hydrographic Service and the Canada Centre for Remote Sensing to support nautical-charting missions in the Arctic during the few weeks a year the region is ice free. Based on surveys performed in the Canadian Northwest Territories, it became the world's first operational ALH system. Additional surveys were conducted in the Great Lakes, the St. Lawrence River, and off the East and West coasts of Canada. Testing of the Australian WRELADS II was successfully completed (Billard and

Wilsen, 1986); design and construction was begun at BHP/Vision Systems on the operational version called LADS (Penny et al., 1989) for the Royal Australian Navy. Design and testing of a number of systems such as the U.S. Navy HALS (Curran et al., 1988) and the Swedish FLASH (Steinvall et al., 1992) continued. Three multi-purpose research systems (GOI, Chaika, and Makrel-II) were actively tested in the Soviet Union (Feigels and Kopilevich, 1993), and work was also conducted in China on their BLOL (Liu, 1990). In 1988, the U.S. Army Corps of Engineers began the SHOALS program with Optech (Pope and Lillycrop, 1988).

In the 1990s, LADS became operational in Australia (Setter and Willis, 1994), SHOALS became operational in the U.S. (Lillycrop et al., 1997), and Hawk Eye became operational in Sweden (Steinvall et al., 1997). LADS was flown in a dedicated Fokker F-27 fixed-wing aircraft. SHOALS originally operated from either of a pair of NOAA Bell 212 helicopters, while two Hawk Eye systems were borne in Bell 212 and Boeing Vertol helicopters. The Canadian LARSEN-500 system continued to perform in a variety of fixed-wing aircraft. Late in the decade, SHOALS added the capability of using kinematic GPS (KGPS) as an optional vertical reference (Guenther et al., 1998); this permitted topographic mapping over land to be conducted in conjunction with bathymetric missions and also permitted operation without concurrent tide data. The pulse repetition rate of SHOALS was doubled, and the system was transitioned from the helicopters to a deHavilland Twin Otter (Dash 6) fixed-wing aircraft (Irish and Lillycrop, 1999). The LADS Mk II system, with many added capabilities such as a much higher pulse repetition rate and improved computer and navigation functions, was built by Tenix LADS Corp. to perform contract surveys worldwide and became operational in a faster deHavilland Dash 8 200-series aircraft (Sinclair, 1999). ALB surveys were proving to be accurate, rapid, flexible, and significantly less expensive than waterborne surveys in appropriate locations. The Joint Airborne Lidar Bathymetry Technical Center of Expertise (JALBTCX) was formed in 1996 as a nexus for operations, research, and development in airborne lidar bathymetry and complementary airborne coastal mapping and charting technologies and still remains a broad collaboration among a number federal organizations, industry (sensor manufacturers and surveying and mapping services providers), and academia with an interest in advancing the technology in a collaborative manner (Wozencraft and Lillycrop, 2003).

The first decade of the 2000s was a time of busy surveying, of the development of new systems, of a breakthrough into the commercial arena, and of thoughtful planning for a bright future. The pace-setting LARSEN, the world's first operational ALB, active for an impressive 16 years, was retired in 2001. Topographic ability was added to LADS Mk II. The SHOALS system was upgraded to 1,000 pulses per second (pps) and 3000 pps. USACE obtained a SHOALS-3000TH system called "CHARTS" with new, high-rate topographic and hyperspectral capabilities (Wozencraft and Millar, 2005). Fugro Pelagos became the first commercial surveyor to purchase an ALB system when they received a SHOALS-1000T (Lockhart et al., 2005). The NASA Experimental Advanced Airborne Research Lidar (EAARL) of unique design was fielded in 2001 (Wright and Brock, 2002; Nayegandhi et al., 2009) to provide topographic and bathymetric surveys combined with imagery from a down-looking RGB digital camera and a high-resolution multi-spectral color infrared (CIR) camera, often in support of U.S. Geological Survey projects. The EAARL sensor was the first narrow-beam topobathymetric lidar sensor. A consortium called Admiralty Coastal Surveys AB was formed in Europe to develop and provide surveys with the new Hawk Eye II system built by Airborne Hydrography AB (AHAB) in Sweden. The LADS Mk-III was fielded in 2009 with a decreased size that enabled deployment into aircraft of opportunity, and performance improvements included greater depth penetration and laser repetition rate.

The second decade of the 2000s is best characterized as a decade of new systems. Nine new systems have been fielded since 2010. Optech's Aquarius was the first of the commercial shallow water systems on the market and was developed in collaboration with the National Science Foundation National Center for Airborne Laser Mapping (NCALM) at the University of Houston. This green-laser variant of the topographic system Gemini was fielded in 2010. Riegl teamed with University of Innsbruck on a research project for the Bavarian Water Authority to develop and assess the capabilities of a new shallow water bathymetric lidar system (Pfennigbauer and Steinbacher, 2012). This system (the Riegl VQ820G) was designed specifically to meet the inland water (riverine) mapping requirements set forth by the European Water Framework Directive and was made commercially available in 2011. The Coastal Zone Mapping and Imaging Lidar (CZMIL) was developed through a government, industry, academic partnership among the USACE, The University of Southern Mississippi, and Optech, Inc. to support the USACE National Coastal Mapping Program. CZMIL was fielded in 2012 as the follow-on system to the USACE SHOALS and NAVOCEANO CHARTS systems, and was designed for better performance in shallow turbid waters, and to exploit fusion of the lidar and imagery data for environmental mapping (Tuell et al., 2010). AHAB built the Chiroptera system for the University of Texas, Bureau of Economic Geology. Like EAARL, this shallow water system was designed in the tradition of bathymetric lidar systems. Chiroptera was delivered in 2012.

Four systems were fielded in 2014 alone. Chiroptera II, which boasts greater depth penetration capability than its predecessor was launched with HawkEye III by Leica AHAB as a modular approach to coastal zone mapping. Chiroptera II surveys over land and shallow water, while the HawkEye III surveys deeper water, typically from the same platform. Titan is the currently available Teledyne Optech shallow-water system first fielded in 2014. Like Aquarius, Titan was also developed in collaboration with NCALM, and is the first bathymetric lidar system to emit three wavelengths of light (green, near-infrared, and infrared (Fernandez-Diaz, 2016). Riegl released the VQ 880-G in 2014, which improves on the depth penetration and measurement rate capability of the VQ 820-G (Martin Pfennigbauer, personal communication). A significantly modified EAARL-B was also field tested in 2014 (Wright et al., 2016). More recently, Teledyne Optech released CZMIL Nova in 2015, which offers a significant size, weight, and power reduction over the original CZMIL, and improved data processing. The Fugro LADS-HD and government counterpart Royal Australian Navy LADS-HD, were fielded in early 2016 with faster laser repetition rate than the predecessor LADS Mk-III, and improved data processing capabilities (Mark Penley, personal communication). In 2018, Fugro announced it developed the RAMMS (Rapid Airborne Multibeam Mapping System) in collaboration with Arete Associates. RAMMS utilizes push-broom laser technology, based on military mine-detection techniques and has the potential to be deployed on fixed-wing unmanned aircraft.

System manufacturers have also developed lidar bathymeters for the small, light-weight UAVs (drones). However, these sensors are generally profiling systems with limited performance; applications include generation of profiles of inland water bodies (rivers, lakes, channels) and repeat survey of water reservoirs. In 2016, Riegl introduced the BathyCopter, a compact and lightweight bathymetric depth finder comprising of tilt compensator, an IMU/GNSS unit with antenna, a control unit and up to two external digital cameras. The device sends out laser pulses at a rate of 4 kHz. The echo signal of each laser pulse is digitized and recorded for the entire range gate of 50m. In 2018, ASTRALiTe successfully demonstrated the use of a drone-ready topo-bathy lidar system that uses a unique polarization technique to determine the water surface (Mitchell and Thayer, 2014).

BASIC CONCEPTS

The ALB technique relies on propagation of green laser pulses through the water to reflect from the seabed, and then to make the return trip back through the water to receivers in the aircraft. Green is selected for sea bottom detection because that is the wavelength which penetrates typical coastal waters with the least attenuation (Jerlov, 1976). Figure 10.2a on page 280 depicts all the interactions of the laser pulse with the water, including a specular interface reflection, refraction (change in light speed and direction) at the air/water interface, forward scattering of the light through the water column into an asymmetric cone, backscatter by particles in the water column as well as water molecules themselves, absorption by particles in the water column, and finally a diffuse bottom reflection. Figure 10.2b is a conceptual green waveform, which is the signal recorded by ALB sensors. The waveform represents the amount of light returning to receivers in the aircraft per unit of time, typically in nanoseconds. In the ideal case, the waveform captures three of the interactions listed above: the surface return captures the specular interface reflection at the water surface, the bottom return captures the diffuse bottom reflection at the seafloor, and the volume backscatter captures light returning from within the water column. The waveform is analyzed by ALB processing software to determine the timing of the surface and bottom returns. The time difference between these returns is the

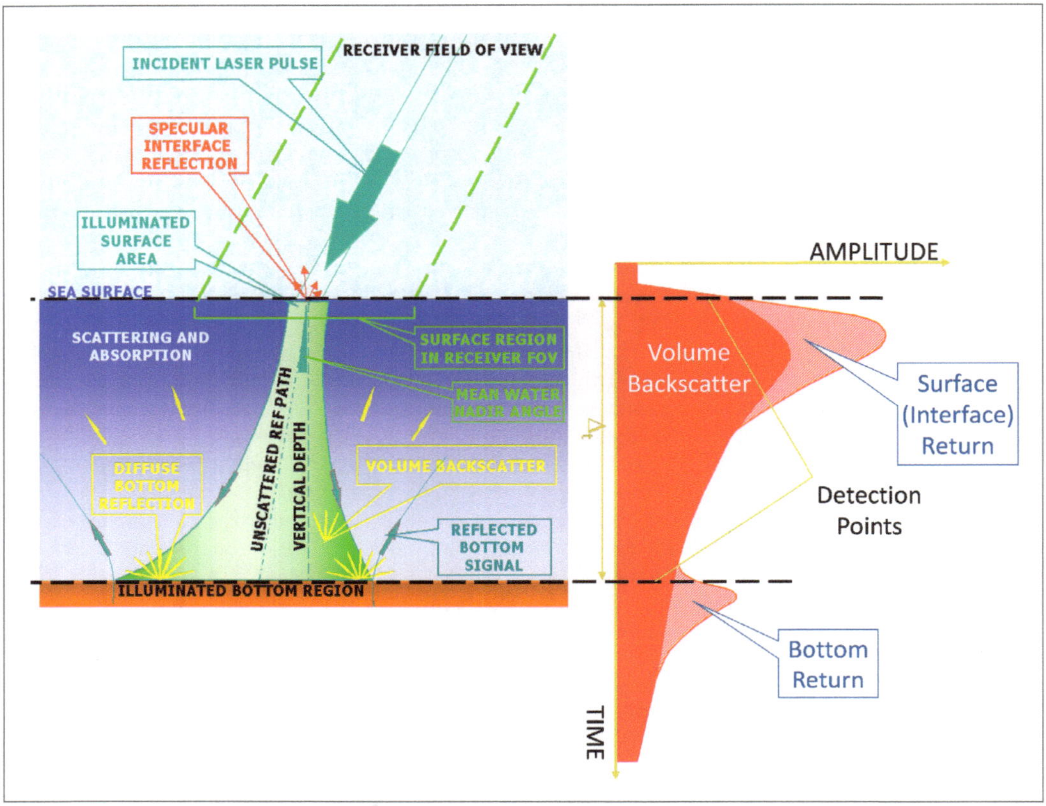

Figure 10.2. Basic concepts of ALB; (a) Left: Schematic illustrating the interaction of a laser pulse as it meets the water surface, traverses through the water column, and is reflected off the illuminated bottom. (b) Right: Schematic illustrating the return waveform received and recorded by the ALB sensor.

round-trip flight time of the laser pulse through the water column, typically on the order of tens or hundreds of nanoseconds. Combining this time difference with the speed of light in water and the beam angle in water gives the water depth for each laser pulse. The following sections detail characteristics of the surface, water column, and seafloor that impact airborne lidar bathymetry.

Water Surface

Waves, or the absence thereof, are the primary water surface characteristic that impact reflection and refraction of ALB laser pulses. A flat, glassy, calm water surface acts like a mirror and reflects a laser pulse away from the aircraft. According to the law of reflection, the angle of reflected light is equal to the angle of incidence measured from the normal to the water surface (Figure 10.3). The only way that surface-reflected light returns to receivers in the aircraft is for light to reflect from water surface wave faces that are perpendicular to the angle of incidence. In general, these wave faces are provided by capillary waves, which are very small wind-generated ripples sometimes called cat's paw waves. Only about 2% of incident light reflects from the sea surface (Guenther, 1985), and only some of that reflects in the direction of aircraft receivers.

Accurate depths can only be determined for pulses with enough light returning from the water surface to be detected by receivers in the aircraft. The timing of the water surface is important because light refracts, or

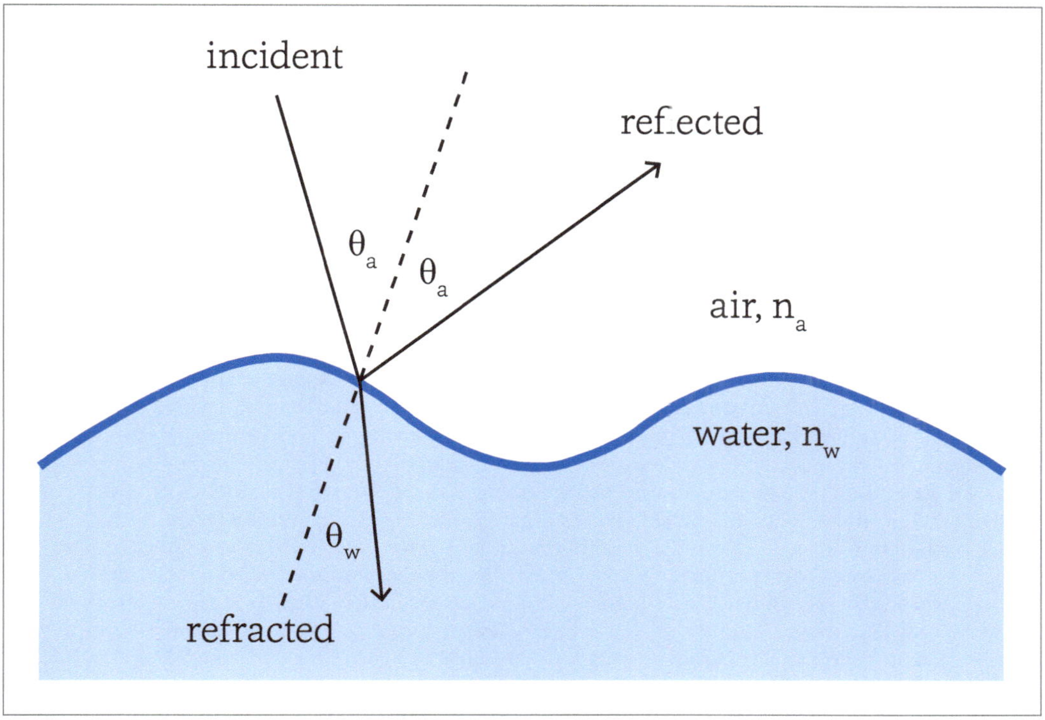

Figure 10.3. Reflection and refraction at the air-water interface.

changes speed and direction at the water surface. Light travels about 30 cm/ns in air and ~22cm/ns in water, so each nanosecond in a lidar waveform represents 15 cm in air, or 11 cm in water, considering the two-way travel time for each pulse. Each nanosecond inappropriately identified as traveling in air or water results in a plus or minus 3.74 cm error in the slant range measurement. Although the interface return component is illustrated in Figure 10.2 as being roughly equal in magnitude with the volume return to which it is added, this is not generally true. The amplitudes of the green interface returns vary statistically with wind/wave conditions over many orders of magnitude because they depend on surface wave-slope statistics. The interface component can be significantly stronger or weaker on a pulse-to-pulse basis (Guenther and Mesick, 1988b). The amplitudes of the near-surface, green volume backscatter peaks are generally relatively constant because they depend on water clarity. If the green interface return is significantly weaker than the volume backscatter return, the latter may be incorrectly detected as the surface (Guenther et al., 2000).

The change in direction of the laser pulses is governed by Snell's law, which can be approximated as $\sin(\theta_w) = \sin(\theta_a)/1.333$ for ALB. Shown in Figure 10.3, θ_w is the refracted angle in the water, θ_a is the incidence angle in air, and 1.333 is the index of refraction for seawater at the laser wavelength of 532 nm. For a typical ALB incidence angle of 20 degrees on the water surface, the refracted angle of propagation is around 15 degrees ("UNSCATTERED REF PATH" in Figure 10.2). Unlike the situation with sonar where the speed of sound is a sometimes elusive quantity and can lead to large depth errors, the speed of light in water is only a very weak function of salinity and temperature and can be easily handled, for example, with three crudely differentiated values representing "fresh", "brackish", and "salty".

Waves impact identification of the surface return, positional accuracy of the resulting depths, and overall performance of ALB in other ways. Waves near shore and breaking waves suspend sand from the seafloor into the water column, increasing the scattering properties of the water column. In some cases, the amount of entrained sediment can be so high as to preclude receiving a bottom return from the seafloor. Breaking waves also create bubbles and foam on the sea surface that increase the amount of light reflected from the sea surface or increase scattering such that there is little to no transmission of the laser pulse into the water column, again resulting in no bottom return. Surface waves affect the direction of beam entry through the water surface on a pulse-to-pulse basis. The mean direction is the same as for a flat surface, and the RMS variation is a function of the beam nadir angle arriving at the surface and RMS wave slopes which vary with wind speed and fetch. There are also factors related to the relative size of the laser spot diameter at the surface, compared to the wavelengths of the water waves, which make it difficult to achieve a precise error estimate. Based on airborne measurements of RMS gravity-wave slopes by Cox and Munk (1954), a reasonable horizontal error estimate for a 20-degree nadir angle and 10-knot winds is around 0.36 m RMS per ten meters of depth. These are random errors from pulse to pulse. Large wave heights and wave lengths pose problems, both directly and indirectly, for several reasons. As wave heights grow, the accuracy of the measurements of the mean water level and the local wave heights degrade, reducing the accuracy of the resulting depth or elevation measurements that rely on these measurements. Wavelengths that are large relative to the size of the laser footprint create water surface slopes that also affect the direction of the beam as it passes through the interface and add an additional small random depth error. Anecdotally, ALB has demonstrated capability to operate accurately in the presence of large wave heights and wavelengths in Portugal when the winds and waves were so great that they forced the survey boats to remain in port (Lillycrop et al., 2000).

Sun glint is the reflection of sunlight off the water surface in the direction of the ALB receivers. The additional light, if strong enough, may blind the airborne receiver, or may add noise to the recorded waveforms that increases uncertainty in the depth detection. For this reason, operations are sometimes scheduled to avoid flight times around local noon. LADS, for example, flew its missions in the late afternoon and early evening (Sinclair and Spurling, 1997). A properly designed receiver, however, can work successfully in moderate glint if the bottom return signal strength is sufficient.

Water Column

Light that is not reflected from the water surface enters the water column where it is subject to absorption and scattering by particulates in the water column like suspended sediments, bubbles, or organic material, and the water molecules themselves. Although laser beams are commonly envisioned as being highly collimated with a small cross section (as they are in space or over short distances in air), this is not the case for ALB. Depending on the sensor and its flying height, the area illuminated by each laser pulse on the water surface varies from a few tens of centimeters up to two meters. As the beam enters the water column scattering and absorption within the water column spreads the beam as depicted in Figure 10.2a. Scattering causes even the narrowest beam to expand into a cone whose interior angle and cross section increase significantly with depth. It should be noted that Figure 10.2a is a cartoon, and the beam spreading depicted has been exaggerated for clarity. For most operating conditions, the 3-db (half energy) beam diameter at the sea bottom will range between ten and thirty percent of the depth; for extreme conditions near extinction, it could be as great as half the depth. The resulting net expansion in irradiated bottom area is beneficial to the detection probability for significant bottom features (Guenther et al., 1996) but, as with broad sonar beams, can be detrimental to depth accuracy when very high-relief features are present.

Water clarity is a general term describing the combined scattering and absorption properties of the water column, and is the primary limitation for depth performance (Guenther and Goodman, 1978). If the water is too turbid, or has a high fraction of suspended sediments, bubbles, or organic material, the volume backscatter will be greater than the bottom return and no depth can be determined. The maximum surveyable depth is the greatest depth, at a given time and location, for which depth measurements can be obtained whose accuracy meets obligatory standards. This requires that the bottom-return signals be reasonably strong and free from excessive noise so that their arrival times can be accurately estimated. This depth will be somewhat less than the greatest depth from which weak, noisy lumps of bottom-return energy are barely detectable in the signal waveforms. The maximum surveyable depth depends on a number of system hardware, software, and logistical parameters in addition to environmental conditions. Of the environmental factors, water clarity is by far the most important because it enters as a negative exponential factor.

Performance of ALB systems is often described in terms of Secchi depth. The Secchi depth is an old and intuitive water clarity measure which is the depth at which a standard white (or black and white) disc, deployed over the side of a boat, is no longer visible to the human eye. A rule of thumb for ALB systems (higher power, larger beam divergence systems) is that successful operations can be expected to depths between 2 and 3 times the Secchi depth. The factor of two applies where the water has a significant amount of absorption (which reduces energy), while the factor of three is appropriate for waters dominated by scattering (which redistributes energy). Topobathy systems generally specify performance of 1-1.5 times the Secchi depth, without specifying scattering or absorption. The Secchi depth is not a particularly good predictor of performance, however, because its relationship to the proper optical parameter, the diffuse attenuation coefficient, varies with the scattering-to-absorption ratio of the water (Gordon and Wouters, 1978).

In the more specific terms of ocean optics, the water property which most nearly dictates the received bottom-return pulse energy in a well-designed, full-capability ALB system is the "diffuse attenuation coefficient for downwelling irradiance", "$K_d(\lambda)$", where the subscript "d" indicates downwelling and "λ" indicates the green laser wavelength of 532 nm. The concepts surrounding various definitions and measures of $K_d(\lambda)$ are far too complex to describe here (Gordon, 1989), but in simple terms, $K_d(\lambda)$ is the exponential factor by which the downwelling vector irradiance of the incident light field, at a given wavelength, decreases with increasing depth. The downwelling irradiance is all photons heading in all downward directions. Both the Sechhi depth and $K_d(\lambda)$ are classified as apparent optical properties of the water column since they depend on the directional structure of the ambient light field in the water column.

The value of K is very different from the so-called "beam attenuation coefficient", (c(λ)) which is defined in terms of the radiant power lost from a single, narrow, collimated beam of photons. (c(λ)) is the sum of the scattering and absorption coefficients. The absorption coefficient describes how light is absorbed and either released as another form of energy or contained in a chemical bond. The volume scattering function describes how light is scattered, or changes direction and/or energy and continues to travel through the water column. The absorption coefficient and volume scattering function are fundamental inherent optical properties of the water column. Inherent optical properties are independent of incoming radiation such as from the sun or an ALB, so they can be measured in the laboratory on a water sample, as well as *in situ* in the ocean (Mobley, 1994). The ratio K/c is always less than unity and for green light typically ranges between one-sixth and one-half for coastal waters. It depends strongly on the scattering-to-absorption ratio of the water, often expressed in terms of the so-called "single-scattering albedo", and also, to a lesser extent, on the scattering "phase function" (Guenther, 1985).

The maximum surveyable depth for a given water clarity can be expressed roughly as n/K, where "n" is a constant. For ALB systems, under customary operational circumstances, the value of n will be around 3 to 4 for daytime operation and perhaps 5 at night (Guenther, 1985) but with smaller values for the detection of small objects on the bottom (Engstrom and Axelsson, 2001). In other words, if, for example, the water clarity can be expressed in terms of a value of K = 0.1 m^{-1}, then one would expect to be able to survey to a depth of 3,040 meters during the daytime. The daytime value of n depends on the extent of solar background and sun glint present during operations and on the optical filter bandwidth of the system. Nighttime operation is preferred from a performance standpoint, because there is no sun glint to avoid, and the shot noise associated with the ambient, reflected solar background in the optical filter bandwidth is absent. Flying at night for extended periods, at low altitude over water, however, may be taxing on pilots, and can be more dangerous, particularly near land. For safety and practicality, most operations are consequently conducted during the daytime, with concomitant reduction in performance. The value of n for operational topobathy systems ranges from 0.5 to 1.5 for daytime operations.

Seafloor

Photons that finally reach the seafloor are reflected back into the water column for a return trip to receivers in the aircraft (Figure 10.2a, "DIFFUSE BOTTOM REFLECTION"). These photons make up the bottom return, which is used to identify the arrival time of the laser pulse on the seafloor. Because the photon paths in the region undercutting the unscattered ray (Figure 10.2a, "UNSCATTERED REF PATH") from the surface to the bottom are shorter, they encounter less total path attenuation. This causes the spatial energy peak (Figure 10.2b "BOTTOM RETURN"), and hence the effective sounding location, to be located in the undercutting region. The undercutting region is located between the "UNSCATTERED REF PATH" and "VERTICAL DEPTH" in Figure 10.2a. The magnitude of the shift of the effective sounding location from the unscattered ray for a flat surface depends on various water clarity parameters (discussed above, such as the scattering phase function and the single-scattering albedo) and depth. Because these water clarity parameters cannot be known from the air, the exact sounding location is a bit uncertain. For constant or slowly-varying water clarity conditions, this horizontal error will be a bias. The magnitude of this shift has been predicted by Monte Carlo simulation for various water clarity conditions. The value of the predicted mean shift, averaged over bounding water clarity conditions, can be applied as a horizontal bias corrector leaving the variation about the mean as a residual bipolar uncertainty of reduced magnitude. Depending on circumstances, recommended mean shift magnitudes are on the order of 10%-30% of the distance from the unscattered ray to the nadir point. Residual horizontal errors after the bias correction, which increase linearly with depth, are estimated at ±0.32 m RMS per ten meters of depth. These are bias errors that are constant from pulse to pulse and vary only with water clarity properties. The quoted variation relates to the statistical uncertainty of the water clarity.

The beam spreading also has a temporal component that affects the arrival time of the bottom return at the receiver. The basis for timing measurements for the slant path to the bottom is the unscattered ray. Scattering causes simultaneous "long" biases due to increased photon path lengths and "short" biases due to the fact that a significant amount of energy is scattered into the "undercutting" region in the direction closer to the aircraft than the unscattered ray. The results of Monte Carlo simulation studies of underwater light propagation (Guenther, 1985) indicate that net propagation-induced depth measurement biases vary with nadir angle, depth, and water clarity parameters and exhibit larger magnitudes, and larger variations versus water clarity parameters, near nadir. The resulting net depth-measurement biases must be predicted by such modeling. These predicted biases, after their accuracy has been confirmed by field intercomparisons, are then applied in software as correctors to the raw measured depths.

The cartoon in Figure 10.2a depicts the ideal case of a flat seafloor. Steep slopes represent a geometry problem and a vertical and horizontal measurement problem for ALB as they do for other active sensing technologies. Because of the finite beam width, the tendency is to bias the depth toward the shoalest point in the beam, while horizontally the slope moves slightly away from its true location. Results will also vary somewhat depending on the direction of flight with respect to the slope because of the off-nadir beam angle (Steinvall and Koppari, 1996). Extremely sophisticated algorithms would be needed to improve performance for this case. Perhaps a better way of looking at the situation is to consider that, for the most part, the error is not really a vertical error, but just a slight horizontal displacement of an otherwise accurate vertical measurement. The horizontal accuracy requirements are generally forgiving enough to accept this shift without a problem. This can, however, cause a problem for navigation channels with steep sides and with the apparent size and location of engineering structures. These errors can be compensated for with appropriate software.

The impact of complex bottom morphologies like coral reefs, rocky substrates, and other submerged objects depends on the size of these features relative to the laser footprint size and the position of the feature in the water column relative to the surrounding depths. In general, light returning from objects smaller than the laser footprint on the seafloor will not be differentiable from bottom-returning light. Exceptions are tall objects that reflect enough light to generate a peak in the bottom return independent of the bottom return (Guenther et al., 1996). Complex morphology of the sea bottom will tend to scatter the signal and may also result in multi-path, where the light bounces off more than one surface within the receiver field of view before returning to the receiver. Depth measurements resulting from such scattering are erroneous and are removed in automated or manual editing.

Bottom material and submerged aquatic vegetation also have potential to impact ALB performance. Typical coastal sand is a relatively bright substrate with a reflectivity near 15% that returns strong signals to the receivers. Darker substrates absorb light and have lower reflectivity resulting in weaker bottom return. In some cases these weaker returns are not distinguishable from system noise and no bottom return is detected. Bottom reflectivity is a linear factor in calculations of ALB performance, and so has a smaller impact on ALB performance than water clarity. Underwater masses of kelp growing from the bottom toward the surface provide strong returns throughout the water column and, when dense, are nearly impossible to penetrate to the bottom. It is possible at least to delineate the boundaries of the kelp beds. Less obtrusive but nevertheless dense bottom vegetation may be detected and may cause a slight shoal bias depending on their height, density, and reflectivity.

SYSTEM DESIGN

ALB depends on state-of-the-art engineering in areas of lasers, optics, electro-optics, and electronics. Although detailed specifications and implementations vary from system to system, there is a fair amount of generic commonality because the designs are driven by the same physical principles and available engineering, and by similar user requirements. Systems are generally installed in fixed-wing aircraft, though helicopters were used in the early days when laser pulse repetition rates were low. Helicopters are occasionally used today when the survey conditions require it, as in river surveys in steep canyons. Typical aircraft altitudes are 300-600 m. Each system contains three major system components: the laser transmitter, scanner, and receiver, along with several ancillary components required for an operational system. This section will briefly describe each of these components.

Laser

Lasers produce and emit laser pulses in ALB systems. Diode-pumped Nd:YAG (neodymium-doped yttrium aluminum garnet) lasers are the predominant laser type for ALB. Nd:YAG lasers produce radiation at 1064 nanometers (nm), an infrared (IR) wavelength. For ALB, some or all of the IR output is frequency doubled to produce light at the second harmonic mode at 532 nm, which is the desired green wavelength for penetration of the water column. Some systems emit IR and green light simultaneously from a single laser source, using the IR as an aid in detection of the water surface.

Laser characteristics that define performance of an ALB system are laser power, laser pulse repetition frequency (PRF) and beam divergence. Laser power may be described by the peak power per pulse in millijoules (mj) or total laser power in Watts (W = peak pulse power * PRF). ALB systems with higher laser power can measure deeper depths than those with low laser power. Laser pulse repetition frequency is the number of laser pulses emitted per second. This is sometimes expressed as pulses per second (pps) or in kilohertz (kHz). ALB systems with higher PRF have more points per square meter (ppsm) than those with lower PRF. In laser manufacturing it is very difficult to produce a laser with both high PRF and high peak pulse power. In general, lasers with higher peak pulse power will have lower PRF than those with lower peak pulse power. This is evident in the lasers used for ALB. Lasers with peak pulse power ranging from 3-7 mJ, have PRF from 1.5-10 kHz Lasers with peak pulse power ranging from 0.02 to 0.1 mJ, have PRF from 10 to 500 kHz.

Beam divergence is an angular measure that describes the increase in laser pulse diameter as a pulse gets further away from the laser emitter. Beam divergence is typically expressed in milliradians (mrad). Laser pulses are intentionally expanded for the eye safety of casual observers in the survey area. The maximum permissible exposure (MPE) for direct ocular exposure in an outdoor environment is determined by averaging the incident power of the beam over an area defined by a 7-mm diameter aperture that is equal to a fully dilated pupil of the human eye. In the case of a green laser pulse (532 nm) the MPE is 5×10^{-7} J/cm^2 for an exposure time that is less than 104 ns. To stay under the MPE limit, lasers with higher peak pulse power must have larger beam divergence than lasers with lower peak pulse power. Lasers with high peak pulse power ranging from 3-7 mJ, have beam divergence from 3-11 mrad. Lasers with low peak pulse power ranging from 0.02 to 0.1 mJ, have beam divergence from 1-3 mrad. Beam divergence controls the size of the laser spot on the ground or sea surface. The size of the laser spot on the ground sea surface is 2-4 m for high power systems and less than a meter for low power ALB systems at nominal flight altitude. Operational ALB systems have added operational controls to ensure eye safety. The pilots have a physical pause switch that can be pressed to pause emission of laser pulses. The system operator has a physical button that can be pressed to stop the laser from firing, and can control laser firing for the ALB system operator interface. Controls built into ALB system firmware automatically power off the laser when the aircraft altitude (or measured range) decreases below the eye-safe altitude (or distance).

Scanner

Scanners direct laser pulses into a pattern on the ground or water surface below the aircraft. The main design goal for scanning mechanisms is to produce a uniform pattern of laser pulses on the ground or water surface (Guenther, 1985), for a given laser pulse repetition frequency and anticipated forward aircraft speed and altitude. Mechanical scanners in use today (Wehr, 2009) include oscillating mirror(s), Palmer scanner, polygonal scanner, and rotating Fresnel prism. Oscillating mirrors with a single axis of motion produce a zig-zag pattern beneath the aircraft, sweeping through angles from about 20 degrees port or starboard of the aircraft to nadir (0 degrees, directly below the aircraft). Using gimbals or galvanometers, the motion of the oscillating mirror can adjust for the forward motion of aircraft, resulting in a parallel line pattern, i.e. parallel scan lines/arcs (Axelsson & Alfredsson, 1999). Oscillating mirrors with two axes of motion can produce nearly any scan pattern, but are used primarily in ALB to produce arcs that maintain a 20 degree angle between the water surface and incident laser pulse, which has been shown to be advantageous in terms of accuracy, surface detection, object detection, and minimizing signal dynamic range (Guenther, 1985). Palmer and polygonal scanners are both rotating mirror-type scanners. The mirror in a Palmer scanner is wedge-shaped, and positioned such that laser pulse incidence angle on the mirror is 45 degrees, and the mirror rotates around the incidence point. The Palmer scanner makes an elliptical pattern below the aircraft. The polygonal scanner is a rotating 6-sided polygon with mirrored faces. This scanner is similar to the single-axis oscillating mirror in that it sweeps through about 40 degrees of arc below the aircraft, but differs in that the scan lines are parallel. The rotating Fresnel prism (Fuchs and Mathur, 2010) is designed to refract outgoing laser pulses at a 20 degree angle. The rotation of the prism produces a circular scan pattern. With maximum scan angles near 20 degrees for all ALB scanner types, and typical operating altitudes from 200-600 m, swath widths range from 140-420 m beneath the aircraft.

Receiver

Receivers collect and record laser pulse energy returning from the ground, water surface, water column, and seafloor. A typical lidar receiver consists of a telescope, various optical filters and field-of-view controls, light detectors, amplifiers, analog surface detection logic for real-time system control, a digitizer, and removable hard drives for data storage. Both the transmitter and receiver are operated under detailed computer control. Light at different wavelengths is split into separate channels for independent detection and timing. Two types of light detectors, photomultiplier tubes (PMT's) and avalanche photodiodes (APD's), are in common use, and a typical receiver may contain both. Details of receiver design vary widely between systems because various approaches are taken in solving the problems associated with handling the extremely large amplitude dynamic range between strong water surface reflections and weak bottom returns. Some systems utilize multiple channels and, possibly, logarithmic amplifiers for this purpose, while others use gain-controlled PMTs.

Along with the number and wavelength of receiver channels, receiver telescope field of view is another characteristic of ALB systems that has an impact on depth performance. The receiver telescope field-of-view (FOV) is typically expressed in milliradians and defines the area from which photons returning in the direction of the airborne receiver are collected. As shown in Figure 10.2a, the receiver FOV in most ALB systems is usually several orders of magnitude bigger than incident laser pulse footprint on the water surface. For these systems, the system attenuation coefficient for pulse energy, "k", will approach K_d (Steinvall et al., 1992). The maximum surveyable depth for a given water clarity can be expressed roughly as n/K_d, where "n" is a constant. For example, if water clarity can be expressed in terms of a value of K_d = 0.1 m^{-1} and the sensor is characterized as being able to operate at $3/K_d$, then the maximum survey depth is 30 m. If that same sensor is operated in more turbid waters where K_d = 0.6, then the maximum survey depth is 5 m. A large receiver telescope FOV (40-100 mrad, 16-40 m diameter on water surface assuming 400 m flight altitude) captures a larger percentage of photons returning in the direction of the airborne receivers and results in greater depth performance capability. Smaller receiver telescope FOVs (1-3 mrad, 0.4 to 1.2 m diameter on water surface) to that may not capture a major fraction of the returning bottom-reflected energy that is scattered beyond the FOV of the receiver telescope. For these systems, k tends in the direction of the larger value of c (Gordon, 1982), and a potentially severe penalty in maximum surveyable depth results. However, a smaller FOV increases the spatial resolution of the ALB system, limits solar background noise, and increases precision of timing the surface and bottom return by minimizing geometrically-induced temporal return pulse broadening. Some ALB systems use a combination of large and small receiver telescope FOVs to capture the benefits of both across a large depth range.

Receivers in operation today record ALB data in three different ways. Some ALB sensors digitize returning pulse waveforms (like the cartoon in Figure 10.2.) in 1 nanosecond or less intervals. Some of the newer topobathy sensors record data more like topographic lidar sensors, recording the timing of discrete returns using either constant fraction discriminators or online waveform processing to identify the timing of a "return" onboard the aircraft. The waveforms or return time information is recorded on hard drives with a great deal of supporting data, for post-flight data processing.

Ancillary System Components

In addition to these three main components, ALB systems are integrated with GPS, IMU, power distribution units (PDU), thermal management systems (TMS), computer controllers, operator interfaces, additional lidar systems, and cameras. GPS and IMU provide positioning and orientation information for the aircraft that is combined with the lidar returns to produce a point cloud. PDUs manage power and deliver it to subsystems as required. A TMS may be required to regulate the temperature of the laser and on-board computers. Air cooling may be sufficient for some systems, but high power systems may require liquid cooling to maintain a constant laser temperature and dissipate heat from other system components like computer boards. System level computer controllers configure and operate each of the subsystems as required, monitor system status, perform error recovery, and communicate with the operator interface. The operator interface provides realtime feedback to the operator and pilots to assist in data collection, including flightline location, data quality, sub-system diagnostics, etc. ALB are sometimes integrated with standalone topographic lidar systems for collection of higher-resolution data on land and to provide additional information on the water surface. All ALB systems are integrated with cameras. Some cameras are strictly for QA/QC, while others are designed to provide high-quality imagery to accompany the lidar data. Hyperspectral cameras are sometimes integrated with ALB systems for the development of deliverables like landcover and seafloor classifications, and water column color products. See Chapter 12 for more information on enabling technologies.

DATA PROCESSING

The goal of ALB data processing is to produce accurate depth measurements, accurately positioned. Accomplishing this goal requires an understanding of ALB accuracy standards and a robust system calibration. Data processing algorithms, software, workflows, and QA/QC are equally important with hardware design for ALB system performance and accuracy. A detailed description of post-flight data processing software and procedures is beyond the scope of this document, but a brief summary is presented here to emphasize important features and typical characteristics. An automated processing step produces a lidar point cloud by combining lidar returns with aircraft position and orientation data. These points may require some level of filtering erroneous or outlier returns, or reprocessing for certain environmental conditions, based on the level of skill contained in the automated processing step. Powerful visualization tools enable detailed QA/QC of the point cloud in 2-D, quasi-3-D, and 3-D environments. These tools may also have capability for interactive manual editing and reprocessing to finalize the point cloud. The final point cloud data are then passed along to production software packages for the generation of standard and specialized deliverable end products. Metadata and reports of survey accompany the final deliverables. This section will briefly discuss each of these critical steps in delivering accurate ALB survey deliverables.

Accuracy Standards for ALB

Accuracy standards for ALB have historically been defined by two primary groups: the International Hydrographic Organization (IHO) and the US Army Corps of Engineers (USACE). These groups represent the main drivers for development of ALB systems, nautical charting and coastal engineering. USACE ALB survey standards are published in Engineer Manual 1110-1-1000 "Photogrammetric and Lidar Mapping" (2015). This manual prescribes accuracy in terms of Root Mean Square Error (RMSE) of 15 cm accuracy for depths and 1.5 m for positional accuracy. IHO standards are published in Special Publication No. 44, "IHO Standards for Hydrographic Surveys" (2008). The special publication prescribes accuracy in terms of total vertical or horizontal uncertainty at the 95% confidence interval. The Total Vertical Uncertainty (TVU) underwater is specified as, $\sqrt{(a^2+(b^*d)^2)}$ where a represents that portion of the uncertainty that does not vary with depth (such as related to system timing, GPS, IMU, etc.), b is a coefficient which represents that portion of the uncertainty that varies with depth (such as propagation-induced depth bias corrections, surface-related bias corrections), and d is water depth in meters. Total Horizontal Uncertainty (THU) IHO has three "Orders" of survey with decreasing uncertainty tolerances that are presented with their intended usage in Table 10.1.

IHO Order	TVU ($\sqrt{(a^2+(b^*d)^2)}$)		THU	Usage
	a	b		
Special	0.25 m	0.0075	2 m	Underkeel clearance is critical, e.g., berthing areas, harbours and critical areas of shipping channels
1	0.5 m	0.013	5 m + 5%*d	Areas shallower than 100 m where a general depiction of the seabed is sufficient
2	1.0 m	0.023	20 m + 10%*d	Areas deeper than 100 m where a general depiction of the seabed is sufficient

Table 10.1. IHO TVU coefficient values, THU, and usage for the IHO survey orders (IHO, 2008)

More recently in the U.S., the National Science and Technology Council has combined the nautical charting and engineering requirements for ALB into a set of Bathymetric Lidar Quality Levels in the draft National Coastal Mapping Strategy 1.0: Coastal Lidar Elevation for a 3-D Nation from the National Science and Technology Council (NSTC, 2016). The ALB quality levels are shown in Table 10.2. Like IHO S-44, the ALB quality levels are specified as uncertainties at the 95% confidence level, with constant and depth-dependent uncertainty components. The vertical accuracy specification for QL0B and QL1B are equivalent to the IHO

Special Order standard for vertical accuracy. The vertical accuracy specification for QL4B is equivalent to the IHO Order 1 standard for vertical accuracy. The National Coastal Mapping Strategy recommends bathymetric lidar data collection to at least QL2B, which is commensurate with recognized bathymetric lidar accuracy performance since 1994, and is the vertical accuracy that meets most requirements for hydrographic survey in USACE (Engineer Manual 1110-2-1003).

Bathymetric Lidar Quality Level	Source	Vertical accuracy coefficients a,b in ($\sqrt{a^2+(b*d)^2}$)	Nominal Pulse Spacing	Point Density	Example Applications
QL0B	Bathymetric Lidar	a = 0.25 b = 0.0075	≤0.7 m	≥2.0 pt/m^2	Detailed site surveys requiring the highest accuracy and highest resolution seafloor definition; dredging and inshore engineering surveys; high-resolution surveys of ports and harbors
QL1B	Bathymetric Lidar	a = 0.25 b = 0.0075	≤2.0 m	≥0.25 pt./m^2	
QL2B	Bathymetric Lidar	a = 0.30 b = 0.0130	≤0.7 m	≥2.0 pt./m^2	Charting surveys; regional sediment management; general bathymetric mapping; coastal science and management applications; change analysis; deepwater surveys, environmental analysis
QL3B	Bathymetric Lidar	a = 0.30 b = 0.0130	≤2.0 m	≥0.25 pt./m^2	
QL4B	Bathymetric Lidar	a = 0.5 b = 0.0130	≤5.0 m	≥0.04 pt./m^2	Recon/planning all general applications not requiring higher resolution and accuracy

Table 10.2. Quality Level Definitions for Bathymetric lidar. (Applicable for areas submerged at the time of survey).

The ALB quality levels also include a nominal pulse spacing, or point density requirement, in order to align with an existing topographic lidar collection standard. The vertical uncertainty requirement is the same for QL0B and QL1B, but the point density requirement differs. The point density requirement is also the only difference between QL2B and QL3B, which share the same vertical accuracy requirement. It is noted in the National Coastal Mapping Strategy that in ALB and topobathymetric lidar, a planned point density is not always achievable due to water conditions. As water gets optically deep for ALB systems, valid returns per square meter decrease. Therefore the nominal pulse spacing and point density listed in Table 10.2 reference these parameters for the final point cloud, not the sensor capability. To further complicate matters, many bathymetric lidar projects include a deeper water component that will have one point density in shallow water, with a much smaller point density in deeper water. So it is possible that a single coastal lidar survey will have one QL for the topographic portion, another for shallow water, and yet another for deep water, based on the vertical accuracy and point density requirements.

The TVU formula is applied to water depths from 0 to 30 m in Table 10.3 on page 290. The results demonstrate that the constant portion of the uncertainty dominates the allowable depth error in depths from 0 to 10 m. For depths greater than 10 m, the depth-dependent portion of the uncertainty begins to add more significantly to the allowable depth error.

Bathymetric Lidar Quality Levels					
	QL0B	QL1B	QL2B	QL3B	QL4B
Pulse Spacing	≤0.7m	≤2.0 m	≤0.7 m	≤2.0 m	≤5.0 m
Point Density	≥2.0 pts/m²	≥0.25 pts/m²	≥2.0 pts/m²	≥0.25 pts/m²	≥0.04 pts/m²
Depth (m)	Bathymetric Accuracy RMSE (cm)				
	QL0B	QL1B	QL2B	QL3B	QL4B
0	25.0	25.0	30.0	30.0	50.0
2	25.0	25.0	30.1	30.1	50.1
5	25.3	25.3	30.7	30.7	50.4
10	26.1	26.1	32.7	32.7	51.7
20	29.2	29.2	39.7	39.7	56.4
30	33.6	33.6	49.2	49.2	63.4

Table 10.3. Bathymetric accuracy at varying depths.

System Calibration

ALB system calibrations have a laboratory and a flight component. In the lab, optical and electronic time delays through the system, both fixed and variable, must be carefully determined, and errors must be either eliminated or calibrated and corrected in software. Timing calibration must be measured and corrected to sub-nanosecond accuracy, and it should demonstrate good stability. Systems with multiple receiver channels have the added benefit of continuous, full-time timing inter-comparisons as a consistency check. Typical examples are signal transit time variations in a photomultiplier tube (PMT) as a function of high voltage and in a logarithmic amplifier as a function of signal amplitude. Computer latencies must be well understood and carefully handled. The stability of laser pulse characteristics such as pulse rise-time, width, and energy is very important and must be monitored. An internal system alignment procedure sets internal geometry of optical elements to the desired output and receive paths. Radiometric calibration establishes the relationship between receiver output in electronic counts to the input optical power in Watts (Tuell et al., 2005a).

Following laboratory calibration, the ALB system is installed into an aircraft. Once the GPS antenna, IMU, and ALB system are securely mounted into the aircraft, the distances between reference points on each of these components is measured in 3 dimensions (x, y, and z). This can be accomplished to within a few centimeters either by using a total station or by geodetic GPS observations from an arbitrary baseline. System installation angles, aircraft orientation angles, and the resulting beam nadir angle in the world frame must be known to high accuracy for every pulse because their effect on distance measurements is magnified by the aircraft altitude. For example, at a 400-m altitude and with a nominal 20-degree nadir angle, a system angle error of 0.05 degrees (<1 mrad), which equates to a nadir angle error of 0.10 degrees after reflection off a misaligned mirror, would yield a 25-cm error in the estimated vertical distance between the water surface and the aircraft. This would be unacceptably large for many applications. It is desirable to limit system error components to about 5 cm and thus system angle errors to about 0.01 degrees. The measurement of system alignment angles in the aircraft to such a tolerance is very difficult.

Instead, a flight calibration procedure known as boresighting is used to define the rotational, or angular offsets among the ALB sensor, IMU, and GPS antenna system components, and the scanner offset angle, which is the angle between zero on the scanner and the long axis of the aircraft (Habib et al., 2010). The boresighting procedure utilizes a prescribed set of flight lines over a well-surveyed target in an optimization routine that minimizes the difference between the airborne data and ground truth. In addition to the angular offsets described above, this procedure also refines the distances measured between the components after system installation. Flight lines are often required to be flown in reciprocal directions, at different altitudes, over both flat ground and complex building geometries in order to completely resolve the positional and angular offsets required for direct-georeferencing of the lidar measurements. ALB systems also require an in-water calibration over well-surveyed hard bottom to determine appropriate correctors for propagation-induced depth biases. Redundant calibration flights over a number of days can help eliminate contributions of ancillary systems like GPS and IMU to the determination of altitude-related range biases, and can help average contributions of environmental factors like water surface and water clarity to environmental-related correctors.

Because field operations take place in harsh environments, the precision, repeatability, and absolute accuracy of the system must be verified on a daily basis. The most reliable method to validate the precision quality and stability of a calibration is comparing processed data from adjacent flight lines, flight blocks, and flight days. Comparisons can be performed on production data alone, or on cross lines flown perpendicular to the predominant flight direction for the specific purpose of QA/QC. In systems with multiple receiver channels, agreement between channels may also be monitored. Occasional inter-comparisons against independent standards such as ground truth measurements, or comparisons with data collected by another lidar or multibeam system, or comparisons to prior validated surveys of an area can give a quantitative measure of the absolute accuracy of the system calibration.

Generating the ALB Point Cloud

Data processing techniques vary considerably from system to system, but they have a number of aspects in common. Major functions of ALB post-flight processing software are: identifying the timing of laser returns, discriminating surface and bottom returns, syncing time stamps of laser pulses and ancillary data streams, and calculating the water depth for each return.

The following list highlights some of the major functions performed by the post-flight waveform processor and depth determination algorithms. Flexible operator interaction is provided to handle special cases through the use of a set of software control parameters. Timing latencies in measured scanner angles, attitude, and altitude data are de-skewed. Low-rate sensor data are interpolated for each pulse. For example, GPS position data are typically measured at a rate of 1-2 Hz and IMU aircraft orientation data are typically measured at a rate of 200 Hz, while lasers fire from 10-500 pps. Each pulse must have an associated aircraft position and orientation to determine accurate positioning of each point. A tracking algorithm may be applied to surface times. Surface and bottom returns are discriminated using signal-to-noise ratio and possibly near-neighbor confidence as criteria. Algorithms containing heuristic rules (Guenther and Mesick, 1988a) are helpful in rejecting noise, system artifacts, and false targets in the environment such as fish and scattering layers. The algorithms must handle complex waveforms associated with a variety of circumstances from extremely strong shallow returns to extremely weak deep returns as well as difficult returns in turbid waters (Perry, 1999). For some applications, it is vitally important to have a reliable capability to identify small objects on the sea bottom. Recorded waveforms, which have been nonlinearly processed in hardware, are linearized, and precise pulse arrival times are estimated. Timing calibration correctors are applied for hardware and environmental time delays. This should include delay versus amplitude tables because transit times through nonlinear components tend to acquire a certain amount of amplitude dependence.

Several types of automated channel priority logic may be available for selection of the optimal surface return for each pulse. The local mean water level is calculated. The depths are determined in a manner consistent with the quality of available surface data and the goals of the survey. Multiple bottom returns per waveform are saved in order to permit valid depths to be calculated in the presence of fish and other biota in the water column. This is also critical for small-object detection. Depths from multiple detections, if available, can be presented for possible manual inspection (Guenther et al., 1998). A predicted corrector for propagation-induced bias is applied as a function of depth and nadir angle (Guenther, 1985). The positions of the depths are calculated taking into consideration parameters such as aircraft position, altitude, attitude, scanner angles, and depth. The time difference between the surface and bottom returns is converted to a slant range (UNSCATTERED REF PATH, Figure 10.2a) using the speed of light in water. The slant range is converted to "vertical depth" using the beam incident angle on the water surface, which is calculated from measurements of scanner angle and the aircraft orientation. A geographic position is assigned to each depth based on aircraft position and orientation, and the slant range between the aircraft and the laser spot on the water surface. A number of internal consistency checks are conducted during processing, and various confidence and accuracy statistics are computed. For each return, an overall level of "confidence" in the result is sometimes provided as a key parameter.

Some topobathy systems use a technique called refraction correction to determine depths. These systems record timing of returns onboard the topobathy sensor using techniques like constant fraction discriminators or online echo-digitization. Constant fraction discriminators (Fernandez-Diaz et al., 2016) trigger recording of a return time when received light energy is a specified fraction of each detected peak. Online echo-digitization analyzes waveform snippets temporarily stored in topobathy system memory using Gaussian decomposition to identify the peak time, amplitude and width of returning laser energy (Ullrich and Pfennigbauer 2011). The times of all returns, both surface and bottom, are converted to slant ranges using the speed of light in air. The result is a point cloud that has land returns, sea surface returns, volume returns and seafloor returns. Using manual, semi-automated, or automated processes the points are classified as land or water. A mean water surface is derived from the surface points, or from surface data collected with another sensor. Points classified as volume or seafloor returns are then corrected in two steps. One corrects the position of each point for the angle of refraction at the water surface, and the other corrects the length of the slant path for the difference in the speed of light in air or water.

Example Waveform Processing Algorithm

Most automated processing of waveforms is performed in sensor-specific software provided by ALB and topobathy manufacturers. Many of these waveform detection and extraction methods and software are proprietary. System manufacturers use a variety of different signal processing methods that rely on the properties and characteristics of the transmit pulse, laser scanning pattern, as well as the water surface and water column environment being mapped. Some of the information used to process waveforms is a priori, whereas other information is determined from the waveforms captured for that mission using heuristic rules. We present one such waveform detection technique as an example so that the reader can understand the complexities related to extracting reliable surface and bottom reflections from a laser pulse. The Airborne Lidar Processing System (ALPS) is waveform processing software based on algorithms originally proposed by C. Wayne Wright for the EAARL system and used to process data from both EAARL and EAARL-B (Nagle and Wright, 2016). The waveform detection algorithm relies on estimating decay of the laser pulse as it traverses through the water column in different types of water (clear, moderately turbid, turbid, extremely turbid, etc.). The resulting inflection in the waveform can then be determined to be a valid bottom using signal processing methods that investigate a neighborhood of bottom returns (Nayegandhi et al., 2009).

The approach is based on processing the waveforms to better define the inflection in bathy bottom by suppressing the effect of the turbidity in the water column. First, the part of the signal from peak of surface return to noise level in waveform is extracted. The waveform is then scaled and normalized to remove any bias. Three exponential decay constants are estimated:

1. Laser Decay: Estimates decay of the laser pulse due to its system properties. This is typically determined using the transmit (T0) pulse and describes the surface return laser event
2. Water Decay Factor: Estimates decay of the laser signal as it traverses through the water column. This is the exponential decay of the backscatter from within the water column
3. Automatic gain control (AGC) factor between 0 and 1 used to equalize returns with depth so effect of near surface water column backscatter can be minimized

The laser decay exponential constant is typically determined from sample return waveforms acquired on a flat surface such as a road. The laser decay exponent is also empirically derived from a group of waveforms. The AGC and water decay factor are based on the turbidity of the water column. Several waveforms are investigated to determine the water decay factor rate and the AGC constant. In the examples below, ALPS is implemented on sample waveforms from the CZMIL sensor. After determining these values, the curve representing the exponential laser decay (red) and exponential water decay (green) is subtracted from the original scaled waveform (See Figure 10.4b). The x-axis (time) originates from the peak reflection of the surface return. This results in the waveform shown in magenta. After applying the AGC equalizer, the water column return is further suppressed resulting in the waveform shown in blue. The resulting blue waveform has a better defined bottom inflection that can be used to describe a potential bottom return.

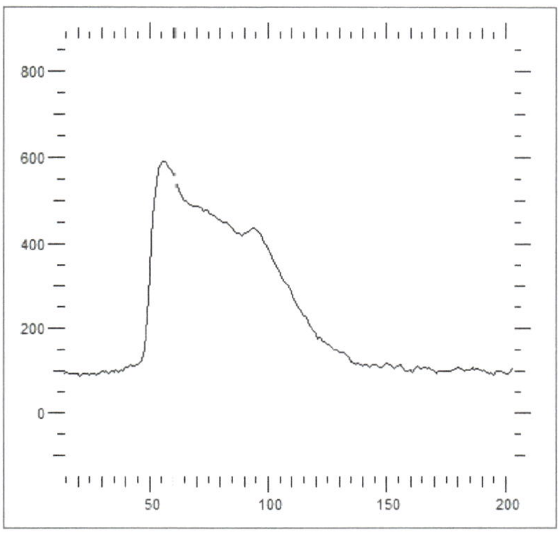

Figure 10.4a. CZMIL waveform from Kings Bay near Tampa Bay, Florida; depth ~3 m. The backscattered signal from the near-bottom layer falls below the noise level.

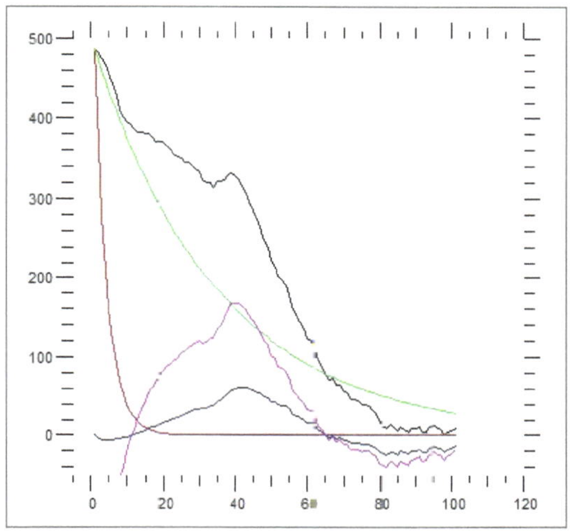

Figure 10.4b. The waveform processed by determining the exponential decay of the laser system and water column to create a resulting waveform that had a better defined bottom inflection. This algorithm was implemented in an open-source programming environment.

Figure 10.5 on page 294 depicts another return waveform from the deep channel of the CZMIL sensor. The original waveform (left) that has a potential bottom inflection (shown by the blue arrow) would not be picked by any signal processing method because the inflection is within the system noise. However, following the application of the algorithm described above, the inflection in the bottom return (blue waveform on the right) is more pronounced and can be picked by most signal processing methods.

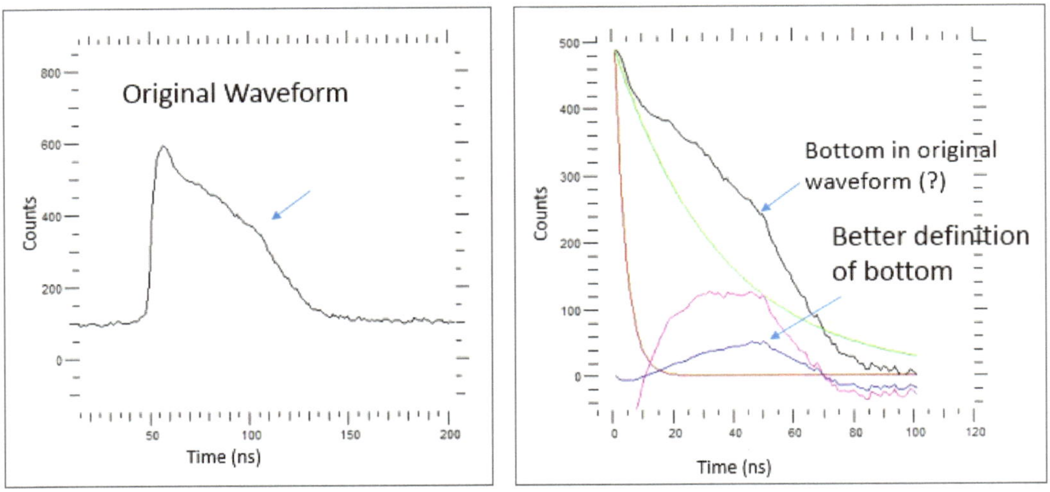

Figure 10.5. A sample CZMIL waveform (left) depicting a hint of a bottom reflection (indicated by the blue arrow). The blue waveform (shown in the image on the right) was processed using Dewberry's turbid water algorithm. This blue waveform depicts a more distinct inflection that can be picked by most signal processing methods.

Filtering and Manual Editing

After automated processing, the amount and type of manual editing required to generate the ALB point cloud varies from system to system and may be carried out in a 2-D or 3-D visualization environment. In general, manual editing is not carried out in manufacturer software, but rather in software packages such as TerraSolid, Quick Terrain modeler, LP360, RapidLasso LAS Tools, or the public domain PFM ABE. Interactive data displays and automatically calculated statistics are available to facilitate quality control and validation of results. Calculated depths are typically color-coded and may be presented in a variety of display formats. The user may "zoom" in to view selected geographic regions of interest. Pulses not returning depths may need to be examined. Areas of indicated land might need to be verified, particularly in surf zones. Camera images taken at the time of the lidar data points may be consulted for ancillary geographic information. Difficult or questionable data segments can be reprocessed, within limited geographic boundaries, with parameters optimized for the situation. Multiple or false depths due to environmental effects such as schools of fish or turbidity layers can be investigated by the analyst to ensure the use of the underlying true bottom depths. The automated decision may be reversed if necessary. In some cases, areas with very little to no depth soundings are marked as voids or low-confidence areas.

For ALB systems, recorded raw waveforms, along with important parameters, can be accessed for selected soundings and displayed for careful evaluation by the analyst (Sinclair et al., 1999). According to the laws of statistics, assuming a normal distribution, 99.73% of data are within the three-sigma level. If a relatively short 13 km flight line contains, say, 100,000 or more pulses, then that flight line will contain 270 pulses whose depths are outside the three-sigma bound. A few of such statistically inevitable errors may leak through automated quality tests and need to be studied, and possibly removed, manually. Waveforms with multiple bottom returns may need to be examined, particularly for the presence of small objects when invoking hazard-detection procedures.

Especially rigorous hazard detection techniques and quality control procedures are used for nautical charting data (Perry, 1999). The waveforms of questionable soundings are examined and compared to those from neighboring points to assist interpretation. The distinction between fish and coral heads or a wreck is not always clear. For nautical charting work, if no conclusive evidence exists to discredit an apparent anomaly, additional flight data are collected to confirm or disprove its existence. Some surveys are flown with 200% original coverage to improve the detection probability of small objects. In this case, the data are examined to determine if the anomalies exist in both flight lines. The job of data processing is not an easy one. The survey manager must be aware of problematic data and maintain control over these procedures. For less-stringent general bathymetry not for navigation purposes, such as reconnaissance or "condition" surveys or modeling work, similar examinations of the anomalies and neighboring soundings are made, but additional data are not normally collected; the decision to keep or discard is made based on the existing data.

Powerful visualization software tools have become commercially available. Manufacturers are willing to customize these tools to meet the unique needs of specific users. This permits the migration of much of the editing previously done in ad hoc two-dimensional manual processing schemes to be accomplished more efficiently in a pseudo-three-dimensional visualization environment.

Data Products

Products, procedures, and deliverables vary far too much from system to system, customer to customer, and over time, to be described in any but the most general terms. Deliverables may be paper or digital; often, both are required. Typical digital products from a survey, provided to the customer in agreed-upon formats, would be the cleaned and verified full xyz point cloud data set, a decimated xyz data set, TIN's, Computer-Aided Design (CAD) and Geographic Information System (GIS) files, plot files, and a metadata file compliant to FGDC standards. Paper products may include two-dimensional forms such as charts, maps, smooth sheets, contour plots, or profiles and pseudo three-dimensional, color-coded, shaded displays. Some customers may desire higher-order products, based on further analysis of the data, such as bathymetric differences from previous years or dredge volumes.

All raw data and processed results should be permanently recorded for possible later review during chart verification and production, as well as for archiving. If decimation has not already been done, it would be done at this point to the cleaned xyz files.

Output Formats and Deliverables

Raw data acquired by lidar sensors are processed in software provided by the system manufacturer. The acquired data can be in the form of digitized waveforms for each laser pulse or many discrete returns of objects, water surface, water column, bottom, and noise encountered by each laser pulse. In either case, and as described above, the system manufacturer's software processes these ranging data along with GPS/IMU data to produce a 3-D point cloud. In order to load the data into other commercial software packages for further editing and distribution, the laser measurements need to be converted from a proprietary system-specific format into a conventional exchangeable binary format.

There are many deliverables that can be generated from ALB data and simultaneously-collected sensor data. The most basic is a lidar point cloud in which points on bare ground are identified. Derived data products include Digital Surface Models (DSMs), Digital Elevation Models (DEMs), lidar reflectance images, shorelines, and aerial photo and hyperspectral image mosaics. These deliverables are generated using automated processes and add little extra cost to the project. More advanced deliverables like landcover and benthic classification, elevation and volume change comparisons to previous data. The Report of Survey deliverable contains a summary of the survey activity and results of QA/QC performed as part of the project. The following sections outline recommended specifications for the most common ALB survey project deliverables.

Most projects are processed and delivered in a specified tiling scheme. This can be provided in shapefile format, if there are specific project-related boundaries that can be used to delineate the survey area. Existing tiling scheme may also be used; for example, the NCMP has historically used 5 km alongshore boxes for data delivery, or the USGS digital ortho quarter quarter-quad boundaries.

LAS is the industry-standard format for exchange of lidar point cloud data. These files are binary and contain a specific structure that is defined by the American Society for Photogrammetry and Remote Sensing (ASPRS, 2013). Documentation for the LAS format is publically available allowing vendors and customers to export and import the data between different software. The LAS 1.4 format provides a public header block, any number of optional variable length records (VLR), any number of optional extended variable length records (EVLR), and a full point data record (Table 10.4). The public header block contains generic data such as point numbers and point data bounds; the data content is often referred to as "payload". The VLRs contain variable types of data including projection information, metadata, waveform packet information, and user application data. They are limited to a data payload of 65,535 bytes. The EVLRs allow a higher payload than VLRs, and have the advantage that they can be appended to the end of a LAS file. This allows adding, for example, projection information to a LAS file without have to rewrite the entire file. The point data record indicates the type of point data records contains in the file. LAS 1.4 defines types 0 through 10. The length of each point data record is also stored in the header block. Details of the LAS format can be found in the LAS v.1.4 specifications (ASPRS, 2013).

LAS 1.4 Format Definition (ASPRS, 2013)
Public Header Block
Variable Length Records (VLR)
Point Data Records
Extended Variable Length Records (EVLR)

Table 10.4. LAS 1.4 Format Definition (ASPRS, 2013)

LAS v1.4 supports the creation of a LAS Domain Profile that adds (but does not remove or alter existing) point classes and attributes. The Topo-Bathy Lidar Domain Profile, published in July 2013, adds point classification values for bathymetric point (e.g., seafloor or riverbed; also known as submerged topography), water surface, derived water surface, submerged object, IHO S-57 object, and bottom-not-found depth. Extra Byte Variable Length Records (EXTRA_BYTES or Extra Byte VLRs) are added for pseudo-reflectance, uncertainty, water column depth, figure of merit, and processing specific flags (See Table 10.5).

Classification Value	Meaning
40	Bathymetric point (e.g., seafloor or riverbed; also known as submerged topography)
41	Water surface (sea/river/lake surface from bathymetric or topographic-bathymetric lidar; distinct from Point Class 9, which used in topographic-only lidar and only designatres "water," not "water surface")
42	Derived water surface (synthetic water surface location used in computing refraction at water surface)
43	Submerged object, not otherwise specified (e.g., wreck, rock, submerged piling)
44	International Hydrographic Organization (IHO) S-57 object, not otherwise specified
45	No-bottom-found-at (bathymetric lidar point for which no detectable bottom return was received)

Table 10.5. LAS domain profile for topobathy lidar.

When paper products are required, the cleaned or decimated data are typically passed into commercial chart or map production software. The scale is determined by customer needs and the size of the project. The complete data set would generally only be used to create very large-scale charts (1:1200 or larger) for small projects, because the large file size is not justified by the minimal increase in information in the final product. The basic element of the mapping process is the triangulated irregular network (TIN) that represents the depths or elevations. Each individual sounding is a mass point in the TIN. TIN's are used as intermediate products in the generation of a variety of deliverables such as color-coded elevation grids, contours, and profiles. Hard breaklines are rarely needed or used when dealing with high-density bathymetric data. Along the shoreline or on land, however, breaklines may be used along road centerlines, toes and crowns of dikes, levies, jetties, and walls, and anywhere they are needed to prevent the software from connecting data points that fall on opposite sides of a physical discontinuity.

Individual CAD and/or GIS files may contain single data representations such as color-coded elevations, contours, profiles, geo-referenced photographs, and sounding elevation text (individual soundings shown on the final chart) which can be overlaid in various combinations. Contours are usually generated on intervals ranging from 0.5 m to 5.0 m depending on the chart scale and intended use of the data. The elevation text is produced by a method that reads the complete or decimated data file. The overlaid data files must be on, or converted to, the same datum and the same projection but need not all cover identical geographic areas. Composite products can be generated by retrieving all data within a selected overlay frame and writing it into a "sheet" file. The sheet file will also contain items such as the title block, legends, notes, index maps, scale bars, and borders. Plot files can be created from these sheet files. Charts and maps may be printed on site in the field, but, more often, cleaned complete and decimated files will be sent to a centralized production and distribution facility for final validation and printing. The customer may perform additional data verification measures according to their standard procedures.

Digital GIS products can provide access to the valuable information contained within the bathymetric lidar and accompanying datasets, especially in cases where customers do not have the knowledge or hardware required to deal with large datasets. These products include raster images, or grids, built from TINs of the combined topographic and bathymetric elevation data. From these grids, a shoreline or other contour of interest is extracted and stored as a shapefile. For topographic data, images of first and last returns are analyzed to extract building footprints and bare earth models. Topographic lidar intensity and bathymetric seafloor reflectance data are used to discriminate between discrete land use areas and bottom types, respectively. These classifications are improved by combining lidar elevation and intensity/reflectance with RGB or hyperspectral image data in complex data fusion algorithms (Tuell et al., 2005b).

Two elevation models are recommended deliverables, one of all the valid data and one of the bare ground identified points. The first, a DSM, will include all vegetation, buildings, other ground features and water surface elevations. Sometimes, a DSM is created with above-ground surface features and the bathymetry (no water surface). The second, a DEM, represents ground topography (no above ground features) and bathymetry. An example DSM and DEM from the Lake Ontario shoreline are shown in Figure 10.6. The DEMs are typically

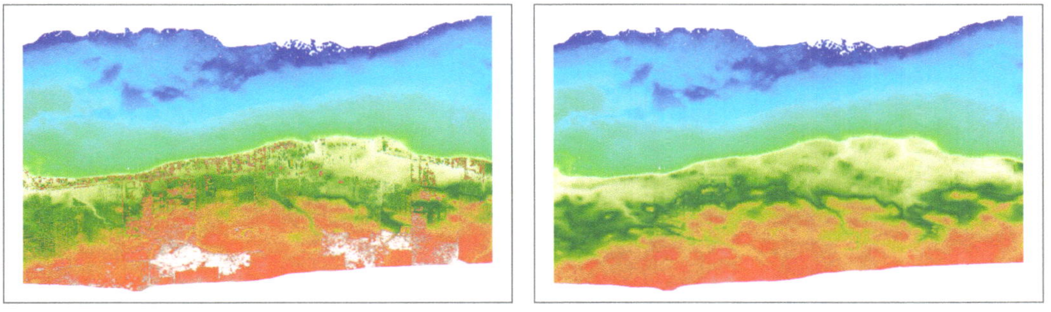

Figure 10.6. Example digital elevation models from the shoreline of Lake Ontario. In both the left and right panels, land is at the bottom of the image and water is at the top. The left panel is a DSM that include trees and building in the elevation model. The right panel is a DEM, where those and other above ground features have been removed, and only bare ground remains. Bathymetry is included in both elevation models.

provided as elevation rasters in 32-bit geo-tiff format. The raster will have 1-meter pixel resolution with the lower-left corner as the reference point. The elevation of each grid cell is interpolated from a TIN surface of the data points. Interpolation across data gaps typically do not exceed 10 meters and elevations will not be interpolated more than 5 m away from actual data points. The exception would be for data gaps resulting from the removal of buildings where full interpolation across the data gap is required. In areas of no data, the raster cells are assigned an elevation value of "No Data". In some cases, such as areas of very dense vegetation over land or at the land water interface (such as mangroves), or in turbid waters, the density of the 'ground' points are very low and may have gaps over 10 m. Depending on the application of the DEM data, it may be worthwhile to interpolate across the larger gaps over 10 m. In such cases, a low-confidence polygon could be added to the output products that clearly identify the areas that have lower point density in the DEM.

DEM rasters are created such that the edges of adjacent tiles have matching elevations. Further, the origin point of the DEM is evenly-divisible by the DEM resolution, such that cell boundaries for subsequent surveys will align. DEMs are generally used in ArcGIS. When the raster is loaded into ArcGIS, the elevation range for the tile is indicative of the actual values observed in the data. There should be no gaps or overlaps between adjacent grids. Metadata in an FGDC-endorsed format is typically provided per geo-tiff.

Aerial Photography: If aerial photography is collected during data acquisition, orthomosaics at 25 cm (1 ft) resolution are the recommended deliverable for aerial photography. Images over navigation structures and other features of interest may be specified for a 5 cm-resolution product. The lidar elevation data is the best elevation source for creation of the orthomosaics. The orthomosaics can be provided in GeoTIFF format and referenced to the coordinate system and vertical reference for the project. Metadata in an FGDC-endorsed format is typically provided per orthomosaic.

Reflectance Images: Laser reflectance images are gray-scale images of the survey area at the laser wavelength of 532 nm. The topographic analog of this project is an intensity image. If the ALB sensor is radiometrically calibrated, the value of each pixel is the absolute reflectivitiy of the object. If the ALB sensor is not radiometrically corrected, the reflectance values are called pseudo-reflectance (Wozencraft and Park, 2013). The laser reflectance image is generated at the resolution of the lidar data. Values for the image shall range from 0 to 1, with 0 representing total absorption and 1 representing total reflection. The image is provided in GeoTIFF format. Metadata in an FGDC-endorsed format shall be provided per reflectance image.

Hyperspectral Imagery: Hyperspectral image strips, if acquired, are processed into 1-m resolution at-sensor radiance images using sensor-specific lens and radiometric correction files. The images may be further corrected to reflectance images using atmospheric correction techniques. The images are typically tiled in the same tiling scheme used for DEMs produced by the lidar data. The coordinate system is typically the appropriate UTM zone for the survey area. Metadata in an FGDC-endorsed format is provided.

Report of Survey: The Report of Survey provides documentation of the survey processes, procedures, collection conditions, and results of the QA/QC tests of the data. The report should include, at a minimum, airborne collection logs, calibration reports, accuracy check reports, daily reports, GPS logs, NGS control sheets, photos of control points, processing logs, list delivered files, description of survey, dates of collection, and a list of problems encountered. Full documentation of all control used, existing or established, is typically provided in the Report of Survey. Results from QA/QC tests, which are the basis for evaluating the final quality of the ALB data collection, processing, and product generation, are summarized in the report. Procedures for and results of testing are included in the Report of Survey.

CURRENT SENSOR TECHNOLOGY

Current sensor technology, known to be operational as of December 2017, can be broadly classified into two categories:

1. Broad beam systems, such as the CZMIL, HawkEye II / III, SHOALS, and the LADS-MK3
2. Narrow beam systems, such as the Chiroptera II, EAARL-B, VQ 820-G / 880-G, and Titan

The single-photon sensor, known as HRQLS, has some bathymetric capability as it uses a green-wavelength lidar. However, specific design efforts towards bathymetric application have not yet been undertaken (Lidtka, 2016).

Airborne lidar bathymetry is an established operational technique which has been proven to be an accurate, efficient, highly cost-effective, safe, and flexible method for rapidly charting large and small projects in near-shore waters, navigation channels, coral reefs, coastal engineering structures and adjacent beaches. There are seven variants of airborne lidar bathymeters in operation today, each with a unique development history and resulting unique design. Fugro operates a modified Optech SHOALS system and LADS-HD. These broad beam sensors are typically flown tandem with narrow-beam Riegl sensors. The Royal Australian Navy operates a customized LADS-HD. NOAA's National Geodetic survey owns and operates a Riegl 880G system, whereas Dewberry and Quantum Spatial have leased Riegl 820G and 880G sensors for various projects in coastal and riverine environments. USACE and NAVOCEANO operate four CZMIL systems out of the JALBTCX and the Japanese Coast Guard operates another. Teledyne Optech offers one CZMIL as a rental system and partners with surveying and mapping companies such as Dewberry to perform contract surveys. University of Texas Austin operates a Chiroptera I, while Geomatics Data Solutions, Nova Scotia Community College and Leading Edge Geomatics, LLC each operate Chiroptera II. For deep water projects Chiroptera is often paired with HawkEye III, which is offered as a rental by Leica Geosystems, along with additional Chiroptera II sensors. University of Houston operates a Teledyne-Optech Titan. Each of these systems is briefly described below.

The venerable SHOALS and LADS families of systems have been providing operational hydrographic and bathymetric surveys virtually full-time in the field for over 20 years, and CZMIL was developed based on knowledge cultivated in those many years of SHOALS operations. These systems, based on differing technical design philosophies, have all proven to be highly effective, and they demonstrate excellent accuracy with good reliability. On the other hand, it is important to note that ALB remains a youthful technique that utilizes state-of-the-art technology and requires knowledgeable implementation. It is far from mature, and the new generation of systems and techniques being implemented now and in the future will improve performance and availability even more with careful application and consideration of basic principles developed over the past 50 years.

SHOALS (Scanning Hydrographic Operational Airborne Lidar Survey)

Fugro Pelagos has been operating commercially and continually evolving the SHOALS-1000 they purchased in 1998. Prior to the CZMIL, USACE and NAVOCEANO operated a SHOALS system and called it CHARTS (Compact Hydrographic Airborne Rapid Total Survey). In its current form, the SHOALS system has a laser pulse repetition rate of 3,000 pulses per second (pps). Altitude can range from 200 to 500 meters, with swath widths as large as 510 m, and on-ground spacing of laser shots (spot spacing) as small as 1 m (Bible, 2016). All survey scan patterns are of the form of a section of a circular arc, and the beam is maintained very nearly at a constant 20-degree nadir angle. At survey speeds from 125-180 knots, and 4 meter sounding spacing, coverage rate of 131 km^2/hour is achieved. A ProSilica digital camera with metric calibration has been integrated into the system and provides decimeter-resolution imagery for QA/QC of the bathymetric lidar data and as an additional product. The lidar and RGB imagery are positioned using an Applanix Pos AV V6 510 integrated into the SHOALS hardware box. Fugro SHOALS is often flown in tandem with a topographic lidar, shallow water topo-bathymetric lidar, and/or hyperspectral imager according to the survey specification.

The SHOALS system maintains the collinear laser design that puts both green and IR beams at the same location on the surface. SHOALS has the unique capability of measuring Raman backscattering from the water, in conjunction with the infrared air/water interface reflection, to provide the most sophisticated and accurate water surface detection strategy of any ALB system (Guenther et al., 1998). Green surface returns are never used for depth calculation due to the inherent ambiguity of their origin. The receiver has four independent channels in which to measure surface and bottom returns: two green, one IR, and one red. Fugro SHOALS system data are processed using the heuristic, leading-edge detection technique with appropriate prediction, modeling, and application of appropriate correctors developed for the original SHOALS system (Guenther et al., 1998) and modified for the SHOALS-1000. A special software algorithm permits acquisition of depths from the problematic region of less than one meter of water (Wozencraft & Millar, 2005). This permits seamless operation across the land/water boundary. The green channels are radiometrically calibrated to enable extraction of bottom reflectance from the lidar waveforms. Fugro operates the SHOALS to meet both USACE "Class 1" and IHO "Order 1" hydrographic accuracy standards for most applications.

Figure 10.7 illustrates an application of data acquired by the SHOALS sensor for USACE to examine coastal environments influenced by Mining (Kerfoot, et al., 2014). Grand (Big) Traverse Bay is an embayment along the southern coastline of Lake Superior, off the Keweenaw Peninsula. Lidar-derived bathymetry of Grand Traverse Bay, in Lake Superior, showing the positions of the original Gay tailings pile, above-water (deep blue region) stamp sands along the beach stretch from the pile to the Coal Dock and southwestward to the Traverse River seawall and the central Buffalo Reef area. The "trough" region, an old riverbed, is in the center. To the northeast, note the underwater stamp sand bars moving across bedrock (Jacobsville Sandstone) to deposit in the "trough". To the southwest, natural sands and silts move out of the bay across the bottom into Lake Superior. Red horizontal contour lines are at 5-m depth intervals. ROV and side-scan sonar sites (1–5) are indicated in red (Kerfoot, et al., 2014)

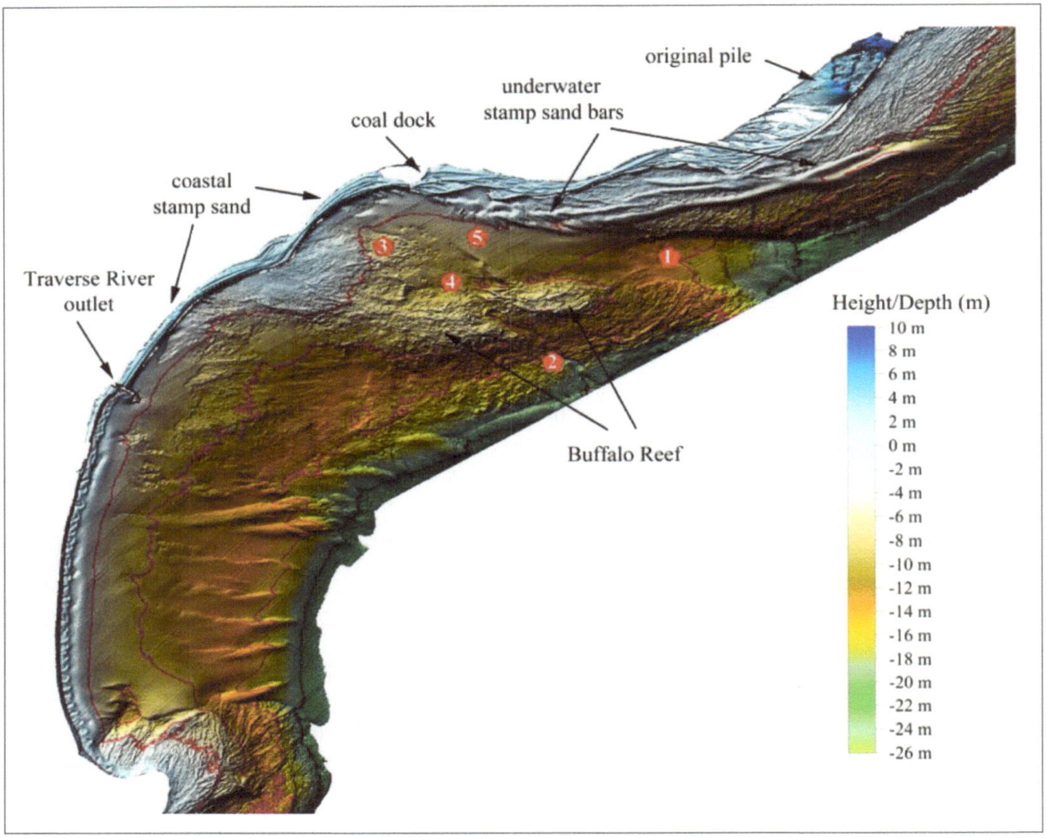

Figure 10.7. Coastal application of SHOALS data to better understand the coastal environment influenced by mining.

CZMIL (Coastal Zone Mapping and Imaging Lidar)

CZMIL was fielded in 2012 to meet the requirements of the USACE National Coastal Mapping Program and NAVOCEANO Airborne Coastal Surveys: high-resolution, high-accuracy bathymetric lidar, topographic lidar, aerial photography, and hyperspectral imagery. The CZMIL laser emits pulses at a rate of 10,000 per second, each with 2-ns pulse width. The pulses are scanned using a spinning Fresnel prism. The resulting scan pattern is a circle on the water surface, with 2.8-m diameter laser footprints spaced 2-m apart. CZMIL is not motion-compensated so angle of incidence varies from the optimal 20 degrees depending on attitude of the aircraft. Operational altitude is 400 m and speed is 140 knots. CZMIL has three receivers, one infrared Avalanche Photo Diode for surface detection, one deep field-of-view (40 milliradians) Photo Multiplier Tube (PMT) for deep water, and a shallow water receiver (6 milliradians) comprised of seven PMT segments (2 milliradians each) for shallow water and topographic measurements. For shallow water (up to 10 m depths depending on turbidity) and topography, measurement spacing is 70 cm. Lidar waveforms are digitized for each receiver and receiver segment, a total of nine waveforms per laser pulse. CZMIL is integrated with a medium format digital camera and uses Applanix POS AV for positioning. CZMIL is operated to meet both USACE "Class 1" and IHO "Order 1" hydrographic accuracy standards for most applications (Tuell et al., 2010).

CZMIL Hydrofusion data processing software evolves the SHOALS empirical heuristic waveform processing technique to a more modern signal processing approach that utilizes signal smoothing and detection of inflection points. Using this technique, CZMIL Hydrofusion can extract up to 32 returns on each waveform, up to 256 per laser shot, with separation along the slant range as little as 10 cm on land. CZMIL Hydrofusion takes advantage of CZMIL radiometric calibration and produces advanced lidar products like lidar reflectance on land and in water, water column attenuation, and water-leaving reflectance. USACE and NAVOCEANO further integrate CASI-1500 hyperspectral imagers into their CZMIL systems to support their missions. The CASI-1500 is a visible/near-IR pushbroom hyperspectral instrument whose design was specified for integration with the CHARTS system. The CZMIL and CASI-1500 occupy a single rigid platform and view window in the aircraft. As such, the hyperspectral imagery may be georeferenced using the position/orientation data stream from CZMIL. The advanced lidar products and hyperspectral imagery are input to CZMIL Hydrofusion spectral optimization technique that produces land and seafloor reflectance at multiple wavelengths by fusing lidar and hyperspectral imagery.

Figure 10.8 illustrates a unique example of mapping the Springs coast in Florida using the CZMIL system (Karlin et al., 2017). In peninsular Florida, where the soluble limestones of the Upper Floridan aquifer occur close to the land surface and are not covered by confining sands, terrains of the Southwest Florida Water Management District (SWFWMD) have developed into karst landscapes, characterized by sinkholes, sinking streams, underground caverns, and springs. The King's Bay Springs group (aka Crystal River Springs), in

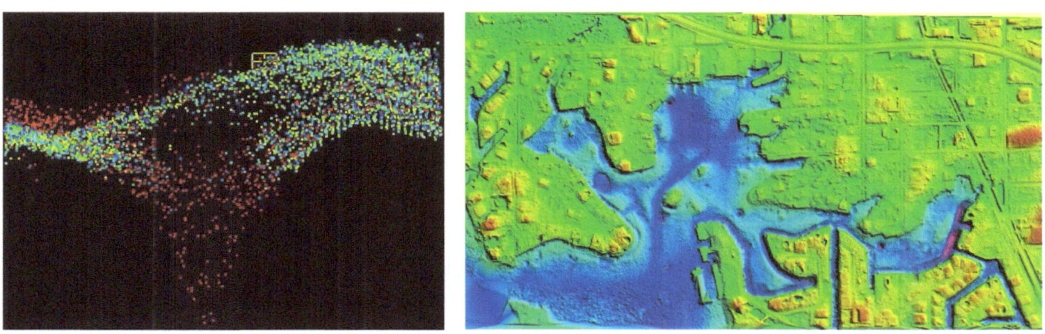

Figure 10.8. Left: A profile across a spring from the CZMIL sensor in King's Bay. Red points are returns from the deep 40-mrad channel that profiles the depth in the spring vent, whereas the other colors represent returns from the seven shallow 2-mrad channels that illustrates the returns from the shallow sea-floor. Right: A topobathymetric DEM produced from the CZMIL sensor with maximum depths of 18 ft.

Citrus County, Florida, has a cumulative discharge of approximately 567 million gallons per day of mostly fresh, warm water that attracts both tourists and the West Indian Manatee population. The nutrients, algae, and debris interact to fill karst depressions with organic-rich materials which require removal to insure the bay's aquatic health. Hence, the District engaged in a survey to determine the volume of material, algae, muck, debris, etc. to be removed from King's Bay. In the Spring of 2015, through a cooperative pilot project with JALBTCX and Dewberry, the CZMIL sensor was used to collect topobathymetric lidar data for a portion of the Kings Bay/Three Sisters springs system. The District and Dewberry discussed the various commercial sensors available to determine the most appropriate sensor to be used for this mission. Deriving reliable bathymetric measurements for this project required a sensor that was capable of mapping shallow bathymetry at high resolution as well as the deeper bathymetry in the spring vent. The system also had to be capable of mapping the depth of the unconsolidated layer to determine the volume of material, algae, muck, debris, etc. to be removed from King's Bay. This requires a relatively short-width pulse that has the ability to reflect off multiple targets over a short distance between the unconsolidated layer and the bottom of the sea floor. The CZMIL sensor was selected for this mission because it has characteristics of high resolution narrow-beam systems as well as broad-beam systems for deeper bathymetry. Figure 10.8 (left) elucidates the benefits of using the CZMIL sensor for this project—it depicts a profile cross-section across a sample spring in King's Bay. The shallow water bathymetry is determined using the seven PMT segments (various colors except red) whereas the deeper bathymetry is obtained using the deep field-of-view PMT (red dots).

LADS (Laser Airborne Depth Sounder)-HD

LADS-HD is the latest generation Fugro LADS technology fielded in 2016. It uses a 3000-pps laser to provide a sounding spacing from 2-4.6 meters in a swath up to 600 m wide from a 400 to 900-m altitude and aircraft speeds from 125-175 knots. Its basic design configuration is the same as LADS with green and IR channels, and the design involves the use of a scanning green beam and a vertically fixed IR beam, both transmitted from a gyroscopically stabilized platform (Penny et al., 1989). The scan pattern is rectangular with a widely-varying nadir angle whose maximum value is 15 degrees. The IR beam is used for surface detection, while the green beam is used for bottom detection and to augment surface detection. The laser footprint size on the water surface is 2.7 m. The LADS-HD system can operate by day and night; operations at night are enhanced by the removal of the day-filter from the receiving optics to improve the optical efficiency. LADS-HD is integrated with a Redlake Mega Plus II High Speed digital camera.

LADS-HD takes advantage of Back 2 Base data processing, a capability for packaging survey data and delivering it to a data processing center. The airborne lidar and imagery data may be positioned using Applanix POS AV or Fugro MarineSTAR WADGPS. LADS-HD has been routinely integrated with the Riegl VQ820 G and HySpex hyperspectral cameras to meet customer requirements for high-resolution topography and shallow water, and hyperspectral imagery. The depth accuracy meets IHO Order 1 standards (Sinclair, 1999). In November 2016, the Fugro LADS sensor was deployed in response to the New Zealand earthquake. At the request of the New Zealand Government, the Royal Australian Navy (RAN) Laser Airborne Depth Sounder (LADS) flight is to conduct a rapid hydrographic survey of the seafloor in the coastal margins of the north east coast of the South Island. The data were primarily acquired to reveal what has happened below the waterline, and identify any shifts in the ocean floor which mariners need to be aware of.

Chiroptera II/Hawk Eye III

Chiroptera II and Hawk Eye III are the recent shallow and deep water bathymetric lidar systems fielded by Leica AHAB in 2013 and 2015, respectively. Chiroptera comprises a green laser for bathymetry that pulses 35,000 time per second, and an infrared laser that pulses 500,000 times per second. The lasers have separate transmit and receive paths. The green laser has pulse energy of 0.1 millijoules, which is an order of magnitude less powerful than bathymetric lidar sensors like SHOALS, CZMIL, and LADS. With a beam divergence of 3 milliradians the on-water laser footprint size is 1.2 m at operational altitude of 400 m. The infrared laser beam divergence is 0.5 milliradians with an on-ground laser footprint size of 0.1 m. Maximum depth penetration expected for this system is 15 m in moderately clear water. The laser pulses are scanned over the ground and water surface using a Palmer scanner which results in an elliptical scan pattern that is 40 degrees across track and 26 degrees along track, and a swath width that is 70% of the operating altitude. Chiroptera II is optionally gyro-stabilized using a PAV-100 mount and maintains a near-constant scan pattern along the flightline.

Hawk Eye III is a Chiroptera II with the addition of a separate laser transceiver for deep water depths. It has characteristics similar to CZMIL in terms of laser power (3 millijoules) and pulse repetition rate (10,000 pulses per second), but has a slightly larger beam divergence so the size of the laser footprint is also larger (4 m). Chiroptera II and HawkEye III both use Novatel SPAN with LCI-100 IMU for positioning and orientation, and are operated to meet both USACE "Class 1" and IHO "Order 1" hydrographic accuracy standards for most applications. Chiroptera II is integrated with a 5 MP camera for QA/QC and may be integrated with an 80-MP RCD30 for higher-resolution imagery. Leica MissionPro and FlightPro provide mission planning and flight operations support, while Leica Survey Studio and FramePro provide lidar processing and image processing. Leica Survey Studio provides automated calibration for all three lasers (Chiroptera infrared and green, HawkEye III green), and uses all three in combination to apply the appropriate processing algorithm based on automated land/water and turbid water detection.

In 2016, Dewberry acquired and processed 205 square miles of airborne topographic and bathymetric, lidar data and developed lidar-derived elevation products for the Beaver Islands Archipelago and South Manitou Island in Upper Lake Michigan for NOAA. These data are being used to enhance coastal forecasting for inundation, seiches, and flooding and provide accurate digital elevation models for modeling, engineering, and planning needs. The lidar intensity data are being used to map aquatic vegetation to support the NPS inventory and monitoring program. The Leica AHAB Chiroptera II was selected for lidar acquisition based on its nearshore depth penetration in moderately clear waters. Figure 10.9 illustrates the Chiroptera II data acquired on and around South Manitou island; note the detail of the geomorphology on the island acquired by the NIR lidar and the greater than 35 ft depth penetration in this lacustrine environment.

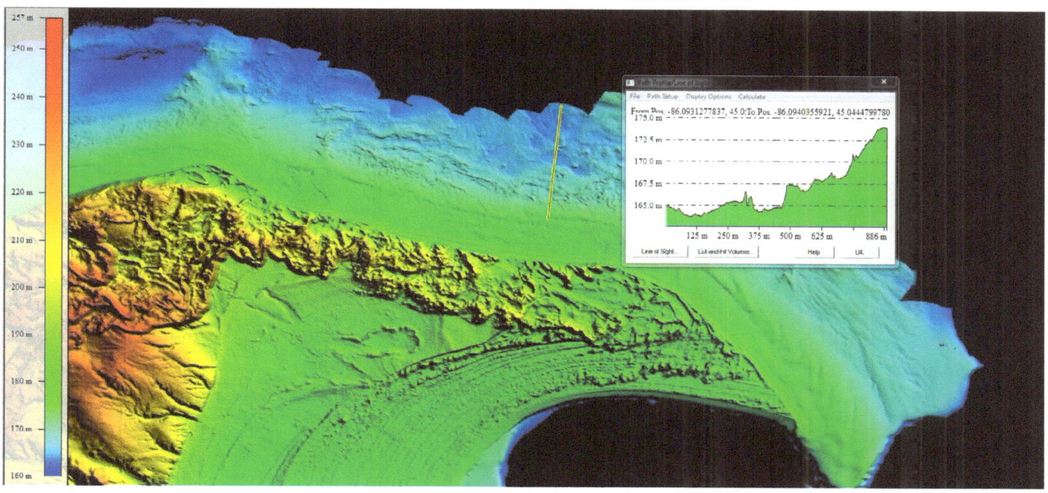

Figure 10.9. DEM produced from the Chiroptera II sensor in South Manitou Island, Upper Lake Michigan.

EAARL (Experimental Advanced Airborne Research Lidar)–B

EAARL-B is an extensive modification of the original NASA EAARL system, both developments led by C. Wayne Wright (Wright et al., 2016). The EAARL-B laser pulses 10,000 times per second, with 0.07 millijoules of power in each 1.3 nanosecond pulse width. Each pulse is split into 3 beamlets that result in 3-30 cm diameter laser footprints on the water surface. The footprints are separated 1.6 m in the along track direction. EAARL-B scans laser pulses across the land and water surface in a raster pattern with 2 m spacing between the beamlets across track and variable incidence angle on the water surface. The scanner is tilted slightly forward to avoid Fresnel reflections at nadir that would saturate receivers in the aircraft. At operational altitude of 300 m, the swath width is 240 m and pulse spacing is nominally 2 m. EAARL-B has four receivers for each laser pulse, one 60-cm FOV PMT receiver for each beamlet and another 5-m FOV PMT receiver that encompasses all three beamlets, and enables EAARL-B to see maximum depths in excess of 40 m in clear water. EAARL-B uses Novatel for positioning and is operated to meet IHO "Order 1" hydrographic accuracy standards for most applications.

EAARL-B data are processed in the open-source Airborne Lidar Processing System (ALPS) developed at NASA and USGS (Nayegandhi et al., 2009). ALPS provides capability to geometrically calibrate EAARL-B data, and apply appropriate depth-induced propagation biases to the shallow water beamlets and the deep channel. ALPS requires the user to interrogate representative waveforms in order to define a water column model for each dataset. After the model is subtracted from waveforms digitized by the system, potential bottom peaks identified based on user-defined thresholds are accepted or rejected based on expected bottom return characteristics like pulse width, rise time, and fall time (Wright et al., 2016). ALPS allows the user to run an Iterative Random Consensus Filter to remove noise points, and to manually delete any remaining noise points (Nayegandhi et al., 2009).

Figure 10.10 illustrates the EAARL-B submerged topography DEM acquired in Fort Lauderdale in April 2014. Depths ranging from 0 to 35 m were reached using a combination of the three shallow beamlets and the 5-m-FOV deep channel. These data were used to calibrate and assess the EAARL-B data.

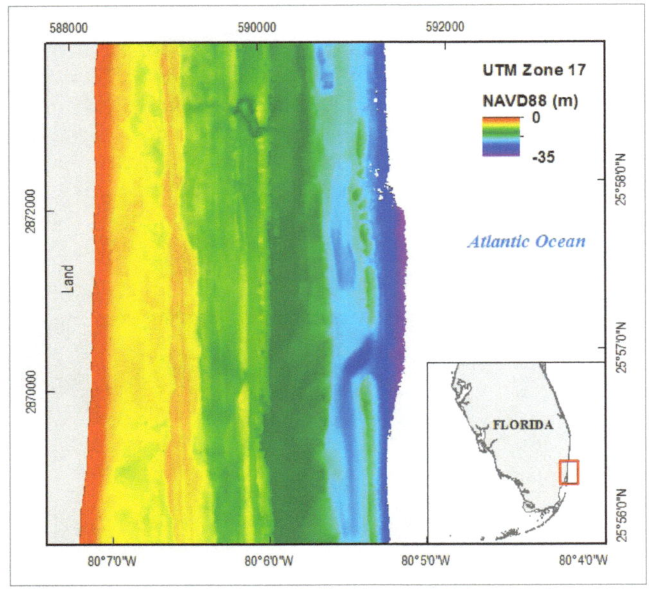

Figure 10.10. EAARL-B submerged topography extent for Fort Lauderdale, FL; data were collected in April 2014 (Wright, et al., 2016).

VQ 820/880-G

The VQ 820-G and VQ 880-G are topo-bathy systems fielded by Riegl in 2011 and 2014, respectively. Both are in operation today. The VQ-820-G was one of the first commercial topo-bathy systems characterized by a high repetition rate fiber laser (485,000 pulses per second), short laser pulse (@ 1 ns), small power per pulse (0.02 millijoules), small beam divergence (1 milliradian), and small receiver aperture (50 mm), (Mandlburger, et al., 2013). The laser pulses are scanned by a multifaceted mirror that describes an elliptical section across the land and water surface. The size of the laser footprint on the water surface at 400m altitude is 0.4 m and the footprint spacing is 0.2 m, up to 10 points per square meter. The VQ 820-G operates solely at the green wavelength and maximum depth penetration is 10 m in clear water. The VQ 820-G utilizes "onboard echo digitization" (reference) techniques for real-time detection of returns on the lidar waveform, rather recording and storing waveforms for post-processing. As such, a refraction correction technique was developed using the lidar point cloud for location of the mean water surface to provide the angular and speed-of-light corrections to lidar point geolocation.

The VQ 880-G adopts a circular scan pattern with an off-nadir angle of 20 degrees and adds an IR channel for improved surface detection. The IR is scanned by a rotating polygonal mirror in a slight arc that falls in the center of the green channel, which is scanned in a circular pattern using a rotating prism. The laser has a slightly higher pulse repetition frequency, up to 550,000 pulses per second, but the pulse power and pulse width are similar to the VQ 820-G. The laser beam divergence is selectable from 0.7 to 2 mrad, as is the scan speed from 10 -200 scans/sec for the green channel. The VQ 880-G also utilizes onboard echo digitization, but has an add-on capability for recording of all waveforms for post-processing, and is integrated with an APC 29 megapixel camera for concurrent aerial photography collection.

The VQ 820-G and VQ 880-G are both delivered with a software suite that includes RiAcquire for acquisition tasks like flight line planning and real-time feedback to the operator, RiWorld for georeferencing, RiProcessing for strip adjustment, visualization, classification, and data export, and RiHydro for water classification, water surface modeling, and refraction correction.

Figure 10.11 depicts a high-resolution DEM acquired along the Eastern Atlantic Seaboard acquired by the VQ 820G sensor for NOAA National Geodetic Survey Remote Sensing Division's Coastal Mapping Program to enable accurate and consistent measurement of the national shoreline following Hurricane Sandy's landfall. The collected data provided seamless topobathymetric data for various applications within the entire coastal community, as well as supported other mapping, charting, geodesy services, marine debris surveys and coastal shoreline modeling for coastal states impacted by Hurricane Sandy. The high-resolution DEM characterized small sand ripples on the sea bottom.

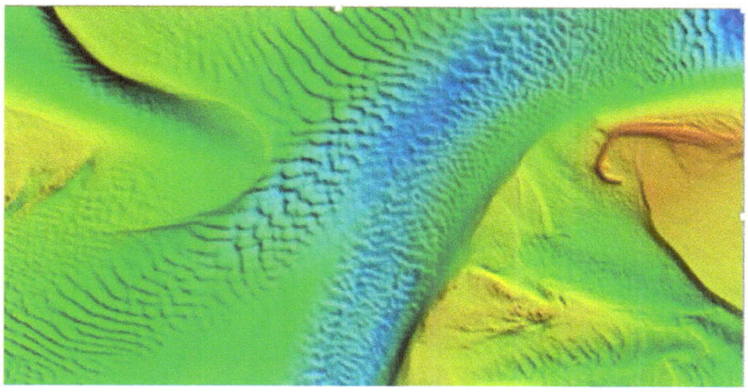

Figure 10.11. Seamless topobathy DEM produced by the VQ820G.

Titan

Titan is the second Teledyne-Optech entrant to the "topo-bathy" market segment following Aquarius. Titan represents the next generation of shallow water mapper, made possible by advances in fiber lasers. Titan is unique in the field of bathymeters as it adds a wavelength in the short-wave infrared region, so it emits and receives light at 532, 1064, and 1550 nanometers. Each channel has a pulse repetition rate from 50,000-300,000 pulses per second, for a measurement rate up to 900,000 measurements per second. The laser beam divergence is 0.35 milliradians for the green and infrared channels and 0.7 milliradians for the short-wave infrared. Operational altitude is 300-600 m for bathymetry and 300-2000 for topography. Each of the three lasers are raster scanned, with the 1064 nanometer beam scanning at nadir, the 1550 nanometer beam scanned at a 3.5 degree forward angle, and the 532 nanometer beam scanned at a 7 degree forward angle. The resulting scan pattern is overlapping sawtooths, with approximately 15 points per square meter per channel, for a total of up to 45 points per square meter. Titan operates primarily in discrete-return mode, but has an add-on capability for recording waveforms for post-processing. Titan depth penetration capability is 15m in relatively clear water, and because of the use of fiber lasers, this capability does not diminish with increases laser repetition rate (similar to the Riegl VQ820G/880G sensors).

Titan is delivered with a set of software modules: Optech FMS for flight planning and execution, Optech LMS for automated calibration, processing and accuracy assessment, and an extension that supports processing of bathymetric data. Titan may be integrated with a medium format 29 Megapixel camera for passive imagery support, or large format 80 Megapixel camera for orthometric aerial photo collection with forward motion compensation.

OPERATIONAL CONSIDERATIONS

Successful ALB surveys require an in-depth understanding of the sensor technology and its operational limitations, sophisticated processing algorithms that provide accurate depth measurements regardless of environmental conditions like the sea surface, sea floor, and water column, and a broad understanding of both topographic and hydrographic surveying and mapping.

Mission planning can be thought of in terms of three epochs: the basic survey plan, detailed scheduling and logistics, and on-site survey management. Given a potential mission requirement, the first agenda involves the development of a plan consisting of a feasibility study, the estimation of logistical considerations, the determination of data requirements, and a cost assessment. The success or failure of a survey can depend on how well it is planned, how those plans are executed, and the ability to deal with unforeseen circumstances.

The feasibility study involves both theoretical and practical aspects. The first determination that has to be made is expectation of success based on depth requirements and expectations of water clarity. For many areas, the water clarity and meteorological factors may vary seasonally and dictate the month(s) appropriate for the survey. In some areas, daily variations in tides and turbidity dictate the most appropriate time within the day for the survey. The next matter is ascertaining the availability of the system at that time and location. Availability conflicts may need to be resolved. Planning a geographical "Grand Tour" can be very important in reducing costs. Customer priorities and requirements must be taken into consideration. If the survey is judged feasible, the next considerations are logistical. The distance from the expected system location to the new survey is determined. The size of the survey area is considered with respect to projected travel costs. The nearest appropriate airport to the survey site is located and contacted. The availability of suitable facilities and equipment for aircraft and personnel are evaluated. Special features such as mountains and potential air space restrictions or conflicts will be factored in.

Discussions with the potential sponsor about data requirements will have indicated whether the survey is to be a bathymetric survey or a topo-bathymetric survey and defined the associated data density, vertical and horizontal accuracy, and small target requirements. Commercial sensors that are available today have varied performance characteristics; a "horses for courses" approach is often used to determine the best sensor for a particular application. The need for surf-zone, shoreline, or topographic products will have been identified. Known features of particular emphasis will have been denoted. For a DGPS survey, the source of differential correctors (wide area or local) must be determined. The number of base stations needed is determined, availability of control monuments and locations for base stations is identified, and the possible need for a control survey must be considered. Experience has proven that things are not always where they were thought to be. Finally, a cost assessment is made based on the logistical considerations, data requirements, and perceived risks.

The next step is to determine schedule and detailed logistics for items such as aircraft and equipment needs, environmental considerations, transportation requirements, personnel assignments, office space, and housing arrangements.

The third and final process can best be categorized as survey management. This involves the implementation of the plan, mobilization of equipment and personnel, establishment and manning of control and water level stations, conducting needed system calibrations, day-to-day selection and planning of specific survey areas and laying out flight lines consistent with data requirements, monitoring weather and water clarity, management of flight times to avoid sun glint and meet the needs of the air crews, coordination with tidal cycles if desirable, reacting to winds, weather, and wave conditions, selection of alternate sites, planning reflights, modification of plans as necessary, selection of data processing protocols, evaluation of data quality, production of the required output products, maintenance of the equipment, conducting necessary remedial actions to ensure the successful completion of the survey, and demobilization. This is a complex and demanding task that requires a great deal of both expertise and experience.

Costs are important at two times: when a job is being planned and bid, and when the job is done. A cost estimate prior to a lidar survey is based on two generic factors—logistics and risk. Once a price bid or cost forecast is accepted, the ultimate survey cost depends on planning, logistics, survey management, and luck. Contract surveys are typically bid at fixed prices; the price must also include factors for profit and recovery of initial investment. Contract surveys with higher environmental risks must be bid at higher rates with the understanding that actual profit margins will vary from mission to mission according to uncontrollable circumstances.

Risks which must be considered in survey planning and bids involve the statistical consideration of the uncontrollable factors such as the expected effects of water clarity, wave heights, local and regional weather, reflight probability, and time on site. Water clarity varies with location, season, tidal phase, and weather. Water clarity problems can be managed by choosing the optimum month for a given location and, once on site, by selecting the best time of day and by having alternate targets in neighboring areas with different characteristics—such as two sides of an island, or protected bay areas compared to exposed offshore areas. Weather varies on both daily and seasonal time scales and, as with water clarity, can be managed by selecting optimum seasons and times of day, and having alternate targets. Some days will be lost to weather. Those days are rarely wasted because they are generally well spent in performing needed system maintenance, planning future surveys, and catching up on data processing. For hydrographic surveys, areas known to have coral heads and fish schools or high winds will require a higher percentage of reflights to resolve anomalies and holidays. Time on target may be affected by winds. Flight lines may have to be flown in only one direction, for example. In some congested airspace near urban areas, time can also be lost to the aircraft "loitering" while awaiting clearance from the air traffic controller to access the survey area. In some cases, daytime flights may be proscribed altogether, requiring a shift to nights where, in some cases, increased dangers near land could reduce production rates.

Lidar survey operation and maintenance (O&M) costs depend strongly on project characteristics. Survey scenarios and costs can vary widely. The characteristics with the largest effects on costs are project location, size, horizontal density requirements, survey accuracy "order", the possible requirement to resolve all questionable points by reflying, the physical shape of survey area and if it includes surf zone or high cliffs, the positioning method employed, and extant environmental conditions on site. Many of these same factors apply to sonar costs.

Two aspects of project location are important: the distance of the airfield to be used from the preceding mission site and the distance of the project area from the airfield being used. Transit (mobilization) costs can be a significant fraction of the total. Greater costs are obviously incurred by flying halfway around the world than by flying to the next county or state. For this reason, it is important to be able to plan ahead as far as possible and plot a "grand tour" route in which surveys fall in logical geographical and environmental progression. On a single continent, survey crews can drive to their next location, but for travel over water, regardless of the size of the survey aircraft, many of the survey crew fly to their destinations by commercial air carriers. A great deal of computer and positioning hardware, spares, and other equipment must also be transported. This greatly increases the costs for distant locations. The size of the survey must be large enough to amortize these costs. Simply put, large areas are cheaper to survey per unit area than small areas. Interestingly, however, increased costs at remote areas are sometimes nevertheless a cost advantage, because some sites may be so remote or dangerous that they cannot practically be accessed other than from the air. In such a case, greater costs may be justified.

Once on site, the project location with respect to the aircraft's base of operations has a large influence upon the survey cost. Each aircraft type has specific flight endurance, and the time spent in transit to reach and return from the survey area reduces the time available for data collection. Aircraft time in the air is money, whether it is in transit to the project location, to the survey site, or in actual surveying.

For a given system, the spatial data density required is another strong contributor to the project cost. Most current lidar systems offer a means of decreasing or increasing "spot" density from their standard value, in order to deal with special circumstances. Spot spacing is typically controlled by varying the coverage swath width of the aircraft flight line and the ground speed of the aircraft while collecting data. There is a direct correspondence between required survey density and the required aircraft time and cost to complete a survey.

Economics dictate that surveys should be performed at the lowest density acceptable to achieve the desired products. If surveys are required to prove or disprove the existence of small objects with high probability, the data processing time and refly percentage will also increase.

The physical shape of the project is also a large factor in cost determinations. Generally speaking, large, rectangular shaped, open-water areas are the most efficient for ALB surveys, and irregular shorelines tend to be least efficient. Data collection plans over a specific survey area can be broken into two parts: Collection lines and line changes (time in turns). The most efficient survey areas incorporate the longest possible collection lines, within reason, and the least number of lines. The efficiency of the survey is best indicated by the ratio of time "on-line" to time "off-line"; the higher this number the more efficient the survey. Another consideration for near-shore areas is whether the flight lines are flown parallel with the shore or perpendicular. Various logistical concerns may dictate one or the other. Flights perpendicular to the shore are less efficient because they spend more time over land. This may not necessarily be true, however, if topographic heights over land are part of the product. Areas with large surf zones are less efficient because the large areas of white foam can require reflights. High cliffs adjacent to the shore can disrupt desired flight patterns for shoreline surveys, and the winds associated with them can be dangerous to the safe operation of the aircraft. It is noteworthy that a factor that is not very important to ALB cost considerations is depth, because the lidar swath is constant. This is very different from the situation with multibeam sonar for which the swath decreases as the depth decreases, and the shallow-water sonar costs soar.

Additional items affecting costs are the positioning system used and the environmental conditions. The use of wide-area differential GPS (WADGPS) or Precise Point Positioning (PPP), if available, is less expensive than placing and manning differential stations on the ground. The use of kinematic GPS, if needed, requires more equipment, more work on the ground, and more data processing. Finally, as noted above under risk management, it is well known that a certain number of days are going to be lost to rain, turbid water, high waves, and high winds. The associated costs are included as part of risk assessment. Savvy survey management and the availability of alternate survey areas can reduce the associated losses.

CONSIDERATIONS FOR PLANNING AN ALB SURVEY

This sections provides a guide to planning an ALB survey. It is intended to give the reader an insight into the complexities involved with bathymetric and topobathymetric lidar surveys and the various factors that must be considered for a successful survey and ensuing data products.

- Define the survey area and intent. A discussion about an ALB survey project begins with an outline of the desired survey area. ESRI shapefiles, Google Earth kml/kmz files, web-based collaboration tools, or even hand-drawn boxes on a map or aerial photo are effective means of providing this information. The flight time required to cover the survey area is one of the major cost drivers for any airborne survey, and ALB is no exception. A common mistake is to draw a very general box or outline, and then find out the cost of the survey, and then go through iterations to get to the final survey area. Though some iteration is expected, starting with a more specific outline of the survey area will reduce the amount of time required to finalize the survey area. A key point to remember is that each flight line, or pass of an ALB sensor, will cover an area about 300 m (1,000 ft) wide under the aircraft at the nominal flying altitude, so that is the level of granularity the project outline can accommodate. A brief description of the intended use of the final data set is also valuable information and may shape some unexpected aspects of the data collection
- Determine the schedule: Several factors will determine when an ALB survey will be performed, primary among them weather, water turbidity, ALB sensor availability, mobilization costs, and airspace restrictions. Best practice is to start discussion early, as soon as ALB is considered to meet project survey requirements. This will ensure quality data are collected and available when needed, and can result in significantly reduced cost for the project

- Weather: Knowledge of general weather patterns for the project area can help identify the optimal time for aircraft operations. Frequency of low pressure storm systems, and daily patterns like afternoon thunderstorms, fog, and marine layers can all impact aircraft operations. Targeting times of the year when these are lowest decreases overall risk of the project, and therefore reduces cost.
- Turbidity: The single most important consideration for success of an ALB survey project is turbidity, or the amount of suspended sediment and organics in the water column. If the water has a high content of sediment and/or organic particles, the photons in each laser pulse will be scattered and/or absorbed to such an extent that an insufficient amount of light returns to receivers in the aircraft to make an accurate depth determination. Local knowledge of turbidity and its drivers in the survey area is key to scheduling an ALB survey for the greatest chance of success. Answers to the following questions help determine the best time to perform an ALB survey:
    - When is the water clearest?
    - What time of year has the lowest winds? Does a certain wind direction increase/decrease suspended sediments in the water?
    - What time of year has the lowest waves, both wind waves and swell?
    - What time of year has lowest rainfall?
    - Are algal blooms a concern? At what time of year?
    - When does submerged aquatic vegetation have the highest biomass?
    - What time of year is river discharge lowest?
    - When is tide range the largest, or when are spring and neap tides?
    - In areas impacted by riverine discharge, do incoming tides push cleaner water nearshore?
    - For river surveys, what water levels and flow conditions result in clearest water?
    - Can river flow be controlled to improve water clarity?
    - What time of year are runoff from rivers, river discharge volumes, effect of flooding and farming activities lowest?
    - Are there ongoing O&M or construction activities that impact water clarity, like channel dredging or sand placement in the nearshore?

- Availability and selection of ALB sensors: Though ALB technology is not new, the entry cost and operational risk of the ALB market are high. As a result, there are not many systems available to perform survey work. Depending on the project requirements, sensor availability and cost, a "horses-for-courses" approach should be used to determine which sensor is most appropriate for the project. Providers with airborne coastal mapping and charting experience that have well-established standard operating procedures and processing pipelines are lowest risk in terms of cost, schedule, and data quality.
- Mobilization cost: For a typical project-scale survey, the cost of mobilizing the ALB sensor and crew to the survey area is often higher than the cost of the ALB survey. To reduce cost of mobilization, it is advantageous to leverage other ALB surveys that may be planned in the area. For example, if several survey requirements in a geographic area can be combined into a single survey effort, the mobilization cost can be shared among the requiring projects
- Airspace restrictions. Survey areas near large airports, military bases and power plants may require coordination in advance of airborne operations. Designated wildlife and sanctuary areas also often require permits for aircraft operations. Managers of these areas should be included early in the discussions to ensure aircraft operations are permitted and occur according to the defined survey schedule. Finally, it is always best practice to engage local officials in the area in which the survey

will occur, as low-flying aircraft occasionally excite the general public. ALB operations can occur in the day or night which can help in navigating through some airspace restrictions; informing the local officials may be needed as the green laser is visible in the night from the aircraft

COMPARISON WITH OVERLAPPING TECHNOLOGIES

One of the most useful features of ALB is to bridge the topographic and marine survey worlds. Traditionally, the hydrographic community and the topographic community have operated in their own way—the hydrographic community has focused on sonar/acoustic based methods whereas the topographic community has used lidar and even photogrammetry to derive land elevations. The hydrographic community has traditionally based its elevations on the tidal datum whereas the topographic community has used orthometric datums (see Chapter 2). The difference in reference datum manifests most at the land/water junction, often due to the varying definition for "shoreline" which can have significant scientific, engineering, social, economic, legal, and safety implications. ALB technology is the capability to fill in the coastal gap between topographic and bathymetric data collection (Figure 10.12). With the advent of VDatum, a free software tool developed by NOAA to vertically transform geospatial data among a variety of tidal, orthometric and ellipsoidal vertical datums - users can convert their data among different horizontal/vertical references into a common system, enabling the fusion of diverse geospatial data in desired reference datums. In this section, we explore the various other technologies that are used for mapping submerged topography.

Figure 10.12. Airborne lidar bathymetry is more efficient for mapping nearshore areas and is complementary to boat-based acoustic technology that can map in deeper waters.

Bathymetry Derived from Imagery

Multispectral satellite and aerial imagery can provide estimates of bathymetry that may be suitable for certain environmental and scientific applications over near-shore areas where sub-meter accuracy is not needed. This method may not meet IHO hydrographic accuracy standards, but it can provide a meaningful bathymetric model if lesser accuracy can be tolerated. Deriving bathymetry from imagery is an inferred technique based on the principle that different wavelengths of light are attenuated by water to differing degrees depending on water depth. Bathymetry is estimated by two main approaches: (1) Using a physical approach that requires data related to water composition, nature of the sea bottom, and atmospheric conditions to determine the physical processes affecting pixel values, or (2) An empirical approach that uses control data (known bathymetry) to calibrate the image. Often, both approaches are used together to derive water depth from imagery. The number of wavelengths in the blue-green spectrum of the source imagery enables better depth penetration and more accurate bathymetry. Low-resolution satellite imagery such as LANDSAT and MODIS produce bathymetry at an accuracy of 25-250 m whereas the higher resolution satellite imagery such as GeoEye1 and WorldView2 produce bathymetry at an accuracy of 3.5-6.5 m at 95% confidence level. WorldView2 is the most suitable for deriving detailed near-shore bathymetry as it contains more spectral bands in the blue-green wavelength and a high spatial resolution. Satellite derived bathymetry has a significant cost and acquisition advantage over airborne or boat-based platforms. Substantially large areas can be mapped in much less time and costs, albeit at poorer resolution and accuracy. The quality and accuracy of satellite derived bathymetry is highly dependent on depth and seafloor type, and the presence of clouds in the imagery precludes depth extraction entirely.

In many coastal areas, video cameras are used to measure coastal processes for scientific research. Pioneered by Oregon State University, a team of universities and research institutes around the work created a global network of Argus monitoring stations. An Argus station usually consists of 3-6 synchronized cameras that cover a stretch of beach up to 5 kilometers centered at the station. Cameras are typically mounted on a tall building or other structure. Argus imagery can be used to determine shorelines, bathymetry, wave period and direction, and long-shore surface currents. Not all measurements are possible in all conditions. The major advantage of Argus is its long-term presence on the coast. Some sites have been collecting data for almost 20 years. For many areas of the world, the bottom is not visible either due to turbidity or bubbles in the surf zone so methods that rely on "seeing" the bottom from imagery do not work and depth must be estimated through the use of ocean surface observables such as those acquired by Argus. Using signal processing techniques, the cBathy algorithm has been tested to provide robust bathymetry maps based solely on long time series observations of surface wave motions (Holman et al., 2013). The cBathy algorithm, based on studies conducted at Duck, North Carolina, estimated errors less than 0.5 m. The bias and RMS error of the cBathy estimate, computed over the entire area from shoreline to 500 m offshore were 0.28 and 0.44 m, respectively. Bathy performance was found to vary with depth. The mean bias and RMS error for the entire data set, binned by depth. Both bias and RMS error were largest in the shallowest depth bin of 0–1 m and rapidly improved in intermediate depths before worsening in deep water.

Maritime Vessel-Based Acoustic Bathymetry

Waterborne acoustic ranging technology has been commercially developed and deployed since the 1970s and its advances have paralleled the advances in ALB. See Chapter 11 for greater details.

Single-Beam Echo Sounder (SBES)

SBESs provide data along the profile delineated by the vessel's track. Surveying using a SBES requires large amounts of ship time to provide dense enough data to characterize an area, and they always include large gaps in the areas between the profiles. A SBES is typically deployed on small vessels in the near-shore areas. SBESs are used extensively for navigation channel surveys, where cross-sections are collected at prescribed stations along the channel length, and for beach profile surveys, or cross-section from the beach out into the water where the land and wading depths are collected using land based survey techniques (rod and level, total station, KGPS), and the deeper depths are collected using SBESs on small boats or sometimes jet skis. SBESs are increasingly being used to provide spot heights to calibrate other bathymetry acquisition techniques such as satellite-derived bathymetry, or provide a comparison dataset for ALB. SBESs are also used to fill in gaps from surveys conducted with ALB caused by lack of water clarity or depths limits of the lidar sensor.

Multi-Beam Echo Sounder (MBES)

MBESs are the most efficient and reliable sensor to acquire bathymetry from a maritime vessel platform. MBES systems have the advantage of producing a higher measurement density than airborne sensors. The increased point density makes MBES more suitable for small object detection, especially when used in conjunction with a side-scan sonar. Although a side-scan sonar only measures seabed backscatter, it provides a superior method to all other sensors for detecting small, off-nadir objects with a high level of confidence. MBES systems are very inefficient in shallow water as swath width is highly dependent on water depth (Figure 10.1 on page 276). The degradation in performance in very shallow waters is further accentuated by the need to compensate for smaller footprints (resulting from the narrow beam angle) by slowing the speed of advance to ensure overlap of footprints along track. The accuracy of MBES systems is dependent on the correction applied for the vessel and sensor position and motion, especially in higher sea states. As with airborne Lidar, the errors associated with the inertial sensor (vessel motion and orientation) and DGPS sensor (vessel location) contribute significantly to the overall accuracy of the acoustic data. MBES surveys can generate bathymetry to the same or better resolution and accuracy as ALB and is often used to complement ALB (Figure 10.12). For a large-area survey, MBES is often used following a bathymetric Lidar acquisition to fill voids in the ALB data due to water turbidity, seabed reflectivity, and maximum depth limitations. ALB provides data in areas that MBES is unable to map due to vessel draft (shallow areas <2 m) and can identify hazards to survey vessels.

GPS Profile Surveys

In the very near-shore coastal and riverine areas, GPS profile surveys are often conducted to measure the topography under water. It is the most localized technique for acquiring bathymetry. A pole-mounted GPS system is typically used by wading through the water to acquire data along profiles that are spaced several hundred meters apart. A nearby GPS base station is deployed by the operator for differential GPS corrections. GPS profile surveys are commonly used to map shallow rivers and mountain streams for fisheries management. However, there are several limitations to using GPS profile surveys for mapping rivers. Private property creates access restrictions along the river in several areas. There can be substantial safety issues due to high water velocities and suspended sediment causing lack of visibility of the stream bed during the time of year the field surveys are performed. Areas of heavy tree canopy limit the accuracy of GPS measurements. The topography of the stream bed is determined by interpolating across profiles that can be separated by several hundred meters. Also, the water depths are limited by the wade-able area and the length of the GPS pole. ALB and topobathy lidar are good alternatives to GPS profiling in many instances, especially when the water column is clear. Figure 10.13 illustrates the comparison of GPS profiling surveys with airborne topobathymetric lidar data acquired along a section of the Sandy River in Oregon using the Riegl VQ820G sensor. The topobathy bare-earth DEM produced from lidar (upper right) shows a continuous and seamless surface that would not be possible to create using any other remote sensing technique. The comparisons of the three GPS profiles with the lidar data within 2 meters of the GPS points show the lidar points (in black) following the GPS profile data points (in magenta) very closely, and the lidar points are also denser along the profile than the GPS profiles. The lidar profile

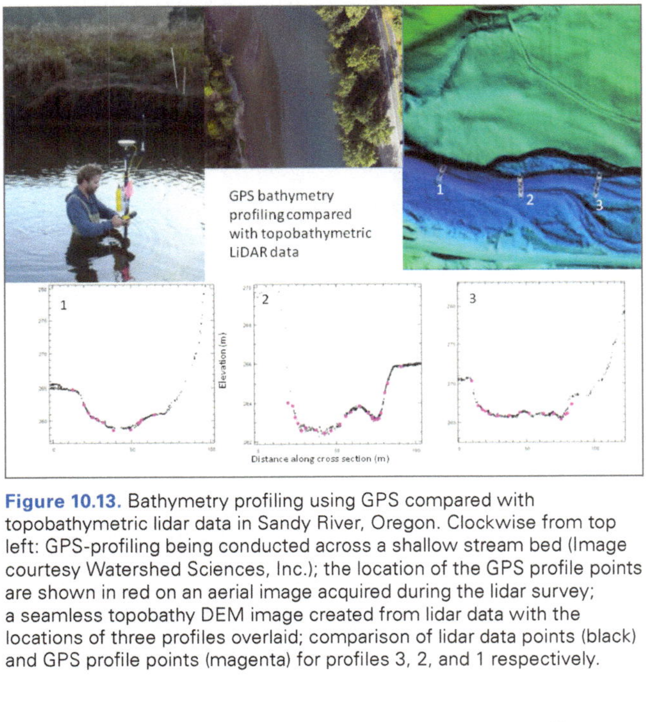

Figure 10.13. Bathymetry profiling using GPS compared with topobathymetric lidar data in Sandy River, Oregon. Clockwise from top left: GPS-profiling being conducted across a shallow stream bed (Image courtesy Watershed Sciences, Inc.); the location of the GPS profile points are shown in red on an aerial image acquired during the lidar survey; a seamless topobathy DEM image created from lidar data with the locations of three profiles overlaid; comparison of lidar data points (black) and GPS profile points (magenta) for profiles 3, 2, and 1 respectively.

also extends over land seamlessly. This example illustrates how airborne topobathymetric lidar is a very good tool for mapping shallow stream beds in relatively clear waters.

SUMMARY

There are now a wide variety of bathymetric sensors to choose from, the broad-beamed, deep water SHOALS, LADS-HD, HawkEye III, and CZMIL, or the narrow-beamed, shallow water Chiroptera, VQ 820/880-G, EAARL-B, and Titan. In practice, the deep water systems SHOALS and LADS-HD are often flown together with the VQ 820/880-G to provide high-resolution topography and nearshore bathymetry, in much the same way that HawkEye II combines the capability of a deep water sensor with the high-resolution topography and shallow nearshore of Chiroptera. CZMIL uses a segmented detector to increase spatial resolution on land and in shallow water. Most of the sensors have integrated cameras for QC, and are integrated with production camera and hyperspectral imagers on a routine basis. Titan provides a unique active-imaging capability that can be used for multispectral classification independently of a passive image source.

Every service provider, whether government agency or private sector company, is addressing a specific client or constituent group, and the performance expectations of a given ALB system are defined by the product specifications of the client. Although the primary product for most ALB surveys is bathymetry, the environmental conditions may vary and survey configurations may differ among service providers. Technical and economic considerations limit the capacity of the ALB system to perform all the expected tasks at the same level of performance. Thus, prioritizing the specifications of the ALB system according to the specific mission is important.

As an example, two U.S. government agencies that use ALB systems for coastal mapping have different requirements. The mission of the USACE National Coastal Mapping Program (NCMP) is to provide repetitive, regional, high-resolution, high-accuracy, seamless bathymetric-topographic data to support regional sediment management among coastal navigation, flood damage risk reduction, and ecosystem restoration projects. The program focuses on data collection where sediment is moving the most: Along sandy coasts, and from the onshore dune system to the depth of closure (an offshore depth beyond which the depths do not change with time). USACE has broader data requirements on other types of coasts with a more extensive cross-shore expression than are addressed by other programs. These data, and a series of information products derived from them, are used to characterize engineering, environmental, and economic conditions along the shoreline, and their changes over time (Wozencraft, 2014). USACE's mission also includes a requirement to monitor changes that occur naturally or due to manmade construction, which is typically over sandy areas. The USACE survey requirements are that the ALB system should be able to acquire data at a spot spacing of 2 m × 2 m with 100% coverage (i.e., no gaps in the survey flight plan) and achieve 30 cm accuracy at the 95% confidence level (Wozencraft, 2014) so that data can be compared from one survey to the next (nominally 5 years between surveys) to produce meaningful shoreline and volume change products.

On the other hand, the goals of the NOAA ALB survey missions are to provide accurate, consistent, and up-to-date bathymetry and shoreline around the coasts of the U.S. and its territories. The survey products need to meet international hydrographic survey standards, such as the IHO S-44 survey standards (IHO, 2008). Accordingly, ALB survey data for hydrographic charting in NOAA are expected to: 1) fill in the data gap shoreward of the navigable area limit line (NALL) (0 to 4 m), and 2) overlap with surveys collected using sonar systems (e.g. multibeam, side scan or single beam) from 4 m to 10 m with a 50 cm accuracy at the 95% confidence level (Imahori et al., 2013). Many of coastal areas surveyed by NOAA include muddy, sandy and rocky coastal regions. The NOAA survey requirements are that the ALB system should be able to acquire data up to a depth of 10 m (where possible) at a minimum spot spacing of 2 m × 2 m with complete coverage along the shoreline and often during two separate tide cycles.

Theoretically, ALB systems can perform tasks mentioned above successfully. However, variations in environmental conditions and sensor characteristics limit the ALB performance. A narrow-beam ALB will not provide successful and continuous bottom detection at depths greater than 10 m and some broad-beam

ALB systems alone may not provide spot spacing smaller than 2 m × 2 m. Thus, the service provider needs to identify and prioritize the tasks that the ALB system is expected to perform. Selecting the most appropriate sensor for each job is a very important decision that is typically based on project requirements, intended application, and sensor availability. Cost is also often a factor. The cost of bathymetric lidar acquisition and processing can vary significantly depending on the type of aircraft and sensor used for each project.

The future is bright for consumers of bathymetric and topo-bathymetric lidar. Researchers are focused on rigorous total propagated uncertainty (TPU) models for the new systems, and on improved automated processing that reduces time to get to final product. Current bathymetric lidar sensor providers are constantly upgrading and improving their systems. Prototype sensors for UAS include the Riegl BDF-1, a bathymetric profiler for low-altitude operation, and the Arete PILLS sensor, which is a streak-tube imaging lidar designed for typical bathymetric lidar operational parameters. Capabilities of single photon-counting lidar for bathymetric application are under assessment by several organizations.

AUTHOR BIOGRAPHY

Jennifer Wozencraft is a Research Physical Scientist in the Coastal and Hydraulics Laboratory of the U.S. Army Corps of Engineers (USACE) Engineer Research and Development Center, and Director of the Joint Airborne Lidar Bathymetry Technical Center of Expertise (JALBTCX). At JALBTCX, she coordinates operations, research, and development in airborne coastal mapping technologies among USACE, Naval Oceanographic Office, National Oceanic and Atmospheric Administration, and U.S. Geological Survey. She also manages the USACE National Coastal Mapping Program, which provides regional scale, engineering-accuracy elevation, depth, and imagery data to support USACE regional sediment management, navigation, environmental restoration, regulatory enforcement, asset management and emergency response activities in the coastal zone. Ms. Wozencraft represents USACE as co-chair of the Interagency Working Group on Ocean and Coastal Mapping. Ms. Wozencraft has worked in the field of lidar bathymetry for 25 years, beginning with the USACE SHOALS program while attending college. She earned her Bachelor's degree from the University of Alabama in Mathematics and Dance, her Master's in Marine Sciences from the University of South Alabama, and her International Hydrographic Office Category "A" Hydrographic Surveyor Certificate from the University of Southern Mississippi and Naval Oceanographic Office Joint International Hydrographic Applied Science Research Program.

Amar Nayegandhi is Vice President and Director of Remote Sensing at Dewberry. He oversees the geospatial and technology services line for Dewberry's contracts with federal, state, and commercial clients. With over 19 years of experience, he is a recognized expert in topographic and bathymetric lidar data acquisition and processing. He has authored 15 refereed manuscripts and over 65 reports for the U.S. Geological Survey. Amar has a Bachelor's degree in Electrical Engineering from University of Mumbai and a Master's degree in Computer Science from University of South Florida. He is the Director of the ASPRS Lidar Division (2018–2020) and is an ASPRS Certified Photogrammetrist and Certified Mapping Scientist—Remote Sensing.

ACKNOWLEDGEMENTS

The authors acknowledge significant contributions from the "Airborne Lidar Bathymetry" chapter, authored by Gary Guenther, in the DEM Users Manual 2nd Edition. We also acknowledge contributions made by Shachak Pe'eri, and Viktor Feygels who are authoring chapters in the new edition of the Airborne Laser Hydrography Blue Book (currently in review).

REFERENCES

ANSI (2005). *IEEE standards interpretations for IEEE Std C95.1™*-2005 IEEE standard for safety levels with respect to human exposure to radio frequency electromagnetic fields, 3 kHz to 300 gHz.

ASPRS (2013). *LAS specification Version 1.4—R13;* Bethesda, Md: American Society for Photogrammetry and Remote Sensing, 15 July, 2013.

Axelsson, R. & Alfredsson, M. (1999). Capacity and capability for hydrographic missions. In *Proc. U.S. Hydrographic Conference,* April 26-29, Mobile, AL, (paper 9-4 on CD), 9 pp.

Banic, J., Sizgoric S., & O'Neil, R. (1986). Scanning lidar bathymeter for water depth measurement. In *Proc. SPIE Laser Radar Tech. and Appl.,* Vol. 663, Quebec City, Quebec, 187-195.

Bible, R. (2016). *The Evolution of SHOALS-1000 into the next general ABL: lidar, yesterday and today.* JALBTCX Annual Conference, 2016.

Billard, B. & Wilsen, P.J. (1986). Sea surface and depth detection in the WRELADS airborne depth sounder. In *Appl. Opt. 25,* 13, 2059-2066.

Cottin, A.G., Forbes, D.L., & Long, B.F. (2009). Shallow seabed mapping and classification using waveform analysis and bathymetry from SHOALS lidar data. In *Canadian Journal of Remote Sensing* Vol. 35, Iss. 5.

Cox, G. & Munk, W. (1954). Measurement of the roughness of the sea surface from photographs of the sun's glitter. In *JOSA,* 44(11): 838-850.

Crawford, G.B. & Farmer, D.M. (1987). On the spatial distribution of ocean bubbles. In *J. Geophys.* Res. 92, 8231-8243.

Cunningham, G., Lillycrop, W.J., Guenther, G.C., & Brooks, M.W. (1998). Shallow water laser bathymetry: accomplishments and applications. In *Proceeding Oceanology International: The Global Ocean,* March 10-13, Brighton, England, Vol. 3, pp. 277-288.

Curran, T., Keck, T., Contarino, V.M., Harris, M.M., & Haimbach, S.P. (1988). Digital ABS laser sounder bathymetry. In *Proc. SPIE Ocean Optics IX,* Vol. 925, 242-249.

Enabnit, D.B. (1981). An evaluation of airborne laser hydrography. *In Proc. 4th Laser Hydrography Symposium, ERL-0193-SD,* Defence Research Centre Salisbury, Sept. 30–Oct. 3, 1980, Salisbury, South Australia, 85-95.

Engstrom, R. & Axelsson, R. (2001). Laser bathymetry and its compliance with IHO S44. In *Proc. Hydro 2001,* The Hydrographic Society, Special Pub. 42, March 27-29, Norwich, England, Paper 18, 10 pp.

Feigels, V.J. and Kopilevich, Y.I. (1993). Russian airborne lidar for oceanography. In *Proc. Symp. on Russian airborne geophysics & remote sensing, SPIE Vol. 2111,* Sept. 13-17, 1992, Golden, CO, 127-141.

Fernadez-Diaz, J.C., Carter, W.E., Glennie, C., Shreshtha, R.L., Pan, Z., Ekhtari, N., Singhania, A., Hauser, D., & Sartori, M. (2016). Capability assessment and performance metrics for the Titan Multispectral Mapping Lidar. In *Remote Sensing,* Vol 8, Issue 8, 936; doi:10.3390/rs8110936.

Feygels, Viktor I., Kopilevich, Yuri I., Surkov, Alexey I., Yungel, James K. & Behrenfeld, Michael J. (2003). Airborne lidar system with variable-field-of-view receiver for water optical properties measurement. In *Proc. SPIE 5155, Ocean Remote Sensing and Imaging II,* November 10, 2003; doi:10.1117/12.506976.

Fuchs, E. & Mathur, A. (2010). Utilizing circular scanning in the CZMIL system. In *Proc. SPIE 7695, Algorithms and Technologies for Multispectral, Hyperspectral, and Ultraspectral Imagery XVI,* 76950R (2010/05/13); doi: 10.1117/12.851905.

Goodman, L.R. (ed.) (1976). Laser hydrography technical review workshop minutes, August 25-26, 1976, Rockville, MD, National Oceanic and Atmospheric Administration, Rockville, MD, 127 pp.

Gordon, H.R. & Wouters, A.W. (1978). Some relationships between secchi depth and inherent optical properties of natural waters. In *Appl. Opt., 17(21):* 3341-3343.

Gordon, H.R. (1982). Interpretation of airborne oceanic lidar: effects of multiple scattering. In *Appl. Opt., 21(16):* 2996-3001.

Gordon, H.R. (1989). Can the Lambert-Beer law be applied to the diffuse attenuation coefficient of ocean water? In *Limnol. Oceanogr., 34(8):* 1389-1409.

Guenther, G.C. (1985). *Airborne laser hydrography, system design and performance factors.* Presented at the NOAA Professional Paper Ser. NOS1, Rockville, MD, 1985.

Guenther, G. C. (2007). Airborne lidar bathymetry. In: Maune, D. F. (ed.). *Digital Elevation Model Technologies and Applications: The DEM Users Manual,* 2nd edition. Bethesda, Maryland, American Society for Photogrammetry and Remote Sensing. pp. 253–320.

Guenther, G.C., Brooks, M.W., and LaRocque, P.E. (1998). New capabilities of the SHOALS airborne lidar bathymeter. In *Proc. 5th Int'l. Conf. on Remote Sensing for Marine and Coastal Environments,* ERIM International, October 5-7, San Diego, CA, Vol. I, 47-55. (Reprinted in 2000, Remote Sens. Environ. 73, 247-255.)

Guenther G.C., Cunningham, A.G., LaRocque, P.E. & Reid, D.J. (2000). Meeting the accuracy challenge in airborne lidar bathymetry. In *Proc. 20th EARSeL Symposium: Workshop on Lidar Remote Sensing of Land and Sea,* European Association of Remote Sensing Laboratories, June 16-17, Dresden, Germany, (paper #1 on CD), 28 pp.

Guenther, G.C. & Goodman, L.R. (1978). Laser applications for near-shore nautical charting. In *Proc. SPIE Ocean Optics V,* Vol. 160, 174-183.

Guenther G.C. & Mesick, H.C. (1988a). Automated lidar waveform processing. In *Proc. U.S. Hydro. Conf. '88, Spec. Pub. 21,* April 12-15, Baltimore, MD, 52-59.

Guenther G.C. & Mesick, H.E. (1988b). Analysis of airborne laser hydrography waveforms. In *Proc. SPIE Ocean Optics IX,* Vol. 925, 232-241.

Guenther G.C. & Thomas, R.W.L. (1981). Error analysis of pulse location estimates for simulated bathymetric lidar returns. In *NOAA Tech.* Rpt. OTES 01, National Oceanic and Atmospheric Administration, Washington, D.C., 51 pp.

Guenther, G.C., Thomas, R.W.L. & and LaRocque, P.E. (1996). *Design considerations for achieving high accuracy with the SHOALS bathymetric lidar system.*

Habib, A., Bang, K.I., Kersting, A.P. & Chow, J. (2010). Alternative methodologies for lidar system calibration. In *Remote Sensing,* 2010, 2, 874-907.

Hickman, G.D. & Hogg, J.E. (1969). Application of an airborne pulsed laser for near-shore bathymetric measurements. In *Remote Sens. of Env.,* 1, Elsevier, New York, 47-58.

Hoge, F.E., Swift, R.N. & Frederick, E.B. (1980). Water depth measurement using an airborne pulsed neon laser system. In *Appl. Opt. 19,* 6, 871-883.

Holman, R., Plant, N., & Holland, T. (2013). cBathy: A robust algorithm for estimating nearshore bathymetry. In *Journal of Geophysical Research, Oceans.*

IHO (International Hydrographic Organization) (2008). *IHO standards for hydrographic surveys, 5th edition.* Monaco: International Hydrographic Bureau, 36p.

Imahori, Gretchen; Ferguson, Jeff; Wozumi, Toshi; Scharff, Dave; Pe'eri, Shachak; Parrish, Christopher E.; White, Stephen A.; Jeong, Inseong; Sellars, Jon; & Aslaksen, M, (2013). *A procedure for developing an acceptance test for airborne bathymetric lidar data application to NOAA charts in shallow waters.* National Oceanic and Atmospheric Administration (NOAA), National Ocean Survey (NOS), NOAA Technical Memorandum NOS CS 32, Silver Spring, MD, p.53.

Irish, J.L. & Lillycrop, W.J. (1999). Scanning laser mapping of the coastal zone: the SHOALS system. In *ISPRS Journal of Photogrammetry & Remote Sensing,* Vol. 54, Nos. 2-3, 123-129.

Jerlov, N.G. (1976). *Marine optics,* Elsevier Scientific Pub. Co., Amsterdam, 231 pp.

Karlin, A., Nayegandhi, A., & Owens, J.F. (2017). Topo-bathymetric lidar on the Springs coast of Florida. In *Lidar Magazine,* Vol. 7, No. 3.

Kerfoot, W.C.; Hobmeier, M.M.; Yousef, F.; Green, S.A.; Regis, R.; Brooks, C.N.; Shuchman, R.; Anderson, J. & Reif, M. (2014). Light detection and ranging (lidar) and multispectral scanner (MSS) studies examine coastal environments influenced by mining. In *ISPRS Int. J. Geo-Inf.* 2014, 3(1), 66-95.

LaRocque P.E. & West, G.R. (1999). Airborne laser hydrography: an introduction. In *Proc. ROPME/PERSGA/IHB Workshop on Hydrographic Activities in the ROPME Sea area and Red Sea,* October 24-27, Kuwait, 16 pp.

Lidtka, K. (2016). Single photon lidar. In *xyHt Magazine.*

Lillycrop, W.J., Irish, J.L., & Parson, L.E. (1997). SHOALS system: three years of operation with airborne lidar bathymetry - experiences, capability and technology advancements. In *Sea Technology,* June, Vol. 38, No. 6, 17-25.

Lillycrop, W.J. Irish, J.L; Pope, R..W. & West, G.R. (2000). GPS sends in the marines: rapid environmental assessment with lidar. In *GPS World,* 11(11): 18-28.

Liu, Z.S. (1990). Estimate of maximum penetration depth of lidar in coastal water of the China sea. In *Proc. SPIE Ocean Optics X,* Vol. 1302, 655-661.

Lockhart, C., Arumugam, D., & Millar, D. (2005). Meeting hydrographic charting specifications with the SHOALS-1000T airborne lidar bathymeter. In *Proc. U.S. Hydro 2005,* The Hydrographic Society of America, March 29-31, San Diego, CA, paper 7-3, 8 pp.

Mandlburger, G.; Hauer, C.; Wieser, M.; & Pfeifer, N. (2013). Topo-bathymetric lidar for monitoring river morphodynamics and instream habitats—a case study at the Pielach River. In *Remote Sens.* 2015, 7, 6160-6195.

McKean, J.; Nagel, D.; Tonina, D.; Bailey, P.; Wright, C.W.; Bohn, C.; & Nayegandhi, A. (2009). Remote sensing of channels and riparian zones with a narrow-beam aquatic-terrestrial lidar. In *Remote Sensing* 2009, 1, 1065-1096.

Mitchell, S.E. & Thayer, J. (2014). Ranging through shallow semitransparent media with polarization lidar. In *Journal of Atmospheric and Oceanic Technology.*

Mobley, C.D. (1994). *Light and water: radiative transfer in natural waters,* Academic Press.

Nagle, David B. & Wright, C. Wayne (2016). Algorithms used in the Airborne Lidar Processing System (ALPS). *U.S. Geological Survey Open-File Report,* 2016–1046, 45 p.

Nayegandhi, A., Brock, J.C., & Wright, C.W. (2005). Classifying vegetation using NASA's Experimental Advanced Airborne Research Lidar (EAARL) at Assateague Island National Seashore. In *Proc. ASPRS Annual Conference,* March 7-11, 2005, Baltimore, MD, Session 25, paper ID 500001, 15 pp. (CD-ROM).

Nayegandhi, A.; Brock. J.C.; Wright, C.W.; & O'Connell, M.O. (2006). Evaluating a small-footprint, waveform-resolving lidar over coastal vegetation communities. In *Photogrammetric Engineering and Remote Sensing* 2006-12:1408-1417.

Nayegandhi, A., Brock, J. C., & Wright, C. W. (2009). Small-footprint, waveform-resolving lidar estimation of submerged and sub-canopy topography in coastal environments. In *International Journal of Remote Sensing* 30.4 861-878.

Nordstrom, G. (2000). The Swedish Hydrographic Service on the eve of a new millennium. In *Integrated Coastal Zone Management,* Spring Edition, 37-40.

NSTC (2016). *National coastal mapping strategy (draft),* National Science and Technology Council.

Pe'eri, Shachak & Philpot, William D. (2006). Increasing the existence of very shallow-water lidar measurements using the red-channel waveforms. In *IEEE Transactions on Geoscience and Remote Sensing,* Volume 75, Number 5, p.1217-1223.

Penny, M.F. & Phillips, D.M. (editors) (1981). *Proc. 4th Laser Hydrography Symposium, ERL-0193- SD,* Defence Research Centre Salisbury, Sept. 30-Oct. 3, 1980, Salisbury, South Australia, 554 pp.

Penny, M.F., Billard, B. & Abbot, R.H. (1989). LADS—the Australian Laser Airborne Depth Sounder. In *Int'l. J. Rem. Sens.,* 10, 9, 1463-1479.

Perry, G.J. (1999). Post-processing in laser airborne bathymetry systems. In Proc. *ROPME/PERSGA/ IHB Workshop on Hydrographic Activities in the ROPME Sea Area and Red Sea,* October 24-27, Kuwait, 13 pp.

Pfennigbauer, M. & Steinbacher, F. (2012). Hydromapping—high-resolution shallow-water mapping from an airborne platform. In *Hydro International,* v. 16, No. 1.

Pope, J. & Lillycrop, W.J. (1988). Development of a helicopter lidar bathymeter system. In *Proc. U.S. Army Corps of Engineers Surveying Conf.,* Fort Belvoir, VA, 213-216.

Reif, M.K., Wozencraft, J.M., Dunkin, L.D., Sylvester, C.S. & Macon, C.L. (2012). U.S. Army Corps of Engineers airborne coastal mapping in the Great Lakes. In *Journal of Great Lakes Research, Supplement 39,* 194–204.

Setter, C. & Willis, R.J. (1994). LADS—From development to hydrographic operations. In *Proc. U.S. Hydro. Conf. 1994,* The Hydrographic Society, April 18-23, Norfolk, VA, Special Pub. No. 32, 134-139.

Sinclair, M. (1999). Laser hydrography—commercial survey operations. In *Proc. U.S. Hydrographic Conference,* April 26-29, 1999, Mobile, AL, (paper 9-2 on CD), 10 pp.

Sinclair, M.J. & Spurling, T. (1997). Operational laser bathymetry in Australia. In *Proc. XVth International Hydrographic Conference,* International Hydrographic Organization, April 21-22, Monaco, Session IV, 4.1-4.17.

Skogvik, J. & Axelsson, R. (2001). Experience and results from Swedish laser surveys and post-processing of laser bathymetry data. In *Proc. Hydro 2001,* The Hydrographic Society, Special Pub. 42, March 27-29, Norwich, England, Paper 20, 13 pp.

Sorenson, G.P., Honey, R.C., & Payne, J.R. (1966). Analysis of the use of airborne laser radar for submarine detection and ranging, SRI Report No. 5583, Stanford Research Institute.

Steinvall, O., Koppari, K., & Karlsson, U. (1992). Experimental evaluation of an airborne depth sounding lidar. In *Proc. SPIE Lidar for Remote Sensing,* Vol. 1714, 108-126.

Steinvall, O. & Koppari, K. (1996). Depth sounding lidar—an overview of Swedish activities and future prospects. In *Laser Remote Sensing of Natural Waters: From Theory to Practice,* V. I. Feigels, Y. I. Kopilevich, Editors, SPIE, Vol. 2964, 2-25.

Steinvall, O.; Koppari, K.; Lejdebrink, U.; Winell, J.; Nilsson, M.; Ellsen, R.; & Gjellan, E. (1997). Theories and experience of the Swedish airborne laser system. In *Proc. XVth International Hydrographic Conference,* International Hydrographic Organization, April 21-22, 1997, Monaco, Session IV, 3.1-3.23.

Sutherland, M. (2013). *Weaker without bathymetry.* Hydro International.

Tuell, G., Barbor, K., & Wozencraft, J. (2010). Overview of the coastal zone mapping and imaging lidar (CZMIL): a new multisensor airborne mapping system for the U.S. Army Corps of Engineers. In *Proc. SPIE 7695, Algorithms and Technologies for Multispectral, Hyperspectral, and Ultraspectral Imagery XVI,* 76950R (2010/05/13); doi: 10.1117/12.851905.

Tuell, G. H., Feygels, V. I., Kopilevich, Y., Weidemann, A. D., Cunningham, A. G., Mani, R., Aitken, J. (2005a). Measurement of ocean water optical properties and seafloor reflectance with scanning hydrographic operational airborne lidar survey (SHOALS): II. Practical results and comparison with independent data. In R. J. Frouin, M. Babin, & S. Sathyendranath (Eds.), *SPIE 5885: Remote Sensing of the Coastal Oceanic Environment,* July 31- August 1, 2005 (Vol. 5885, p. 58850E–58850E–13). San Diego, CA: SPIE.

Tuell, G.H., Park, J.Y., Aitken, J., Ramnath, V., Feygels, V.I., Guenther, G.C., and Ko-pilevich,Y.I. (2005b). SHOALS-enabled 3-D benthic mapping. In *Proc. SPIE Vol. 5806, Algorithms and Technologies for Multispectral, Hyperspectral, and Ultraspectral Imagery XI,* S. Chen and P. Lewis Ed., 816-826.

Ullrich, Andreas & Pfennigbauer, Martin (2011). Echo digitization and waveform analysis, in Airborne and Terrestrial Laser Scanning. In *Photogrammetric Week 2011.*

UN (1989). Status of hydrographic surveying and nautical charting world-wide. In *Proc. 4th United Nations Regional Cartographic Conference for the Americas,* January 23-27, 81/INF/9.

US Army Corps of Engineers (2015). Airborne bathymetric lidar. In USACE EM 1110-1-1000, *Photogrammetric and LiDAR Mapping,* Chapter 7, 24 pp.

Wang, C-K. & Philpot, W.D. (2007). Using airborne bathymetric lidar to detect bottom type variation in shallow waters. In *Remote Sensing of Environment* 106:123–135.

Wehr, A. (2009). LiDAR systems and calibration. In *Topographic Laser Ranging and Scanning: Principles and Processing* (J. Shan and C.K. Toth, editors), CRC Press, Taylor and Francis Group, Boca Raton, Florida.

Wellington, M. (2001). The Laser Airborne Depth Sounder (LADS)—a broad range of applications. In *Proc. Hydro 2001,* The Hydrographic Society, Special Pub. 42, March 27-29, Norwich, England, Paper 19, 11 pp.

White, S. A., Parrish, C. E., Calder, B. R., Pe'eri, S., & Rzhanov, Y. (2011). Lidar-derived national shoreline: empirical and stochastic uncertainty analyses. In *Journal of Coastal Research.* West Palm Beach, FL, USA: Coastal Education & Research Foundation, Inc. (CERF).

Wozencraft, J.M. & Lillycrop, W.J. (2003). SHOALS airborne coastal mapping, past, present, and future, In *J. Coastal Research, Special Issue,* 38, 207-215.

Wozencraft, J.M. & Lillycrop, W.J. (2006). JALBTCX coastal mapping for the USACE. In *Int'l. Hydro. Rev.*

Wozencraft, J.M. & Millar, D. (2005). Airborne lidar and integrated technologies for coastal mapping and charting. In *Marine Technology Society Journal,* 39, 3 (Fall), 27-35.

Wozencraft, Jennifer & Park, Joong. (2013). *Integrated lidar and hyperspectral.* 175-191. 10.1007/978-90-481-9292-2_7.

Wozencraft, J.M. (2014). *Remote sensing for the MTS, innovative technologies for a resilient MTS,* 24 June 2014.

Wright, C.W. & Brock, J.C. (2002). *EAARL: a lidar for mapping shallow coral reefs and other coastal environments.*

Wright, C.W., Kranenburg, C.J., Troche, R.J., Mitchell, R.W., & Nagle, D.B. (2016). *Depth calibration of the experimental advanced airborne research lidar,* EAARL-B: U.S. Geological Survey Open-File Report 2016–1048, 23 p.

Chapter 11

SONAR

Guy T. Noll and Douglas Lockhart

TECHNOLOGY OVERVIEW

SONAR is an acronym for (So)und (Na)vigation and (R)anging. Sound and acoustic are terms that can be used interchangeably, although the latter is a broader term and now more common. The former is most often used to describe acoustic frequencies within the range of human hearing, which leads to terms like ultrasonic and subsonic for acoustic frequencies that are respectively, above and below the range of human hearing.

Sound waves have a physical character that differs from that of other types of electromagnetic propagating waves, i.e., light and radio waves. Acoustic waves are based on vibrations of the actual material of the medium and are manifested as periodic variations of pressure in the medium. As a result of this physical nature of acoustic waves, the exact composition of the material through which an acoustic wave travels will impact the energy that is necessarily lost as the wave propagates through the material. When a propagating acoustic wave encounters a sudden change in the properties (sound speed and/or density, but specifically the product of sound speed and density) of the actual material of the medium, a portion of the acoustic wave will change its direction of propagation. That portion of the acoustic wave that reverses its propagation direction is the echo which echo sounders are designed to exploit for distance measurements. Figure 11.1 shows the basic principle of an acoustic wave reflecting from the ocean floor and the simple formula for reducing the time of flight to a depth value.

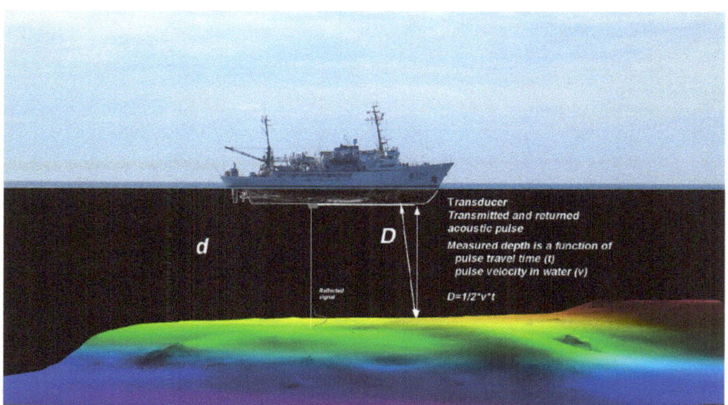

Figure 11.1. Echosounding fundamentals.

If the transmission and reception of acoustic energy can be confined to a unique narrow angular sector, the detection of an echo at some time after a pulse is transmitted provides both the range and bearing to the point in space where the echo was generated. Measuring the local configuration of the seabed with acoustics begins this simply: transmit acoustic energy toward the bottom and precisely detect the arrival times and directions of the acoustic energy that returns from the bottom. The measured ranges and 3-D directions to points where the echoes were generated can be converted into 3-D locations, relative to the transducer through trigonometric calculations. Finally it is necessary to geometrically transfer the echo generation locations from the transducer frame of reference into the ship's frame of reference and finally into the appropriate reference frame for presenting the survey results.

The accuracy standards for hydrographic surveys are set forth in Special Publication 44 (SP 44). SP 44 is issued by the International Hydrographic Organization (IHO) in Monaco and was updated to the 5th edition in February 2008. Chapter 3 of this book discusses the various Orders of IHO vertical and horizontal charting specifications along with issues concerning detection of features on the bottom. While reading this chapter on sonar, it is sufficient to keep in mind that shallow water hydrographic surveys, which are typically conducted to "Order-1" specifications, must meet the requirement that depths, corrected relative to chart datum, are within 50 cm of the true depth, with 95% confidence. Thus, much of the final accuracy is dependent upon the precision and accuracy of the corrections applied through a robust process.

Developmental History

The earliest methods for sounding water depths were simple and direct, and a natural extension of the forms of propulsion in use, such as poles and oars. A weight attached to a rope served the needs of early man to measure depths that were too deep to measure with a pole. In approximately 450 BC, Herodotus, a Greek chronicler, reported a water depth of 11 fathoms at a distance of one day's sailing distance off the Egyptian coast (Harre, 1992). In the middle part of the nineteenth century when the advent of oceanic submarine cables accelerated the need for soundings in deep water, the measurement tool was naturally a similar technology to a weight attached to a rope. It was a lead weight attached to a wire line. As an aside, the action of measuring depth is known as "taking a sounding", based on the old French verb *sonder,* meaning to probe the unknown, and was not derived from the English "sound." The use of sound to take a sounding was an invention made in the early part of the twentieth century. Nevertheless, in the twenty-first century, a graduated wire line with an attached lead weight remains an accepted means to confirm the correct operation of modern echo sounders.

In 1490, Leonardo da Vinci wrote, "If you cause your ship to stop and place the head of a long tube into the water and place the other extremity to your ear, you will hear ships at a great distance from you." This is generally accepted as the earliest documented reference to acoustics in the sea (Bell, 1962). In 1827, documented investigations into the propagation of sound in water (Lake Geneva) were conducted. In 1877, Lord Rayleigh described the basic mathematics of acoustic waves in "The Theory of Sound." In 1912, one month after the RMS Titanic collided with an iceberg, a patent application was filed with the British Patent Office for echo ranging with underwater sound. That device was intended to detect submerged objects, like for example, the underside of an iceberg. In 1914, the outbreak of World War I provided the impetus for developing the capability to detect submerged objects, such as an enemy submarine.

Echo sounders were commercially available for use in deep water by 1925 (Klein, 1968). Depths were determined by measuring the time interval required for a short sequence of sound waves (acoustic pulses) to travel from the ship to the ocean bottom (where the sound waves are reflected, producing an echo) plus the time for the subsequent echo to return to the ship. One-half of that time interval multiplied by the speed of sound gives the depth. These early echo sounding designs were limited by the technology for making precise measurements of short time intervals. Consequently, the measurement of acoustic time of flight was unsuitable for application to measuring shallow depths until a decade later. It is interesting to note, that in this same era, acoustic travel times were also employed in the determination of horizontal positions (Hawley 1931). In 1935, the U.S. Coast and Geodetic Survey began employing shallow water echo sounders for hydrographic surveys. Note as an aside that these improvements in technology and techniques precede by sixty years, and are paralleled by, the recent evolution of lidar.

The development of acoustic echo sounders advanced considerably as a result of acoustics research in World War II that was directed toward the detection of submarines and radar research for the development of electronic wave forming. That research resulted in improved electronic circuitry and improved materials for use in constructing acoustic transducers, which are necessary to convert between electrical energy and mechanical (acoustical) energy, and vice versa (Tucker, 1966). Those improvements led to scientific investigations to better understand the transmission of sound in the oceans (Horton, 1957) and the physical mechanisms involved in generating echoes through the interactions of sound with the seabed (Urick, 1956). These investigations demonstrated that the interactions at the seabed depended on the acoustic frequency, the angle of incidence, and physical properties of the seabed, like roughness, porosity and grain size. (Owaki, 1963) (Hamilton, 1971). In 1958, acoustic waves were first used to purposefully map geologic surface features of the seabed utilizing acoustic backscatter (Chesterman, Clynick, and Stride, 1958). As electronic systems have become more sophisticated, the ability to develop more advanced calculators based on analytical models has created new systems such as the camera-like Blueview (Thompson, 2003), simultaneously shrinking the size of accompanying electronics.

BASIC PRINCIPLES OF ACOUSTIC MAPPING SYSTEMS

All sonars must have the capability to generate electrical signals that are converted to acoustic energy via a transmit transducer thereby emitting into the water column a short burst or pulse of acoustic energy at a particular frequency. All sonars must also have the capability, via a receive transducer, to convert the acoustic energy of the returning echoes into electrical energy and separate those echoes from the background sound. Additionally, all sonars require the ability to accurately associate time, particularly time relative to the time of transmission, with the returning echoes. The details of the transmit transducer, receive transducer, echo processing, and the interpretation of the echo returns are what distinguishes a particular sonar product as vertical beam, multibeam, or side scan, etc.

Acoustic Sources

If a short pressure fluctuation were to be inserted at a point in the water column that is a large distance from either the surface or bottom of the sea, that pressure fluctuation would spread equally in all directions as a spherical wave. This is termed omni-directional, because the pressure fluctuations are the same in all directions projected from the source, as is shown in Figure 11.2. As the radius of the spherical wave increases, the cross-sectional area over which the pressure fluctuation is spread increases according to the squared distance from the point of origin. This is the basis for the well-known "inverse-square law" which describes the decrease in the pressure fluctuation of an acoustic wave as it moves through the water. Assuming that the seabed below the point of the initial pressure insertion is horizontal, the intersection of the expanding spherical shell of the pressure fluctuation starts as a point, which expands to a circle of radius "R" and then transforms into an expanding annulus of width "R." Echoes are generated where the spherical shell is intersecting with the bottom, or any other discontinuity such as the surface. The vertical angles between the intersection points and the initial insertion point are continually increasing from an initial value of zero. Echoes resulting from the interaction between the pressure fluctuation and the seabed will propagate back toward the insertion point of the initial pressure fluctuation and over time will arrive from a continually changing vertical direction.

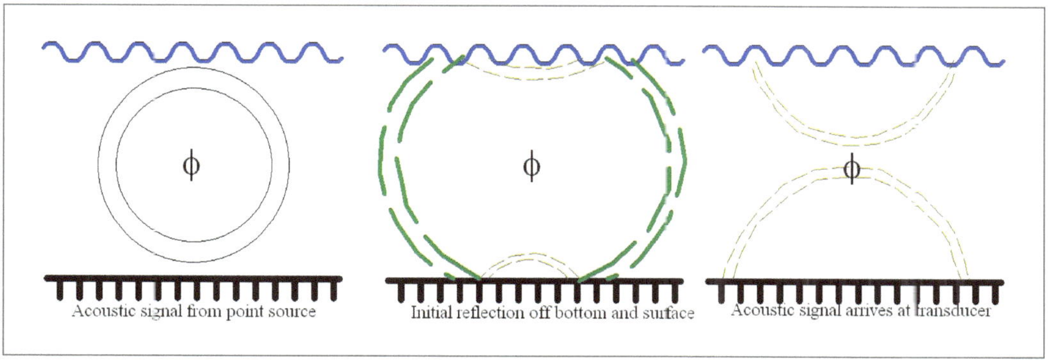

Figure 11.2. Signal reflection off surface and bottom from point source.

All of the echoes from any given annulus will arrive at the insertion point at the same time. In order to identify which specific section of the seabed generated a particular echo, it is necessary to estimate the vertical and azimuthal angle of arrival of any particular echo. Those angular estimates are not possible to make if the echoes are received using an omni-directional receiver. In order to estimate the vertical and azimuthal arrival angle of an echo, it is necessary to use receive transducers that are directional and it is also preferable to use a directional transmit transducer.

Sensitivity of the transducer elements drives the ability of the system to determine directionality, and is the primary factor determining cost. This directionality of the transmit and receive transducers and the techniques of achieving that directionality are the major factors that distinguish one sonar type from another.

Directional Transmit/Receive Transducers

It has already been stated that acoustic energy will spherically radiate from a point source in a homogenous medium. Let us consider what would happen if two point sources, in close proximity to each other, were activated at a particular frequency (local pressure caused to alternate periodically between being above and below the local ambient pressure) using the same driver. There would be a set of spherical waves expanding around each of the point sources, as seen in Figure 11.3. Because the sources are driven periodically, the spherical waves might be visualized as a set of expanding concentric spherical shells where adjacent spheres alternate between being an increase in pressure and a decrease in pressure. Since there is a set of expanding spheres centered on each of the point sources, the shells must intersect at a number of places. If at the intersection between two shells, each shell is an increased pressure shell, the pressure at that point in space will be doubly increased relative to the ambient pressure. Likewise, if at the intersection between two shells, each shell is a decreased pressure shell, the pressure at that point in space will be doubly decreased relative to the ambient pressure. The third possible combination of two intersecting shells is the combination where one is an increased pressure shell and the other is a decreased pressure shell, resulting in the pressure at that intersection point being the ambient pressure. Once that mental picture has been formed, imagine that the two sources were activated by two different drivers operating at the same frequency such that the pressure at one point source is being increased relative to ambient pressure while the other point sources is decreased relative to ambient pressure. The radiating concentric shells will be essentially as before, but with one important difference. At the shell intersections where both shells were previously either increased or decreased pressure relative to ambient pressure, the resultant pressure will now be ambient rather than a pressure that is doubly increased or decreased relative to ambient pressure.

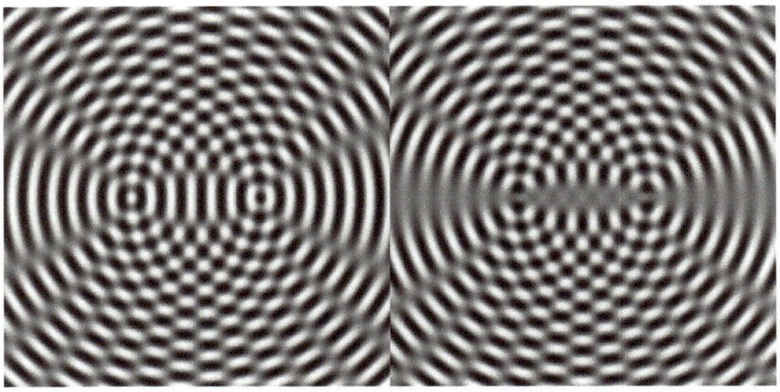

Figure 11.3. Interference pattern from two wave sources, right-hand side is phase-shifted.

Figure 11.3 illustrates this point in two, rather than three, dimensions. In the right half of the figure, the phase of one of the point sources has been shifted by one-half a cycle relative to the cyclic pressure of the other point source. The darker areas represent increased pressure and the white areas represent decreased pressures, relative to ambient pressure. The figure demonstrates how the simple act of changing the relative phase angle between the two point source drivers has completely changed the two-dimensional spatial pattern of pressure fluctuations relative to ambient. If this two point source example were expanded to a series of point sources regularly spaced in a straight line, there would still be concentric shells of pressure increases and decreases around each point source. There also would be a complicated set of points in space where all of the shells would intersect with different combinations of pressure increases and decreases relative to ambient. As in the two point source example, the relative phases with which the individual point sources are driven can dramatically alter the three-dimensional spatial pattern of pressure fluctuations relative to ambient. Selecting the proper combination of relative phases can cause the pressure to be zero or near zero everywhere in space except along a particular angle from the long axis of a line drawn through each of the point sources. In this manner a transmit transducer can be caused to direct a spherically spreading acoustic pulse into a particular direction, rather than into all directions as was the case with the omni-directional point source. Since echoes can only be generated from sections of the seabed where the spherically spreading pressure fluctuations from the transmitter have intersected with the seabed, it is possible to reduce the uncertainty in which a specific section of the seabed generated a particular echo.

The main lobe of a transmit/receive transducer is defined as having a beam width equal to the angular sector between the half-power points of the transducer response. The characteristics of the main lobe and associated side lobes of a transmit/receive transducer depend on the actual frequency and physical size of the transducer. For example, the width of a transducer's main lobe in a given plane will be reduced as the dimension (measured in acoustic wavelengths) of the transducer in the perpendicular plane increases. When discussing the physical dimension of a transducer like its length, width or diameter, it is common practice to express the dimension in multiples of the acoustic wavelength (Urick, 1967, 1975, 1983). In the case of a horizontal looking, rectangular shaped transducer the vertical beamwidth is controlled by the vertical height of the transducer, whereas the horizontal beamwidth is controlled by the (horizontal) length of the transducer. The angular width of a line shaped by the transducer's main lobe in a particular direction is roughly 50 degrees divided by the extent of the transducer in that direction, expressed in number of wavelengths. In the case of a circular transducer the beamwidth in the direction perpendicular to the transducer face is controlled by the diameter of the transducer. The angular width of a circular transducer's main lobe is roughly 60 degrees divided by the diameter of the transducer, expressed in number of wavelengths.

A receive transducer, like the transmit transducer above, can be caused to respond only to acoustic energy that arrives at the receive transducer from a particular direction relative to the long axis of a line drawn through each of the linearly arranged point receivers. To accomplish this feat, it is necessary that the electrical signals from each of the point receivers be phase shifted (a.k.a. time-shifted) relative to each of the other point receivers and then summed to provide one composite electrical signal. If a directional transmitter and a directional receiver are mounted such that their main lobes are co-aligned, then it is possible to determine with considerable assurance which specific section of the seabed generated a particular echo.

Acoustic survey systems are generally constructed such that the transmit/receive transducers have negligible response in one hemisphere and finite response in the other hemisphere. They are mounted in such a manner that the responsive hemisphere is aimed in the direction where echoes of interest might be generated, like the seabed. The basic physical shapes of the transducers used in acoustic survey systems are circles and lines, both curved and linear. The lengths of the line transducers are usually ten to twenty times their width.

The transducer design and its associated beam characteristics fundamentally enable or restrict the ability of a particular type of sonar to associate a specific and unique direction, relative to both nadir and the heading of the vessel, with each echo that the system receives. An acoustic survey system is designed to measure echoes that are generated in the region on the seabed where the projection of the main lobe of the transmit beam onto the seabed and the projection of the main lobe of a receive beam onto the seabed coincide. Object detection is also directly linked to the extent of differentiation of bottom return to the other sources of sound in the water column. An object must typically comprise at least one-half the "volumetric" measurement of the sonar ping, measured by sound duration of intersection of the first and last part of the ping with the bottom and object, to be recognized by the sonar processor as distinct from the bottom alone. However, this is also significantly affected by the bottom type, the amount of signal returned, and the frequency and dynamic range of the sonar used. Higher frequency sonars with shorter ping periods and smaller beams typically have higher resolution.

TYPES OF SONARS

Six types of sonar survey systems are discussed. Vertical beam sonar, multibeam sonar, side scan sonar, interferometric sonar, focusing sonar, and Doppler sonar systems broadly represent the types of sonars that are used to obtain information about the configuration and condition of the seabed. There are important differences and obvious similarities in the functional designs of these six types of sonar.

Vertical Beam Sonar

Vertical beam sonars are primarily designed to produce quantitative information about nadir water depths. Vertical beam sonars have one, and sometimes two, transducer(s) that are each used for both transmitting and receiving acoustic energy at a given frequency. The vertical orientation of the beam(s) means the transmitted acoustic waves will most likely interact with the bottom at near vertical incidence, which will maximize the energy in the echo returns. The received echoes are processed to determine the onset time of the first echo's arrival defined by the "leading edge" of the echo envelope waveform. The time measured by a vertical beam echo sounder is associated with the shortest distance from the ship to a point on the seabed. Depending on characteristics of the transducer and the configuration of the local seabed, that distance may not be the depth directly beneath the survey vessel, also referred to as nadir. However, it is generally assumed to be nadir because in shallow depths the error in position is likely not great, and in deeper depths the footprint is so large as to include the nadir at nearly all realistic vessel attitudes.

In detecting the vertical beam sonar echo return, one is looking for a significant rise in voltage level above the mean level of the noise fluctuations that are also present in the output of the receiving transducer. The ability to distinguish one arrival time from another is limited by the bandwidth of the receiver. Wider bandwidths are associated with greater ability to precisely measure the arrival time of an echo. The amount of noise present in the output of the receive transducer depends on both the amount of acoustic noise in the water and the bandwidth of the receiver. The engineering aspects of designing and building a vertical beam sonar deal with the tradeoff between time resolution and rejection of noise which allows weaker echoes to be reliably detected. Other factors that the sonar engineer may change to improve the vertical beam sonar performance for a particular application, like measuring into the seabed (sub-bottom profiling), are acoustic frequency, transmit power, transducer beam width, and the signal processing technique (Urick, 1967, 1975, 1983), (Chramiec and Morton, 1970) and (Mayer and LeBlanc, 1983).

Vertical beam echo sounders point in the direction of the main lobe, which is directed vertically toward the seabed except when the vessel is experiencing high values of roll. Vertical beam echo sounders generally have main lobe angular widths between five and twenty-five degrees. Since vertical beam sonars are characterized as looking in only one direction, it is therefore evident that for each transmitted pulse, there is only one area on the seabed where transmit and receive beams coincide. This provides a high degree of confidence, but not an absolute guarantee, that the echoes measured in a vertical beam echo sounder originate from a point that was located in the direction that the main lobe of the transmit/receive transducer was pointing. If a narrow main lobe is not orientated vertically, there will be a horizontal displacement on the seabed between where the leading edge of the echo is generated and where the leading edge of the echo would have been generated if the beam were orientated vertically. The result is that the measured depth is vertically offset from true depth. The vertical offset associated with a non-vertical beam will nearly always make the measured depth appear to be deeper than it is and the offset will scale with true depth. Figure 11.4 illustrates an example where a depth offset is produced as the result of a horizontal displacement caused by vessel roll. This case is more often observed in areas of extremely steep bottom relief, such as channel boundaries or fjords, in which small changes in pointing angle have a high impact on the measured angular response. A change of 10% of the depth is possible in steep channels; very significant in fjords.

Because the reduction to half-power in the transducers angular response occurs on both transmit and receive, the half-power point specified for the transducer is actually the quarter-power point for the system throughput. As a word of caution, relentless pursuit of narrower and narrower beamwidths comes at the cost of greatly increasing the size of the transducer or greatly reducing the operating range of the sonar by greatly increasing the frequency to the point where the acoustic energy suffers increased attenuation by absorption. Furthermore, if a very narrow beam width is achieved, then even small values of vessel roll and pitch can negate the basic assumptions of vertical beam echo sounders, which are that the received echoes were generated directly beneath the vessel and consequently that the time after transmit of the leading edge of the echo relates to the depth directly beneath the vessel (MacPhee, 1979).

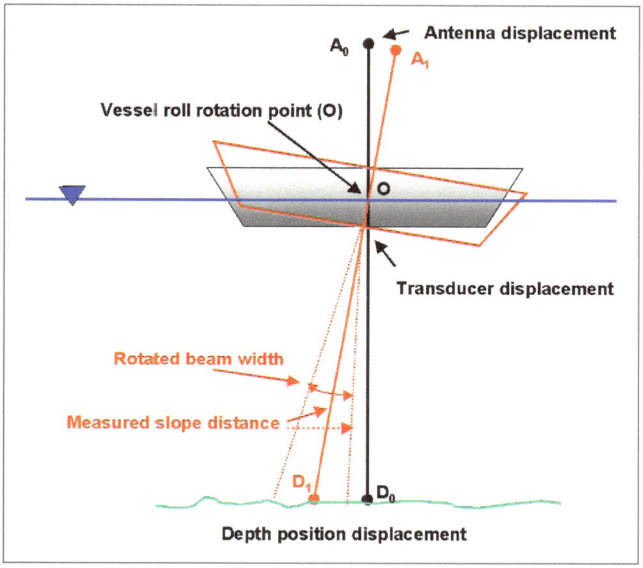

Figure 11.4. Horizontal displacement and depth offset caused by roll rotation of the vertical beam.

To achieve additional confidence in the accurate delineation of the sea bottom with vertical beam echo sounders, a narrow beam transmitting/receiving at one frequency may be operated in conjunction with a broad beam transmitting/receiving at another frequency. There are numerous echo sounders on the market that incorporate two simultaneous vertical beam depth measurements at different frequencies (for example 50 and 200 kHz). In such dual frequency vertical beam echo sounders the beamwidths may be the same, but that is the exception rather than the rule. Usually the lower of the two frequencies has a broader beamwidth. These systems can increase the robustness of the solution, both as a "check" and by bringing additional information to the measurement, such as density differences in the sediment or excluding vegetation returns from the primary bathymetry.

The received echoes in a vertical beam depth sounder may be subjected to various signal processing schemes, which provide information from phase integration that allows the user to infer variations in the interaction of the transmitted acoustic pulse and the seabed that might, in turn, imply spatial variations in the composition of the seabed.

A notable variation of the vertical beam echo sounder is the multiple vertical beam echo sounder, or sweep sonar. Sweep sonar is specialized system comprised of a parallel logging of standard single beam transducers vertically mounted on a boat, barge, or other stable platform. The bottom coverage of a sweep sonar is a function of transducer spacing, beam width, and water depth. Bottom coverage is nearly independent of water depth and is directly related to the mechanical form factor of the sweep arms. That fact makes a sweep sonar attractive for surveys in very shallow water. Sweep systems may use any number of transducers, though typically at least three per side comprise a sweep sonar. Two or more transducers may be mounted permanently in the vessel hull in the simplest configuration. Additional transducers may be mounted on "over-the-side" outriggers or, more commonly, from hinged, retractable booms deployed to port and starboard. In 1985, the Detroit District of the U. S. Army Corps of Engineers was the first in the U.S. to install a 32-channel, vertical-looking multi-channel echo sounder (Rougeau, et.al. 1992). The more common systems deploy between 3 and 12 transducers on combinations of hull and retractable boom mounts. Figure 11.5 on page 328 illustrates a sweep sonar with two hull-mounted transducers and 12 transducers mounted on booms.

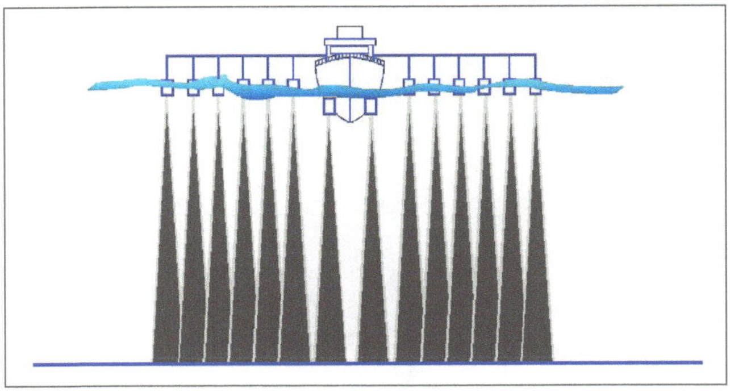

Figure 11.5. Sweep sonar with two hull-mounted and 12 boom-mounted vertical beam transducers.

Sweep systems are normally used on shallow surveys where sea state conditions are calm, and can achieve enough resolution to discriminate very small features when operated at their typically slow survey speed. Thus, full X-Y-Z inertial motion sensors, as discussed in Chapter 12 of this book, are rarely added to a sweep system unless sea states, usually swell, require compensation for heading or heave. Roll correction may be required to correct for vertical motion at the outer transducers, or a mechanical dampening system may be used to reduce the effects of roll on the booms. Regardless of roll, to maintain the full positional accuracy of the navigation system for each of the multiple soundings, it is essential that corrections be made for eccentricities due to yaw.

Multibeam Sonar

Multibeam sonars are primarily designed to produce quantitative information about water depth. Multibeam sonars are first characterized as having significant system response and the ability to measure depths at angles that are non-vertical to the seabed, as well as, at nadir, earlier defined as the term used to describe "perpendicular to the seabed," or simply "vertical." Secondly, multibeam sonar is typically characterized as having separate transducers for transmit and receive. All multibeam sonars measure travel times between the echo sounder transducer and the seabed using a transmitted acoustic pulse. One of the major differences between multibeam and the other types of sonars is the manner in which the sonar processes/interprets the echo waveforms that are received subsequent to the pulse transmission (Heald and Pace, 1996). Conventional multibeam sonars measure the acoustic time of flight to the seabed as a function of angle from nadir. Through the use of trigonometric functions, the travel times are converted to a set of points, each with a vertical and horizontal coordinate, relative to the multibeam transducer (depth and position). Seemingly minor errors in the beam angle relative to nadir can result in unacceptably large depth errors. Because of the non-vertical measurement geometry, it is absolutely essential that full roll-pitch-yaw motion sensors, as discussed in Chapter 12 of this book, be installed and operated on the survey platform along with the multibeam sonar, with precisely measured accurate time-linkage between the system components. Too often, very serious investments are made in sonar technology, training, and software without understanding that the time-tagging of all measurements must be accurately and precisely maintained for the resulting depths to be combined into a high-quality DEM surface.

The first multibeam sonars were designed in the mid-1960s and deployed for military applications, they were known as SASS, for Sonar Array Sounding System (Glen, 1970) (Satriano, 1991). In 1977, the first non-military multibeam bathymetric sonar was placed into service (Farr, 1980) (Renard and Allenou, 1979). Being deep water systems, the frequencies of SASS and other early multibeam sonars were necessarily low, and the transducers were correspondingly large and expensive to construct. The operating frequencies of subsequent shallow water sonars were more than 50 times higher than the frequencies of the early deep-water sonars. By 1990, there were multibeam sonars operating at frequencies as low as 12 kHz and as high as 455 kHz (Harre, 1992) (Steenstrup, 1992).

A conventional multibeam sonar has a single linear transmit transducer oriented along-track. Its width and length are such that the entire width of the depth measurement swath is ensonified by the same acoustic transmit pulse. The main lobe of the transmit transducer is narrow in the along-track direction (horizontal plane) and broad in the cross track direction (vertical plane). The receive transducer is orientated cross-track and is formed by multiple cells, each with its own main lobe that is relatively broad in the along-track direction (horizontal plane) and narrow in the cross track direction (vertical plane). For each transmitted pulse, there are a large number of locations on the seabed where the projected main lobe of the transmit pulse and the projected main lobe of a computed receive "beam" coincide. Collectively the multiple locations comprise the entire measurement swath, as shown in Figure 11.6.

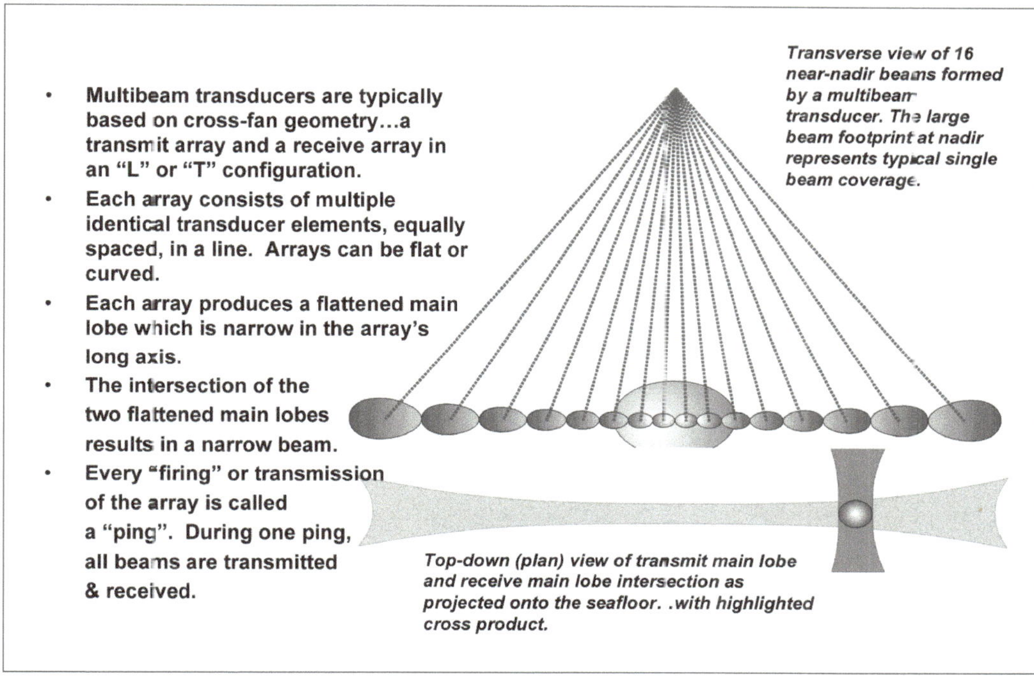

Figure 11.6. Individual beams in multibeam transducer form a swath.

The technique used in multibeam sonar to measure the acoustic time of flight may employ the amplitude of the echoes received in a given transducer cell or the relative phases between the echoes received on adjacent transducer cells. The amplitudes of the echoes received in each of the several cells are also measured and associated with horizontal coordinates relative to the transducer (Talukdar and Tyce, 1991). With suitable processing the along-track and cross-track variations in the amplitudes of the received echoes provide information that may allow the user to infer spatial variations in the composition of the seabed. The variations in echo amplitudes may be presented in a planar view or presented as an overlay on the apparent surface of the seabed established by the multibeam sonar depths. Fundamentally, most sonar processing solutions use the area and shape of the amplitude-time series curve to determine seabed characteristics.

Side Scan Sonar

Side scan sonar is a widely-used tool for qualitative observations and supplements other quantitative measurement tools. The objective of side scan sonar is to provide a detailed presentation of the seabed features and manmade objects that may lie on the surface of the seabed, in the form of a raster image. The first side scan sonar was developed in 1960 at the Institute of Oceanographic Sciences (IOS) in England (Tucker and Stubbs, 1961). The first side scan sonar was a shallow water system, and most current versions of this technology are still used in estuaries, lakes, and bays. In 1969, ISO developed the Geological Long Range Inclined Asdic (GLORIA), side looking sonar for surveying in the deep ocean (Laughton, 1981). The fundamental physics remain the same as with other sonars, so higher resolution requires higher frequency and thus decreased range, so even in deep water a towed side scan sonar must remain relatively close to the bottom to be effective.

As in multibeam sonar and interferometric sonar, a side scan sonar ensonifies the entire measurement swath with the same acoustic transmit pulse. Actually there are two pulses, one transmitted from a continuous line array transducer looking to port and one from a continuous line array transducer looking to starboard. The main lobe of the (port and starboard) transmit transducer is narrow in the along-track direction (horizontal plane) and broad in the cross track direction (vertical plane), again due to the physics of the array size. Like vertical beam echo sounders, conventional side scan sonars use the same transducer for receive and for transmit. This provides a high degree of confidence, but not an absolute guarantee, that the echoes received by the side scan sonar originated from points that are located in the direction that the transducer is pointing. One of the greatest challenges facing the side scan sonar processor is removing multi-path acoustic returns from the primary return. These can be reflections from the underside of the surface-air interface, objects in the vicinity, or even other vessels.

The spatial resolution capabilities of side scan sonars are different in the cross- and along-track directions. Furthermore, the cross- and along-track resolutions each vary with the cross-track distance from nadir and the character of those variations differs between the two directions. The along-track resolution is determined by the horizontal beamwidth of the transmit/receive transducer and changes linearly with slant range. The cross-track resolution is determined by the sonar's basic range resolution, largely dependent upon the duration period of the acoustic transmit pulse, and by geometric effects that vary nonlinearly with slant range, based on the height of the tow fish above the bottom and the cross-track distance from nadir. The cross-track resolution may be 50 to 100 times worse directly under the tow fish (zero cross track distance) as it is at the maximum cross-track range. Thus, although the measured returns may be gridded, the returning acoustic signal varies considerably across the track, complicating the processing.

The basic range resolution of a side scan sonar, as well as a vertical beam echo sounder and a multibeam sonar, is determined by the bandwidth of the transmit pulse. In most systems the bandwidth of the transmit pulse is determined by the time duration of a short transmitted tone burst of a given frequency. However, in some designs the bandwidth is defined by the bandwidth of a long frequency-modulated transmit pulse. These systems are commonly referred to as "chirp" side scan sonars. The basic range resolution in meters is one-half the speed of sound divided by the bandwidth.

In the design of a side scan sonar a high premium is placed on achieving transmit/receive beams that are narrow in the along-track direction to maximize resolution and thus the discrimination of features. Side scan sonars tend to use high frequencies in order to achieve narrow beamwidths with transducers of moderate length. Due to the typically high frequencies of side scan sonar, 455–900 kHz, the useful operating range of a side scan sonar is less than 200 meters to either side of the tow fish. A notable exception is GLORIA II (Figure 11.7) which operates at a frequency of 6.5 kHz and has a maximum imaging range of 60 km (Mitchell, 1991). The experimental 100 kHz side scan interferometer by L-3 Klein has nominal range of 1 km when towed appropriately (Huff, McConnaughey, 2011) as proven on NOAA missions in the Bering Sea for Fisheries Habitat studies.

If the objective of a particular side scan sonar survey is to detect small objects, then the operating range may be limited by the spatial divergence of the beam in the horizontal plane as the acoustic pulse travels farther from the tow fish. For maximum ability to detect specific small targets at any given range, the along-track width of the transmit/receive beam, at that range, should be less than the least cross section of that specific target, and the range resolution should be much less than the largest cross section of that specific target. Delineation of small details on large complex targets requires both narrow beamwidths and wide bandwidths.

The raster image of the seabed obtained via side scan sonar will almost always contain spatial variations in the intensity of the received backscatter signals. There are several possible effects that explain the majority of the spatial variations observed in the intensity of the received signals. The first possible cause of the spatial variation is an actual spatial change in the various materials that comprise the seabed. The relative surficial and

Figure 11.7. GLORIA II wide-swath towed side scan sonar system as deployed in 1970s.

volumetric backscatter characteristics differ for different materials. For example, rock and gravel will reflect more of the incident acoustic signal than sand will backscatter. Sand will backscatter more of the incident acoustic signal than silt, which will backscatter due primarily to volumetric properties, and silt will create stronger returns than decomposed organic matter. A second possible cause of the spatial variation is a change in the angle between the propagation direction of the outgoing (transmitted) acoustic pulse and the seabed, which is designated the incident angle. The incident angle varies systematically from ninety degrees directly below the tow fish to approximately five degrees at the maximum slant range expressed in the side scan imagery. Portions of the seabed with slopes that face toward the transducer on the tow fish (incident angle is closer to ninety degrees) will reflect more of the incident acoustic signal than surfaces with slopes that face away from the transducer (incident angle is closer to zero degrees), with resultant impacts on signal-to-noise and overall backscatter results.

If the seabed is level and flat, the vertical beam pattern of the transmit/receive transducer approaches omni-directional across the track. Assuming further that water column effects are homogenous, rarely true (see Colbo, et. al., 2014), then any variations in the intensity of the side scan sonar received signals that are observed along the survey track line can readily be attributed to variation in the material composition of the bottom. However, in general, most of those conditional statements are not simultaneously true and the side scan imagery contains spatial variations that are due to local changes in both the incident angle, the water column, and the composition of the seabed. Consequently, manual interpretation of side scan sonar imagery requires considerable experience and first-hand knowledge about the particular backscatter characteristics of various rock types, gravels, sands, and characteristic bed forms associated with them such as bedding, jointing, ripple marks, sand waves, etc. (Flemming, 1976) (Fish and Carr, 1990) (Lurton and Lamarche, 2015). Automated processing that includes interpretation is even more difficult to achieve, though software is continuously being improved with the intent of increasing the repeatability of the classification result.

A side scan sonar is typically towed close to the seabed to control the geometry of the interactions between the acoustic transmit pulse and the seabed. When a tow fish is close to the bottom, the incident angle will change quickly from ninety degrees into a range of incident angles where the variation in backscatter is less sensitive to small changes in incident angle. The higher the tow fish is above the bottom the longer the incident angle is in the range where the variation in backscatter is more sensitive to small changes in incident angle. The geometry associated with towing the side scan sonar close to the bottom also enhances the generation of "shadows" behind areas of the seabed that distinctly rise above the surrounding area and consequently may represent a feature that warrants further investigation. Side scan sonar information is missing from these shadows, however, so full coverage is only achieved by combining two or more tracks that are offset. This allows the operator to also recover the nadir region under the tracks with improved incident angular coverage.

Interferometric Sonar

Interferometric sonar is another type of sonar that is primarily designed to produce information about depths and exhibits significant acoustic response and the ability to measure depths at angles that are non-vertical to the seabed, as well as, at nadir. Consequently, it sometimes is grouped under the general term of multibeam sonar. The measurement geometry of interferometric sonar is similar to that of side scan sonar. Consequently, it is sometimes called interferometric side scan sonar and grouped under the general term of side scan sonar. Interferometric sonar differs from multibeam sonar in that not only is the travel time of the echo measured but the vertical angle of arrival for each time sample of the echo must also be measured (Cloet and Edwards, 1986) (Denbigh, 1989). Interferometric sonar differs from side scan sonar in that not only is the amplitude of an echo measured for each time sample of the echo, but the vertical angle of arrival for each time sample of the echo is also measured.

Interferometric sonars have a single transmit transducer whereby the entire width of the depth measurement swath is insonified by the same acoustic transmit pulse. Interferometric sonars have multiple receive transducers, each with a relatively narrow main lobe response in the along-track (horizontal) direction and a broad main lobe response in the cross-track (vertical) direction. There is a small vertical separation between the several receive transducers which provides the physical geometry required to determine the vertical angles of arrival from the time samples of the received echoes. Because the geometry associated with echoes from near nadir is unfavorable for the accurate resolution of vertical angle of arrival, many interferometric sonars incorporate a vertical beam echo sounder to measure the depths directly below the vessel. As with multibeam sonar, it is absolutely essential that full roll-pitch-yaw motion sensors, as discussed in Chapter 12 of this book, be installed and operated on the survey platform along with the interferometric sonar.

Interferometric signal processing techniques include determination of the amplitudes of the backscatter from the seabed, as well as, determination of depths based on vertical angle of arrival. Those processing techniques provide a three-dimensional surface on which to view the along track and cross track variations in the amplitudes of the received echoes. Those spatial variations in echo amplitudes provide qualitative information that may allow the user to infer spatial variations in the composition of the seabed (Green, Hewitt, and Adams, 1993).

Due to the simplified design of the transducers and electronics, interferometric sonars are less costly than conventional multibeam sonars. The typically small transducer size means that they are quite portable and can be rapidly mounted onto small survey platforms, as well as, be configured for towing. However, if the transducer is towed, the roll/pitch and azimuth angles of the tow body must also be measured, as well as, the depth of the tow body.

Manufacturers of interferometric sonars state that their systems can typically measure depths over a swath that is up to ten times the depth, or more particularly, the height of the transducer above the bottom. The data density in a beam-formed swath system (multibeam), and thus cross-track resolution will halve if depth is doubled. The data density of a phase measurement swath system (interferometric) is far higher, typically hundreds of times higher, and stays roughly constant with depth. Therefore, the interferometric sonar can afford to average or statistically filter through many real data points and still provide soundings in a high spatial resolution grid. However, they also tend to provide depth measurements that are less accurate and precise than depths measured with beam-formed multibeam sonar. Factors contributing to the uncertainties in depths from shallow water interferometric sonar are: unraveling the true phase (time) differences in light of the ambiguities in differential phases, interference from multipath arrivals, and interference due to simultaneous arrival of echoes with different vertical angles of arrival (such as echoes from both the sea surface and seabed). All of these factors are exacerbated in nearshore areas, but these are also the areas of greatest interest for interferometry due to the higher potential efficiency vice coverage obtained with directional multibeam technology.

Focusing Sonar

The latest sonar development was originally developed in the late 1990s for underwater inspection in low visibility. These techniques can be applied in lower frequencies to develop additional resolution at greater distance. Focusing sonars utilize virtual or actual multiple transducer arrays to create a larger sensing surface and calculate over a larger physical distance. This creates the ability to solve for sonar returns to create:

1. Resolution that cannot be seen using a single array
2. Multiple views from varying angles
3. Redundant measurements to calculate backscatter variance due to water column path effects

The research and application of focusing sonar continues to be actively pursued across academia, defense, and commercial entities. Key technical issues in virtual sonar arrays, sometimes referred to as Synthetic Aperture Sonar or SAS, are ultra-precise tracking of the sonar attitude in time and space, discriminating between main lobe and multi-path sonar returns, and water column scattering effects in shallow waters (Sternlicht, Hagen, 2009).

Doppler Sonar

Doppler sonars are designed to measure the velocity of a moving platform relative to the bottom, profiles of the water currents relative to the sonar, or both. They were originally designed to measure a vessel's speed over ground as an aid to navigation, which has proven particularly useful when operating in tight quarters and/or significant currents. The basic principle is that the echo of the transducer's transmitted signal from the bottom will have its frequency shifted from the transmitted frequency proportionally to the relative velocity between the sonar and the bottom—one of the many applications of the well-known "Doppler Shift" named for the Austrian physicist Christian Doppler after his first published description of the phenomenon observed in light (Doppler, 1842). Velocity can only be measured along the raw path of an acoustic beam. In order to measure horizontal velocity, the beams of a Doppler sonar need to be inclined and organized in a geometry that allows for the vector decomposition of the beam velocities into 3-D vector components. Typically three or four transducers are angled at twenty to thirty degrees off of the vertical and at symmetrical azimuth. The independent measurements of the velocity along each beam are then combined to obtain the true vector velocity over ground (under the assumption that the bottom is not moving). Use of an internal compass or integration with the vessel's gyroscopes then allows rotation of the velocity field from a coordinate frame based on the instrument attitude to a coordinate frame based on East-North-Up coordinates independent of the vessel. On most vessels an internal compass is not feasible because of the many potential sources of magnetic bias on the vessel, so the ship's gyroscopes are used. However, when using the ships gyroscopes it is important that any offset in the instrument's as-mounted orientation from the vessel's coordinate system be measured and factored into the coordinate rotation; otherwise a velocity bias that is proportional to the vessel's speed will occur.

The bottom is the strongest, but not the only, source of acoustic backscatter. Most water is filled with zooplankton and/or sediment that will also reflect sound back to the transducer, though at a much lower intensity. If this smaller signal can be teased from the noise (Pinkel, 1980) then the returned signal from the water can be range-gated into bins and the Doppler shift of the aggregate of backscattering sources within each bin along each beam can be determined. Combining the profiles along three or four beams as above yields a profile of the three dimensional water velocity extending away from the transducer. Two critical assumptions are implicit: that the backscattering sources are passively following the flow (as opposed to swimming in a particular direction); and that the flow is relatively uniform through all of the beams at a given range. If the first assumption is violated, then the measured velocity will be strongly biased toward the swimmer's velocity rather than the water velocity. This fact has proven problematic with fish (which reflect very strongly and tend to congregate under any shadow (such as a buoy or a vessel) and special considerations have to be taken to remove the bias. However, this fact has also proven useful in the study of the daily vertical migration of zooplankton (they swim up at night to eat and down during the day to avoid being eaten) where the vertical velocity of the aggregate can be measured. The second assumption is required to combine beams pointing in different directions into horizontal and vertical velocities. If the flow is not uniform across the beams, then the combination has little skill in resolving the true flow field.

Doppler sonars on vessels can measure the speed over ground when the bottom is in range, and can also measure a profile of the relative water velocity beneath the vessel. To obtain true water velocity profiles, a mechanism to remove the vessel motion must be employed. If the bottom is in range, then the true velocity can be found simply be subtracting the bottom velocity from the measured profiles. One very nice side effect of using the Doppler sonar to measure both the speed over ground and the water velocity is that any attitude errors are applied equally to both measurements and are automatically removed from the final true current measurement in the subtraction. When the bottom is out of range, the vessel's GPS fixes can be used to measure the vessel velocity to subtract from the relative velocity profiles. However, it is again important to measure any offset between the Doppler sonar's as-mounted orientation relative to the ship's coordinate frame to ensure that no velocity-dependent biases are introduced in the process.

Each velocity measurement made by a Doppler sonar is an aggregate measurement over an area (in the case of the bottom) or of a volume of scattering centers passively following the flow. In addition, the independent beams of the Doppler sonar are deliberately pointed away from the vertical in order to measure velocities that are primarily horizontal. The leading edge of the sonar pulse reflecting from the bottom reflects more strongly than does the trailing edge, so for Doppler sonars this means there is a potential for a velocity bias toward zero because the more vertical return from the leading edge has a smaller measured horizontal component of velocity. Also, when measuring profiles of water velocity, the computed Doppler shift of aggregate return from all scattering centers in the volume can be strongly influenced high or low by a single strong scattering center located off the beam's centerline. For these reasons and others, round transducers designed with narrow main lobes and significantly reduced side lobes are the norm for Doppler sonars.

PRESENT OPERATING STATUS

Platforms and Installation

The obvious concerns in planning a sonar installation for a particular survey begin with the type of information that is required, along with any accuracy and precision requirements placed on that information. This gives the first indication as to what type of sonar(s) may be applicable, as well as the acoustic frequencies and consequently the transducer sizes that are required. Next, one must reconcile the primary engineering considerations of weight, size, volume, and power required by the sonar with the capabilities of a particular deployment platform. The next level of concern includes definition of ancillary requirements, such as additional sensors beyond the sonar which are necessary components of a survey system, to include attitude sensors, heading device, sound speed profiler, navigational control, and whether to use a proprietary or open data acquisition format. Data processing, quality control and data visualization and presentation must be considered, particularly the extent to which those functions must be performed in near real time in the field for immediate delivery, or later in a shore facility.

Sonars of the types discussed above are deployable on a surface ship, a submarine, a small surface craft including Surface Autonomous Vessels (SAV), a tow body, and sub-sea vehicles including remote-operated vehicles (ROVs) or fully autonomous underwater vehicles (AUVs). Virtually all combinations of sonars with all types of platforms are possible. Interferometric sonar may either be mounted to the hull of a vessel or incorporated into a tow fish (Tyce, 1986). Side scan sonars are generally incorporated into a tow body, or in one of the types of sub-sea vehicles. However, when mounted to the hull of a small surface craft, side scan sonars are also capable of providing quite satisfactory imagery in water depths less than 20 meters, (Huff, 1996).

Compact sized multibeam sonars are commonly mounted on small craft and used in shallow water to provide high spatial resolution while maintaining accurate positioning. Compact sized multibeam sonars are also sometimes operated on a sub-sea vehicle, as a means of achieving high spatial resolutions of small areas of the seabed (Steenstrup and Luyneburg, 1992). The transducers for multi-beam sonar that are applicable to deep ocean surveys are necessarily larger than those for application to shallow water surveys. The operating frequencies for deep water are necessarily lower than frequencies that are useable in shallow water. It is basically the difference in frequency that dictates the difference in size.

Calibration Procedures

In a sonar there are only a few things to calibrate, namely the operating frequency, the pulse length, the transmit power, the main lobe amplitude of the transmit pulses, the main lobe sensitivity of the receive transducer, beam patterns of the transmit/receive transducer(s), and the accuracy of the time base. A user typically does not have the resources to calibrate many of these factors, and therefore relies on the manufacturer for the information. Even there, the normal practice is for manufacturers to provide typical or nominal response data, as opposed to the response function of a particular set of transducers and electronics. As transducers age, this information will vary from the nominal response data, impacting final measurements.

Sound speed within the water column between the transducer and the seabed is an important parameter that affects the accuracy of sonars. It could be viewed as an extension of the accuracy of the sonar's time base. In a vertical beam echo sounder the measured distance (depth) is the product of the total travel time, which transpires between the transmit pulse and reception/detection of an echo, and the average sound speed along the travel path from the transducer to the seabed and back to the transducer. In multibeam and interferometric sonars the distance measurement is further complicated by gradients in sound speed, which will cause the acoustic energy to travel along paths that are curved lines rather than straight lines. Errors due to sound speed are typically addressed by regularly measuring vertical profiles of sound speed at intervals in space and time throughout a survey.

Calibration becomes an issue when a particular sonar is incorporated into a survey system. In the instance of multibeam sonar and interferometric sonar, or fixed-mount side scan sonar, that means interfacing with additional equipment such as an attitude reference unit, heading device, navigational unit, and a data acquisition computer. The parameters that need to be calibrated are the locations, relative to the roll/pitch center of rotation of the platform, of the sonar transducer (acoustic center), the navigational (GPS antenna phase center) system, and the attitude reference unit (rotational center of roll-pitch-heave sensor), as seen in Figures 11.8 and 11.9 and 11.10 on page 336. Inaccuracies in these parameters impact the accuracy with which the measured acoustic travel times are converted to vertical and horizontal positions.

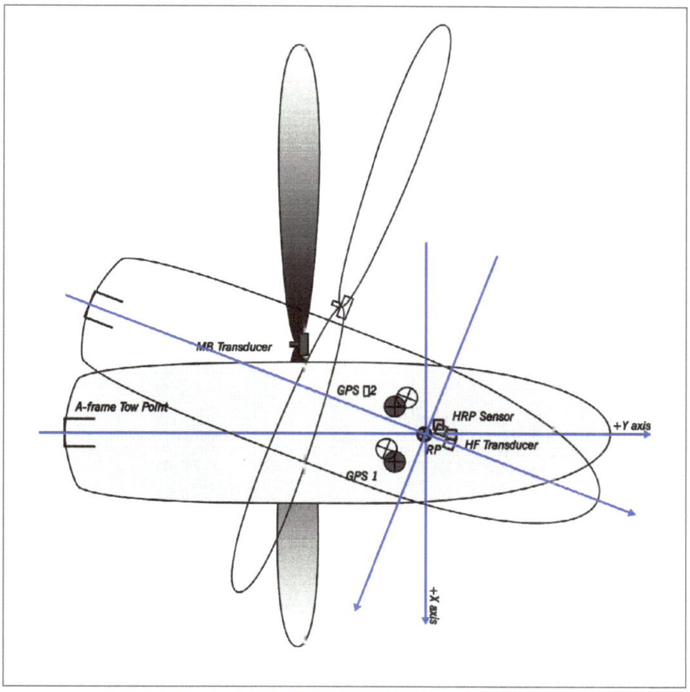

Figure 11.8. Yaw effects on offsets of sensors.

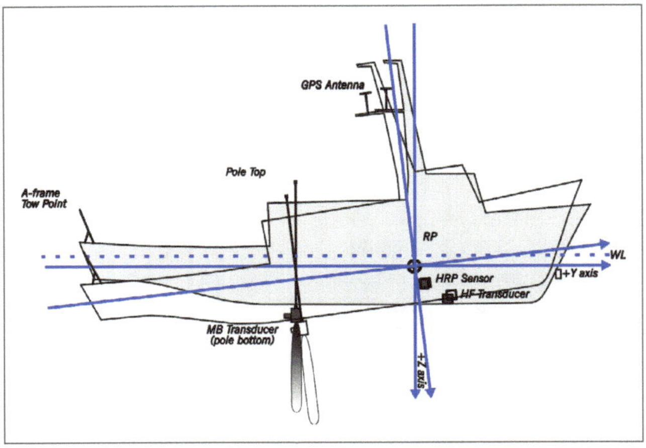

Figure 11.9. Pitch effects on offsets of sensors.

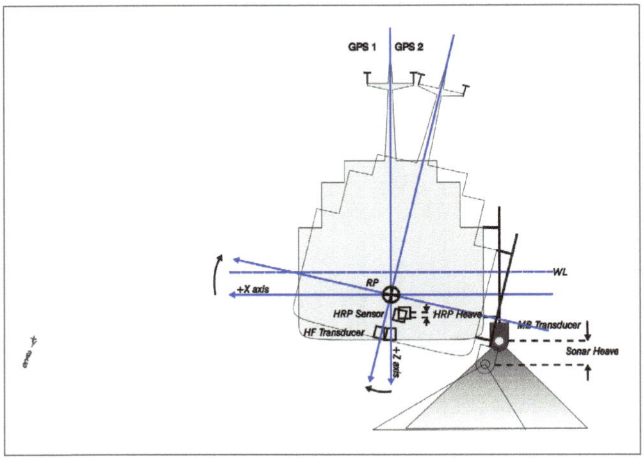

Figure 11.10. Roll effects on offsets of sensors.

There are also three angles that must be determined when the sonar and associated instrumentation are installed on the survey platform; the reduction of the data to compute these angles is more complex if the transducer is not installed squarely and in a recoverable geometry in the vessel frame of reference.

First, there is a calibration procedure, called a patch test, which is used to calibrate the offset angles. The exact offset angle between the azimuthal mounting of the transducer and the heading unit must be known in order to accurately convert acoustic travel times into horizontal positions relative to the transducer. As shown at Figures 11.9 and 11.10, the exact offset angle between the vertical angles (roll and pitch) of the transducer and the reference plane of the vertical reference unit must be known to accurately convert the acoustic travel times into vertical and horizontal positions relative to the transducer. Seemingly minor errors in the beam angle relative to nadir can result in unacceptably large depth errors. The patch test includes provisions for determining the offset angle between the vertical angles (roll and pitch) of the transducer and the reference plane of the vertical reference unit (Wheaton, 1988) (Godin and Ramalho-Marreiros, 1998).

The final calibration parameter (also determined during the patch test) is the time latency, which is the difference between the actual times when a measured parameter is reported from the depth sensor (sonar), the navigational unit (GPS), and heading unit is true and the times which the data acquisition computer associates with those reports. For example if the survey vessel is running 5 meters per second and there was a one second time lag between when a GPS measured position is true and the computer time that is assigned to the GPS measurement, then the horizontal positions that are assigned to any particular sounding will be in error. The assigned positions will be displaced from the true positions by 5 meters in the direction the ship was traveling. In many gyroscopes, flux gate compasses and GPS heading devices, the output of the sensor is smoothed with a low pass filter to reduce unrealistic short-term fluctuations. The low pass filtering causes the output heading values to be presented at times after they were the true values for vessel heading. Timing latency in heading values will impact the horizontal positions assigned to any (especially outer beam) soundings that are taken while the heading is changing. A patch test calibration should be done each time the installation of the survey system is modified. For example if the vertical reference unit were to be replaced, that would be cause to perform a patch test to determine the roll/pitch offset angles between the transducer and the vertical reference unit. Effects on timing of data through changes to software and firmware should also be tested to ensure offsets are consistently applied after new software has been installed.

As shown in Figure 11.11, when a vessel's speed increases, it generally settles (sinks relative to overall water surface) and squats (changes pitch trim), causing errors in depth measurement that must be corrected. In multibeam sonar the former is corrected by using settlement correctors and the latter is corrected by using the attitude (roll/pitch) sensor. In a vertical beam echo sounder, the effects of both settlement and squat are typically corrected through a tabular lookup. A settlement/squat test should be performed at least annually to determine the relation between boat speed and transducer height above or below the static sounding reference plane. It is important that the positions of the GPS antenna and the sonar transducer relative to the pitch center of the vessel are known accurately. Otherwise the effects of settlement and squat may cause depths to be reported that differ significantly from their actual values. Kinematic (dual frequency) GPS solutions that provide direct (absolute) antenna-transducer elevation eliminate the need for the settlement corrections, as the antenna height will record the settlement in real-time. However, if the Kinematic GPS system is set up to provide only the antenna height and is not configured to resolve the transducer elevation, then the squat correction must still be applied. Also, the final result must still be corrected for water level (tidal) changes.

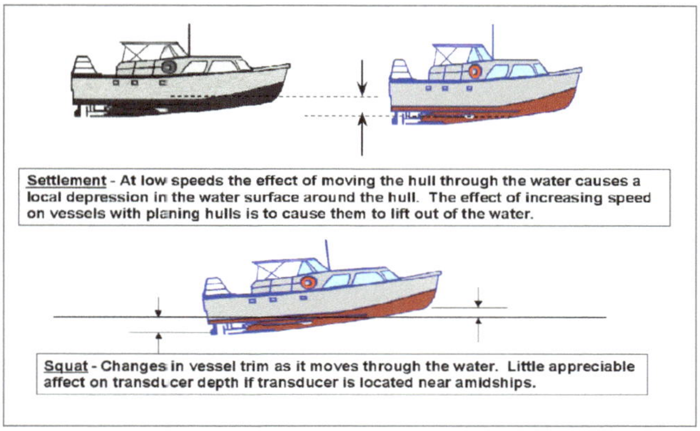

Figure 11.11. Effects of speed through water on sensor attitude and height.

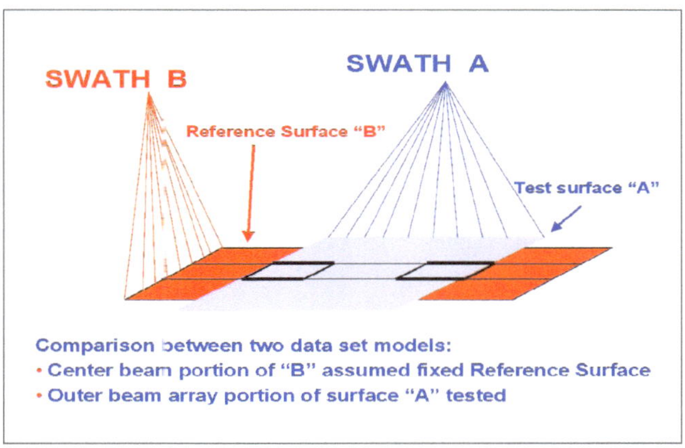

Figure 11.12. System test using comparisons between surfaces.

Often it is desirable to establish a performance measure for each beam in a multibeam sonar. That performance measure can readily be accomplished by surveying a flat section of seabed in two perpendicular directions followed by comparing the data for internal consistency. First the area is surveyed using closely spaced survey lines like Swath B on Figure 11.12. Those swaths are processed into a gridded surface using primarily the central beam depths from the several swaths. Such a surface is designated as Reference Surface "B" in Figure 11.12. The same area of seabed is also surveyed using track lines like Swath A, which are perpendicular to the survey lines used to develop the Reference Surface B. In this configuration the depths measured with each beam in Swath A can be compared to the spatially corresponding depths of the Reference Surface B. The differences between the Reference

Surface B and depths measured in a particular beam of Swath A are expressed as the root mean square of the differences (RMSD). The Swath B and Swath A data sets should be acquired in a span of time that is too short for there to be changes in the water level and/or sound speed profile. With environmental factors relatively homogenous, the depth differences are attributable to the slant range measurements of the bathymetric sonar, which are processed to depths using the vertical reference data. Low RMS values are desirable and necessary to satisfy the IHO depth measurement requirements.

If the RMS differences in any particular beam(s) are not less than seventy percent of the required specification, then it is doubtful that a hydrographic survey containing depths measured in those particular beams will meet the required survey accuracy. The seventy percent rule of thumb is based on equally dividing the total allowable uncertainty between the measurement of depth and the knowledge of water level that is necessary to translate the measured depths, acquired relative to the actual water level, into depths relative to the MLLW chart datum.

During actual survey operations, the essence of this specific test can be accomplished through the use of cross line checks during the survey. However, in an actual survey the constraint for the same water level and sound speed profile are rarely met. Consequently, cross line checks exhibit elevated values of RMS depth differences, compared to those observed in a specific test between comparison surfaces. This shows that there is no substitute for understanding how the real-world constraints of operational sonar usage are deduced from the test measurements used in preparation for surveying after the equipment is installed.

Even when all of the factors which impact on the accuracy of a sonar's distance measurements are known, the measured depths reported by the system will undoubtedly be biased, relative to the actual depth reduced to charting datum. That is because the measured depths are relative to the sea surface at the times when each of the measurements were made and the depths for charting must be relative to the chart datum. In the U.S. the chart datum is mean lower low water (MLLW), whereas internationally it might be MLW or lowest astronomical tide (LAT). The water level at a given time and location of a depth measurement is a function of a predictable astronomical (tidal) component and a meteorological component, which is much less predictable. Understanding how these components relate to the sonar measurements made throughout the survey area is critical to the task of creating a completely accurate bathymetric surface or grid.

Depths measured with respect to the local mean water level must be reduced to the chart datum by applying either predicted tides or, ideally, measured tides across expertly-derived zones of equal time (phase) and height relative to controlling gages of the world water level network. In the USA, this network is called National Water Level Observation Network (NWLON). Alternately, kinematic GPS with rapid ambiguity resolution (RTK-GPS) may be used to determine the depth with respect to the WGS-84 ellipsoid (Krabill and Martin, 1987), and later after applying the ellipsoidal height of MLLW using a tool like VDATUM (*http://nauticalcharts.noaa.gov/csdl/vdatum.htm*), determine the depth relative to chart datum.

Planning Considerations

Good survey products begin with proper planning of the data acquisition. Planning begins with understanding the survey requirements and the characteristics of the area to be surveyed. Planning culminates in the determination of the procedures, equipment, and software, which are linked and should always be considered carefully, so as to optimize the survey efficiency and effectiveness for the production of both foreseen and unpredictable uses of the survey results.

Surveyors have significant capital equipment costs, so are often constrained to conduct a particular survey with the equipment on hand, rather than rent or purchase new equipment that might be particularly suited to a given survey. In such cases, consideration must be given to the capabilities and limitations of a given system and how to design the survey to maximize the quality of the results. The design of the survey will require a choice of line orientation, line spacing, vessel speed, and sonar ping duration and rates. Those will in turn depend on the distribution and configuration of depths in the survey area, expected types of objects or relief to be found, and the characteristics of the particular system that will be used to conduct the survey. Line orientation and line spacing may also depend on the relative importance of imagery and bathymetry in meeting the objectives of the survey. These factors are obviously regressive, so that there are many different "right" solutions to the problem of conducting an accurate survey. The main differences in modern surveys often will be in the efficiency with which the survey is completed to meet the desired data goals.

Any particular hydrographic sonar survey might be for one and often more than one of the following reasons: nautical charting, regional bathymetric mapping, engineering applications, geologic studies, military applications, habitat mapping, etc. The type of equipment used will often be driven by the primary objective, and influenced by the interest of other customers in the data types potentially collected. In fact, there is a growing interest in "survey once, use many times" (NOAA Integrated Ocean & Coastal Mapping motto) to optimize the information for multiple users in the future who are unknown at the time of survey operations.

Surveys for nautical charting are conducted for the purpose of safety of navigation and require very accurate bathymetry. However, the detection of all hazards to surface navigation such as wrecks and other obstructions is critical. This type of surveying is most often conducted with multibeam sonar and/or a sweep sonar, which might be augmented with towed side scan sonar to ensure full coverage of the seabed. If towing a side scan is not possible, many multibeam systems now collect reasonable side scan sonar backscatter data. However, such data have not been as reliable as towed systems for object detection in the past, so areas "swept" by multibeam are not necessarily free of isolated hazards (Wetzler, VanDenAmeele, Noll, 2001).

Regional bathymetric surveys are conducted to determine the location and orientation of depth contours in water depths and areas where safety of surface navigation is not a major concern. This type of surveying might be accomplished with lidar systems, interferometric sonar, vertical beam echo sounder, or multibeam sonar, in order of speed of survey. Because this type of survey is most often coastal areas and possibly at considerable distance from a port, a sweep sonar is generally inappropriate, and backscatter imagery is acquired only if that parameter is an output of the specific survey system. These surveys are commonly used in support of water level modeling applications such as tsunami run-up models, storm surges from hurricanes, and flooding studies.

Surveys conducted for engineering applications include: pipe line and cable routing, dredging, and site selection for offshore platforms. This type of surveying must acquire information about bathymetry, sediment type, and the presence of abrupt features like rock outcrops. Systems most likely to be employed are multibeam sonar, vertical beam echo sounders (for sediment analysis), and side scan sonar with or without interferometry.

Geologic studies, both for mineral exploration and research, require bathymetry as well as identification of characteristics of the surface and subfloor seabed that can potentially convey information about the geological processes that may have occurred in the past as well as geological processing that might be currently active. This type of survey might be accomplished with towed interferometric sonar or multibeam sonar combined with sub-bottom vertical beam, either of which will provide a level of detailed imagery to supplement the bathymetry. The limiting factor in these studies is often the *in situ* observations, so supplemental ground truth data acquired with remote vehicles is often necessary. Autonomous vehicles are becoming more prevalent in this market, such as the diamond mining off South Africa using the Maridan AUV (2001).

Mine counter measures, also euphemistically referred to as "environmental assessment" is one military application that requires survey data. In surveys of this type, the ability to identify different sediment regimes and to detect targets of appropriate sizes and shapes throughout the water column are paramount in determining the selection of a sonar system. Side scan sonar, because of its superior ability to provide detailed imagery of the seabed and objects on the seabed, is a prevalent sensor for this type of survey. Integrated sensors for water characteristics are a prerequisite, and processing of these data are often accomplished in real-time from AUV data acquisition.

Habitat mapping is a rapidly emerging application for sonar and lidar survey systems. This interest is based on requirements for fisheries management that relate geographic knowledge to population data acquired over huge areas, and these are still being defined. However, it is clear at present that habitat mapping requires information about sediment types, the prevalence and extent of plant cover on the seabed such as eelgrass and kelp, as well as bathymetry, both general trends and local rugosity, or surface roughness. These types of surveys might employ the entire suite of sonar systems discussed above. Vertical beam echo sounders and side scan sonars are widely employed in habitat mapping based on their ability for the identification/characterization of sediments by analysis of the individual echo waveforms or statistical analysis of "spatial patterns" in the backscatter imagery. Integration of information to develop valid and robust habitat solutions is a current area of development in the survey industry.

Capabilities and Limitations

Acoustic techniques for bathymetry are fundamentally limited by the ability of the sonar instrumentation to determine angle of return and time of flight. Multipath, the return of multiple echoes to the acoustic receiver, creates noise in the digital or analog recording. The uncertainty of the velocity of sound in the water column, and the actual source of the returning echo from the sediment-water interface, also lead to uncertainty in the result. These problems can, to some extent, be eliminated through increasing the density of the acoustic measurements and doing inter-comparisons among neighboring echoes. Note that performing a trending analysis without independent check measurements can cause undue confidence in one's results. Therefore, using another source of data, if only for occasional comparison with the normal survey techniques, can prevent blunders in equipment or procedures from invalidating the survey product. In general, weather is responsible for the most variability in the operating conditions that limit effective work being performed. Weather, unfortunately, cannot be controlled. The continual requirement for the correct sound speed profile can also restrict the amount of acceptable data that can be acquired during a given period in a specific area. This is particularly true if there are several different water masses in the survey area and the survey vessel must stop often to update the sound speed profile information because it is not possible to reorient survey lines to increase the survey time in a water mass. Using towed undulating sound velocity instruments is becoming an increasingly common method to mitigate this problem and increase efficiency of the survey.

Comparisons with Complementary and Competing Technologies

As indicated above, many survey tasks require specific approaches and equipment. Complete, high-resolution bathymetric surface models can, and have been for decades, developed using vertical beam echosounder techniques. However, a modern multibeam echosounder does make quicker work of the job in water depths deeper than about twenty meters (60 feet). In calm waters shallower than ten meters, the use of a sweep sonar is very effective. Interferometric sonar sensors are quite effective at bathymetry between one and 50 feet deep. As object size decreases, side scan sonar should become a more prominent method of the survey, and should be used whenever object detection is the primary reason for surveying and environmental factors allow it to be accomplished safely. There is competition between the different sonar types with regard to which is best to use for a particular survey requirement, but they still represent the same basic acoustic technology.

Airborne lidar bathymetry technology is a survey technology which is distinctly different from sonar technology. Airborne lidar bathymetry is often thought to be a competing technology to that of sonar. However, the fact is that airborne lidar bathymetric systems were initially designed with the intent of complementing sonar survey systems. Chapter 10 of this publication discusses airborne lidar bathymetry in detail. It is important to point out that when airborne lidar bathymetric systems were initially designed, vertical beam echo sounders and side scan sonars were the primary survey tools for use in shallow water. It was initially envisioned that airborne lidar bathymetry would provide greater spatial density of soundings and be more efficient than sonar in water depths less than approximately 12 meters (36 feet). That vision was based on the standard vertical beam sonar procedure, which was to survey perpendicular to the depth curves, at a nominal spacing of 50 meter (150 feet) between lines. This vertical beam echo sounder procedure required a large amount of ship time to safely survey the areas between the 12-meter depth curve and the beach.

The technologies of multibeam sonar and airborne lidar bathymetry were developed at about the same time. With the change from vertical beam to multibeam sonar, there was a change in the sonar survey procedure, which switched from surveying perpendicular to the depth curves to surveying parallel to the depth curves. That switch was made because the cross-track angular coverage of multibeam sonar provides swath widths that vary proportionally with depth, narrower swaths are therefore associated with shallower depths. Another change in sonar survey procedure that accompanied the change from vertical beam to multibeam sonar was to adjust the survey speed to ensure along track overlap of footprints on successive pings. In shallow water this is an important factor because as the depths become shallower, the footprints become smaller and smaller, necessitating further reductions in survey speed to ensure the required overlap. Airborne lidar bathymetry is not subject to speed limitations in shallow water, though care must be taken to avoid loss of object detection using lidar, and there can be safety issues with nearshore survey regardless of platform.

Airborne lidar bathymetry has satisfied its design intent of complementing sonar survey systems by virtue of being an accurate, cost-effective, rapid, safe, and flexible method for surveying in shallow water and on coastlines where sonar systems are less efficient, such as for bathymetric modeling. The authors have extensive experience in effectively combining data from sonar and lidar for efficient production of final surveys. Figure 11.13 illustrates an example of an area that benefited from combined lidar and sonar surveys in an outer coast environment in which nearshore sonar data were dangerous to acquire.

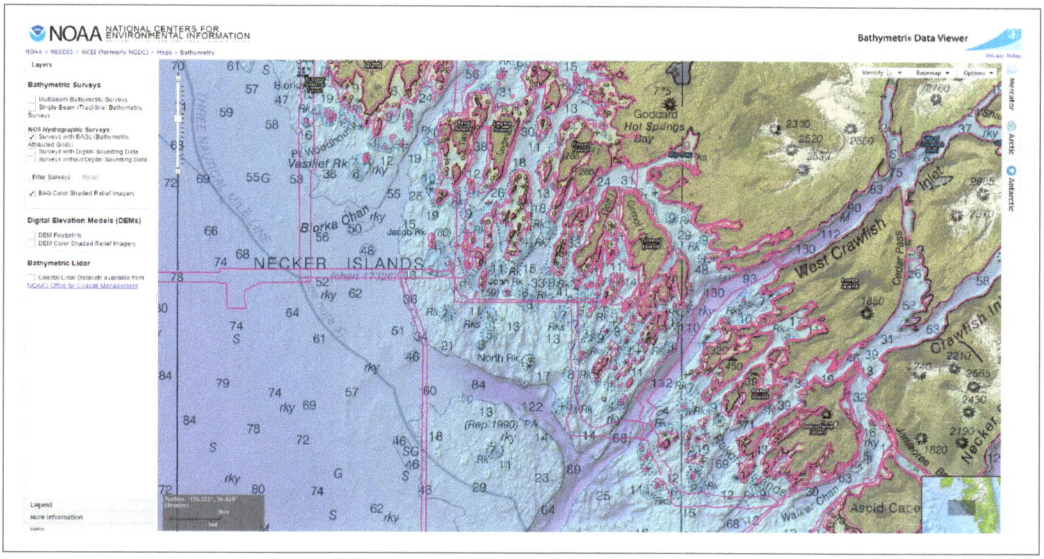

Figure 11.13. Digital Elevation Model BAG Color Shaded Relief Imagery combined with NOAA Alaska nautical chart and Esri Imagery base map using National Center for Environmental Information (NCEI) Interactive Bathymetric Data Viewer.

Post-Processing

Processing hydrographic survey sounding data is highly tailored to the goal of the survey. The marine bathymetric modeler may desire a smooth gridded surface, the nautical charting authority needs the more conservative shoal-biased binned data, and the dredge operator may want only high-density cross-sections. These and other different survey products, which can be produced by a variety of software packages, form a basis for differentiation of processing software. Each of several software vendors has optimized the features that the vendor's customers deem to be the most useful. There is industry-wide pressure to increase the speed and usability of the software that, in turn, provides further differentiation between the various vendors as they attempt proprietary data storage and automated processing techniques. The basic task of any processing software remains the same; take the acquired data from one or more types of sonar acquisition packages, convert it to a common processed data format, remove incorrect bottom detection points, and output the cleaned data for visualization. Another important step is the re-application of corrector data based on updated knowledge of the vessel, sensors, or water column parameters, though this is not always performed.

Data density within a hydrographic survey has increased by orders of magnitude each decade in concert with improvements in computer technology. Similar strides have not been made in the human interface to the computer. The application of manual methods to cleaning high-density multibeam data is time-consuming and inefficient when faced with the volumes of data acquired in modern multibeam surveys. Fortunately, robust algorithms are being developed (Hou, et. al., 2001) (Calder and Mayer, 2001) that promise significant improvement. Until these are implemented, the surveyor must educate the client with respect to the true cost of accurate high-density sounding data, and make particular note of the cost of processing the data, especially coastal data. As acoustic sensors improve and data become denser, it is likely that manual methods will succumb to mostly automated ones because operators will no longer be able to review the full data density at common screen resolutions, let alone edit it. However, there will always be a need for review of odd acoustic returns that the algorithms will flag as aberrant. These realities are reflected in the IHO standard S-102 based on Bathymetry Attributed Grids (BAGs) (Mayer, Calder 2002) that support standard metadata for bathymetric grids that tie final products to the underlying measurements. BAG data management is thus the emerging best choice for survey visualization and reuse. In the end, the choice of processing software should reflect the surveyor's goals, including the type of product required by the client, the degree of automation with which the client is comfortable, and the desired overall confidence in the survey products.

Quality Control

The higher the confidence with which the hydrographer can state the survey results, the more likely that the survey has been performed correctly. A quality control plan, or a control process, ensures that the final data will meet the desired specification when designed, implemented, and followed. No specification will magically create good data; the old maxim of "Garbage In-Garbage Out" will apply to hydrographic survey data collected with the intent of good quality if the plan does not provide for the necessary controls.

One vital factor, which has considerable influence on the quality of the sonar data, is the sonar operator. The operation of the echo sounder is carried out by hydrographers, who in the course of a hydrographic survey must review the incoming sonar data and set echo sounder system parameters like range scales, pulse lengths, transmitted power levels, and receiver gains in order to maximize the quality of the survey data that are acquired. Since hydrographers are expected to make the correct decisions regarding the operation of the echo sounder, they must be given specific training. It is not sufficient that the hydrographers simply understand the instrumentation. If the hydrographers are to collect meaningful data, then they must know which data are meaningful. Their training must provide some background in underwater acoustics, marine geomorphology and perhaps even a bit of marine biology, as well as extensive training in the operation of the sounding system and the on-line data quality control tools.

The quality control of hydrographic survey data requires an understanding of basic seabed conditions in the survey area and whether or not to expect the existence of man-made objects. The more complete that understanding, the higher the resultant confidence will be that features, with a certain size and shape, will be discovered and mapped correctly. Thus, a hydrographer should reasonably expect that sand waves are found off a rocky sandstone headland, and that man-made objects generally are not found in a recently glaciated region. Extending this reckoning, the hydrographer can estimate the likelihood of a given type and/or size of feature being detected, given the survey procedures and survey equipment that were employed. When absolute knowledge of the presence or absence of hydrographically significant objects is needed, then the hydrographer can focus resources on those areas which reflect the largest concern, and apply the appropriate procedures to locate the objects.

In addition to the extra attention that is necessary for potential problem areas in a survey, it is important to maintain the practice of acquiring data on survey lines that cross the main scheme of the survey (cross line checks) and comparing the resulting depths with those acquired on the main scheme survey lines. A patch test calibration, as discussed above, should be done when the character of the depths on either the main scheme survey lines or cross lines indicate that there might be something unnatural about the results. For example if the survey data always indicate that depths systematically decrease from port to starboard, regardless of heading of the survey line, that would be unnatural and indicate a problem with the roll sensor or the relative angle between the transducer and the vertical reference unit. Another example might be when the position of a particular localized feature, like a rock on the seabed that is measured on one survey line, does not agree with the position of the same localized feature that is determined on another survey line. This and the occurrence of a disjoint mismatch in a large-scale bottom feature, when viewed on different survey lines, may indicate one of several parameters about the installation of the sonar and its associated equipment have changed Conducting a patch test calibration is just about the only way to resolve the root cause of such changes in performance of the survey system.

An important component of controlling the quality of survey data is for the sonar operator to assure that sound speed profiles in the survey area are adequately sampled in space and time. As part of this practice, it is necessary to maintain operator review of the swath data to quickly detect evidence that the sound speed profile being used to acquire/process the swath data differs from the actual sound speed conditions. In general, it is a fairly safe assumption that the seabed is comprised of large flat facets that may be tilted in one direction or another. Therefore, when the seabed seems to curve upwards or downwards to both port and starboard of the survey track, the sound speed profile of the water column is definitely not the same as the sound speed profile that is being used in the conversion of travel times at particular angles (beams) to vertical and horizontal positions. This situation should be remedied as soon as possible by measuring the local sound speed profile and using the new profile for converting acoustic travel times into depths and positions.

Modeling of the Total Propagated Uncertainty (TPU) is a key part of modern remote sensing, including quantitative analysis of errors of measurement in multibeam sonars. This work has extended the traditional step-wise correction to echosoundings to incorporate near-real-time modeling of uncertainty of measurements in sonar, including the physical properties of the equipment itself. Storage of the TPU within the data acquired will allow utilizing the TPU in both metadata and potentially in re-processing of the recorded data.

It is difficult to impossible with many of the current acquisition systems to re-process, in an efficient way, the raw acoustic information into new bathymetric solutions after data have been collected. Thus, primacy of quality control during acquisition must be maintained. Normally, the use of equipment and software operating manuals, augmented by the hydrographer's notes from past experience with the equipment, form the basis for this control. There are increasingly stringent requirements to maintain quality control at the operator's console and minimize the first-time usage of new technology on riskier projects. Training, improved software, mentored data acquisition operation, and management-by-results will provide users with the requisite tools and incentives to ensure adequacy of nearly all data, and all of these measures prove their worth over time.

Once collected, data may be analyzed in several ways. Traditionally, the manual methods discussed in the post-processing section above have been built around a dual-operator approach similar to the land surveying method of observing-recording. One person performs the tasks, one person checks the values acquired, and the initial operator double-checks the final results. Denser data sets now provide more measurements of the seabed, enabling the automation of the data inter-comparison and trending of statistical results, or stochastic analysis. These methods, still under development, will greatly simplify and generalize the quality control process in hydrography and are a key step towards sonar sensors becoming part of the Internet of Things (IoT).

Data Deliverables

The traditional view of a complete survey, which is a drawing representing the area at a specific scale, is still produced for nearly all hydrographic projects today. However, provision of a digital data set that represents the survey area, with at least as much resolution as the plotted sheet, is becoming standard practice. The usefulness of this digital data, when combined with other Geographic Information System (GIS) data, is reliant upon its accuracy, so it must be scrupulously compared with interim products, including plotted sheets, to ensure the signed survey and the digital data are equivalent at the required scale. A gridded product must similarly have requisite metadata that links back to the original source with relevant quality information. The new S-102 standard (BAGs, above) store metadata to meet International standards for broad use of geographic data, and this assists the industry to evolve away from scale-based products. Modern positioning and sonar methods have made the use of survey scale somewhat quaint, since accuracies and precision of measurements are now more closely linked to the depth of the survey than the scale of the survey. Per IHO Standard SP-44 (see Chapter 3 of this volume for more information) the typical allowable error for a modern shallow water survey, at the 95% confidence level, would be expressed mathematically as:

$$\pm\sqrt{[\,a^2 + (b^*d)^2\,]}$$

Where "a" is equal to 0.5 meters and represents the sum of all uncertainties invariant with depth, (b*d) represents the sum of all depth dependent errors, "b" is equal to 0.013 and is the factor for depth dependent uncertainty, and "d" is depth in meters (IHO, 2008).

Seabed Survey Data Model (SSDM) is an approach to develop a single deliverable for as-built bathymetry surfaces survey including pipelines, drilling structures, and other man-made items.

Cost Considerations

The costs for most hydrographic projects are driven primarily by the availability of personnel and platforms. When there are large numbers of petroleum and other high-cost deep-ocean projects in progress, qualified hydrographic personnel become scarce and expensive. Vessel time is similarly tied to the ocean industry, where daily vessel lease costs are several hundred to several thousand dollars per day, dependent on vessel size and equipment. Another major factor is the working area; remote regions require higher mobilization costs and strict adherence to vessel availability schedules, whereas surveys of harbors near major metropolitan areas are easier to both equip and man. Finally, the type of survey, as indicated above, has a very large effect on the potential costs of the work. Higher accuracy and confidences require more people with more knowledge to be involved in the work, thus increasing the project cost, and the length of the project affects the ratio of mobilization costs to overall survey production costs. The equation is not simple, and there are other factors that could affect final survey costs in measurable ways, but not in all situations. Surveying is a high-risk industry because it is closely linked to capital expenses, weather, and schedules mandated by these availability restrictions.

Costs for a hydrographic survey can easily range from $1,000 to over $100,000 per square nautical mile, a substantial range to be interpreted by an inexperienced surveyor bidding on contract work or a potential customer interested in procuring services. Proposing an expensive, technically challenging solution to the survey task is not always required. Often the smaller one- or two-man survey firms, with only a vertical beam echosounder and analog side scan sonar, can do a job well and efficiently. Matching experience to the task at hand, these smaller firms can focus on one aspect of the hydrographic survey industry. They optimize their processes to the fullest extent, maintain their equipment through complete amortization, and thus maximize their profits while providing their customers with good survey products. Larger firms with higher values of investments in personnel and equipment can perform more remote and extended survey tasks.

Technology Advancements

The next major technology advancement in sonar systems, both bathymetric and imaging, will be in the manner in which beams are electronically formed. The present techniques will be modified to include the principle of focusing. Through the use of focusing, it is possible to achieve angular resolutions at ranges very close to the transducer that are as spatially acute as the resolutions that current technology achieves at ranges that are very far from the transducer. One benefit of focusing is readily apparent in a side scan sonar design that can simultaneously form several focused beams which lie parallel to each other in the along-track direction. In such an instance, focusing provides the ability to considerably increase the speed at which side scan imagery can be obtained without having gaps in the along-track coverage. That translates into a considerable reduction in ship time for a particular side scan survey, thus increasing survey efficiency. Another benefit of focusing is evident in a multibeam bathymetric system which forms focused beams. In such an instance, focusing can provide exceptional definition of bottom features and man-made objects on the seabed that have a vertical expression which is significant compared to the horizontal expression, for example a thin pile rising proudly from the bottom of a lake. This also translates into better definition of small details on large complex targets, which in turn provides improvement in the quality of range (depth) measurements that may translate into a cost benefit on many surveys.

Improving bandwidth, especially in lower-cost systems, has also recently improved the ability of sonars to differentiate the returns from objects with small phase (time) differences. The fisheries-targeted echo sounders application requires pelagic (water column) returns very close to the bottom to identify benthic-habitat based fish. As these systems continue to improve, their technology bleeds over into the bathymetry-based systems, resulting in ever-increasing resolution. Display and analysis of these data products are a growing interest in the GIS community. By leveraging NetCDF and other open formats to display complex three dimensional data with time in several automatically-clustered layers, the development of new digital elevation models for data types beyond bathymetry or sea surface is expected to transform oceanography.

In addition to beam forming and bandwidth improvements, mapping sonar electronics are being reconsidered in order to offer more predictable backscatter products. The words, 'calibrated' or 'documented' are being used to describe backscatter that is more repeatable and lends itself to quantitative analysis. While truly calibrated backscatter will be difficult to achieve, improvements are being made. These include the design of robust linear electronics that minimize the variability in backscatter within the sonar and a trend toward higher dynamic range analog to digital conversion and digital time varied gains. A few multibeams now offer 24-bit analog-digital (A/D) conversion and digital time varied gain (TVG). These acquisition parameters have been used for a long time in oil field seismic exploration and are now making their way into high frequency sonar design.

Processing of side scan sonar data to account for the incident angle of the echo on the underlying bathymetry in near real time is a promising area of development. Using this and other techniques to improve the inter-comparison of data from different sensors should lead to faster surveys and increased confidence in the survey results, especially for object detection. Further research into the oceanographic properties of the littoral ocean, specifically the local variability of sound propagation, will also lead to operational efficiencies due to increased allowable swath widths in multibeam echosounding. Automation of the processing algorithms used to clean multibeam echosounding depths in near-real time, such as the Combined Uncertainty Bathymetric Estimation (CUBE) algorithm (Calder, CCOM/UNH, 2002) will lead to rapid acquire-review-submit cycles, and improve overall throughput from the surveyor. All of these advancements will enable hydrographic survey data to be an increasingly important facet of the GIS databases used for habitat preservation, municipal planning, and ecosystem development in the future.

AUTHOR BIOGRAPHIES

Captain Guy T. Noll, NOAA Corps, retired, is currently the Maritime Principal Consultant and Database Services Department head at the Redlands, California location of Geographic System Information software provider Esri. While on active duty, he was the Commanding Officer, NOAA Marine Operations Center-Pacific, and also served as Chief of Party/Commanding Officer of NOAA Ship RAINIER, a nautical charting-tasked hydrographic ship in the National Oceanic & Atmospheric Administration Office of Marine and Aviation Operations. He was formerly the Chief of the Hydrographic Systems and Technology Programs and Deputy Chief of the Hydrographic Surveys Division for the NOAA Office of Coast Survey in Silver Spring, Maryland. Captain Noll was the lead systems integrator for the five multibeam echosounder hydrographic systems installed aboard the NOAA Ship RAINIER from 1998 through 2000 and deeply involved in documenting best-practices and specifying sonar acquisition systems for NOAA's fleet of fisheries and mapping ships. Captain Noll has a Bachelor of Science degree in Civil Engineering from Columbia University and a Bachelor of Arts degree in Physics/Mathematics from Whitman College.

Mr. Doug Lockhart is General Manager/VP of Teledyne Caris. Previously, Mr. Lockhart served as Chief Scientist at Fugro Pelagos and as Senior Staff Engineer at the Marine Mineral Technology Center at the University of Mississippi. Mr. Lockhart holds bachelor and master's degrees in Petroleum Engineering from the University of Alaska, Fairbanks.

ACKNOWLEDGEMENTS

The initial two editions of this chapter were co-authored with Dr. Lloyd Huff, who has more than 50 years of experience in designing and using acoustic instrumentation for marine surveys of various parameters, including: tides, currents, surface waves, sediment transport, and depths. From 1985 to 1999, he was Chief of the National Ocean Service Hydrographic Technology Programs. Starting in 2000, he became a senior hydrographic researcher at the Center for Coastal Ocean Mapping/Joint Hydrographic Center, University of New Hampshire, in Durham, New Hampshire. Dr. Huff has a Bachelor of Science degree in Physics from Southwestern University, a Master of Science degree in Electrical Engineering from the University of Rhode Island, and a Doctorate of Philosophy degree in Ocean Engineering from the University of Rhode Island. His work from these previous editions has been incorporated into this chapter, with modification to update technology references from 2008.

The authors also express gratitude to Al Rougeau of Odom Hydrographic Systems, Inc., who provided considerable background material for the prior editions of this chapter. They acknowledge extended use of the U.S. Army Corps of Engineers Publication EM 1110-2-1003, Hydrographic Surveying, and the NOAA Hydrographic Survey Specifications and Deliverables without specific citation.

REFERENCES

Bell, T. A. (1962). Sonar and submarine detection. In *U.S. Underwater Sound Lab Report 545.*

Calder, B. R. and Mayer, L. A. (2001). Robust automatic multi-beam bathymetric processing. In *Conference Proceedings, US Hydrographic Conference, 2001,* May 21–24, 2001, Norfolk VA.

Chesterman, W.D., P. R. Clynick, and A. H. Stride (1958). An acoustic aid to seabed survey. In *Acustia,* 8, (pp. 285–290).

Chramiec, M. A. and Morton, R. W. (1970). High resolution near sub-bottom profiling. In *Proceedings, Second Annual Offshore Technology Conference.*

Cloet, R.L. and C. R. Edwards (1986). The bathymetric swathe sounding system. In *The Hydrographic Journal,* 40, (pp. 9–17).

Colbo, Keir, Ross, Tetjana, Brown, Craig, Weber, and Thomas, A. (2013). A review of oceanographic applications of water column data from multibeam echosounders. In *Estuarine, Coastal and Shelf* Science 145 (2014) 41e56.

Denbigh, P. N. (1989). Swath bathymetry: principles of operation and analysis of errors. In *IEEE, Journal of Oceanic Engineering,* 14(4) (pp. 289–298).

Doppler, C. J. (1842). *Über das farbige licht der doppelsterne und einiger anderer gestirne des himmels* (About the colored light of the binary stars and some other stars of the heavens). Publisher: Abhandlungen der Königl. Böhm. Gesellschaft der Wissenschaften. V. Folge, Bd. 2, S. (pp. 465–482) [Proceedings of the Royal Bohemian Society of Sciences (Part V, Vol. 2)]; Prague.

Farr, H. K. (1980). Multibeam bathymetric sonar: sea beam and hydrochart. In *Marine Geodesy*, 4, (pp. 77–93).

Fish, J. P. and H. A. Carr (1990). *Sound underwater images: a guide to the generation and interpretation of side scan sonar data.* Orleans, MA, Lower Cape Pubs.

Flemming, B. W. (1976). Side-scan sonar: a practical guide. In *International Hydrographic Review LIII* (1), (pp. 65–92).

Glenn, M. F. (1970). Introducing an operational multi-beam array sonar. In *International Hydrographic Review,* XLIII, (pp. 35–39).

Geen, M. F., P. D. Hewitt, and A. R. Adams (1993). The ISIS interferometric seabed inspection sonar. In *Proceedings of the Institute of Acoustics, Acoustic Classification and Mapping of the Seabed,* Vol. 15 Pt 2, University of Bath.

Godin, A. and J. P. Ramalho-Marreiros (1998). Attitude and squat assessment for hydrographic launches using GPS positioning. In *Proceedings of the Canadian Hydrographic Conference,* March 10–12, 1998. Victoria, BC, (pp. 291–308).

Hamilton, E. L. (1971). Elastic properties of marine sediments. In *Journal of Geophysical Research* Vol. 76, No 2, Jan. 1971.

Harre, I. (1992). Multi-beam echosounders for inshore to rough-seas applications and the related data processing—an equipment survey. In *Proceedings of The Eighth Biennial International Symposium of the Hydrographic Society, HYDRO'92,* IHO Special Publication No 29.

Hawley, J. H (1931). *Hydrographic manual, special publication No. 143, Coast and Geodetic Survey,* U S Department of Commerce, United States Government Printing Office, Washington, D.C.

Heald, G. J. and N. G. Pace (1996). Implications of a bi-static treatment for the second echo from a normal incidence sonar. In *Proceedings of 3rd European Conference on Underwater Acoustics,* June, 1996 (pp. 649–554).

Horton, J.W. (1957). Fundamentals of Sonar, United States Naval Institute, Annapolis, MD.

Hou, T., L. C. Huff, and L. Mayer (2001). Automatic Detection of Outliers in Multibeam Echo Sounding Data, U.S. Hydrographic Conference, May 21-24, 2001, Norfolk VA.

Huff, L. C. (1993). High-Resolution Multi-Beam Focused Side Scan Sonar, Proceedings of the Institute of Acoustics, Acoustic Classification and Mapping of the Seabed, Vol. 15 Pt 2, University of Bath.

Hughes Clarke, J.E. et al. (1999). Data Handling Methods and Target Detection Results for Multibeam and Side Scan Data Collected as Part of the Search for Swiss Air Flight 111, Proc. Shallow Survey-99, Australian Defense Science and Technology Organization, Oct 18-20, Sydney, Australia, Paper 6-2, 11 pp.

Klein, E. (1968). Underwater Sound and Naval Acoustical Research before 1939, *Journal of the Acoustical Society of America,* 43:931.

Krabill, W.B., and Martin, C.F. (1987). Aircraft Positioning Using Global Positioning System Carrier Phase Data, Navigation: Journal Institute of Navigation, Vol. 34, Spring, pp. 1-21.

Laughton, A. S. (1981). The first decade of GLORIA, *Journal of Geophysical Research, 86,* pp. 11511-11534.

Lurton, X. and Lamarche, G. (2015). *Backscatter measurements by seafloor-mapping sonars*—Guidelines and Recommendations: A collective report by members of the GeoHab Backscatter Working Group.

MacPhee, S. B. (1979). Underwater Acoustics and Sonar and Echo Sounding Instrumentation, Canadian Hydrographic Service Technical Report.

Mayer, L. A. and L. R. LeBlanc (19983). The CHIRP Sonar: A New Quantitative High-Resolution Profiling System, Pace, N. G. ed., Acoustics and the Sea-Bed, Bath University Press.

McConnaughey, R. and Lomnicky, J. (2008). *Integrating Charting and Acoustic Habitat Research, Sea Technology* (ISSN 0093-3651). Compass Publications Inc. with additional links through *NOAA Office of Coast Survey blog*.

Miller, J. E., J. E. Hughes Clarke, and J. Paterson (1997). How effectively have you covered your bottom? *The Hydrographic Journal,* No. 83, pp. 3-10.

Mitchell, N. C. (1991). Improving GLORIA Images Using Sea Beam Data, *Journal of Geophysical Research,* 96, pp. 337–351

NOAA (2000). *NOS Hydrographic Surveys Specifications and Deliverables,* p. 43, originally published in 2000.

Owaki, N. (1963). A Note on Depth When the Bottom is Soft Mud, *International Hydrographic Review.,* XL, No. 2, pp. 41–43.

Pinkel, R. (1980). Acoustic Doppler Techniques. In Dobson, F., L. House, & R. Davis (Eds.), *Air Sea Interactions* (pp. 171–199). New York, NY: Plenum Press.

Renard, V., and J-P Allenou (1979). Sea Beam, Multi-Beam Echo-Sounding in "Jean Charcot" Description, Evaluation, and First Results, *International Hydrographic Review,* LVI (1), January, 1979.

Rougeau, A., L. Rosenbalm, and C. Sonnier (1992). Integrated Acoustic Technologies Form New Shallow Water Survey System, *Marine Geodesy* 15, pp. 187–198.

Satriano, J. H. (1991). Wide Swath Bathymetry for Multibeam Sonar Systems, *Proceedings of OCEANS'91, Ocean Technologies and Opportunities in the Pacific for the 90's,* Honolulu, HI, October 1991, Vol. 2, pp. 733–736.

Steenstrup, J. R., and R. W. E. Luynenburg (1992). Multibeam Echosounder—On an ROV, *Sea Technology,* 33 (6), pp.17–21.

Steenstrup, P. R. (1992). New cost effective multi-beam echo sounder systems for shallow water: 0 to 50 meters., *Proceedings of the Eighth Biennial International Symposium of the Hydrographic Society, HYDRO'92,* IHO Special Publication No 29.

Sternlicht, D.D. and Hagen P.E. (2009). Special Issue on Synthetic Aperture Sonar—Guest Editorial in *IEEE Journal of Oceanic Engineering,* Vol. 34, No. 3, p 205, July 2009.

Talukdar, K. K. and R. C. Tyce (1991). Digital Processing of Sidescan Images from Bottom Backscatter Data Collected by Sea Beam, *Marine Geodesy,* 14, 81–100.

Tyce, R. C. (1986). Deep Seafloor Mapping Systems—A Review, *Marine Technical Society Journal,* 20, pp. 4–16.

Tucker, D. G. (1966). Underwater Observations using Sonar, Fishing News Books Ltd, London.

Tucker, M. J. and A. R. Stubbs (1961). A narrow beam echo-ranger for fisheries and geological investigations, British Journal of Applied Physics, V. 12, pp. 103–110.

Urick, R.J. (1956). Processes of Sound Scattering at the Ocean Surface and Bottom, *Journal for Maritime Research* 15:134 1956.

Urick, R.J. (1967, 1975, 1983). Principles of Underwater Sound, 1st, 2nd, and 3rd editions, McGraw-Hill, Inc.

Wetzler, M.A., Van den Ameele, E.J. and Noll, G.T. (2001). Shallow Water Multibeam System Testing for Object Detection over a Defined Reference Surface, *Proceedings of US Hydrographic Conference 2001.*

Wheaton, G.E. (1988). "Patch Test", A System Check for Multibeam Survey Systems, *Proceedings of the U.S. Hydrographic Conference '88,* Baltimore, pp. 85–90, 1988.

Chapter 12

ENABLING TECHNOLOGIES

Bruno M. Scherzinger, Joseph J. Hutton, and Mohamed M.R. Mostafa

This chapter provides an overview of the georeferencing technologies that have enhanced or in some cases enabled DEM data acquisition. These include:

- Precise GNSS positioning
- GNSS-Aided Inertial Navigation System
- Direct Georeferencing

PRECISE GNSS POSITIONING

A satellite based positioning system is referred to as a Global Navigation Satellite System (GNSS). Today there are four such systems either in development or in operation: the Global Positioning System (GPS) (U.S.), GLONASS (Russian Federation), Galileo (European Union), and the Beidou Navigation Satellite System (BDS) (Peoples Republic of China). At the time of publication GPS and GLONASS are fully operational. Galileo reached initial operational capability at the end of 2015 and is expected to be fully operational by 2019. Beidou is fully operational in China and surrounding countries and is expected to be globally operational by 2020. Japan has deployed the Quasi-Zenith Satellite System (QZSS) comprising four satellites in geostationary orbit to supplement GPS in Japan and surrounding regions.

Most high-accuracy survey grade applications use a combination of signals from satellites in view belonging to any of the fully or partially operational GNSS's to improve the reliability and accuracy of the overall solution, primarily through enhanced satellite coverage. Thus a GNSS receiver designed to support such applications can track signals from GPS, GLONASS, Galileo, and Beidou.

Technology Overview

A GNSS is fundamentally comprised of a constellation of satellites carrying atomic clocks that broadcast time over radio frequency signals, and an arbitrary number of receivers which receive the signals from visible satellites. Each receiver then computes its position on the Earth by measuring the time it takes for the signals to propagate from the satellites to its location. The currently operational GNSS's; namely: GPS and GLONASS, have been extensively documented, and each operate in approximately the same way. Every satellite transmits its time encoded in a scheduled pseudo-random bit stream on each of two or more L-band (1-2 gHz) frequencies, and a navigation message stream that contains the information required to compute a satellite's position and clock correction. At each sample time or epoch a GNSS receiver does the following:

- Determines the approximate signal transit times between multiple satellites and the receiver
- Computes the positions of the satellites at their respective times of transmission
- Using the above mentioned data it computes the receiver's position on the surface of the Earth

The actual mechanizations of the GNSS's differ somewhat. This overview starts with GPS to describe the basic characteristics of a GNSS, and then adds the key features that differentiate the other GNSS's. It then describes how a GNSS can be used to compute a position fix with centimeter accuracy.

GPS comprises three major segments: Space, control, and ground. The GPS space segment comprises 24 to 32 satellites that orbit the Earth in six orbital planes with an approximate 12-hour orbital period (see Figure 12.1). A complete constellation comprises 24 satellites for global coverage. Up to 32 satellites may be active if older satellites and their newer replacements are both active, as has been the case during the past GPS replenishment program and current GPS modernization program. The orbital planes are inclined 55 degrees with respect to the equatorial plane. The design of the space segment with the minimum 24 satellites puts at least four satellites in view at any time anywhere on the Earth. Typical visibility in most parts of the world ranges from 5 to 12 satellites.

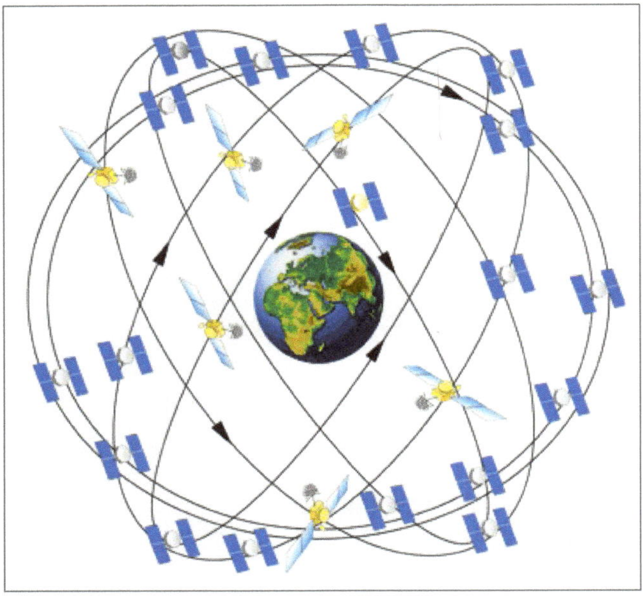

Figure 12.1. GPS Satellite Constellation (Image courtesy of www.gps.gov).

The satellites transmit spread-spectrum modulated signals centered at 1575.42 mHz, known as the L1 signal, and 1227.6 mHz, known as the L2 signal. The code division multiple access (CDMA) modulation used by GPS (also used in cellular telephony) is a form of spread-spectrum modulation that allows multiple satellites to transmit signals on the common L1 and L2 bands. The modulated signal contains one or more pseudo-random codes that are unique to each satellite.

GPS satellites are identified by the pseudo-random number (PRN) in the range 1 to 32 used to initialize a GPS satellite's pseudo-random code generators. The legacy signals that have been transmitted since GPS's advent in 1970s are the Coarse Acquisition (CA) code and the Precise (P) code.

The CA code has a repetition period of one millisecond and a bit or chip length of 1 microsecond. The P code has a repetition period of 267 days and a chip length of 100 nanoseconds. It is encrypted to form a Y code, the purpose of which is to deny unauthorized access to precise positioning and to prevent spoofing of the P-code by an adversary. Each satellite broadcasts a unique P(Y) code on the L1 and L2 signals that enables authorized users to calibrate and remove signal delays induced by the ionosphere. Each bit of the pseudo-random codes is transmitted on a published time schedule as determined by the satellites' on-board atomic clocks. In this sense the pseudo-random codes encode their respective times of transmission. Figure 12.2 shows GPS signals and their associated frequencies.

The current operational GPS Control Segment includes a master control station located at the Schriever Air Force Base in Colorado Springs, Colorado, an alternate master control station at Vandenberg Air Force Base, in California, 11 command and control antennas, and 15 monitoring sites as shown in Figure 12.3.

The control segment has the task of controlling the operation of the satellites and calibrating their orbits. The control station periodically uploads navigation and clock correction parameters that the satellites then broadcast in their navigation messages. GPS time is coordinated universal time (UTC) plus the number of leap seconds since January 6, 1980 when GPS time was initialized. GPS time increments continuously

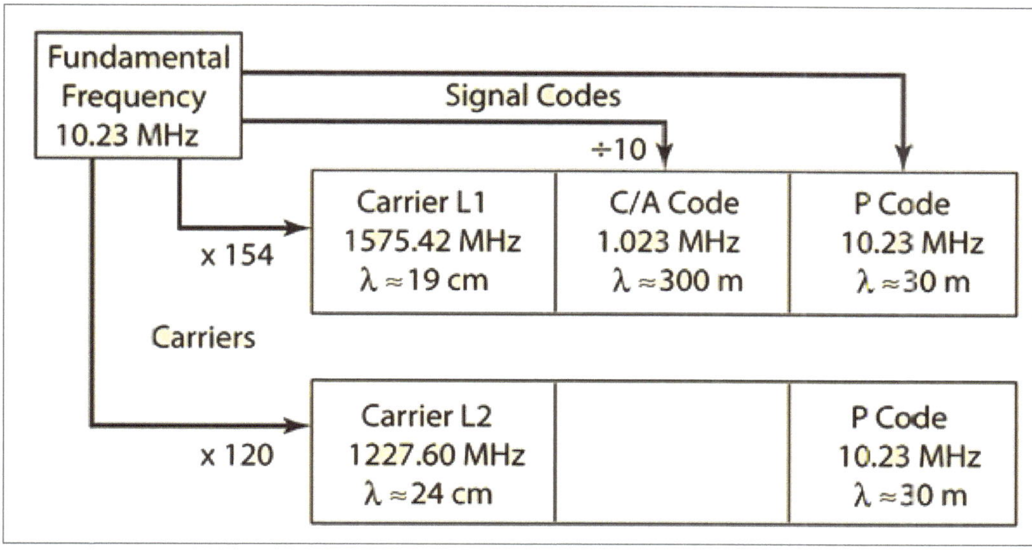

Figure 12.2. GPS Signals.

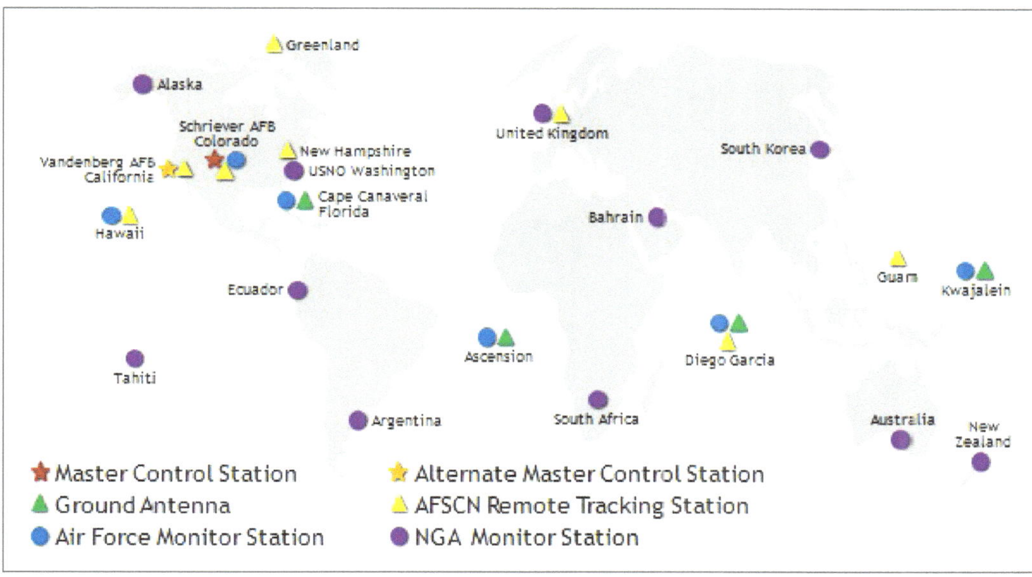

Figure 12.3. GPS Control Segment (Image courtesy of www.gps.gov).

without leap second correction, and consequently has, as of June 30, 2015 advanced 17 seconds against UTC since initialization. It is maintained by the Master Clock at the United States Naval Observatory (USNO) in Washington D.C.

The GPS modernization program was started in 2008 and is currently ongoing with the replacement of older satellites with Block IIR(M) shown in Figure 12.4 on page 354, Block IIF and Block III satellites. It adds several new signals to the legacy L1 and L2 signals described previously. The civilian L2 or L2C signal

Figure 12.4. A GPS IIR(M) satellite (courtesy of Lockheed Martin).

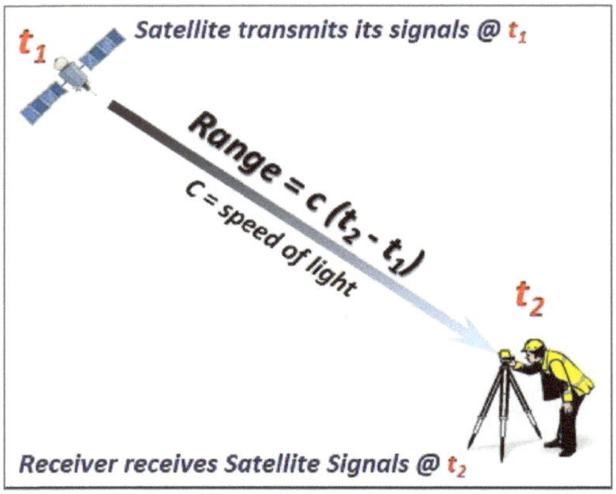

Figure 12.5. GPS ranging concept.

appearing on Block IIR(M) and subsequent satellites is a Coarse Acquisition code signal modulated on the L2 carrier that permits commercial users to remove ionosphere-induced signal delays. The L5 signal appearing on Block IIF and subsequent satellites is a civilian-use signal with a center frequency of 1176.45 mHz, and is intended for safety-of-life applications such as civilian aircraft terminal guidance and landing.

The GPS user segment comprises all user equipment, commonly called GPS receivers, which perform positioning or time transfer using the GPS signals. A GPS receiver performs two fundamental functions, namely: Generating observables and computing user position and velocity. The observables for a given frequency (L1, L2, L5) and a given signal type (CA, P(Y), L1C, L2C) from a given GPS satellite comprise the pseudorange, the carrier phase and the Doppler frequency. Figure 12.5 shows a schematic of the GPS ranging concept.

A GPS receiver contains a radio frequency (RF) section that converts the received signal from the antenna to a digital sample stream and a signal processing section that processes the digitized signal. The signal processing section contains high-speed hardware organized into multiple tracking loops. Each tracking loop comprises a delay-lock loop (DLL) and a phase-lock loop (PLL). Each DLL locks onto one of the modulated codes in the GPS signal. Each DLL computes the correlation between a received code and a receiver generated version of the same code delayed in time with respect to the published time of transmission so as to maximize the correlation.

The receiver has a crystal clock that provides the time to the DLL's. The amount of delay is the estimated signal transit time, which multiplied by the speed of light yields the estimated range. This comprises the true range plus range errors, the largest of which is the receiver clock error. For this reason it is called the pseudorange observable. The corresponding PLL locks onto the carrier phase of the signal and yields an integrated Doppler frequency called the carrier phase observable. The tracking loops are organized into channels, each of which generate the observables from one satellite. An all-in-view GPS receiver traditionally had 12 channels to track all GPS satellites in view and generate their observables. Nowadays, GNSS receivers have a minimum of 60 and up to hundreds of channels for tracking all GNSS satellites in view. A single frequency GPS receiver generates the CA pseudoranges and carrier phases on the L1 carrier from each satellite. A high-performance multi-frequency receiver generates the pseudoranges and carrier phases from the CA, P(Y) and L2C signals modulated on the L1 and L2 frequencies. This requires significantly more signal processing hardware, consequently such a receiver is more expensive than a single frequency receiver.

GLONASS, Galileo, and Beidou are similarly structured to GPS in that they comprise space, control and user segments. Their respective space segments comprise satellites that generate signals on L-band frequencies containing satellite time and position information that a receiver uses to compute its position. They mainly differ in transmitted frequencies, modulation schemes and pseudo-random codes.

The GLONASS space segment comprises a constellation of 24 satellites in three orbital planes with 64.8 degrees inclination to provide better coverage at higher latitudes. The satellites use frequency domain multiple access (FDMA) modulation to transmit CA code and P codes on one of 20 dual-frequency channels. The G1 frequencies comprise 1602 mHz plus or minus an integer multiple of 562.5 kHz. The G2 frequencies comprise 1246 mHz plus or minus an integer multiple of 437.5 kHz. The second generation GLONASS-M satellites broadcasting only FDMA signals are fully deployed. Third-generation GLONASS-K1 satellites are being deployed to replace aging M satellites. The K1 satellites broadcast both legacy FDMA signals and CDMA signals on GPS L1 and L5. Later generation satellites will transmit on a new G3 band centered at 1207.14 mHz. The eventual goal of the GLONASS modernization program is to completely replace the legacy FDMA signals with CDMA signals that are inter-operable with GPS and Galileo.

The GLONASS control segment is located in Moscow and performs a similar task of satellite ephemeris and clock management. The GLONASS user segment comprises GLONASS receivers that track GLONASS signals and compute the position of the antenna. The hardware configuration of a GLONASS receiver is very similar to that of a GPS receiver, the only difference being the demodulation and tracking of signals on multiple frequencies. A GLONASS receiver can contain frequency-dependent inter-channel biases that do not appear in a GPS receiver in which all channels operate on the same L1, L2, and L5 frequencies. Consequently GLONASS receivers are inherently less accurate than GPS receivers, which is another reason for the intended transition to CDMA modulation.

The Galileo space segment when completed in 2019 will comprise 30 satellites in three orbital planes with 56 degree inclinations, thereby providing better coverage of northern and southern hemispheres than GPS. As of May 2016 the system has 14 of 30 satellites in orbit. The orbital radius of 23,222 km provides a 14 hour orbit period, which is slightly higher and longer than those of GPS. The satellites transmit signals on four bands called E1, E5a, E5b, and E6. The E1 and E5a bands are centered on the GPS L1 and L5 frequencies thereby providing inter-operability with GPS. The E5b band is centered on 1207.14 mHz, which is the new GLONASS frequency. The E6 band with center frequency 1278.75 mHz is adjacent to the GPS L2 band. Galileo offers three services on these frequencies. The Open Service (OS) provides positioning, velocity and timing information that can be accessed free of direct user charge. The Commercial Service (CS) provides higher performance than offered by the Open Service plus other value-added services for a subscription fee. The Public Regulated Service (PRS) is limited via encryption to use by civil institutions such as police coast-guards and customs.

The Galileo Ground Segment comprises the Galileo Control System and the Galileo Mission System. The Galileo Control System (GCS) is responsible for satellite constellation control and management of Galileo satellites. It provides the telemetry, command and control functions for the whole Galileo satellite constellation. The Galileo Mission System (GMS) is responsible for the computation and uplink of navigation and integrity data messages needed to provide navigation and UTC time transfer services. The Galileo User Segment comprises the Galileo or GNSS receivers that receive Galileo signals, determine pseudoranges and possibly other observables, and solve the navigation equations to obtain a position-time solution. Galileo system time (GST) is expected to be within a few nanoseconds of GPS time, thereby making GPS and Galileo inter-operable.

The Beidou Navigation Satellite System is a second-generation navigation system designed to provide global navigation services via 35 satellites comprising 27 satellites in medium earth orbits, 5 satellites in geostationary orbits over China and surrounding regions and three satellites in inclined geosynchronous orbits also over China and surrounding regions. The satellites transmit on four bands: B1 centered at 1561.098 mHz,

B1-2 centered at 1589.742 mHz, B2 centered at 1207.14 mHz, and B3 centered at 1268.52 mHz. Beidou provides an Open Service available to all users and an Authorized Service limited to authorized users via encryption. Beidou has reached regional operation status in China and surrounding regions, and is scheduled to reach global operational capability in 2020.

The Beidou ground segment comprises a master control station, two upload stations and 30 ground monitor stations that collectively perform satellite control, orbit and clock monitoring, and navigation data upload functions similar to the other GNSS's. The Beidou user segment again comprises Beidou-capable receivers that receive Beidou navigation signals, determine observables and solve the navigation equations to obtain a position-time solution.

GNSS Single Point Positioning Concept

A high-accuracy GNSS receiver is designed to receive, demodulate and track signals from all GNSS satellites. The signal processing hardware contains over 100 channels that can be configured for any supported GNSS satellite. Such a receiver becomes capable of generating observables for all GNSS satellites in view. This becomes advantageous when visible sky occlusion by buildings or trees blocks the signal paths to some satellites. A GNSS receiver typically uses GPS time to schedule observables sampling and therefore needs to handle the differences between GPS and other GNSS times.

The pseudorange to each GNSS satellite is the computed time of signal propagation from the satellite to the GNSS antenna multiplied by the speed of light as shown in Figure 12.5 and Figure 12.6.

Each GNSS signal includes the navigation message (shown in Figure 12.7) that contains the parameters required by the receiver to compute the satellite position and system time when the received signal left the satellite. The receiver maintains a local inaccurate estimate of time in a crystal clock in order to compute the time of propagation. A computed pseudorange thus comprises the true range between a satellite and the receiver antenna plus the receiver clock error plus several smaller range errors due to atmosphere-induced signal propagation delay and satellite orbit and time errors. The receiver clock error is common in all pseudoranges and can be as large as one millisecond, which is an equivalent range error of 300,000 meters. Other range errors include ionosphere delay, troposphere delay, orbital error, multipath error and receiver noise. These are typically less than 10 meters, and hence negligibly small compared with the receiver clock error.

Figure 12.6 shows the vector position relationship between a GNSS satellite and a receiver on the Earth's surface. If the receiver has access to the satellite position and the relative satellite to receiver position, then it can compute its position with respect to the

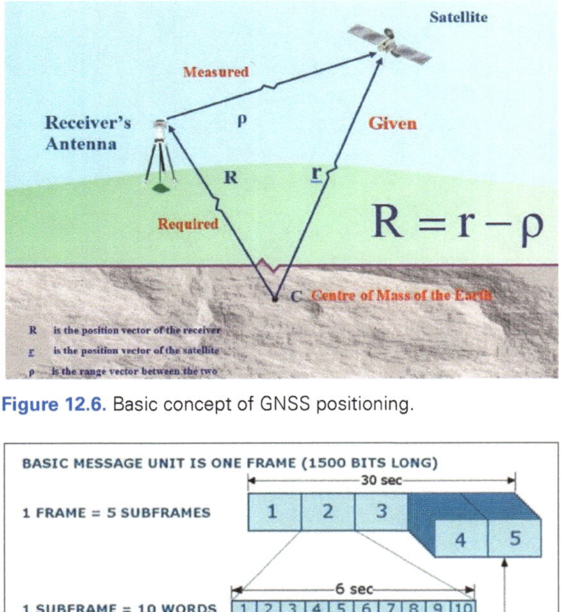

Figure 12.6. Basic concept of GNSS positioning.

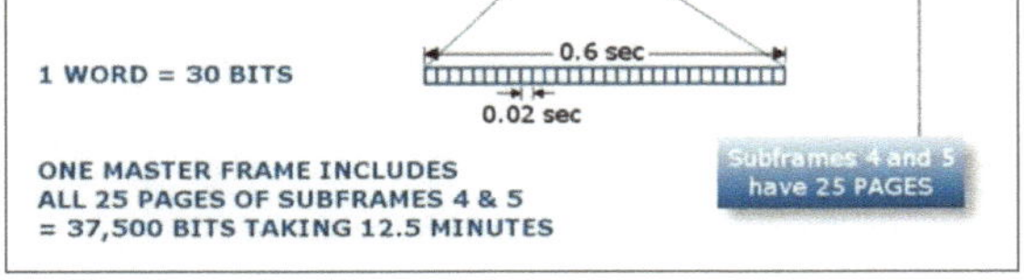

Figure 12.7. GPS Navigation Message (Guide to GPS Positioning, 1987).

Center of Mass of the Earth. In reality, the receiver measures the pseudoranges to the satellites and computes the satellite positions from the navigation data messages (shown in Figure 12.7) that the satellite includes in its transmitted signal. Using those instantaneously computed satellite positions, the receiver computes its position using trilateration of the pseudoranges.

Figure 12.8 shows the basic geometry between a GNSS antenna and four satellites. The lines between the satellites and the antenna indicate the line-of-sight ranges. The receiver requires a minimum of four pseudoranges to solve for four unknowns: three components of position on the Earth's surface and the receiver clock error.

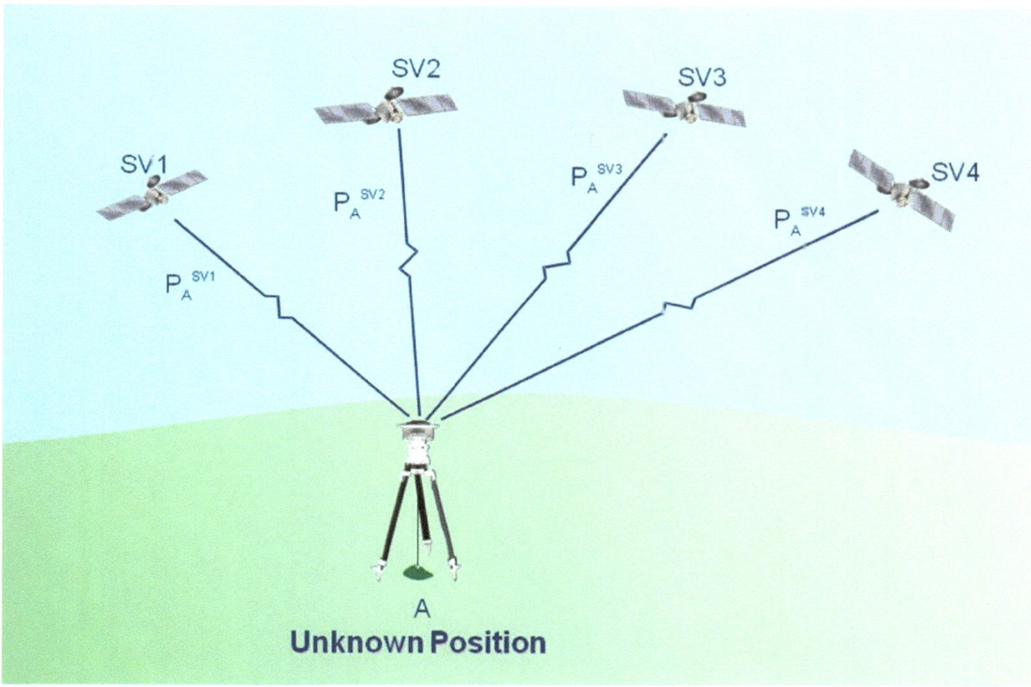

Figure 12.8. GNSS positioning by trilateration.

The geometry of the satellite sky distribution impacts the positioning and timing accuracy. Positioning and timing accuracy is higher when the satellites are uniformly distributed across the visible sky (as shown in Figure 12.9). Accuracy degrades if the satellites are located in one sector of the sky, which frequently happens in land applications in urban canyons. This is captured in a figure of merit called the Dilution of Position (DOP).

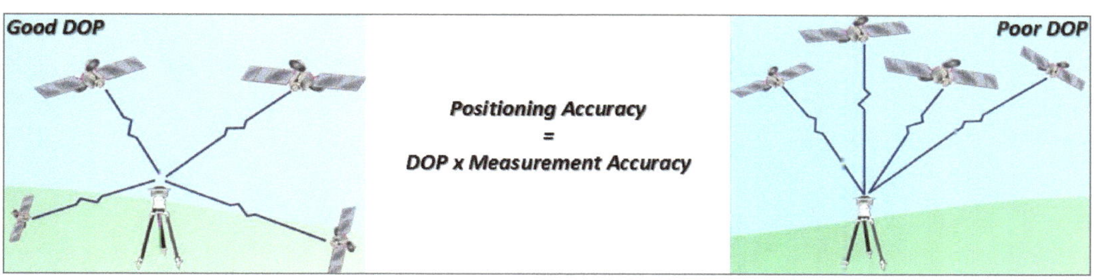

Figure 12.9. GPS DOP.

In simplest terms, the position and time solution error standard deviation is given by the pseudorange error standard deviation multiplied by DOP. The DOP is typically used as a figure of merit for coverage of a position on the Earth by the GNSS satellites. A position solution obtained from CA pseudoranges exhibits an error of 1.5 to 5 meters RMSE, depending on the satellite geometry.

Differential GNSS (DGNSS)

The key technology advancement that moved GPS into the commercial market was differential GPS (DGPS). With the advent of the other GNSS's, this has become differential GNSS (DGNSS). A DGNSS system comprises one or more reference receivers (also called base receivers, master stations, or base stations) at known positions and any number of roving receivers that move within an area covered by the reference receivers.

Figure 12.10 shows the geometry between two GNSS antennas and four satellites. The lines between the satellites and each of the two antennas indicate the ranges. The baseline vector between the ground antennas is the DGNSS solution. If the position of one antenna is known, then the position of the other antenna can be computed from the baseline vector.

The base receivers both individually or collectively provide observations of the errors in the code and phase observables and the broadcast navigation message. If the separation between a roving receiver and a reference receiver is within 50 km, then these errors are sufficiently correlated between the rover and base stations to allow for reducing or removing those errors. Errors increase as the baseline separation increases.

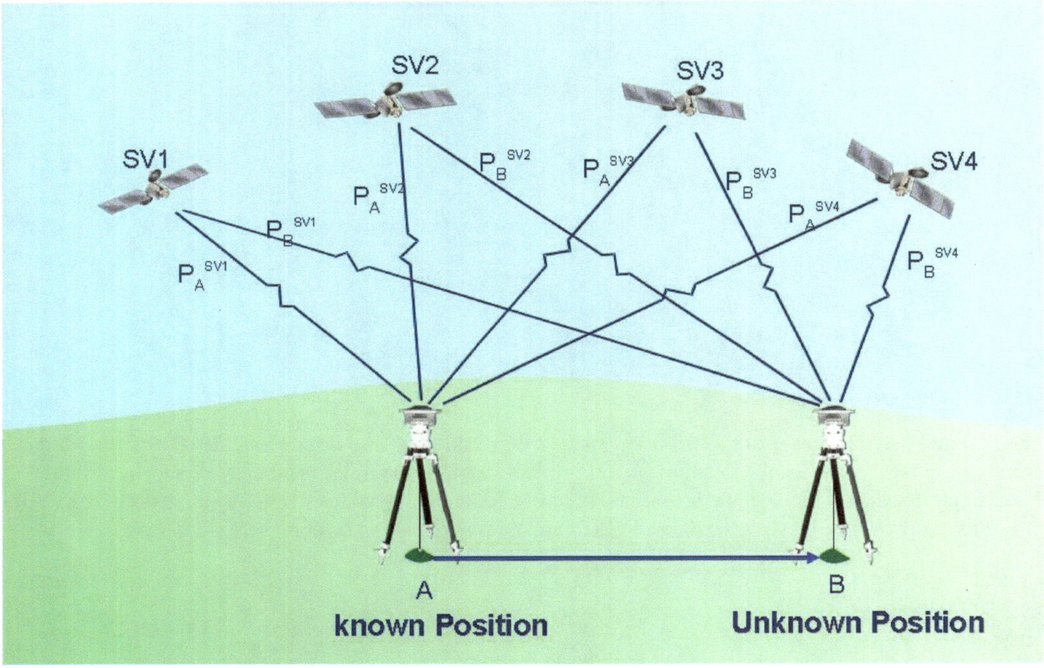

Figure 12.10. Differential positioning.

In a real-time DGNSS system, also called real-time kinematic GNSS (RTK GNSS), a reference station comprising a reference receiver plus a radio transmitter broadcasts the corrections to the roving receiver over a data channel so that the roving receiver can correct its observables and achieve sub-meter accuracy using the CA code. DGNSS correction services that provide these corrections have local, regional, continental or even global coverage.

Precise Phase Interferometry Positioning

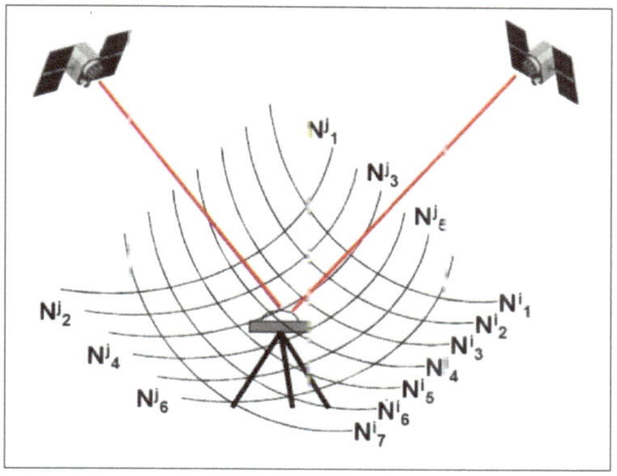

Figure 12.11. GNSS Carrier Phase Ambiguity Resolution.

Precise DGNSS positioning at the centimeter accuracy level is made possible using carrier phase interferometry. This technique uses the fact that a DGNSS receiver can track the different carrier phases modulo one wavelength coherently across the receiver channels with high accuracy. The receiver typically contributes phase noises on the order of a few millimeters. Each reported carrier phase contains among other errors an unknown integer number of wavelengths. A DGNSS system with integer ambiguity resolution combines the phase data from two receivers so as to eliminate all significant errors other than the integer ambiguities. An integer ambiguity resolution algorithm is a statistical search for the set of unknown integer ambiguities among all feasible combinations of integers as shown in Figure 12.11.

In this simple example, the intersecting carrier phases from two satellites define a grid of feasible positions in a search space. Each feasible point corresponds to one candidate set of two integer ambiguities. The carrier phase from a third satellite allows one of the feasible positions to be selected. Redundant phase observations on multiple frequencies from two or more receivers each tracking five or more satellites provide the information for rapid unambiguous resolution. This method of signal processing can be thought of as coherent multi-frequency phase interferometry between two GNSS receivers. Successful ambiguity resolution for baseline lengths between roving and reference receivers of 10 kilometers or less typically requires a few seconds of signal processing. Once the ambiguities are resolved, the corrected phases for each observed satellites become precise ranges that allow the computation of the baseline vector between the receivers with a typical accuracy of 1 to 2 centimeters, depending upon the baseline.

Integer ambiguity resolution was first used to compute baseline vectors between stationary receivers as an alternative to conventional theodolite or total station survey methods. The receivers recorded the phase data, and the surveyor computed the baseline vectors in post-processing. Each receiver was required to occupy a station for 10-30 minutes so that phase errors could be averaged. The achievable baseline accuracy after averaging is less than one centimeter, which is within the requirements for geodetic survey applications.

A subsequent advancement in this methodology is kinematic ambiguity resolution (KAR) resolution where the rover receiver is moving. This allows the relative positioning of the rover receiver with respect to the base receiver with accuracies on the order of a few centimeters. Hence if a fixed base receiver broadcasts the data requirement for KAR, then a roving receiver can compute its position in real time relative to the position of the base receiver to centimeter accuracy. This is called real-time kinematic (RTK) positioning. Due to its efficiency, RTK has become the preferred method of real-time positioning for most types of surveying applications with minimal baseline separation of up to 5 km. Therefore, it's been widely acceptable in construction surveying applications and small land surveying and mapping projects.

Another application of RTK is attitude determination. The relative positioning vector(s) between two or more antennas rigidly mounted with respect to each other on a platform are used to compute the roll, pitch and heading angles of that platform as shown in Figure 12.12 on page 360. A GNSS attitude measurement system

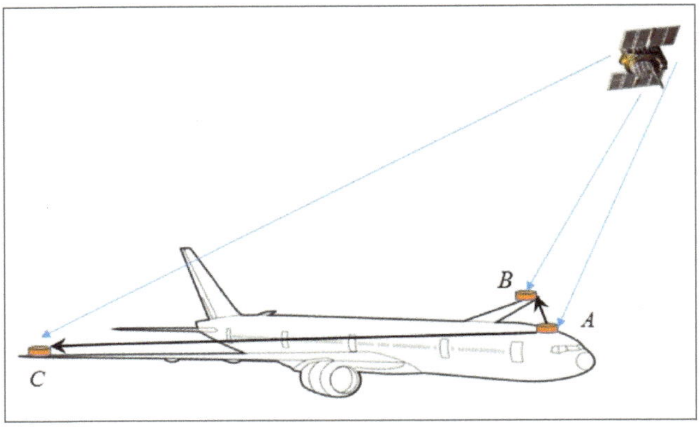

Figure 12.12. GNSS-based attitude determination (courtesy of Technical University of Munich).

thus becomes a possibly cost-effective alternative to an inertial system. The accuracy of the attitude array is dependent on the separation distance between antennas in the array. The attitude error can be computed as the RTK position error between antennas divided by the baseline length. Thus a one-centimeter position error divided by a one-meter baseline length yields a 10-milliradian or 0.5 degree attitude error. GNSS attitude arrays by themselves are typically not suitable for continuous and accurate attitude measurement. However they can be used to shorten the alignment time and enhance the performance of a GNSS aided inertial navigation system.

Precise Point Positioning (PPP)

Precise point positioning (PPP) is a methodology by which a user can obtain centimeter level position accuracy without a local reference receiver. PPP services are usually provided by online Public Sector websites, such as the US NGS, the Canadian CSRS, or the Japanese GEONET, or globally using the International GPS Service (IGS). The IGS is further addressed subsequently. Recently, the Private Sector started to provide similar services, an example of which is Trimble's CenterPoint© RTX. Both public and private sector PPP services use a global network of GNSS reference receivers to calibrate satellite-generated range errors to 10 centimeters or better. It is typically done in either post-processing or in real time where the corrections are broadcasted to users via satellite links, cellular networks or the internet.

The calibrated errors include satellite clock errors, orbital position and velocity errors and signal delay biases. A user's receiver that is enabled for this kind of positioning is then able to compute a more precise position than would be possible using only the GNSS broadcast navigation data. Position accuracies of better than 4 centimeters horizontal, 95%, can be achieved using carrier phase ambiguity resolution techniques. Atmospheric range errors that are canceled or attenuated in a DGNSS must now be averaged over time in a PPP solution. Consequently full PPP accuracies can be obtained only after a convergence time of 5 to 30 minutes, depending upon geographic location. This is the main disadvantage of PPP over other forms of DGNSS.

Local, Regional and Global Differential GNSS

Differential GNSS in its simplest form relies on a strong correlation of atmospheric and satellite orbital and clock errors between the roving receiver and a single reference receiver at a known position. This allows the cancellation of these common-mode errors at various levels of accuracy, the most accurate being the phase interferometry methods described previously. Almost exact cancellation of common mode errors can be had on baseline lengths of 10 kilometers or less. A local DGNSS system achieves this using a single reference receiver and a radio link to the rover to relay the encoded reference receiver data. An example of a local Survey setup using RTK GNSS is shown in Figure 12.13. Such short baselines may however be inconvenient or impossible to implement, for example in survey applications that cover larger distances. Examples are any type of airborne survey, large area road mapping and marine hydrography.

Figure 12.13. RTK GNSS (Image courtesy of Topcon).

A regional DGNSS system uses a network of reference stations to provide a DGNSS service throughout a given region. In North America the U.S. and Canadian Coast Guards provide meter-level DGNSS service using differential beacon networks. The networks modulate the differential corrections on existing 500 kHz marine direction-finding beacons along coastal regions and inland waterways. Other parts of the world have their own services.

The wide area augmentation system (WAAS) is another form of regional DGNSS system. It was implemented and is maintained by the U.S. Federal Aviation Administration and the Department of Transportation for the purpose of en route and terminal guidance navigation for commercial aviation. It provides differential corrections with 1-3 meter positioning accuracy to users on an L-band channel that is compatible with the GPS signal structure, so that a GPS receiver can receive and decode the signal without requiring a separate antenna and RF section. It uses 25 reference receivers located across the United States and two master stations that collect reference receiver data and compute the corrections. Europe has an equivalent Satellite-based Augmentation System (SBAS) called the European Geostationary Navigation Overlay System (EGNOS).

A regional DGNSS system for precise positioning uses a network of reference receivers to calibrate spatial models of the atmospheric signal delay errors that are applicable over large areas. These regional networks typically use four or more reference receivers separated by up to 50 kilometers to provide accurate spatial error models capable of calibrating atmospheric delay errors in the GNSS observables to the decimeter-level accuracy need for integer ambiguity resolution. An example of a network DGNSS product is the Virtual Reference Station (VRS). It comprises a dedicated set of four or more reference receivers and a server that computes and transmits differential correction information to a roving receiver via radio modem, cell phone or wifi service. It allows the roving receiver to implement RTK positioning accuracy over a large operational area instrumented by VRS reference receivers as if a single reference receiver were always close by (hence the word virtual in VRS). The concept is illustrated in Figures 12.14 and 12.15.

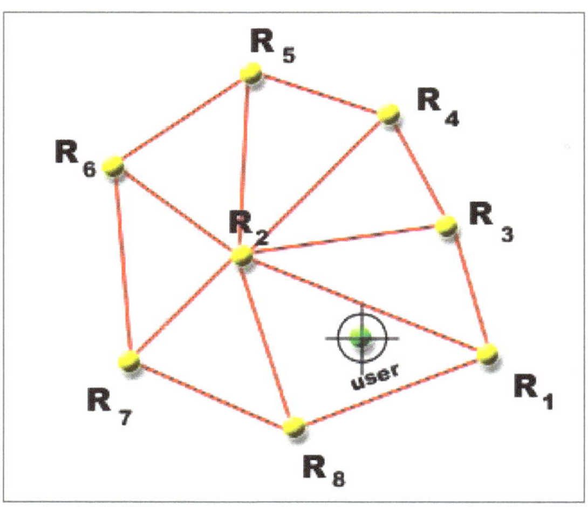

Figure 12.14. The concept of a Regional DGNSS Network.

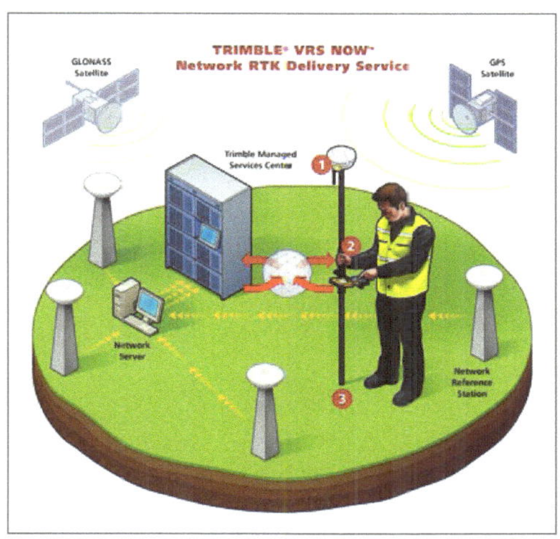

Figure 12.15. Real Time Regional GNSS Service Network (Image courtesy of Trimble).

The size of the operational area depends on the reference receiver network. Complete countries in Europe and Asia as well as U.S. states and Canadian provinces have been instrumented with reference receivers to provide RTK positioning accuracies to users within these jurisdictions on a subscription basis. Figure 12.15 shows a schematic of the Trimble VRS concept as an example of these regional GNSS Network Services.

A global network expands the regional GNSS model to a global scale. It uses a global network of reference receivers to continuously collect GNSS data at precisely surveyed reference receiver locations. This allows for computing the precise orbits, satellite clock corrections and atmospheric errors. The first global network was International GNSS Service for Geodynamics (IGS), which was established by the International Association of Geodesy (IAG). The IGS officially started its activities on January 1, 1994 after a successful pilot phase of more than a year. It collects, archives, and distributes GNSS observation data sets for a wide range of applications mainly via the internet. For details on the IGS, see www.igs.org. Figure 12.16 shows the IGS network distribution around the globe while Table 12.1 lists the IGS products together with their associated accuracy and availability.

Figure 12.16. The IGS Network (Image courtesy of the IGS).

Product	Availability	Accuracy
Precise Orbits		
Ultra-Rapid (Predicted half)	Real Time	0.05 m
Ultra-Rapid (Observed half)	3–9 hours	0.03 m
Rapid	17–41 hours	< 0.025 m
Final	12–18 days	< 0.025 m
Satellite Clock Corrections		
Ultra-Rapid (Predicted half)	Real Time	3 nanoseconds
Ultra-Rapid (Observed half)	3 hours	150 picoseconds
Rapid	17 hours	75 picoseconds
Final	13 days	75 picoseconds

Product	Availability	Accuracy
Atmospheric Parameters		
Final Tropospheric Zenith Path Delay	4 Weeks	4 mm
Rapid Ionospheric TEC Grid	24 hours	2-9 TECU
Final Ionospheric TEC Grid	11 days	2-8 TECU

Table 12.1. IGS Combined Product Accuracy and Timelines (Data courtesy of IGS).

PPP is also advantaged by post-processing because the forward and backward processing leading to a smoothed solution removes or attenuates the convergence time. The precise orbit and clock parameters can be downloaded over the internet from several sources such as the IGS. The IGS is a service established by the IAG, officially started its activities on January 1, 1994 after a successful pilot phase of more than a year. It collects, archives, and distributes GNSS observation data sets for a wide range of applications mainly via the internet.

Governmental Continuously Operating Reference Stations (CORS)

CORS is one of the examples of a regional GNSS differential network. The U.S. CORS shown in Figure 12.17, or the Australian Regional GNSS Network (ARGN) shown in Figure 12.18 on page 364, both exemplify national efforts in different countries to establish a permanent tracking GPS network for different applications. U.S. CORS and ARGN networks have permanent tracking GNSS receivers around the country, from which data are accessible for free via the internet. Earlier analysis showed that using multiple base receivers achieve significant improvement in positioning accuracy (c.f., Mostafa, 2004).

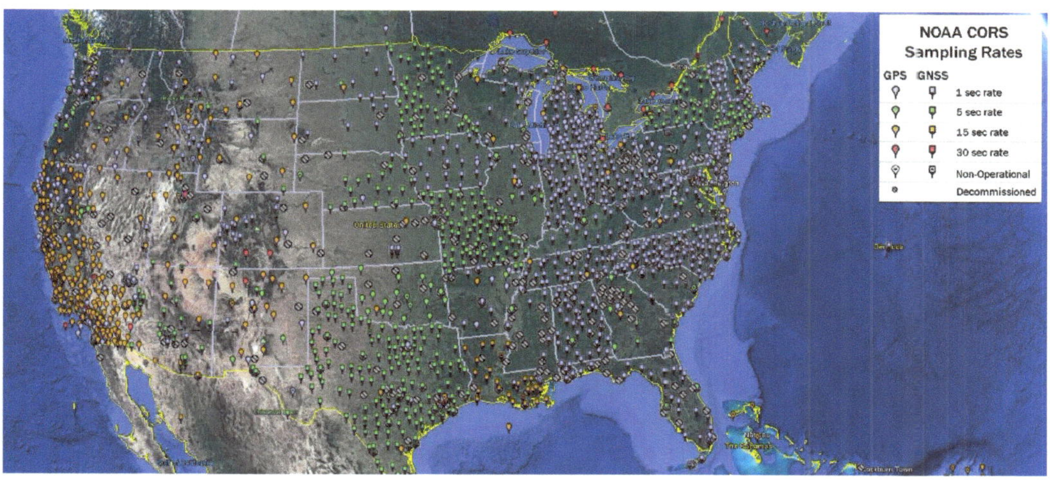

Figure 12.17. The Continental U.S. CORS—Alaska and Hawaii are not shown (Image courtesy of NOAA).

Among a number of GNSS and geodetic local, regional and global differential GNSS networks, is the CORS concept. The IGS produces high accuracy GNSS satellite ephemeris, satellite clock corrections, and ionospheric and tropospheric information. These products can be used to refine the recorded GNSS data from an airborne survey mission. GNSS data refinement can be done on different levels. Firstly, the IGS data can be used to minimize orbital and atmospheric errors and thereby improve the accuracy of a DGNSS solution when the reference receiver is more than 100 km away from the mapping area. Secondly, the IGS data can be used to compute a PPP solution when no reference receiver is available in some remote regional mapping applications.

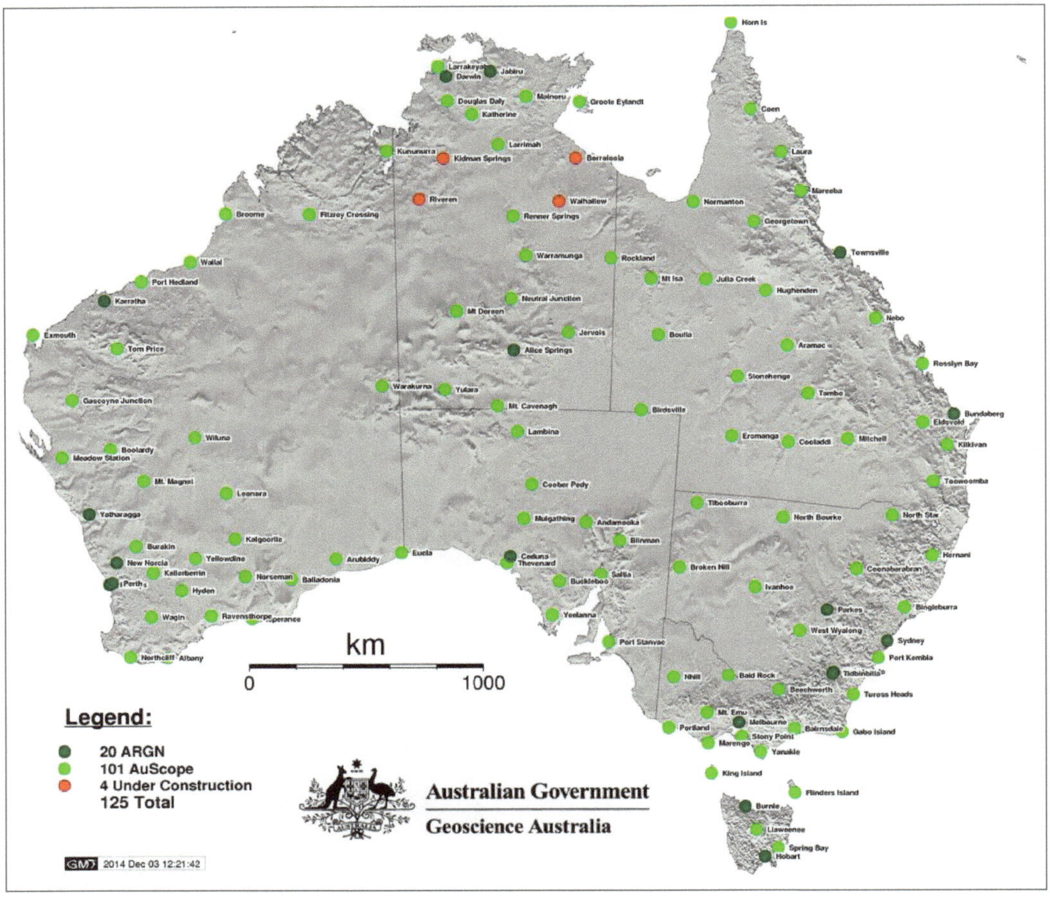

Figure 12.18. Australian Regional GNSS Network ARGN (Image courtesy of ARGN).

Another important advantage of post-processing is the ability to use a Post-processed Virtual Reference Solution (PP-VRS). Using the same concept as the real-time VRS described in the previous section, a PP-VRS can significantly improve the overall quality of the differential corrections by using the IGS corrections and doing a forward-reverse processing of the stations in a network to better estimate and correct for the atmospheric errors within the network. Furthermore a PP-VRS does not require a dedicated real-time network of stations. Any existing Continuously Operating Reference Stations (CORS) can be added to the network and used in the solution. An example of a PP-VRS is the Applanix SmartBase™.

Development History

GPS first started development in 1973 with the formation of the U.S. Department of Defense NAVSTAR Joint Program Office (JPO) under the direction of Dr. (Col.) Brad Parkinson. An experimental constellation of 11 Block I navigation development satellites was used to test the concept and explore operational issues. The first Block I was launched in February 1978, the last on September 9, 1985. Thereafter, 24 Block IIA operational satellites were launched between 1989 and 1994 to replace the Block I satellites and achieve a fully operational constellation of a minimum of 24 satellites. In 1996, the first of 21 Block IIR satellites was launched to replace the aging Block IIA satellites as part of a GPS constellation replenishment program. The GPS modernization program was started informally with the May 2000 Presidential Decision Directive ordering the shutdown of selective availability (SA). This was part of an ongoing campaign to make GPS more accessible to commercial users and to respond to the emerging competitive challenge raised by the nascent European Galileo program. The GPS modernization program also includes an L2 Civilian (L2C) signal to allow commercial users access to dual frequency pseudoranges, and an L5 signal that provides commercial users with three carrier frequencies. The first Block IIR satellite broadcasting the L2C signal was launched in January 2006.

Commercial GPS activity began in 1984 when Trimble Navigation released its first GPS product. Since then Trimble and other companies have delivered GPS receivers and GPS-based products into the marketplace for a variety of applications. These include precision surveying, aircraft and ship navigation, commercial positioning for hiking and fishing and time transfer. In fact, most commercial transactions over electronic media such as wire transfers or the internet are tied to GPS time.

Councilman first developed precise GPS positioning using carrier phase processing in 1980 (Councilman and Gourevitch, 1981). Hatch (Hatch, 1981) and Remondi (Remondi, 1985) conducted further developments in ambiguity resolution and decimeter level positioning. Their positioning methods required a static initialization of the roving receiver to identify the phase ambiguities. The first on-the-fly (OTF) ambiguity resolution was demonstrated in 1990. The first RTK receivers from Trimble, Leica and Novatel emerged during the early 1990s. Since then RTK GPS positioning has found extensive use in precision positioning applications such as land survey, structure and crustal motion monitoring and control of construction and paving machines.

Types of Sensors

GNSS receivers can be categorized as low cost and moderate performance for mass-market applications, and high performance for precision positioning applications that include surveying. The low-end receivers are single frequency (L1) receivers configured as original equipment manufacturer (OEM) components or chip sets for use in applications such as cell phones, car navigation systems and hand-held personal navigators. They are designed to deliver a position and time solution at minimum cost.

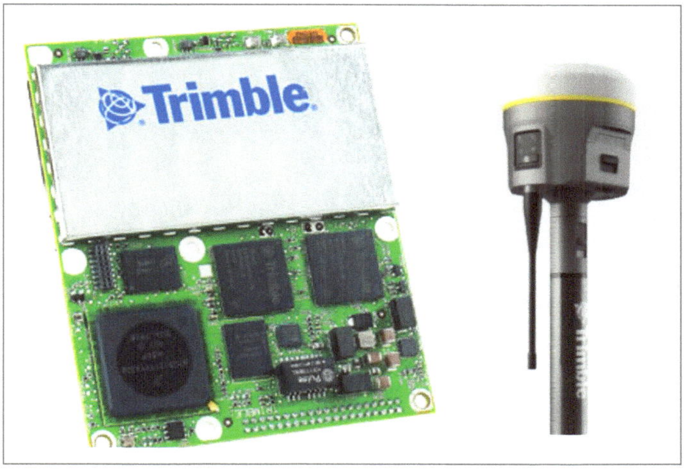

Figure 12.19. Examples of an OEM GNSS receiver board (left) and a ready-to-use survey instrument (right) (Image courtesy of Trimble).

High performance receivers are multi-frequency GNSS receivers configured either as OEM boards for integration into high-end positioning and navigation systems or as self-contained instruments for a particular application (Figure 12.19). Each of these requires a survey-grade antenna such as those shown in Figure 12.20 on page 366 that provides a stable phase center and a reception pattern that rejects multipath ground-reflected signals.

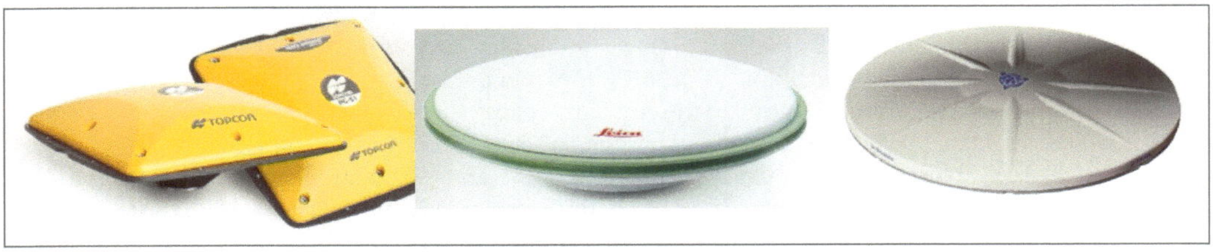

Figure 12.20. Examples of survey quality GNSS antennas by Topcon (left), Leica (center), and Trimble (right).

Post-Processing

An alternative to real-time processing is to post-process the GNSS observables and satellite ephemeris data recorded by GNSS receivers during a mission. This is referred to as Post-Processed Kinematic (PPK). Several GNSS manufacturers and third-party software developers offer post-processing software.

A GNSS post-processing software package takes advantage of the non-causal post-processing environment to compute a more accurate and more consistent ambiguity resolution solution than is possible in real-time. The simplest such technique is to compute position time histories forward and backwards in time and then combine them in some fashion to ensure maximum position accuracy at each time point.

A simple data quality assurance method compares the separation between the forward and backward position time histories and thereby identifies ambiguity resolution failures or wrong fixes in either solution when the difference exceeds a few centimeters.

Capabilities and Limitations

GNSS positioning has become a significant utility to surveyors and navigators. It is capable of providing baseline position accuracies comparable with traditional survey methods, and requires significantly less expertise and training to operate. It is however sensitive to its environment and can, without proper planning and quality control, give poor or misleading data. The two key limitations that a user of precise GNSS positioning must plan for are signal shading and multipath reflection and residual GNSS errors. These are described briefly below and apply to any GNSS.

Signal Shading and Multipath Reflections

The L-band signal from any GNSS satellite propagates along the line of sight from the satellite to a GNSS antenna, and thus is prone to shading by obstacles. Also, the signal can be reflected off of fixed surfaces or refracted by foliage, both causing signals with multiple path lengths to impinge on the antenna. This is caused multipath error. Multipath signals are always delayed compared to line-of-sight signals because of the longer travel paths caused by reflection. Code phase multipath can be as large as several meters if the antenna is stationary or slowly moving. For kinematic applications, the GNSS antenna motion with respect to the signal reflectors tends to randomize the multipath error and thereby diminish its impact on positioning accuracy. Typical phase multipath does not exceed 0.05 centimeters. Consequently a GNSS receiver must be operated with a clear view of the sky and away from reflecting surfaces. This is the key limitation to performance that operators of GNSS equipment must be aware of when installing a GNSS receiver or planning a survey mission.

Residual GPS Errors

GPS errors can be classified into three major groups as shown in Figure 12.21. DGNSS processing can either eliminate or attenuate these errors provided that a mission has been planned to account for these error sources. Such planning requires the separation between the base and roving receivers to be less than a maximum distance over which the residual common errors are small.

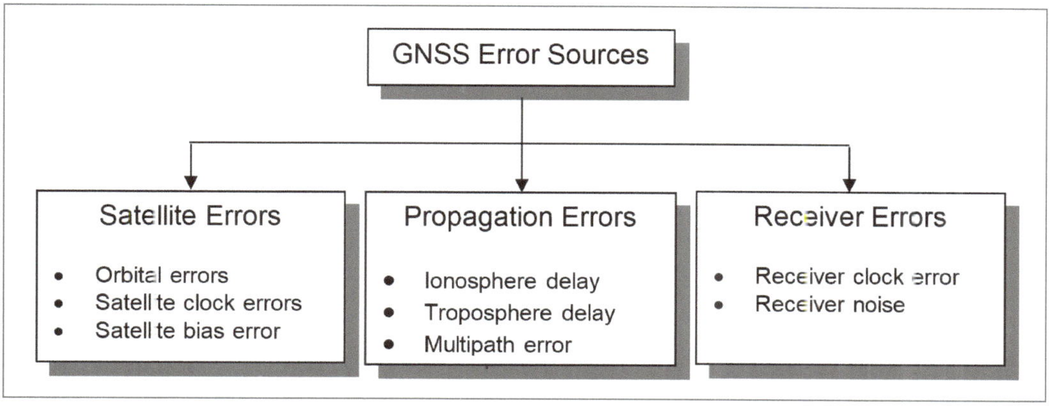

Figure 12.21. GNSS error sources.

Satellite Orbital Errors

A number of gravitational and non-gravitational forces affect each satellite in its orbit. The actual satellite orbit therefore deviates from its predicted elliptical form. The navigation message received by a GNSS receiver includes the predicted (broadcast) ephemeris, which is a number of parameters from which the satellite trajectory is computed with respect to an Earth-Centered Earth-Fixed (ECEF) frame such as the World Geodetic System 1984 (WGS84) used by GPS. The satellite trajectory is required to be able to compute the position of the receiver's antenna. GPS broadcast ephemerides allow for satellite position computation with 1-5 m accuracy depending on the age of an ephemeris. Galileo and Beidou have similar characteristics. A GLONASS satellite navigation state has a maximum age of 15 minutes after which its usage can lead to some large errors. Currently, the precise ephemeris freely available online are accurate to 2.5 cm.

Satellite Clock Errors

Although GNSS satellite clocks are very precise, small biases or drifts result in large positioning errors (0.3 meters per nanosecond). Therefore, each satellite-transmitted navigation message carries coefficients that a GNSS receiver or GNSS processing software uses to compensate for satellite clock errors. In post-mission, DGNSS processing removes the satellite clock error. When using single point positioning (without a reference receiver) in either static or kinematic positioning modes, the broadcast clock correction parameters are used with a residual error of 10 nanoseconds or 3 meters. Using single point positioning together with the IGS-derived satellite clock corrections (accurate to 0.025 nanoseconds) improves the satellite-to-receiver remaining range error to about 10 to 15 cm.

Propagation Media Errors

Another limitation that must be considered in planning GNSS operations is the effect of the atmosphere. The ionosphere and troposphere introduce errors in the observables, which can contribute several meters of position error. GNSS satellites transmit spread spectrum signals that propagate through about 20,000-26,000 km of media to be received by a GNSS receiver. Signal transmission–to-reception time is about 0.07 seconds, during which the satellite signals pass through different layers of the Earth's atmosphere. Two major atmospheric layers, the ionosphere and the troposphere, affect the propagation of satellite signals.

Ionospheric Errors

The ionosphere has lower and upper limits of approximately 50 and 2000 km, within which ions and electrons are available in quantities sufficient to affect the propagation of L-band signals. The ionospheric distortion is due to the total electron count (TEC) along the signal path between a satellite and the receiver. It is affected by the daily radiation of the sun and by the 11-year solar cycle. The ionosphere is particularly active at equatorial and northern latitudes. The year 2000 was the most recent "solar max" at which the sun's electromagnetic radiation reached a peak and thereby caused a corresponding peak in ionospheric activity. The ionosphere's effect on a GNSS signal, among other things, is a bending of the signal path, time delay of arriving code signals, time advance of arriving carrier phase signals, and scintillation. In some extreme cases of high solar activity, ionospheric scintillation causes rapid fluctuation of the signal strength, which may prevent a GNSS receiver from computing a position fix. Ionospheric errors can be dealt with using different approaches. Using DGNSS significantly reduces the effect of the ionospheric error, especially when using dual-frequency receivers. In single point positioning, the IGS-derived ionospheric grid help reduce the ionospheric errors. In real time applications, either modeling or compensating for the errors using a differential GNSS service is necessary.

Tropospheric Errors

The troposphere is the non-ionized portion of the atmosphere, and extends from the Earth's surface to about 50 km. It introduces a non-dispersive (i.e., frequency-independent) propagation delay of the GNSS signals. The total effect can be divided into two main effects, the effect of the dry and the wet components. The dry component contributes about 90% of the total error and can be modeled to about 2% accuracy using meteorological surface measurements. The wet component is more difficult to deal with because of the variations of the water vapor content that cannot be correlated over time and space and, thus, can only be modeled to about 10-20% accuracy. Tropospheric errors can be reduced using DGNSS. Modeling tropospheric errors is also used in either real time or post-mission.

Receiver Noise

A GNSS receiver noise comprises the code noise and the carrier phase noise. Code noise is currently at the level of 0.1 to 0.5 meters depending on the received signal to noise ratio (SNR). Carrier phase noise is usually at the level of one centimeter or less. The extremely low carrier phase noise is a key reason for the high accuracy of a kinematic ambiguity resolution position solution. DGNSS processing combines the observables from two receivers, and typically increases the single receiver noise variance by a factor of 2.

DGNSS Residual Errors

Generally, DGNSS reduces atmospheric (ionospheric and tropospheric) and orbital errors, eliminates satellite and receiver clock errors, and increases the single receiver noise. Atmospheric and orbital errors are correlated with distance; the shorter the separation baseline between receivers, the greater the correlation between these errors at each receiver. These errors are almost the same at two receivers separated by short baselines (1 to 30 km), so that differential processing of the receiver data will affect the almost complete cancellation of the errors. Since these errors also distort phase data used in kinematic ambiguity resolution, the reliability of ambiguity resolution on short baselines is better than on longer baselines. Table 12.2 briefly describes the DGNSS remaining errors.

Error	Error Characteristics
Orbital	• Correlated between satellites • Significantly reduced by between-satellite differencing (DGNSS) • Using precise orbits and satellite clock corrections improves positioning accuracy for long baselines
Ionospheric	• Frequency-dependent, thus, dual frequency data eliminates the error for long baselines. • Broadcast model reduces the error by 50% • In double difference airborne kinematic case error is typically 1-2 PPM for mid-latitudes between sunspot highs

Error	Error Characteristics
Tropospheric	• Frequency-independent, thus, cannot be removed by dual-frequency data • Dry component can be modeled and removed • Wet component needs meteorological data and more difficult to model because of variable nature of water vapor • Over long baselines the wet component effect on positioning can be estimated for airborne applications
Multipath	• Site-dependent and, thus, cannot be removed using differential GPS • In kinematic applications, the multipath signature has a strong correlation with vehicle speed. Therefore, multipath gets random (and less) for higher speed

Table 12.2. DGNSS residual errors.

Planning Considerations and Quality Control

Use of GNSS for precision positioning requires mission planning to avoid or diminish its limitations discussed previously, and thereby ensure the quality of the data. The following aspects of a mission must be planned:

- The time of day of the mission
- The expected mission trajectory
- The baseline length(s) between receivers

The ever-changing satellite constellation of an operational GNSS such as GPS provides different levels of coverage throughout a 12-hour period, ranging from 5 to 12 satellites. The combined coverage of two or more GNSS's can provide 10 or more visible satellites at all times in almost all locations. Consequently mission planning does not need to consider satellite coverage unless the receiver is limited to GPS reception. Mission planning software or services can be used to predict satellite coverage at a specified project area during a specified mission time.

Ionospheric activity during years of high solar activity can also compromise the quality of GNSS data if not planned for. Some parts of the world experience ionosphere scintillation that can completely disrupt GNSS reception. Where feasible, a mission-planning option is to conduct a survey at night when the ionosphere is quieter.

The mission trajectory should be assessed for sources of signal shading and multipath reflections. An L-band GNSS signal travels along its line of sight and is easily shaded by structures and vegetation. Buildings and trees also can reflect the signal so as to generate multipath reflections to the antenna. Multipath error is the hardest error to control because it is local to the receiver antenna and therefore cannot be attenuated in a differential GNSS solution. The best method to date of attenuating multipath is with an antenna ground plane or choke ring.

In order to achieve a KAR solution having centimeter-level accuracy with low risk, the baseline length between two receivers used to generate the KAR solution should be limited to a maximum of 10 kilometers during high solar activity and 50 kilometers during low solar activity. It has therefore been considered a rule of thumb that the reference to rover separation is acceptable if it is within 25-30 kilometers. This ensures that common mode errors in the observables from the two receivers, most notably ionospheric and tropospheric delays, cancel almost exactly in the differential solution. If a longer baseline is contemplated, then two or more base receivers should be installed so that the roving receiver is within the maximum radial distance of any one receiver.

Technological Advances

GPS was originally designed as a military global navigation system, but has become a core capability in many commercial products and services. Positioning accuracy using differential phase interferometry techniques have made GPS into a powerful survey tool. In order to extract this accuracy, the commercial GPS industry has designed around the intended and unintended limitations that the original GPS configuration had. The simplest circumvention of the original Selective Availability (SA) degradation of the CA signal was the basic DGPS configuration. Commercial receiver manufacturers developed codeless L2 phase tracking techniques to get around the L2 signal access limitation and extract dual frequency phases needed for reliable RTK positioning.

The U.S. government has recognized the importance of GPS as an important source of commercial economic activity. To improve GPS's accuracy and reliability in the commercial sector, the GPS JPO launched the GPS modernization program that coincides with the ongoing satellite replenishment program. It implements L2C and L5 signals that supplement the legacy L1 and L2 signals. L2C is a commercially accessible code modulated onto the L2 frequency that allows a commercial precision receiver to generate L2 pseudoranges so that the user can calibrate and remove the effects of the ionosphere in a position solution. It allows the receiver to track the L2 phase with lower noise than current semi-codeless phase tracking algorithms are capable of.

L5 has a center frequency 1176.45 mHz and is transmitted with approximately 3 db more power than the L1 signal. It was designed primarily for civil aviation. The L5 carrier falls into the Aeronautical Radionavigation Services band, which allows the aviation community to manage interference in this band more effectively than is possible in the L1 and L2 bands. The implication to survey applications is a third frequency for constructing combinations with longer wavelengths than the currently used L1-L2 wide lane phase. A longer wavelength allows faster and more reliable phase interferometry with ambiguity resolution to be performed.

Galileo was conceived by the European Union (EU) as a civilian-controlled GNSS because the United States would not guarantee the reliability of GPS service since it is a military-controlled asset. Also the satellite orbital planes have higher inclinations than this of GPS and so provide better coverage of northern European countries than GPS is capable of. Galileo also provides three levels of service to address the different needs of individual users, commercial aviation and government agencies.

An important attribute of the different GNSS's is inter-operability. This is defined to be the ability of a user to combine the signals from the different GNSS's to compute a navigation solution. Inter-operability becomes important when the satellites visible to a receiver from any one GNSS are insufficient to compute a reliable position fix, such as might occur when the user is surrounded by buildings or trees that occlude part of the sky. Inter-operability allows the receiver to use a sufficient number satellites from two or more GNSS's to compute a position fix. Inter-operability is enhanced when two GNSS's use signal bands with the same center frequencies, as then the receiver's frequency-dependent biases are common among the tracked signals from the two GNSS's and can be cancelled or lumped with the receiver clock error. Inter-operability is also enhanced when two GNSS's use time bases whose time differences can be accurately calibrated and broadcast to users. For example, GPS time is computed from UTC time maintained by the U.S. Naval Observatory (USNO), whereas Galileo system time (GST) is maintained by the Galileo mission segment. They differ by a few nanoseconds, which has an equivalent distance of 30 to 100 centimeters. Precise knowledge of the GPS-GST time difference is required to do precise positioning using GPS and Galileo observables.

GNSS-AIDED INERTIAL NAVIGATION SYSTEM

Technology Overview

Direct Georeferencing (DG) is the direct measurement of the position and orientation parameters of a remote sensor for the purpose of georeferencing the data acquired by the sensor. It is an enabling technology for quantitative data acquisition and mapping applications where precise orientation of the sensor in addition to its position is required. Such mapping applications include aerial photogrammetry, airborne multi-spectral scanners, airborne lidar and multibeam sonar bathymetry. In all of these applications, the sensors have a wide field of view of the terrain being mapped. A direct georeferencing system provides the position and orientation of the sensor that is required to stabilize and register the acquired data in geographic coordinates. The state-of-the-art technology in DG systems is the GNSS-aided inertial navigation system (GNSS-AINS) which combines GNSS positioning with inertial navigation. The basic concept of inertial navigation, as illustrated in Figure 12.22 and Figure 12.23, is to make measurements of acceleration and angular rate and integrate them into changes of position and orientation with respect to known initial conditions.

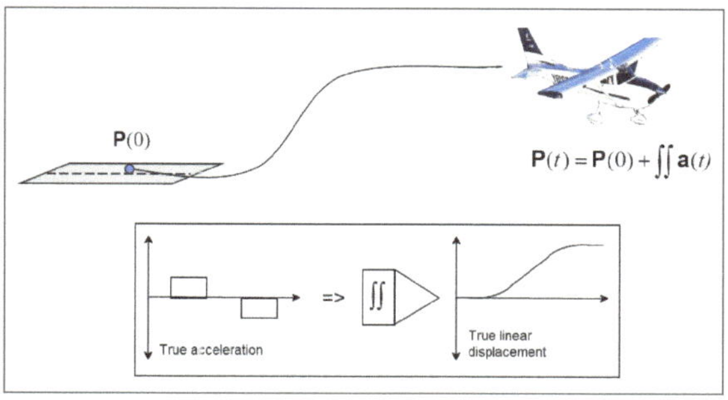

Figure 12.22. Relative positioning concept by inertial navigation.

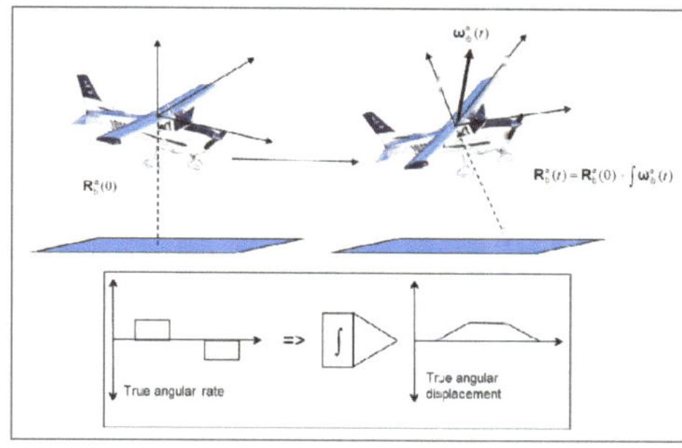

Figure 12.23. Attitude determination concept by inertial navigation.

Inertial navigation is done by an INS. In a GNSS-AINS, errors in the inertial navigation solution are estimated and controlled by the GNSS position and velocity measurements.

Complete descriptions of inertial navigation concepts and technology can be obtained in Britting, 1971; Kayton and Fried, 1997; and Siouris, 1993. An overview of GNSS-aided inertial navigation can be found in Parkinson and Spilker, 1996 (Volume 2). For a more detailed description, see Siouris, 1993.

Inertial Navigation System (INS)

An inertial navigation system is comprised of two components:

1. Inertial Measurement Unit (IMU)
2. Inertial Navigation Processor

An IMU contains three accelerometers, three gyros and the electronics to provide digitally encoded samples of the accelerometer and gyro data on a serial interface. The accelerometers are arranged into an orthogonal triad that measures the specific force vector experienced by the IMU. The specific force vector is the vector sum of the local gravity vector and the acceleration vector experienced by the IMU. The gyros are likewise arranged into an orthogonal triad that measures the angular rate vector experienced by the IMU. The angular rate vector comprises the sum of the earth rotation vector plus the angular rate vector experienced by the IMU with respect to the Earth. A typical IMU outputs digitized samples at a 50-1000 Hz sampling rate of the total specific force and angular rate of the IMU with respect to a hypothetical stationary inertial reference system. High data rates are typically required to adequately sample short-term vibration dynamics as well as long-term vehicle dynamics. An IMU can be a stand-alone device or integrated directly into an INS. Figure 12.24 on page 372 shows examples of IMU's.

The inertial navigation processor is a computer that solves Newton's equations of motion on the rotating Earth based on the measured accelerations and angular rates from the IMU. In order to do so, the inertial navigation processor must establish the local North, East and Down directions of a navigation coordinate

Figure 12.24. Example IMUs (Image courtesy of Applanix and Seasonor).

frame using a procedure called alignment. Determining the Down direction is called leveling, and is based on the assumption that the local gravity vector is the vertical reference. Once leveling is completed, the horizontal axes of the navigation frame define a plane that is tangent to the local geoidal surface and hence said to be locally level. Aligning a locally level navigation frame to the true North direction is called heading alignment. A mechanism called gyrocompassing uses the horizontal component of the earth angular rate vector as the heading reference.

A stationary or otherwise non-accelerating INS uses the earth rate vector measured by the gyros to gyrocompass its navigation frame to true North. A typical aircraft INS must undergo a stationary alignment lasting 10 to 30 minutes to accurately align its navigation frame. Thereafter it navigates free-inertially, which implies that it computes its navigation solution using only the acceleration and angular rate data from the IMU without any subsequent corrections to its alignment or position solution.

The INS technology from the early 1950s to the late 1970s comprised the gimbaled platform. In a gimbaled platform INS, the accelerometer and gyro triads were located on a mechanical 4-axis gimbaled platform. The alignment process rotated the platform into alignment with the North, East and Down directions. Thereafter during navigation, the gyros measured the platform angular rate and provided the feedback signal to the gimbal axis motors to null the platform rotation with respect to the Earth's surface, thereby maintaining the platform locally level and aligned to true North in spite of the vehicle's angular rate.

The accelerometers thus reported North, East and Down accelerations at all times, which the inertial navigation processor integrated to compute North, East and Down velocity and from this updated latitude, longitude and altitude. The gyro technology used in these mechanical INS's was the floated spinning mass gyro.

The platform gimbal axes were instrumented with synchro resolvers that measured the gimbal shaft angles. The mechanical design of the gimbaled platform caused the shaft angles to be the vehicle roll, pitch and heading, which formed a complete orientation solution for the vehicle. The INS thus computed a complete navigation solution for the vehicle comprising geographic position, North-East-Down velocity and orientation with respect to the North-East-Down directions as shown in Figure 12.25.

The strapdown INS is the current state-of-the-art in inertial navigation. It emerged as the dominant INS technology during the late 1970s with the development of the ring laser gyro (RLG) and a class of avionics

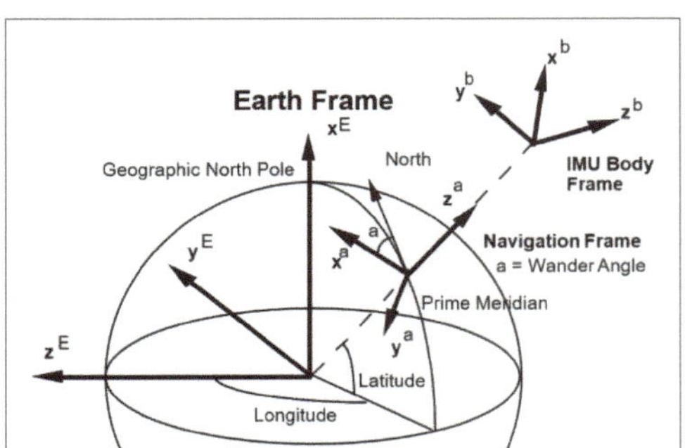

Figure 12.25. Frames of reference for inertial navigation.

computers that could perform the numerically intensive strapdown navigation computations. It is so named because the accelerometers and gyros are hard-mounted, or "strapped down," to the INS case, and hence rotate with the vehicle. The accelerometers and gyros thus report the acceleration and angular rate of the INS resolved in the INS body frame.

Using the angular rate data from the gyros, the navigation processor updates the direction cosine matrix (DCM) that describes the orientation of the INS body frame with respect to a mathematical navigation frame that is locally level and aligned to true North. The alignment process in the strapdown INS initializes this DCM and thereby establishes the alignment of the mathematical navigation frame. The body-to-navigation DCM is thus the mathematical equivalent of the mechanical gimbals in the platform INS.

The strapdown INS then transforms the measured acceleration vector to the navigation frame and thereafter computes velocity and position in the same way as does the platform INS. The vehicle orientation parameters are derived from the components of the body-to-navigation DCM. The complete mechanization of a strapdown INS is shown in Figure 12.26.

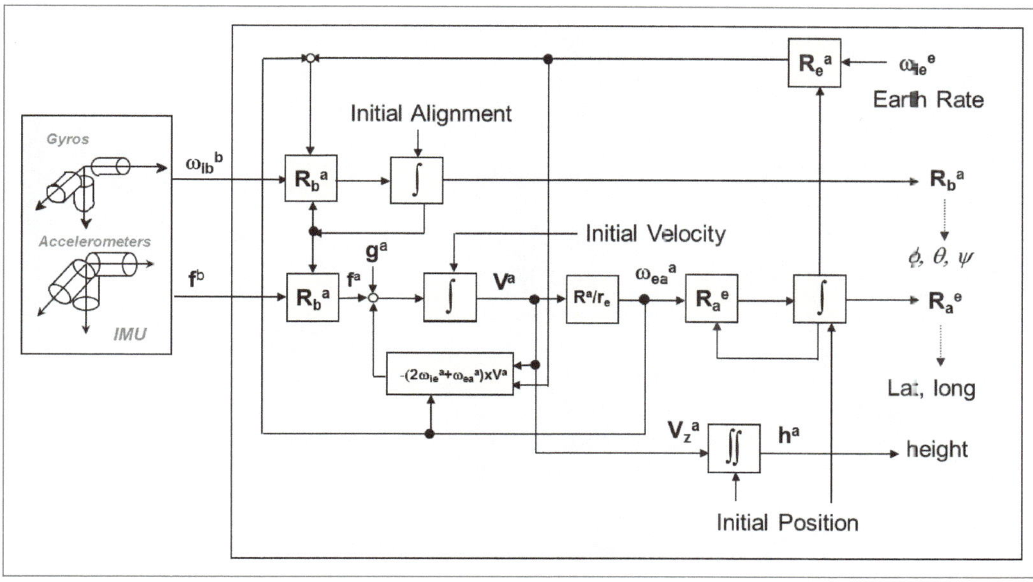

Figure 12.26. Strapdown inertial mechanization.

An INS is thus a sophisticated dead reckoning navigation system that navigates from an initial known position to its current position using only the measured accelerations and angular rates as measured by the inertial sensors. As with any dead reckoning method, it is prone to position drift, in this case caused by errors in the measured inertial data. A free-inertial navigator exhibits a position drift that on average grows linearly with time. An INS is typically classified in terms of the number of nautical miles that it drifts in one hour following a full ground alignment.

A so-called medium accuracy INS that is used for aircraft navigation uses RLG's and typically exhibits an approximate one nautical mile per hour drift. Superimposed on this average position drift is the Schuler oscillation. This is a sinusoidal position error with an approximate period of 84 minutes, and is a consequence of the INS being constrained to navigate on the Earth's surface. It was first documented by Schuler, 1923. Higher accuracy INS's such as are used on ships and submarines are capable of better long-term position accuracies.

The dominant errors in an INS that cause a position drift are the alignment errors and the inertial sensor errors. The alignment errors are the deviation of the navigation frame alignment from locally level and true North in azimuth. A navigation frame tilt error from locally level causes the gravity vector to be projected into the horizontal components of the acceleration vector transformed into the navigation frame. The dominant inertial sensor errors are the accelerometer and gyro biases, which are the non-zero outputs of the accelerometers and gyros when the INS is "stationary" in inertial space (for example floating in deep space with no specific forces acting on it). The gyro biases integrate into additional tilt errors which increases the gravity projection error. The INS integrates the projection errors and accelerometer errors as if they are real accelerations, and outputs a position that typically grows erroneously as time squared (see Figure 12.27).

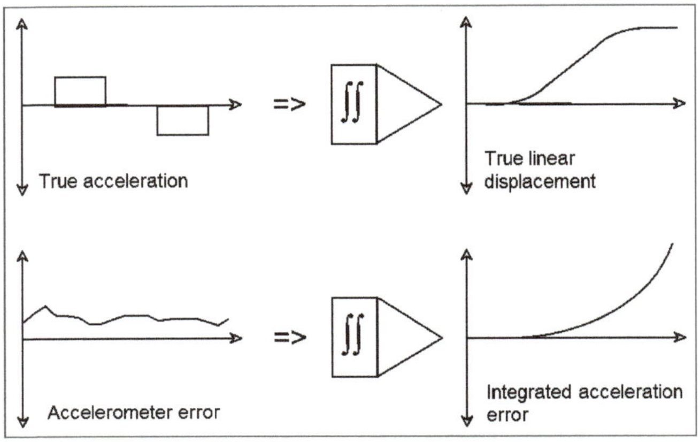

Figure 12.27. Simplified concept of inertial navigation position error growth.

The other major inertial errors are the accelerometer and gyro random noises. These are uncorrelated noises caused by a variety of mechanisms that includes analog amplifier noise and digital quantization error. Smaller inertial sensor errors include scale factor error, input axis non-orthogonality and vibration rectification. All of these further increase the velocity and position errors in the navigation solution as the INS performs its integration process.

The inertial sensor errors are the residual errors following a factory calibration of the IMU. An uncalibrated IMU will exhibit large biases, scale factor errors and input axis non-orthogonalities that vary with the internal temperatures of the inertial sensors. The calibration firmware in an IMU implements temperature-dependent models of these errors. A typical model is a 3rd to 6th order polynomial interpolation of an error as a function of temperature. An IMU calibration procedure estimates the model parameters at different test temperatures across the specified operational temperature range. Ideally, an IMU calibration will model the mean inertial sensor errors at any temperature, so that the residual errors are random from one IMU power-up to the next. A typical IMU performance specification thus quotes the standard deviations of these inertial sensor errors about their calibrated mean values.

GNSS-Aided Inertial Navigation System (GNSS-AINS)

A GNSS-AINS combines the components of an INS with GNSS and other sensors for the purpose of regulating the INS errors to an improved level of accuracy. Hence the GNSS and other sensors are said to "aid" the INS. Figure 12.28 shows a typical GNSS-AINS architecture which uses a Kalman filter to estimate the errors in the INS. The Kalman filter uses measurements of the INS position and velocity differenced with measurements from the GNSS to observe and estimate the errors in the inertial solution. This is possible when the aiding sensor error characteristics are different from the INS errors. A GNSS position solution contains errors that can be characterized as noisy and statistically bounded within an error ellipse of fixed volume. By contrast, the INS position errors are characterized as smooth but unbounded. These error characteristics are said to be complementary, and hence amenable to Kalman filter error estimation. The Kalman filter processes measurements that are the differences of the inertial navigator and GNSS positions, and thus comprises the noisy but bounded GNSS position errors plus the growing INS position error. The Kalman filter contains a model for these errors, and thereby is able to suppress the GNSS short-term errors and estimate the INS errors.

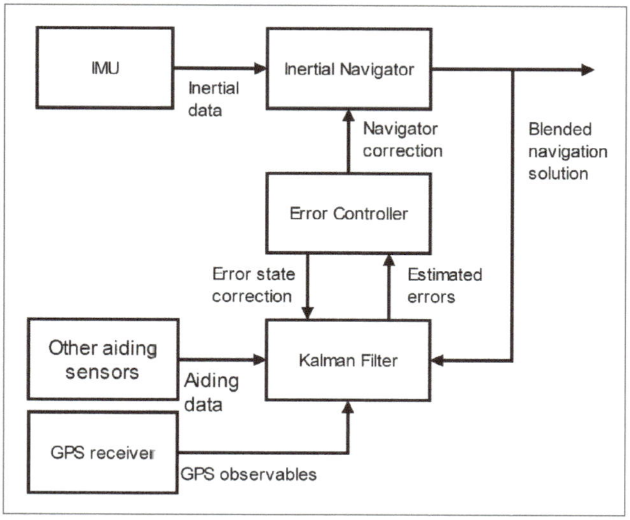

Figure 12.28. GNSS-Aided INS architecture.

The accuracy of the INS error estimates depends on the magnitude of the GNSS position error. It will obtain a better error estimate when the GNSS solution is a RTK solution with centimeter accuracy than when the GNSS solution is a non-differential solution.

The GNSS-AINS uses the Kalman filter's estimated inertial navigation errors to correct the position, velocity and attitude computations in the inertial navigation algorithm. This error correction closes the loop around the INS and thereby regulates the INS errors to be consistent with the aiding sensor errors. In a GNSS-AINS, the inertially derived position and velocity errors are regulated by the error controller to be consistent with the smoothed GNSS position and velocity errors. It furthermore regulates the inertial attitude errors and thereby continuously improves the inertial solution alignment. Thus a GNSS-AINS is able to align from a cold initialization while the vehicle is moving. This is called a mobile alignment, or in the case of an airborne system an "air-start".

A key capability of the GNSS-AINS is an attitude solution whose accuracy is better than the INS attitude solution on its own. This is because its Kalman filter continuously calibrates the inertial sensor errors as part of its overall error estimation function. For example, an INS will integrate a constant accelerometer bias to a time-linear velocity error. A gyro bias will rotate the navigation frame out of locally level at a constant angular rate. The gravity vector will project increasingly into the horizontal accelerometer input axes, and the INS will integrate the resulting time-linear acceleration error to a time-quadratic velocity error. The GNSS-AINS Kalman filter recognizes these velocity errors when it compares the INS and GNSS positions. The Kalman filter has a mathematical model that describes the velocity error as a function of the various sources of acceleration error, including the accelerometer and gyro biases. Using this error model, the Kalman filter is able to estimate these biases from the observed velocity error, effectively doing an "on-line" and continuous calibration of the IMU, which in turn provides for pitch and heading accuracies that are significantly better than the INS could achieve without aiding. This means a GNSS-AINS that uses a relatively inexpensive IMU can achieve roll and pitch accuracies on the order of a few hundredths of a degree and a heading accuracy on the order of a tenth of a degree or better.

Components of a GNSS-AINS for Direct Georeferencing

A GNSS-AINS system for survey and remote sensing applications typically comprises the following major components (shown in Figure 12.29 on page 376):

- IMU
- Roving GNSS receiver
- Navigation computer system (NCS)

The following are optional components for obtaining differential GNSS (DGNSS) or real-time kinematic (RTK) position accuracies:

- Differential corrections via a radio modem, cell modem, internet or satellite
- RTK subsystem: a dedicated base receiver complete with radio modem transmitter, or a virtual reference station network
- Distance Measurement Indicator (DMI)
- Secondary GNSS receiver and antenna for computing heading from GNSS to aid the inertial solution

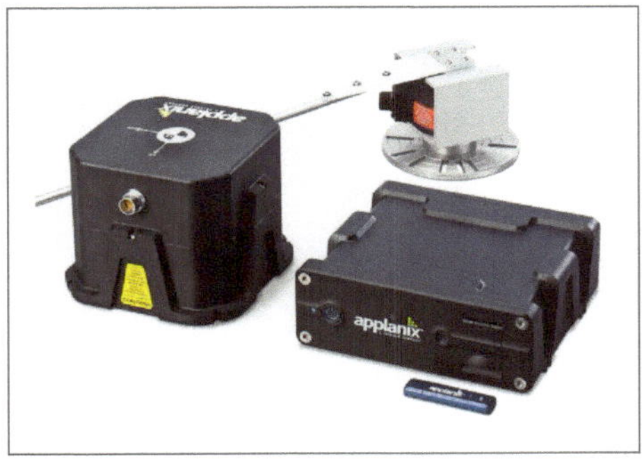

Figure 12.29. An Example of a DG System for Land Applications showing the external IMU, NCS and optional DMI (GNSS antenna not shown). (Image courtesy of Applanix).

Those applications requiring the best achievable accuracy in post-processing will also include the following components:

- Mobile data acquisition system (DAS)
- Reference (base) receivers with self-contained data logging
- Data retrieval and post-processing computer

The IMU containing the accelerometers and gyros is ideally compact and unobtrusive so that it can be located at or close to the desired instrumentation point, for example an aerial camera perspective center. At the same time, its inertial sensors must be sufficiently accurate so that the overall system delivers the required orientation accuracy. These requirements typically conflict; high quality inertial systems using optical gyros are quite large, whereas micro-electrical machined sensor (MEMS) based IMU's having the desired small size are often not sufficiently accurate. A best compromise is achieved by using a small IMU with low noise inertial sensors in a GNSS-AINS designed and manufactured specifically for Direct Georeferencing.

The NCS runs a software mechanization of the GNSS-AINS algorithm described previously and computes a position and orientation solution in real-time. For some applications such as sonar bathymetry, the real-time solution is the primary data product of interest. For other applications requiring a post-processed solution for georeferencing such as the case for airborne and land-based mobile mapping, the real-time solution is used as input to Flight Management Systems (FMS) to guide the pilot and control the mapping sensor, and as input to stabilized mounts to provide levelling control and correct for crab. A properly designed system will initialize and align while the vehicle is moving. The alignment and transition to full navigation performance is automatic and hence transparent to the operator after he has pushed the power switch.

The roving GNSS receiver is typically embedded in the NCS. It must be capable of outputting observables of sufficient accuracy and quality for computing a KAR position solution. The antenna must also be multi-frequency capable and of sufficient quality to reject most multipath interference, and be able to track L-Band frequencies if a satellite based correction service is to be used.

Without any source of differential corrections, the computed position accuracy is on the order of 5 meters (one sigma). Satellite based corrections services (such as WAAS, Omnistar, StarFire and Trimble RTX) reduce this error down to a range of 0.05-5 meters, depending upon the type of correction service. When using PPK or RTK coupled with a dedicated base station or a local, regional, or global network of GNSS receivers, the NCS computes position with an accuracy of 0.02–0.1 meters (one sigma). Typically, a PPK solution is more accurate and easier to QC than an RTK solution.

The data acquisition system (DAS) is also typically embedded in the NCS. It records time stamped IMU and GNSS data to a non-volatile storage media such as a hard disk or tape. The IMU time stamp accuracy must be 50 microseconds or better to ensure that the position and orientation solution is accurate at the sensor sampling time on a possibly fast moving platform such as an airplane. It is also required to ensure a good inertial-GNSS integration in the GNSS-AINS.

A few hundred microseconds of time misalignment between the IMU and GNSS data can cause the GNSS-AINS to estimate incorrect accelerometer biases and hence compute inaccurate orientation data.

The base receivers with data logging are required to compute a precise position solution in post-processing. These are also dual frequency receivers, and are located around or within a project area so that the distance from any one base receiver to the roving receiver is within 10-50 kilometers. This tolerable distance depends on the ionospheric activity in the project area at the time of the survey. Ionospheric interference with the GNSS signals causes signal delay and scintillation. The data logging capability is either built into the receiver or is provided by a personal computer connected to the receiver.

A DMI provides along-track velocity measurements when the GNSS-AINS is mounted on a vehicle. The most common DMI's are wheel mounted, and produce a "pulse" each time the wheel rotates a specific angle. The pulses are summed to compute the distance travelled, or converted into a measurement of vehicle speed. When processed in the Kalman filter, the DMI measurements help to constrain the position error growth of the GNS-AINS solution, especially when the GNSS signals are blocked when in a tunnel or under a bridge, or by other obstructions such as buildings or trees.

The data retrieval and post-processing computer is used to retrieve and process the recorded data. The end product of post-processing is either the desired position and orientation solution formatted to the particular application, or parameters derived from the solution. For example, the end product for aerial photogrammetry is the set of exterior orientation parameters for each photo. These are essentially the position and orientation solution at the time of exposure in a particular industry standard format

Development History

The aided INS concept emerged shortly after the development of the Kalman filter, a core component of the AINS. This technology was first developed for and deployed in military aircraft, beginning in 1968. More recently, the U.S. Air Force Embedded GNSS-Inertial (EGI) program has spawned a standard GNSS-AINS configuration.

The direct georeferencing concept for remote sensing has been postulated in technical research publications for several years (for example Schwarz et al., 1993). It has been put into practice in several experimental systems such as airborne multi-spectral scanners and laser altimeters. The orientation sensor typically was an aircraft INS that navigated free-inertially during the data acquisition. These systems demonstrated the DG concept, but were impractical because of the large size and high cost of the INS. An example is the early work performed by the Canada Centre for Remote Sensing (CCRS) led to the development of the Multi-spectral Electro-optic Imaging Sensor (MEIS).

A practical DG system became feasible with the advent of GNSS as a commercially available precise positioning system and the arrival of strapdown IMU's for tactical and motion sensing applications in military weapons systems. A "tactical" IMU is designed for short-term missile guidance. It is required to be small and light in order to fit into an air-to-air missile case, and relatively inexpensive in order to be affordable in an expendable weapon. It uses low cost inertial sensors such as fiber-optic gyros (FOG) or Micro Electromechanical Sensor (MEMs) based gyros to meet a 1-10 degree/hour performance requirement. Such an IMU is impractical in a free-inertial navigation application. Its gyro errors would drive the position error drift to upwards of 100 nautical miles per hour. It is however suitable in a GNSS-AINS which calibrates the inertial sensor errors on an ongoing basis. A "motion sensing" IMU is designed to measure the short-term motion of a platform such as a radar antenna for the purpose of signal error compensation. A typical application is a synthetic aperture radar (SAR) motion compensation system, which is required to measure the radar phase center displacement along the line of sight with an accuracy standard deviation of less than one centimeter over a 10-30 second integration time. Its inertial sensors exhibit stable and low noise errors that can be accurately calibrated on-line. These same characteristics become important for high accuracy orientation measurement such as required in the aerial photogrammetry application.

Applanix was the first organization to offer for sale a GNSS-AINS specifically for commercial surveying applications. The Position and Orientation System for Land Vehicles (POS LV) first appeared in 1993. The POS MV for multibeam sonar bathymetric survey was made available in 1994 (Scherzinger, Woolven and Field, 1997). Shortly thereafter, the POS AV for airborne remote sensing and POS TG for railroad track geometry measurement were released (Hutton, Savina, and Lithopoulos, 1997).

DIRECT GEOREFERENCING SYSTEMS FOR AIRBORNE DEM GENERATION

This section overviews how DG systems are used with airborne sensors for the production of Digital Elevation Models.

Development History

Direct Georeferencing has been the subject of fundamental research for many years. The first use of Inertial/ GPS systems in airborne mapping was proposed by Schwarz et al., (1984). Experimental systems that explored the feasibility of the technology have been developed by several university and government agencies (c.f., Schwarz, et al., 1993; Toth and Grejner-Brzezinska, 1998; Mostafa and Schwarz, 2000). Early DG systems used in the airborne survey market were adaptations of the INS products. For an overview, see Schwarz et al., 1993; Toth and Grejner-Brzezinska, 1998; Mostafa and Schwarz, 2000; Mostafa and Schwarz, 2001; and Grejner-Brezezinska, 2001.

The first commercial development of a DG system for airborne photogrammetric mapping took place in 1996 for use on existing film cameras such as Leica RC30 or Zeiss LMK (Hutton et al., 1997). The photogrammetric community at that time used the DG-measured position and orientation parameters of each 'photo' as an auxiliary source of data to aid the aerotriangulation (AT) process. The photogrammetric community gradually realized that DG systems allowed for more productivity, flexibility, and relaxed the need for ground control points (GCP) which resulted in gradually using the DG systems with a lighter AT and less GCP. This progress resulted in significant cost savings for photogrammetric mapping service providers. Over time, photogrammetric mapping service providers started to use DG for more challenging applications such as corridor mapping, oblique photogrammetric mapping, and rapid response.

In 2001, The OEEPE, a European Union Funded Research Program led by the University of Hannover, Germany started to extensively research the use of DG in photogrammetric mapping (c.f., Heipke, et al., 2001; Jacobsen and Wegmann, 2001). The OEEPE organization designed and executed detailed test flights using two DG systems; namely Applanix POS and IGI Aerocontrol. Then, the OEEPE made all test flight data publicly available which allowed a large group of academic, public and private sector entities from around the globe to participate in the analysis. The OEEPE held an international conference in Hannover on September 12, 2001 where the test results and analyses were presented by different organizations from around the world. The conference came to a conclusion that DG is here to stay and will be used in different flavors and recommendations have been also made on different technical aspects.

In parallel, from 2001 to 2003, digital airborne camera systems came into being; such as Leica ADS40 and Zeiss DMC. Shortly after, Vexcel UltraCam and Applanix DSS came into being. The abovementioned digital systems shared a few common grounds one of which was the inclusion of a DG system as an integral part of an airborne camera system (c.f., Sandau, et al., 2000; Madani and Mostafa, 2004). This was a clear statement by the professional photogrammetric community that DG became a procedure in the airborne mapping field. Academic research never stopped to analyze DG systems (c.f., Ip et al., 2004; Casella et al., 2006). Further integration of DG systems together with flight management systems took place shortly after for system integration and data fusion improvements (c.f., Ip et al., 2008). Theoretical and practical accuracy assessment efforts have also been made (c.f., Mostafa and Hutton, 2010).

Types of Airborne DG Systems

Direct Georeferencing systems used in the commercial airborne mapping market tend to be split into two categories:

- Custom OEM solutions built for a specific type of imaging sensor
- Generic "ready-to-use" or "stand-alone" GNSS-Aided Inertial systems

A custom OEM solution is usually comprised of an electronics board set and separate IMU, or a single board with combined GNSS and inertial components that are embedded directly into the airborne sensor to be georeferenced. The IMU and GNSS-AINS HW and SW are developed specifically to match the requirements of the imaging sensor and the platform that it will be flown in, whether this be a manned platform or an Unmanned Aerial Vehicle (UAV). A generic solution is typically a pre-packaged boxed product with standard connectors and cables that has been designed to provide a specific level of performance when installed on a typical aerial platform. The advantages of an OEM solution include a reduction in size, weight and power, and improved performance that is "tuned" to the imaging sensor.

Present Operating Status

Direct Georeferencing systems are now an integral component of most if not all airborne mapping systems flown on manned platforms, and are quickly becoming adopted for use on UAV platforms. Imaging sensors used for DEM measurement that routinely use direct georeferencing include Light Detection and Ranging (lidar), photogrammetric cameras, and Interferometric Synthetic Aperture Radar (IfSAR). A partial list of companies that specialize in supplying GNSS-AINS solutions specifically for DG include: Applanix, IGI GmbH, Novatel, OxTS, SBG, IxBlue, and Advanced Navigation.

Technology Overview

Lidar

In an airborne lidar installation, whether on a manned aircraft or UAV, a DG system is used to measure the position of the laser reference point and the orientation of the laser range at the exact time of measurement. In a scanning lidar system, the laser reference point is either the scan mirror or the detection element. For flash lidar, the reference point would typically be a location at the center of the array, from which the relative location of every other detection element in the array is known. The orientation is given in Euler angles with respect to the North, East and Down directions. The IMU is mounted to the lidar housing so that it is perfectly rigid with respect to the laser reference point, and remains so over temperature changes and when exposed to shock and vibration. During a mission, the DG system records the IMU and GNSS data and the time of each laser scan, all in a common time base such as GPS time. The DG post-processing software computes the time tagged position and orientation of the laser reference point at a high data rate, typically at 200 Hz. The lidar post-processing software then interpolates the DG position and orientation data to the exact time of scan. With this data and the range measured by the laser, it computes the 3-dimensional ground spot coordinates of each laser range. Typically, the software first computes the Earth-Centered-Earth-Fixed (ECEF, also called Cartesian) coordinates of each point, and then converts these to the desired mapping frame. The resulting georeferenced point cloud is then ready for exploitation into data products such as Digital Surface Models (DSMs).

When a calibrated frame camera is also attached to the lidar, the DG system can be used to measure the position and orientation of each image at the exact time of exposure. This allows the imagery to be directly projected onto the point cloud from the lidar, enabling it to be colorized to produce photo-realistic 3D products. A DEM can also be extracted from the lidar and used to directly orthorectify each image, all without the need of running Aerial Triangulation (AT).

The angular accuracy requirements for a DG system for use with lidar is typically a function of its field of view (FOV) and flying height. Low altitude systems, such as those used on UAV's, require only moderate roll and pitch measurement accuracy, while high altitude aircraft based systems require angular accuracy at the milli-degree level. Furthermore, a system with a wide FOV (such as with a 360 degree scanner) will require a much higher heading accuracy measurement vs a scanner with a narrow FOV. Most lidar manufactures today produce their systems with built-in DG that has been optimized to meet accuracy requirements over the operating range of the sensor.

Calibration parameters for the DG system include the relative positional offsets (also called Lever Arms) of the IMU sensing center with respect to the lidar reference point and the GNSS antenna phase center, and the fixed rotational offsets of the IMU with respect to the scanner housing (boresight angles).

IfSAR

In an Interferometric Synthetic Aperture Radar (IfSAR) system, the DG System is used to compute the relative displacement of the phase center of one SAR antenna with respect to another. The relative displacements are then used to correct the radar phase. This is the equivalent of stabilizing the IfSAR platform against aircraft motion due to maneuvers and air turbulence. The accuracy of the DG system is selected based upon both the absolute orientation accuracy required to produce a specific ground accuracy from a given flying height, and the relative orientation accuracy (sample to sample noise) required to produce a sharp radar image.

Photogrammetric Cameras

Airborne photogrammetric cameras used for DEM measurement on manned platforms or UAV's are typically comprised of a single or multiple cameras mounted looking downward (nadir), or multiple cameras mounted in an oblique configuration. Regardless of the platform or configuration, a DG system integration and the georeferencing workflow remains the same.

The IMU is attached rigidly to the cameras structure (usually embedded directly inside) so it can measure the exact motion the camera experiences during flight. The DG system records the IMU and GNSS data, plus the mid-exposure pulses of all images in a common time base such as GPS time. The DG post-processing software then computes the position and orientation (also referred to as the Exterior Orientation) of each image at its perspective center at the time of exposure, with respect to a local mapping frame and datum. These data are then used as inputs into the photogrammetric mapping process to generate a DEM without any or with just a few ground control points (GCP's).

Calibration parameters include the lever arm offsets of the IMU to each camera perspective center, and the GNSS antenna phase center, and the IMU boresight angles to each camera. If the camera system is mounted on a stabilized mount, the IMU and camera will be rotating with respect to the GNSS antenna that is mounted on the airframe. In this case the angles of the mount (gimbal angles) with respect to the airframe are required as input to the DG system so that the GNSS measurements can be accurately translated from the antenna phase center to the center of rotation of the mount.

Boresight Calibration Requirements

An IMU usually does not provide an orientation reference for its inertial sensor input axes to the accuracy required by direct georeferencing, which means the physical mounting offset angles between the IMU and image sensor reference frame axes must be calibrated. This procedure is commonly called boresighting of the IMU. The methods used to compute the boresight angles depends on the type of imaging sensor used.

Lidar

The mounting misalignment or boresight angles between the IMU and lidar reference frame are usually computed though a combination of laboratory and airborne calibration, or sometimes solely through an airborne calibration.

The laboratory calibration typically involves collimating the laser beam to achieve a perfect vertical reference, against which the roll and pitch output of the DG system are compared to determine the roll and pitch misalignment angles. Such a calibration requires a true horizontal reflecting surface. The laser beam is directed towards the horizontal reflecting surface and its scan angle is set to nadir. The complete lidar instrument is then rotated in roll and pitch until the incident and reflected beams are coincident and hence perpendicular to the horizontal reflecting surface. The roll and pitch angles reported by the DG system are then the roll and pitch misalignment angles of the DG IMU with respect to the laser beam. This procedure is repeated a number of times to average down the errors in the DG orientation.

The airborne calibration is usually done using a reference surface test. Here a series of opposing flight lines with 100% overlap are flown over a flat bare surface (usually a runway or parking lot). Cross sectional heights of opposing flight lines are then compared with each other. A roll misalignment angle will cause the cross sections to look like an "X" when they are overlaid, and hence can be estimated directly from the height differences. The pitch misalignment can similarly be determined by flying opposing flight lines over a bare sloped surface (e.g., a treeless hill). Along track sections are then compared to determine the pitch misalignment angle, which has the effect of offsetting the slope horizontally. Finally the heading misalignment is determined by flying opposing flight lines with 100% overlap over the corner of a building. The flight lines are then overlaid, and any heading misalignment will have the effect of the corner of the building to be translated horizontally between the opposing lines.

A variation of the reference surface test is to fly a block of opposing flight lines with 50% overlap and several cross strips. Then the overlapping surfaces can be adjusted in a least squares fashion to solve for the boresight angles that minimize the height separation of the surfaces.

Some lidars also have the ability to provide intensity for each return. This allows specific points to be identified in a lidar "image" for which the ground coordinates can be surveyed, thus allowing traditional photogrammetric techniques to be used to determine the boresight angles.

No matter which method is used, the boresight flight must be designed such that the number of flight lines, flying height and overall geometry of the setup are enough to compute the misalignment angles to an accuracy significantly better than the specified accuracy of the overall system. This can often be challenging, especially for the high-altitude systems that require better than 0.005 degrees of roll and pitch accuracy.

IfSAR

For an IfSAR system, the mechanical structure of the SAR antenna is usually machined to such a high level of accuracy that the INS or IMU can be mechanically aligned to the antenna in the laboratory to a sufficient accuracy. Then an airborne test using corner reflectors is conducted as a final check.

Photogrammetric Cameras

Two approaches are followed to calibrate the boresight angles between the IMU and the camera image plane (see Figure 12.30). The traditional absolute approach is to use aerial triangulation by flying over a test field having many ground control points. The boresight angles are computed by constructing the mean differences of the orientation angles from the bundle adjustment with the angles from the DG system. The other approach (the relative approach) is to directly and relatively solve for the boresight angles in the least squares filter without ground control. The relative approach is more or less the typical approach nowadays implemented by most photogrammetric

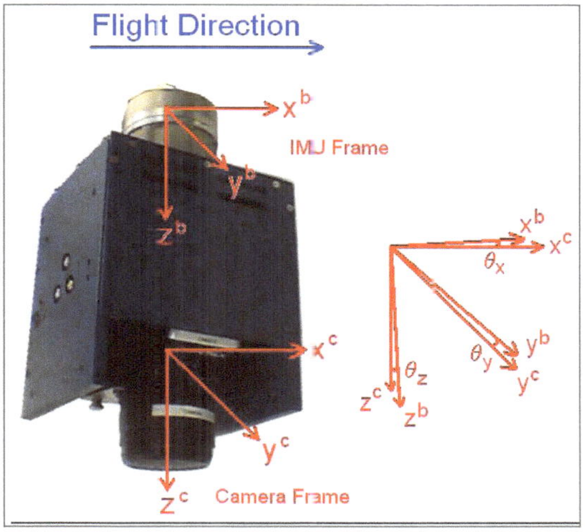

Figure 12.30. Boresight between camera and IMU frames of reference.

software packages. The relative approach allows for computing the boresight angles as a quality check using a small image block from any flight mission data and does not necessarily require flying over a test field, unlike the case of the absolute approach where ground control was always needed. For details, see Mostafa and Schwarz (2001).

Empirical approaches have been developed for boresight calibration for a variety of digital camera systems where only one or two ground control points are needed for datum calibration only if the relative boresighting approach is used. One of the advantages of the relative approach is that the boresighting could be done anytime using a part of the flight mission data, as depicted in Figure 12.31, where only a small block of images is chosen at either the center or the corner of the mapping area to be used to calibrate the boresight, camera, and datum as a quality check (Mostafa, 2002).

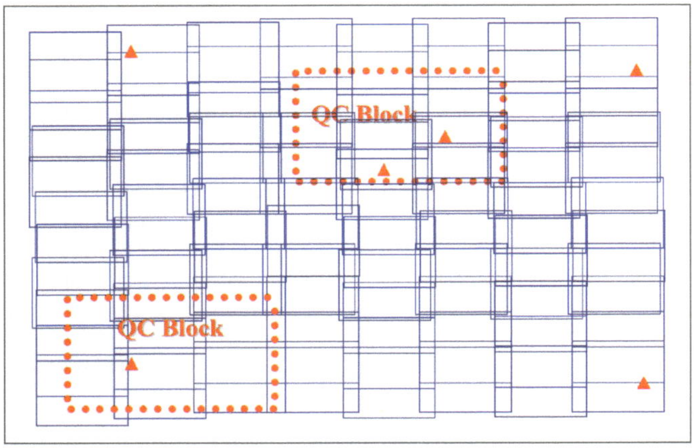

Figure 12.31. Quality Control using real mission data.

Lever Arm Calibration Requirement

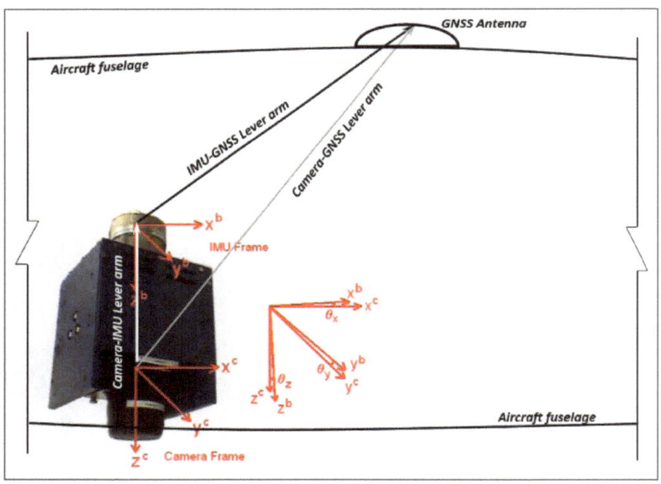

Figure 12.32. Lever arm calibration.

A lever arm is a relative position vector of one sensor with respect to another. It is usually convenient to define a sensor coordinate frame in which lever arms are defined and measured, whose origin is the sensor reference point and the origin for all lever arm measurements. In a photogrammetric camera installation, the sensor reference point is the camera perspective center. In a DG system installation, there are typically two lever arms that must be measured. The IMU lever arm is the relative position of the inertial sensing center of the IMU with respect to the sensor reference point. The GNSS lever arm is the relative position vector of the phase center of the GNSS antenna with respect to the sensor reference point, as shown in Figure 12.32. These two lever arm vectors characterize the installation of the DG system with respect to the sensor, and by vector subtraction provide the relative position of the GPS antenna with respect to the IMU required by the inertial-GPS integration software. As with the boresight angles, these lever arms must be measured to a better accuracy than the DG positioning accuracy in order for it not to contribute to the overall positioning error budget.

Post-Processing

The final product of an airborne sensor such as a lidar or a camera is generated by a post-processing software. This implies that the airborne sensor essentially stores its raw data in some form during a mission. Likewise, a DG system for airborne applications logs its raw IMU and GNSS data during a mission, and the position and orientation solution is computed post-mission with specialized GNSS/Inertial processing software. The software first implements a Kalman filter based algorithm (see Figure 12.28 on page 375) that generates the position and orientation time history. It then performs a smoothing operation backward in time on the data to generate a time history with improved accuracy. This method of post-mission processing generates the best achievable accuracy that can be had from the recorded inertial and GNSS data. It is an important component of a high accuracy DG system capable of 0.0025 degree orientation accuracy for airborne Digital Elevation Model extraction data process flow.

Planning Considerations

The absolute accuracy provided by a DG system is limited to the absolute positional accuracy of the GNSS. Hence it is important that proper mission planning be done to ensure that the best possible GNSS accuracy is achieved. To this end, the generic GNSS mission-planning considerations described previously apply here.

The best GPS positioning accuracy (1 to 10 cm) is achieved using carrier phase PPK DGNSS techniques described earlier. To obtain this accuracy, a mission must be planned to provide conditions for reliable ambiguity resolution throughout the mission.

Error sources that can prevent maintenance or re-fixing of integer ambiguities include ionospheric delays, multipath, and poor satellite geometry. Even if the correct ambiguities are found and maintained for the entire mission, if these error sources are not properly managed, they can still degrade the accuracy of the solution. Airborne mission planning should therefore include the following components.

Multipath Mitigation

Multipath reflected signals can be a major source of position error and cause for integer ambiguity resolution failures. It is therefore recommended that base station GNSS receivers use antenna choke rings or ground planes to attenuate low elevation signals, and should be mounted at least 100 m away or above all reflecting surfaces.

Inertial Navigator Alignment

A DG system can align itself while stationary or in motion. In fact, the in-air alignment is accelerated and the quality of the alignment improved if the aircraft performs an accelerating maneuver such as take-off or a turn. An in-air alignment requires a period of nominally straight and level flight to allow the DG system to compute an initial roll and pitch, followed by a series of turns to align the heading. Thereafter the alignment and hence accuracy of the DG solution improves with every maneuver. On manned platforms of fixed wing UAV's a typical survey pattern provides the maneuvers required by the DG system to maintain a high quality alignment. On rotor based UAV's typically a fast linear acceleration/deceleration maneuver is required before and after the survey area to ensure alignment.

Quality Control

The quality of the data generated by a DG system does not become directly apparent until it is combined with a lidar or a camera data. Consequently quality control for a DG system becomes a process of managing each step in the data acquisition and post-mission processing phases to achieve a consistent and reliable quality assessment.

Proper mission planning as described in the previous section goes a long way towards obtaining repeatable results. Once the mission begins, the DG system must be monitored frequently for GNSS dropouts or other data acquisition failures. A severe failure such as loss of GNSS data for an extended time period may be grounds for aborting the mission. Once the aircraft or UAV has landed, the recorded data should be checked in the field for outages and other immediate indications of bad or missing data such as cycle slips. This allows the mission to be re-flown possibly the same day.

If the recorded data are deemed to be acceptable, then the data are handed over to post-mission processing. The DG post-processing software typically has several quality assessment indicators. The most basic of these are the inertial-GNSS residuals. These are the corrected differences between the inertial and GNSS position solutions at each GNSS epoch, and indicate the consistency between the solutions. The residuals will appear to be random in a successful inertial-GNSS integration, indicating that the integration process has removed all sources of bias errors in the data. The processing software will typically perform a statistical analysis on the residuals and report a simple quality indicator to the user.

Once all the data have been processed and the georeferenced data assembled, the final quality is given by the separation between computed and true positions of ground control points in the georeferenced data. This is the first direct measurement of quality, and comes at the end of the mapping process. If the previous steps have been managed properly, then the final quality measurement will be consistent, reliable and reproducible.

MOTION SENSING SYSTEM FOR MULTIBEAM SONAR BATHYMETRY

Overview

A multibeam sonar generates a fan of listening beams that spans 70-90 degrees. The bottom swath covered by the sonar fan has a width in excess of 7 times water depth. The multibeam sonar provides raw ranges from return echoes along each listening beam, which an on-line processing computer translates into georeferenced depth images. Figure 12.33 shows an example of a sea bottom DEM generated with a multibeam sonar. This process requires accurate measurement of the position and orientation of the sonar head at high data rates so that the sonar system can interpolate the sonar position and orientation to the time of echo reception.

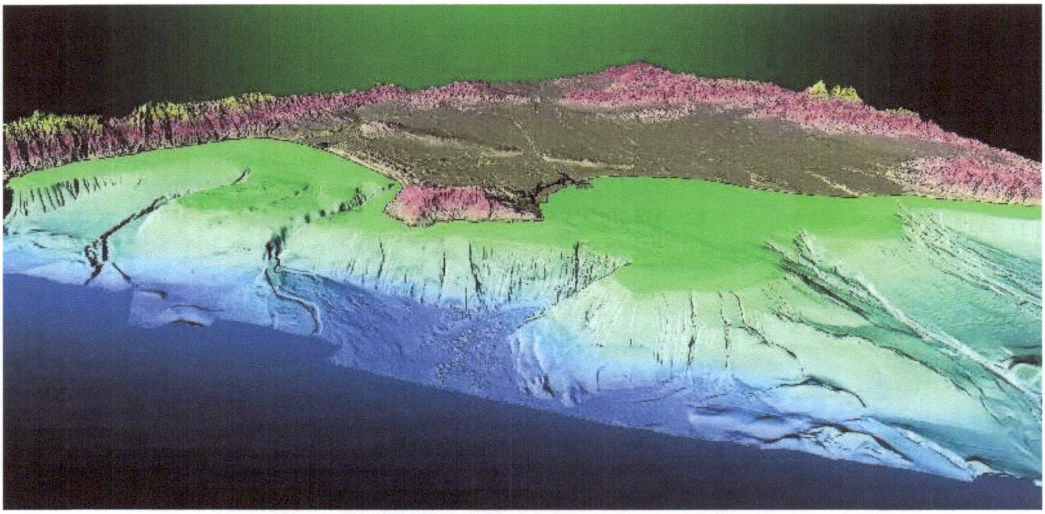

Figure 12.33. Greater Los Angeles continental shelf DEM generated with a multibeam sonar (Image courtesy U.S. Geological Survey).

The International Hydrographic Organization (Special Publication No. 44, 1987) specifies an accuracy requirement of 1 percent of water depth greater than 30 meters and 0.1 meters for shallower depths. The depth error γ as a ratio of water depth incurred in the outer beam of a multibeam sonar with fan angle Θ is: $\gamma = \delta\Theta \tan(\Theta)$ where $\delta\Theta$ is the error in the measurement of roll. Figure 12.34 shows the multibeam geometry. In order to achieve full utilization of a multibeam sonar, i.e., $\gamma = 0.01$ or 1 percent in the outer beam of a 75 degree fan, the error standard deviation of the measured orientation must be better than 0.05 degrees. Heading error impacts the position error of each pixel in the bathymetry data. In order to obtain a 0.5 meter pixel position error in the outer beam of a 75 degree fan at a depth of 100 meters, the heading error must be better than 0.05 degrees. Consequently the requirement for a high accuracy multibeam bathymetry system is 0.5 meters horizontal position error and 0.05 degrees orientation error in all three axes.

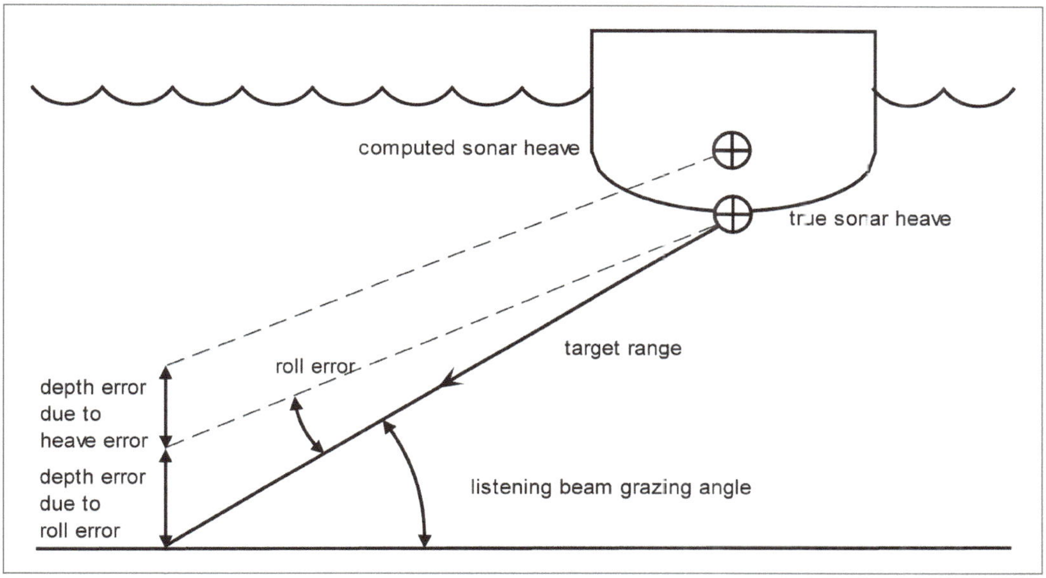

Figure 12.34. Bathymetric depth error in a multibeam sonar.

In addition, the heave of the vessel must also be measured to an accuracy of better than 0.1 meters to be consistent with the IHO standard. Absolute measurement of height with this accuracy is currently not feasible in a typical marine survey scenario that is far from the nearest shoreline where a GPS reference receiver or other form of positioning reference can be installed. Furthermore, the datum for mean sea level does not necessarily correspond to a GPS zero altitude measurement. GPS uses the WGS-84 ellipsoid as the reference for positioning and an undulation model to compute height with respect to an approximate mean-sea level reference called the geoid. Consequently, vessel heave for marine survey is specified as a relative displacement with respect to an assumed vessel waterline in calm water. Any offset between the assumed heave datum and true mean sea level is removed during the final map generation. A heave sensor is thus required to measure the non-constant or high frequency vertical displacement of the vessel, which is an approximately sinusoidal displacement with frequency given by the wave encounter frequency of the vessel. A simple heave sensor comprises a vertical accelerometer and a heave filter that performs a combined double integration and a high-pass filtering action on the measured vertical acceleration. The high-pass filter blocks the nearly constant accelerometer bias from reaching the double integrator, which in turn computes the vertical displacement of the vessel in the filter pass band. The heave filter must be tuned so that it provides the required accuracy in a pass band that includes the lowest expected wave encounter frequency when the vessel is moving downwind.

A roll-pitch-heave (RPH) sensor provides basic measurements of vessel roll, pitch and heave. It is typically a single unit that contains triads of accelerometers and gyros and a processor that implements a vertical gyro (VG) algorithm and the previously described heave filter. The VG algorithm computes roll and pitch with respect to an apparent vertical reference assumed to be the gravity vector. If the vessel does not accelerate, then the accelerometers will measure only the gravity vector that defines the true vertical. The VG algorithm computes the short-term roll and pitch motion with respect to the apparent vertical using the gyro data. It uses a complementary filter to block high frequency accelerations due to wave motion in the apparent vertical estimation and block low frequency gyro biases in the roll and pitch propagation. If the vessel experiences a sustained horizontal acceleration during a turn, then the apparent vertical will shift away from the true vertical and the computed roll and pitch will contain offsets. Once the sustained acceleration ends, the VG algorithm will exhibit a transient that decays within the complementary filter's settling time. A vessel that uses a RPH sensor must therefore allow a settling time (typically less than 5 minutes) at the beginning of a survey line following a turn before using the RPH data.

The state-of-the-art in high accuracy marine motion sensors is a GNSS-AINS. This technology provides all of the required motion data at the required accuracy to provide full utilization of a wide-swath multibeam sonar. The GNSS-aiding provides for initialization, alignment and full accuracy motion sensing independent of vessel motion, as well as continuous refinement of the alignment and calibration of the inertial sensors. The GNSS-AINS delivers full accuracy in any vessel dynamics, including sharp turns and severe accelerations and decelerations. It has no requirement to maintain a straight trajectory or to allow for a settling time after a turn, as does a VG-based RPH sensor. The IMU provides the vertical acceleration measurement for the heave filter. The heading accuracy is achieved with a dual-antenna azimuth aiding system. The Applanix POS MV exemplifies the state-of-the art in high accuracy marine motion sensing.

Development History

The Canadian Hydrographic Service (CHS) first recognized the possibility of marrying the multibeam sonar with a sufficiently accurate motion measurement system for quantitative bathymetry in 1992. Until then, the multibeam sonar, at that time a relatively new device, was used by oceanographers for seafloor imaging. The advantage of vastly improved production and data density over the traditional single-beam sonar became immediately apparent, and assured the eventual dominance of multibeam sonars for bathymetric survey. Applanix worked in partnership with the CHS to demonstrate the concept, and then developed the Position and Orientation System for Marine Vessels (POS MV) for this new application (Woolven, and Scherzinger, 1996). Since then the POS MV has become one of the dominant choices for high accuracy multibeam applications.

Types of Sensors

Motion sensors available for bathymetry range from simple roll-pitch-heave (RPH) sensors such as the iXBlue Octans, TSS DMS, or Kronsberg Seatex MRU series, to a full GNSS-AINS such as the SBG Apogee or the Applanix POS MV (see Figure 12.35). The RPH sensors must be supplemented with a positioning system such as a GNSS receiver and a heading sensor such as a gyrocompass. A GPS-AINS provides a complete position, orientation and heave solution in a single system and therefore requires no supplementary data.

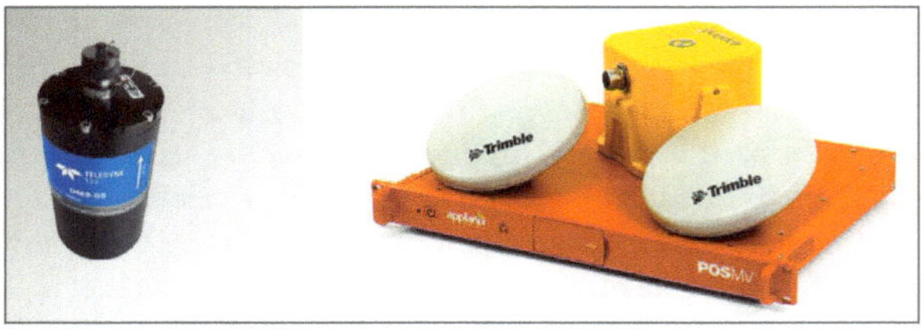

Figure 12.35. Examples of motion sensors for marine bathymetry, TSS DMS (courtesy TSS) and Applanix POS MV (courtesy Applanix Corporation).

AUTHOR BIOGRAPHIES

Dr. Bruno Scherzinger is the Chief Technology Officer Trimble Applanix, in Toronto Canada, where he is responsible for advanced navigation technology development and the core navigation technology in the Applanix product line. He has over 30 years of experience in GNSS-INS integrated navigation, and is a member of the Institute of Navigation (ION) and the IEEE. Dr. Scherzinger obtained the B.Eng. degree from McGill University in 1977 and the M.A.Sc. and Ph.D. degrees in System Control Engineering from the University of Toronto respectively in 1979 and 1983. He is also a Trimble Fellow.

Mr. Joe Hutton is the Director of Inertial Technology and Airborne Products at Trimble Applanix, and is responsible for the management of the airborne product line. He received a BASc degree in Engineering Science, Aerospace Option, from the University of Toronto in 1989, and a Master of Applied Science degree in Aerospace Controls from the University of Toronto Institute for Aerospace Studies in 1991. He has been with Trimble Applanix since its inception, and has over 25 years of experience in Integrated GNSS/Inertial technology as applied to airborne mapping.

Dr. Mohamed M.R. Mostafa is the Chief Technology officer at Navmatica Corporation. He has 17 years of experience in the field of multi-sensor systems for geomatics applications. He co-authored over 160 publications including the Manual of Photogrammetry. He has served as the chair of the ASPRS Direct Georeferencing Committee from 2001 to 2010 and currently serves as the chair of the ASPRS committee on High Precision Mapping by UAS. He completed his Ph.D. in geomatics engineering from The University of Calgary in 1999.

REFERENCES

GNSS

Ackermann, F., and H. Schade (1993). Application of GPS for aerial triangulation. In *Photogrammetric Engineering & Remote Sensing,* 59(11), (pp. 1625–1632).

Brown, R.G. and Hwang, P.Y.C. (2012). *Introduction to random signals and applied kalman filtering,* 4th Edition, John Wiley & Sons.

Councilman, C.C. and S.A. Gourevitch (1981). Miniature interferometer terminals for earth surveying: ambiguity and multipath for the global positioning system. In *IEEE Transactions on Geoscience and Remote Sensing,* GE-19(4), (pp. 244–252).

de Jonge, P.J., Y. Bock and M. Bevis (2000). Epoch-by-epoch™ positioning and navigation. In *Proceedings of ION-GPS-2000,* Salt Lake City, September 21–25.

Gleason and Gebre-Egziabher (2009). *GNSS Applications and Methods, Artech House.*

Hatch, R. (1981). The synergism of GPS code and carrier measurements. In *Proceedings of the 3rd International Geodetic Symposium on Satellite Doppler Positioning,* Las Cruces, New Mexico.

Hoffmann-Wellendorf, B., H. Lichtenegger and J. Collins (2001). *Global positioning system theory and practice, 5th edition,* New York, NY, Springer-Verlag.

Hofmann-Wellenhof, Lichtenegger, and Wasle (2007). *GNSS: global navigation satellite systems,* Springer-Verlag.

Parkinson, B.W. and J.J. Spilker (1996). *Global positioning system theory and applications, Volumes I and II,* Progress in Astronautics and Aeronautics Volume 163, American Institute of Aeronautics and Astronautics.

Radio Technical Commission for Maritime (RTCM) Services (2013). *Differential GNSS (Global Navigation Satellite Systems) services—Version 3.0, RTCM 10403.2,* RTCM Special Committee No. 104, February 1, 2013.

Remondi, B.W. (1985). Performing centimeter-level surveys in seconds with GPS carrier phase: initial results. In *Journal of the Institute of Navigation,* Vol. 32, No. 4, Winter 1985–86.

European Space Agency. *Navipedia.*

Federal Space Agency (Russia). *Information and analysis center for positioning, navigation and timing.*

GPS.gov. *The official U.S. government information about GPS.*

International GNSS Service (IGS).

Omnistar.

Trimble CenterPoint RTX.

United States Naval Observatory Time Service Department.

Inertial Navigation System

Britting, K.R. (1971). *Inertial navigation system analysis.* New York, NY: Wiley-Interscience.

Chatfield, A.B. (1997). *Fundamentals of high accuracy inertial navigation, progress in astronautics and aeronautics,* Volume 174, American Institute of Astronautics and Aeronautics.

Jekeli, C. (2001). *Inertial navigation systems with geodetic applications.* New York, NY: Walter de Gruyter.

Schuler, M. (1923). The disturbance of pendulum and gyroscopic apparatus by the acceleration of the vertical. In *Physikalische Zeitschrifte,* Vol. 24, (pp. 334–350). English translation (1952) by W. E. Frye.

Aided INS

Brown, R.G. and Hwang, P.Y.C. (2012). *Introduction to random signals and applied kalman filtering, 4th Edition,* John Wiley & Sons.

Farrell, J.A. and M. Barth (1999). *The global positioning system and inertial navigation.* New York, NY: McGraw-Hill.

Gelb, A. (editor) (1984). *Applied optimal estimation.* Cambridge, MA, MIT Press.

Kayton, M. and W.R. Fried (1997). *Avionics navigation systems, Second Edition.* New York, NY, Wiley and Sons.

Rogers, R.M. (2000). *Applied mathematics in integrated navigation systems.* AIAA Education Series, 2000.

Siouris, G.M. (1993). *Aerospace avionics systems, a modern synthesis.* San Diego, CA, Academic Press.

Airborne Georeferencing

Casella, V., K. Jacobsen, M.M.R. Mostafa, and M. Franzini (2006). A European project on direct georeferencing. In *Proceedings, ASPRS Annual conference, Reno, Nevada, May 1–5, 2006.*

Grejner-Brezezinska, D.A. (2001). *Direct sensor orientation in airborne and land-based mapping application, Report No. 461,* Geodetic GeoInformation Science, Department of Civil and Environmental Engineering and Geodetic Science, The Ohio State University, Columbus, Ohio.

Heipke, C., Jacobsen, K., Wegmann, H. (2001). *Analysis of the results of the OEEPE test "integrated sensor orientation".* OEEPE Workshop, Integrated Sensor Orientation, Hannover, Germany, Sept 17–18.

Hutton, J., Savina, T., and Lithopoulos, L. (1997). *Photogrammetric applications of Applanix's position and orientation system (POS).* ASPRS/MAPPS Softcopy Conference, Arlington, Virginia, July 27–30, 1997.

Ip, A.W.L., El-Sheimy, N., Hutton, J. (2004). Performance analysis of integrated sensor orientation. In *International Archives of Photogrammetry and Remote Sensing,* ISPRS Comm. V, 35(B5), (pp. 797–802), Istanbul, Turkey, July 12–23, 2004.

Ip, A.W.L., M.M.R. Mostafa, J. Hutton and J.P. Barriere (2008). An optimally integrated direct georeferencing and flight management system for increased productivity of airborne mapping and remote sensing. In *International Archives of the Photogrammetry, Remote Sensing and Spatial Information Sciences.* Vol. XXXVII. Part B1: 579–584, Beijing, 2008.

Jacobsen K., Wegmann, H. (2001). *Dependencies and problems of direct sensor orientation.* OEEPE Workshop, Integrated Sensor Orientation, Hannover, Germany, Sept 17–18, 2001.

Madani, M., M.M.R. Mostafa (2004). Georeferencing the DMC images—data flow and performance analysis. In *Proceedings, ASPRS Annual Conference, Denver, CO, May 22–28, 2004.* CD-ROM.

Moffit, F., and E.M. Mikhail (1980). Photogrammetry. New York, NY, Harper and Row.

Mostafa, M.M.R. and J. Hutton (2010). Theoretical ground accuracy analysis derived from today's airborne digital frame cameras and direct georeferencing. In *Proceedings of the ASPRS Annual Conference, San Diego, California, April 26–30, 2010.*

Mostafa, M.M.R. (2004). Airborne GPS positioning using continuously operating reference stations for mapping applications. In *Proceedings of the 4th International Symposium on Mobile Mapping Technology (MMT2004), Kuming, China, March 29–31, 2004.*

Mostafa, M.M.R. and J. Hutton (2004). A fully integrated solution for aerial surveys: design, development, and performance analysis. In *Photogrammetric Engineering and Remote Sensing,* 71 (4), (pp. 391–399).

Mostafa, M.M.R. (2002). Camera/IMU boresight calibration: new advances and performance analysis. In *Proceedings, ASPRS Annual Meeting, Washington, D.C.,* April 21–26, 2002.

Mostafa, M.M.R., J. Hutton, and E. Lithopoulos (2001). Airborne direct georeferencing of frame imagery—an error budget. In *Proceedings, The Third International Mobile Mapping Symposium, Cairo, Egypt, January 3–5, 2001.*

Mostafa, M.M.R. and K.P. Schwarz (2001). Digital image georeferencing from a multiple camera system by GPS/INS. In *ISPRS Journal of Photogrammetry & Remote Sensing* 56 (2001), (pp. 1–12).

Mostafa, M.M.R. and K.P. Schwarz (2000). A multi-sensor system for airborne image capture and georeferencing. In *Photogrammetric Engineering and Remote Sensing,* 66 (12), (pp. 1417–1424).

Reid, D.B., E. Lithopoulos, and J. Hutton (1998). Position and orientation system for direct georeferencing (POS/DG). In *Proceedings, Institute of Navigation 54th Annual Meeting, Denver, Colorado, USA, June 1-3, 1998* (pp. 445–449).

Sandau, R., Braunecker, B., Driesher, H. Eckardt, A., Hilbert, S., Hutton, J, Kirchhoffer, W., Lithopoulos, E., Reulke, R., Wicki, S. (2000). *Design principles of the LH Systems ADS40 airborne digital sensor.* IAPRS, Vol. XXXIII, Part B1. Amsterdam, 2000.

Scherzinger, B. (1997). A position and orientation post-processing software package for inertial/GPS integration (POSProc). In *Proceedings of the International Symposium on Kinematic Systems in Geodesy,* Geomatics and Navigation (KISS 97), Banff, Canada, June 1997, (pp. 197–204).

Schwarz, K.P., M.A. Chapman, M.E. Cannon, and P. Gong (1993). An integrated INS/GPS approach to the georeferencing of remotely sensed data. In *Photogrammetric Engineering and Remote Sensing,* 59(11), (pp. 1167–1674).

Schwarz, K.P., C.S., Fraser and P.C., Gustafon (1984). Aerotriangulation without ground control. In *International Archives of Photogrammetry and Remote Sensing,* 25 (Part A1), Rio de Janeiro, Brazil, (pp. 237–250).

Shi, J and E.M. Cannon (1994). Precise airborne DGPS positioning with a multi-receiver configuration: data processing and accuracy evaluation. In *Proceedings of the International Symposium on Kinematic Systems in Geodesy,* Geomatics and Navigation, Banff, Canada, (pp. 393–402).

Škaloud, J. (1999). Problems in direct-georeferencing by INS/DGPS in the airborne environment. In *ISPRS Comm. III, WG III/1 Barcelona, Spain,* Nov. 25–26.

Toth, C. and D.A., Grejner-Brzezinska (1998). Performance analysis of the airborne integrated mapping system (AIMSTM). In *International Archives of Photogrammetry and Remote Sensing,* 32 (2), (pp. 320–326).

Shipborne Bathymetry

International Hydrographic Organization (IHO) (1998). *Special Publication No. 44,* IHO Standards for Hydrographic Surveys, 4th Edition.

International Hydrographic Organization (IHO) (2008). *Special Publication No. 44,* IHO Standards for Hydrographic Surveys, 5th Edition.

Loncarevic, B.D. and B.M. Scherzinger (1994). Compensation of ship attitude for multibeam sonar surveys. In *Sea Technology,* June 1994.

Scherzinger, B.M., S. Woolven and M. Field (1997). Seafloor mapping solution. In *Sea Technology,* March 1997.

Woolven, S. and B.M. Scherzinger (1996). An integrated position, attitude and heading solution for seafloor mapping with multibeam sonars. In *Proceedings of the IEEE Oceans 1996 Conference, Fort Lauderdale, FL,* September 1996.

Chapter 13

DEM USER APPLICATIONS

David F. Maune

INTRODUCTION

This chapter explains how elevation data are used and why elevation data are so valuable. Combined with the DEM User Requirements and Benefits in Chapter 14, this chapter provides credible information to assist in justifying the cost of national., state and local community initiatives to acquire enhanced elevation data. This chapter includes major DEM user applications documented in the final report of the National Enhanced Elevation Assessment (NEEA) which the U.S. Geological Survey (USGS) sponsors called "the most comprehensive benefit/cost analysis ever performed for any layer of The National Map." The author of this chapter was also the primary author of the NEEA, prepared by Dewberry in 2012 for USGS.

The NEEA was based on 27 Business Uses (Table 13.1) from which 602 Functional Activities were defined by users with mission-critical requirements for enhanced elevation data in one of five Quality Levels and one of five update frequencies. Extensive benefit/cost analyses were performed initially for eight national implementation scenarios, subsequently expanded to ten scenarios from which USGS selected the scenario for Quality Level 2 (QL2) lidar of 49 states and U.S. territories and QL5 IfSAR of Alaska as having the highest return on investment (ROI).

	DEM Business Uses
1	Natural resources conservation
2	Water supply and quality
3	River and stream resource management
4	Coastal zone management
5	Forest resources management
6	Rangeland management
7	Wildlife and habitat management
8	Agriculture and precision farming
9	Geologic resource assessment, hazard mitigation
10	Resource mining
11	Renewable energy resources
12	Oil and gas resources
13	Cultural resources preservation and management
14	Flood risk management
15	Sea level rise and subsidence
16	Wildfire management, planning, and response
17	Homeland security, law enforcement, and disaster response
18	Land navigation and safety
19	Marine navigation and safety
20	Aviation navigation and safety
21	Infrastructure and construction management
22	Urban and regional planning
23	Health and human services
24	Real estate/banking/mortgage/insurance
25	Education K-12 and beyond
26	Recreation
27	Telecommunications

Table 13.1. DEM Business Uses Documented in the NEEA Study.

This chapter consolidates DEM user applications into eight major categories: (1) General Mapping Applications, (2) Coastal Mapping Applications, (3) Transportation Applications, (4) Underwater Applications, (5) Other Technical Applications, (6) Military Applications, (7) Commercial Applications, and (8) Individual Applications.

GENERAL MAPPING APPLICATIONS

Planimetric Maps

A planimetric map shows the 2-D horizontal location of features, but without contour lines or other means for displaying the 3-D topography. Planimetric maps are commonly produced from aerial imagery and stereo photogrammetry. Outside the mapping profession, it is not widely understood that digital elevation data are critical in the production of planimetric maps. The reason for this is relief displacement—the displacement of an image, outward from the center of an aerial image, caused by the elevation (relief) of features above an established base elevation.

A single aerial photograph or image is a perspective view of the terrain as viewed from above. Whether or not a photograph is exactly vertical., i.e., photographed with the camera pointed straight down, elevated features on the photo will have a different scale than features closer to sea level. The magnitude of the relief displacement for a feature photographed with aerial imagery is a function of the camera's focal length, the elevation of the feature mapped, and the feature's distance from the center of the photo. Relief displacement is corrected by stereo photogrammetric procedures that account for these variables. Furthermore, if the aerial images are not exactly vertical, they also have tilt displacement caused by the roll and pitch of the aircraft at the instant a photograph is taken. Stereo photogrammetry corrects for tilt displacement in addition to relief displacement.

Planimetric mapping requirements are normally satisfied by one of the stereo photogrammetric mapping procedures described in Chapter 6, perhaps in combination with additional land surveys of parcel and administrative boundaries and other features not visible on the aerial images. However, aerial lidar is also popularly used for mapping some planimetric features, especially when high density lidar data are available with four points per square meter or better. The mapping of building footprints from high-density lidar is often more accurate than mapping building footprints from digital orthophotos that include building lean.

Planimetric features at ground level can also be digitized from digital orthophotos that have already been corrected for relief displacement. However, elevated features, including buildings, still lean outward from the center of the photos from which produced; therefore, rooflines are normally displaced outward from building foundations at ground level. Taller features like towers and smokestacks may have very large displacements on digital orthophotos.

Digital Orthophotos

Digital orthophotos include no elevation data, but they cannot be produced cost-effectively without DEMs because of relief displacement and tilt displacement. Aerotriangulation yields the six exterior orientation parameters for each image, as explained in Chapter 6. Computers can compensate for tilt displacement, but relief displacement is still present in ortho images. By projecting aerial images over a DEM, the "draped" image compensates for most of the relief displacement and provides an approximate orthographic projection, as though theoretically looking straight down from space. But this is true only for features at the ground elevation of the DEM used. Elevated features, such as buildings, towers and trees are still displaced outward, and the sides of buildings are shown, both as a function of their elevation and as a function of their distance from the center of the photograph "draped" over the DEM. Furthermore, elevated bridges may be distorted if orthorectified to the elevation of the terrain beneath the bridge, as shown in Chapter 6, Figure 6.9 on page 143.

A digital orthophoto serves as the base map for overlay of planimetric and topographic features. Because the sides of buildings are shown (Figure 6.11 on page 144), most standard digital orthophotos produced today are not "true digital orthophotos." A true digital orthophoto would show the roof of buildings, even skyscrapers, directly over their foundations, and there would be no pixels showing the sides of a skyscraper, for example. Normally, the sides of buildings are recorded on the perspective aerial images, and there are no simple solutions for deleting image pixels that have no place to go. Although several software programs are designed to produce "true digital orthophotos," the process is still complex and expensive. For now, most digital orthophotos are not the perfect "straight down" views of the earth from infinity that we wish we had in an ideal GIS. For the foreseeable future, we will continue to rely upon DEMs to produce the forms of digital orthophotos in common use today. The only other alternative is to fly higher, with longer focal length lenses, and acquire imagery with heavy forward overlap (e.g., 80%) and heavy side lap (e.g., 60%) so that only the "sweet spot" at the center of images are used for mosaicking images together that minimize building lean.

Orthophotomaps

An orthophotomap is an orthophoto or mosaic of multiple orthophotos to which other optional information may be added, such as a map grid, north arrow, scale bar, planimetric and/or topographic features, names, text, marginalia and/or other information characteristic of a map but missing on the orthophoto or orthophotos from which produced. An example orthophotomap is at Figure 13.1.

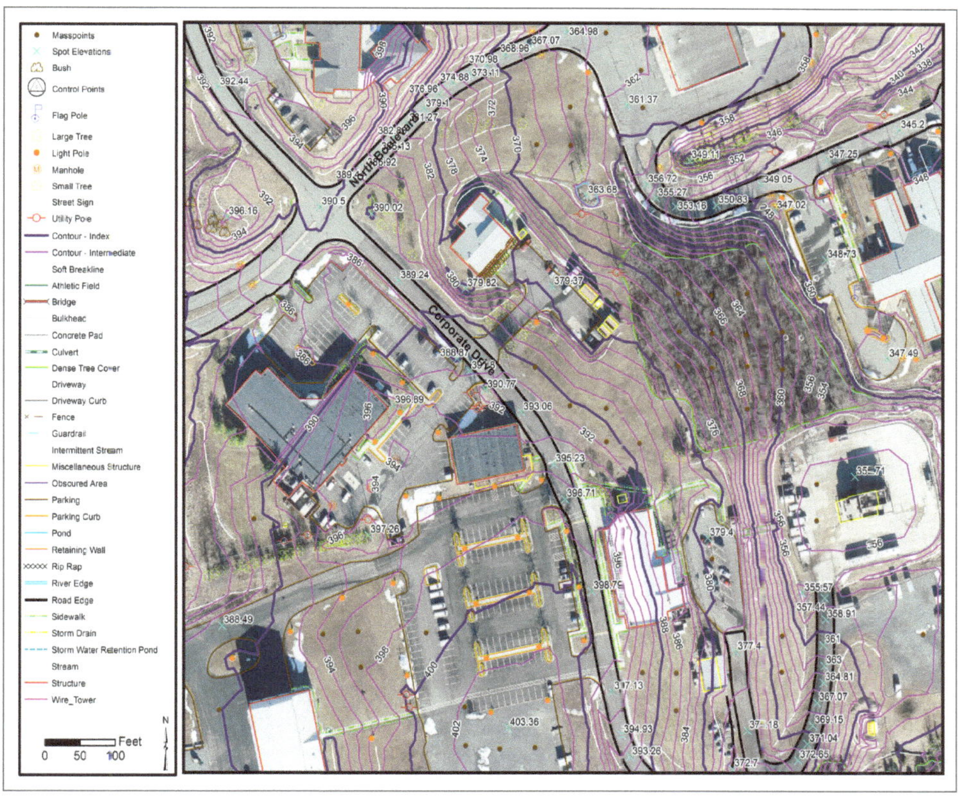

Figure 13.1. Large-scale orthophotomap with addition of planimetric and topographic features.

Topographic Maps

Topographic maps normally include planimetric features as well as topographic contours and spot heights. It goes without saying that modern topographic maps must include digital elevation data of some sort in order to produce the critical elevation component of a topographic map.

The topographic mapping requirement can be satisfied by stereo photogrammetry (see Chapter 6) or in combination with digital elevation data produced by IfSAR (see Chapter 7) or topographic lidar (see Chapter 8). Bathymetric contours are produced from bathymetric lidar (see Chapter 10) or sonar (see Chapter 11).

Figure 13.2 shows what USGS 7.5-minute topographic quadrangles maps looked like in the past. These paper maps were almost impossible to maintain up-to-date, and their traditional color-separation processes were very expensive and archaic.

U.S. Topo Maps

The current "U.S. Topo" maps are produced by USGS' National Geospatial Program (NGP). The term "U.S. Topo" refers specifically to quadrangle topographic maps published in 2009 and later, modeled on the familiar 7.5-minute quadrangle maps of the period 1947–1992 (see Figure 13.2) but mass-produced from national GIS databases on a repeating cycle. US Topo maps repackage GIS data in traditional map form; this benefits non-specialist map users, as well as applications that need traditional maps.

Layered PDF files allow users to turn data layers on and off. The maps include layers not present on most traditional topographic maps, such as aerial photo and shaded relief images. For the same 7.5-minute quad area shown in Figure 13.2 for the traditional 7.5-minute topographic paper quad map, Figure 13.3 shows a new U.S. Topo map with selected digital layers turned on including the contour layer. Figure 13.4 shows a U.S. Topo map of that area with selected digital layers turned on including the shaded relief layer, and Figure 13.5 shows a U.S. Topo map of that area with selected digital layers turn on including the orthophoto layer. All four images are courtesy of USGS.

In 2001, the USGS published a new vision for topographic mapping of the Nation: The National Map. The objective of The National Map is to ensure current, consistent, seamless, and integrated geospatial data for the Nation through Federal, state, local and tribal partnerships.

In 2009, the USGS NGP, working with partners in the public, private and academic sectors, developed capabilities to rapidly produce U.S. Topo maps from The National Map data. These new maps are vital for homeland security, research, industry, disaster response, recreation, land and resource management, and many other applications.

As shown in the figures, U.S. Topo quadrangles are digital topographic maps from the USGS, arranged in the familiar 7.5-minute quadrangle format and designed to look and feel much like the traditional paper topographic maps from USGS, while providing modern technical advantages that support faster, wider public distribution and enable basic, on-screen geographic analyses for users. Important characteristics include the following: (1) Made from nationally consistent data that are quality assured to high standards; (2) Richer content, multiple layers of data; more than a street map; (3) Continuously evolving, incorporating additional data layers and features; (4) Look and feel like legacy paper USGS topographic maps but with technical advantages; (5) Can be used on the computer or printed to scale; (6) Users can select the reference systems, e.g., latitude-longitude, UTM, or military grid reference system; (7) Free downloadable user tools; and (8) Interactive capabilities with Google Maps®.

U.S. Topo maps are available free on the Web. Each map quadrangle is constructed in GeoPDF® format from key layers of geographic data, including orthoimagery, topographic contours, hydrography, transportation, structures, boundaries and geographic names found in The National Map. Newer features include Public Land Survey System (PLSS) in 12 western states and woodland tint.

U.S. Topo quadrangles can be printed from personal computers or plotters as complete, full-sized maps or in customized sections, in a user-desired specific format. Paper copies of the maps can also be purchased from the USGS Store. Download links and a user's guide are featured on the U.S. Topo website.

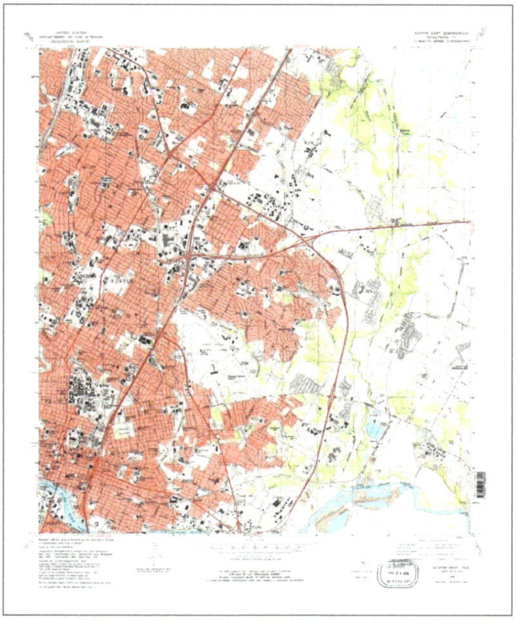

Figure 13.2. Traditional hardcopy USGS 7.5-minute topographic quadrangle map.

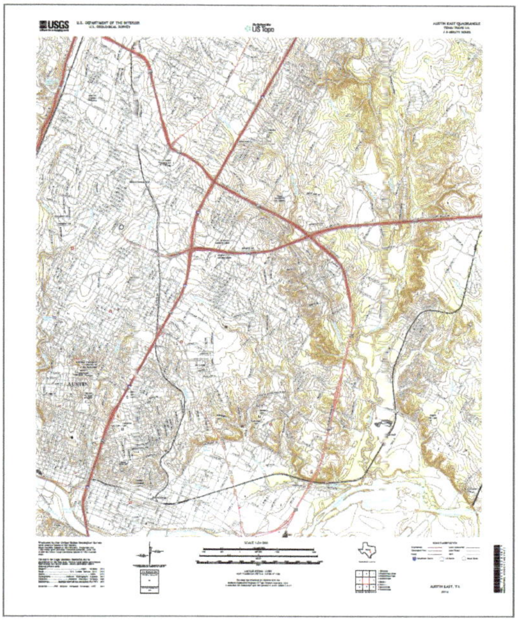

Figure 13.3. U.S. Topo map with selected digital layers turned on, including the contour layer.

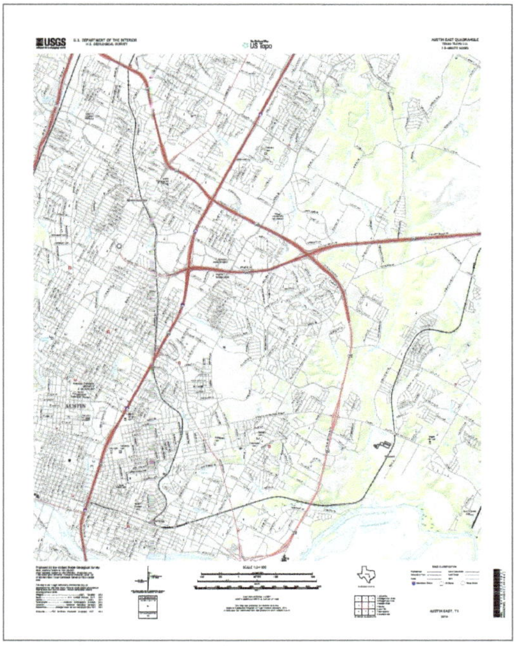

Figure 13.4. U.S. Topo map with selected digital layers turned on, including the shaded relief layer.

Figure 13.5. U.S. Topo map with selected digital layers turned on, including the orthophoto layer.

U.S. Topo users can turn geographic data layers on and off as needed; they can zoom in and out to highlight specific features or see a broader area. File size for each digital 7.5-minute quadrangle, about 15-20 megabytes, is suitable for most users. Associated electronic tools for geographic analysis are available free for download.

The first 3-year production cycle for the conterminous 48 states was completed in September 2012, and the second in September 2015. USGS is currently in the third cycle. Hawaii, Puerto Rico, and U.S. Virgin Islands also have U.S. Topo coverage. Alaska has been started, and should be complete by 2020.

Flood Insurance Rate Maps (FIRMs)

Flood Insurance Rate Maps (FIRMs), produced by the Federal Emergency Management Agency (FEMA), are combinations of the above planimetric and topographic data and orthoimagery. The older FIRMs are essentially planimetric base maps overprinted with Base Flood Elevations (BFEs) and Special Flood Hazard Area (SFHA) boundaries (computed from hydrologic modeling of elevation data for entire watersheds, and hydraulic modeling of elevation data in floodplains), but they don't include actual contours.

If contour lines were printed on FIRMs, users would see that SFHA boundaries cross contour lines as the flood waters slope downstream in accordance with computer models for standard flood events (10%, 2%, 1% or 0.2% annual chance floods commonly referred to as 10-, 50-, 100-, or 500-year floods). A river doesn't simply flood to a specified elevation, but to variable elevations along different reaches of a river. In fact, FIRMs regularly show BFEs that are relatively close to other BFEs (upstream or downstream) that are, respectively, at least one foot higher or lower. Sometimes, a BFE immediately upstream of a bridge or culvert is several feet higher than a BFE immediately downstream of that bridge or culvert. This does not result from abrupt changes in surface topography, but from the bridge or culvert being undersized to carry the flow expected to result from a 1% annual chance flood (100-year flood), causing that bridge/culvert to act as a dam that increases the flood risk to buildings upstream.

FEMA's digital FIRMs (DFIRMs) often use digital orthophotos as the base map for overlay of georeferenced digital flood data. Figure 13.6 shows the concept behind the DFIRMs. It shows that the DFIRM results from the merger of (1) A digital orthophoto base map; (2) Digital elevation data used to perform hydrologic and hydraulic (H&H) modeling and analyses; and (3) The digital flood data (BFEs and SFHAs) that result from such H&H modeling. The resulting DFIRM is the map designed to predict locations that have a 1% or greater chance of flooding during any given year.

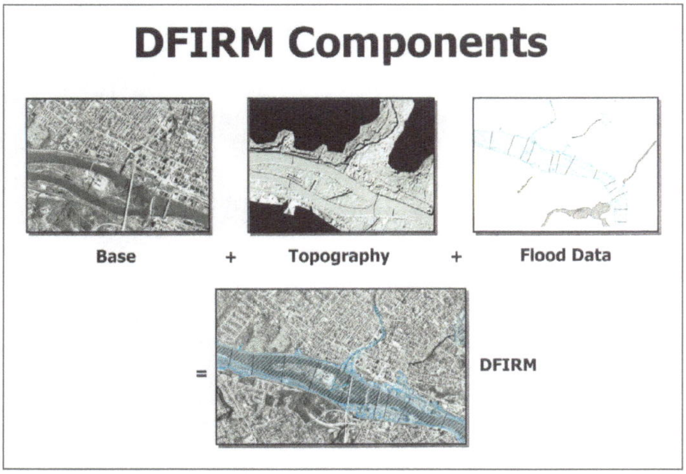

Figure 13.6. Components of a Digital Flood Insurance Rate Map (DFIRM). Images courtesy of FEMA.

FIRMs and DFIRMs are often obsolete because the risk of flooding is ever increasing, even for the same predicted amounts of rainfall, because land development activities (cutting down trees, and covering the land with concrete and asphalt) causes less water to be absorbed into the ground and more water to run off into the stream with each passing year, increasing the flood risk. Thus, although the shape of the topography may not change that much, the changing land cover on that topography continuously worsens the flood risks.

Hydrologic modeling can be performed on less-accurate DEMs available from virtually any source described in this manual, including most data in the National Elevation Dataset (see Chapter 4). The slope of the terrain (derived from DEMs) is one of the factors in determining the volume and speed of surface water run-off.

Hydraulic modeling is performed on more-accurate TINs, including various breaklines that define channel cross-section geometry and ditches, for example. FEMA now utilizes QL2 lidar or better, consistent with the 3D Elevation Program (see Chapter 5). The demanding requirement for hydraulic modeling requires TINs that are provided either by conventional stereo photogrammetric procedures described in Chapter 6, or topographic lidar described in Chapter 8. However, breakline generation is best performed by stereo photogrammetric compilation procedures, so lidar and photogrammetry may combine to deliver the optimal solution for hydraulic modeling necessary for FIRM production. As explained in Chapter 9, lidargrammetry is gaining acceptance for such breakline generation.

Soils Maps

Hugh Hammond Bennett, the first Chief of the Soil Conservation Service (SCS), maintained a consistent focus on the slope of the land and the impact of slope on drainage, erosion, and soil and water conservation. When the SCS was founded, steep slopes were clear cut, dust storms were carrying soil from the nation's midsection to Washington, D.C., and Hugh Hammond Bennett was campaigning for Congressional action to stop erosion. Congress acted and passed the Soil Conservation Act in 1935, creating the Soil Conservation Service, now known as the Natural Resources Conservation Service (NRCS), a major component of the U.S. Department of Agriculture (USDA).

One of the founding principles of SCS is that "science matters." Nearly a century after the creation of the SCS, NRCS managers agree that "science still matters," drawing on a tradition of scientific principles in working with private landowners that is as relevant today as when it was a dream to Hugh Hammond Bennett.

As documented in (NRCS, 2011), terrain influences, especially slope, aspect and curvature, are the most universal factors determining the pattern of soil distribution. Software available today provides the potential for more objective determination of the terrain characteristics that are used to differentiate soils. Elevation presented as a raster array (gridded DEM) is the universal format used by soil scientists. Data delivered as a TIN or Terrain is converted to a raster format prior to use. Terrain features used by soil scientists include: elevation, hillshade, slope, aspect, planform curvature, profile curvature, tangential curvature, flow accumulation, soil wetness index, relative position, various landform classification schemes, and sinks. Elevation is used to map flooding frequency and parent materials. Elevation is also used to estimate soil temperature, soil moisture regime and precipitation. Temperature and moisture regimes are part of the classification scheme used by soil scientists.

Soil scientists have traditionally used a combination of science and art to make a soil map. Soil borings and soil pits reveal the soil properties used to classify soils at selected locations. Elsewhere, the tone of aerial photos, the landform and slope breaks provide logical places to delineate soil boundaries; these decisions are dominated by experience, art, and judgment. Soil line placement has been largely subjective because by necessity the soil scientist must project out from the soil borings, interpret the landscape, and sketch the boundary lines. Many of these lines are intended to occur at predetermined slope breaks (0-2%, 2-5%, etc.). Accuracy in delineating lines that separate slope groups has been based on a relatively few readings with clinometers, stereoscope interpretations, and interpolation based on field experience. In addition, vegetation (trees or crops) on imagery can obscure slope readings when trying to create the best line placement. Line placement between soil scientists has also varied with the level of experience and skill in landform interpolation. Because of these limitations, the soil scientist's best efforts at slope breakline placement have not always been as accurate as desired for modern applications such as precision farming.

Figure 13.7 on page 400 demonstrates how lidar data, used with the Lidar Enhanced Soil Survey (LESS) process, can improve NRCS' Soil Survey Geographic Database (SSURGO). The LESS report, which describes capabilities and limitations, is accessible via an Internet search engine such as Google. The LESS report concludes: "Until now, line placement between different sloping units has been mostly a subjective decision. Soil scientists have done their best to represent the landscape as they saw it and sketched the lines. With lidar

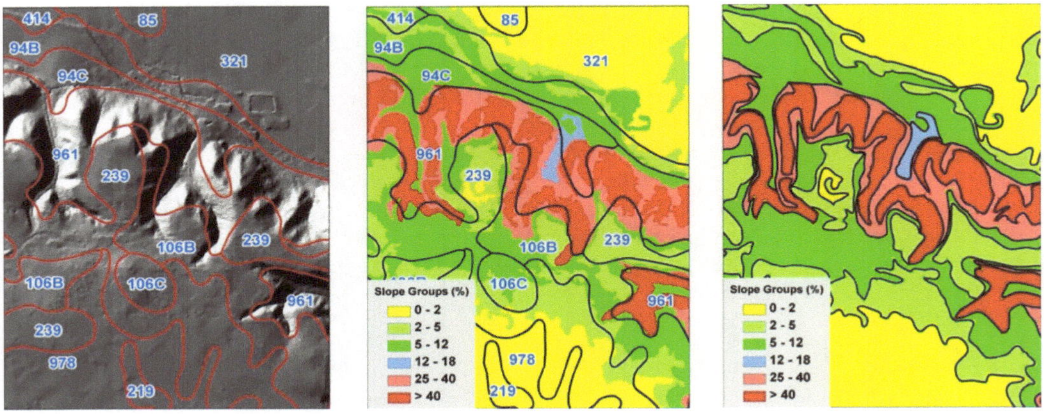

Figure 13.7. (Left) published SSURGO soil layer overlying a hillshade; line placement is inconsistent with the image. (Center) published SSURGO soils layer overlying the lidar based slope group map; the slope designations in the SSURGO map units are also inconsistent with the slopes generated by lidar data—better line placement is needed. (Right) automated slope group lines produced by the LESS model overlying the lidar based slope map; the position of the lines much more accurately fit the lidar data. The automated lines provide most of the line work to update/correct the SSURGO lines—lowering time and cost while improving accuracy and efficiency and promoting science-based decision-making. Images courtesy of NRCS.

data as a tool, it is possible to automate most of the soil line placement with speed and consistency, reducing the amount of dissimilar slope inclusions, and adding more objective science to soil line placement. The slope map can be tailored to fit the desired mapping scale. The accuracy of lidar enables the soil scientist to produce a soil map at a scale of 1:4000 or larger with excellent line placement or at a scale as small as that of a general soil map. The LESS model and slope map that it generates provides an update tool that is ready to use when lidar data becomes more readily available."

Were he still alive today, Mr. Bennett would be pleased that the key organization he founded still relies heavily on science in general, and lidar in specific, for soils mapping, soil and water conservation, and best engineering practices within NRCS.

Geologic Maps

In explaining how geologic maps are produced, largely from knowledge of slopes and outcrops, a brochure from the Pennsylvania Geologic Survey states: "Lidar data provides a wealth of information not discernible on photography or even in the field." Digital elevation models have a variety of uses in geology, geomorphology, geophysics, and related fields. Typical applications include:

- Generation of shaded relief images with variable illumination angles, topographic contour maps, and other derivative products to support mapping of land forms and geologic hazards such as landslides and faults
- Empirical and process-based geologic hazard modeling based on topographic variables such as slope angle, curvature, soil shear strength, and pore water pressure
- Surface water or rainfall/runoff modeling
- Generation of topographic profiles for use in groundwater modeling, slope stability, or earthquake hazard studies

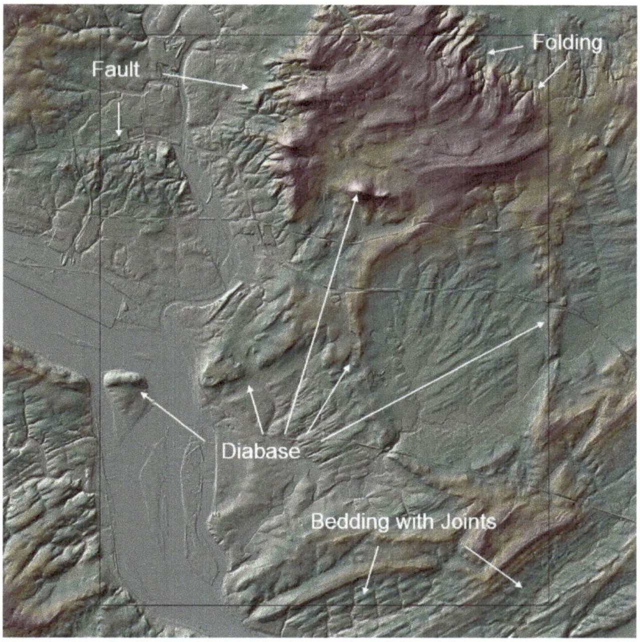

Figure 13.8. Mapping of geomorphological features. Image courtesy of the Pennsylvania Geologic Survey.

As shown at Figure 13.8, derivative maps depicting variables such as slope angle, residual topography, and various measures of topographic roughness can be particularly useful for identifying landforms when layered over shaded relief images and covered with topographic contours. The National Enhanced Elevation Assessment (NEEA) (see USGS, 2012) identified Geologic Resource Assessment and Hazard Mitigation as one of the top three Business Uses with potential annual benefits of $1B or more. During the NEEA, geologists were the users who most consistently specified that they needed the higher density QL1 lidar to perform their mission critical activities.

Wetland Maps

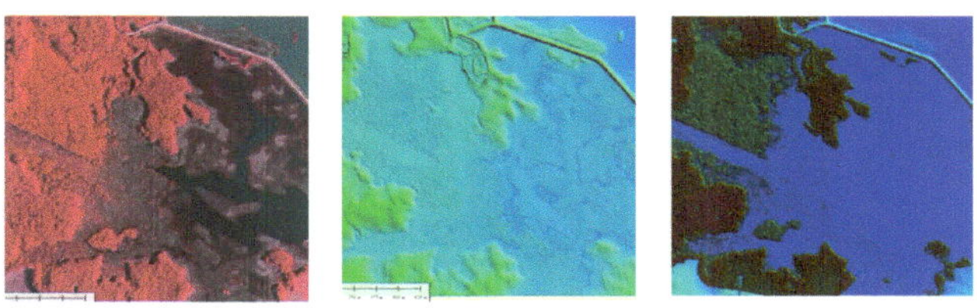

Figure 13.9. Color infrared imagery (left), lidar DTM (center) and lidar DSM (right) all work together for wetland identification. DTMs help differentiate wetland water regimes (flooding duration). DSMs help differentiate vegetation classification (e.g., trees vs. shrubs). Lidar point clouds help determine vegetation structure (e.g., pines vs. oaks).

In the U.S., EPA, FWS, USGS, NPS, BLM, USFS, NRCS, NOAA, NASA, USACE and other federal agencies have long been working with state Departments of Natural Resources and various ecosystem research centers to improve wetland maps, recognizing the cumulative effects of wetland loss and degradation on entire landscapes and regions. Remote sensing technologies are crucial for mapping and monitoring of wetlands. As shown in Figure 13.9 color infrared (CIR) imagery, combined with DTMs and DSMs from airborne lidar, have been standard research and common operational tools of the past decade for mapping and monitoring of wetlands. There is still a strong reliance on aerial photography for operational fine resolution wetland mapping. Operational wetland mappers understand that lidar data are very helpful; however, they are only used when available and standard protocols are needed. In Lang, McCarty et al., 2012, the authors report the results of studies that indicate that lidar-based topographic metrics (e.g., slope, aspect, and topographic wetness indices) have the potential to improve accuracy and automation of wetland mapping.

In Lang and McCarty, 2009, the authors report how lidar intensity images have significantly improved the mapping of inundation below the forest canopy relative to CIR aerial photography, with an accuracy of 97 percent from lidar intensity imagery versus 70 percent from CIR photography, but intensity data need to be calibrated to improve their utility for broad-scale operational use.

In McCarty and Lang, 2010, the authors report their use of satellite synthetic aperture radar (SAR) and airborne lidar for mapping forested wetlands in the Chesapeake Bay watershed. The SAR allows for more frequent image collections over large areas, maps through clouds, and is best for spotting short-term changes in water levels, such as flooding or short-lived pools of water. Airborne lidar, though flown less frequently, often has greater spatial resolution, can penetrate the forest canopy better than shorter wavelength SAR data, is better for producing highly accurate maps of flooding, and is better for understanding the surface-water flow paths in landscapes or recording changes over long time periods.

In Lang et al., 2010, the authors provide a detailed overview of how discrete return and waveform lidar provides new information for improved wetland mapping and monitoring. They explain how lidar is used to determine the topographic wetness index and how lidar data provide information not only on height of the ground surface, but also on the height of vegetation and other plant size/structural characteristics, such as biomass and plant canopy closure—information used to improve the ability of wetland maps to discriminate between different types of wetlands or to characterize key wetland services, such as carbon sequestration. Structurally unique plant species are often identifiable using lidar data, furthering the ability of managers to locate sensitive or invasive species.

In Lang, McCarty et al., 2012, the authors report on the enhanced detection of wetland-stream connectivity using lidar data, concluding with the following: "The importance of wetland-stream connectivity has been heightened due to the current dependence of wetland regulatory status on this connectivity, although the importance of wetland function to adjacent stream health has been and will continue to be substantial regardless of government policies and regulations. Lidar based DEMs can be used to significantly improve the accuracy of wetland-stream connectivity determinations, at multiple scales, where lidar data are available. The degree of accuracy gained through the inclusion of lidar based information is likely to vary across the landscape based on a number of variables including physiographic region, finer scale geomorphology, human alteration, and meteorological conditions. If not available, users of aerial photography based data sets (e.g., NHD) should be aware of the limitations of these data and communicate these limitations to data users."

In Snyder and Lang, 2012, the authors report on the significance of the National Enhanced Elevation Assessment (USGS, 2012) which documented the benefits of lidar data for wetland mapping, and which led to the USGS' 3D Elevation Program (3DEP). The authors' major conclusions are quoted as follows: "Although the benefits of elevation data for wetlands mapping are well known, the current operational use of lidar data to improve wetland mapping represents a small portion of what could be gained in the future with an operational 3DEP. Current use of lidar data provides a foundation upon which to develop even more advanced applications so that the wetland mapping community is better prepared to take full advantage of the significant benefits that 3DEP would enable. In addition to supporting the improved accuracy and timeliness of relatively fine-spatial-scale wetland maps, such as NWI (National Wetlands Inventory), 3DEP has the potential to support moderate-spatial-scale wetland mapping programs, like the NOAA Coastal Change Analysis Program, through the fusion of multispectral satellite data and lidar-derived topographic metrics. This possibility is only one of many. We have yet to fully comprehend the potential of 3DEP to the wetland mapping and management communities."

Forestry Maps

Forest Resources Management is one of the 27 major Business Uses documented in the NEEA (USGS, 2012). Among Federal agencies, for example, the mission of the U.S. Forest Service (USFS) is to sustain the health, diversity, and productivity of the nation's forests and grasslands to meet the needs of present and future generations. This mission includes forest inventories (inventory metrics include hardwood/conifer discrimination, canopy height, mean diameter, volume, biomass, basal area, canopy density) and forest assessments (including forest health) of USFS lands as well as forested areas in non-USFS lands.

Other agencies and commercial timber companies assess forest resources in different ways for different purposes, e.g., BIA for timber harvests on Indian lands, NPS for monitoring changes in forests in national parks, FWS for analyzing changes in wildlife habitat, of BLM for wildfire management, for example. State agencies have similar interests.

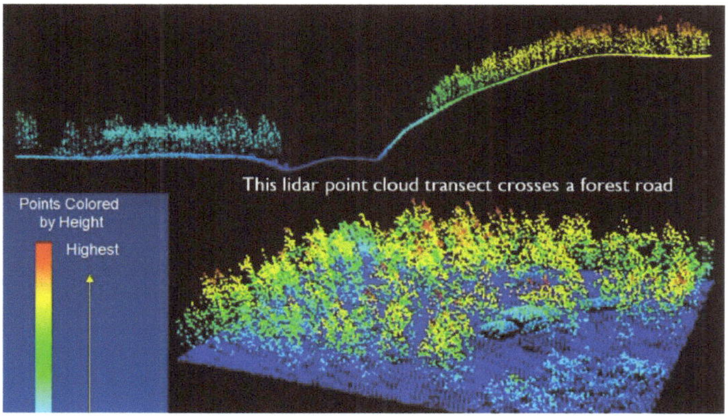

Figure 13.10. Lidar point clouds map the forest canopy, understory and bare-earth terrain beneath the canopy; transects, as shown at the top, enable the measurement of many forms of forest metrics. Image courtesy of the USFS

Clearly, the commercial timber industry has major interests in sustainable forestry for the $20B/year timber, paper and forest products industry in the U.S. A timber tract is a parcel of land used for the purpose of growing timber for harvesting on a rotation cycle. According to *manta.com*, there are over 5,000 timber tracts companies in the U.S., categorized by company revenue.

Lidar is the only technology that penetrates through openings in the forest canopy to map the terrain beneath. Lidar Quality Level 1 was added as a Quality Level in the NEEA specifically because of the need to penetrate dense forest canopy where studies have concluded that high-density lidar data, with 8 points per square meter, are required for forest inventories and topographic mapping in dense forests. Figure 13.10 shows an example of a lidar point cloud and transect used to determine forest metrics.

Lefsky et al., 2002, provides an excellent overview of the use of discrete return and waveform lidar remote sensing for ecosystem studies, especially as pertains to forest mapping. The authors conclude that lidar is "an extremely accurate tool for measuring topography, vegetation height, and cover, as well as more complex attributes of canopy structure and function. In addition, the basic canopy structure measurements made with lidar sensors have been shown to provide highly accurate and nonasymptotic estimates of important forest stand structure indices, such as leaf area index and above ground biomass. Because the basic measurements made by lidar sensors are directly related to vegetation structure and function, we expect that these findings will continue to be corroborated in a variety of biomes, with similar results".

Rangeland Maps

Rangeland Management is another one of the 27 major Business Uses documented in the NEEA (USGS, 2012). Rangelands are vast natural landscapes in the form of grasslands, shrub lands, woodlands, and deserts. Types of rangelands include tall grass and short grass prairies, desert grasslands and shrub lands, woodlands, savannas, chaparrals, steppes and tundra. Rangelands are not: Barren desert, farmland, closed canopy forests, or land covered by solid rock and/or glaciers. Rangelands are important for America's livestock and wildlife habitat.

The Federal Land Policy Management Act (FLPMA) of 1976 formally recognized what the Bureau of Land Management (BLM) had been doing for many years—managing the public lands under principles of multiple use and sustained yield. This includes rangeland management. Livestock grazing, an important use of public lands, is central to the livelihood and culture of many local communities; a significant portion of the cattle and sheep produced in the West graze on public lands managed by BLM.

The Natural Resources Conservation Service (NRCS) assists private landowners with rangeland management and considers lidar as an essential tool. Lidar data provide an efficient method to map and monitor the various forms of rangelands, vegetation, and hydrographic features that nourish them; to assess changes in topography; to map erosional features such as gully cross-sections; and to assess changing morphology resulting from a shift from grassland to shrub land vegetation, for example.

Figure 13.11 shows one such example where the DTM (bare-earth) surface is subtracted from the DSM (top reflective surface) to quantify and convert acres to percent for different vegetation types per Planning Land Unit (PLU). This can be used with other terrain derivatives and soil information for brush management planning and improved estimation of potential stocking rates.

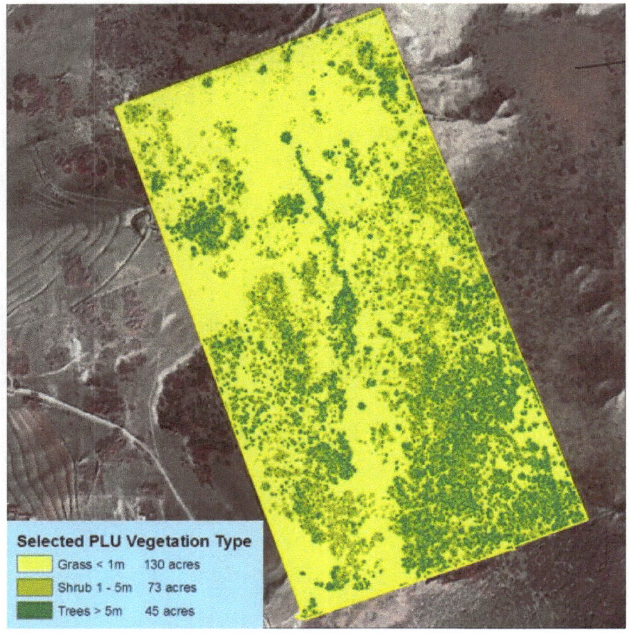

Figure 13.11. Example of how NRCS uses lidar elevation data for quantification of vegetation types in rangeland. The number of acres of shrub and tree vegetation types can be estimated by spatial analysis map algebra functions. Image courtesy of the NRCS.

Wildlife Habitat Maps

Wildlife and Habitat Management is another one of the 27 major Business Uses documented in the NEEA (USGS, 2012). Figure 13.12 shows how the U.S. Fish and Wildlife Service (FWS) used lidar datasets from multiple years to map the rapid gain (bright red) and loss (bright blue) of Loblolly pine forest habitat in Craven County, NC. Loblolly pines are the most important commercial timber in the southeastern U.S. They are planted in mine reclamation areas; they stabilize soil and reduce erosion; their biomass is being studied as a possible alternative source for energy; and they provide excellent wildlife habitat for many animals and birds.

Various state agencies, as well as non-governmental agencies such as The Nature Conservancy, have similar interests in mapping and tracking changes to wildlife habitat.

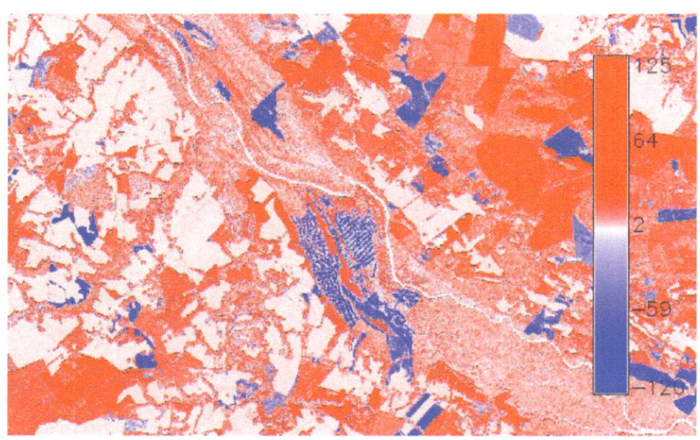

Figure 13.12. Lidar-derived differences in Loblolly pine canopy heights of ±120 feet over a 6-year period in Craven County, NC. New growth (red) and timber harvest (blue) map additions and deletions to wildlife habitat for animals and birds. Image courtesy of the FWS.

Cultural Resources Maps

Cultural Resources Management is yet another one of the 27 major Business Uses documented in the NEEA (USGS, 2012). Figure 13.13 shows a brochure from the National Park Service (NPS), explaining how lidar was used to map the Sny Magill Mound Group Effigy Mounds National Monument in Iowa.

The most famous use of lidar for cultural resource mapping is currently ongoing in Guatemala where lidar has mapped an ancient megalopolis of over 60,000 structures and a superhighway system over 150 miles long, built over 2,000 years ago. Lidar essentially removed the dense forest canopy to show that Maya cities were much larger than previously believed. Readers are encouraged to Google “Lidar mapping of Mayan ruins” to see how lidar is revolutionizing archaeology much as the Hubble telescope revolutionized astronomy.

Figure 13.13. National Park Service brochure that explains the lidar mapping of prehistoric Indian mounds in Iowa. Photo courtesy of the NPS.

Corridor or Right-of-Way Maps

Airborne lidar mapping allows rapid, cost-effective, accurate mapping of linear corridors such as power utility rights-of-way or gas pipelines, for example. Transportation corridor mapping will be discussed separately in a subsequent section. Aerial access with highly detailed information has proven to be a major advantage that lidar offers over traditional methods to clients interested in engineering projects along these corridors.

In the case of power line surveys, many of these projects cover terrain with few roads and require the definition of conductor catenaries, crossing catenaries, accurate attachment point elevations, DTM, vegetation encroachment and structure locations.

In the past, engineers would have to budget months for survey and mapping for critical projects. Topographic lidar significantly reduces the time and costs necessary to satisfy such engineering requirements along narrow and meandering corridors.

The North American Electric Reliability Corporation (NERC) regulates the bulk power system, the facilities and control systems necessary for operating an interconnected electric energy supply and transmission network, including over 450,000 miles of bulk transmission lines but excluding smaller lines used for local distribution of electricity. Poor vegetation management was a key contributor in the August 2003 blackout in the Northeast. Three high-voltage lines in Ohio relayed out in the span of approximately an hour, acting as the trigger for a wide-area cascade of the system and the largest blackout in North American history. This blackout was one of the motivating factors in the certification of the NERC by the Federal Energy Regulatory Commission (FERC) to serve as the electric reliability organization in the United States, under the provisions of the Energy Policy Act of 2005. NERC is actually an independent corporation with a mission to ensure the reliability of the bulk power system of North America, including the United States, Canada and a small portion of Mexico.

To address the issue of poor vegetation management practices the NERC in 2006 issued a mandatory management standard known as FAC-003. One year later, in August of 2007 a flashover involving a 230-kv transmission line caused Duke Energy to make a proposal to the NERC requesting that they use lidar to better determine the actual field conditions of their facilities and to accurately calculate the positions of the wires under projected electrical and meteorological loadings. There was concern on Duke’s part that not only could there be issues with vegetation, but that the as-built geometry of their lines was not matching their facility design files.

In a paper published by the NERC in 2010, the agency reported, “The results from Phases one and two validated Duke Power’s assertion that the use of lidar would provide enhanced reliability by developing a precise 3-D model to confirm conductor-to-ground clearance issues throughout the entire right-of-way of the system.”

Figure 13.14. Lidar data enables the accurate mapping of power lines relative to vegetation that could damage these power lines.

Just a few months later out of concern that transmission line as-built conditions were still not being properly considered in most of North America, the NERC issued a recommendation to the industry, published in a press release as "Consideration of Actual Field Conditions in Determination of Facility Ratings." This, in addition to the Duke Energy findings, has led to the increasing use of lidar by transmission line owners to cost effectively comply with these NERC regulations.

As shown at Figure 13.14, lidar is now the de facto standard for accurately documenting the 3-D as-found condition of power transmission lines and their rights of way. Lidar can support a number of applications, but the two most widely used are for vegetation management and power line re-rating. Both of these applications require an accurate 3-D model of the facilities. Lidar is the most cost effective technology for acquiring this data.

Urban and Regional Planning Maps

As documented in the NEEA (USGS, 2012), elevation data are critical in urban and regional planning, often because of the need to address potential drainage issues when considering sites for future development. Figure 13.15 shows color-coded elevations in Baltimore, MD where the brightest blue shows the lowest elevations and the brightest red shows the highest elevations. Figure 13.16 shows a potential economic development site in Richland County, SC, where a technology company had narrowed its search for a billion dollar production facility to two locations and asked the county to pay for a topographic survey to assess the elevation, slope, and drainage characteristics prior to making a site selection. Because the county already had lidar data of the area, slope, aspect, shading, two-foot contours, and potential drainage were all calculated from the lidar data and provided to the client within 24 hours, saving the county $140,000 for the land survey. This example was documented in (ESRI, 2004).

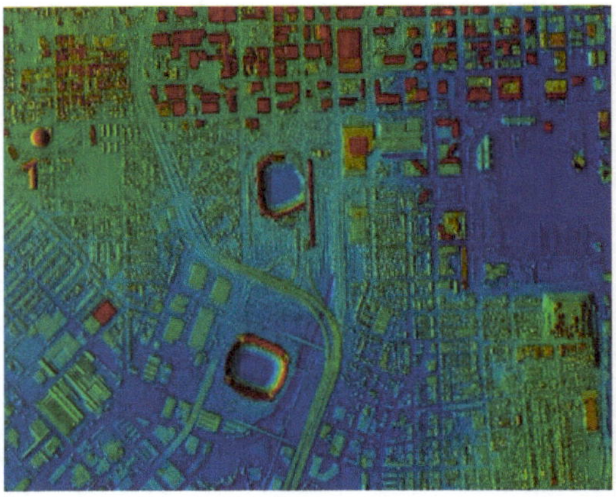

Figure 13.15. Color-coded lidar 3-D model of Baltimore, MD used for urban planning. Image courtesy of the U.S. Army Geospatial Center (AGC).

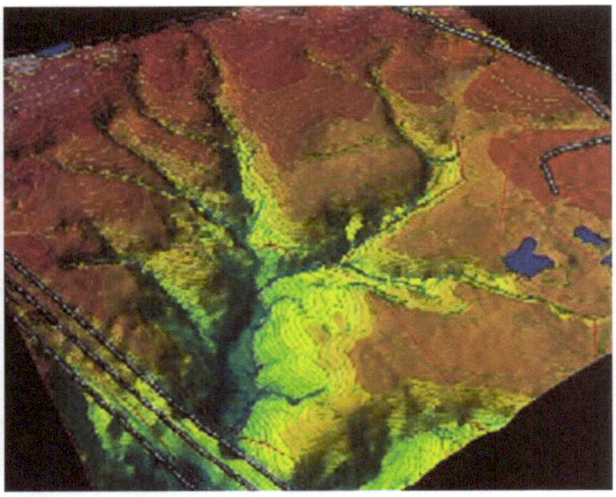

Figure 13.16. A 150 acre site under consideration for development in Richland County, SC where lidar saved the county $140,000 for a traditional topographic survey. Image courtesy of Esri.

3-D City Models

Examples abound where 3-D virtual city models are developed for diverse applications, to include presentations on the TV news reports, traffic reports, etc. seen daily by average citizens. Two innovative examples will be presented here in Figures 13.17 and 13.18.

Figure 13.17. Example of a Philadelphia urban landscape developed with City Engine, used for conceptual 3-D design.

Figure 13.18. Example of lidar data used for modeling of population distribution and dynamics, using LandScan U.S.A. developed by the DOE for the DHS. Image courtesy of DOE.

As shown in Figure 13.17, Esri's City Engine allows users to efficiently create 3-D urban landscapes using their existing GIS data, and perform conceptual design in 3-D. Scenarios can be brought back into ArcGIS for further analysis or exported to high-end simulation engines or web clients. Building facades are added from a catalog of standard choices, or actual facades could be added from oblique imagery, for example.

As documented in USGS, 2012, many agencies responsible for homeland security use lidar data for viewshed analyses to determine "what can be seen from where" because many threats are based on intervisibility between points with different elevations. Furthermore, 3-D city models are used for homeland security, law enforcement and disaster response and are relevant to most aspects of infrastructure protection, border protection, search and rescue, and special security events. As shown at Figure 13.18, LandScan U.S.A. city models are used for disaster response planning and execution for major cities of the U.S. High-resolution population distribution data are critical for homeland security because operational activities and policy decisions are significantly influenced by the number of people impacted at different times of the day when incidents occur. LandScan USA data, produced from lidar and census data as well as ancillary datasets (e.g., land cover, roads, slope, urban areas, village locations, and image analysis) are used to estimate the number of people in residential and non-residential structures at different hours of the day, with phased increases in populations in morning hours and phased decreases in the evening hours.

COASTAL MAPPING APPLICATIONS

Over half of the U.S. population lives in coastal zones where communities are at risk from climate change, sea level rise, subsidence, hurricanes, tsunamis, coastal flooding, oil spills and other natural and manmade disasters. Population increases cause conflicts between human development (e.g., need for new roads, homes, commercial development and public infrastructure) and the environment (e.g., protection of wetlands, beaches and dunes). A common denominator between each of the applications in this section is that accurate and up-to-date topographic and bathymetric data of coastal zones are essential for all coastal communities.

NOAA's Digital Coast

Information in this subsection is taken from (NOAA, 2015) and *coast.noaa.gov/digitalcoast/*. NOAA's Digital Coast provides a delivery system with data and information most needed by those who manage coastal resources and work to protect coastal communities from coastal hazards. In addition to data, the Digital Coast also provides: (1) Training, so people can fully understand how to best use the data, (2) Data tools, so users can see the important answers and information data are capable of providing, and (3) Information, so users can see how others in similar situations are using these resources to better their communities. The primary components of Digital Coast include:

- Data: Topographic and bathymetric lidar; land cover; economic data; navigational charts
- Tools: Sea Level Rise viewer; coastal inundation toolkit; county snapshots; marine cadastre
- Training: Technical skills; effective projects; working with people
- Information: Webinars; blogs; stories of how organizations are using Digital Coast content

The four images at Figure 13.19 help users imagine how accurate elevation data would be critical in helping coastal managers estimate the impact of a predicted sea level rise this century, 50-cm for example, or areas to be inundated by a predicted hurricane storm surge.

Figure 13.19. Elevation data are vital for assessing the impact of predicted sea level rise or hurricanes on coastal communities. Images courtesy of NOAA's Digital Coast.

Shoreline Delineation

NOAA's National Geodetic Survey (NGS) has the responsibility to conduct surveys of coastal regions of the United States and its possessions for demarcating the nation's legal coastline. The national shoreline provides the critical baseline data for demarcating America's marine territorial limits, including its Exclusive Economic Zone, and for the geographic reference needed to manage coastal resources as well as many other uses. The National Shoreline is fundamental to the growth of the nation's shipping, export, manufacturing, coastal development, and insurance industries. Therefore, accurate, consistent, up-to-date national shorelines are needed for incorporation onto NOAA nautical charts to facilitate safe marine transportation and navigation. The shoreline represented on NOAA nautical charts is based on a Mean High Water (MHW) tidal datum. This MHW tidal datum is defined as the average of all the high water heights observed over the National Tidal Datum Epoch (explained in Chapter 2). The utilization of remote sensing sensors that allow for derivation of high resolution DEMs allow the possibility for extraction of datum based shorelines (see Figure 13.20).

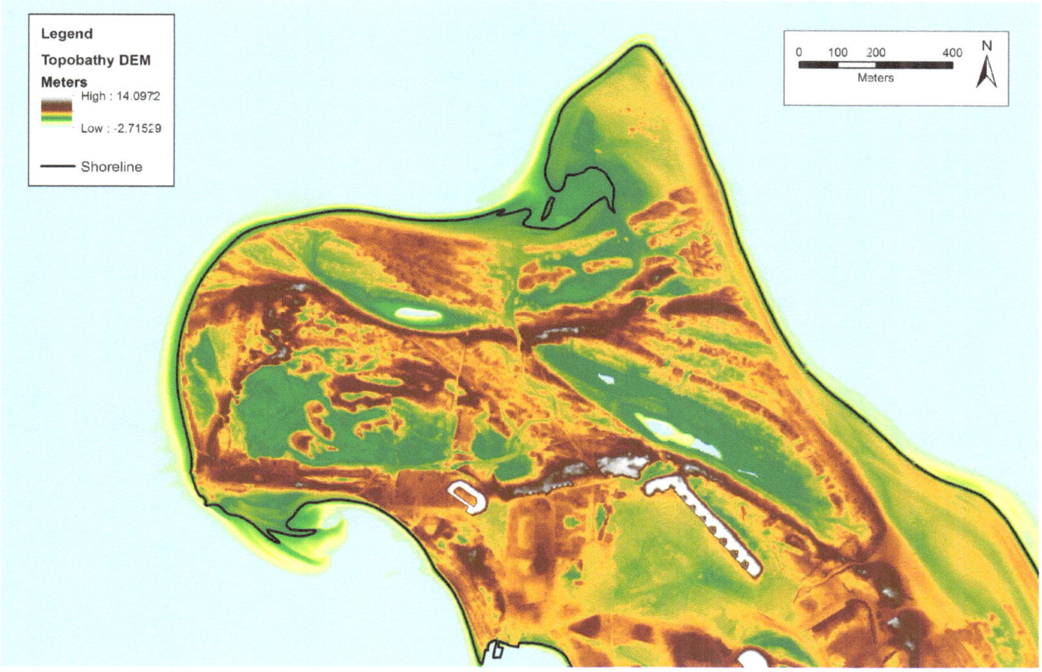

Figure 13.20. Extraction of a Mean High Water (MHW) shoreline at Sandy Hook, NJ from topobathymetric lidar. For most coastal states, the MHW level defines the official boundary between state-owned and privately-owned lands.

Projects are currently being performed in conjunction with NOAA/NGS' vertical datum transformation tool (VDatum) to allow for consistent, non-interpreted tidally based coastal lines to be extracted from elevation data. VDatum allows for transformation of elevation data expressed between any two differing vertical datums. VDatum currently supports 29 vertical datums that can be placed into three categories: 3-dimensional (realized through space-borne systems), orthometric (defined relative to a form of mean sea level), and tidal (based on a tidally derived surface). VDatum's geoid grids represent the departure of NAVD88 from the NAD83/GRS80 ellipsoid. The grids from hydrodynamic modeling depict the variation of local mean sea level from NAVD88 and the other tidal datums from local mean sea level.

NOAA/NGS now has an operational workflow for lidar-based mapping of the National Shoreline using VDatum. There is still some manual work as automated extraction does not produce a cartographically sufficient line for charting. Nevertheless, extracting shorelines utilizing VDatum along with elevation datasets such as highly accurate topobathymetric lidar data has several advantages over traditional methods.

- Shorelines derived from highly accurate DEMs are accurately referenced to a tidal datum, which provides a routine that maximizes consistency with minimal interpretation
- High resolution elevation data allows for deriving consistent tidal datum based demarcation lines across varying coastal environments, when compared to others such as the high water line, vegetative lines, and berms
- The possibility of extracting multiple shorelines may be achieved from a single elevation data set, providing that data acquisition is properly coordinated with a specified tidal stage

Sea Level Rise (SLR)

It has long been known that global sea levels are rising because of warming temperatures, melting glaciers and ice sheets and because the ocean expands as it warms. Digital elevation data give users the ability to visualize the impacts and consequences of sea level rise (SLR), regardless of actual levels of rise during the coming centuries. More importantly, digital elevation data enables coastal managers to plan ahead in order to mitigate future impacts of SLR, generally considered to be inevitable.

In 1995 (see Gornitz, V., 1995), NASA estimated that sea levels could rise by 40 to 65 cm by the year 2100, due to predicted greenhouse-gas-induced climate warming, threatening coastal cities, ports and wetlands with more frequent flooding, enhanced beach erosion, and saltwater encroachment into coastal streams and aquifers.

In 2007 (see IPCC, 2007), the Intergovernmental Panel on Climate Change (IPCC) predicted that sea levels would rise, worldwide, between 18 and 59 cm by the year 2100.

In 2011, scientists at NASA's Jet Propulsion Laboratory (JPL) and the University of Colorado in Boulder reported that NASA satellites had detected a (temporary) lowering of sea levels when El Nino of 2010 became La Nina in 2011. See "NASA Satellites Detect Pothole on Road to Higher Seas" and Figure 13.21 which enables users to see both the "pothole" and where the extra water went in 2011.

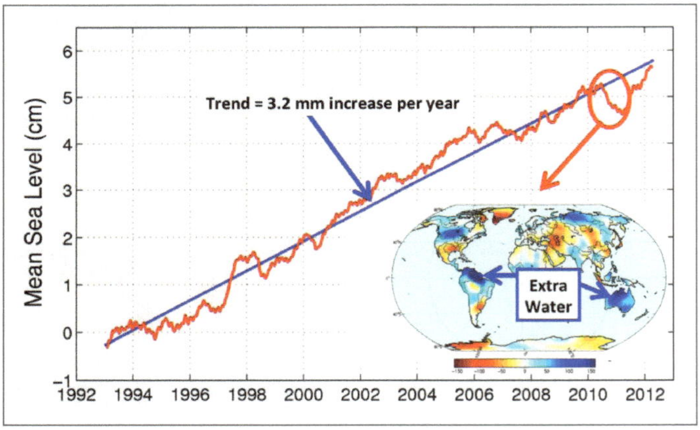

Figure 13.21. Graph showing 3.2 mm annual increase in mean sea level during recent decades. Image courtesy of NASA/JPL-Caltech/CNES.

In 2012, a new NASA study found that global sea level, which dipped sharply in 2010-11 due to a strong La Niña event, had recovered and resumed its long-term upward climb, also shown at Figure 13.21, returning to the two-decade upward trend of 3.2 mm per year. See "What Goes Down Must Come Back Up."

Sea levels are determined from NASA and European Topex/Poseidon, Jason-1 and Jason-2 satellites that track the gradual rise of the world's oceans in response to global warming.

There is no best estimate of sea level rise, but there are ranges. The climate program office in 2012 showed a range from 20-cm to 200-cm *(cpo.noaa.gov/Home/Home/AllNews/Tabid/315/ArtMID/668/*

ArticleID/80/Global-Sea-Level-Rise-Scenarios-for-the-United_States-National-Climate-Assessment aspx). More recent estimates eliminate the low end of that and have gone even higher with the faster than expected ice sheet melting.

NOAA Sea Level Rise Viewer

As documented in (NOAA, 2011), the NOAA Office for Coastal Management, working with partner agencies, has developed a new mapping tool and techniques for visualizing SLR and coastal flooding impacts. This reference should be read in order to understand the various caveats that accompany each view. The SLR Viewer is accessed at *coast.noaa.gov/digitalcoast/tools/slr*. It has five major buttons for mapping (1) SLR, (2) Confidence, (3) Marsh migration, (4) Social and economic vulnerability, and (5) Coastal flood frequency. Images 13.22 through 13.26 are courtesy of NOAA.

Mapping Sea Level Rise: The SLR feature, accessed by clicking on the "Sea Level Rise" button, enables the user to use a slider bar to see how various levels (0-6 feet) of SLR will impact the zoomed-in area. These levels represent inundation at MHHW. Areas that are hydrologically connected, according to the DEM used, are shown in shades of blue that represent depth of inundation. Low-lying areas, displayed in green, are hydrologically "unconnected" areas that may flood. They are determined solely by how well the elevation data capture the area's hydraulics. A more detailed analysis of these areas is required to determine susceptibility to flooding. The user can click on visualization locations to view simulation photos of selected inundation levels on selected landmarks (see Figure 13.22).

Figure 13.22. The Sea Level Rise button allows users to see the impact of 4 foot SLR above MHHW in Galveston, TX. Local impacts of this amount of water at local landmarks can be seen in simulation photos. This is one of five features of the tool. Users can choose any level of SLR between 0 and 6 feet.

Mapping Confidence: The Mapping Confidence feature, accessed by clicking on the "Confidence" button, illustrates that the inundation areas depicted in the "Sea Level Rise" feature are not as precise as they may appear. Levels of confidence are depicted in this feature, with blue areas denoting a high confidence of inundation, orange areas a low confidence of inundation, and unshaded areas a high confidence that these areas will be dry given the chosen water level (see Figure 13.23). There are many unknowns when mapping future conditions, including evolution of the coastal landforms (e.g., barrier island overwash and migration, man-made alterations), as well as the data used to predict the changes. The presentation of confidence only represents the known error in the elevation data and tidal corrections and not the uncertainty associated with these other physical processes or the selected SLR projections.

Figure 13.23. Mapping confidence depicting uncertainty in the elevation tidal surface used to depict 4 feet of SLR in the viewer for Galveston, TX.

Mapping Marsh Migration: The Mapping Marsh Migration feature can be accessed by clicking on the "Marsh" button. The user can use a slider bar to see how various SLR scenarios may impact marsh distribution. Maps represent the potential distribution of 12 marsh and wetland types (Cowardin et al., 1979) based on their elevation and how frequently they may be inundated under each scenario (Figure 13.24). As sea levels increase, some marshes may migrate into neighboring low-lying areas, while other sections of marsh will be lost to open water. There are advanced options where users can select an appropriate accretion rate and time interval for their area of interest. The wetland data portrayed as the initial condition within the viewer are derived from NOAA's Coastal Change Analysis Program (C-CAP) *(coast.noaa.gov/dataregistry/search/collection/info/ccapregional)*.

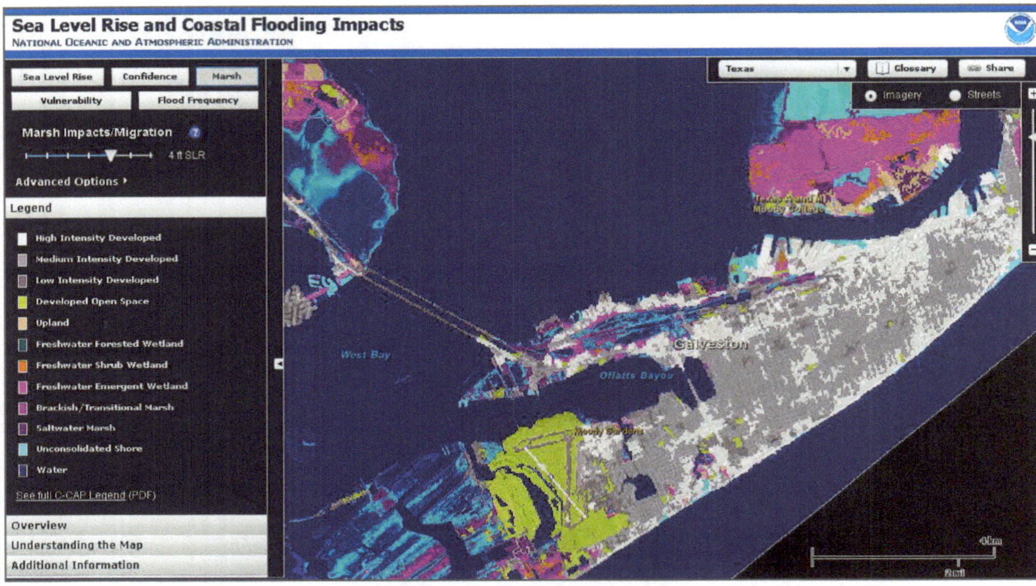

Figure 13.24. Potential distribution of marsh and wetland types based on their elevation and how frequently they may be inundated under different SLR scenarios (showing 4 feet). As sea levels increase, some marshes may migrate into neighboring low-lying areas, while other sections of marsh will be lost to open water.

Mapping Social and Economic Vulnerability: By overlaying social and economic data on a map that depicts SLR, a community can see the potential impact that SLR can have on vulnerability of people and businesses. This feature of the tool, accessed by clicking on the "Vulnerability" button, shows sea level inundation layers over social and economic data from the University of South Carolina (USC) and U.S. Bureau of Labor Statistics (BLS). By looking at the intersection of potential SLR and vulnerable population, one can get an idea of how vulnerable populations might be affected by SLR (Figure 13.25). The Social Vulnerability Index (SOVI) (Cutter et al., 2003) shows areas of high human vulnerability to hazards based on population attributes (e.g., age and poverty) and the built environment (31 variables in all). Census block group analysis of SOVI data is shown in this tool, based on methods developed by USC to downscale the SOVI analysis from county to block group level.

Mapping Coastal Flood Frequency: The Mapping Coastal Flood Frequency feature, accessed by clicking on "Flood Frequency," demonstrates that everyday coastal flooding from tides will become more frequent as sea level rises. In a sense, today's flood will become tomorrow's high tide because SLR will cause flooding to occur more frequently and last for longer durations. The red layer in Figure 13.26 represents areas currently subject to shallow coastal flooding as determined by NOAA National Weather Service (NWS) criteria. These areas were mapped using a single-value threshold for each weather forecast office warning area. The user can click

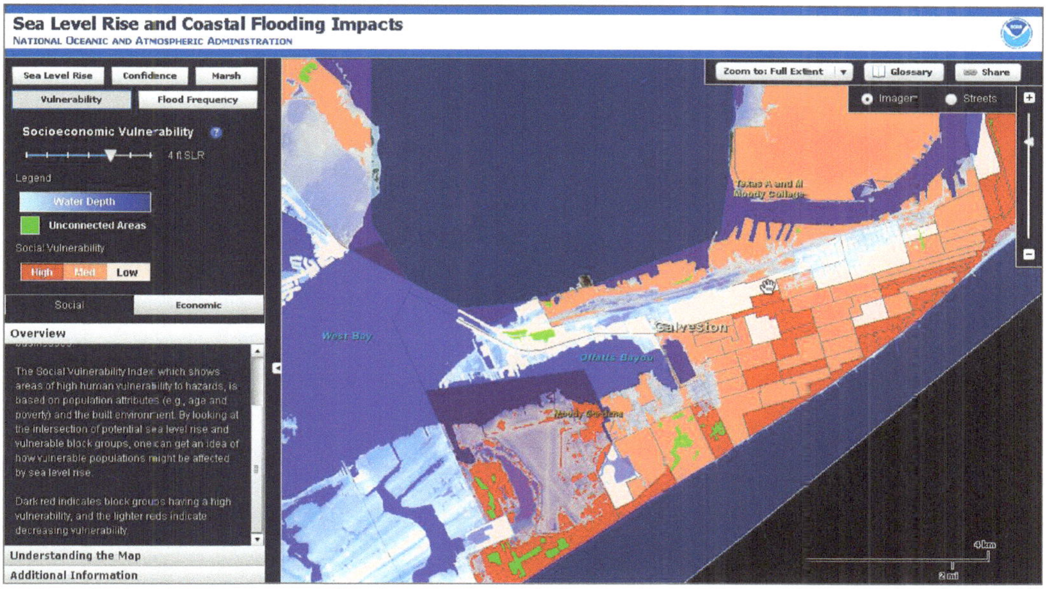

Figure 13.25. Sea level inundation of 4 feet overlaid on Social Vulnerability Index data. Dark red indicates Census block groups having a high vulnerability, and the lighter reds indicate decreasing vulnerability.

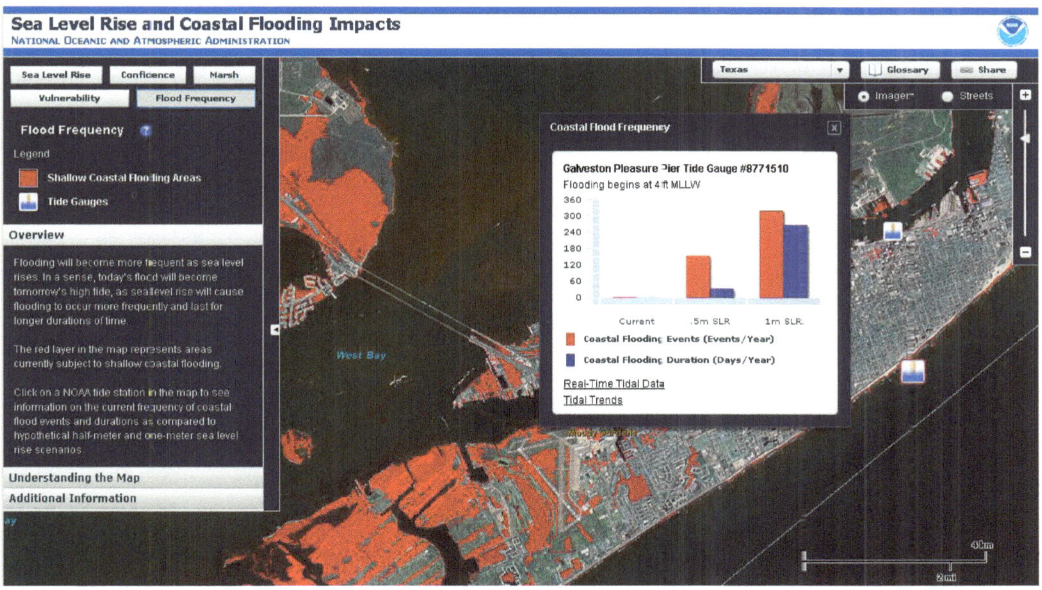

Figure 13.26. Areas currently exposed to shallow coastal flooding are depicted in red. The graph shows the number of flooding events and duration for 0, 0.5 and 1 meter of projected SLR at a specific NOAA tide gauge. Analysis is based on three years of observed tide data.

on a NOAA tide station to see information on the current frequency of coastal flood events and durations as compared to hypothetical half-meter and one-meter SLR scenarios (see inset in Figure 13.26).

Federal., state, county and local officials require accurate DEMs of coastal areas to gather a good understanding of sea level rise impacts and to develop mitigation strategies for local communities. The hyperlink at *businessinsider.com/us-city-sea-level-rise-maps-nickolay-lamm-2013-4* also gives good examples of what water levels would look like under various sea level rise scenarios.

To assess the impacts of sea level rise, simulations can be developed by combining a finite element hydrodynamic model with the consistent, continuous elevation dataset. The scenarios can simulate tidal response, synoptic wind events, and hurricane storm surge propagation in combination with sea level rise. Accurate prediction of inundation patterns can be accomplished by merging the high resolution, continuous topobathymetric data with an accurate wetting/drying algorithm. Shoreline migration can then be dynamically computed from the algorithm's output as a function of sea level rise, and coupled to characterize the effects of sea level rise on coastal ecosystems. The availability of both high resolution bathymetric and topographic datasets holds an enormous possibility for helping understand the impact of sea level rise to the coastal environment.

Coastal Flood Hazard Mapping

In addition to mapping flood risks for riverine areas as discussed above, FEMA's FIRMs and flood studies also identify and map coastal flood hazards that allow residents, businesses and community officials to make informed decisions about reducing their coastal flood risk and proactively prepare for the next hurricane landfall. By integrating accurate digital elevation data with state-of-the-art storm modeling techniques, the National Flood Insurance Program (NFIP) has traditionally depicted coastal flood hazards in two different zones on its FIRMs:

- Zone VE, where the delineated flood hazard includes wave heights equal to or greater than three feet
- Zone AE, where the delineated flood hazard includes wave heights less than three feet

However, it is known that wave heights as low as 1.5 feet can cause significant damage to structures when constructed without consideration to the coastal hazards. Additional flood hazards associated with coastal waves, including floating debris, high velocity flow, erosion, and scour, can cause damage to Zone AE-type construction in these coastal areas.

To help communities and individuals to understand their coastal flood risk even more, a new informational layer—the Limit of Moderate Wave Action (LiMWA)—is depicted on FIRMs. The LiMWA maps the 1.5-foot wave height line and alerts property owners on the coastal side of the line that although their property is in a Zone AE, their property may be adversely affected by waves as low as 1.5 feet high. See Figure 13.27. For more details, see the FEMA LiMWA Fact Sheet, dated May 2012, at *txchart.com/Documents/ResourcesDocument/LiMWA%20Factsheet_062912.pdf*.

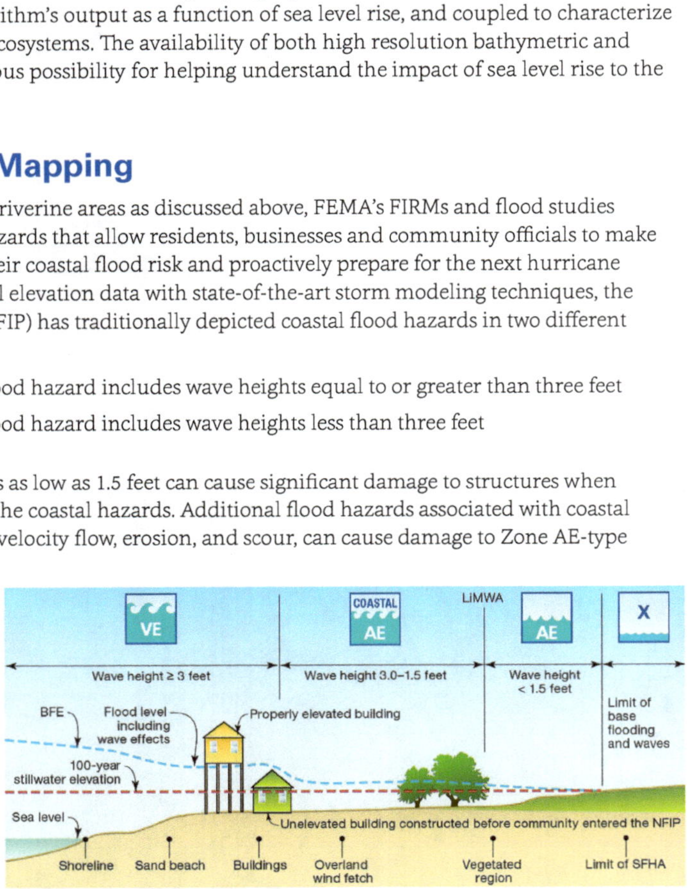

Figure 13.27. The LiMWA identifies the limit of the coastal portion of the AE zone where wave heights between 1.5- and 3-feet can cause significant damage to structures, even though they are not in the coastal VE zone. Image courtesy of FEMA.

As shown at Figure 13.28, the LiMWA is depicted on the FIRM as two black dots and three white dash lines in a sequential pattern.

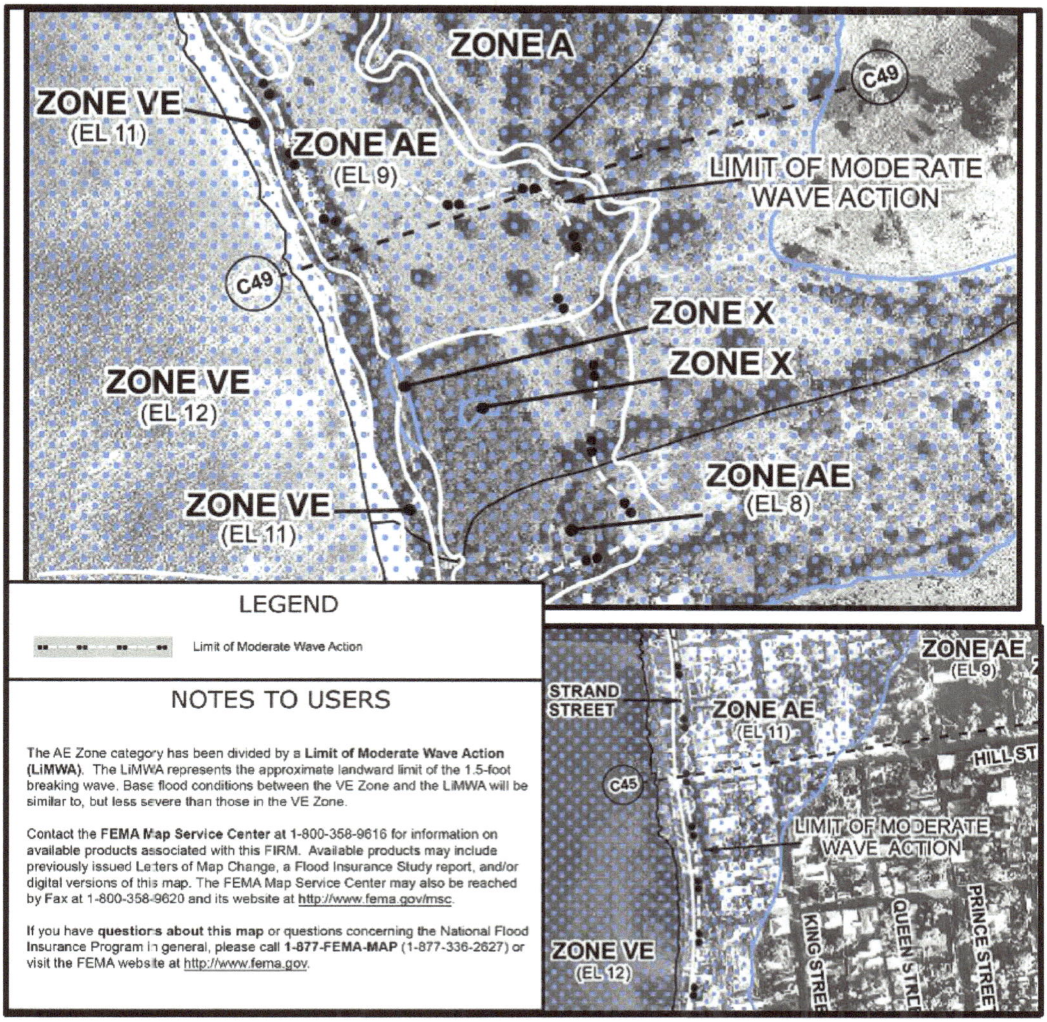

Figure 13.28. The LiMWA (line with two black dots and three white dashes), as well as other flood zone boundaries, are determined from the best available digital elevation data and storm modeling techniques. Image courtesy of FEMA.

Coastal Management

Coastal management organizations are regularly relying on lidar remote sensing data to provide high accuracy elevation models and contours for various scientific and regulatory applications. The Maine Geological Survey (MGS) uses lidar topographic data to help determine whether or not coastal development projects are located within Erosion Hazard Areas (EHAs), as defined by the state's revised Coastal Sand Dune Rules. In Maine, an EHA is an area that may become a coastal wetland in the next 100-years through the combined effects of historic shoreline change, short-term erosion resulting from a 100-year storm, and flooding in a 100-year storm after 2-ft of sea level rise is taken into account, or any area mapped as an AO-zone by the effective FIRM. (An AO zone is a Special Flood Hazard Area with sheet flow, ponding, or shallow flooding.) To examine this, MGS uses the position of the frontal-and-back dune boundaries, historic shoreline change data, short-term erosion data, FEMA FIRM data, highest annual tide levels, and lidar topographic data. This data is compiled within a GIS and overlain onto orthorectified aerial photographs. The topographic lidar data set is then color-gridded and draped over the orthophotographs. ArcGIS is used to examine topographic transects along a project site in reference to published 100-year FEMA base flood elevations (BFE's), in addition to projected future shoreline, washover, and highest annual tide (after 2-ft of sea level rise) positions and elevations. MGS uses EHA determinations to recommend whether or not a proposed structure should be placed on posts, acceptable flow-through foundation, or may use a full foundation.

In the state of South Carolina, lidar data are used by the Department of Health and Environmental Control, Office of Ocean and Coastal Resource Management (OCRM), for determining the primary dune crest which determines setback regulations for development in the coastal zone. In the past, this important line was determined by either aerial photo interpretation or by traditional survey methods, both costing considerably more time and money than to analyze quality lidar DEMs or contours derived from the data. To update the state's setback lines in 1997, the lidar elevation points were processed into elevation contour lines. From these contours, the crest of the dune was delineated, and the setback line was established a minimum of 20 feet landward. It was estimated that one technician at a computer with the data could map an average of 10 miles of beach a day. That amounts to about 20 days of work per year, as compared to the 132 manpower days required using the GPS technology. Today, OCRM continues to use these beachfront jurisdictional lines, developed using lidar.

Coastal Engineering

Nautical charting provides raw data for a broad assortment of coastal engineering applications. Shoreline dynamics and the movement of sand are central to many such uses. Areas of interest are sediment transport, sediment budgets, sand as a local resource, and the design, evaluation, and monitoring of structures such as breakwaters and jetties for shoreline stabilization and dredging. Other interests are mapping anchorage areas, support for harbor engineering and construction projects, port projects, pipelines, cables, and gas and oil production.

According to a FEMA study (FEMA, 1997b) that relied on study data from the U.S. Army Corps of Engineers, approximately 20,500 miles of the 84,240 miles of U.S. shoreline experience "significant" erosion, while 700 miles are subject to "critical" erosion. There are 260 coastal counties in the U.S. FEMA inventoried 26 of these counties (10%) in 1997, including GPS elevation surveys of 45,000 buildings within coastal high hazard zones (V-zones, subject to velocity wave action). Thousands of buildings are vulnerable to coastal erosion and/ or velocity wave action from hurricane tidal surges. If other coastal counties are similar, then 450,000 buildings nationwide could be in V-zones. Projected erosion rates along the Gulf and Atlantic coasts are typically two feet per year, and one foot per year on the Pacific Coast. Elevation surveys are needed for all buildings in V-zones for multi-hazard mitigation. Decades after this FEMA study was conducted, these statistics remain essentially correct to this date.

Modeling and Mapping Coastal Inundation

Coastal inundation is the flooding of coastal lands by ocean waters. Severe inundation is an infrequent event and is normally of short duration; nevertheless, it can cause significant threat to life and damage to public and private property. Severe inundation can be caused by hurricanes, tsunamis, or large coastal storms such as nor'easters. Many areas within the coastal zone can be affected by a combination of coastal inundation and freshwater flooding during an event. The extent and behavior of such flooding is complex and depends upon the magnitude, variation and relative timing of freshwater inflows and coastal water levels. Identification of the level of coastal inundation is an essential starting point in the management of this hazard.

One of the most valuable parameters to coastal inundation modeling and mapping is a continuous topobathymetric DEM of coastal regions. Computer models and GIS mapping utilize DEMs and simulated coastal water levels to determine the extent of coastal inundation as well as visualize the impacts of the flooding. High-resolution data sources such as lidar are extremely useful for providing accurate elevation data in upland, coastal., and intertidal areas and for defining the geographic boundary of flooding. The need for bathymetric lidar is particularly acute in the intertidal region, since there is generally a gap in conventionally obtained elevation data where collection systems cannot gain access.

Many computer models exist to model coastal inundation. Some models take the form of a structured grid (grid cells are the same size) where a DEM will naturally transfer easily, and some have unstructured grids (grids cells vary in size) where a DEM will have to be translated to the model grid. In either case, the DEM serves as the bottom boundary for the model both offshore and on land. Most coastal inundation models are 2-D models that use the bottom boundary to calculate the forces of bottom friction that ultimately affects the propagation of the water towards the land. Bottom roughness and morphology affect the amount and morphology of this inundation propagation greatly and is considered to be equally important to coastal inundation modeling as atmospheric forcing. Wave effects on top of still water levels are also an important factor that many inundation models do not account for. Wave heights are directly related to bottom depth and morphology as well. The height of wave run-up on beaches depends upon many factors that include wave height and period, the slope, shape and permeability of the beach, the roughness of the foreshore area, and wave regularity. Many of these parameters are resolved in a high resolution DEM.

Appropriate grid size of a DEM must balance the resolution of the data needed for modeling, the file size best utilized by the model, and the grid size supported by the available data. A higher resolution DEM will resolve ground features that could potentially obstruct water flow across the land; however, these features tend to disappear in the gridding process, which reduces file size. Similarly for larger regions, a lower resolution DEM may be needed to model coastal inundation, where feature details are sacrificed for larger coverage.

The use of coastal DEMs is becoming an increasingly important part of coastal inundation modeling to better predict and prepare for coastal inundation events. Seamless integrated bathymetric and topographic DEMs are even more valuable to modelers for treating the offshore and onshore as one continuous surface. Because topography is generally referenced to an ellipsoidal or orthometric datum and bathymetry is generally referenced to a tidal datum, tools have been created to convert among the different datums. One of these tools is VDATUM which can convert bathymetry and topography to a common reference, and, in this way, bathymetry outside of the intertidal area can be incorporated into the DEM with the high-resolution near-shore bathymetric and coastal topographic elevation data. However, topographic to tidal only works very close to the shore as tidal grids don't go far inland.

The 2005 hurricane season illustrated the devastating hazards of coastal inundation, with Hurricane Katrina causing a higher storm surge on the Northern Gulf Coast than Hurricane Camille (1969), the benchmark storm for that region for over 30 years. To quickly show the inundation footprint of Katrina, FEMA let contracts for the rapid collection of high water marks and mapping of the coastal inundation that occurred. The maps, called Coastal Flood Recovery Maps also displayed advisory base flood elevations and were intended to help state and local officials, as well as homeowners, to identify existing and increased flood hazards caused by Hurricane Katrina and other storms that have struck this region in the last 25 years, and to use this information during recovery and redevelopment to avoid future flood damages. The creation of the maps was possible in such a short amount of time due to the availability of lidar derived DEMs for three Mississippi coastal counties. High water mark elevations were mapped using the high resolution DEMs, allowing for the most accurate portrayal of storm surge effects from Katrina as possible in a short amount of time. The maps were distributed via the FEMA website. Figure 13.29 is an example of a Coastal Flood Recovery Map for Waveland, MS.

Figure 13.29. FEMA Coastal Flood Recovery Map of Waveland, MS. Photo courtesy of FEMA, State of Mississippi, and NOAA. Photo courtesy of FEMA.

TRANSPORTATION APPLICATIONS

Land Navigation and Safety

Highway design and construction requires high accuracy land surveys, normally equivalent to contours with 6" contour intervals. However, no land survey project is more dangerous than surveying next to speeding traffic. Warning signs to drive slowly often go unheeded, and serious accidents and deaths result. For this reason, aerial., mobile and terrestrial lidar, combined with aerial imagery (see Figure 13.30), have become popular for generating high accuracy topographic maps and DEMs on which highways and interchange construction plans are based.

During highway construction, it has long been standard practice to perform construction stake-out operations to guide the operators of earth-moving equipment. For major construction and mining projects, Computer-Aided Earthmoving Systems (CAES) now integrate differential GPS (DGPS) with real-time kinematic (RTK) positioning and control of construction vehicles. Construction vehicles are equipped with DGPS receivers, linked to a GIS with DTMs updated "on-the-fly" as earth-moving machines make their cuts and fills. As each vehicle's onboard computer radios GPS-based position information back to a dispatch computer, the dispatcher monitors the location and status (full or empty, heading and velocity) of each vehicle in the fleet, monitors where trucks are waiting to be loaded, and redirects them for maximum efficiency. Using the in-cab display, the earthmoving equipment operator views the design grade with cut or fill requirements. Using a moving cut/fill isopach map as a guide, the operator can minimize push distances and increase efficiency. After the CAES machine has started its cut, the heading and a long section is displayed on the monitor, providing the operator with a graphical display of the current topography in relation to the design

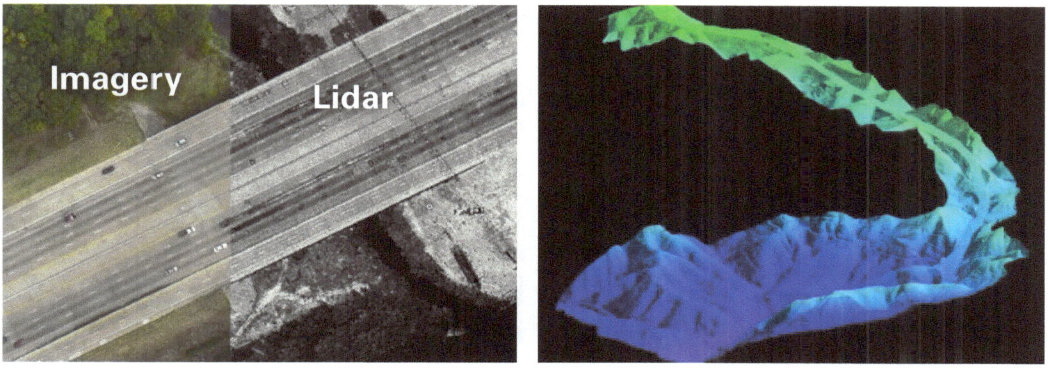

Figure 13.30. These images show how imagery and lidar are used to efficiently and safely survey and map existing roads (left) and proposed roads (right) to determine cut/fill requirements and plan efficient construction of safe roads by conventional methods or modern machine control. Images courtesy of Tuck Mapping Solutions.

surface. As the operator make the cuts, the onboard processor updates the current DTM in real time. This on-the-fly updating enables the operator to assess excavation progress for each individual pass and maximize productivity by receiving immediate feedback. All of these features have combined to produce a dramatic increase in the amount of useful work that construction and mining machines can accomplish each day.

As reported in the National Enhanced Elevation Assessment (NEEA) report, cars and trucks are now under development that will utilize elevation data on road surfaces for intelligent transportation systems, including self-driving cars. In anticipation of increased fuel efficiency standards, and based partly on research performed by TomTom engineers in Germany, they have estimated that fuel consumption can be reduced between 4% and 12% by building vehicles that use accurate elevation and slope data, combined with transmission-control technology and in-vehicle location and navigation products, to down-shift and up-shift transmissions in anticipation of changing gradients ahead. The same elevation datasets used to reduce fuel consumption, saving billions of dollars annually for consumers, would also be used to reduce accidents and deaths.

Aviation Navigation and Safety

Digital Surface Models (DSMs) support air navigation and safety in two principal ways:

- Terrain avoidance during en route flight operations
- Airspace analysis for runway, takeoff and landing operations

Terrain avoidance is especially serious in the state of Alaska which has few roads and relies heavily on aviation to fly to towns and villages outside of major cities; even the state capital in Juneau is inaccessible by roads. Thousands of small aircraft in Alaska are not powered to fly over mountains, but instead fly between mountains and through valleys to get from point A to point B. As shown at Figure 13.31, Alaska has a high rate of controlled flight into terrain (CFIT) accidents (yellow) and deaths (red), shown over a 5-year period. CFIT accidents are not caused by mechanical or electrical failures but by human errors or

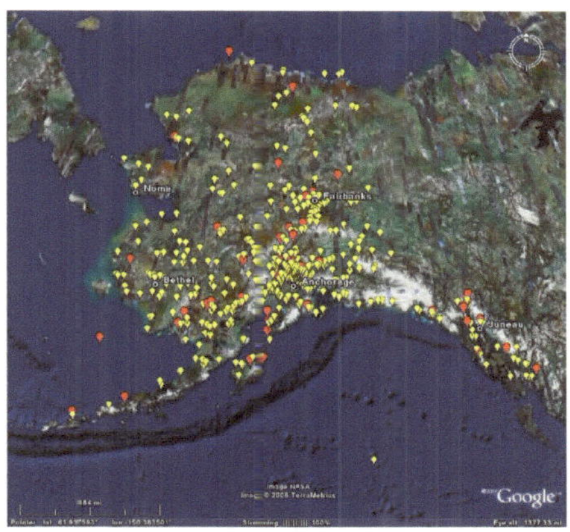

Figure 13.31. Controlled Flight Into Terrain (CFIT) accidents, shown for 2001-2005, are frequent in Alaska. Accurate elevation data are critical for flying through valleys using Instrument Flight Rule (IFR) procedures when valleys are dark or obscured by clouds or fog. Image courtesy of Alaska Department of Transportation.

errors in elevation or other data supporting aerial navigation under Instrument Flight Rules (IFR) used when flying under poor visibility conditions. For this reason, accurate elevation data in Alaska are especially important, the main reason why the 3D Elevation Program (3DEP), explained in Chapter 5, starts in Alaska. In Chapter 1, Figure 1.61 on page 36 shows a flight simulator which uses imagery and elevation data to train pilots how to fly more safely under dangerous flying conditions in Alaska. Chapter 7 of this manual also discusses Alaska's elevation problem and its IfSAR solution for aviation safety.

Airspace analysis is critical near airports where lidar data are used to detect potential obstacles that present hazards to air navigation and safety. The Federal Aviation Administration (FAA), supported by the National Geodetic Survey (NGS), utilizes lidar data for preparation of aeronautical surveys, Airport Obstruction Charts (AOCs) and related products used in development of instrument approach and departure procedures at airfields. Lidar DSMs are ideal for digital airspace obstruction analysis, which includes modeling the mathematical airspace surfaces defined by FAA FAR Part 77. These surfaces include: (1) Approach and departure surfaces, (2) Transitional zones, (3) Primary zones, and (4) Horizontal surfaces above the airport. These surfaces must be free of trees, towers, and other obstructions. High-density lidar data are used to detect potential obstacles that penetrate Obstruction Identification Surfaces (OISs) as shown at Figure 13.32. AOCs are graphics that depict OISs as well as aircraft movement and apron areas, navigational aids, prominent airport buildings, and a selection of roads and other planimetric detail in the airport vicinity. Also included are tabulations of runway and other operational data.

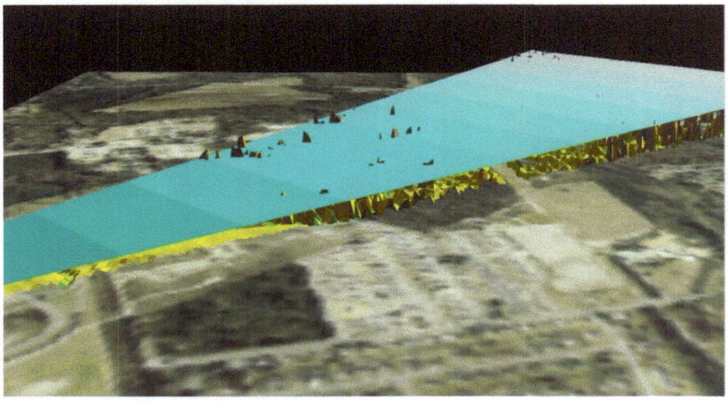

Figure 13.32. Lidar data are used for digital airspace obstruction analyses which include modeling the mathematical airspace surfaces defined by FAA FAR Part 77 that depicts airspaces surrounding and emanating from airports. Image courtesy of the FAA.

Marine Navigation and Safety

The safety and efficiency of maritime transportation depends upon the acquisition of accurate, up-to-date hydrographic data disseminated through authoritative nautical charting products and services. Local and national economies benefit from unencumbered passage of freight and people, built upon prudent investments in navigational support infrastructure and safety. Margins of safety are challenged as ship cargo capacities are pushed to the limit and cruise travel routes expand into previously inaccessible areas. A major accident within the marine industry has the potential for disastrous consequences, with loss of life and significant economic damages. Enhancements to marine navigation infrastructure and improvements to safety and efficiency are synergistic. A key component of safe navigation is the successful management of the risk associated with the available charting information through prudent mariner decisions and actions. Rich datasets in modern electronic navigation systems can help meet the demands for greater protection of life, property, and the environment, and significantly improve the efficiency of maritime commerce.

The Marine Transportation System (MTS) is the primary system by which goods enter and leave the United States. According to the U.S. Department of Commerce, U.S. Census Bureau, Foreign Trade Division, FT920, U.S. Merchandise Trade Selected Highlights, 2015, during 2015, waterborne trade accounted for over 72% of all international trade by tonnage and 44% by value, moving $2.9 trillion of goods. Annually, over 600 commercial vessels are involved in accidents on the nation's waterways and major seaports. Over one hundred of the largest ports in the U.S. annually draw many tens-of-thousands of deep draft vessels, and the traffic has been growing year after year. In 2015, over 82,000 ocean vessels called on U.S. ports; the Port of Houston averaged nearly 22 calls per day that year, with the Port of New Orleans not far behind. Vessel traffic for the group of twenty or so next-largest ports in the U.S. ranges between fifteen and sixty percent of the topmost busy ports (SOURCE: U.S. Department of Transportation (DOT), Maritime Administration, Vessel Calls, available at *marad.dot.gov/*).

The North American cruise industry and U.S. recreational boats each supply several tens of billions of dollars annually to the U.S. economy. Cruise lines are capacity driven—setting fares to fill ships. Each quarter around one hundred cruise ships conduct more than one thousand cruises, carrying approximately ten million passengers per year on stays that average over six nights (SOURCE: U.S. DOT, Maritime Administration, *North American Cruise Snapshot*, publication discontinued after 2012-Q1). The number of registered recreational power boats in the U.S. is in the vicinity of eleven million (SOURCE: U.S. DOT, U.S. Coast Guard, 2014 Recreational Boating Statistics, available at *uscgboating.org/*).

The United Nations International Maritime Organization (IMO) promulgates the Safety of Life at Sea (SOLAS) Convention; specifically, the SOLAS Chapter V publication covers safety-of-navigation regulations for all vessels at sea. Warships, naval auxiliaries, and other ships owned or operated by a country on government non-commercial service are exempt to IMO directives, but United Nations member states are otherwise compelled to develop products which meet the SOLAS definition into local legislation. The concept of hydrographic services is outlined in SOLAS Chapter V, prescribing that an adequate surveying program be carried out to prepare reliable nautical charts and publications to satisfy the needs of safe navigation. More explicit inter-governmental coordination of hydrographic surveying and nautical charting standards is within the purview of the International Hydrographic Office (IHO). The IHO standards and guidelines are adopted by the government-authorized hydrographic offices presiding in the maritime regions around the world.

SOLAS Chapter V includes carriage requirements for shipborne navigational systems and equipment, including statutes concerning nautical charts and nautical publications; for a good summary, see IHO Publication S-66—*Facts about Electronic Charts and Carriage Requirements*. The U.S. Code of Federal Regulations (CFR) legally establishes the National Ocean Service (NOS), National Oceanic and Atmospheric Administration (NOAA) as the provider of official nautical charts and related information in U.S. waters. The CFR also establishes the standard for the U.S. Army Corps of Engineers (USACE) to maintain navigation channels nationwide and to regularly report up-to-date conditions in harbors and waterways to NOAA for charting. U.S. carriage requirements for the use of official nautical charts are established by the U.S. Coast Guard (USCG). USCG responsibility also includes the establishment and maintenance of a system of aids to navigation (ATON), requiring collaboration with USACE and NOAA. NOAA's charting responsibilities near shore and port regions encompass approximately 95,000 miles of coastline. Both paper and electronic charts are produced to fulfill official carriage requirements and the official chart content for a given area is subject to weekly updates compiled by the governing hydrographic office. To see how to get NOAA nautical charts, visit *oceanservice.noaa.gov/facts/find-charts.html*.

Traditional paper nautical charts have moved into the realm of print-on-demand (POD). NOAA POD charts are purchased through certified agents and orders are printed at time-of-sale to include all available updates. As summarized in IHO Publication S-66, electronic charts may be divided into two types: raster charts and vector charts. The Raster Navigational Chart (RNC) follows IHO Product Specification S-61 and may be regarded as simply a digital "copy" of the official paper chart. In actuality, the production lines used by hydrographic offices for the paper chart and the RNC may be in fact unified.

A vector chart consists of a hydrographic database of all the objects to be represented within a region; coupled with an official product specification from the governing hydrographic office, this database is called the Electronic Navigational Chart (ENC). Electronic charts are the greatest advancement in maritime safety since the introduction of radar to ships. A sophisticated navigation system using an ENC "knows" that the charted symbol on an otherwise paper chart is a dangerous wreck, as well as an estimate of how much safe water may present. Such a system may check planned routes over ENCs to see if they cross or pass too near isolated dangers or shoal areas, sound alarms if a vessel's projected course will carry it close to charted hazards, and inform the mariner about various warnings and regulations that relate to areas that the vessel is transiting. Most marine mishaps are due to human error. Vector chart data with proper software applications will enhance safe navigation and provide the mariner with advance electronic warnings of any impending dangers. The mariner may control the display of ENC data, to declutter the screen by hiding features that are not important to the safety of the vessel. This promotes navigational safety by enhancing situational awareness and improving resource management within the pilot house of a vessel.

The structure and format of the current ENC standard is defined in the IHO Special Publication S-57—*Transfer Standard for Digital Hydrographic Data;* specifications of how to display the ENC—feature symbolization that includes portrayal rules for object types and attributes, and associated point, line, and area topology—are contained in IHO Publication S-52. Electronic chart carriage requirements are in fact established according to the definition of the Electronic Chart Display and Information System (ECDIS) in conjunction with ENC data. The ECDIS follows strict performance standards set by IMO to comply with the concept of having the required up-to-date "chart"; often, the modifier "type approved" is used in reference to ECDIS equipment to clarify that official chart carriage requirements are being met. The ECDIS specification includes fail-safe requirements (including the ability to display RNC data), and "system ENC" (SENC) portrayal is influenced by mariner settings in the well-defined manner according to S-52. For example, the mariner's ship safety depth settings changes the precision display of shoal soundings; the location and number of depth area shades are controlled through shallow, safety, and deep contour mariner settings. The ECDIS standard also includes aspects related to the managed-display of additional navigation-related information. Modern ECDIS equipment can integrate data from several sources; e.g., own-ship and radar contacts position, speed and heading, radar vector contacts, etc. Electronic navigation using vector chart data provides the mariner trained in ECDIS with the most powerful tools to achieve a safe passage.

Another term encountered in the industry is the Electronic Chart System (ECS). ECS is developed by the Radio Technical Commission for Maritime Services (RTCM) and, in contrast to the stringent meaning of ECDIS, does not identify specific chart content and associated performance standards to satisfy official carriage requirements for maritime safety. An ECS is perhaps best regarded as a special-purpose geographic information system (GIS); many ECS can display ENC data, just not in the rigorous type-approved sense dictated by ECDIS. An IMO SOLAS regulation was put into place in 2012 for the mandatory installation of ECDIS on certain ships engaged in international voyages. The implementation of this ECDIS carriage requirement follows a 6-year phase-in plan, with new-build passenger ships, tankers, and cargo ships atop the list and completed in 2014. Existing vessels of the same types listed above follow in schedule, in order of decreasing gross tonnage, to full-implementation of type-approved ECDIS carriage in 2018 (SOURCE: IHO Publication S-66).

Of special interest to DEM users is the "chart of the future" initiative that is emerging from work within the IHO S-100 Universal Hydrographic Data Model. S-100 builds upon International Organization for Standardization (ISO) geospatial standards to define the next-generation ENC product (IHO Publication S-101) and to provide a framework to develop additional products required by maritime-related hydrographic, cartographic, and other GIS communities. The DEM environment plays an important role in many of these emerging S-100 compliant product specifications; S-100 Part 8 addresses imagery and gridded data, and IHO Publication S-102 describes the Bathymetric Surface Product Specification that may be used alone or as an auxiliary layer of data with an ENC. Other IHO publications involving such bathymetric DEM work are S-104—Tidal Information for Surface Navigation, and S-112—Dynamic Water Level Data.

Most marine port facilities—certainly those in the U.S.—are affected by ocean swells and tidal conditions. Significant swells make navigation with large container ships and tankers difficult and hazardous (see Figure 13.33). Large vessels in particular need to be acutely aware of under-keel clearance, as well as overhead clearance of the ship's superstructure (relative to bridges, suspended cables & pipelines, etc.). Depending on the type of cargo, every additional inch of draft may be worth tens of thousands of dollars per voyage, so shippers are tempted to load their vessels to the maximum. Ports with shallow channels lose business to competing ports with deeper channels or otherwise inaccurate or outdated data on actual conditions. Because timing of arrival and departure is critical in all shipping operations, it is important to have reliable

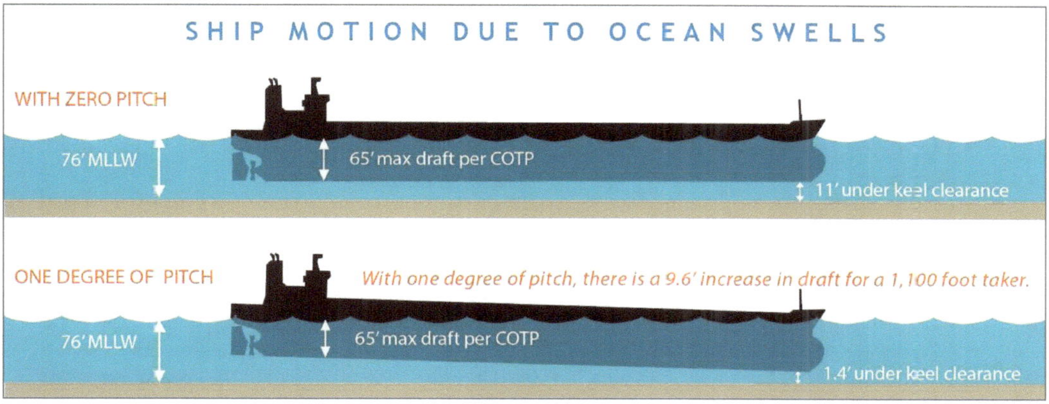

Figure 13.33. Ultra large crude carriers (ULCC) entering the Long Beach channel must contend with potential winter storm swells where a single degree of pitch may increase a 1000-foot ULCC's draft by as much as ten feet. The channel at the Port of Long Beach is dredged to 76 feet. The Captain of the Port sets maximum draft to enter port at 65', leaving 11' clearance under-keel. However, one degree of pitch increases draft for a 1,100 foot tanker by 9.6', leaving a hazardous 1.4' of actual clearance. Image courtesy of NOAA.

bathymetry information on a real-time basis: a DEM adjusted for *in situ* water level to indicate actual depth conditions into navigation channels. Ships that touch the seafloor, however slight, are stopped for several days for mandatory safety inspections. Accurate bathymetry can minimize the use of (de)ballasting operations with ships; e.g., lightering operations to remove cargo prior to entering port, or pumping operations to take-on seawater in holding tanks to lower the ship below overhead obstructions. These operations are not only time consuming, but are also very expensive. Accurate bathymetry can also limit dependency on tidal conditions; excessive waiting for conservative rising/failing tide conditions to achieve safe passage translate to tighter schedules or unnecessary delays.

Discussions about the "chart of the future" usually include requests to display high resolution data. A future S-100 based ECDIS might permit the mariner to make fluid adjustments to symbolized contour intervals and depth areas based on an S-102 DEM, adjusted for tidal and dynamic water level per S-104 and S112 data; certainly, automatic generalization of high-resolution hydrographic (source) data to cartographic portrayal-scales with rigor and repeatability remains a challenge. Cartographically-symbolized point, line, and area features must be derived from source data in a manner that enhances navigation and safety, providing the mariner with an unambiguous level of detail on the nautical chart.

UNDERWATER APPLICATIONS

Resource Management

Nautical charts and associated DEMs are also used for assessment and management of coral reefs and other critical environmental resources such as coastal wetlands and ecologically sensitive areas. The combination of multispectral data with bathymetry is a particularly powerful tool which lends itself to DEM representations. Because of the declining health of many coral reefs, this topic has been raised to one of national importance.

Seafloor Morphology

Seafloor morphology is the study of the form and structure of the seafloor, i.e., the physical geography of the seafloor. As demonstrated in the prior section pertaining to marine navigation and safety, complex hydrographic data sets can be represented by a DEM in a natural and intuitive manner. A DEM allows the integration of multiple components or data sets. The integration can be accomplished without compromising the quantitative nature of an individual data set, provided that each of the data sets is properly georeferenced. Artificial sun-illumination and shading of DEM's can be used to form natural looking and easily interpretable, yet quantitative underwater scenes. Color can be used to represent depth or other parameters, like acoustic backscatter or sediment properties that may be draped over a DEM. The 2nd edition of this manual included four examples of seafloor morphology that will not be duplicated in this 3rd edition.

Underwater Archeology

Many important archeology sites are underwater, making the task of inspection, monitoring, and conservation of those sites significantly more difficult, particularly as depth increases. Fortunately, it is possible to use high-resolution multi-beam sonar and other bathymetric instruments to rapidly acquire information about such marine archeological sites that can be rendered into a DEM that provides a basic map of the area of interest. The underwater DEM also provides 3-D spatial information that significantly improves the efficiency of the dive work and is useful for optimizing the bottom time of divers that might be employed in direct inspection of the sites.

The 2nd edition of this manual included two examples of how underwater DEMs from sonar were invaluable for this purpose. Those examples will not be repeated in this 3rd edition.

OTHER TECHNICAL APPLICATIONS

Natural Resources Conservation

Working in partnership with others, the Natural Resources Conservation Service (NRCS), within the U.S. Department of Agriculture (USDA), is an agency that has many activities that fall into the category of Natural Resources Conservation, including Lidar Enhanced Soil Surveys (LESS) described previously in this chapter. Approximately 70% of the land in the U.S. is privately owned, making stewardship by private landowners absolutely critical to the health of our nation's environment. NRCS works with private landowners through conservation planning and assistance designed to benefit the soil, water, air, plants and animals that result in productive lands and healthy ecosystems. Other federal., state and local agencies have similar but different environmental stewardship responsibilities for federal., state, local., tribal and privately owned lands, and they all require enhanced elevation data, primarily lidar, for this purpose.

Figure 13.34 compares a low-resolution (10-meter), low-accuracy DEM, currently available in the National Elevation Dataset (NED) shown on the left, with higher-resolution (1-meter), high-accuracy lidar data shown on the right. In this example, the elevations and slopes of farm terraces can be measured and reviewed in the office by NRCS staff, helping them avoid the time and cost associated with expensive on-site field visits or field surveys, and enabling the NRCS to support many more customers with the same budget. High resolution DEMs are especially helpful for control of soil erosion as well as farm run-off into streams.

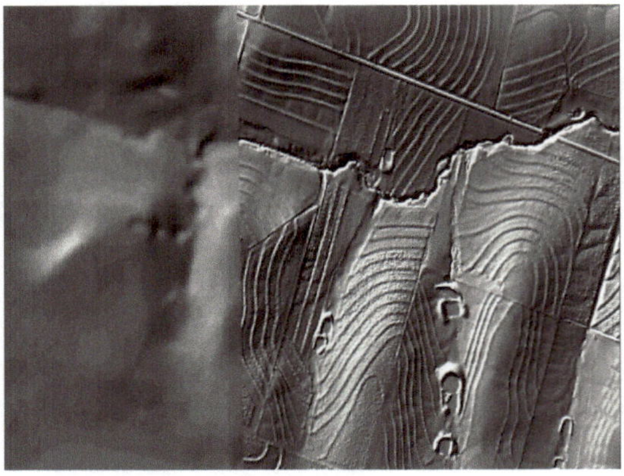

Figure 13.34. Low resolution DEM from the NED (left) compares poorly with the high resolution lidar DEM from the 3DEP (right) vital for natural resources conservation. Image courtesy of NRCS.

Hydrologic and Hydraulic Modeling

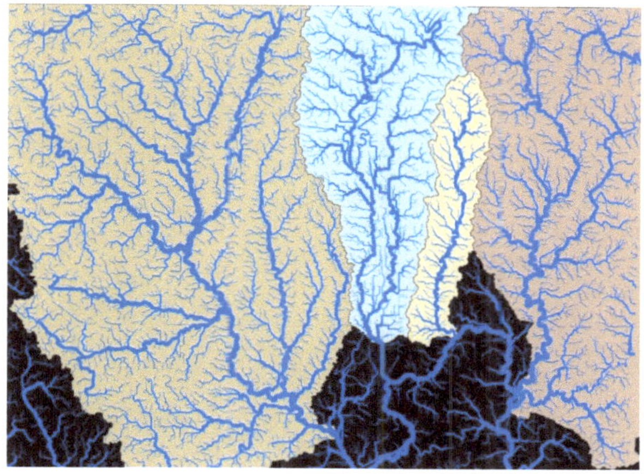

Figure 13.35. Lidar DEMs are ideal for hydrologic modeling to determine where water flows. Such streams need to be hydro-enforced to allow water in the model to pass beneath bridges (that appear as dams in a lidar dataset) and to enforce the continuous, downstream gradient of river/stream shorelines. Image courtesy of USGS.

Figure 13.35 shows an example of a DEM used for delineation of watersheds. DEMs are extensively used for hydrologic and hydraulic (H&H) modeling.

Hydrologic modeling is the computer modeling of rainfall and the effects of land cover, soil conditions, and terrain slope to estimate rainfall runoff in streams and lakes. Hydraulic modeling of rivers and streams is typically the use of rainfall runoff data from hydrologic models, surface roughness data, and information of hydraulic structures (e.g., bridges, culverts, dams, weirs, sewers) to predict flood levels and manage water resources.

Lidar and IFSAR data help improve the National Hydrography Dataset (NHD), the surface water component of The National Map. The NHD is a digital vector dataset used by geographic information systems (GIS), and includes features such as lakes, ponds, streams, rivers, canals, dams and stream gages. These data are designed to be used in general mapping and in the analysis of surface-water systems.

In mapping, the NHD is used with other data themes such as elevation, boundaries, and transportation to produce general reference maps. The NHD is often used by scientists using GIS. GIS technologies take advantage of a rich set of attributes imbedded in the NHD to generate specialized information. These analyses are possible because the NHD contains a flow network that allows for tracing water downstream or upstream. It also uses an addressing system based on reach codes and linear referencing to link specific information about the water such as water discharge rates, water quality, and fish population. Using basic NHD features like flow network, linked information and other characteristics, it is possible to study cause and effect relationships, such as how a source of poor water quality upstream might affect a fish population downstream.

Water Supply and Quality

Everyone knows that a sustained supply of safe drinking water is mandatory worldwide, but few stop to realize the importance of elevation data for wellhead protection, water supply and quality, flood protection, and management of natural systems that support hydrological and ecological functions.

As indicated in EPA, 1989, the 1986 Amendments to the Safe Drinking Water Act (SDWA) established a Wellhead Protection Program (WHPP) to protect ground waters that supply wells and well fields that contribute drinking water to public water supply systems serving 50% of all Americans and 95% of rural America. The wellhead protection area (WHPA) is the surface and subsurface area surrounding a water well, or well field, supplying a public water system, through which contaminants are reasonably likely to move toward and reach such water well or well field. To comply with the SDWA, each State's WHP must:
(1) Delineate the WHPA for each wellhead, (2) Identify sources of contaminants within each WHPA,
(3) Develop management approaches to protect the water supply within WHPAs from such contaminants,
(4) Develop contingency plans for each public water supply system to respond to well or well field

contamination, and
(5) site new wells properly to maximize yield and minimize potential contamination. Without high-resolution, high-accuracy DEMs, compliance with this mandate is essentially impossible.

Digital elevation data are required to help ensure: (1) the availability of water where and when required; (2) that drinking water is safe; and (3) that aquatic ecosystems remain healthy and reduce causes for water pollution.

Figure 13.36. Withlacoochee River Wastewater Treatment Plant, Valdosta, GA. Lidar data are vital for proper locating of water and wastewater facilities, or, in this case, the relocation of the plant. Image from Georgia DNR.

America's sources of safe drinking water are constantly vulnerable to contamination, as demonstrated with the example at Figure 13.36. The Withlacoochee River Wastewater Treatment Plant in Valdosta, GA serves 70% of the city. This plant is located outside the FEMA 100-year floodplain. However, during the March/April 2009 flood event, ⅓ of the plant was under water and an emergency berm had to be constructed to save the main pump station. Various options were considered regarding repair and future flood prevention, but the only viable option was to relocate the plant at a conservative cost of $94.5 million.

Many federal., state and local governmental agencies, as well as private sector companies, are responsible for water supply and quality, to include reductions in farm runoff and/or establishment of wetlands that filter runoff into streams that supply us with water that sustains life, and accurate elevation data are mission-critical for planning for the delivery of safe drinking water and control of water pollutants.

Stormwater Management

Of all utility system managers, stormwater managers have the greatest need for DEMs for efficient hydrologic and hydraulic modeling of watersheds, streams, and channels. Hydrologic models predict volumetric concentrations of water from peak rainfall events, and hydraulic models compute where those waters will go and how flood waters will back-up behind undersized culverts and bridges. Diverse computer models, with DEMs and detailed breaklines, are used for stormwater management.

Subsidence Monitoring

Land subsidence, the loss of surface elevation due to removal of subsurface support, occurs in nearly every state in the United States. Subsidence is one of the most diverse forms of ground failure, ranging from small or local collapses to broad regional lowering of the earth's surface. The major causes of subsidence include: (1) subterranean mining and withdrawal of fluids (petroleum, geothermal., and ground water), (2) dewatering of peat or organic soils, (3) dissolution in limestone aquifers, (4) first-time wetting of moisture deficient low density soils (known as hydro-compaction), and (5) the natural compaction of soil, liquefaction, and crustal deformation.

Figure 13.37 demonstrates the magnitude of the problem in the San Joaquin Valley in California. The sign near the top of the electric pole shows the position of the land surface in 1925; the sign in the middle of the pole shows the elevation in 1955; and the sign on the ground shows the elevation of the ground in 1977 when this photo was taken. Subsidence has continued into the 21st century. This entire valley has subsided for many miles in all directions.

During a five year period of drought, the California Department of Water Resources estimated the state's aquifers were being over-drafted at the rate of 10 million acre-feet per year. Unfortunately, the results of over-drafting of aquifers has led to many problems caused by land subsidence, including:

- Changes in elevation and gradient of stream channels, drains, and other water transporting facilities
- Damage to civil engineering structures—weirs, storm drains, sanitary sewers, roads, railroads, canals, levees, and bridges
- Structural damage to private and public buildings
- Failure of well casings from forces generated by compaction of fine-grained materials in aquifer systems
- In some coastal areas, subsidence has resulted in tidal encroachment onto lowlands

The National Research Council (NRC, 1991) conservatively estimated the annual subsidence costs due to increased flooding and structural damage to be in excess of $125 million. These estimates do not include loss of property value due to condemnation, and they do not consider increased farm operating costs (re-grading of land, replacement of pipelines, and replacement of damaged wells) in subsiding areas. The NRC estimates annual subsidence costs may be about $400 million nationally, including over $180 million per year for the San Joaquin Valley, California, over $30 million per year for Santa Clara County, California, over $30 million per year for the Houston-Galveston, Texas area, $30 million per year for New Orleans, Louisiana, and $10 million per year for the State of Florida. The Louisiana coast line is undergoing constant coastal change, and it is subsiding at an alarming rate—as much as 1.5" per year is some areas. Combined with the predicted sea level rise of 1" every 30 months, millions of people now living in south Louisiana will see this land area and population living at and below sea level by the end of the current century.

Figure 13.37. Subsidence in California's Central Valley near Mendota. Image courtesy of USGS.

Homeland Security, Disaster Preparedness and Response

Many agencies responsible for homeland security use lidar data for viewshed analyses to determine "what can be seen from where" because many threats are based on intervisibility between points with different elevations; this application is relevant to most aspects of infrastructure protection, border protection, search and rescue, and special security events.

In the section above on 3-D City Models, it was shown how LandScan U.S.A. data are used by DHS for planning and execution of emergency response plans for major urban areas.

Floodplain Management

As documented in (USGS, 2012), the National Flood Insurance Program (NFIP) was established to reduce future flood damage through hazard identification and mapping, effective community floodplain management, and insurance protection for property owners. FEMA's Risk MAP (Mapping, Assessment, and Planning) approach integrates risk assessment, mitigation planning, risk communication, and actionable mitigation. It emphasizes updating the flood hazard data and maps of the nation's coastal areas; a reevaluation of the level of protection provided by levees; and watershed-based updates to reflect changes since current Flood Insurance Rate Maps (FIRMs) were produced (see Figure 13.6 on page 398).

FIRMs are utilized 20-30 million times a year by communities issuing permits for building in high risk flood zones and mitigation planning; lenders in their enforcement of the insurance purchase requirement; Federal agencies under EO 11988 (Floodplain Management); developers who build in and near the high risk areas; emergency responders and those making decisions on where and how to recover and rebuild after disasters; and individuals who rely on insurance offered by the NFIP as a financial backstop to their most valuable investment, their homes.

Two studies by the National Research Council (NRC) emphasized the high value of lidar data for the NFIP. The first study (NRC, 2007), entitled "Elevation Data for Floodplain Mapping," documented the fact that existing elevation data were inadequate and that *Elevation for the Nation,* based on lidar with 2-foot contour accuracy or better, is needed for accurate mapping of floodplains and coastal areas. The second study (NRC, 2009), entitled "Mapping the Zone: Improving Flood Map Accuracy," documented 19 benefits of improved map accuracy—benefits that reduce loss of life, property and business, and reduce issues pertaining to the purchase of flood insurance by improved public confidence in the accuracy and legitimacy of FIRMs. Finding 1 from NRC's 2009 study states: "Topographic data are the most important factor in determining water surface elevations, base flood elevation, and the extent of flooding and, thus, the accuracy of flood maps in riverine areas;" and Recommendation 1 from NRC's 2007 study states: "*Elevation for the Nation*" should employ LiDAR as the primary technology for digital elevation data acquisition." (Note: Elevation for the Nation (EFTN) is now known as the 3D Elevation Program (3DEP)).

In part as a result of these recommendations, the FEMA Risk MAP program has implemented new elevation standards for flood map updates and increased coordination with other federal agencies on elevation acquisition and planning. According to FEMA, the decisions and actions taken by communities that participate in the NFIP contribute to an estimated $1.6 billion in losses avoided each year. This would not be possible without sound and accurate elevation data (primarily lidar) used for production of FIRMs. In addition to the benefits from FEMA's use of lidar for Risk MAP, many other federal and state organizations have flood risk management responsibilities. For example, the International Boundary and Water Commission (IBWC) estimates $14.3 million in damages avoided annually along the Rio Grande River alone because of lidar.

FEMA has identified many areas throughout the country where the flood hazards shown on the older FIRMs understate the true risk of flooding. Those who have built to standards on those older maps are in fact subject to a higher probability of flooding. FEMA has also found that in some areas flood hazards on the older maps overstated the true risk (see Figure 13.38a) which means those properties insured under the NFIP are paying more than they should be paying. Figure 13.38b illustrates the improvements in map accuracy attained by the use of lidar data; over 300 structures, some of which are outlined in red, were removed from the Special Flood Hazard Area (SFHA) in Towns County, GA through the use of higher accuracy lidar data. In other locations nationwide, additional structures are added to SFHAs as a result of more-accurate data and modeling. Whether flood risks are currently understated or overstated, the NRC concludes that higher accuracy from lidar yields the following benefits: (1) structures insured at appropriate levels; (2) most consistent insurance ratings through better information about risk; and (3) more insurance purchased because of improved understanding of risk.

Upon publication of (FEMA, 2016), FEMA has now aligned its requirements with the USGS Lidar Base Specification V1.2 for QL2 lidar, consistent with the goals of the 3DEP explained in Chapter 5.

Figure 13.38a. Old SFHA (shown under water), pre-lidar. This image shows homes that were previously required to purchase flood insurance. Images from the Georgia Geospatial Advisory Council.

Figure 13.38b. New SFHA, post-lidar. This image shows homes (outlined in red) that are no longer required to purchase flood insurance, though still recommended. Most lidar datasets demonstrate more homes require insurance.

Geological Applications

As documented in (USGS, 2012), lidar data are mission-critical for the mapping, inventory and assessment of geologic resources and for mitigation of geologic hazards such as earthquakes, tsunamis, landslides and volcanoes.

In explaining how geologic maps are produced, largely from knowledge of slopes and outcrops, a brochure from the Pennsylvania Geologic Survey states: "Lidar data provides a wealth of information not discernible on photography or even in the field." The same is true for seismic fault detection where many geologic faults have been detected from lidar, though not discernible when walking over these faults on the ground. The discovery of over a dozen new faults in western Washington over the last decade through the use of lidar (see Figure 13.39) identified surface ruptures from past earthquakes previously hidden from view by thick vegetation. Lidar has detected previously unknown earthquake faults near a $12 billion nuclear waste treatment plant and a $735 million suspension bridge then under construction across the Tacoma Narrows, both in Washington. Lidar is also widely used for mapping of landslide hazards nationwide. Both lidar and IfSAR are used for mapping of active volcanoes in the U.S.

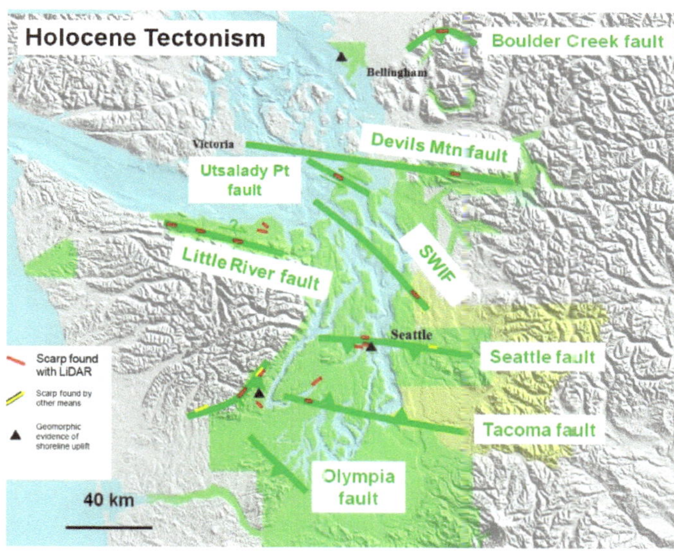

Figure 13.39. Lidar has revolutionized geologic science worldwide, by mapping geologic formations and seismic faults not detected from imagery or field surveys, including these faults near Seattle, WA. Image courtesy of USGS.

Wildfire Management

As documented in (USGS, 2012), lidar data are mission-critical for wildfire management, planning and response because lidar data are used to determine forest fuel and fire susceptibility for wildfire modeling and suppression activities. Wildfire modeling software requires: (1) DEMs, (2) Slope, (3) Aspect, (4) Canopy cover, (5) Fuel models, (6) Weather, (7) Wind, and (8) Fuel moisture. Items (1) through (5) all come from lidar which is well known for determining fuel biomass statistics. In addition to the obvious damages and threats to homes, communities, human life and wildlife, wildfires also leave the terrain more susceptible to landslides.

MILITARY APPLICATIONS

At *globalsecurity.org/military/library/report/2005/050100-bergman2.pdf* (Bergman, 2005) provides a good summary on military uses of digital geospatial data, entitled: "Integration of Topographic Engineering Skills and Tools—Providing Assured Mobility with C2 Systems."

In the U.S. Department of Defense (DOD), Digital Terrain Elevation Data (DTED) has been used for several decades for weapon systems as well as command and control (C2) systems. Produced by the National Geospatial-Intelligence Agency (NGA), DTED has many military applications including the following:

- Intervisibility (line-of-sight) for optimal positioning of command, control, communications and intelligence (C3I) and weapon systems
- Cover and concealment planning, i.e., using the terrain to best advantage so that enemy ground forces cannot see friendly forces and engage them with direct line-of-sight weapons
- Cross country movement analyses, using DTED as one of several factors in determining the ease or difficulty in off-road movement of military vehicles
- Near shore bathymetry as needed for amphibious landings and logistics-over-the-shore support
- Terrain contour matching for guidance of cruise missiles (no longer used)
- Terrain avoidance for military aircraft
- Battlefield "fly-throughs" and simulation for mission planning and rehearsal

Many DOD research and development (R&D) projects of past decades, involving different levels of DTED, have had direct benefits to the civilian community. Many of the GIS technologies in use today had their start in DOD. Furthermore, digital photogrammetry, IfSAR, topographic lidar, airborne lidar bathymetry, and sonar all had roots in DOD. GPS technology alone, on which all modern DEM technologies are dependent to some degree, is totally a result of DOD funding and ingenuity in using satellite technology to enable many of the DEM applications used today. GPS technology became so indispensable to the civilian world that civilian applications came to overshadow many of the military applications for which GPS was originally funded.

COMMERCIAL APPLICATIONS

Precision Agriculture

As documented in (USGS, 2012), precision agriculture has the second highest potential dollar benefits, i.e., with over $2B in benefits annually from nationwide lidar.

- Lidar data are required for all agricultural land areas of the U.S. for analysis of slope, aspect, curvature and potential soil wetness
- Lidar data are used for site-specific application of seed, fertilizer, lime, pesticides and water to optimize farm yields

- Lidar yields major dollar benefits where applied
- Lidar data are used to reduce farm and pasture runoff that pollutes streams

Modern farmers now manage their lands by the square meter. Slope data, applied water and rainfall statistics, seed, fertilizer, lime, and pesticide application rates as well as farm yields this year will vary application rates next year. Figure 13.40 shows a tractor with GPS antenna and row planter. The upper inserts show the display and controller. The lower inserts show elevation and yield maps.

Subtle elevation changes from accurate lidar data are especially relevant when farm fields are nearly flat because the lower areas are wetter than the higher areas.

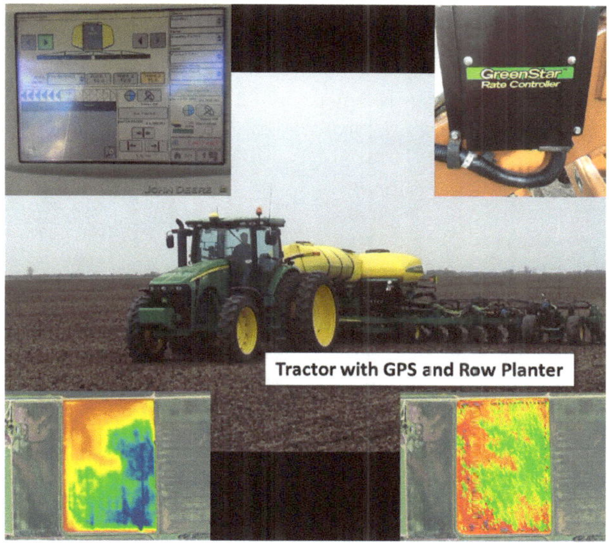

Figure 13.40. Row planter pulled by tractor with GPS, shown with precision agriculture tools, including elevation data and yield maps. Images from the J.R. Simplot Company.

Resource Mining

As documented in (USGS, 2012), in addition to mapping of geologic and other resources to be mined, elevation datasets are used by the DOI Office of Surface Mining Reclamation and Enforcement (OSM) and state agencies for the following:

- Existing pre-mining contours and volumetric analyses are needed as the baseline against which to measure permit proposals, set bond amounts, and measure restoration activities
- As mining operations continue, ongoing data collection is used to monitor backfill and grading activities and determine whether incremental reclamation activities have met permit requirements, thereby being eligible for release of bond monies (see Figure 13.41)

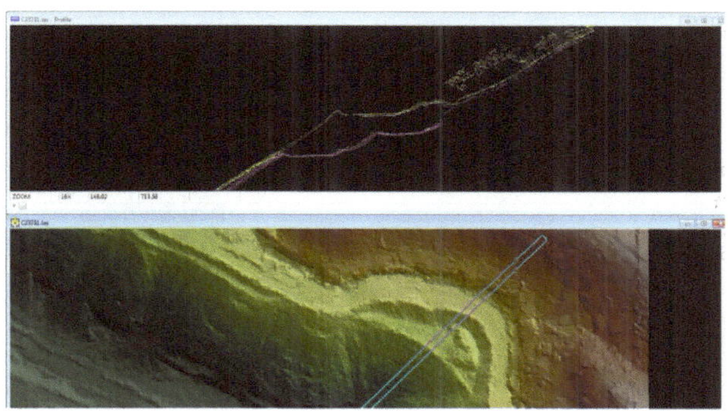

Figure 13.41. This image shows a lidar cross-section in Pennsylvania after only one month of mining. Lidar is ideal for evaluation of slopes for safe mining operations, environmental protection, and for comparisons between baseline surfaces (prior to mining) and final surfaces (after reclamation). OSM and state partners conduct thousands of inspections annually for which lidar data are needed for effective and efficient monitoring of compliance with the Surface Mining Control and Reclamation Act.

- At the end of mining activities, elevation data are used to determine if reclamation is adequate and the landscape has been restored to its "original" condition within the requirements of the legislation
- Thousands of annual inspections can be performed more accurately and cost-effectively by "virtual inspections" performed with before/after elevation datasets. For mines, periodic lidar resurveys are typically performed with terrestrial lidar scanners rather than airborne sensors

Renewable Energy Resources

Digital elevation data are essential for analysis of solar energy potential and placement of solar farms and individual solar panels, wind energy potential and placement of wind farms, and hydroelectric energy potential for placement and/or safe operation of hydro-electric power plants. Figure 13.42 shows an example home selected at random from the Los Angeles County Solar Map. Individuals can zoom in and see exactly which parts of their roof are good for solar—then print the results in a report. The LA County Solar Map is based upon a large-scale solar model with solar radiation calculated every 5 feet.

Figure 13.42. Example of a Los Angeles County Solar Map, available online, for which home owners can print a report for their home/property explaining its potential for solar energy and/or solar water heating, and potential cost savings and carbon savings as well as estimated installation costs. Image courtesy of Los Angeles County.

Oil and Gas Resources

Lidar data are used by the oil and gas industry for: (1) Seismic programs and exploration; (2) Locating well sites, facilities and pipelines based on slope data; (3) Selection of well locations and pipeline routes; (4) Location and classification of buildings and other objects within special protection zones; (5) Identification of land cover and timber removal calculations to minimize tree cutting; (6) Asset inventories; (7) Encroachment control; (8) Construction planning; and (9) Fast checking of critical heights of ground coverage when combined with existing sub-surface data.

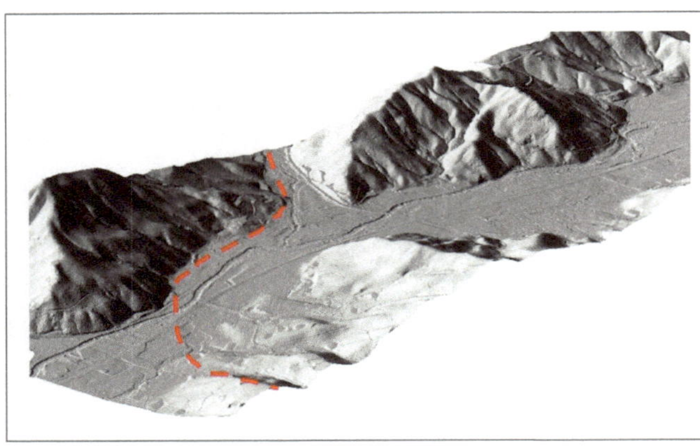

Figure 13.43. Lidar slope data are essential for pipeline routing across mountain ranges and beneath rivers as shown here, construction planning, encroachment control and asset inventories. Lidar data are especially good for estimation of timber removal requirements along alternative pipeline routes. Image from Dewberry.

Lidar is used to pre-select suitable locations in the office, as shown at Figure 13.43, which is quicker, safer and more cost-effective than sending survey crews to the field to search for suitable well locations and pipeline routes.

Telecommunications

Elevation data are needed to determine line-of-sight conditions between transmit and receive locations for broadcast, microwave, cellular, WiFi, and WiMAX users. Elevation data are also used as inputs to automated propagation prediction software. Both lidar and IFSAR data are ideal for this purpose because they can provide both the Digital Terrain Model (DTM) and Digital Surface Model (DSM) that could interfere with wireless telecommunications. Figure 13.44 demonstrates how elevation data can be used to determine where line-of-sight exists and doesn't exist for antenna viewpoints simulated at different heights; this example demonstrates a short antenna, only 9.5-feet high; as the antenna is elevated by simulation, much broader areas become visible.

Figure 13.44. Using a DSM, a 9.5 foot antenna would have line-of-sight to the areas shown in red. The simulated antenna viewpoint can be elevated to any level necessary to achieve line-of-sight coverage to a large percentage of the total service area desired. Image courtesy of USGS.

Recreation

Lidar data have long been used for design of lakes and ski slopes; but lidar is now being used for design and mapping of golf courses as well as professional auto racetracks and trails used by hikers, bikers, snowmobiles, or all-terrain vehicles (ATVs). If elevation data or derivative products (e.g., slope, aspect, and curvature) are important, or if detailed information regarding forests or vegetation is needed, lidar has become the technology of choice. DTMs are often required with centimeter level precision for design of professional golf courses and subsequently for golf course mapping, navigation and simulation systems.

Real Estate, Banking, Mortgage, and Insurance

Whereas most homeowners have mandatory fire insurance, there is a greater risk of severe damage to the average home from other natural disasters than from fires. In recent decades, there has been a 4-fold increase in natural disasters caused by thunderstorms and tornadoes, winter storms, floods, wildfires, earthquakes and hurricanes. Hurricanes normally cause the greatest losses, with four of the five most damaging hurricanes occurring since 2005. According to NOAA statistics, the five worst hurricanes, from a property damage perspective, were as follows:

1. Hurricane Harvey (2017): $180B
2. Hurricane Katrina (2005): $108B
3. Hurricane Maria (2017): $85B
4. Hurricane Sandy (2012): $65B
5. Hurricane Irma (2017): $65B

The 1994 earthquake in Northridge, CA caused insured losses of over $17B, and many areas of the U.S. are in close proximity to seismic faults that are now more-readily detectable with lidar. The risk of buildings to geophysical events (earthquakes, tsunamis, volcanic eruptions), meteorological events (storms), hydrological events (floods, landslides), and climatological events (drought and forest fires) are partly impacted by parameters determined from high resolution digital elevation data.

The author predicts that it is only a matter of time before the real estate, banking, mortgage and insurance industries determine the actuarial risks to properties from these multi-hazards, and high resolution elevation data will play a significant role in quantifying those risks.

INDIVIDUAL APPLICATIONS

Elevation Certificates

FEMA Elevation Certificates have long been used to compare the surveyed elevations of the lowest floor, lowest adjacent grade (LAG) and highest adjacent grade (HAG), compared with the Base Flood Elevation (BFE) on Flood Hazard Maps. Rather than spending hundreds or even thousands of dollars, per home, for a complete Elevation Certificate, the potential exists for the very same lidar data, used for Flood Insurance Studies to estimate BFEs, to also be used to estimate LAG and HAG elevations to determine who needs flood insurance and who doesn't. Thus, elevation criteria would be used to make this determination, rather than using the horizontal placement of controversial Special Flood Hazard Area (SFHA) boundaries.

Parcel Topographic Maps

It is known that local land surveyors are the most common users who download lidar datasets where available. One reason is that there are over 3,000 counties and county equivalents in the 50 States and the District of Columbia, and each of these counties make zoning decisions weekly that typically require some form of topographic map to identify (and resolve) potential drainage issues. The simple topographic map at Figure 13.45 would cost over $10,000 if produced by traditional land surveys; however, it was produced with only two hours of labor because of access to lidar data that had been previously certified. If landowners in a typical county saved just $1,000 per week from use of lidar data as demonstrated in this example, Americans would save over $150M per year in costs avoided, exceeding the total annual costs of lidar data acquisitions throughout the U.S. for the 3DEP.

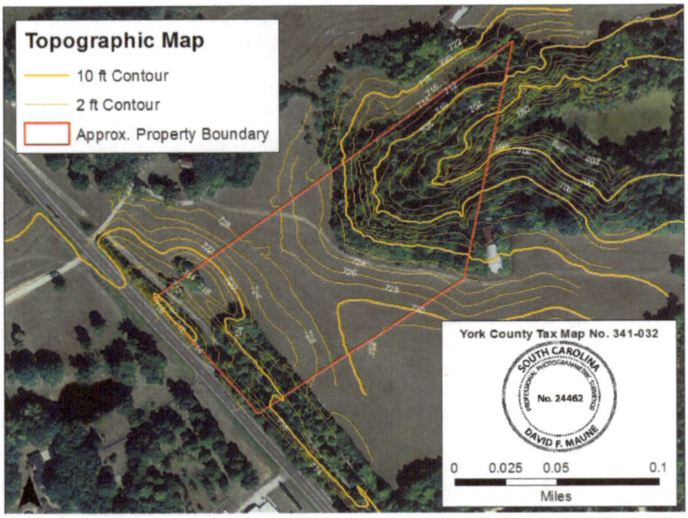

Figure 13.45. Lidar data were used to produce this simple topographic map of an 8-acre parcel that otherwise would have cost over $2,000 by traditional land surveys. Such maps are required weekly for zoning decisions in counties and communities throughout the U.S.

AUTHOR BIOGRAPHIES

Dr. David F. Maune is a Senior Project Manager for Mapping and Remote Sensing at Dewberry in Fairfax, VA and principal author of the NEEA. See Chapter 1 for his biography.

Dr. Kirk Waters of NOAA's Office for Coastal Management (OCM) is acknowledged for his contribution to the section on coastal mapping applications.

Mr. Jack Riley of NOAA's Office for Coast Survey is acknowledged for his contribution to the section on Marine Navigation and Safety.

Ms. Megan Lang of the U.S. Department of Agriculture (USDA) is acknowledged for her contribution to the section on Wetland Mapping.

REFERENCES

Bergman, Ken (2005). *Integration of topographic engineering skills and tools*—providing assured mobility with C2 Systems. In Engineer, January–March 2005.

Cowardin, L.M., V. Carter, F.C. Golet, and E.T. LaRoe (1979). *Classification of wetlands and deepwater habitats of the United States.* U.S. Department of the Interior, Fish and Wildlife Service, Washington, D.C. Jamestown, ND: Northern Prairie Wildli fe Research Center Home Page. (Version 04DEC98).

Cutter, S.L., B.J. Boruff, and W.L. Shirley (2003). Social vulnerability to environmental hazards. In *Social Science Quarterly.* Volume 84, Number 2, (pp. 242–261).

Esri (2004). *Measuring up: the business case for GIS.* Esri Press.

FEMA (2016). *Guidance for flood risk analysis and mapping: elevation guidance,* May, 2016.

Gornitz, V. (1995). A comparison of differences between recent and late Holocene sea-level trends from eastern North America and other selected regions. In *J. Coastal Res., Special Paper No. 17: Holocene Cycles: Climate, Sea Levels, and Sedimentation,* (pp. 287–297).

IPCC (Intergovernmental Panel on Climate Change) (2007). Climate Change 2007: The Physical Science Basis. Contribution of Working Group I to the *Fourth Assessment Report of the Intergovernmental Panel on Climate Change* [S. Solomon, D. Qin, M. Manning, Z. Chen, M. Marquis, K.B. Averyt, M. Tignor and H.L Miller (eds.)] Cambridge UK, and New York, Cambridge University Press.

Lang, Megan & Greg McCarty (2009). Lidar intensity for improved detection of inundation below the forest canopy. In *Wetlands* 29, (pp. 1166–78).

Lang, Megan, Greg McCarty, Bill Wilen & Jane Awl (2010). Light detection and ranging: new information for improved wetland mapping and monitoring. In *National Wetlands Newsletter* (September–October, 2010).

Lang, Megan, Owen McDonough, Greg McCarty, Robert Oesterling & Bill Wilen (2012). Enhanced detection of wetland-stream connectivity using lidar. In *Wetlands* 32, (pp. 461–473).

Lang, Megan, Greg McCarty, Robert Oesterling & In-Young Yeo (2010). Topographic metrics for improved mapping of forested wetlands. In *Wetlands,* published online, DOI 10.1007/s13157-012-0359-8.

Lefsky, Michael A., Warren B. Cohen, Geoffrey G. Parker & David J. Harding (2002). Lidar remote sensing for ecosystem studies. In *Bioscience* (January 2002), Vol. 52, No. 1, (pp. 19–30).

McCarty, Gregory & Megan Lang (2010). Forested wetlands in the Chesapeake Bay watershed. In *Agricultural Research Magazine,* (August, 2010).

NOAA (2011). *New mapping tool and techniques for visualizing sea level rise and coastal flooding impacts.* NOAA Coastal Services Center, Charleston, SC.

NOAA (2015). *Projected benefits and costs of the digital coast.* Charleston, SC: NOAA Office for Coastal Management.

NRC (1991). *Mitigating losses from land subsidence in the United States.* National Research Council, Washington, DC, National Academy Press.

NRC (2007). *Elevation data for floodplain mapping.* National Research Council, Washington, DC, National Academy Press.

NRC (2009). *Mapping the zone: improving flood map accuracy.* National Research Council, Washington, DC, National Academy Press.

NRCS (2011). *Digital elevation model (DEM) whitepaper: NRCS high resolution elevation data.* Natural Resources Conservation Service. Fort Worth, TX: NRCS National Geospatial Center.

Snyder, Gregory & Megan Lang (2012). Significance of the 3D elevation program to wetland mapping. In *National Wetlands Newsletter* (September–October, 2012), Vol. 34 (5), (pp. 11–15).

USGS, 2012, *Final Report of the National Enhanced Elevation Assessment.*

Chapter 14

DEM USER REQUIREMENTS AND BENEFITS

David F. Maune

MINIMUM DEM USER REQUIREMENTS

In prior editions of this manual, the author provided a "User Requirements Menu" that provided many options for the elevation surface(s) to be mapped, the elevation type, data model types, data source, vertical and horizontal accuracy requirements, accuracy reporting, surface treatment factors, horizontal and vertical datums, geoid model, coordinate system, units, data formats, file size, tile size, and metadata.

Since the conclusion of the National Enhanced Elevation Assessment (NEEA) in 2012, summarized below, and the resultant 3D Elevation Program (3DEP) summarized in Chapter 5 of this manual, prior variable choices should no longer be considered because the nationwide minimum standard is for QL2 lidar (or better) which promotes data standardization, consistency, interoperability and usability at the most affordable cost for all.

Simply stating that you need QL2 lidar answers the majority of questions pertaining to DEM user requirements. To acquire any lidar poorer than QL2 simply means the data will have limited usage by others and may not be considered as an authoritative source for elevation data that could impact public welfare and safety. QL2 lidar has been established as the minimum required Quality Level for new lidar data collections to be made available to the public for standardized usage of authoritative data, tested for relative and absolute accuracy consistent with the ASPRS Positional Accuracy Standards for Digital Geospatial Data (Chapter 3) which tests absolute vertical accuracy of lidar by two new terms: Non-vegetated Vertical Accuracy (NVA) and Vegetated Vertical Accuracy (VVA).

The following sections will also include common data upgrades, all also acceptable to USGS and others, above the minimum requirements for QL2 lidar described below.

3DEP QL2 Lidar Standard Deliverables

The USGS Lidar Base Specification (LBS) V1.3 (Heidemann, 2018) describes standard deliverables as follows for any project area, normally with a small surrounding buffer and called a buffered project area (BPA). For data in the USGS National Geospatial Program (NGP), including the 3D Elevation Program (3DEP), the USGS requires unrestricted rights to all delivered data and reports, which will then be placed in the public domain. USGS places no restrictions on the rights of the data provider to resell data or derivative products. Because V1.3 was designed for traditional linear mode lidar, exceptions may be made on a case-by-case basis for Geiger mode lidar, single photon lidar and other emerging technologies that deliver QL2 or QL1 lidar for the 3DEP to newer specifications not yet published by USGS as of early 2018 when this manual was published.

These deliverables are very technical and are most reliably and cost-effectively delivered by specialized prime contractors under USGS' Geospatial Products and Services Contract (GPSC3 or later).

Metadata

The term "metadata" refers to all descriptive information about the project, and metadata includes text reports, graphics, and supporting shapefiles. Product metadata files shall comply with the Federal Geographic Data Committee (FGDC) standards which facilitate the development, sharing, and use of geospatial data. Metadata deliverables shall include the following:

- A collection report detailing mission planning and flight logs
- A survey report detailing the collection of all ground control, including the following:
    - Control points used to calibrate and process the lidar and derivative data
    - Check points used to validate the lidar point data or any derivative product
- A processing report detailing calibration, classification, and product generation procedures including methodology used for breakline collection and hydro-flattening. In the USGS LBS V1.3, Appendix 2 supplements the section on Hydro-Flattening below
- A QA/QC report, detailing procedures for analysis, accuracy assessment and validation of the following:
    - Point data (absolute Non-vegetated Vertical Accuracy (NVA) and relative vertical accuracy)
    - Bare-earth surface (absolute vertical accuracy (NVA and VVA)
    - Other optional deliverables as appropriate
- A georeferenced, digital spatial representation of the detailed extents of each delivered dataset:
    - The extents shall be those of the actual lidar source or derived product data, exclusive of TIN artifacts or raster void areas
    - A union of tile boundaries or minimum bounding rectangles is not acceptable
    - For the point clouds, no line segment in the boundary will be further than four times the aggregate nominal pulse spacing (ANPS) from the nearest lidar point
    - Esri polygon shapefile or geodatabase is required
- Product metadata (FGDC-compliant, XML format metadata). Metadata files for individual data files are acceptable but not required. FGDC-compliant metadata shall pass the USGS Metadata Parser (MP) without errors. One XML file is required for each of the following datasets:
    - The Overall Project: Describing the project boundary, the intent of the project, the types of data collected as part of the project, the various deliverables for the project, and other project-wide information
    - Each Lift: Describing the extents of the lift, the swaths included in the lift, locations of GPS base stations and control for the lift, preprocessing and calibration details for the lift, adjustment and fitting processes applied to the lift in relation to other lifts, and other lift-specific information
    - Each Deliverable Product Group: Including classified point data; bare-earth DEMs; breaklines; and any other datasets delivered (digital surface models (DSM), intensity images, height above ground surfaces, and others)

A block of lidar-related metadata tags specified by the USGS shall be included in FGDC metadata files for all lidar point data deliverables. All tags are required. This block was developed so information often provided in reports or in free-text metadata fields can be made machine-discoverable in a predictable location in a single file. The descriptive template of this lidar metadata block and a completed example are provided in the USGS LBS V1.3, Appendix 3, "Lidar Metadata Example" and Appendix 4, "Lidar Metadata Template."

Raw Point Cloud

Raw point cloud deliverables are required and shall include or conform to the following procedures and specifications:

- All collected points, fully calibrated, georeferenced, and adjusted to ground, organized and delivered in their original swaths, one file per swath, one swath per file. (This is one example of differences with linear mode lidar in that Geiger mode lidar does not collect data in swaths.)
- If production processing requires segmentation of the swath files, the requirements listed in the section "Swath Size and Segmentation" shall be met
- Fully compliant LAS Specification version 1.4, Point Data Record Format 6, 7, 8, 9, or 10
- If collected, waveform data in external auxiliary files with the extension .wdp. See ASPRS LAS Specification V1.4
- Correct and properly formatted georeference information as Open Geospatial Consortium (OGC) well known text (WKT) in all LAS file headers
- GPS times recorded as Adjusted GPS Time at a precision sufficient to allow unique timestamps for each laser pulse
- Intensity values, normalized to 16-bit. See ASPRS LAS Specification V1.4
- A report of the assessed relative vertical accuracy of the point cloud (smooth surface repeatability and overlap consistency). For QL2 lidar, the smooth surface repeatability is ≤6 cm; the swath overlap difference, RMSDz, is ≤8 cm; and the swath overlap maximum difference is ±16 cm. Raw swath point cloud data shall meet the required accuracy levels before point cloud classification and derivative product generation
- A report of the assessed absolute vertical accuracy (NVA only) of the unclassified lidar point data in accordance with the guidelines set forth in the *ASPRS Positional Accuracy Standards for Digital Geospatial Data* (ASPRS, 2014). The required NVA for QL2 lidar is ≤19.6 cm at the 95-percent confidence level, which is equivalent to RMSEz of 10 cm in non-vegetated terrain. Raw swath point cloud data shall meet the required accuracy level before point cloud classification and derivative product generation

Classified Point Cloud

Delivery of a classified point cloud is a requirement for USGS-NGP lidar projects. Specific research projects may be exempt from this requirement. Classified point cloud deliverables shall include or conform to the following procedures and specifications:

- All project swaths, returns, and collected points, fully calibrated, adjusted to ground, and classified, by tiles. Project swaths exclude calibration swaths, cross-ties, and other swaths not used and not intended to be used, in product generation
- Fully compliant LAS Specification version 1.4 Point Data Record Format 6, 7, 8, 9, or 10
- If collected, waveform data in external auxiliary files with the extension .wdp. See the LAS Specification V1.4
- Correct and properly formatted georeferenced information as OGC well-known text (WKT) included in all LAS file headers
- GPS times recorded as Adjusted GPS Time at a precision sufficient to allow unique timestamps for each laser pulse
- Intensity values, normalized to 16-bit. See the LAS Specification version 1.4 for additional information.
- Tiled delivery, without overlap, using the project tiling scheme

- Minimum classified point cloud classification scheme for each topographic lidar point:
    - Class 1: Processed, but unclassified
    - Class 2: Bare earth
    - Class 7: Low noise
    - Class 9: Water
    - Class 17: Bridge decks
    - Class 18: High noise
    - Class 20: Ignored ground (near a breakline)
- With bathymetric or topobathymetric lidar, additional LAS codes may apply for each lidar point:
    - Class 40: Bathymetric point (e.g., seafloor or riverbed; also known as submerged topography)
    - Class 41: Water surface (sea/river/lake surface from bathymetric or topobathymetric lidar, distinct from point Class 9 which is used with topographic lidar to designate "water," not "water surface")
    - Class 42: Derived water surface (synthetic water surface location used in computing refraction at water surface)
    - Class 43: Submerged object, not otherwise specified (e.g., wreck, rock, submerged piling)
    - Class 44: International Hydrographic Organization (IHO) S-57 object, not otherwise specified
    - Class 45: No-bottom-found-at (bathymetric lidar point for which no detectable bottom return was received)

Other LAS codes may apply for specialized lidar projects on a case-by-case basis. See Chapter 3 for a listing of additional LAS codes

Bare-Earth Surface (Raster Digital Elevation Model)

Delivery of a hydro-flattened bare-earth DEM is a requirement for USGS-NGP lidar projects. Specific research projects may be exempt from some or all these requirements. Bare-earth surface deliverables shall include or conform to the following procedures and specifications:

- Bare-earth DEM, generated to the limits of the buffered project area (BPA)
- DEM with minimum cell size of 1 meter (or 2 feet) for QL2 lidar. For optional QL1 lidar, the DEM minimum cell size is 0.5 meter (or 1 foot)
- An industry-standard, GIS-compatible, 32-bit floating point raster format (ERDAS .IMG preferred)
- Georeference information in or accompanying each raster file
- Tiled delivery without overlap
- DEM tiles with no edge artifacts or mismatch. A quilted appearance in the overall DEM surface will be cause for rejection of the entire DEM deliverable, whether the rejection is caused by differences in processing quality or character among tiles, swaths, lifts, or other non-natural divisions
- Void areas (for example, areas outside the BPA but within the project tiling scheme) coded using a unique "NODATA" value. This value will be identified in the appropriate location within the raster file header or external support files, e.g., .aux
- Hydro-flattening as outlined in the USGS LBS V1.3 section: "Digital Elevation Model Hydro-Flattening," described in the next section of this chapter. Depressions (sinks), whether natural or man-made, are not to be filled (as in hydro-conditioning and hydro-enforcement). The methodology used for hydro-flattening is at the discretion of the data producer (refer to Appendix 2, "Hydro-Flattening Reference" of the USGS LBS V1.3 for more information on hydro-flattening)

- Bridges removed from the surface. See Appendix B of this manual for the definition of bridges
- Road or other travel ways over culverts remain intact in the surface. See Appendix B of this manual for the definition of culverts
- QA/QC analysis materials for the absolute vertical accuracy assessments
- A report on the assessed absolute vertical accuracy (NVA and VVA) of the bare-earth surface in accordance with the *ASPRS Positional Accuracy Standards for Digital Geospatial Data* (ASPRS, 2014). For either QL2 or QL1 lidar, the RMSEz in non-vegetated terrain must be ≤10.0 cm; the Non-vegetated Vertical Accuracy (NVA) must be ≤19.6 cm at the 95-percent confidence level (RMSEz x 1.9600); and the Vegetated Vertical Accuracy (VVA) must be ≤29.4 cm at the 95th percentile

DEM Hydro-Flattening

Hydro-flattening pertains only to the creation of derived DEMs. No geometric changes are to be made to the originally computed lidar points. Breaklines developed for use in hydro-flattening may be used to support classification of the point data.

Bare-earth lidar points that are near the breaklines shall be classified as Ignored Ground (LAS code 10; will be changed to LAS code 20 in the revised USGS LBS V1.3) and excluded from the DEM generation process. This process prevents unnatural surface artifacts from being created between mass points and breakline vertices. The proximity threshold for reclassification as Ignored Ground is at the discretion of the data producer, but in general will not exceed the aggregate nominal pulse spacing (ANPS).

The goal of the USGS-NGP is not to provide accurately mapped, geographically corrected water-surface elevations within the NED or 3DEP—it is to produce topographic DEMs that, with respect to water surfaces, resemble DEMs derived from traditional photogrammetric methods and to the degree practical are free of unnatural triangulation effects. Best professional judgment should be used to achieve this traditional smooth water-surface effect.

The requirements for hydro-flattening are listed below. These requirements also define the minimum features for which breaklines shall be collected and delivered.

Inland Ponds and Lakes

- Water bodies of 8,000 m^2 (2 acres) or greater surface area at the time of collection shall be flattened
- Flattened water bodies shall present a flat and level water surface (a single elevation for every bank vertex defining the water body's perimeter)
- The entire water-surface edge shall be at or below the immediately surrounding terrain (the presence of floating water bodies will be cause for rejection of the deliverable)
- Long impoundments—such as reservoirs, inlets, and fjords, whose water-surface elevations decrease with downstream travel—shall be treated as streams or rivers

Inland Streams and Rivers

- Streams and rivers of a 30-m (100-ft) nominal width (or greater) shall be flattened
- Streams or rivers whose width varies above and below 30 meters will not be broken into multiple segments; data producers will use their best professional cartographic judgment in determining when a stream or river has attained a nominal 30-m width
- Flattened streams and rivers shall present a flat and level water surface bank-to-bank (perpendicular to the apparent flow centerline)
- Flattened streams and rivers shall present a gradient downhill water surface, following the immediately surrounding terrain

- In cases of sharp turns of rapidly moving water, where the natural water surface is notably not level bank-to-bank, the water surface will be represented as it exists while maintaining an aesthetic cartographic appearance
- The entire water-surface edge shall be at or below the immediately surrounding terrain
- Stream channels shall break at culvert locations leaving the roadway over the culvert intact
- Bridges in all their forms shall be removed from the DEM
- Streams shall be continuous at bridge locations
- When the identification of a structure as a bridge or culvert cannot be made definitively, the feature shall be regarded as a culvert

Non-Tidal Boundary Waters

- Boundary waters, regardless of size, shall be represented only as an edge or edges within the project; collection does not include the opposite shore
- The entire water-surface edge shall be at or below the immediately surrounding terrain
- The water-surface elevation will be consistent throughout the project
- The water surface shall be flat and level, as appropriate for the type of water body (level for lakes, a gradient for streams and rivers)
- Any unusual changes in the water-surface elevation during the course of the collection (such as increased upstream dam discharge) shall be documented in the project metadata
- In the event of an unusual change in water-surface elevation, the water body shall be handled as described in Tidal Waters below

Tidal Waters

Tidal water bodies are defined as any water body that is affected by tidal variations, including oceans, seas, gulfs, bays, inlets, salt marshes, and large lakes. Tidal variations during data collection or between different data collections will result in lateral and vertical discontinuities along shorelines. As it is the USGS-NGP's intent for the DEM to represent as much ground as the collected data permits, lidar ground points shall not be removed for the sake of adjusting a shoreline inland to match another shoreline. Likewise, adjusting a shoreline outland will create an equally unacceptable area of unmeasured land in the DEM. It is recommended that, to the highest degree practical, collections be planned to minimize tidal differences at the land-water interface. In addition to meeting the requirements for inland water bodies listed above in Inland Ponds and Lakes, and Inland Streams and Rivers, the treatment of tidal water bodies shall also meet the following requirements:

- Within each water body, the water surface shall be flat and level for each different water-surface elevation
- Vertical discontinuities within a water body resulting from tidal variations during the collection are considered normal and shall be retained in the final DEM
- Horizontal discontinuities along the shoreline of a water body resulting from tidal variations during the collection are considered normal and shall be retained in the final DEM

Long tidal water bodies that also exhibit downhill flow (such as a fjord) can present unusual challenges; data producers are to exercise their best professional judgment in determining the appropriate approach solution to meet the overall goal of hydro-flattening as described in this section. For projects located in coastal areas, cooperating partners may impose additional requirements for tidal coordination.

Islands

Permanent islands 4,000 m^2 (1 acre) or larger shall be delineated within all water bodies.

Breaklines

Delivery of the breaklines representing all hydro-flattened features in a project, regardless of the method used for hydro-flattening, is a requirement for USGS-NGP lidar projects. Specific research projects may be exempt from these requirements. Breakline deliverables shall include or conform to the following procedures and specifications:

- Breaklines developed to the limit of the BPA
- Breaklines delivered in shapefile or file geodatabase formats, as PolylineZ and PolygonZ feature classes, as appropriate to the type of feature represented and the methodology used by the data producer
- Breaklines in the same coordinate reference system and units (horizontal and vertical) as the lidar point delivery
- Properly formatted and accurate georeferenced information for each feature class, stored in that format's standard file system location. Each shapefile shall include a correct and properly formatted .prj file

Breakline delivery may be in a single layer or in tiles, at the discretion of the data producer. In the case of tiled deliveries, all features shall edge-match exactly across tile boundaries in both the horizontal (x, y) and vertical (z) spatial dimensions. Delivered data shall be sufficient for the USGS to effectively re-create the delivered DEMs using the lidar points and breaklines without substantial editing.

Cooperating partners may require collection and integration of breaklines representing single-line streams, rivers, culverts, and other features within their lidar projects. Although the USGS does not require these breaklines to be collected or integrated into the DEMs, the USGS does require that if collected and incorporated into the DEMs, the following requirements are met:

- All vertices along single-line stream breaklines shall be at or below the immediately surrounding terrain
- Breaklines representing single-line streams, culverts, or other hydrographic features shall not be used to introduce hydrologic flow paths through road crossings (culverts), dams, or other similar topographic features
- All additional breaklines developed for the project shall be delivered to the USGS
- The final DEM shall be a hydro-flattened (not hydro-enforced) topographic DEM suitable for integration into the NED or 3DEP

COMMON DATA UPGRADES

The most common data upgrade above QL2 lidar is QL1 lidar which is nearly the same as QL2 lidar but with higher point density. QL1 lidar has 8 or more points/m^2 compared with 2 or more points/m^2 for QL2 lidar. Absolute and relative vertical accuracy requirements are the same for both QL1 and QL2 lidar.

The following is a partial list of other common upgrades which is neither comprehensive nor exclusive, covering topographic and/or bathymetric lidar:

- Independent third-party QA/QC by another contractor experienced with QL2 lidar specifications
- Full waveform collection and delivery
- Additional environmental constraints, e.g.,:
    - Tidal coordination, flood stages, crop or plant growth cycles
    - Shorelines corrected for tidal variations within a collection

- Top-of-Canopy (first return) Raster Surface (tiled); the raster representing the highest return within each cell is preferred
- Intensity images (8-bit gray scale, tiled)
    - Interpolation based on first returns
    - Interpolation based on all-returns, summed
- Detailed classification (additional LAS classes for topographic lidar), e.g.,
    - Class 3: Low vegetation
    - Class 4: Medium vegetation (use for single vegetation class)
    - Class 5: High vegetation
    - Class 6: Buildings, other man-made structures
- Additional LAS classes for bathymetric or topobathymetric lidar, e.g.,
    - Class 40: Bathymetric point (e.g., seafloor or riverbed; also known as submerged topography)
    - Class 41: Water surface (sea/river/lake surface from bathymetric or topobathymetric lidar; distinct from Point Class 9 which is used in topographic lidar and only designates "water," not "water surface")
    - Class 42: Derived water surface (synthetic water surface location used in computing refraction at water surface)
    - Class 43: Submerged object, not otherwise specified (e.g., wreck, rock, submerged piling)
    - Class 44: International Hydrographic Organization (IHO) S-57 object, not otherwise specified
    - Class 45: No-bottom-found-at (bathymetric lidar point for which no detectable bottom return was received)
    - Class n: additional classes or features as agreed upon in advance
- Hydrologically enforced (Hydro-Enforced) DEM as an additional deliverable
- Hydrologically conditioned (Hydro-Conditioned) DEM as an additional deliverable
- Breaklines (PolylineZ and PolygonZ) for additional hydrographic and topographic features
    - Narrower double-line streams and rivers
    - Single-line streams and rivers
    - Smaller ponds
    - Culverts and other drainage structures
    - Retaining walls
    - Hydrologic areas, e.g., swamp or marsh
    - Appropriate integration of additional features into delivered DEMs
- Extracted building footprints (PolygonZ): footprints with maximum elevation or height above ground as an attribute
- Other products as defined by requirements and agreed upon before a funding commitment

DEM USER BENEFITS

The author of this chapter was the primary author of the 800+ page report of the National Enhanced Elevation Assessment (NEEA) completed in 2012 that the U.S. Geological Survey (USGS) called "the most comprehensive Benefit-Cost Analysis ever performed for any layer of The National Map." This chapter will summarize the

NEEA process for assessing DEM user requirements and benefits based on five DEM Quality Levels (QLs) and five update frequencies, and explains the benefit-cost analysis process that resulted in USGS' 3D Elevation Program (3DEP) summarized in Chapter 5 of this manual. The 3DEP is based on QL2 lidar for 49 states and all U.S. territories as well as statewide QL5 IfSAR for Alaska.

Quality Levels and Update Frequencies

The five pre-defined topographic data Quality Levels (QL's) used in the NEEA are in Table 14.1, and the five pre-defined update frequencies are: annually, 2-3 years, 4-5 years, 6-10 years, and >10 years. Note: For consistency with the *ASPRS Positional Accuracy Standards for Digital Geospatial Data* (ASPRS, 2014) which uses incremental round numbers, the USGS LBS V1.3 specifies RMSEz of 10 cm for QL1 and QL2 lidar, and RMSEz of 20 cm for QL3 lidar. V1.3 also includes a new QL0 with the same point density as QL1 lidar but with tighter standards for both absolute vertical accuracy (RMSEz ≤5.0 cm, NVA ≤9.8 cm, and VVA ≤ 14.7 cm) and relative vertical accuracy (smooth surface repeatability ≤3 cm, swath overlap difference RMSDz ≤4 cm, and swath overlap maximum difference ±8 cm).

Elevation Quality Levels (QL)	Source	Horizontal Resolution Terms			Vertical Accuracy Terms	
		Point Density	Nominal Pulse Spacing (NPS)	DEM Post Spacing	Vertical RMSEz	Equivalent Contour Accuracy
QL1	Lidar	8 pts/m^2	0.35 m	1/27 arc-sec ~1 meter	9.25 cm	1-ft
QL2	Lidar	2 pts/m^2	0.7 m	1/27 arc-sec ~1 meter	9.25 cm	1-ft
QL3	Lidar	1-0.25 pts/m^2	1-2 m	1/9 arc-sec ~3 meter	≤18.5 cm	2-ft
QL4	Imagery	0.04 pts/m^2	5 m	1/3 arc-sec ~10 meter	46.3 cm-139 cm	5-15 ft
QL5	IfSAR	0.04 pts/m^2	5 m	1/3 arc-sec ~10 meter	92.7 cm-185 cm	10-20 ft

Table 14.1. The five pre-defined topographic data Quality Levels used in the NEEA. Note: RMSEz values have subsequently changed.

Chapter 5 of this manual explains the 3D Elevation Program (3DEP) that resulted directly from the NEEA; Chapter 8 on airborne topographic lidar summarizes several of the emerging lidar technologies introduced in the NEEA; and Chapter 13 on DEM user applications summarizes the 27 major Business Uses documented in the NEEA study. In 2016, Dewberry completed Part A of a similar National Hydrography Requirements and Benefits Study (NHRBS) for the USGS, but without the benefit-cost analysis for different implementation scenarios. The NHRBS documented that the accuracy of hydrographic data is highly dependent on the accuracy of elevation data. As this manual goes to press, Dewberry is currently executing Part B of the NHRBS, developing tools to analyze the data collected in Part A to summarize the benefits reported by the agencies and to provide USGS the ability to generate reports accessing program implementation scenarios.

DEM Business Uses and Functional Activities

The NEEA was based on 27 major Business Uses (BU's) from which 602 Functional Activities (FA's) were defined by users with mission-critical requirements for enhanced elevation data in one of five Quality Levels and one of five update frequencies. These FA's are documented in the NEEA Appendix B for federal agency FA's; Appendix C for state, territory and local FA's; Appendix D for nongovernmental (not-for-profit and private companies); and summarized by BU in Appendix E of the NEEA.

BU#1, Natural Resources Conservation, was defined in terms of FA's that preserve the health of soil and vegetation, minimize soil erosion and runoff into streams, and preserve wetlands.

BU#2, Water Supply and Quality, was defined in terms of FA's that ensure the availability of water where and when required, ensure that drinking water is safe, ensure that aquatic ecosystems remain healthy, and reduce causes for water pollution.

BU#3, River and Stream Resource Management, was defined in terms of FA's that ensure America's rivers and streams sustain economic recreational and subsistence activities and their other beneficial functions.

BU#4, Coastal Zone Management, was defined in terms of FA's that ensure America's coastal zones sustain economic, recreational and subsistence activities and their other beneficial functions.

BU#5, Forest Resources Management, was defined in terms of FA's that ensure America's forests sustain economic, recreational and subsistence activities and their other beneficial functions.

BU#6, Rangeland Management, was defined in terms of FA's that ensure America's rangelands are managed and sustained for their beneficial functions.

BU#7, Wildlife and Habitat Management, was defined in terms of FA's that sustain economic, recreational and subsistence activities of wildlife habitats.

BU#8, Agriculture and Precision Farming, was defined in terms of farming systems that use new technology to allow a closer, more site-specific management of the factors affecting crop production and farm run-off.

BU#9, Geologic Resource Assessment and Hazards Mitigation, was defined in terms of FA's that map, inventory and assess geologic resources, and mitigate geologic hazards such as earthquakes, tsunamis, landslides, and volcanoes.

BU#10, Resources Mining, was defined in terms of FA's that ensure that surface mines are operated in a manner that protects citizens and the environment during mining, and assures that the land is restored to beneficial use following mining.

BU#11, Renewable Energy Resources, was defined in terms of FA's that analyze solar energy potential and placement of solar farms and individual solar panels, wind energy potential and placement of wind farms, and hydroelectric energy potential for placement and/or safe operation of hydro-electric power plants.

BU#12, Oil and Gas Resources, was defined in terms of FA's that select sites for oil/gas wells and facilities, for pipeline routing and construction planning, and for environmental impact assessments and mitigation.

BU#13, Cultural Resources Preservation and Management, was defined in terms of FA's that identify evidence of past human activity, to include pioneer homes, buildings or old roads; prehistoric village sites; historic or prehistoric artifacts or objects; or earthworks such as battlefield entrenchments, prehistoric canals or mounds.

BU#14, Flood Risk Management, was defined in terms of FA's that mitigate flood risks, including dam, dike, and levee safety.

BU#15, Sea Level Rise and Subsidence, was defined in terms of FA's that model and/or mitigate the effects of sea level rise (SLR) and subsidence that cause the loss of land.

BU#16, Wildfire Management, Planning and Response, was defined in terms of FA's that determine forest fuel and fire susceptibility for wildfire modeling and suppression activities.

BU#17, Homeland Security, Law Enforcement and Disaster Response, was defined in terms of FA's that promote homeland security, minimize threats from terrorism and criminal activities, and respond to natural or manmade disasters.

BU#18, Land Navigation and Safety, was defined in terms of FA's that promote safe and economical navigation and commerce on America's roads and railroads.

BU#19, Marine Navigation and Safety, was defined in terms of FA's that promote safe and economical navigation and commerce on America's waterways.

BU#20, Aviation Navigation and Safety, was defined in terms of FA's that promote safe terminal and en route aerial navigation and commerce.

BU#21, Infrastructure and Construction Management, was defined in terms of FA's that support all forms of infrastructure and construction management needs of Federal, state, local and Tribal governments, not-for-profit, private industry and individuals.

BU#22, Urban and Regional Planning, was defined in terms of FA's that support diverse forms of urban and regional planning and development.

BU#23, Health and Human Services, was defined in terms of FA's that identify unhealthy environments, improve air quality, mitigate the spread of diseases and contaminants, and promote occupational health and safety.

BU#24, Real Estate, Banking, Mortgage, Insurance, was defined in terms of FA's that identify and mitigate risks from natural disasters other than floods (BU#14), i.e., other risks that need to be known by home owners and business owners as well as representatives of the real estate, banking, mortgage and/or insurance industries.

BU#25, Education K-12 and Beyond, was defined in terms of FA's that educate the public about the world in which they live or train them to exploit the virtual earth to solve human problems. This includes basic research in sciences that may not clearly align with any of the other BU's.

BU#26, Recreation, was defined in terms of FA's that support recreational activities for which elevation data and/or derivatives (e.g., slope, aspect and curvature) are design factors.

BU#27, Telecommunications, was defined in terms of FA's that support telecommunications spectrum management and frequency coordination.

Estimated Costs by Quality Levels

For the 48 conterminous states, USGS provided average cost estimates by Quality Level from its Geospatial Products and Services Contract (GPSC2) contractors for Quality Levels 1 through 4. These estimates, in 2011 dollars, are in column B in Table 14.2. Columns C and D include the 15 percent estimated costs of QA/QC to include the survey of QA/QC checkpoints. Column E assumes 5 percent for USGS to manage the processing, storage and distribution of data. Column F includes the total cost per square mile used in the Benefit-Cost Analyses. Dewberry provided cost estimates for Quality Level 5 IfSAR in Alaska and in the other 49 states where acquisition costs were estimated to be about 18 percent lower than costs in Alaska because of improved access to suitable airports and facilities. Future costs are dependent on the changing price of aviation fuel and the Consumer Price Index (CPI), and technological improvements.

A	B	C	D	E	F
Quality Level	$/mi^2	QA/QC	Subtotal	USGS	Total $/mi^2
QL1 Lidar (48 states)	$453.25	$67.99	$521.24	$26.06	$547.30
QL2 Lidar (48 states)	$277.00	$41.55	$318.55	$15.93	$334.48
QL3 Lidar (48 states)	$209.25	$31.39	$240.64	$12.03	$252.67
QL4 1-m image DEM (48 states)	$134.00	$20.10	$154.10	$7.71	$161.81
QL5 IfSAR (Alaska)	$90.00	Included	$90.00	$4.50	$94.50
QL5 IfSAR (49 states)					$80.00

Table 14.2. Estimates of costs per square mile for the five Quality Levels used in the NEEA study. Actual costs today are much lower, meaning that the actual benefit/cost ratio today is much better than the 5.356 estimated in 2012 for QL2 in Table 14.5.

All costs assumed that the same Quality Level of elevation data is acquired in the most efficient manner for entire 1-degree cells (1° latitude by 1° longitude) or other large, rectangular project areas. Most elevation data today are acquired in a less efficient manner, i.e., for smaller, irregularly-shaped areas that cost more per square mile. Nevertheless, the actual cost per square mile are now generally lower nationwide for all the quality levels, even when acquired in smaller and/or irregular shapes, because of technological improvements since the NEEA study was completed in 2012. Today's lower costs, summarized in chapter 5, when compared with higher cost estimates used in the NEEA study, explain why the actual return on investment (ROI) today is higher than estimated in the NEEA.

Tangible Benefits Aggregated by Business Use

For the 602 mission-critical Functional Activities, tangible and/or intangible benefits were documented for operational benefits to the organization, customer service benefits to others, public/social benefits, environmental benefits, and strategic/political benefits. Even when operational and/or customer service benefits were defined as "major," annual dollar benefits were often unknown or were understated for approximately half of the FA's. The NEEA documented "conservative dollar benefits" as well as "potential dollar benefits" considered to be more realistic in many cases.

Table 14.3 summarizes the conservatively estimated benefits as well as potential benefits of enhanced elevation data aggregated for each of the 27 Business Uses. Financial benefits were never double-counted because even when Functional Activities appeared to apply to more than one Business Use, a single Business Use was selected as the primary BU for aggregation of dollar benefits. Only the conservatively-estimated benefits were used in the benefit-cost analyses for the scenarios developed in this assessment and described in Appendix F of the NEEA report. Note: the conservative benefits will not be fully realized unless every FA receives the optimal Quality Level and update frequency determined by the user to be mission critical. For benefit-cost analyses, the master geodatabase, summarized below, documents procedures for degradation of dollar benefits for theoretical implementation scenarios that would provide data of poorer quality and/or currency than specified as mission critical for each of the 602 FA's.

BU	Business Unit (BU) Name	Enhanced Elevation Data Annual Benefits	
		Conservative Benefits	Potential Benefits
1	Natural Resources Conservation	$159.225M	$335,152M
2	Water Supply and Quality	$85.288M	$156.351M
3	River and Stream Resource Management	$38.422M	$86.582M
4	Coastal Zone Management	$23.785M	$41.740M
5	Forest Resources Management	$43.949M	$61.655M
6	Rangeland Management	$0.000M	$0.000M
7	Wildlife and Habitat Management	$1.510M	$4.020M
8	Agriculture and Precision Farming	$122.330M	$2,011.330M
9	Geologic Resource Assessment and Hazard Mitigation	$51.750M	$1,066.750M
10	Resource Mining	$1.686M	$4.864M

BU	Business Unit (BU) Name	Enhanced Elevation Data Annual Benefits	
		Conservative Benefits	Potential Benefits
11	Renewable Energy Resources	$10.050M	$100.050M
12	Oil and Gas Resources	$10.000M	$100,000M
13	Cultural Resources Preservation and Management	$0.000M	$7.000M
14	Flood Risk Management	$294.706M	$501.576M
15	Sea Level Rise and Subsidence	$5.780M	$21 660M
16	Wildfire Management, Planning and Response	$75.700M	$158,950M
17	Homeland Security, Law Enforcement, Disaster Response	$9.975M	$126.469M
18	Land Navigation and Safety	$0.191M	$7,124,875M
19	Marine Navigation and Safety	$0.000M	$0.000M
20	Aviation Navigation and Safety	$35.000M	$56 000M
21	Infrastructure and Construction Management	$206.212M	$941.951M
22	Urban and Regional Planning	$4.197M	$68 569M
23	Health and Human Services	$0.000M	$1.000M
24	Real Estate, Banking, Mortgage, Insurance	$0.000M	$0.000M
25	Education K-12 and Beyond, including Research	$0.264M	$2.264M
26	Recreation	$0.050M	$0.050M
27	Telecommunications	$0.185M	$1.350M
	Total Estimated Annual Dollar Benefits	**$1,180.224M**	**$12,980.707M**

Table 14.3. Conservative and potential annual dollar benefits aggregated by Business Use.

Whereas all of the details are available in the full NEEA report that is downloadable, six of the Business Uses warrant a brief explanation for values unusually low or high:

BU#6, Rangeland Management, shows zero dollar benefits from enhanced elevation data because the Bureau of Land Management (BLM), Bureau of Indian Affairs (BIA), U.S. Fish and Wildlife Service (FWS), National Park Service (NPS), and U.S. Forest Service (USFS) could not quantify their benefits for rangeland management. The Natural Resources Conservation Service (NRCS) indicated that lidar and IfSAR data enables NRCS specialists to perform detailed terrain and vegetation analysis of Planning Land Unit (PLU) rangelands from the office, avoiding time-consuming and costly field visits, but these benefits were included in NRCS' broader FA for Conservation Engineering and Practices with dollar benefits already counted as part of BU#1.

BU#13, Cultural Resources Preservation and Management, shows zero dollar benefits from enhanced elevation data only because the BIA, BLM, NPS, NRCS, state, local and tribal governments, Tennessee Valley Authority (TVA), and The Nature Conservancy all indicated benefits that could not be quantified in dollar terms.

BU#18, Land Navigation and Safety, shows the highest potential dollar benefits for the future from enhanced elevation data because smart cars of the future are expected to automatically upshift or downshift based on changing road gradients ahead (from lidar data) and realize an estimated $7B per year fuel savings.

BU#19, Marine Navigation and Safety, shows zero dollar benefits from enhanced elevation data because the NEEA study was primarily based on topographic data rather than bathymetric data required for marine navigation and safety. Dewberry is currently executing a follow-on to the NEEA study funded by NOAA for the Interagency Working Group on Ocean and Coastal Mapping under the Subcommittee on Ocean Science and Technology that includes five bathymetric lidar Quality Levels (QL0B, QL1B, QL2B, QL3B, and QL4B). An important component of the National Coastal Mapping Strategy, these QLs are intended to provide a standard approach to coastal shallow mapping where bathymetric lidar can perform. When combined with 3DEP topographic data and acoustically acquired bathymetric data, this elevation data will contribute to a seamless 3D Nation from the mountains into the oceans.

BU#23, Health and Human Services, shows zero dollar benefits from enhanced elevation data because the Centers for Disease Control and Prevention (CDC), Environmental Protection Agency (EPA) and various states were unable to quantify benefits from improved water and air quality, health emergency response activities, environmental modeling, vector habitat modeling, and disease prevention planning.

BU#24, Real Estate, Banking, Mortgage, and Insurance, shows zero dollar benefits from enhanced elevation data because the relevant financial institutions have not yet developed DEM or lidar-related actuarial models and insurance programs for risks from earthquakes, wildfires, landslides, tornadoes, and volcanoes—separate from flood risks already covered by BU#14. Risks from earthquakes, wildfires, landsides and volcanoes could all be modeled from lidar, but risks from tornadoes are not yet modeled from lidar.

The Benefit-Cost Analysis Process

Recognizing that benefits are degraded if users do not receive the Quality Level and/or update frequency required for mission-critical activities, Dewberry developed a geodatabase methodology for degrading annual dollar benefits with reduced value multipliers. Table 14.4 shows how the benefits value multiplier is decreased for any Functional Activity that has the most demanding requirement (QL1 lidar with annual updates) and receives something less than that (shown in 24 other alternatives). For other less-demanding requirements, the value multiplier is 1.0 (full benefit value) if the Quality Level and update frequency is equal to or better than required, but decreases by half for every column to the right for Quality Level and for every row beneath for update frequency that is poorer than required for each Functional Activity.

Update Frequency	QL1 Lidar	QL2 Lidar	QL3 Lidar	QL4 DEM	QL5 IfSAR
Annually	1	1/2	1/4	1/8	1/16
2-3 years	1/2	1/4	1/8	1/16	1/32
4-5 years	1/4	1/8	1/16	1/32	1/64
6-10 years	1/8	1/16	1/32	1/64	1/128
>10 years	1/16	1/32	1/64	1/128	1/256

Table 14.4. Benefits value multipliers with poorer Quality Level and/or update frequency

For example, suppose a mission-critical Functional Activity required QL2 lidar updated every 4-5 years in order to receive $100,000 in annual benefits, but only received QL3 lidar updated every 8 years, the benefits would be cut in half once for being of poorer quality and then cut in half again a second time for being updated every 8 years instead of 4-5 years as required. The calculated benefits for that Functional Activity would be $25,000 per year because of the two 50% reductions in the value multiplier.

The NEEA Master Geodatabase

For the benefit-cost analysis of potential implementation scenarios in the NEEA, Dewberry developed a master geodatabase with over 100,000 lines, collected from elevation data users in 34 federal agencies, all 50 states (including local and tribal input) and 13 non-governmental organizations. Each line represented one area polygon that specified a validated mission-critical Functional Activity, defined in each agency's own words that required enhanced elevation data with a specified Quality Level and update frequency for the area within each polygon in order to achieve specified tangible and/or intangible benefits. The geodatabase included extensive additional information about the user stating the mission-critical requirement, explanations of those requirements, and explanations of operational internal benefits, customer service benefits to clients, public/social benefits, environmental benefits, and strategic/political benefits. Figures 14.1 through 14.6 were all extracted from the NEEA study prepared by Dewberry for USGS and other sponsoring agencies.

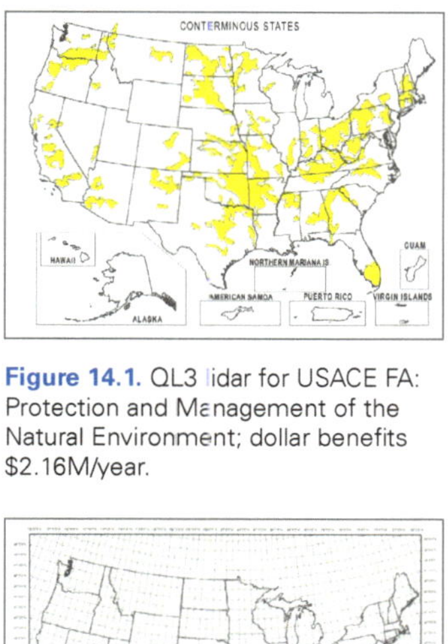

Figure 14.1. QL3 lidar for USACE FA: Protection and Management of the Natural Environment; dollar benefits $2.16M/year.

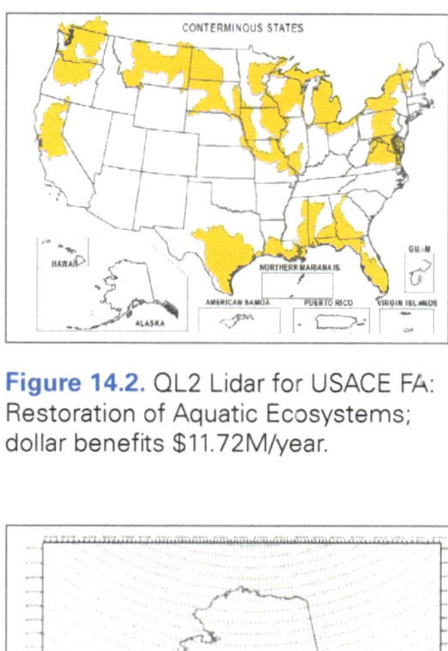

Figure 14.2. QL2 Lidar for USACE FA: Restoration of Aquatic Ecosystems; dollar benefits $11.72M/year.

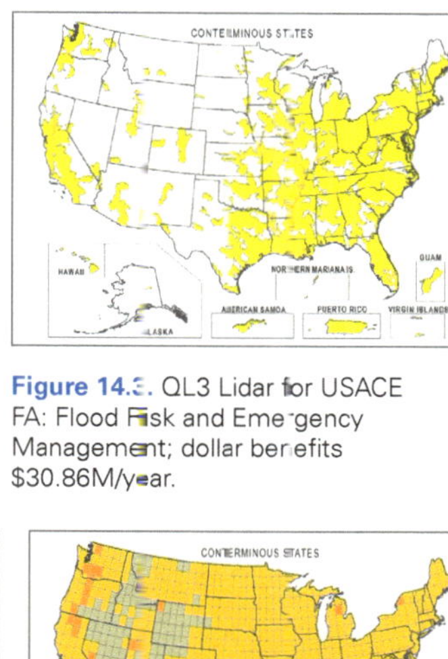

Figure 14.3. QL3 Lidar for USACE FA: Flood Risk and Emergency Management; dollar benefits $30.86M/year.

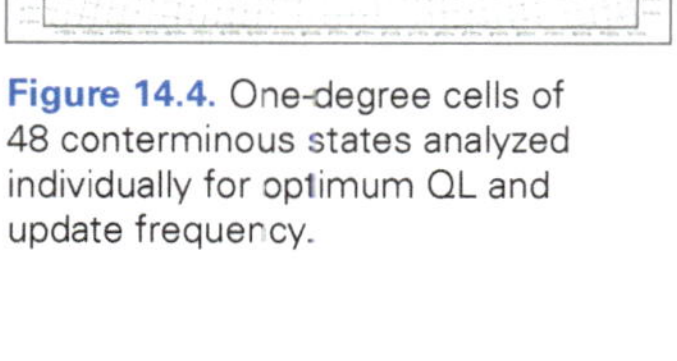
Figure 14.4. One-degree cells of 48 conterminous states analyzed individually for optimum QL and update frequency.

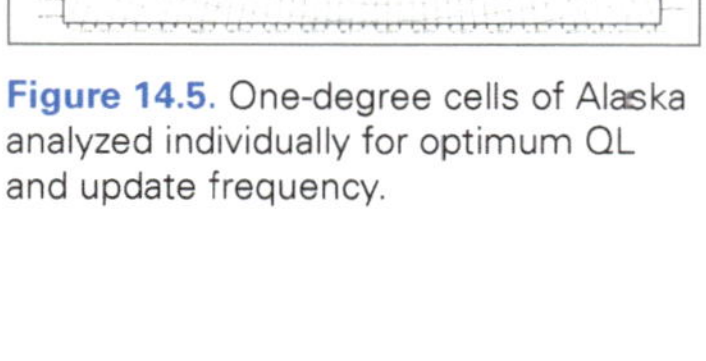
Figure 14.5. One-degree cells of Alaska analyzed individually for optimum QL and update frequency.

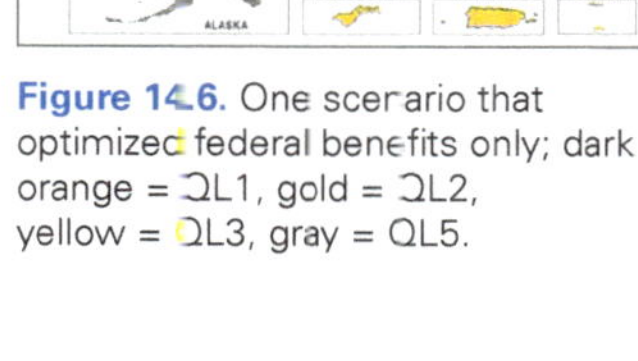

Figure 14.6. One scenario that optimized federal benefits only; dark orange = QL1, gold = QL2, yellow = QL3, gray = QL5.

Figures 14.1, 14.2, and 14.3 were selected as examples of different Functional Activities from the U.S. Army Corps of Engineers (USACE) alone. The purpose for showing these three images is to demonstrate that different FA's from a single agency could have different Quality Level and update frequency requirements for areas that overlap one another; for all three of these FA's, USACE specified required update frequencies of 6-10 years. These are just three of 602 FA's that overlap one another with 100,000 polygons having different user requirements and benefits; USACE gave specific authorization to publicize their Functional Activities (FA's).

A total of 10 implementation scenarios were evaluated by one-degree cells (Figures 14.4 and 14.5) and Figure 14.6 shows one of the implementation scenarios that determined, by cell, the highest return on investments for Federal agencies only. Other implementation scenarios optimized benefits for all users.

In addition to ten implementation scenarios where each 1-degree cells was evaluated individually for the optimum Quality Level and update frequency for that cell, Dewberry used the geodatabase to evaluate all 25 options (five update frequencies for each of five Quality Levels) for collecting elevation data uniformly for the 48 conterminous states, excluding Alaska, Hawaii and U.S. territories. For each of these 25 options, Table 14.5 shows annual total data costs (column D), annual total dollar benefits (column E), the benefit/cost ratio (column F), and the net benefits (benefits minus costs (column G)). The five colors used in this table match the colors for Quality Levels used for all maps in the NEEA report.

A	B	C	D	E	F	G
Option #	Quality Level	Update Frequency	Annual Total Costs	Annual Total Benefits	Benefit/Cost Ratio	Net Benefits (Benefits, Costs)
1	1	Annual	$1,646M	$1,111M	0.674	(-$536M)
2	1	2-3 years	$659M	$1,110M	1.685	$451M
3	1	4-5 years	$366M	$1,066M	2.914	$700M
4	1	6-10 years	$206M	$800M	3.887	$594M
5	1	>10 years	$110M	$403M	3.671	$293M
6	2	Annual	$1,006M	$923M	0.917	(-$84M)
7	2	2-3 years	$402M	$922M	2.291	$520M
8	2	4-5 years	$224M	$888M	3.970	$664M
9	2	6-10 years	$126M	$674M	5.356	$548M
10	2	>10 years	$67M	$339M	5.049	$272M
11	3	Annual	$760M	$697M	0.917	(-$63M)
12	3	2-3 years	$304M	$696M	2.291	$392M
13	3	4-5 years	$169M	$673M	3.983	$504M
14	3	6-10 years	$95M	$501M	5.278	$406M
15	3	>10 years	$51M	$252M	4.970	$201M
16	4	Annual	$487M	$361M	0.741	(-$126M)
17	4	2-3 years	$195M	$360M	1.851	$166M
18	4	4-5 years	$108M	$346M	3.198	$238M
19	4	6-10 years	$61M	$256M	4.204	$195M
20	4	>10 years	$32M	$129M	3.962	$96M
21	5	Annual	$241M	$190M	0.788	(-$51M)
22	5	2-3 years	$96M	$190M	1.970	$93M
23	5	4-5 years	$53M	$180M	3.365	$126M
24	5	6-10 years	$30M	$131M	4.369	$101M
25	5	>10 years	$16M	$66M	4.118	$50M

Table 14.5. Comparison of benefit/cost ratios and net benefits for all 25 Quality Level and update frequency options.

The annual total costs in column D were computed by the costs/mi² in column F of Table 14.2 on page 447. The annual total benefits in column E for each of the 25 options were computed by aggregating all of the conservative dollar benefits from the 602 mission critical Functional Activities, but with benefits degraded for each Functional Activity that, for any of the 24 other options, received data of poorer Quality Level or update frequency than required to achieve full benefits; the reduced value multiplier process was explained with Table 14.4. The benefit/cost ratio (BCR) in column F was computed by dividing the benefit values in column E by the cost values in column D. The net benefit values in column G were computed by subtracting the annual total costs in column D from the annual total benefits in column E.

Major Conclusions from the Benefit-Cost Analysis Process

- All options (see Table 14.5) with annual updates (options 1, 6, 11, 16, 21) have negative net benefits, meaning the annual costs exceed the annual benefits and should not be considered as a viable option
- Option 3 (QL1 lidar with 4-5 year updates) has the highest net benefits ($700M/year) but with a relatively poor benefit/cost ratio of only 2.914
- Option 9 (QL2 lidar with 6-10 year updates) has annual net benefits of $548M and the highest Return on Investment with a benefit/cost ratio of 5.356, meaning there are well over $5 in benefits for every dollar spent on QL2 lidar when the update frequency is 6-10 years. USGS subsequently chose QL2 lidar, updated every eight years, as its standard for the 3DEP program (explained in Chapter 5) for 49 states and all U.S. territories
- Because Table 14.5 pertained only to the 48 conterminous states, separate benefit-cost analyses were performed for Alaska which could justify lidar only for a few cities and high value corridors. QL5 IfSAR was chosen as the Alaska statewide standard for the 3DEP because QL5 IfSAR yielded statewide benefits in excess of $19.3M per year, essentially paying for itself within three years
- Figure 14.6 on page 451 shows just one of the ten implementation scenarios analyzed—all of which indicated that either QL1 or QL2 lidar had the highest return on investment (ROI) for Hawaii and all U.S. territories. However, their benefit-cost analyses were skewed because we did not have reliable numbers for the higher costs of deploying lidar aircraft to islands far from the mainland

CHAPTER HIGHLIGHTS

This chapter has documented the procedures used by USGS and Dewberry to document 602 mission-critical requirements for enhanced elevation data and benefits derived therefrom, and the benefit-cost analysis that concluded that QL2 lidar provides the highest return on investment, i.e., over $5 in benefits for each $1 spent. Readers can look further into this 860-page report to review each of the 602 mission critical requirements submitted by 34 federal agencies, all 50 states, plus non-governmental organizations. To review the full report in greater detail, it is easiest to Google "National Enhanced Elevation Assessment (NEEA)" which will take you directly to the full report which includes an Executive Summary and various appendices that can be downloaded individually.

3DEP Standard Products

As a result of the NEEA, USGS developed the 3D Elevation Program (3DEP), explained in Chapter 5 of this manual. The 3DEP is based on nationwide minimum standards for QL2 lidar for 49 states and all U.S. territories, and QL5 IfSAR for the state of Alaska. Today, lidar acquisitions of large areas at a lower quality than QL2 is probably money not invested well.

It is perfectly acceptable to "buy up" to QL1 or QL0 lidar if the customer is willing and able to pay the higher price. Dewberry has reviewed lengthy Scopes of Work (SOW) for proposed lidar projects that, due to their organization, confused the entire technical staff. In these SOWs, 90-95% of the requirements were identical to the USGS LBS for QL2 lidar, but a few requirements were more or less demanding than the QL2 requirements. Minor variations make it very difficult to identify the subtle changes and data producers may miss some important modifications. A lengthy and confusing 20-page Scope of Work for lidar can often be simplified to a single page by specifying that you need QL2 lidar consistent with the latest USGS LBS, followed by bullets that explain any requirements that are more demanding than QL2, and preferably with a brief explanation as to why these modifications are needed.

Before preparing a Scope of Work for lidar, a responsible manager should first read the 3DEP QL2 Lidar Standard Deliverables section above and determine if USGS has published a newer version of the v1.2 LBS, before articulating what requirements are different. By focusing only on how your needs differ from standard QL2 deliverables, confusion and uncertainty regarding your true requirements will be minimized, increasing the likelihood that the USGS and/or data providers will be able to execute a successful project.

Summary of Benefits from 3DEP Standard Products

- Users benefit from seamless, high-accuracy and high-resolution 3DEP datasets that are consistent nationwide and are maintained up-to-date. Users do not benefit when there is a proliferation of non-standard elevation products
- Users benefit when there is a single authoritative source of elevation data that satisfies multiple requirements, preventing conflicts between competing datasets of questionable quality
- Users benefit when high quality, consistent 3DEP elevation data can be produced at lower costs for broad areas, and they benefit from cost-sharing with other users through USGS' BAA process
- Users benefit when they can use standard 3DEP products and standard training courses without specialized hardware, software or training
- Even when users generate and tailor their own elevation derivative products from 3DEP datasets, users benefit from knowing they can rely on 3DEP source data of consistent quality, data formats, etc.
- Taxpayers benefits when they receive a minimum Return-on-Investment (ROI) of 5:1, and possibly 50:1 or better

AUTHOR BIOGRAPHIES

Dr. David F. Maune is a Senior Project Manager for Mapping and Remote Sensing at Dewberry in Fairfax, VA and principal author of the NEEA. See Chapter 1 for his biography.

Mr. Hans Karl Heidemann is a physical scientist at the USGS and the author of the USGS LBS referenced through this chapter. Mr. Heidemann is acknowledged for his assistance in reviewing this chapter for consistency with the USGS LBS V1.3. See Chapter 1 for his biography.

REFERENCES

Florida (2017). *Florida statewide lidar assessment.*

Heidemann, Hans Karl (2018). Lidar base specification (ver. 1.3, November 2014): *U.S. Geological Survey Techniques and Methods*, book 11, chap. B4, 67 p. with appendixes.

USGS (2012). *Final Report of the National Enhanced Elevation Assessment.*

Chapter 15

QUALITY ASSESSMENT OF ELEVATION DATA

Jennifer Novac

INTRODUCTION

This chapter has two basic goals:

- To explain the best methods for quantitative assessment of digital elevation data, including testing and reporting methods endorsed by the Federal Geographic Data Committee (FGDC), National Digital Elevation Program (NDEP), American Society for Photogrammetry and Remote Sensing (ASPRS), Federal Emergency Management Agency (FEMA), and U.S. Geological Survey (USGS)
- To explain the best methods for comprehensive qualitative assessment of digital elevation data to ensure final products are usable and meet specified requirements and expectations

Quality Assurance (QA): Steps taken: (1) To ensure the end client receives the quality products it pays for, consistent with the Scope of Work, and relevant standards, guidelines, and industry accepted best professional practices; and/or (2) To ensure an organization's Quality Program works effectively. Quality Programs include quality control procedures for specific products as well as overall Quality Plans that typically mandate an organization's communication procedures, document and data control procedures, quality audit procedures, and training programs necessary for delivery of quality products and services.

Quality Control (QC): Steps taken by data producers to ensure delivery of products that satisfy the specifications identified in the Scope of Work and relevant standards, guidelines, and accepted best practices. These steps typically include production flow charts with built-in procedures to ensure quality at each step of the work flow, in-process quality reviews, and/or final quality inspections prior to delivery of products to a client.

Independent QA/QC: Steps taken by a QA/QC specialty firm, hired by the client (e.g., government or data producer) to independently validate the effectiveness of the data producer's quality processes and compliance of the delivered data with the specifications identified in the Scope of Work and relevant standards, guidelines, and accepted best practices.

DEM Quality Assessment: Steps taken to test and report the accuracy of a digital elevation dataset and evaluate its usability. This includes quantitative assessment of data accuracy as well as qualitative assessment of data usability. It is not only possible, but common, for digital elevation datasets to pass vertical accuracy testing requirements and still fail other quality factors that adversely impact the usability of the elevation data.

STANDARDS, GUIDELINES AND SPECIFICATIONS

The following standards, guidelines and specifications are documented in Chapter 3 of this manual and will not be duplicated here. They include the following:

Federal Geographic Data Committee (FGDC) Standards

FGDC Geospatial Positioning Accuracy Standards

Part 1: Reporting Methodology (1998).
Part 2: Standards for Geodetic Networks (1998).
Part 3: National Standard for Spatial Data Accuracy (NSSDA)(1998).
Part 4: Standards for Architecture, Engineering, Construction (A/E/C) and Facility Management (2002).
Part 5: Standards for Nautical Charting Hydrographic Surveys (2005).
FGDC Geographic Information Framework Data Content Standards (2008).

American Society for Photogrammetry and Remote Sensing (ASPRS) Standards

ASPRS Positional Accuracy Standards for Digital Geospatial Data (2014).
ASPRS LAS Specifications V1.4 (2011).

U.S. Geological Survey (USGS) Specifications

USGS Lidar Base Specification, V1.3 (2018).

National Oceanic and Atmospheric Administration (NOAA) Specifications

NOS Hydrographic Surveys Specifications and Deliverables (2010).

National Digital Elevation Program (NDEP) Guidelines

NDEP Guidelines for Digital Elevation Data (2004).

U.S. Army Corps of Engineers (USACE) Guidelines

EM-1110-1-1000, Photogrammetric and Lidar Mapping (2015).

QUANTITATIVE ASSESSMENT

Digital elevation datasets should always be tested both quantitatively and qualitatively. Quantitative assessments determine if the datasets meet specified accuracy standards, both absolute and relative, while qualitative assessments determine the usability of the dataset. Common methods of testing and reporting have been established for the geospatial industry with the various standards and guidelines described in Chapter 3. The most recent standards and specifications have established accuracy thresholds for different types and levels of digital data.

Accuracy Assessment

Relative Vertical Accuracy

Relative accuracy is critical to the quantitative review of a project because it is impractical and cost-prohibitive to have QA/QC checkpoints located in every swath to test absolute vertical accuracy. However, if the relative accuracy between swaths and between points within the same swath can be measured, this consistency can be applied to the final absolute vertical accuracy of the project. Good relative vertical accuracy lends confidence that the final absolute vertical accuracy of the project applies project wide while poor relative vertical accuracy implies the final absolute vertical accuracy may only apply to the locations of the QA/QC checkpoints. There are two types of relative vertical accuracy: Intra-swath and inter-swath relative vertical accuracy.

Intra-Swath Relative Accuracy

Intra-swath relative accuracy is also known as point-to-point or within swath relative vertical accuracy. Intra-swath relative accuracy generally applies to lidar data and is the accuracy from one lidar point to the next lidar point within the same swath. Intra-swath relative accuracy can also be defined as the degree of agreement between nearby points measuring the same surface, or the repeatability of measurements. Intra-swath relative accuracy should be higher than inter-swath relative accuracy as issues between points within the same swath will degrade the inter-swath relative vertical accuracy and overall absolute vertical accuracy of the dataset. Intra-swath relative accuracy issues will also introduce inconsistency in the dataset by causing qualitative issues (such as corn rows) in the final dataset. Intra-swath issues can be caused by sensor issues, such as incorrect laser pulse power, scanning mechanism issues, and range-walk (intensity gain issues caused by bright and dim features located near each other). Issues with the Inertial Measurement Unit (IMU), including drift occurring from very windy conditions, can also contribute to larger than anticipated intra-swath relative accuracy. Intra-swath relative accuracy issues can sometimes be reduced through additional post-processing, but in many cases it may be impossible to fully correct the issue post acquisition. Figure 15.1 shows an example of lidar points with poor intra-swath relative accuracy. Figure 15.2 on page 458 shows the affect this poor intra-swath relative vertical accuracy has on the bare-earth DEM produced from this data.

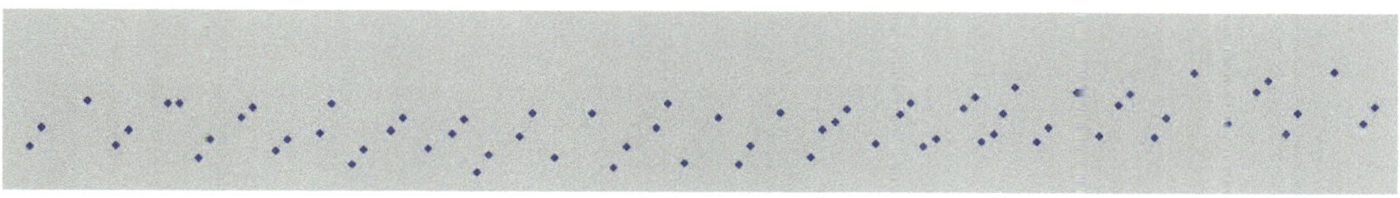

Figure 15.1. Profile view of lidar points (blue) with different elevations from a single swath. All points should be representing bare-earth, void of above-ground features. These points should not be "stacked" on top of each other but should be more horizontal, so that each lidar point represents the bare-earth ground. Instead, there is 1.5 feet of vertical difference between points of the same swath within each grouping when the points in each vertical 'stack' should actually be representing very similar elevations on the ground.

Intra-swath, or within swath, relative accuracy should be tested by measuring the repeatability of single returns on non-vegetated, hard planar surfaces, such as parking lots or large rooftops. Repeatability should be assessed by measuring the differences between single returns on planar surfaces, allowing or normalizing for actual elevation change across the planar surface. USGS recommends creating single swath difference rasters over sample areas where repeatability testing is suitable. According to USGS Lidar Base Specifications v1.3 (2018), all lidar points located within each cell will be used to calculate a precision value per cell using the formula:

Precision = Range - (Slope x Cellsize x 1.414)

Where:

- Range is the difference between the highest and lowest lidar point elevations in each pixel
- Slope is the maximum slope of the cell to its eight neighbors, expressed as a decimal value, calculated from the minimum elevation in each cell
- Cellsize is set to the Aggregate Nominel Pulse Spacing (ANPS), rounded up to the next integer, and then doubled
- 1.414 is the factor to compute the diagonal dimension of the pixel

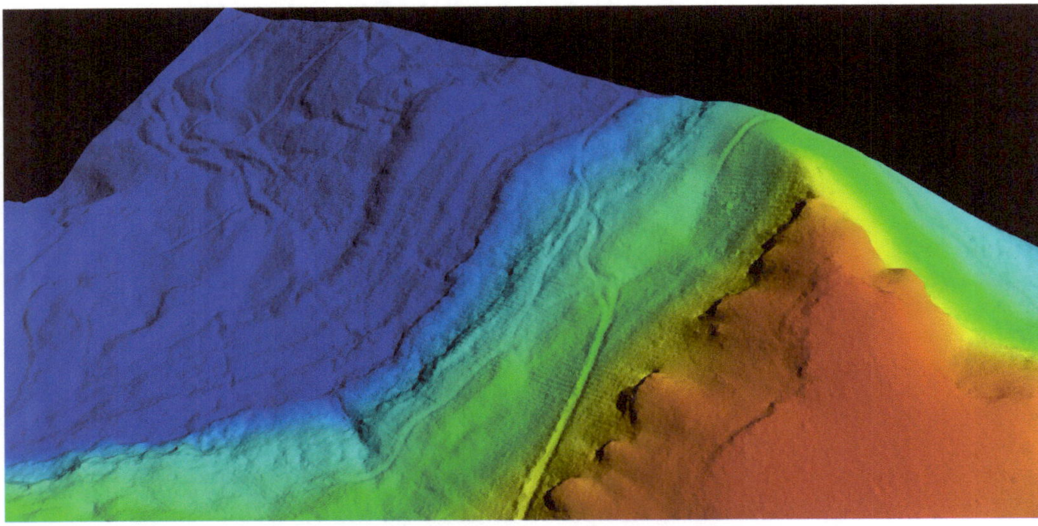

Figure 15.2. This bare-earth DEM was created from data with poor intra-swath relative accuracy. The result is undulations or ridges (sometimes referred to as corn rows) of up to 1 foot in the surface.

The rasters can then be statistically summarized to verify the Root Mean Square Difference in the z direction (RMSDz) for each test area do not exceed project requirements for intra-swath relative accuracy. RMSDz is used rather than Root Mean Square Error (RMSEz), as no absolute "truth" is identified in this calculation. In addition to calculating metrics, the rasters can also be symbolized to visualize areas both within and out of tolerance. Difference rasters can be produced for entire project areas to allow users to visualize relative accuracy for a full dataset, ensuring gross or systematic errors are not missed. Identifying particular sample areas meeting testing suitability (relatively flat and level surfaces) allows reliable metrics to be calculated without the process being overly burdensome. Figure 15.3 provides an example of an intra-swath difference grid. In this example, the difference grid is partially transparent and overlaid on intensity imagery for increased visualization. This example shows a large rooftop and parking lot. All areas where individual cells meet the intra-swath accuracy specifications are symbolized green and all cells exceeding specifications are symbolized as red.

Figure 15.3. This difference grid was created from single returns in a single swath. All cells colored green have values within 6 cm (specifications for this project is 6 cm RMSDz) and all cells with values exceeding 6 cm are colored red. There is some speckling of red in the parking lot, which can be attributed to isolated noise and ignored; the majority of the parking lot is green and within specifications. Red pixels outline portions of the building where lidar returns include the top and sides of the building and represent legitimate elevation change. The rooftop is green and within specifications. Once the difference raster is clipped to the desired testing area, removing the vertical walls of the building from the assessment, the RMSDz can be calculated for each test area. The RMSDz of this test area is within specification.

Inter-swath Relative Accuracy

Inter-swath, or swath-to-swath, relative accuracy is the accuracy of a single swath compared to overlapping or adjacent swaths and applies to all geospatial data collected in flight lines. Inter-swath relative accuracy is the difference in elevation between one or more pairs of overlapping swaths, generally expressed as RMSDz. RMSDz is used rather than RMSEz, as neither swath is identified as an absolute "truth." Airborne GPS and IMU data should be continuously collected during acquisition and final swaths should be tied to ground control. However, considerable differences in the final positioning of one swath compared to overlapping or adjacent data points can occur. Inter-swath relative accuracy issues may occur due to issues with the airborne GPS or IMU, such as losing locks on satellites during acquisition. Issues with the airplane heading, including roll, pitch, and yaw errors, can contribute to inter-swath relative accuracy errors. Roll, pitch, and yaw errors may occur during very windy or other unfavorable flying conditions, if the flying speed is inconsistent, if a consistent Above Ground Level (AGL) flying height is not maintained, or there are other problems maintaining consistent aircraft headings during acquisition. Issues with ground control may cause swath-to-swath relative accuracy issues, especially if different ground control was used for adjacent areas. This may occur if adjacent areas were collected at different time periods and different control was used for each respective area, causing relative accuracy issues when the two adjacent areas are tied together. Lastly, inter-swath relative accuracy issues may also occur from poor calibration and/or boresight techniques, including horizontal alignment issues which can translate to vertical inter-swath differences. Inter-swath relative accuracy can be corrected by re-processing the affected swaths, and in some cases entire lifts may need to be re-processed, with the corrective solution, which will depend on the cause of the original issue. Inter-swath relative accuracy issues left uncorrected will not only reduce the validity of applying tested absolute vertical accuracy values to an entire dataset, but will also cause qualitative issues, such as flight line ridges and inconsistency in the bare earth ground. Figures 15.4 and 15.5 both show examples of lidar points with poor inter-swath relative accuracy but Figure 15.5 is a direct result of horizontal alignment issues. Figure 15.6 on page 460 shows an example of the affect poor swath-to-swath relative vertical accuracy has on the bare-earth DEM produced from this data.

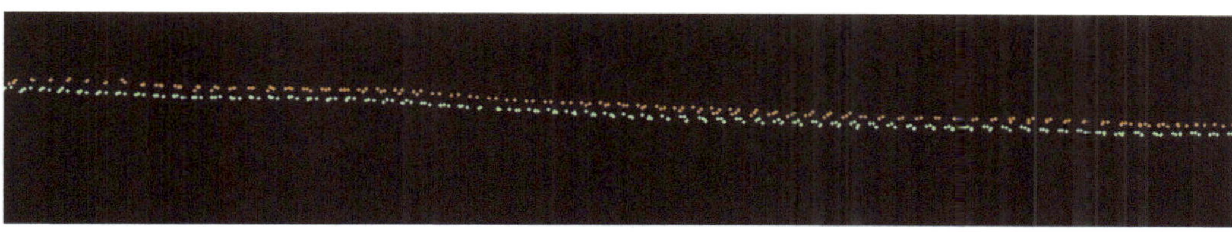

Figure 15.4. This profile shows two different swaths, each colored differently. There is approximately 1.5 feet of vertical difference between the two overlapping swaths due to poor calibration/boresight techniques. These two swaths should have consistent z elevations and should overlap each other.

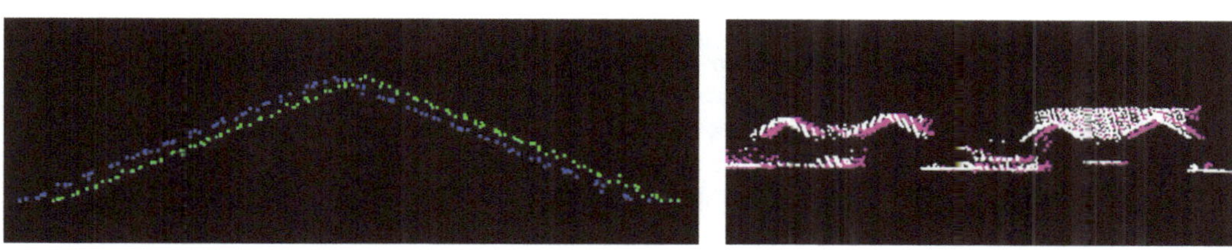

Figure 15.5. The image on the left shows two swaths (blue/green points) from a single rooftop where horizontal misalignments result in vertical offsets of approximately 20 cm between the two swaths. The image on the right also shows vertical offsets (20 cm) between two swaths (pink/white points) resulting from horizontal misalignments between the swaths, but this image shows several rooflines.

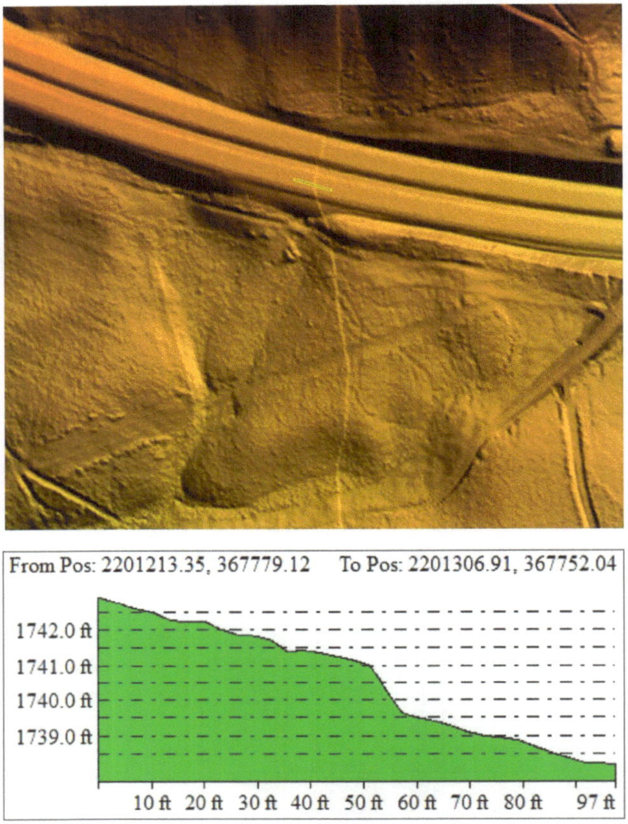

Figure 15.6. This bare-earth DEM was created from data with poor swath-to-swath relative accuracy. A flight line ridge of approximately 1.5 feet is visible in the final DEM, and quantified in the accompanying profile, which was taken on the road at the yellow line.

USGS Lidar Base Specifications v1.3 (2018) recommend testing inter-swath relative accuracy using difference rasters, sometimes also referred to as DeltaZ orthos. DeltaZ orthos (ΔZ) are raster images that compare the differences between data from overlapping swaths on a cell-by-cell basis. Inter-swath relative accuracy testing should be performed in non-vegetated areas (single lidar returns) of less than 10 degree slope where adjacent swaths overlap. The cellsize used for these difference rasters should equal the ANPS, rounded up to the next integer, and then doubled. The rasters can then be statistically summarized to verify the RMSDz for each test area do not exceed project requirements for inter-swath relative accuracy. Identifying particular sample areas meeting testing suitability requirements allows reliable metrics to be calculated without the process being overly burdensome. However, in addition to calculating metrics, the rasters can also be symbolized to visualize areas both within and out of tolerance. Difference rasters should be produced for entire project areas to allow users to visualize relative accuracy for a full dataset, ensuring gross or systematic errors are not missed. Relative accuracy errors are generally not consistent across a project and may not be consistent across the affected overlapping flight lines. Swath-to-swath relative accuracy issues cause both quantitative and qualitative issues. Tested absolute vertical accuracy values do not apply to areas where inter-swath relative accuracy is out of tolerance because using limited checkpoints to calculate absolute vertical accuracy results project wide requires confidence in relative accuracy. Qualitative issues resulting from inter-swath relative accuracy issues, such as flight line ridges, offsets, and misalignments can be quite severe and render data unusable.

In Figure 15.7, ΔZ orthos were created with a 2-meter cell size, which tells the software to compare points from overlapping swaths that are in the same 2-meter grid cell. Grid cells whose greatest differences in elevation were less than 8 cm were colored green and grid cells whose greatest differences in elevation were greater than 8 cm were colored red. The relative accuracy of this project was specified as 8 cm RMSDz. All elevation differences less than 8 cm were well within the RMSDz value and pass relative accuracy standards. Elevation differences greater than 8 cm are colored red because they warrant investigation to ensure they are not issues, especially contiguous red cells following flight line edges or the flight path. Figure 15.8 and Figure 15.9 found on page 462 illustrate how ΔZ orthos correlate to an actual lidar dataset. Both figures show the same location in a dataset but Figure 15.8 highlights an inter-swath relative accuracy issue while Figure 15.9 illustrates the dataset after corrections have been applied.

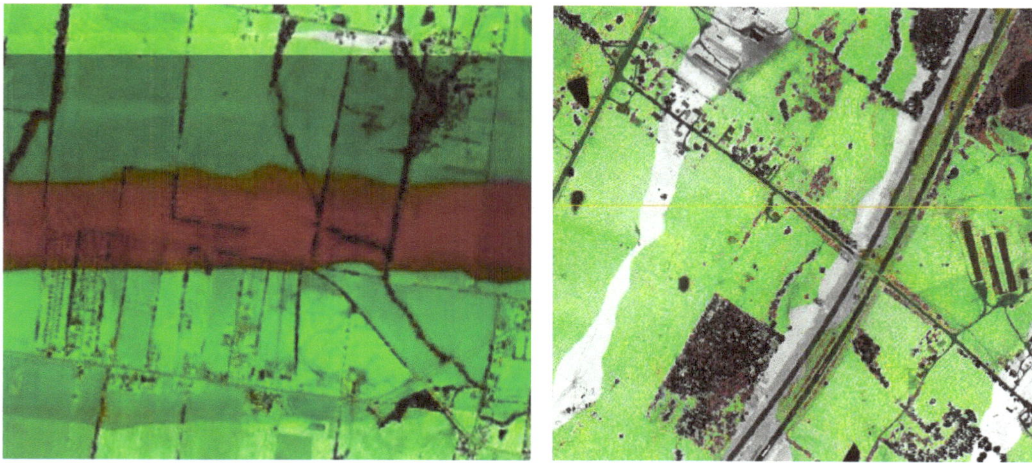

Figure 15.7. The left ΔZ ortho shows a swath exceeding relative accuracy standards, colored red. The right ΔZ ortho shows a portion of a project area with good relative accuracy. Some red pixels are visible along embankments, sloped terrain, and in vegetated land cover, as expected. Open, flat areas that are green indicate the calibration and relative accuracy of the data is acceptable whereas red in open flat areas indicate the relative accuracy of the data is not acceptable. Once the difference rasters are clipped to suitable testing areas (non-vegetated areas on terrain sloped less than 10 degrees), the RMSDz for the entire sample area can be calculated.

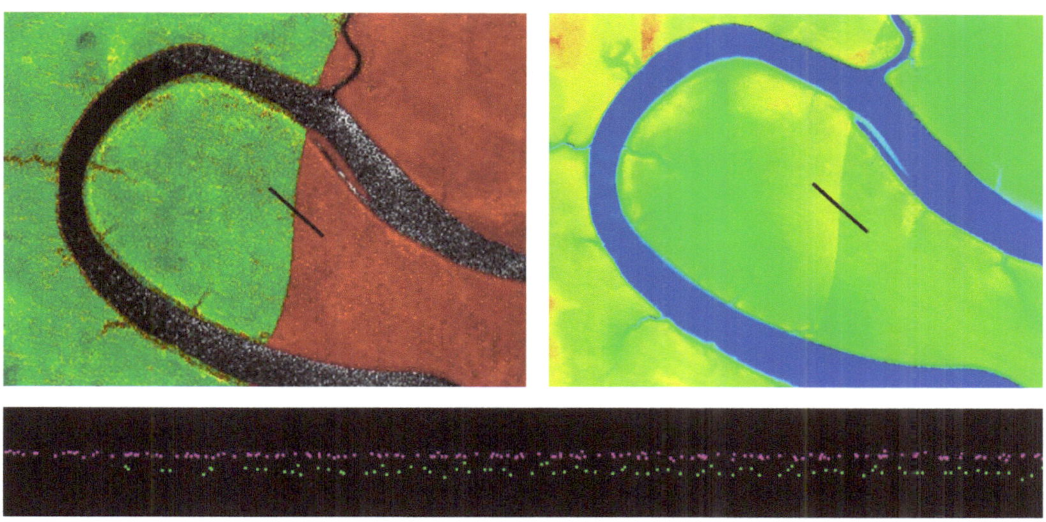

Figure 15.8. The top left image shows a ΔZ ortho created from single returns. The entire swath on the right side of the image is colored red, indicating the inter-swath relative accuracy is larger than allowed by project specifications. The top right image shows a bare earth surface model created from this data. There is a large flight line ridge present in the bare earth ground. The bottom image is a profile of the lidar point cloud at the location of the black line in the top images. The lidar points are colored by flight line. The two flight lines (purple and green) are separated by 15-20 cm when their elevations should be consistent.

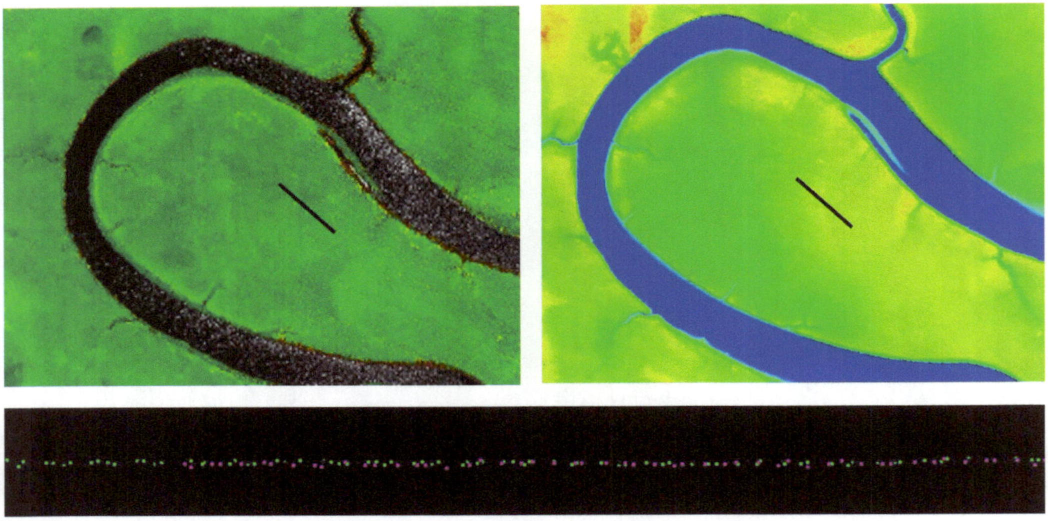

Figure 15.9. The top left image shows a ΔZ ortho created from single returns. The entire ΔZ ortho is now green, indicating the inter-swath relative accuracy has been corrected to be within project specifications. The top right image shows a bare earth surface model created from this data. The flight line ridge no longer exists; the bare earth surface is consistent. The bottom image is a profile of the lidar point cloud at the location of the black line in the top images. The lidar points are colored by flight line. The two flight lines (purple and green) now have consistent elevations and are no longer separated.

As mentioned in earlier paragraphs, roll, pitch, and yaw errors with the aircraft can contribute to inter-swath relative accuracy issues. Even datasets that are calibrated correctly may still have some of these issues present, particularly roll errors which are often visible at the edges of swath data. Roll errors may be present due to inherent or natural roll of the aircraft or from the lidar scanner mechanism, especially oscillating scanner mirrors where the geometric accuracy of lidar points at the highest scan angles (the edges of the swath) is not as accurate as the lidar points at lower scan angles (at or near nadir). As roll errors are often present in datasets, it is common industry practice to flag the points along the outer edges of swaths as withheld so that only the most geometrically usable portion of each swath (typically the center 95%) is used to produce the final bare earth surface or other derived data products. Figure 15.10 shows an example of how roll errors may look in ΔZ orthos and in the lidar point cloud.

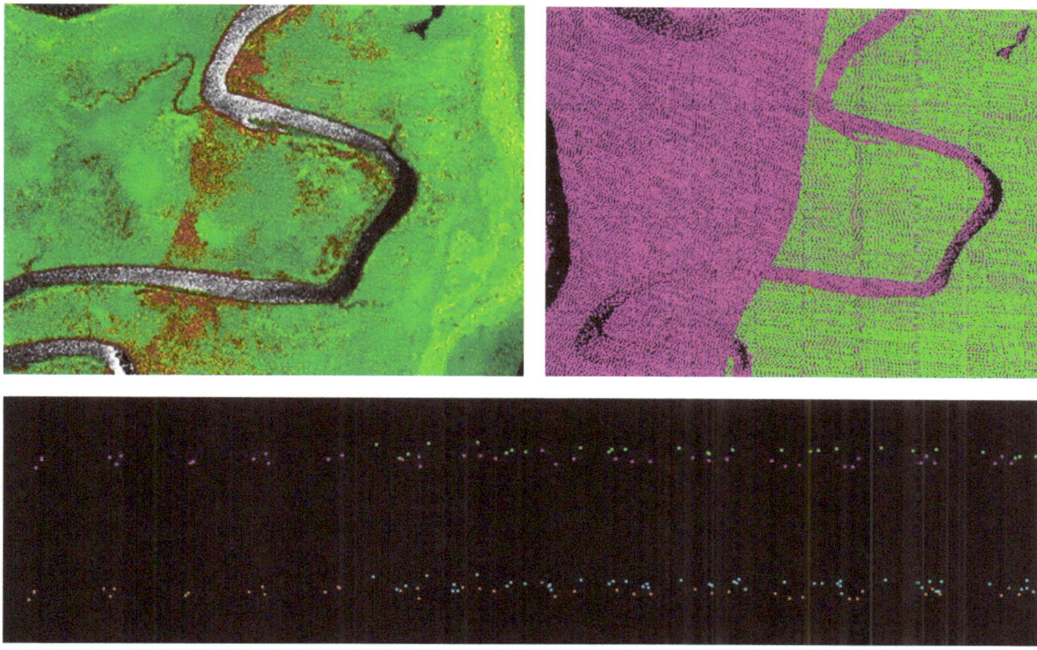

Figure 15.10. The top left image is a single return ΔZ ortho where a diagonal line of red is visible in the middle of the image. The rest of the image is green, indicating the rest of the data is within the project specifications for inter-swath relative accuracy. The top right image shows the lidar point cloud colored by flight line. A comparison of the two top images shows that the line of red visible in the ΔZ ortho occurs at the edge of the green flight line where roll errors may exist. The top profile also shows lidar points colored by flight line. In the profile, points from the green flight line are visibly separated from the purple flight line. These points are approximately ~1 cm above the purple flight line and as the user moves towards the center of the swath, both flight lines will eventually match in elevation. As the rest of the data is calibrated correctly, these points at the edges of flight lines will be flagged as withheld and not used in any of the data processing for bare earth ground or other products. The bottom profile shows the lidar points colored by class where orange points are the final ground and turquoise points are those flagged as withheld.

Absolute Vertical Accuracy Assessment

The absolute vertical accuracy of digital elevation data should be tested with independently surveyed and collected checkpoints that are at least three times as accurate as the dataset to be tested. In order to test vertical accuracy, check point coordinates must be in the spatial reference system, including the coordinate system, datum, and geoid model. Generally, checkpoints should be distributed throughout the project area. However, if certain areas are of higher interest or value, more checkpoints may be concentrated in these areas. NSSDA requires 20 checkpoints per dataset, but NDEP and FEMA Guidelines specify 20 checkpoints (30 preferable) per major land cover type in order to thematically identify areas of lower expected accuracy or reliability within the elevation dataset. Thus, a dataset covering five major land cover categories would require 100 checkpoints. Any land cover category comprising 10% or more of the project area should be considered a major land cover category. Digital elevation data acquired by radar or IFSAR sensors may be differentiated by slope categories rather than land cover categories. Additionally, FEMA requires project areas greater than 2,000 square miles to be divided in smaller blocks no larger than 2,000 square miles and tested as individual areas. ASPRS recommends 100 static vertical checkpoints for the initial 2,500 square kilometer (965 square mile) area within the project, as shown in Table 15.1 on page 464. This ratio recommended by ASPRS provides enough sample measurements to be statistically defensible when used to test vertical accuracy.

Project Area (km^2)	Horizontal Accuracy Testing of Orthoimagery and Planimetrics	Vertical and Horizontal Accuracy Testing of Elevation Data sets		
	Total Number of Static 2-D/3-D Check Points (clearly-defined points)	Number of Static 3-D Check Points in NVA	Number of Static 3-D Check Points in VVA	Total Number of Static 3-D Check Points
≤500	20	20	5	25
501-750	25	20	10	30
751-1,000	30	25	15	40
1,001-1,250	35	30	20	50
1,251-1,500	40	35	25	60
1,501-1,750	45	40	30	70
1,751-2,000	50	45	35	80
2,001-2,250	55	50	40	90
2,251-2,500	60	55	45	100

Table 15.1. Number of horizontal and vertical checkpoints recommended by ASPRS based on project area.

Checkpoints should be located on flat terrain or uniformly sloping terrain for x-units in all directions from each checkpoint, where "x" is the nominal spacing of the DEM or mass points evaluated. If this is not possible, checkpoints should be located on terrain that is less than 20 percent grade so that slope does not translate into vertical errors. Exceptions to this rule may occur, such as with some IFSAR elevation datasets where vertical accuracy is reported by slope categories. When multiple slope categories are being tested, the vertical accuracy specifications will be lower for steeper slopes where greater vertical terrain change over the same "x" horizontal distance is accounted for. Additionally, checkpoints should not be located near breaks in the terrain (such as bridge abutments, road edges, or ditches) where vertical errors may be a result of the interpolation within the elevation data that is caused by the severe terrain break. Checkpoints for vertical accuracy should not be located next to objects that may obstruct the sensor from detecting or recording the ground in that location, such as building overhangs or large tree trunks. Lastly, checkpoints for vertical accuracy should not be located near abrupt, substantial changes in infrared reflectivity.

Vertical accuracy testing does not require well-defined features and is required to be tested by NSSDA, NDEP, and ASPRS. Table 15.1 shows the recommended number of horizontal and vertical checkpoints based on the project area. For vertical testing of areas >2,500 km^2 ASPRS recommends adding 5 additional vertical check points (3 NVA check points and 2 VVA check points) for each additional 500 km^2. The terms NVA (Non-vegetated Vertical Accuracy) and VVA (Vegetated Vertical Accuracy) are described in the Absolute Vertical Accuracy Assessment section in the pages below.

All independent checkpoints should be accompanied by sketches and field photographs of each checkpoint. These field sketches and photographs are essential for horizontal accuracy testing to identify the feature that was surveyed in order to find the same feature in the digital dataset. However, these sketches and photographs are also very important in vertical accuracy testing to provide context for each survey point. For example, field photographs may show that a survey checkpoint labeled as open terrain tested with higher than expected errors because the survey checkpoint is actually located within tall grasses. Survey checkpoints should always be reviewed as issues with the survey checkpoints will cause issues with the accuracy testing. Accuracy testing results should always be reviewed for errors that can represent errors in the dataset or errors with the survey. Reviewing field sketches and notes can identify some potential survey errors prior to testing the elevation dataset. The example in Figure 15.11 shows a field sketch identifying a forest checkpoint that should be located west of the main road. The intensity imagery shows this location as the yellow dot. The orange triangle shows the coordinates of the delivered checkpoint. The coordinates of this checkpoint were processed incorrectly by the surveyor, causing the forest checkpoint to be located east of the main road in open terrain. This error in horizontal location caused errors in the vertical accuracy testing.

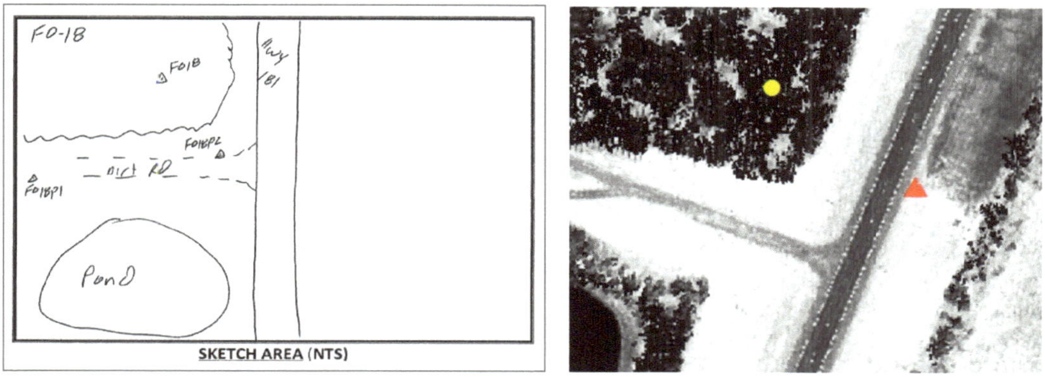

Figure 15.11. Field sketch on the left shows the survey checkpoint should be located in the forest west of the main road (yellow dot on intensity imagery on the right). Incorrect coordinates delivered by surveyor placed this checkpoint east of the main road in open terrain (orange triangle), where the GPS receiver was located, prior to a Total Station offset measurement to a prism reflector in the forest. This survey error resulted in apparent large vertical accuracy errors.

Interpolation of Digital Elevation Datasets

Because elevation data are not horizontally well-defined, vertical accuracy testing does not require checkpoints to be photo-identifiable. Instead, the xy coordinates from the survey are overlaid on the elevation dataset and the elevations at the same checkpoint xy locations are extracted from the elevation dataset and recorded. The extracted z-values are then compared to the survey checkpoint z-values. However, it is unlikely that a survey checkpoint will be located at the precise location of a discrete source masspoint or elevation return. Elevation datasets use interpolation to create continuous surfaces (see Chapter 1) so the extracted z-value at the same location of the xy survey coordinates is an interpolated value. If TIN based surfaces are used, then the irregularly spaced source masspoints become nodes within the TIN structure, and the surface is created by interpolating between these nodes. TINs use contiguous triangle faces and have an underlying topological data structure. DEMs, however, are evenly spaced cells where one elevation value is defined per cell. Vertical accuracy results may slightly differ depending on which type of surface is used for testing since each type of surface uses different creation methods and can use different interpolation methods as well. Several source masspoints may be averaged together to produce one elevation value for each cell within a DEM while TINs would produce an elevation value by triangulating between three source masspoints. While differences between TINs and DEMs should be minimal if rigorous processing methods are used, differences in vertical accuracy results may exist between the two surfaces. Even though differences between the two surface types should be minimal, larger differences may result if best practices are not followed during processing, such as using a coarse cell size in the raster when the elevation data supports a higher resolution raster. The surface type most applicable to each dataset should be used for vertical accuracy testing. Multiple surfaces may be used to test the vertical accuracy within a single project. For example, TINs may be used to test the vertical accuracy of lidar swaths or the classified lidar point cloud while the final vertical accuracy is tested on the derived bare-earth DEMs. The differences between calculated accuracies from TINs and DEMs will be more severe with larger DEM cell sizes where there are more possibilities for one cell to encompass several points at different heights (which will be averaged) due to terrain changes and slope over the horizontal distance of the pixel. Figures 15.12 and 15.13 on page 466 illustrate how the elevations of DEM pixels may be interpolated from several source masspoints, resulting in different values for the DEM surfaces compared to TIN surfaces.

Figure 15.12. The image on the left shows a terrain (TIN based surface) with nodes or masspoints represented as pink dots. The image on the right shows this same terrain with a 1-meter grid overlaid. When producing bare-earth DEMs with a 1-meter cell size from this masspoint data, several source masspoints will be averaged together to interpolate one elevation value for each cell.

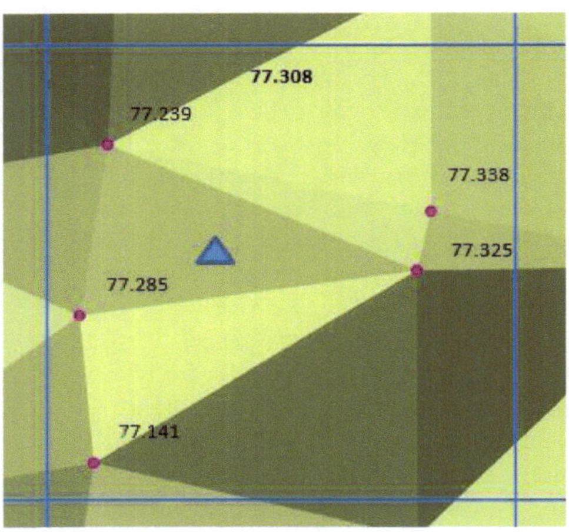

Figure 15.13. Close-up view of one cell (1-meter) overlaid with source masspoints. The TIN triangulates between the source masspoints while the DEM surface will average these five masspoints within this one cell to interpolate one elevation value for the cell. These two surface creation methods can produce slightly different vertical accuracy results. The elevation value of each masspoint is shown to the right of the node (pink dots) while the elevation of the DEM cell is shown at the top of the cell (blue box). A survey checkpoint (blue triangle) located in this cell will have an extracted DEM elevation of 77.308 meters while the same checkpoint interpolated from the TIN will have a TIN elevation of 77.297 meters.

Digital Elevation Datasets Benefiting from Vertical Accuracy Assessment

In addition to testing the final DEMs, source lidar swath data should be tested and verified (using TINs/terrains) prior to any point classification, manual editing, or processing beyond calibration procedures. Testing lidar swath data ensures the elevation data meets project specifications for vertical accuracy prior to processing. This initial testing helps to identify issues at the beginning of a project, promotes or encourages "correct" methods of fixing the entire dataset rather than quick patches, and can help lessen rework, the impact to budgets, and extended time schedules. In order to test lidar swath data, only checkpoints in non-vegetated terrain can be used as the checkpoints will be compared against the full point cloud prior to vegetation being filtered from the ground. Some open terrain checkpoints may even need to be removed from this testing if they coincide with overhanging features, such as power lines, that will distort vertical accuracy results prior to filtering the ground. Once lidar swath data has been verified to meet vertical accuracy specifications, filtering the ground and manual classification can be completed.

If the fully classified lidar point cloud is a deliverable, vertical accuracy testing may also be completed on this product (using TINs/terrains) in addition to the final DEMs and the calibrated swath data. Testing the classified point cloud is beneficial as this is the source data that is used to create all derivatives, including the final DEMs, and derived data will not be more accurate than the source. While vertical accuracy is completed on the final DEMs, future users of the data may need or want to create additional derived products, such as first return DSMs, contours, and height-above-ground maps (see Chapter 1). These additional products should be created from the source lidar point cloud, so knowing the vertical accuracy of the point cloud and not just the final DEM surface will allow users to determine if the data will meet their needs. With the exception of the surface used in vertical accuracy testing (TIN vs DEM), the methodology and reporting is consistent across lidar swath data or lidar point cloud data and DEM datasets. The methodology and reporting described in the following sections can be used for all digital elevation data.

Vertical Accuracy Assessment Methodology

NSSDA standards specify that vertical accuracy of digital elevation data should be tested using RMSE. When errors follow a normal distribution, vertical accuracy at the 95% confidence level ($Accuracy_z$) can be computed for a dataset by multiplying $RMSE_z$ x 1.9600. NSSDA states:

Let:

$$RMSE_z=sqrt[\sum(Z_{data\ I}-Z_{check\ I})^2/n]$$

Where:

$Z_{data\ I}$ is the vertical coordinate of the Ith check point in the dataset

$Z_{check\ I}$ is the vertical coordinate of the Ith check point in the independent source of higher accuracy

n = the number of points being checked

I is an integer from 1 to n

It is assumed that systematic errors have been eliminated to the degree possible. If vertical error is normally distributed, the factor 1.9600 is applied to estimate linear error at the 95% confidence level. As stated in the ASPRS Positional Accuracy Standards for Digital Geospatial Data (2014), "based on empirical results, if the mean error is small, the sample size sufficiently large and the data is normally distributed, 1.9600 * $RMSE_z$ is often used as a simplified approximation to compute the $Accuracy_z$ at a 95% confidence level." Therefore, vertical accuracy, $Accuracy_z$, reported according to the NSSDA shall be computed by the following formula:

$$Accuracy_z = 1.9600 * RMSE_z$$

NDEP recognized that land cover types may affect elevation errors, especially in vegetated terrain where sensors may not be able to penetrate dense vegetation in order to detect the bare-earth ground. NDEP introduced three different vertical accuracy values to be calculated: Fundamental Vertical Accuracy (FVA), Consolidated Vertical Accuracy (CVA), and Supplemental Vertical Accuracy (SVA). Fundamental Vertical Accuracy is tested only with checkpoints in open terrain where there is a high probability that the sensor reached the ground and that the errors will likely follow a normal distribution. Fundamental Vertical Accuracy is tested using the NSSDA method, where vertical accuracy at the 95% confidence level, $Accuracy_z$, is computed as $RMSE_z$ x 1.9600. Supplemental Vertical Accuracy is calculated for each individual land cover category using the 95th percentile. Because these errors, especially in vegetated terrain, may not follow a normal error distribution, SVA is tested as the 95th percentile using the absolute value of all vertical errors in each land cover category. The 95th percentile may be computed with a spreadsheet command and indicates that 95 percent of the errors in the dataset, regardless of distribution or outliers, will have absolute values equal to or smaller than the calculated value. Consolidated Vertical Accuracy is computed with the 95th

percentile method using the absolute value of all vertical errors from all land cover categories combined, included open terrain and urban land cover categories. NDEP allows individual land cover categories to fail target SVA values as long as the overall tested CVA meets specifications. These NDEP categories are now obsolete.

The ASPRS Positional Accuracy Standards for Digital Geospatial Data (2014) are similar to NDEP and NSSDA methods but replace FVA, CVA, and SVA with two categories: Non-vegetated and Vegetated. NVA measures vertical accuracy at the 95% confidence level for non-vegetated terrain by computing $RMSE_z$ x 1.9600. NVA should be computed using survey check points located in open terrain and urban land cover categories where elevation errors typically follow a normal error distribution. VVA measures vertical accuracy at the 95% confidence level for vegetated terrain by computing the 95th percentile. VVA should be computed using survey check points located in all vegetated land cover categories combined, including tall weeds and crops, brush lands, and fully forested, where elevation errors typically do not follow a normal error distribution. The ASPRS Positional Accuracy Standards for Digital Geospatial Data apply accuracy thresholds to data and specify that VVA is allowed to be 3.0 times larger than the $RMSE_z$. Areas that do not meet VVA must be defined by low confidence polygons.

Random Errors, Systematic Errors, and Blunders

Errors computed during vertical accuracy statistics may be random, systematic, or blunders. $RMSE_z$ statistics are used to test vertical accuracy in non-vegetated terrain only because it assumes errors will follow a normal error distribution. These types of errors are random and are produced by irregular causes that cannot be corrected by use of a standardized adjustment. If the number of QA/QC checkpoints is large enough, errors in open terrain or non-vegetated land cover categories will approximate a normal error distribution, as shown in Figure 15.14, which was computed from over 100 checkpoints.

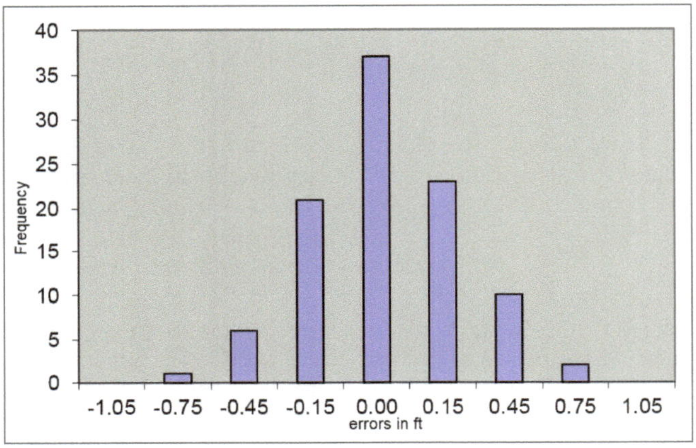

Figure 15.14. Errors in non-vegetated land cover categories do follow a normal error distribution, if a large enough checkpoint sample is used.

If a small number of checkpoints are used to test a dataset, then errors in non-vegetated land cover categories may not follow a normal distribution according to the traditional bell curve, as shown in Figure 15.15. This distribution was computed from only 25 checkpoints. While not following the traditional bell curve as shown in Figure 15.14, this histogram is still fairly close to a normal error distribution. The distribution curve has been exaggerated by only a few checkpoints.

For comparison, Figure 15.16 shows the histogram for checkpoints located in the brush lands and low trees land cover category. These errors do not follow a normal error distribution as more checkpoints have "positive" errors. The positive errors may be a result of the filtering algorithm which did not fully filter all vegetation from the bare earth ground resulting in ground elevations in the dataset that are higher than the survey QA/QC checkpoint elevations. Or, the lidar pulses may never have made it to the actual ground and the lowest points in these areas of the data set are actually representing low vegetation. Both scenarios provide explanations why vegetated land cover categories do not follow a normal error distribution; these categories should be tested using the 95th percentile method.

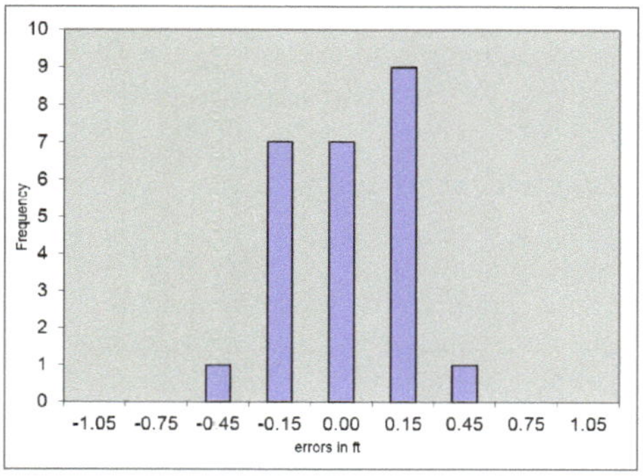

Figure 15.15. Errors in non-vegetated land cover categories may not follow a normal error distribution or traditional bell curve if a small checkpoint sample is used for vertical accuracy testing.

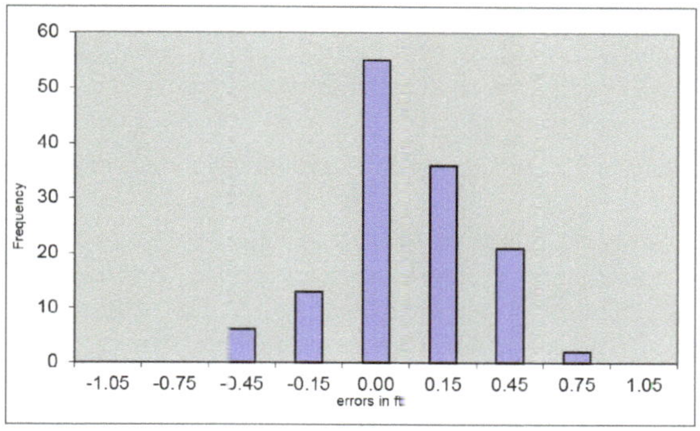

Figure 15.16. Histogram of checkpoints located in brush and low trees shows that vegetated land cover categories do not follow a normal error distribution and should be tested with the 95th percentile method rather than $RMSE_z$ x 1.9600.

Additional statistics, including standard deviation, mean, absolute mean, median, minimum and maximum differences, kurtosis, and skew are very helpful when evaluating the vertical accuracy results and characteristics of a dataset. These statistics are defined below:

- Minimum: The smallest number in a set of (ΔZ) values, i.e., the most negative ΔZ value
- Maximum: The largest number in a set of (ΔZ) values, i.e., the largest positive ΔZ value
- Mean: The result obtained by adding several values together and then dividing by the total number of values to produce a calculated central value. Also known as the average; this function measures central tendency. This calculation is performed with both positive and negative DeltaZ values to evaluate how the central tendency of the tested dataset compares to a normally distributed dataset with a mean of zero
- Mean Absolute: This function is performed the same as Mean except it is calculated from the Absolute DeltaZ values only. Absolute values of numbers are the number without a positive or negative sign. This calculation evaluates the average vertical difference between each checkpoint and corresponding DEM elevation. Using both positive and negative values for this function would "cancel" each other out
- Median: The middle number in a set of values organized from lowest to highest. If there is an even number of values in the set, the median is the calculated average of the two middle numbers
- Skew: Skewness is a measure of symmetry or asymmetry within a dataset. Symmetric data will have a skewness towards zero. Negative values indicate data are skewed left or that the left tail is longer or extended compared to the right tail. Positive values indicate data are skewed right or that the right tail is longer or extended compared to the left tail
- Kurtosis: The measure of relative peakedness or flatness of a distribution compared with a normally distributed dataset. Positive kurtosis indicates a relatively peaked distribution near the mean while negative kurtosis indicates a flat distribution near the mean
- Standard Deviation: Standard deviation measures how much variation or dispersion exists relative to the mean. Standard deviation is used to measure the precision of a dataset. Low standard deviation results indicate the values of the tested dataset tend to be very close to the mean. High standard deviation results indicate values of the tested dataset are spread out over a large range of values

- $RMSE_z$: Root mean square error estimates positional accuracy and is the square root of the average of the set of squared differences between dataset coordinate values and coordinate values from an independent source of higher accuracy for identical points. $RMSE_z$ measures the root mean square error in elevation or the z-direction
- 95th Percentile: The absolute value in which 95% of all measurements fall at or below.

Vertical accuracy is tested using RMSE and the 95th percentile. These additional statistics should also be calculated for each dataset to further describe the nature of the errors. Standard deviation can be used to test a dataset's precision, but not its accuracy. Standard deviation will show how much variation or dispersion exists relative to the mean. Low standard deviation indicates the data points tend to be very close to the mean while high standard deviation indicates the data points are spread out over a large range of values. Generally, good datasets will have standard deviation values and RMSE values that are very close to each other as this indicates high precision and high accuracy, respectively. Standard deviation, however, will not identify systematic errors. Systematic errors follow a regular pattern and can be introduced into a dataset by sensor errors, data collection procedures, or data processing methods. Common sources of systematic errors include the elevation provider using different GPS base stations for airborne GPS control than the GPS base stations used by the independent surveyor for QA/QC checkpoints and the elevation provider not using rigorous procedures for processing airborne GPS data. Figure 15.17 and Figure 15.18 show two histograms; one histogram is an unbiased dataset while the second is a dataset with a 1-foot vertical bias.

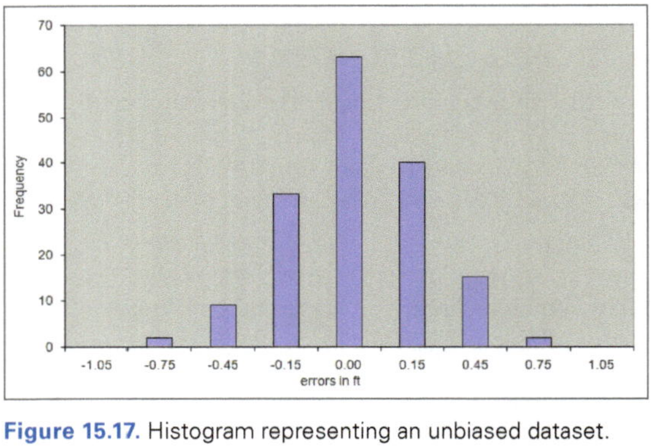

Figure 15.17. Histogram representing an unbiased dataset.

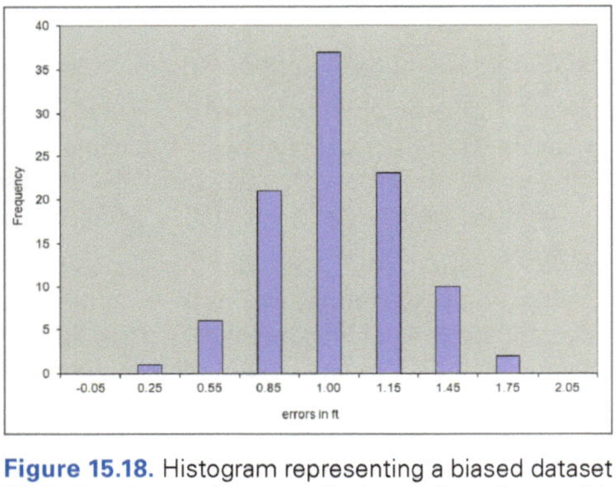

Figure 15.18. Histogram representing a biased dataset (Note the changes in errors displayed along the x-axis).

In addition to random and systematic errors, vertical accuracy results should be reviewed for blunders. Blunders are generally single, large errors that are greater than three times the standard deviation of error. Blunders should always be analyzed to determine if they represent an error within the dataset or within the survey checkpoints. An example of a blunder is shown in Figure 15.19 where incorrect processing by the land surveyor resulted in incorrect horizontal coordinates. The error in horizontal position translated into vertical error. While this blunder could be identified prior to vertical accuracy testing, some blunders may not be apparent until the testing results are reviewed. Table 15.2 shows one forest point that tested with a very high DeltaZ value, compared to other forest checkpoints within the same dataset, which may indicate a potential blunder. Figure 15.19 visually represents the vertical error of this checkpoint. All possible blunders should be reviewed as they may represent an error in the elevation dataset, may represent an error in the survey checkpoints, or may represent a temporal change in ground conditions between aerial and ground surveys. After review, this error was determined to be a result of incorrect surveying methods and was removed from the final vertical accuracy testing. Regardless of the source of the error, all blunders should be excluded from accuracy testing, but included in the project deliverables with explanations of the cause of the error and why it was not used.

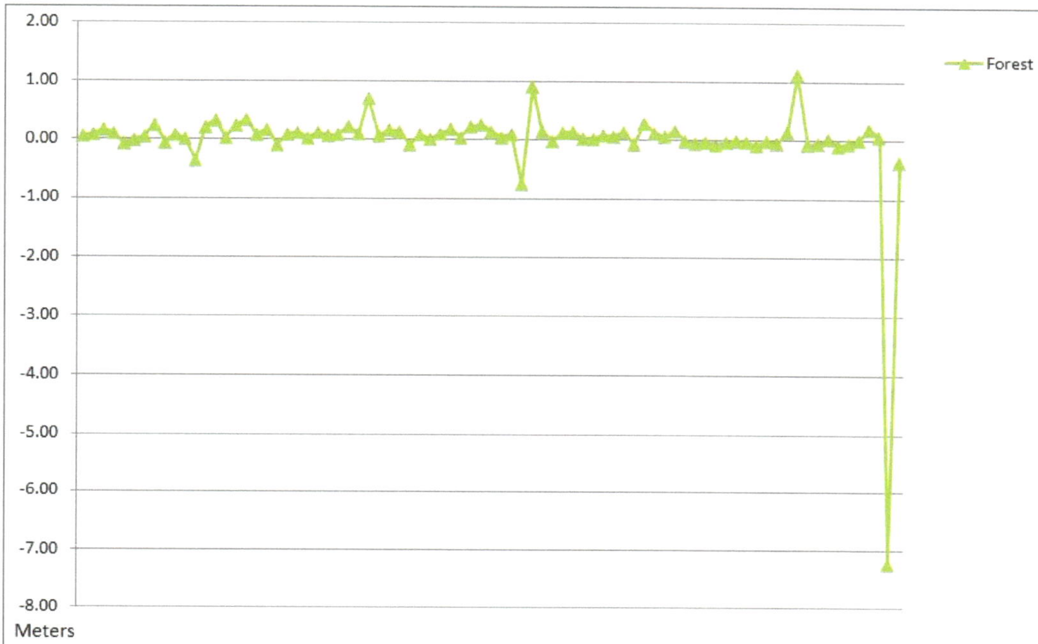

Figure 15.19. This line graph visually represents the errors of all forest checkpoints. One checkpoint has an error much higher than all other checkpoints and represents a potential blunder.

Point ID	NAD83 UTM Zone 15		NAVD88		DeltaZ	Absolute DeltaZ	Land Cover Category
	Easting X (m)	Northing Y (m)	Survey Elevation (m)	Lidar Elevation (m)			
FO-5	365,264.314	3,971,479.695	158.371	151.132	-7.24	7.24	Forest
FO-7	529,300.349	4,716,905.688	634.455	634.670	0.215	0.215	Forest
FO-12	553,516.249	4,729,764.322	255.674	255.650	-0.024	0.024	Forest

Table 15.2. Forest point FO-5 tested with a very high DeltaZ value, indicating a potential blunder, compared to checkpoints FO-7 and FO-12.

Vertical Accuracy Assessment Reporting

Vertical accuracy should be reported according to ASPRS standards. This requires vertical accuracy to be reported at the 95% confidence level and for the method and land cover types to be included in the vertical accuracy statements. Projects using the most recent ASPRS Positional Accuracy Standards for Digital Geospatial Data will report NVA and VVA.

For tested Non-vegetated Vertical Accuracy (NVA), the following statement should be used:

- This data set was tested to meet ASPRS Positional Accuracy Standards for Digital Geospatial Data (2014) for a ___ (cm) RMSEz /Vertical Accuracy Class. Actual NVA accuracy was found to be RMSEz = ___ cm, equating to +/- ___ at 95% confidence level.

For tested VVA, the following statement should be used:

- This data set was tested to meet ASPRS Positional Accuracy Standards for Digital Geospatial Data (2014) for a ___ (cm) RMSEz /Vertical Accuracy Class. Actual VVA accuracy was found to be +/- ___cm at the 95th percentile.

Because the Vegetated Vertical Accuracy is computed with the 95th percentile method, this value should be accompanied by a list of the 5% outliers. Report x/y coordinates and z-error for each QC check point error larger than the 95th percentile when only a few checkpoints are 5% outliers or it is manageable to report each 5% outlier. If there are numerous checkpoints with errors above the 95th percentile or it is unmanageable to report each individual 5% outlier, report only the quantity and range of values.

As IFSAR data are generally reported by slope rather than land cover category, the following statement should be used for IFSAR data:

- Tested __ (meters, feet) at 95 percent confidence level in (specify slope category) using RMSEz x 1.96

Lidar and IFSAR data are generally always tested for vertical accuracy and will be reported with the "tested" statements. However, because photogrammetry is a mature technology with accuracies known from the aerotriangulation solutions, the vertical accuracy of digital elevation data (DEMs, breaklines or contours) from stereo photogrammetry may be documented in the metadata as follows:

- This data set was produced to meet ASPRS Positional Accuracy Standards for Digital Geospatial Data (2014) for a ___ cm RMSEz Vertical Accuracy Class equating to NVA = +/- ___ cm at 95% confidence level and VVA = +/- ___ cm at the 95th percentile.

Breaklines may be used to produce required deliverables, such as hydro-flattened or hydro-enforced DEMs. While there is no current standard for testing or reporting breakline accuracies, the method in which breaklines were derived may impact the accuracy of the breaklines as well as the elevation dataset that the breaklines will be used with. Additionally, the breaklines cannot be more accurate than the source from which they were collected. Therefore, the accuracy of the source along with the method of breakline collection should be documented for all projects using breaklines with the following statement or other similar statement:

Breaklines compiled using (lidargrammetry, terrain drape, 2-D digitization, or other methods). Breaklines compiled from source that was tested to meet ASPRS Positional Accuracy Standards for Digital Geospatial Data (2014) for a ___ (cm) RMSEz /Vertical Accuracy Class. Actual NVA accuracy was found to be RMSEz = ___ cm, equating to +/- ___at 95% confidence level. Actual VVA accuracy was found to be +/- ___ cm at the 95th percentile. The breakline collection method specified should reflect the method or methods used for the project. Common examples of breakline collection methods include, but are not limited to:

- Lidargrammetry
- Terrain drape
- Terrain drape with downhill constraints and flattening constraints
- 2-D digitization

Vertical Accuracy Assessment Sample

Table 15.3 shows an example of how a spreadsheet may be organized and set up in order to perform vertical accuracy assessments. This table was extracted from a larger QA/QC checkpoint sample used to test the vertical accuracy of a lidar dataset by the author's firm. Table 15.4 on page 474 lists the project specifications used during this vertical accuracy assessment. Project specifications should always be clearly listed in final project reports and project metadata. Table 15.5 on page 474 shows the statistics, including RMSE, mean, median, skew, standard deviation, minimum, and maximum that were calculated with the full checkpoint dataset. These three tables, which evaluate and analyze the elevation dataset in different ways, are included here to show the transition from the old way of documenting vertical accuracy statistics for five different land cover categories to the new way of documenting vertical accuracy statistics for only two different land cover categories, i.e., NVA and VVA. Table 15.6 on page 474 compares the final vertical accuracy of the dataset according to NSSDA, NDEP Guidelines, and ASPRS Positional Accuracy Standards for Digital Geospatial Data (2014); this table documents that the NVA of 12 cm greatly exceeded the NVA standard of 19.6 cm for QL2 lidar, and the VVA of 16.7 cm greatly exceeded the VVA standard of 29.4 cm for QL2 lidar.

Point ID	NAD83 Epoch 2010, UTM Zone 15		NAVD88 (Geoid09)		DeltaZ	Absolute DeltaZ	Land Cover Category
	Easting X (m)	Northing Y (m)	Survey Elevation (m)	Lidar Elevation (m)			
OT-402	617,333.684	3,834,231.496	62.965	62.978	0.013	0.013	Open Terrain
OT-403	617,388.762	3,837,427.571	65.299	65.262	-0.037	0.037	Open Terrain
OT-404	609,083.182	3,840,597.054	69.820	69.751	-0.069	0.069	Open Terrain
OT-405	614,173.217	3,845,589.861	68.363	68.342	-0.021	0.021	Open Terrain
OT-406	611,614.800	3,829,519.971	62.147	62.169	0.022	0.022	Open Terrain
OT-407	606,958.200	3,829,517.639	64.754	64.784	0.030	0.030	Open Terrain
OT-408	596,667.440	3,839,976.760	69.650	69.628	-0.022	0.022	Open Terrain
OT-409	598,758.374	3,842,584.564	69.131	69.094	-0.037	0.037	Open Terrain
OT-410	590,922.365	3,845,070.132	72.018	71.972	-0.046	0.046	Open Terrain
F-705	621,678.981	3,829,286.507	59.922	60.105	0.183	0.183	Forest
F-713	638,884.139	3,816,933.829	64.623	64.717	0.094	0.094	Forest
F-714	638,883.609	3,810,096.809	63.375	63.280	-0.095	0.095	Forest
WGC-505	619,995.876	3,826,758.634	59.258	59.744	0.486	0.486	Weeds, Grass, Crops
WGC-514	622,908.626	3,810,554.080	59.238	59.216	-0.022	0.022	Weeds, Grass, Crops
WGC-515	628,040.334	3,819,493.273	61.031	61.120	0.099	0.099	Weeds, Grass, Crops
WGC-516	592,867.201	3,858,610.285	79.316	79.270	-0.046	0.046	Weeds, Grass, Crops
UR-318	585,325.800	3,874,521.760	89.590	89.466	-0.124	0.124	Urban
UR-319	578,952.920	3,873,709.960	180.990	180.841	-0.149	0.149	Urban
BLT-611	629,043.537	3,822,678.364	58.650	58.818	0.168	0.168	Brush and Low Trees
BLT-612	630,721.801	3,813,423.251	63.786	63.814	0.028	0.028	Brush and Low Trees
BLT-613	628,377.261	3,814,526.458	60.905	60.886	-0.019	0.019	Brush and Low Trees

Table 15.3. Subset of a vertical accuracy spreadsheet used to test lidar data. The full set includes 100 checkpoints, 20 each in the five land cover categories listed.

The first column represents the check point name or ID. The next three columns should contain the x and y coordinates and z-elevation of each check point, as defined by the survey or independent source of higher accuracy. The fifth column should contain the z elevations extracted from the dataset to be tested at the same xy locations as the independent check points. The sixth column is the difference in elevation values between the independent source of higher accuracy and dataset to be tested, also called ΔZ. These are always computed by subtracting the independent source of higher accuracy from the dataset to be tested (i.e., tested elevation minus higher accuracy elevation). The seventh column represents the absolute value of the ΔZ for each check point. The eighth column represents land cover type as testing methodologies will vary depending on this attribute. If this dataset was elevation data other than lidar, the eighth column may change. For example, the eighth column

Quantitative Criteria	Measure of Acceptability
Non-vegetated Vertical Accuracy (NVA) at the 95% confidence level in open terrain and urban land cover categories	19.6 cm (based on RMSEz (10 cm) *1.9600)
Vegetated Vertical Accuracy (VVA) at the 95% confidence level in all vegetated land cover categories, including tall weeds and crops, brush lands and small trees, and forested and fully grown trees	29.4 cm (based on 95th percentile for all vegetated land cover categories combined)

Table 15.4. Quantitative Criteria outlining project specifications for QL2 lidar.

Land Cover Category	# of Points	Min (m)	Max (m)	Mean (m)	Mean Absolute (m)	Median (m)	Skew	Kurtosis	Std Dev (m)	RMSEz (m)
Bare Earth-Open Terrain	20	-0.10	0.08	-0.02	0.04	0.00	-0.19	-0.64	0.05	0.05
Urban	20	-0.15	0.11	0.01	0.06	0.02	-0.84	0.22	0.07	0.07
Tall Weeds and Crops	20	-0.13	0.49	0.02	0.08	-0.01	2.68	9.43	0.13	0.13
Brush Lands and Small Trees	20	-0.10	0.17	0.04	0.06	0.04	-0.18	-0.31	0.07	0.08
Forested and Fully Grown	20	-0.13	0.70	0.03	0.10	0.00	3.08	11.46	0.18	0.17
Consolidated	100	-0.15	0.70	0.02	0.07	0.01	3.18	17.12	0.11	0.11

Table 15.5. Vertical accuracy statistics calculated for a lidar dataset.

Land Cover Category	NSSDA's Accuracyz at 95% confidence level based on RMSEz x 1.9600	NDEP's FVA (RMSEz x 1.9600), SVA (95th Percentile), and CVA (95th Percentile)	NDEP Accuracy Term	ASPRS NVA (RMSEz x 1.9600) and VVA (95th Percentile)	DeltaZ ASPRS Accuracy Term
Bare Earth-Open Terrain	0.10 m	0.10 m	FVA	0.12 m	NVA
Urban	0.14 m	0.13 m	SVA	0.167 m	VVA
Tall Weeds and Crops	0.25 m	0.15 m	SVA	0.167 m	VVA
Brush Lands and Small Trees	0.16 m	0.14 m	SVA	0.167 m	VVA
Forested and Fully Grown	0.33 m	0.21 m	SVA	0.167 m	VVA
Consolidated	0.22 m	0.13 m	CVA	N/A	N/A

Table 15.6. Vertical accuracy calculated at the 95% confidence level according to NSSDA, NDEP Guidelines, and ASPRS Positional Accuracy Standards for Digital Geospatial Data. The different accuracy terms are also noted.

would list slope categories rather than land cover categories if IFSAR data was being tested for vertical accuracy. The formula to calculate RMSE by taking the square root of the summed squared differences divided by the number of check points is loaded into the spreadsheet and calculates RMSEz which is multiplied by 1.9600 to calculate NVA according to ASPRS Positional Accuracy Standards for Digital Geospatial Data. The 95th percentile is used to calculate VVA according to ASPRS Positional Accuracy Standards for Digital Geospatial Data. Additional statistics, such as mean, median, skew, kurtosis, standard deviation, minimum, and maximum, should be calculated in order to provide additional information and illustrate characteristics of the dataset.

Table 15.7 shows vertical accuracy and associated statistics that were calculated from swath lidar data only. As only checkpoints in non-vegetated terrain are used for this testing, RMSEz x 1.9600 is used to determine vertical accuracy at the 95% confidence level. Additionally, fewer land cover categories result in less statistics being reported for swath lidar data.

100 % of Totals	RMSEz (ft) Open Terrain Spec = 0.31ft	Fundamental Vertical Accuracy (FVA) (RMSEz x 1.9600) Spec = 0.60ft	Mean (ft)	Median (ft)	Skew	Std Dev (ft)	# of Points	Min (ft)	Max (ft)
Open Terrain	0.09	0.18	0.02	0.02	-0.13	0.09	85	-0.27	0.25

Table 15.7. Vertical accuracy statistics calculated for swath lidar data.

The 5% outliers that are larger than the 95th percentile value for VVA should be reported. If only a few outliers exist, then the individual points along with coordinates and Δz values should be reported. If a large number of outliers exist, it is acceptable to only list the number of outliers and range of Δz values for those outliers. While not required, it is recommended to list the quantity and error range of 5% outliers in the vertical accuracy section of FGDC compliant metadata and to include a full table or listing of all individual outliers in the accuracy assessment or project report. From the total 100 checkpoints, Table 15.8 lists the 5% outliers from the tested lidar dataset shown in Tables 15.5 and 15.6.

Point ID	NAD83 Epoch 2010, UTM Zone 15		NAVD88 (Geoid09)	Lidar Z (m)	DeltaZ	Absolute DeltaZ	Land Cover Category
	Easting X (m)	Northing Y (m)	Survey Z (m)				
BLT-619	578,952.920	3,873,709.960	180.990	181.200	0.210	0.210	Brush and Low Trees
WGC-505	619,995.876	3,826,758.634	59.258	59.744	0.486	0.486	Tall Weeds and Crops
BLT-611	629,043.537	3,822,678.364	58.650	58.818	0.168	0.168	Brush and Low Trees
F-720	638,854.146	3,829,933.259	62.332	63.032	0.700	0.700	Forest
F-705	621,678.981	3,829,286.507	59.922	60.105	0.183	0.183	Forest

Table 15.8. 5% outliers greater than the 95th percentile calculated for VVA.

Table 15.9 illustrates project specifications for IFSAR data where vertical accuracy assessment is categorized by the slope of terrain at the location of the checkpoint rather than the land cover category at the location of the checkpoint. Table 15.10 on page 476 lists the vertical accuracy statistics and results for an IFSAR dataset tested by the author's firm. RMSEz is used for all slope categories, but steeper slopes are allowed more vertical error due to real vertical change across xy horizontal distances and interpolation that will impact vertical accuracy.

Slope Category	RMSEz Specifications (m)	$Accuracy_z$ Specifications (m)
0-10	1.85	3.63
10-20	3.71	7.27
20-30	5.56	10.90
>30	7.41	14.52

Table 15.9. Quantitative Criteria outlining project specifications for an IFSAR project.

100 % of Totals	# of Points	RMSEz	Accuracy$_z$ (RMSEz x 1.9600)	Mean (m)	Median (m)	Skew	Std Dev (m)	Min (m)	Max (m)
Slope 0-10	382	1.69	3.31	1.39	-1.18	0.55	1.37	-4.28	5.00
Slope 10-20	185	2.37	4.65	1.73	-0.26	-0.43	2.18	-6.91	4.41
Slope 20-30	6	3.41	6.69	2.53	0.19	0.26	3.71	-5.32	5.31
Slope >30	31	4.17	8.17	2.99	-1.82	-1.21	3.34	-13.16	1.76

Table 15.10. Vertical accuracy calculated at the 95% confidence level according NSSDA and associated vertical accuracy statistics for an IFSAR project.

Absolute Horizontal Accuracy Assessment

Testing horizontal accuracy requires well defined checkpoints that are identifiable or recoverable in the dataset to be tested. Since elevation datasets may not contain well-defined features, horizontal accuracy is not required to be tested by NSSDA, NDEP, or ASPRS. However, ASPRS does recommend that for lidar datasets at least half of the NVA check points should be located at the ends of paint stripes, 90-degree corners of different reflectivity, or other point features visible on the lidar intensity image, allowing them to double as horizontal check points. See Table 15.1 on page 464 for the recommended number of checkpoints based on the project area.

Horizontal accuracy testing is performed by using field photographs and sketches of each well defined, surveyed feature and identifying the same feature in the digital dataset. The coordinates of the feature are then extracted from the digital dataset, recorded, and compared to the survey coordinates for that feature. While elevation datasets may not always contain well-defined features suitable for horizontal accuracy testing, horizontal accuracy impacts vertical accuracy because horizontal errors usually induce vertical errors. High vertical accuracy requires high horizontal accuracy. Horizontal accuracy should always be tested where possible. While it is difficult to identify well-defined features within the digital elevation data itself, it is common to produce intensity imagery and Orthorectified Radar Imagery (ORIs) from lidar and IFSAR datasets, respectively. Intensity imagery and ORIs are intrinsically georeferenced and aligned to the elevation data, allowing horizontal accuracy of elevation data to be tested by using well defined, photo-identifiable features in the intensity imagery or ORIs.

Horizontal Accuracy Assessment Methodology

In order to test horizontal accuracy, check point coordinates must be in the same coordinate system and datum as the dataset to be tested. Using the survey photos and sketch drawings, the user determines what well-defined feature was surveyed and its general location. While viewing the same area in the dataset, the user then identifies the same well-defined feature and records its coordinates as represented in the dataset being tested. Figure 15.20 shows an example where the northeast corner of a stop bar was surveyed in order to test the horizontal accuracy of 10 cm Vertical Accuracy Class or QL2 lidar data. This same stop bar was located in the intensity imagery and the coordinates, as determined by the intensity image, for the northeast corner of the stop bar were recorded.

This process would be repeated until the coordinates for all horizontal accuracy check points have been determined from the dataset to be tested. Typically, all coordinates are recorded in a spreadsheet so that the necessary computations can be performed easily. RMSE can then be used to calculate radial horizontal accuracy at the 95% confidence level. RMSE is the square root of the average of the set of squared differences between dataset coordinate values and coordinate values from an independent source of higher accuracy for identical points.

Figure 15.20. The field survey photograph of a painted stop bar is shown on the left while the same stop bar is shown in the intensity image on the right. The surveyed coordinates are represented in the right image as the pink dot. By using well-defined features that are photo-identifiable in lidar intensity imagery or IFSAR ORIs, the horizontal accuracy of elevation datasets can be tested.

Horizontal accuracy can be calculated according to the following, as described by NSSDA:

Let:

$$RMSE_x = \text{sqrt}\,(\sum (x_{data\,I} - x_{check\,I})^2/n)$$
$$RMSE_y = \text{sqrt}\,(\sum (y_{data\,I} - y_{check\,I})^2/n)$$

Where:

$x_{data\,I}$, $y_{data\,I}$ are the coordinates of the Ith check point in the dataset

$x_{check\,I}$, $y_{check\,I}$ are the coordinates of the Ith check point in the independent source of higher accuracy

n is the number of check points tested

I is an integer ranging from 1 to n

Horizontal error at point I is defined as:

$$\text{Sqrt}((x_{data\,I} - x_{check\,I})^2 + (y_{data\,I} - y_{check\,I})^2)$$

Horizontal RMSE is:

$$RMSE_r = \text{sqrt}\,(\sum((x_{data\,I} - x_{check\,I})^2 + (y_{data\,I} - y_{check\,I})^2)/n)$$
$$= \text{sqrt}\,(RMSE_x^2 + RMSE_y^2)$$

Computing Accuracy According to the NSSDA when $RMSE_x = RMSE_y$:

$$RMSE_r = \text{sqrt}\,(2*RMSE_x^2) = \text{sqrt}(2*RMSE_y^2)$$
$$=1.4142*RMSE_x = 1.4142*RMSE_y$$

It is assumed that systematic errors have been eliminated as best as possible. If error is normally distributed and independent in each the x- and y-component, and error for the x-component is equal to the error for the y-component, the factor 2.4477 is used to compute horizontal accuracy at the 95% confidence level. As stated in the ASPRS Positional Accuracy Standards for Digital Geospatial Data (2014), "based on empirical results, if the mean error is small, the sample size sufficiently large and the data is normally distributed, 2.4477*$RMSE_{xy}$ is often used as a simplified approximation to compute the $Accuracy_r$ at a 95% confidence level." When the preceding conditions apply, the accuracy according to NSSDA, shall be computed by the formula:

$$Accuracy_r = 2.4477*RMSE_x = 2.4477*RMSE_y = 2.4477*RMSE_r/1.4142$$

$$Accuracy_r = 1.7308*RMSE_r$$

Horizontal Accuracy Reporting

Horizontal accuracy should be reported according to ASPRS and NSSDA standards. For tested horizontal accuracy, the following statement should be used:

- This data set was tested to meet ASPRS Positional Accuracy Standards for Digital Geospatial Data (2014) for a ___ (cm) RMSEx /Horizontal Accuracy Class. Actual positional accuracy was found to be $RMSE_x$ = ___ (cm) and $RMSE_y$ = ___ (cm) which equates to +/- ___at 95% confidence level.

For datasets that were not independently tested, but have been produced according to procedures that have been demonstrated to produce data with particular horizontal accuracy values at the 95% confidence level, the following statement should be used:

- This data set was produced to meet ASPRS Positional Accuracy Standards for Digital Geospatial Data (2014) for a ___ (cm) $RMSE_x$/$RMSE_y$ Horizontal Accuracy Class which equates to +/- ___ cm at a 95% confidence level.
- While horizontal accuracy testing may not always be required for elevation data, an expected horizontal accuracy value is required, using the "produced to meet" statement.

Horizontal Accuracy Assessment Sample

The calculations and methodology described in the sections above may be performed in a spreadsheet, similar to Table 15.11. This report was created by the author's firm for horizontal accuracy assessment performed on a lidar dataset. The project specifications, all inputs, and the final results, are all contained within the table. This dataset tested 0.553 meters horizontal accuracy at the 95% confidence level. This dataset met project specifications.

The Point ID column would list the names or IDs for every horizontal accuracy check point. The next two columns should contain the coordinates for each check point, as determined by the survey or independent source of higher accuracy. The fourth and fifth columns will contain the coordinates for each check point, as determined by the dataset to be tested. The units (meters, feet), datum, and coordinate system must be the same for both the independent source of higher accuracy and the dataset to be tested. The sixth and seventh columns are the differences in coordinate values between the independent source of higher accuracy and dataset to be tested, also called Δx and Δy. These are always computed by subtracting the independent source of higher accuracy from the dataset to be tested (i.e., tested coordinates minus higher accuracy coordinates). The last two columns are the x and y differences squared, per each tested check point. The squared differences are then summed for the x and y component, accomplished with the Sums row towards the bottom of the spreadsheet. The average of the summed squared differences is then computed with the Mean Square Error (MSE) row. Next, $RMSE_x$ and $RMSE_y$ is computed by calculating the square root of the MSE. $RMSE_x^2$ and $RMSE_y^2$ are added together. The square root of this sum equals $RMSE_r$. Lastly, $Accuracy_r$, or the radial accuracy, is calculated by multiplying $RMSE_r$ by 1.7308. It is standard practice to list the acceptance criteria for $RMSE_x$, $RMSE_y$, $RMSE_r$, and $Accuracy_r$, as set forth by individual project requirements, in the horizontal accuracy spreadsheet.

Point ID	QA/QC Survey Checkpoints		Intensity Imagery Coordinates		Intensity Minus Surveyed Coordinates		Discrepancies Squared as required for RMSE calculations	
	NAD83 UTM Zone 18		NAD83 UTM Zone 18		NAD83 Zone 18			
	Easting X (m)	Northing Y (m)	Easting X (m)	Northing Y (m)	Δx (Easting) (m)	Δy (Northing) (m)	Δx^2 (m^2)	Δy^2 (m^2)
UT-001	521,874.403	4,759,203.098	521,874.677	4,759,203.050	0.274	-0.048	0.075	0.002
UT-002	530,981.187	4,753,475.350	530,981.159	4,753,475.671	-0.028	0.321	0.001	0.103
UT-003	534,974.773	4,750,222.414	534,974.963	4,750,222.457	0.190	0.043	0.036	0.002
UT-004	525,745.190	4,747,077.396	525,745.074	4,747,077.562	-0.116	0.166	0.013	0.028
UT-005	551,599.465	4,754,892.144	551,599.480	4,754,892.199	0.015	0.055	0.000	0.003
UT-006	555,891.057	4,749,545.140	555,890.615	4,749,545.206	-0.442	0.066	0.195	0.004
UT-009	554,127.961	4,733,999.462	554,128.115	4,733,999.836	0.154	0.374	0.024	0.140
UT-010	538,247.373	4,734,350.557	538,247.410	4,734,350.719	0.037	0.162	0.001	0.026
UT-014	556,497.530	4,723,915.762	556,497.438	4,723,916.299	-0.092	0.537	0.008	0.288
UT-015	569,566.667	4,727,748.799	569,566.781	4,727,748.743	0.114	-0.056	0.013	0.003
UT-016	554,571.798	4,716,327.775	554,571.794	4,716,327.967	-0.004	0.192	0.000	0.037
UT-017	535,719.312	4,720,119.551	535,719.072	4,720,119.483	-0.240	-0.068	0.058	0.005
UT-018	533,910.887	4,714,450.086	533,911.072	4,714,449.574	0.185	-0.512	0.034	0.262
UT-019	544,376.024	4,701,928.885	544,376.428	4,701,928.888	0.404	0.003	0.163	0.000
UT-021	546,091.953	4,693,561.830	546,092.236	4,693,562.132	0.283	0.302	0.080	0.091
UT-022	533,116.949	4,697,933.296	533,117.137	4,697,933.243	0.188	-0.053	0.035	0.003
UT-023	534,169.310	4,730,265.322	534,169.338	4,730,265.280	0.028	-0.042	0.001	0.002
						Sums	**0.739**	**0.999**
					Horizontal Accuracy Acceptance Criteria:	**MSE**	**0.043**	**0.059**
					$RMSE_x$ and $RMSE_y$ must be ≤0.409 m	**$RMSE_{xy}$ (m)**	**0.208**	**0.242**
					$RMSE_r$ must be ≤0.578 m	**$RMSE_r$ (m)**	**per NSSDA**	**0.320**
Result: PASS			$ACCURACY_r$ must be ≤1 m			**$ACCURACY_r$ (m)**	**per NSSDA**	**0.553**

Table 15.11. Spreadsheet showing tested horizontal accuracy of lidar elevation data using 17 check points that were photo-identifiable in the intensity imagery. This dataset tested 0.553 meters horizontal accuracy at the 95% confidence level.

QUALITATIVE ASSESSMENT

DEM QA/QC involves a qualitative assessment along with the quantitative assessment. Where the quantitative assessment determines if the DEM dataset meets required accuracies, the qualitative assessment determines the overall quality of the dataset regardless of accuracies achieved. DEMs may meet required accuracies while still failing overall quality requirements and rendering the data unusable to the client or end-users. Whereas the quantitative assessment involves comparing DEMs to checkpoints of higher accuracy and reviewing areas of overlap for relative accuracy, qualitative assessments are made by reviewing and looking at the data itself. DEM qualitative assessments can be split into two general categories: Macro level and micro level checks. Macro level or overview checks include verification of completeness and formatting and micro level checks include more detailed and manual visual reviews. Portions of the macro or overview checks can often be automated with various tools and scripts. While producers of data should always review 100% of their data at macro and micro levels, some projects may only require independent QC parties or other reviewers to QC a portion of the data due to various constraints such as budget. When only a portion of data are to be independently reviewed, macro level checks should always be performed on the entire dataset, followed by micro level checks on the required percentage of data. Macro level checks may help to identify potential problem areas where time would be best spent on micro level reviews.

As DEMs are created from source elevation data (lidar, photogrammetrically derived masspoints and breaklines, or IfSAR), and sometimes modified with breakline datasets, the quality of the DEM is a direct result of the quality of all DEM inputs. Many issues identified in the final DEMs often require corrections to the source data. Therefore, many projects will benefit in increased efficiency and reduced re-work from QA/QC performed on the source data prior to creation of the final DEMs. The QA/QC of source data is so important it should always be performed whenever possible. If the source data has issues, then any DEMs or other derived products will also have issues, requiring corrections to be applied to the source data, not just the derived DEMs.

All examples shown are for illustration purposes and do not constitute a final delivered product to the end clients. No one company, entity, or area is identified or implied, and some data has been manipulated for visualization purposes only.

Source Data QA/QC

If the source data are a final deliverable, such as the classified full lidar point cloud or breakline dataset used to enforce the modeled surfaces, then QA/QC should always be performed on that source data. Breakline QA/QC is covered in detail in the section below. Photogrammetry, IfSAR, and lidar technologies and methodologies are covered in chapters 6, 7, and 8, respectively. Generally, the same concepts outlined in this chapter can be applied to reviews of the source elevation data, including completeness, formatting, and qualitative reviews. QA/QC performed on source data does not replace or negate the need for QA/QC performed on the final DEMs. Some source elevation data may be difficult to fully review and requires DEM creation for a complete QA/QC. Some issues occur in the DEMs from processing and are not a direct result of the input datasets. And some issues will occur in the final DEMs due to an issue with how multiple input source datasets, such as lidar and breaklines, or adjacent input elevation datasets, match or fit together in the final product. These mismatches may not be identified when each input source dataset is reviewed independently. Even though a full QA/QC should always be performed on the source elevation data, a full QA/QC of the DEM dataset is always required. As outlined above, some issues are specific to the DEM creation process and some issues result from the combination of inputs. But each product type, i.e., lidar, breaklines, DEMs, required for a project should always be fully reviewed when possible as this provides the most comprehensive QA/QC for the full project. Many products are reviewed and visualized in different software and platforms. Some issues may be more readily identified in one software or product so reviewing all products by various methods ensures the most complete, detailed, and thorough QA/QC.

Breakline QA/QC

Breaklines are commonly used to hydro-flatten and/or hydro-enforce DEMs and can significantly impact the quality of DEMs, for better or for worse. Breaklines may be 2-D, 2.5-D, or 3-D and collected using several different methodologies. 2-D breaklines only contain x/y coordinates and do not contain elevation values. These breaklines only define edges of features, but do not enforce or maintain z-elevations in the surface. 2.5-D breaklines are breaklines that contain x/y coordinates in the internal geometry, but have an elevation assigned to the feature, such as an attribute in the attribute table. 3-D breaklines include x/y and z coordinates in the internal geometry of the breakline, i.e., each vertex contains an x/y/z coordinate. These breaklines will maintain z-elevations in the surface and can be used to define breaks or interruptions in the surface as well. Please refer to chapter 1 for additional breakline definitions. Some issues caused by breaklines are extremely visible in the final DEMs, e.g., floating water where breakline elevations are higher than the surrounding terrain. However, some issues caused by breaklines are not as visible in the final DEMs, e.g., misclassification of hydrographic features where horizontal breakline placement is incorrect but may appear to match the surrounding terrain in the DEM. As not all breakline issues are readily identifiable by just reviewing the DEMs, a separate, full QA/QC should be performed on all breakline datasets used to create the final DEMs. Again, performing QA/QC on all product deliverables will result in the most comprehensive QA/QC for the project. Breaklines should enhance the final DEMs, not detract from the final DEMs. Additionally, some breaklines, such as 2-D polygons representing low confidence areas, are not used to create the final DEMs. However, QA/QC should be performed on all breakline deliverables as they are pertinent to the project and are a deliverable. Issues identified in the breaklines (or any source data or DEM input) should be corrected, even if they do not directly impact the DEMs, again, because they are a deliverable. The sections below outline breakline QA/QC and provide examples of how some breakline issues can impact DEMs. The QA/QC steps below should be used to assess the quality of all input breakline datasets. If adjacent or overlapping breakline datasets are used, the steps below should be used to assess the edge-match quality of the breakline datasets as well. The accuracy and qualitative requirements of topographic lidar projects is for the topographic data, and not the hydrographic data. It is more imperative to retain as much detail and accuracy as possible in the ground data, which is why there is a more stringent review of breaklines collected landward of the land/water interface in all of the breaklines QA/QC steps discussed in the following sections.

Breakline Completeness

Breakline QA/QC should always begin with an overview. Just as with DEM QA/QC, this involves checking the extent of coverage as well as formatting of the breaklines. The number of breakline feature classes or types and naming convention should be verified against project specifications. Formatting, such as shapefile, geodatabase, DGN, should be verified as well. All breaklines should be reviewed to ensure correct spatial reference information, geometry type, attribute definitions, and attribute domains. Lastly, the breaklines should be compared to the project boundary to ensure breaklines cover the full project, including buffer areas if buffered project boundaries are applicable, but do not unnecessarily extend beyond the project boundary. The overview should be performed on all breaklines, 2-D and 3-D, whether or not the breaklines are used in the final DEM datasets.

After the overview, a manual visual completeness check should be performed on all breakline deliverables. The completeness check is a visual review of the breaklines to ensure all features meeting minimum project requirements have been captured with a breakline. This check validates completeness of capture, breakline placement, and breakline type or coding. For this check, breaklines should be compared to imagery that is temporally consistent with the source elevation data. For lidar projects, intensity imagery should be used for this review. IfSAR projects should utilize Orthorectified Radar Images (ORIs) and photogrammetry projects should use the source imagery (orthoimagery is better than stereo-pairs as it is harder to ensure consistency over a larger area using stereo-pairs). These imagery products should always be used over other imagery sources that may not have been acquired at the same time as the elevation data. Temporal differences between reference imagery and elevation data could lead to incorrect breakline collection, missing breakline features, incorrect enforcement of breaklines, and non-enforcement where breaklines are needed. Temporal differences could lead to serious relative accuracy issues between the breaklines and elevation data that could negatively affect the DEMs. If digital imagery was acquired concurrently with the elevation data and is temporally consistent, the digital imagery could be used during the completeness review.

For lidar projects, the intensity imagery used in the completeness check should represent all features, i.e., first return surfaces, so that the breaklines are compared against all available data. If "ground only" lidar intensity images are used, features may be misinterpreted and placement may become distorted, such as where a small patch of trees visually resembles a pond in the ground-only image where vegetation has been removed.

Imagery collected as part of the project, i.e., intensity imagery for lidar projects, ORIs for IfSAR projects, or orthoimagery for photogrammetry projects, should always be used for completeness checks rather than relying solely on the elevation data. Imagery provides additional information, such as if a feature actually contains water at the time of acquisition. Relying on the elevation surfaces alone will not yield the most accurate breaklines. Figure 15.21 shows an example where initial breakline completeness was performed against elevation data only and resulted in the incorrect call to capture and hydro-flatten a dry feature. Two apparent water bodies are shown in the bare DEM on the left in Figure 15.21 but only the southern water body is hydro-flattened. QA/QC was performed, only using the elevation data with no imagery, and an edit call was made to hydro-flatten the northern water body. Intensity imagery, shown in the right image of Figure 15.21 shows that the northern water body was actually dry at the time of acquisition and correctly left un-flattened. Using intensity imagery as part of the completeness review would have prevented this erroneous edit call. Breaklines should only be collected and used to flatten, show gradient downhill flow, and enforce edges for hydrographic features that contain or represent water at the time of data acquisition. Dry hydrographic features should not be captured with breaklines or artificially adjusted in the elevation surface models.

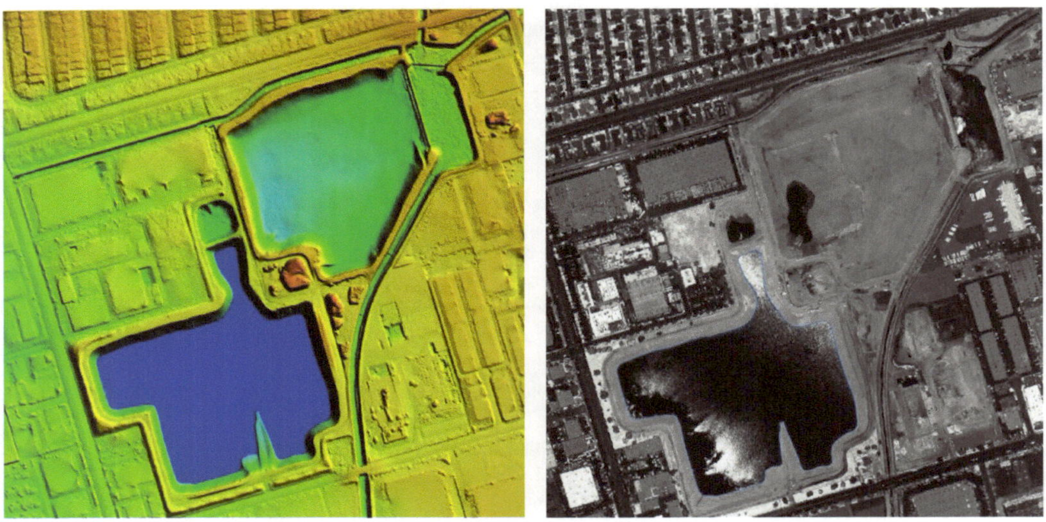

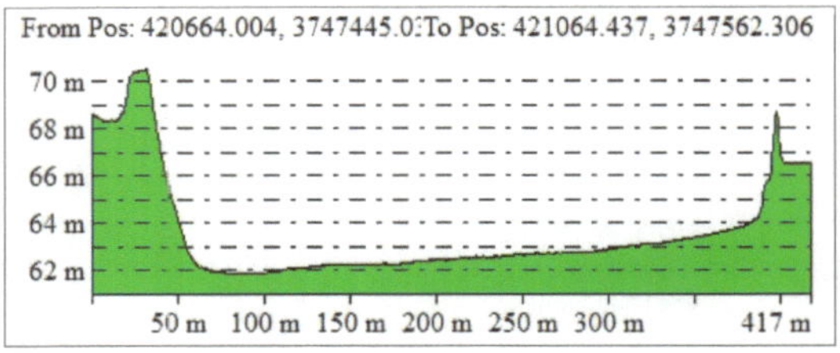

Figure 15.21. The image on the left shows two features in a DEM that appear to be water bodies. The southwestern water body has been flattened, but the northeastern water body is not flattened, shown in the accompanying profile. The right image is the intensity imagery of this same location and shows that the northeastern feature is mostly dry and therefore was not captured with breaklines or flattened in the DEM. Imagery captured during acquisition that is temporally consistent with the elevation data is necessary for accurate breakline collection and QA/QC.

Additionally, the highest resolution imagery supported by the data should be used for the manual visual completeness review when possible. Higher resolution imagery will ensure higher breakline accuracy during collection and will provide a more robust QA/QC of the collected breaklines. The spatial resolution for ORIs and imagery used for photogrammetry is generally determined by project specifications with little room for modification or resampling that would provide additional information during QA/QC. However, because the full point cloud is generally delivered with all lidar projects, highest resolution intensity images for QA/QC can be generated from the point cloud, regardless of what the project specifications require regarding intensity imagery deliverables. For example, a project that requires DEMs with 1-meter post spacing may also require intensity imagery with 1-meter spatial resolution even though the density of the lidar may support intensity imagery with a spatial resolution of 1-foot. The final ANPS for each lidar project will need to be calculated to determine the highest spatial resolution of intensity imagery that can be supported by the data. Intensity imagery with a higher resolution of 1-foot compared to 1-meter cell sizes may provide a more precise and robust breakline QA/QC by providing more clarity along edges of features, such as shorelines. Additional clarity along shorelines where breaklines are collected can help ensure the best possible relative accuracy between the breaklines and source elevation data. Figure 15.22 illustrates the difference between lidar intensity imagery with 1-foot resolution and 1-meter resolution.

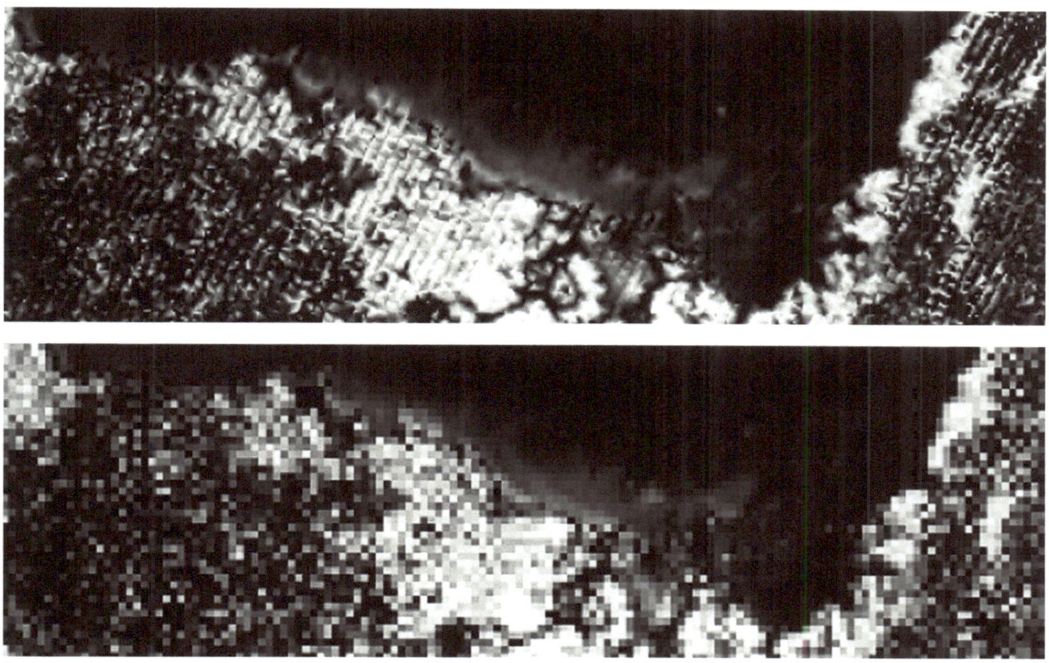

Figure 15.22. The image on the top is intensity imagery with 1-foot resolution while the image on the bottom is the same feature/location but created with 1-meter resolution. The shoreline is more distinct in the 1-foot imagery, especially where the shoreline is more complex or where vegetation starts to overhang and partially obscure the shoreline. Using higher resolution imagery during breakline QA/QC can improve results by ensuring breakline placement truly matches the source elevation data.

The completeness of breaklines should be reviewed by panning through the imagery overlaid with the breaklines at a scale appropriate for each project. A data dictionary, typically included as part of the defined project specifications, should be referenced for breaklines QA/QC. Using the data dictionary, each breakline type should be symbolized appropriately to differentiate between feature types. Each breakline should be reviewed for correct coding (Figure 15.23 on page 484). Missing features, features meeting minimum capture criteria that are present in the imagery but not in the breakline dataset, and incomplete features should be identified (Figure 15.24 on page 484). While checking for complete capture and correct breakline type and coding, the reviewer should also be checking the horizontal placement of the breaklines to ensure the placement aligns with the imagery. The breakline variance check outlined below will also help to identify

Figure 15.23. While both features are water bodies with constant elevations, the feature on the left is coded as a water body (blue) while the feature on the right is coded as a stream (pink). Incorrect coding and feature types should be identified during the breakline QA/QC completeness review.

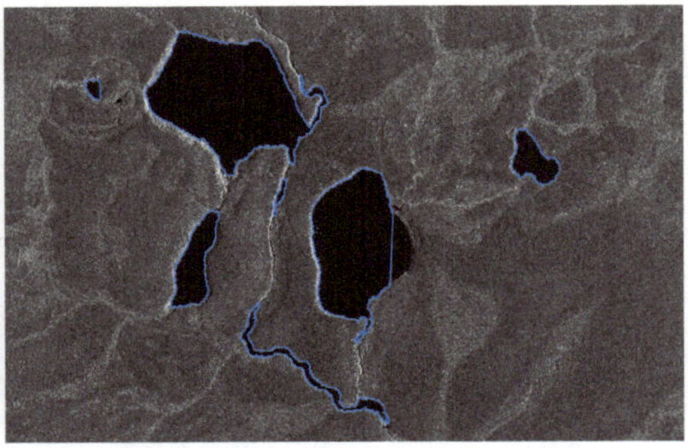

Figure 15.24. There is incomplete breakline capture of the water body in the middle of the image. The breakline does not correctly capture the eastern shoreline. Incomplete breakline capture and missing features should be identified during the breakline QA/QC completeness review.

incorrect horizontal placement of breaklines that are digging into banks/shorelines but the horizontal placement should still be reviewed as part of the completeness check as not all incorrect horizontal placement will result in "digging" features (see Figure 15.31 on page 490 in the breakline variance section below). The various breakline checks are designed to overlap and complement each other to ensure all issues are identified for correction during the breakline QA/QC process. Sometimes the placement of breaklines is hard to interpret against the imagery alone, which is why breaklines should always be compared to both the source elevation data and imagery. A surface model, such as an Esri terrain, created from all ground and water points should be used in conjunction with imagery to verify breakline placement. Breaklines should never contradict the source elevation data and should match the imagery as best as possible.

Breakline Variance

After the overview and completeness checks, all 2.5-D and 3-D breaklines created for lidar projects should be compared to the source elevation data through a process called breakline variance to ensure acceptable relative accuracy between the 3-D breaklines and the lidar point cloud data. As 3-D breaklines are used to enforce, including hydro-enforcement as well as maintaining z-elevations along non-hydro features, and hydro-flatten DEMs, poor relative accuracy between the two datasets can result in degradation of the final DEMs. Only 3-D and 2.5-D breaklines should be used in breakline variance as this process compares the elevations of the breaklines to the source elevation dataset. 2-D breaklines have no elevations to be used in the comparison. Breakline variance is only performed on lidar projects as the point cloud data are generally part of the final deliverables, along with the final DEMs, so that two different elevations, the point cloud and breaklines, are available for comparison. Additionally, most clients and end-users have software with the ability to read or convert the point cloud data into more usable formats. IfSAR projects typically do not include raw source data as part of the deliverables as this data requires specialized software and requires specialized training and experience to interpret, analyze, and process. Traditional photogrammetry uses stereo imagery and aerotriangulation solutions so that analysts can compile 3-D breaklines and masspoints that are used to create the final DEMs. If only the final DEMs, which have been enforced or hydro-flattened with the breaklines, are delivered, two different elevations are not available for comparison. Lastly, breakline variance is typically not performed on bathymetric data, even lidar bathymetric data, as 3-D hydrographic breaklines are not used in these datasets. 3-D hydrographic breaklines would negate or compromise the actual bathymetric data. Breakline variance should still be performed on any 3-D non-hydrographic breaklines that may have been collected for a lidar topobathymetric project, such as road breaklines.

The breakline variance process begins by overlaying all 3-D breakline vertices on a surface model created from the bare earth and water elevations of the source lidar dataset. Water elevations must be included in this surface model because water is typically classified based on the extents of the breaklines, and if the lidar were classified incorrectly, then these issues would not be evident in a surface model created from ground only points. Water point elevations may not be reliable in all areas and this must be taken into account during the manual review of the breakline variance results. While the breakline completeness review uses a combination of both the source elevation data and temporally consistent imagery (lidar intensity, IfSAR ORIs, or orthoimagery for photogrammetry projects), the breakline variance should rely mostly on the surface model elevation data and use imagery only as ancillary information. Too many rely on lidar intensity imagery alone for both QA/QC and the actual breakline collection without fully understanding the intensity images are active, narrow band Infrared (IR) images and not panchromatic images. Ensuring the source elevation data are used in both breakline collection and QA/QC will help avoid many issues, including the example illustrated in Figures 15.25, 15.26 on page 486, and 15.27 on page 486. Figure 15.25 shows two surface models with a hydrographic breakline overlaid. The surface model on the left was created with ground only points while the surface model on the right was created with ground and water points. The hydrographic breakline was used to classify all points within its boundary as water. As the surface model on the right shows, however, the feature in question is not a hydrographic feature but a mounded feature with considerable elevation. Figure 15.26 shows the intensity imagery on the left and a color image on the right. This feature, which is actually a coal pile, looks very dark in the intensity imagery and was mistaken for a water body. If the source elevation data was used during breakline collection, this error would have been avoided. Surface models used in breakline variance QA/QC must always be created with both ground and water points in order to catch classification issues resulting from incorrect hydrographic breakline collection. Figure 15.27 shows the impact of this error in the bare earth DEM in the left image and the corrected bare earth DEM in the right image.

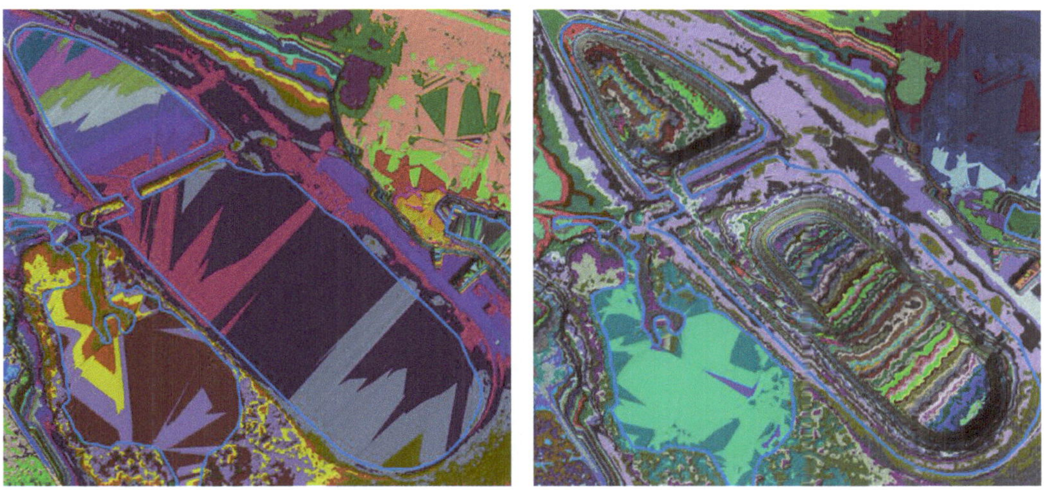

Figure 15.25. Water body breaklines (blue) are erroneously overlaid on terrains in both images where each color band represents 1-foot in elevation change. The terrain in the image on the left was created using only ground points whereas the terrain in the image on the right was created using both ground and water lidar points. If water points were not used in the terrain for the breakline variance process, this water body breakline may not have been flagged as incorrectly classifying ground as water. Only the polygon in the southwest corner is actually water.

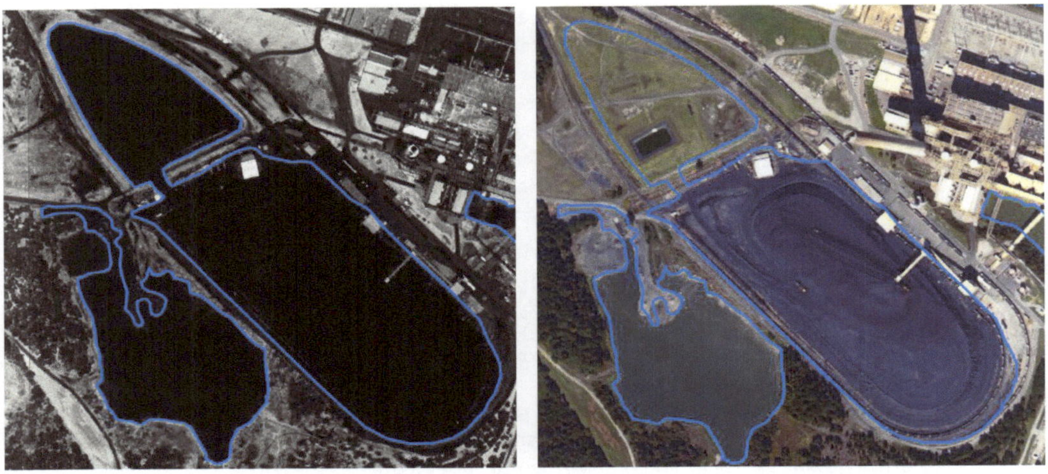

Figure 15.26. Water body breaklines (blue) erroneously overlaid on intensity imagery in the image on the left and Esri World Imagery Layer (Source: Microsoft for this location) on the right. During breakline collection, this coal pile, shown in the right image, was misinterpreted as a water body in the intensity imagery, shown on the left. Intensity imagery should always be used in conjunction with the source elevation data for breakline completeness reviews as intensity imagery will be temporally consistent with the lidar data. However, the source elevation data always takes precedence in all breakline QA/QC and must be the main source of comparison in the breakline variance QA/QC. If the source elevation data had been used during breakline collect, this error would have been avoided. Using the source elevation data in the breakline QA/QC ensures these types of errors are identified.

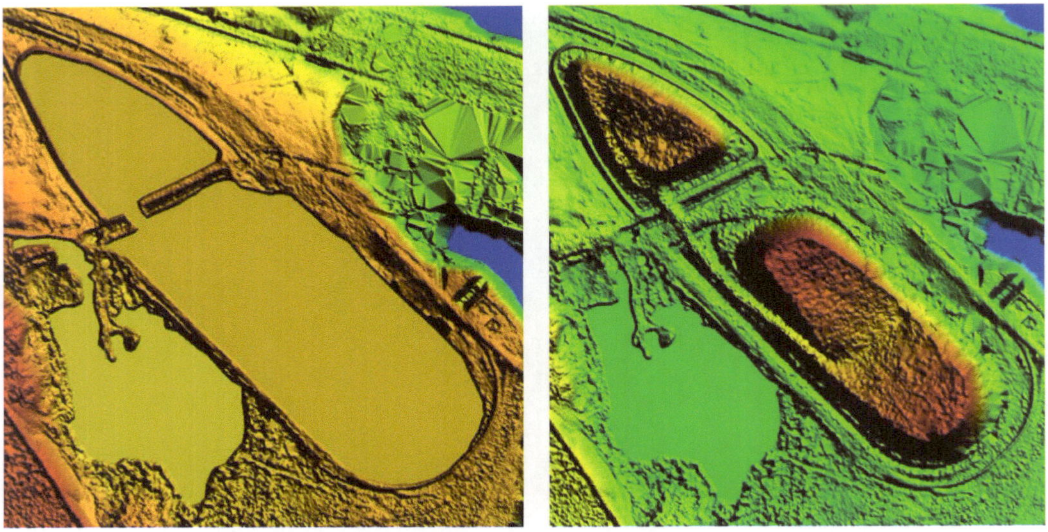

Figure 15.27. These two images show how incorrect hydrographic breaklines can affect the final DEMs. In the image on the left, the hydrographic breakline collected over the coal piles has been flattened. The image on the right shows the corrected DEM with the fully modeled coal piles. Note that there are two coal piles in the source elevation data while there is only one coal pile in the Esri World Imagery Layer in Figure 15.26. Temporal differences will occur between datasets not co-acquired and the source elevation data must always take precedence during breakline collection and QA/QC.

During the breakline variance review, the interpolated lidar elevations at the same xy location of each breakline vertex are extracted and stored in the attribute table of the breakline vertices. The differences between the assigned 3-D breakline elevations and the interpolated lidar elevations are computed. The lidar elevations should always be subtracted from the breakline elevations (breakline z - lidar z) so that positive ΔZ values indicate breaklines are floating above the lidar surface and negative ΔZ values indicate breaklines are digging into the lidar surface. These ΔZ values are then analyzed to determine if high ΔZ values (both positive and negative) are a result of incorrect horizontal or vertical breakline placement, or are a result of normal lidar characteristics, such as increased interpolation in surface models across large water body surfaces or increased interpolation in surface models along vegetated bank lines. There are no absolute thresholds for high, normal, or low ΔZ values that will guarantee a breakline is correct or incorrect. The terms high and low are used in an effort to provide guidance as to where the greatest focus of the breakline variance analysis should be located. In general terms, all positive ΔZ values, which indicate breaklines are floating above the lidar surface, should be reviewed for possible corrections as breaklines should not float under normal circumstances. Negative ΔZ values, which indicate the breaklines are digging into the lidar surface, can generally be considered high and require review when the ΔZ values are -1 meter/-3 feet or more for QL2 or 10 cm RMSEz vertical accuracy class lidar. Floating or high negative ΔZ values do not automatically make breaklines incorrect. There are many exceptions or circumstances which can result in correct breaklines having floating ΔZ values or high negative ΔZ values. For this reason, breakline variance is not simply a straightforward calculation where all breaklines with ΔZ values higher than a specified threshold are automatically considered incorrect. While many exceptions or different circumstances exist that result in a wide range of ΔZ values, the sections below provide consistent guidelines for the collection and QA/QC of breaklines.

Please note that the breakline variance check is testing the relative accuracy of the breaklines compared to the source lidar dataset, not the absolute accuracy.

Horizontal Breakline Variance

Even though z-elevations are being compared during the breakline variance process, this process allows the analyst to review horizontal placement of the breaklines as incorrect horizontal placement can translate into high vertical ΔZ values where there are elevation changes in the bare earth surface along the bank or shorelines. Horizontal placement issues in very flat areas where there may not be associated high ΔZ values must be identified during the manual completeness review.

When possible, breaklines should be placed slightly sea-ward of the land/water interface. A threshold of up to 2 x NPS has been established by some federal agencies for breakline placement inside of a hydro feature. It is unrealistic that a breakline will always be placed exactly at the land/water interface, especially because interpretation can vary depending on the type of product being used for interpretation (intensity raster vs. TIN/TERRAIN) and spatial resolution of that product. While appropriate data types and spatial resolutions are project dependent, it is always better to err on placing the breakline slightly seaward. This placement, up to 2 x NPS, may include small amounts of water in the final bare-earth surface model, but will not detrimentally impact the DEM. Placement of a breakline sea-ward by more than 2 x NPS may result in the request for corrections as now larger sections of water will be included in the final bare-earth surfaces and these points (especially over deeper water) may be unreliable, resulting in undesirable results in the DEM. Breaklines collected sea-ward within the 2 x NPS threshold will avoid all issues identified and discussed below.

There is no acceptable threshold for breaklines collected landward of the land/water interface because this would result in ground being classified as water and flattened in final bare-earth surfaces. While a 2 x NPS threshold may seem minimal, banks, berms, and other elevated terrain features that are located within that threshold could be removed or altered in the final bare-earth surfaces. Figures 15.28 and 15.29 on page 488 show examples of breaklines placed landward of the land/water interface that result in significant impacts on the final bare-earth surfaces.

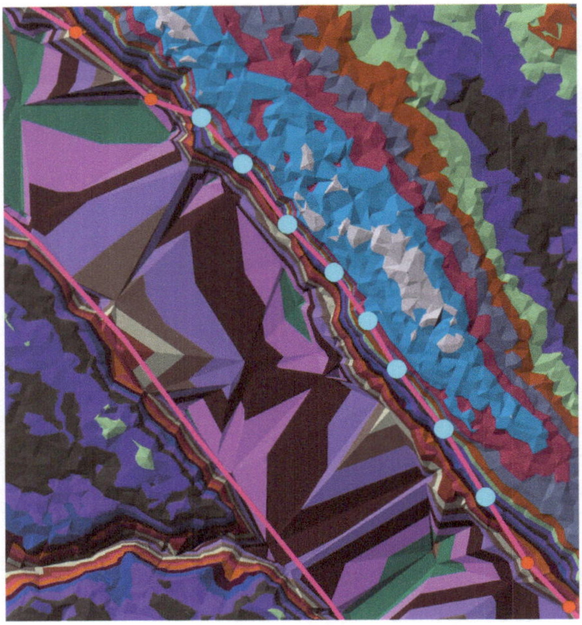

brkl_z	Z	diff_z
516.1972	542.3054	-26.1082
516.2366	542.1661	-25.9295
516.2161	542.1449	-25.9288
516.2273	541.7764	-25.5491
516.1889	541.5448	-25.3559
516.2061	541.4376	-25.2315
516.1774	540.0941	-23.9167
516.1656	537.7023	-21.5367

Figure 15.28. A stream breakline (pink) is overlaid on a terrain where each color band represents 1-foot in elevation change. The vertices highlighted in blue are also shown in the accompanying table. These vertices have ΔZ values ranging between -21 feet and -26 feet, indicating the breakline elevations are lower than the lidar elevations. These vertices are on top of the breakline embankment rather than at water surface level. The high ΔZ values are a direct result of incorrect horizontal breakline placement. Accepting this breakline placement would result in elevated ground features being flattened and removed from the bare-earth surfaces. The bank line would be incorrectly altered in this location, removing features and surface detail from the final bare-earth surfaces.

brkl_z	Z	diff_z
516.1972	542.3054	-26.1082
516.2366	542.1661	-25.9295
516.2161	542.1449	-25.9288
516.2273	541.7764	-25.5491
516.1889	541.5448	-25.3559
516.2061	541.4376	-25.2315
516.1774	540.0941	-23.9167
516.1656	537.7023	-21.5367

Figure 15.29. A stream breakline (yellow) is overlaid on a terrain where each color band represents 1-foot in elevation change. The vertices highlighted in light blue are also shown in the accompanying table. These vertices have ΔZ values ranging between -7 feet and -14 feet, indicating the breakline elevations are lower than the lidar elevations. These vertices are on the bank of the shoreline rather than at water surface level. The high ΔZ values are a direct result of incorrect horizontal breakline placement. Accepting this breakline placement would result in the ground and slope being altered in the final bare-earth surfaces. The bank line would be incorrectly altered in this location, removing slope and surface detail from the final bare-earth surfaces.

To reiterate, the accuracy and qualitative requirements of topographic lidar projects is for the topographic data, and not the hydrographic data. It is imperative to retain as much detail and accuracy as possible in the ground data, which is why there is a more stringent review to identify breaklines collected landward of the land/water interface. While this rule is applicable to all breakline QA/QC steps, the placement of breaklines is scrutinized in the most detail during the variance review.

As Figure 15.30 illustrates, incorrect horizontal breakline placement that goes across shoreline should always be corrected as this type of breakline placement not only alters ground features, but can completely change the geometry of features or remove entire features or portions of features from the final bare-earth surfaces. This type of incorrect horizontal placement will generally be associated with high ΔZ values.

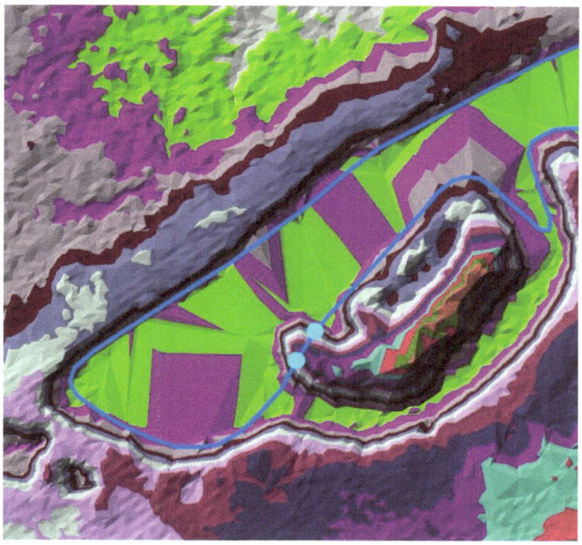

brkl_z	Z	diff_z
851.409845	858.146052	-6.736207
851.409845	858.766138	-3.356293

Figure 15.30. A stream breakline (blue) is overlaid on a terrain where each color band represents 1-foot in elevation change. The vertices highlighted in light blue are also shown in the accompanying table. These vertices have ΔZ values ranging between -3 feet to -7 feet, and cut across the island feature. These vertices are on the bank of the shoreline rather than at water surface level and completely cut off the tip of the island. The high ΔZ values are a direct result of incorrect horizontal breakline placement. Accepting this breakline placement would result in the alteration of the geometry of the island in the final bare-earth surfaces. The shoreline would be incorrectly altered in this location, removing a portion of the island from the final bare-earth surfaces.

Incorrect horizontal placement along shoreline, as shown in Figures 15.28 and 15.29, should always be corrected when the horizontal placement will result in the incorrect modification or change of elevated ground features (embankments, berms, or other significant features), slopes or other terrain breaks as defined in the source lidar. This type of incorrect horizontal placement will generally be associated with high ΔZ values.

Incorrect horizontal placement that does not necessarily coincide with high ΔZ values, but clearly classifies large amounts of ground as water should still be corrected. Although there is some subjectivity associated with this requirement, the goal of the horizontal placement should be that ground must not be classified as water. Large amounts of ground classified as water must be corrected, regardless if upland vegetation is present or not. However, the presence of upland vegetation within hydrographic breaklines is a strong indicator that too much ground is classified as water. Generally, only slivers of ground classified as water (shown in Figure 15.32 on page 491) are allowable and all other ground classified as water must be corrected. Slivers of ground can be defined as thin, narrow strips along the shoreline or parallel to the shoreline. Large amounts of ground can be classified as non-slivers; larger amounts of ground have both width and length and result in greater measurable area being incorrectly classified as water. Figure 15.31 on page 490 illustrates this point. In Figure 15.31, low vegetation is visible in the intensity imagery within the collected breaklines, indicating this area is ground and should be corrected. The terrain shown in Figure 15.31 shows that this feature does not result in high ΔZ values and is not flagged as an issue by the breakline variance process. The breakline variance review cannot be used as the sole method to determine the quality of breaklines. The breakline variance review should always be used as part of a system, which includes visual checks against intensity imagery and automated topology tools for data integrity.

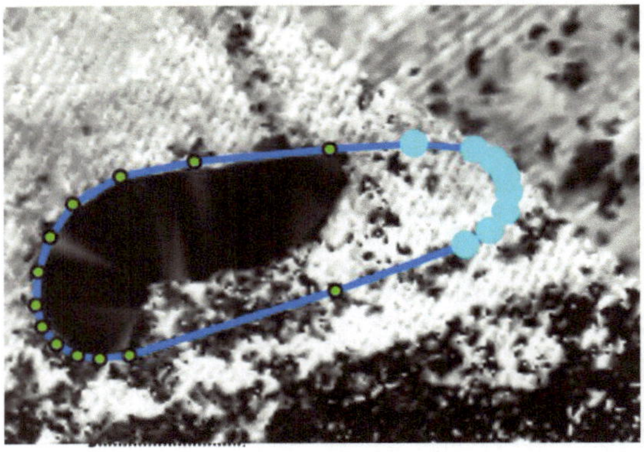

brkl_z	z	diff_z
928.61015	931.943149	-3.332999
928.61015	931.640984	-3.030834
928.61015	931.627533	-3.017382
928.61015	931.311588	-2.701438
928.61015	931.081602	-2.471452
928.61015	931.027796	-2.417646
928.61015	930.966445	-2.356295
928.61015	930.871957	-2.261806
928.61015	930.40411	-1.79396

Figure 15.31. A water body breakline (blue) is overlaid on a terrain in the top image where each color band represents 1-foot in elevation change. The same water body breakline is overlaid on an intensity image with 1-foot pixel size in the bottom image. The vertices highlighted in light blue are also shown in the accompanying table. These vertices have ΔZ values ranging between -1.7 feet to -3 feet, which is within normal ranges for a water body. While the breakline variance review does not flag this feature as having incorrect horizontal breakline placement, a visual review of the breaklines against lidar intensity imagery highlights this feature as capturing too much ground as water. The amount of ground classified as water within this water body is greater than 300 square meters. The horizontal breakline placement must still be corrected, even though the ΔZ ranges of the vertices were acceptable.

As stated previously, there will be more stringent review of breaklines collected landward of the land/water interface because the goal of topographic lidar projects is to collect and acquire topographic data. But breaklines cannot always be placed at the exact land/water interface because that interface may appear slightly different depending on what type of surface model, image, and spatial resolution is used. Different analysts may also interpret the same land/water interface differently, especially in complicated areas. Unfortunately absolute thresholds for ΔZ values cannot be specified as these values will depend on the type of terrain, connectivity, and monotonicity requirements. Guidelines, however, can be established. As shown in Figure 15.30, breakline placement across shoreline must always be corrected. Breakline placement along shoreline that results in significant negative impacts on the final bare earth surface, such as the flattening or removal of elevated features or the classification of large areas of ground as water, must always be corrected (Figures 15.28, 15.29, and 15.31). But breaklines with relatively low ΔZ values whose horizontal placement (a) Does not result in the incorrect modification or change of elevated ground features; (b) Does not classify large amounts of ground as water; (c) Does not capture vegetation within the breaklines; and (d) May classify small slivers of ground as water, will not usually require modification. Figure 15.32 illustrates an example where the horizontal breakline placement classifies a very small sliver of ground as water but does not modify any elevated ground features; no modifications are required to the breakline placement in this instance. However, features and land cover types or habitats, such as wetlands, marsh, and coastal mudflats, deemed significant or important by clients, will result in more stringent review of breaklines inland on the land/water interface on the specified feature types. In these circumstances there will be an even lower threshold or tolerance for allowable ground slivers classified as water in an effort to preserve as much of the identified features of interest for the client as possible. Habitats and land cover types deemed important by a client and requiring

brkl_z	Z	diff_z
590.410039	595.015017	-4.604978

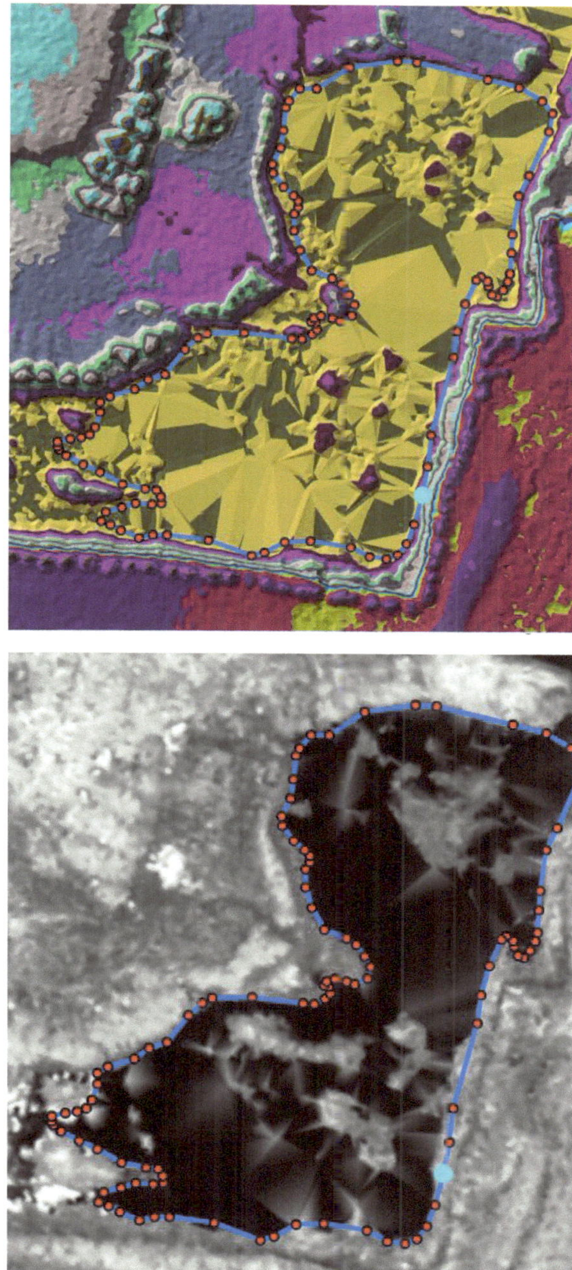

Figure 15.32. A water body breakline (blue) is overlaid on a terrain in the top image where each color band represents 1-foot in elevation change. The same water body breakline is overlaid on intensity imagery with 1-foot pixel size in the bottom image. The vertex highlighted in light blue is also shown in the accompanying table. This vertex has a ΔZ value of -4.6 feet. This Δz value is within normal ranges for a water body, but visually the vertex is slightly elevated up the bank. While this highlighted vertex and a few of the adjacent vertices are marginally placed, they only affect a small sliver of ground (8 feet in width), and minimally affect the slope adjacent to the breakline. A visual review of the breaklines against lidar intensity imagery enforces the notion that only a very small sliver of ground has been captured as water and that this would be within acceptable limits.

special considerations should be outlined in the project specifications. It is important that those performing QA/QC are intimately familiar with the specifics of each project they work, and understand how to appropriately apply their best professional judgment when making error calls.

Breaklines with high ΔZ values that are a direct result of interpolation should not be modified. Due to inherent characteristics of lidar, interpolation is expected on water bodies and hydrographic features. This interpolation may be more severe along banks or other areas that are also densely vegetated. Figure 15.33 on page 492 illustrates an area of breaklines with high ΔZ values as a result of interpolation.

To summarize, the following guidelines should be used when producing and evaluating horizontal breakline placement:

- Emphasis during breakline collection should be on collecting breaklines slightly sea-ward of the land/water interface, up to 2 x NPS
- Incorrect breakline placement across shoreline must always be corrected
- Incorrect breakline placement along shoreline that impacts elevated terrain features, such as embankments, berms, or other significant features, or alters the slope or any terrain breaks as defined by the source lidar must always be corrected

brkl_z	Z	diff_z
770.840125	780.09798	-9.257856
770.849967	778.120294	-7.270327
770.800098	777.513668	-6.713569
770.789928	776.165245	-5.375317
700.820111	776.144576	-5.324464
770.809941	775.332898	-4.522957
770.769915	774694119	-3.924205

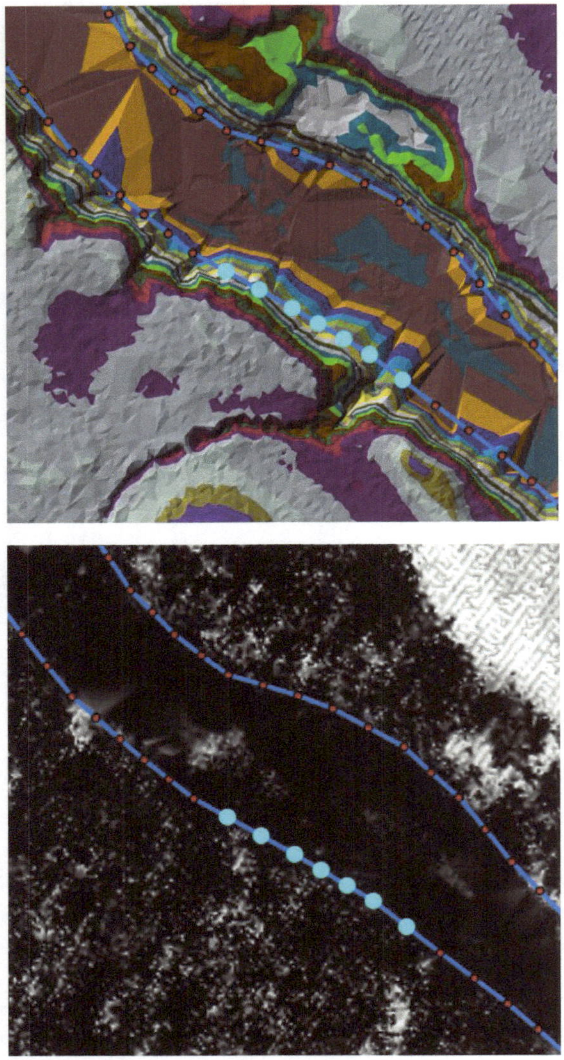

Figure 15.33. A stream breakline (blue) is overlaid on a terrain in the top image where each color band represents 1-foot in elevation change. The same stream breakline is overlaid on intensity imagery with 1-foot pixel size in the bottom image. The vertices highlighted in light blue are also shown in the accompanying table. These vertices have ΔZ values ranging between -4 feet and -9 feet. While these ΔZ values are high, the ΔZ values are a direct result of interpolation. When viewed against the terrain, these vertices are located in very wide, generalized elevation bands that slope downward from the adjacent bank. This area is interpolated in the terrain surface. The breakline follows the actual shore line, represented by the more detailed, narrower elevation bands within the terrain. The breaklines also match the land/water interface visible in the intensity imagery as well. Dense vegetation is visible along the bank line and is contributing to the interpolation visible in the terrain.

- Breakline variance should not be used in isolation to validate breaklines. Breaklines that are within the expected ΔZ value range, but are clearly capturing ground as water based on other methods of review, such as visual completeness checks against intensity imagery, must still be corrected
- Breaklines with high ΔZ values due to interpolation but correct horizontal placement when verified against the terrain and intensity imagery should not be modified
- Breaklines with relatively low ΔZ values whose horizontal placement does not result in the incorrect modification or change of elevated ground features, does not classify large amounts of ground as water, and does not include vegetation within the breaklines, but may classify small slivers of ground as water may be acceptable without modification
- Features and land cover types or habitats such as wetlands, marsh, and coastal mudflats deemed significant or important by clients will result in more stringent review of breaklines landward of the land/water interface on the specified feature types. Lower tolerances for slivers of ground classified as water will be used on these identified feature types in an effort to preserve the fullest extent of the feature as possible in the final DEMs Vertical Breakline Variance

Vertical Breakline Variance

If high Δz values cannot be attributed to incorrect horizontal breakline placement, high ΔZ values may be an indicator of incorrect breakline elevations. High ΔZ values identify breakline vertices that are significantly different from the lidar elevation at the same xy location. These differences must be reviewed as some differences will be legitimate due to normal characteristics of lidar datasets. Generally, all positive ΔZ values are considered high and should be reviewed as breaklines should not float above the lidar surface. Negative ΔZ values of -1 m/-3 ft or more are generally considered excessive and should be reviewed for potential issues that require corrections. Interpolation along stream bank lines, as shown in Figure 15.33 in the Horizontal Breakline Variance section above, is a normal occurrence in lidar datasets and will result in large but acceptable negative ΔZ values. Elevated vertical features, such as seawalls and dams, will usually result in acceptable high positive ΔZ values because the lidar may not fully model the elevated feature, resulting in lower or interpolated lidar elevations compared to breakline elevations that capture the full height or extent of the feature. Acceptable high ΔZ values often occur in areas where breaklines are used to enforce surface models because the lidar itself may not adequately model these features. Figure 15.34 shows an example of a breakline enforcing a dam that results in acceptable high ΔZ values. Figure 15.35 on page 494 shows the intensity imagery and Esri world base imagery for this same location.

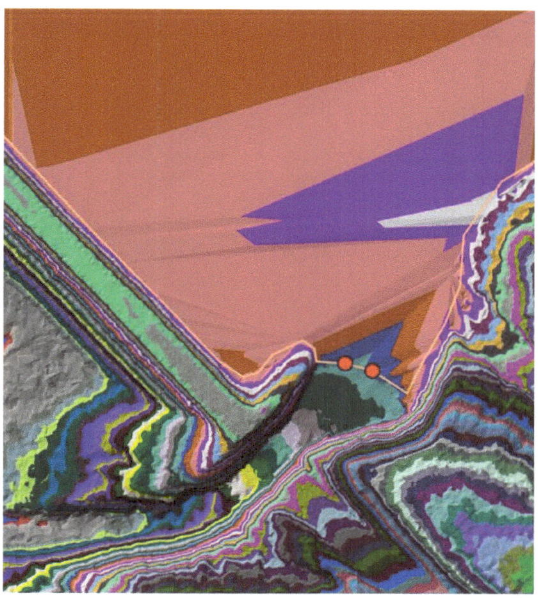

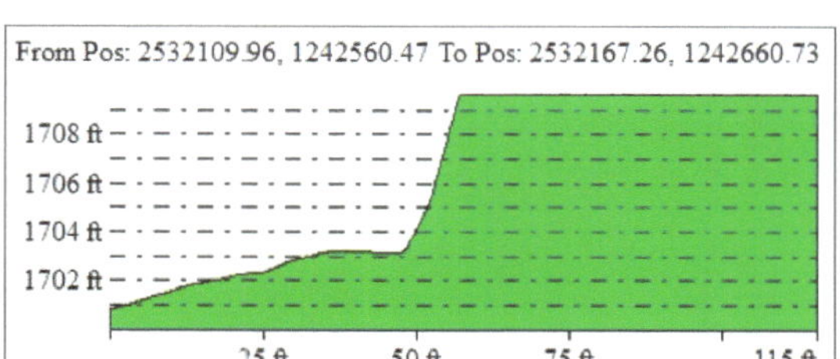

brkl_z	Z	diff_z
521.1001	520.009589	1.0905
521.1001	519.633278	1.46678

Figure 15.34. The image on the left shows a breakline (pink) overlaid on a terrain. The two red circles are breakline vertices floating above the terrain by more than 1 foot. The terrain also shows that there is interpolation occurring at the location of the floating vertices. When this same area is viewed in the DEM (right image), the "floating" area is level and consistently above the terrain directly to the south, but below all other surrounding embankments.

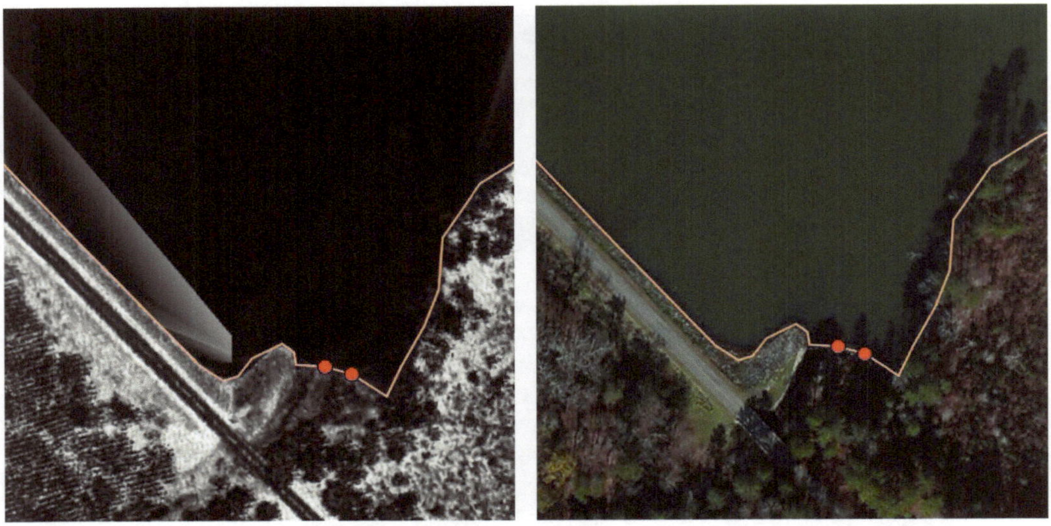

Figure 15.35. Lidar intensity imagery (left) and Esri world base imagery (right) are shown for the same feature identified in Figure 15.34. The imagery for this location shows that the "floating" vertices are located at the spillway for the dam. The breakline is enforcing a consistent elevation for the water body and these vertices (red circles) only appear to be floating because the spillway elevations decrease rapidly and the resulting interpolation in the terrain (when the breakline is not used to enforce this feature) makes the vertices appear to be floating. The high ΔZ values are acceptable in these and other similar circumstances; no corrections are required for this feature.

Breakline variance ΔZ values can either be positive or negative. Subtracting lidar elevations from breakline elevations (breakline z-lidar z), always produces ΔZ's where positive values indicate breaklines are floating above the lidar surface and negative values indicate breaklines are digging into the lidar surface. All floating breakline vertices should be reviewed as these are generally unacceptable and require corrections, but some exceptions are allowed. The term water body generally refers to features such as lakes or ponds that are at constant elevations while the term linear hydrographic feature generally refers to streams or rivers with gradient flow. Water bodies should never have any floating vertices. Floating vertices on these features fail both hydro-enforcement and hydro-flattening requirements. Few exceptions are allowed on water bodies as water bodies are always set to one constant elevation and that elevation can be lowered and adjusted to ensure no vertices or portions of the water body are floating above the lidar surface. Exceptions may sometimes be made when the water body connects to stream features and is part of a flowing network. Depending on the geometry of the network, water bodies can often still be lowered when necessary, but exceptions may occur if lowering the water body elevation would result in unwanted or unacceptable changes along the rest of the hydrographic network. Larger bodies of water that have gradient flow, such as large reservoirs and fjords, should be treated as linear hydrographic features and not water bodies with constant elevations; water body guidelines do not apply to these features. Figure 15.36 shows a water body unacceptably floating and results after correcting the feature.

Continuous segments of stream or river features (dual line and single line) that are floating above the surrounding terrain must always be corrected as these floating vertices fail hydro-enforcement and hydro-flattening requirements. Figure 15.37 on page 496 shows an example of a floating stream that must be corrected along with the results after the correction has been applied.

Whereas water bodies are set to one constant elevation, streams and rivers (both dual line and single line) model gradient flow. As streams and rivers model flow, these breaklines should be monotonic, or flow downhill, rather than have undulating elevations. Only streams and rivers representing actual water flow should be monotonic. Dry streambeds will not be monotonic and should not be artificially adjusted to model

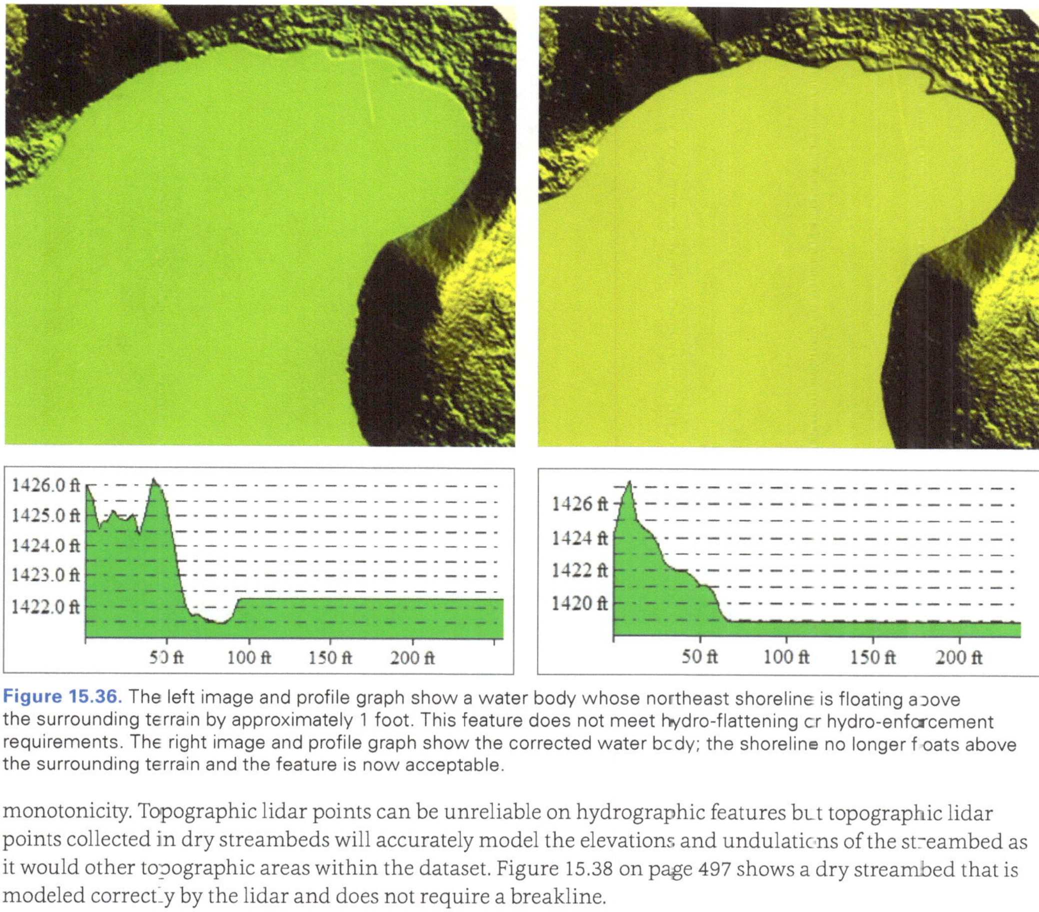

Figure 15.36. The left image and profile graph show a water body whose northeast shoreline is floating above the surrounding terrain by approximately 1 foot. This feature does not meet hydro-flattening or hydro-enforcement requirements. The right image and profile graph show the corrected water body; the shoreline no longer floats above the surrounding terrain and the feature is now acceptable.

monotonicity. Topographic lidar points can be unreliable on hydrographic features but topographic lidar points collected in dry streambeds will accurately model the elevations and undulations of the streambed as it would other topographic areas within the dataset. Figure 15.38 on page 497 shows a dry streambed that is modeled correctly by the lidar and does not require a breakline.

If any vertex on a stream or river representing gradient flow is floating due to undulating elevations along the breakline, or erroneously high and low elevations known as spikes and divots respectively, resulting from incorrect and non-monotonic vertex elevations, then these vertices must always be corrected. However, if a stream is monotonic but a few vertices are moderately floating above the surrounding surface, then these are generally accepted. Corrections applied to the minimal number of vertices may require alterations to large sections of the stream network and in some cases may not be feasible without causing large portions of the stream network to dig into the surrounding terrain. When the negative impacts of corrections may outweigh the benefits of the corrections, the minimal number of vertices that are moderately floating are generally acceptable. While moderately floating is a subjective term, it generally applies when there is a small number of floating vertices and the small number of vertices only float by a few tenths of a meter or less. Vertices that are excessively floating usually result from undulating breakline elevations, or where there are continuous portions of floating vertices. If a stream is monotonic, but excessively floating, then the incorrect elevations will impact a larger number of vertices compared to a monotonic stream where vertices are only moderately or slightly floating which will impact a smaller number of vertices.

Additionally, dual line streams should have consistent elevations from bank to bank, i.e., elevations across banks perpendicular to the centerline should be consistent. Isolated floating vertices on one side of the bank are generally accepted when these elevations are monotonic and are consistent with the elevations on

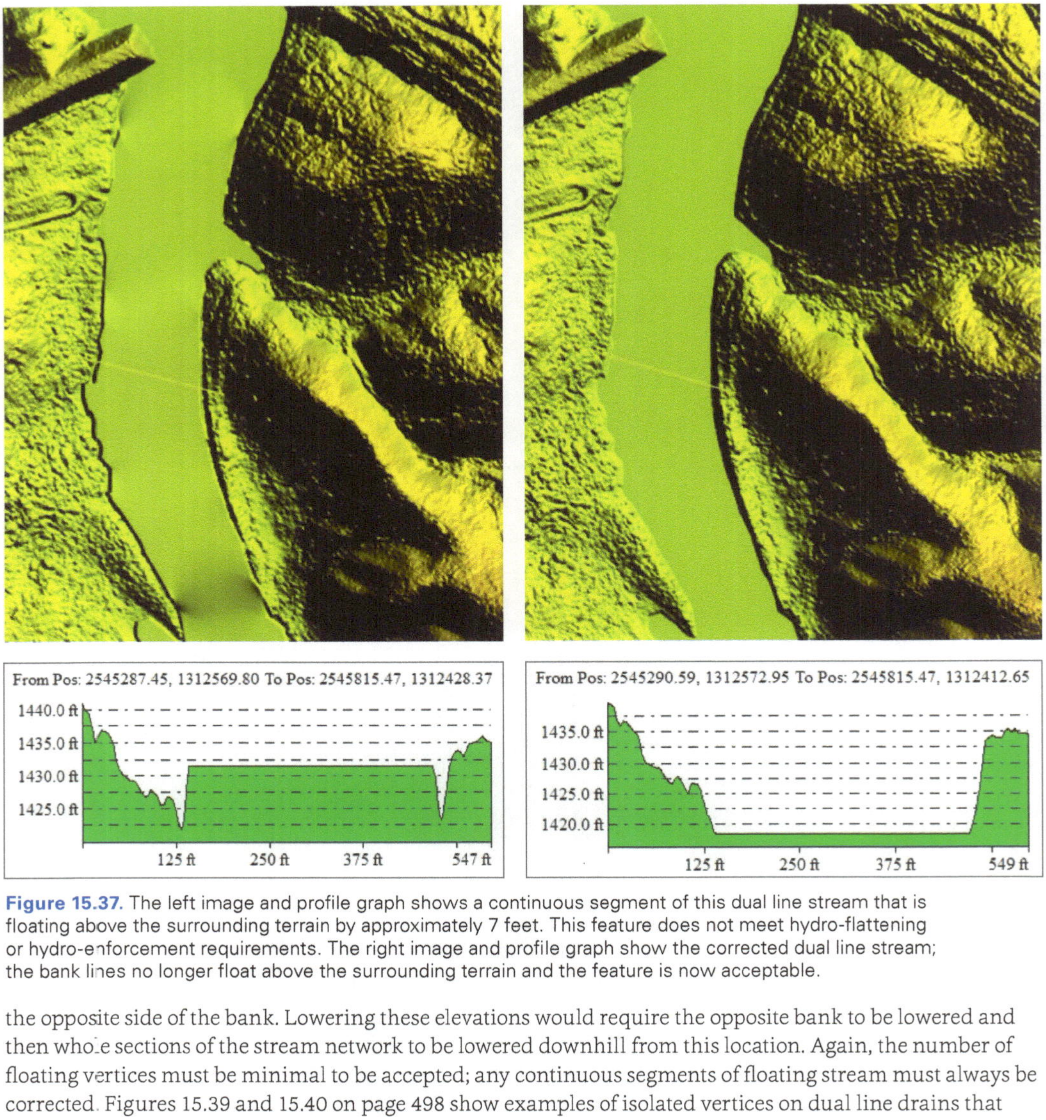

Figure 15.37. The left image and profile graph shows a continuous segment of this dual line stream that is floating above the surrounding terrain by approximately 7 feet. This feature does not meet hydro-flattening or hydro-enforcement requirements. The right image and profile graph show the corrected dual line stream; the bank lines no longer float above the surrounding terrain and the feature is now acceptable.

the opposite side of the bank. Lowering these elevations would require the opposite bank to be lowered and then whole sections of the stream network to be lowered downhill from this location. Again, the number of floating vertices must be minimal to be accepted; any continuous segments of floating stream must always be corrected. Figures 15.39 and 15.40 on page 498 show examples of isolated vertices on dual line drains that are acceptable.

If 3-D stream centerlines are part of the project deliverables, these features must always match the elevations of the dual line streams. There may be small differences between centerlines and bank lines in areas of river bends or complex geometry. However, any differences that cause visible impacts to the DEM, i.e., ridges, spikes, or divots along the centerline, must be corrected. While the centerline must maintain monotonicity, the centerline is an artificial path and should be adjusted to model the correct flow. Generally, the centerline elevations are acquired by extracting the elevation from the monotonic dual line stream banks and applying it back to the centerline or the centerline elevations are collected first and the centerline elevations are then applied to the dual line stream breaklines. These methods help to ensure consistency between dual line stream breaklines and stream centerline breaklines. Figure 15.41 on page 499 shows an

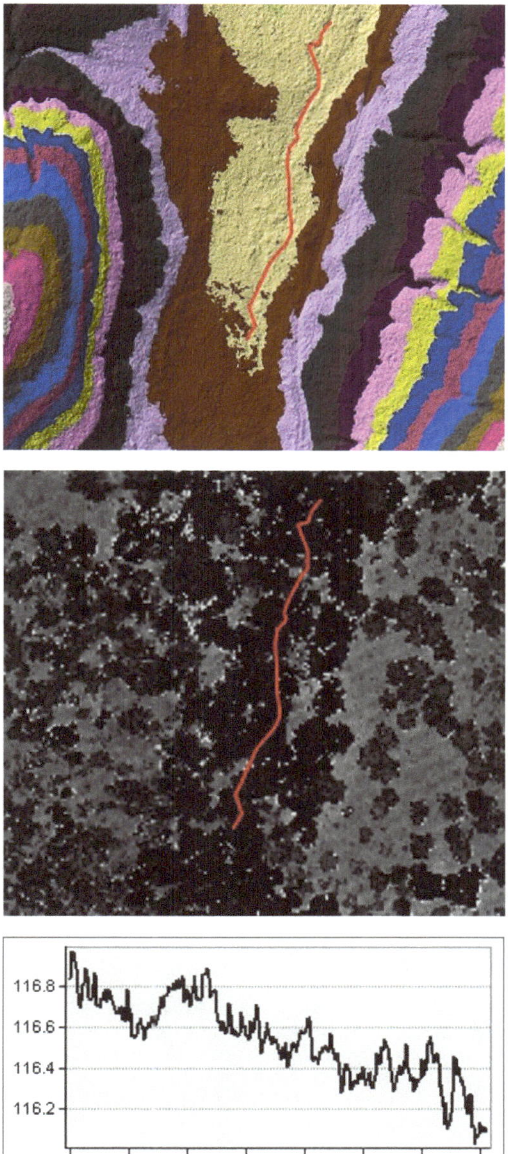

Figure 15.38. The top image shows a terrain where each color band represents 1 meter in elevation change. A profile (red line) has been taken within the dry streambed and the profile graph shows the dry streambed is not monotonic. The intensity image of this feature (with profile location) is shown in the bottom image. Lidar will accurately model dry features; breaklines should not capture dry features or enforce monotonicity in dry streambeds.

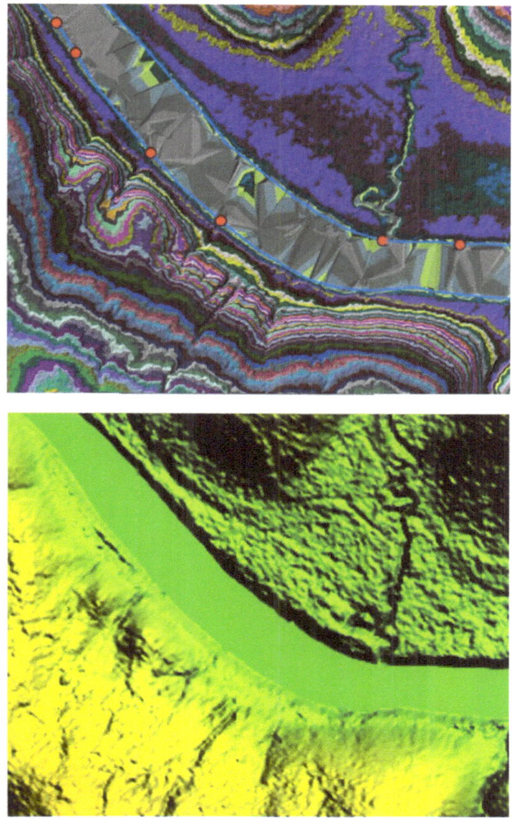

brkl_z	Z	DZ
498.7960	498.71	0.0737
498.0916	497.56	0.5231
498.9925	498.37	0.6131
498.8688	498.12	0.7335
498.9758	498.22	0.7529
498.9148	497.83	1.0730

Figure 15.39. The image on the top shows a terrain (colored by 1-foot elevation bands) overlaid with dual line hydrographic breaklines (blue lines). The red circles identify breakline vertices that are floating compared to the lidar elevations at the same xy coordinate. The table shows that these vertices are floating between 0.08-1 foot. Because the floating vertices are isolated, rather than continuous segments, these floating vertices have little to no impact on the final DEM, shown in the image on the bottom. As these floating vertices have little to no impact on the final DEM, these isolated floating vertices are acceptable.

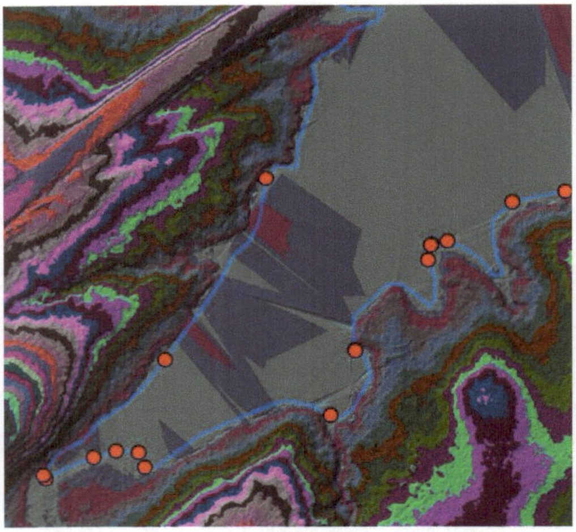

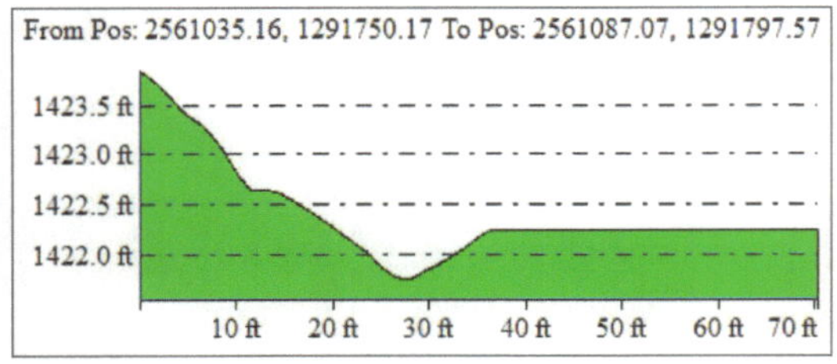

brkl_z	Z	DZ
433.4999	433.496639	0.0032
433.4999	433.490734	0.0091
433.4999	433.466784	0.0331
433.4999	433.394277	0.1056
433.4999	433.393949	0.1059
433.4999	433.392309	0.1076
433.4999	433.376889	0.1230
433.4999	433.371967	0.1279
433.4999	433.367702	0.1322

brkl_z	Z	DZ
433.4999	433.362453	0.1374
433.4999	433.358188	0.1417
433.4999	433.355235	0.1446
433.4999	433.351626	0.1482
433.4999	433.338831	0.1610
433.4999	433.336206	0.1637
433.4999	433.31849	0.1614
433.4999	433.23286	0.2670
433.4999	433.16265	0.3372

Figure 15.40. The image on the top shows a terrain (colored by 1-foot elevation bands) overlaid with dual line hydrographic breaklines (blue lines). The red circles identify breakline vertices that are floating compared to the lidar elevations at the same xy coordinate. The accompanying table shows that these vertices are floating between 0.003 feet to 0.34 foot. Because the floating vertices are isolated, rather than continuous segments, these floating vertices have minimal impact on the final DEM, shown in the image on the bottom. One small area is floating in the DEM (identified with the arrow) by up to 0.5 feet. This small area of floating hydrographic feature is isolated on the end of a branch or channel. Adjusting the feature would require lowering the entire network and may cause issues with monotonicity, connectivity, and other portions of the network excessively digging into the surrounding terrain. This small, isolated area that is minimally floating is acceptable.

example of stream centerline breaklines that were elevated compared to the dual line stream breaklines, causing a ridge down the center of the stream feature.

Because dual line streams and rivers have the monotonicity and consistent parallel bank requirements, more leeway is given to dual line streams and rivers than single line streams and rivers in regards to allowable floating vertices. While some isolated floating vertices may be allowable on single line streams, if the vertices are floating by enough to cause the single line stream to appear as raised or elevated in the final

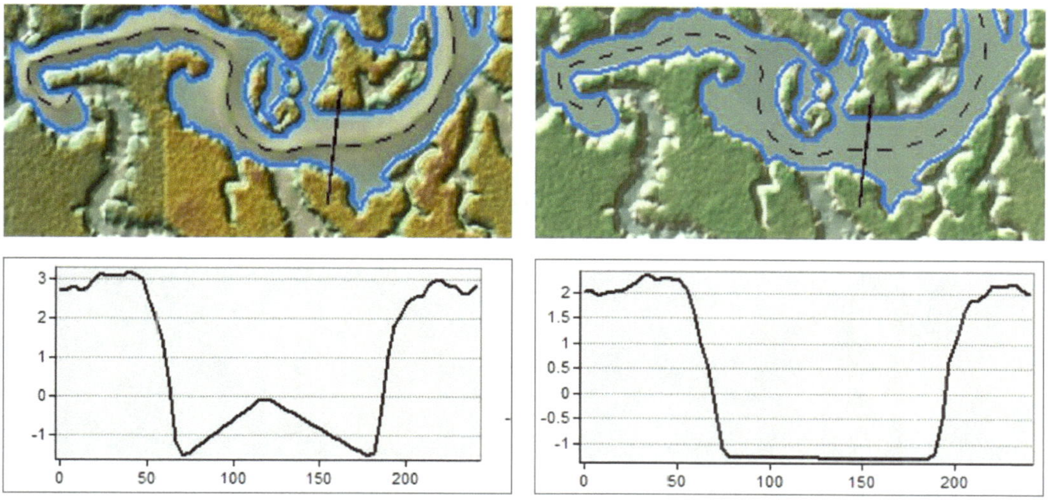

Figure 15.41. The image and graph profile on the left show a stream centerline breakline which is approximately 1 foot higher in elevation than the dual line stream bank breaklines, and results in a ridge down the center of the stream feature. The image and graph profile on the right show the same feature, after corrections have been applied to make the stream centerline elevations consistent with the dual line stream banks; the ridge is removed and the centerline functions as an artificial path.

DEM, then these features should be corrected. While single line streams must be monotonic, corrections only affect a single breakline, rather than dual breaklines for dual banks. Also, isolated vertices on dual line streams will still be identified as water in the final DEM because the dual line stream will be flattened as an area in the final DEM. Single line streams are collected as a single breakline or polyline and not an area polygon or area defined by dual polylines. Single line streams are enforced along the breakline edge in DEMs and are not flattened as an area in DEMs. As only an edge is enforced, floating single line drain vertices act as a ridge or barrier in the final DEM instead of functioning as a hydrographic feature. For this reason, any floating vertex on single line streams that will cause the feature to be elevated in the DEMs must be corrected. Figure 15.42 on page 500 shows an example of floating vertices on a single line stream that must be corrected.

3-D hydrographic breaklines should all be slightly digging into the lidar elevation surface. This ensures that the hydrographic features are at or below the surrounding terrain per hydro-flattening and hydro-enforcement requirements. Breakline vertices that are digging by values up to approximately ½ meter are considered to be at optimal elevations as they are slightly below the surrounding terrain but not digging into the surrounding terrain. Breakline vertices within the optimal digging threshold ensure compliance with hydro-enforcement and hydro-flattening requirements (as long as their horizontal placement is acceptable), but most hydrographic vertices will have much higher ΔZ values than -0.5 meters. Most hydrographic breakline vertices will dig by more than ½ meter as lidar returns along hydrographic features will undulate and vary by several meters. Monotonicity and network connectivity in breaklines remove these undulations to model consistent downhill gradient flow. Across bank consistency and water body consistency where all elevations of a water body are at one constant elevation require flattening in the breaklines, which does not naturally exist in the lidar data along shorelines. The removal of undulations and variability in the breaklines for monotonicity and flattening requirements can result in breakline vertices digging by more than ½ meter. Additionally, areas of interpolation will create even higher ΔZ values, as shown in Figure 15.33 on page 492. Lastly, terrain type can cause features to appear to be digging if they are near elevated features, such as mountains, or have very steep banks, such as some reservoirs. Figure 15.43 shows an example of a water body

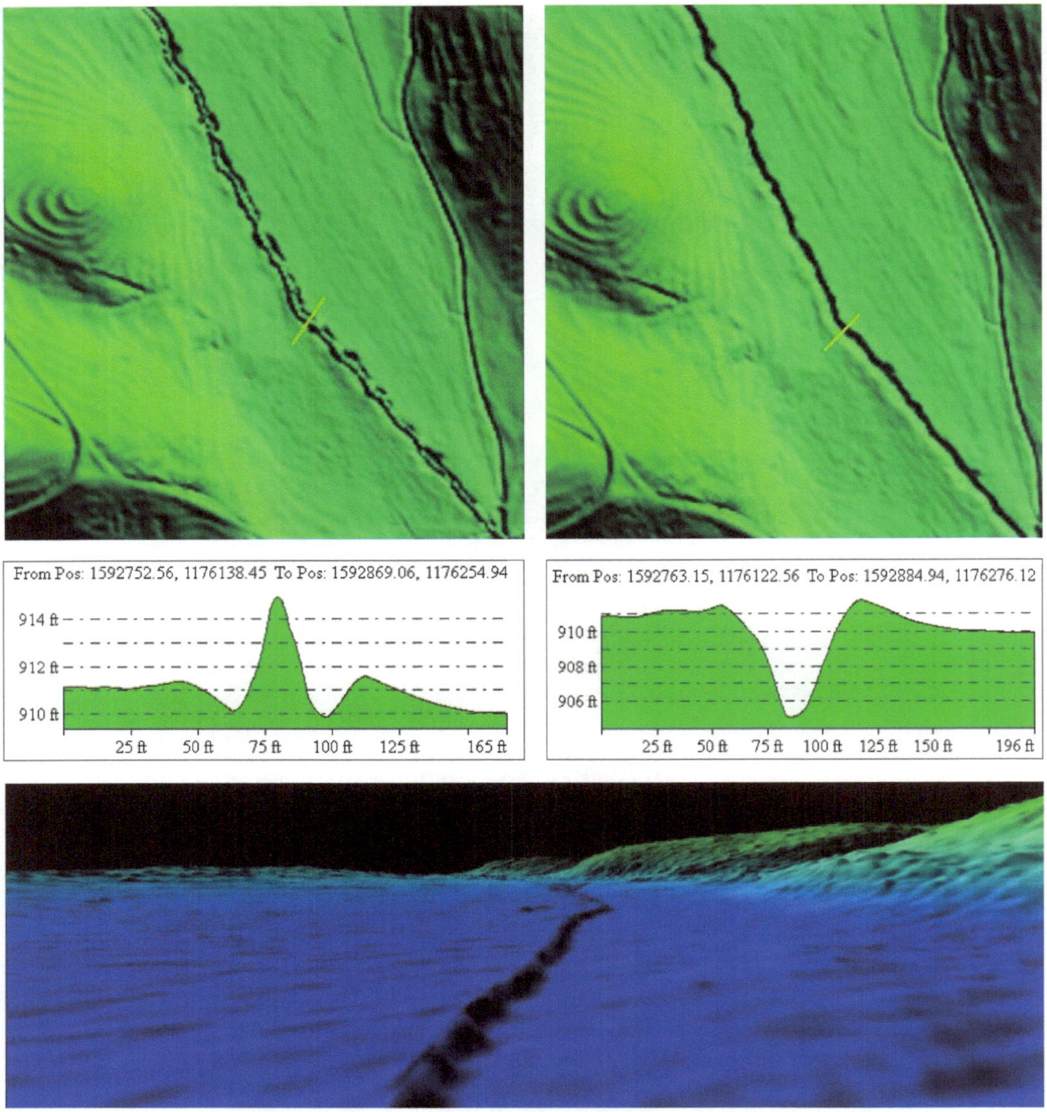

Figure 15.42. The image and accompanying profile on the left show a single line stream that is floating in the DEM and the image and accompanying profile on the right show the same feature after corrections. The bottom image shows a 3-D perspective view of a raised single line stream. Any floating vertices on single line streams that cause the feature to act as a ridge or elevated feature rather than a hydrographic feature must be corrected.

that appears to be digging because it is near a mountain, but is actually within acceptable limits. Because of these different constraints, a maximum digging value cannot be established, but must be reviewed per project or even on a per feature basis.

Water bodies that are digging can almost always be corrected as they are set to one elevation and can be raised, unless they are connected to stream/river features which cannot also be raised. However, stream/river features that are digging may not be able to be adjusted due to monotonicity, connectivity and consistent

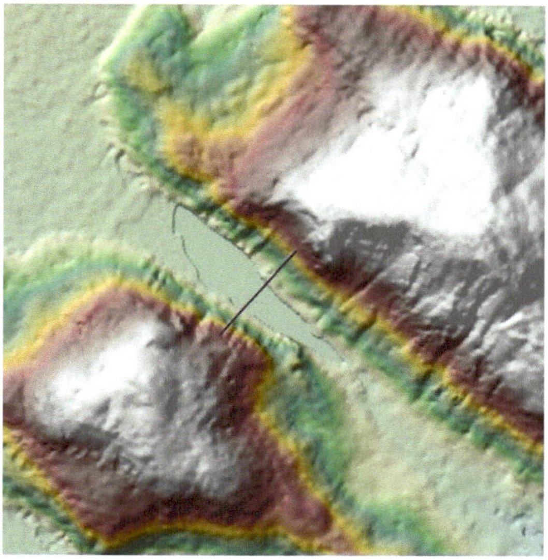

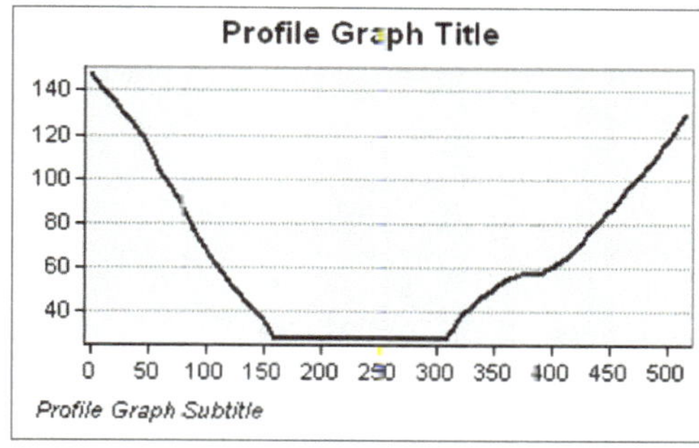

Figure 15.43. When viewing the profile (right image, in meters) for the water body flattened in the DEM (left image), the profile looks as though the water body may be digging and may be too far below the surrounding surface. But when looking at the DEM, this water body is surrounded by mountains and actually fits the surrounding elevation quite well. Different terrain types can cause features to appear to be digging though they may be within acceptable limits.

bank to bank elevation constraints where raising stream/river elevations could cause other portions of the stream network to float, be non-monotonic, or cause bank to bank flatness issues. Water body features that are excessively digging should be corrected. While excessive is a subjective term, excessive can generally be described in two ways: (1) The feature is digging enough that it would cause anomalous sinks in the DEM, or (2) The feature is digging enough that it would cause an erroneous build-up of contours. The smallest appropriate equivalent Class 1 contour interval for each project would need to be known (see Table 3.10 in Chapter 3) as this is dependent on the vertical accuracy of the project. An erroneous build-up of contours rather than just a build-up of contours is suggested as an indication of excessive digging along water bodies because some features may actually have steep banks or slopes that result in a legitimate build-up of stacked contours. An erroneous build-up of contours occurs when at least four or five contours represent a water body due to the water body's low elevation and not valid steep topography. Breaklines represent features as they exist at the time of collection. This means that if a hydrographic feature is drier than normal at the time of collection, it may be large enough to meet breakline capture requirements, but it may also have more exposed banks that could result in stacked contours. Examples of features that may legitimately be represented with stacked contours include reservoirs with lower than normal water levels, water bodies at the bottom of old quarries with steep sides, and streams and rivers with steep banks. Additionally, if a feature is only represented by two or three contours, this would not be a large enough build-up of contours to warrant modification to the dataset. Two or three stacked contours generally do not impact representation or analysis of features negatively and the amount of digging represented by two or three stacked contours will not negatively impact the DEM. Figure 15.44 on page 502 shows an example of a water body excessively digging, resulting in an anomalous sink in the DEM. Figures 15.45 on page 502 through 15.47 on page 503 show examples of stacked contours compared to flattened hydrographic features in bare-earth DEMs.

Breaklines at the boundaries of tidally influenced water bodies may be handled differently for different projects and different clients, depending on the intended uses of the data. Due to tidal variations, the exact land/water interface will change along tidally influenced hydrographic features, such as coastal waters, bays, inlets, and gulfs. The tidal variations can affect the delineation or capture of the land/water interface between different collections, different missions or lifts, and even between different flight lines within the same mission or lift. Some projects may have specific requirements to acquire data along tidally influenced features within a specific tidal range to minimize the tidal variations, but discontinuities along the land/

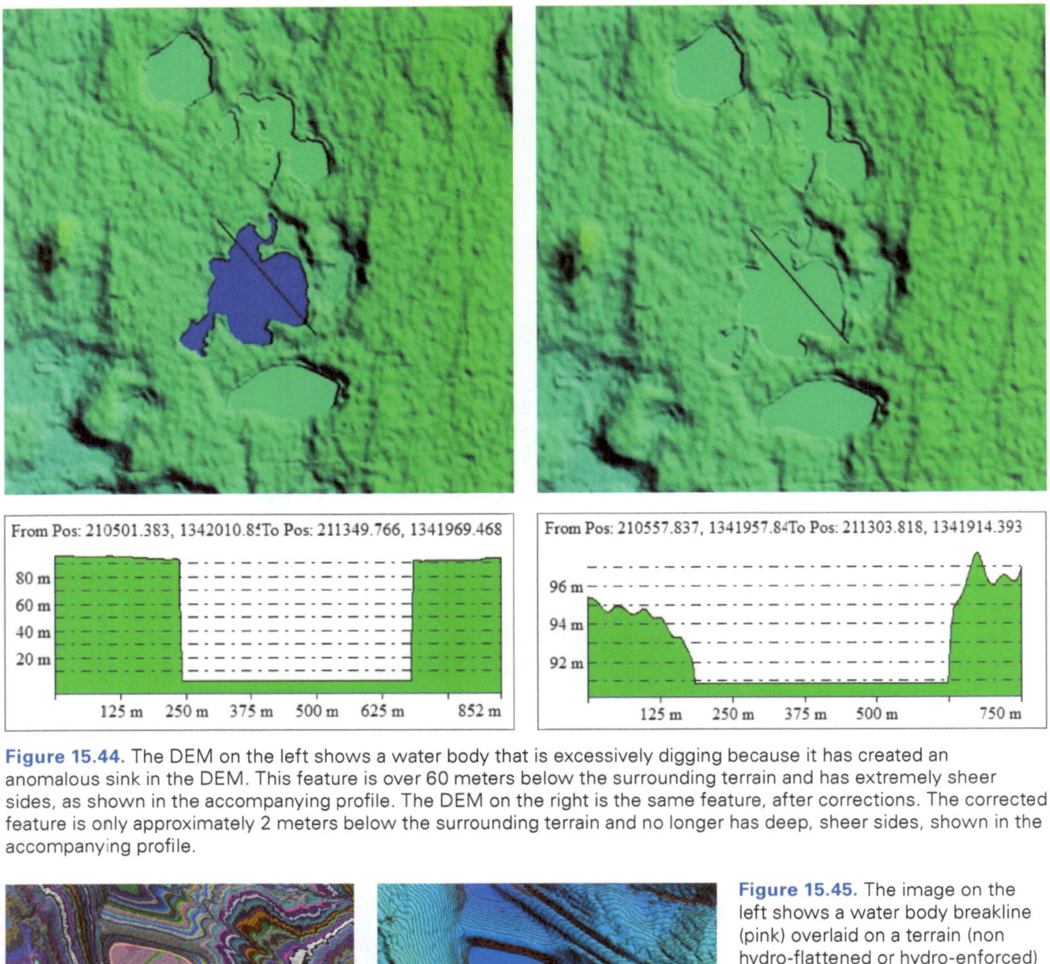

Figure 15.44. The DEM on the left shows a water body that is excessively digging because it has created an anomalous sink in the DEM. This feature is over 60 meters below the surrounding terrain and has extremely sheer sides, as shown in the accompanying profile. The DEM on the right is the same feature, after corrections. The corrected feature is only approximately 2 meters below the surrounding terrain and no longer has deep, sheer sides, shown in the accompanying profile.

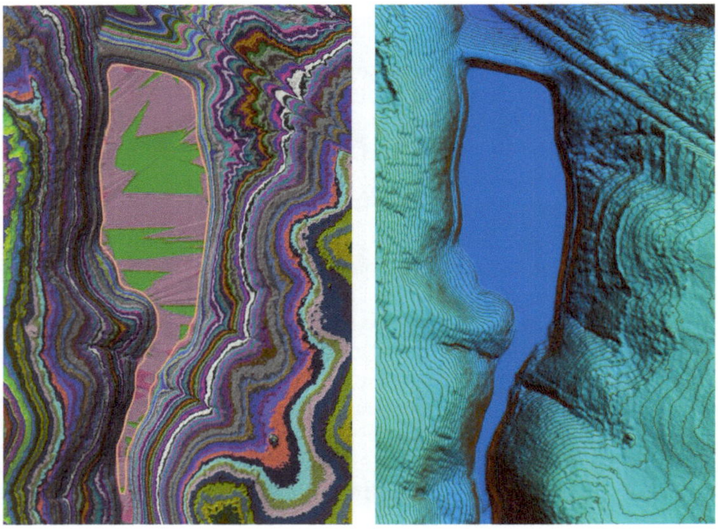

Figure 15.45. The image on the left shows a water body breakline (pink) overlaid on a terrain (non hydro-flattened or hydro-enforced) where each color band represents 1 foot in elevation. The right image displays the same feature in a hydro-flattened DEM with 1-foot contours (brown) overlaid. While there may appear to be stacked contours in the image on the right, the terrain, which is created from source elevation data only, shows that this feature does indeed have steep terrain surrounding it and the stacked contours are legitimate. Breakline elevations at these types of features should not be raised due to stacked contours.

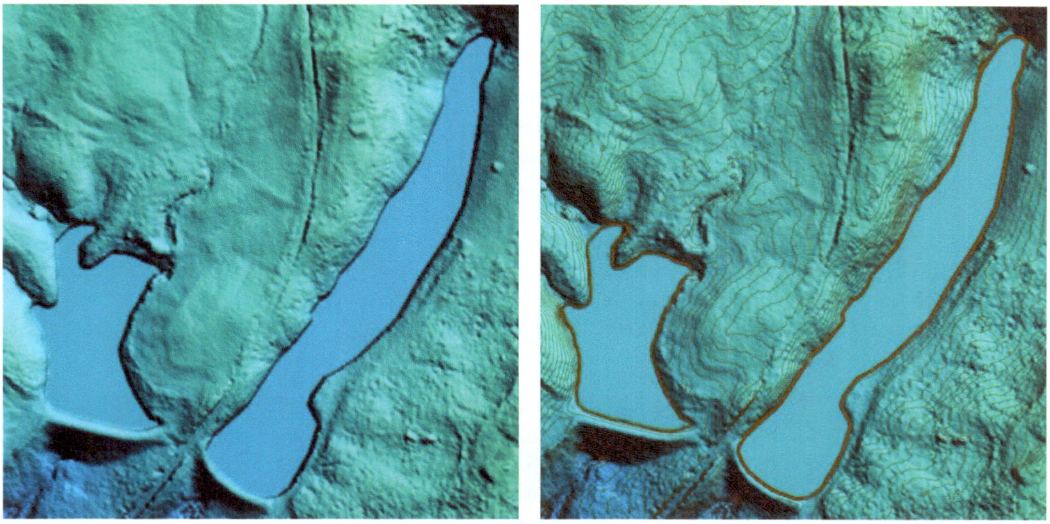

Figure 15.46. The left image shows two water body features in a hydro-flattened DEM. The right image shows the same features with 2-foot contours (brown) overlaid. The build-up of eight stacked contours around these features (the brown lines around the water body polygons) when the surrounding terrain is not that steep indicates that both features are excessively digging and the breakline elevations should be raised. Additionally, there is heavy shadowing in the hillshade around the shoreline of these features and the shorelines are rather sheer. Heavy shadowing in hillshades combined with sheer shorelines that seem unnatural can be further indicators that breaklines may be excessively digging.

Figure 15.47. These images show the same features as Figure 15.46 after corrections have been applied. There is no longer a build-up of stacked contours around these features; only two contours represent each feature. Additionally, heavy shadowing and sheer or steep shorelines are no longer present in the hillshade or DEM.

water interface may still exist. All project requirements should specify how tidally influenced features should be captured in the breaklines and elevation dataset. Method #1 for tidal breakline collection is to set all tidal breaklines to a single elevation. An elevation low enough should be selected so that no portions of the tidal breaklines are floating. However, this will cause some portions of the tidal breaklines to dig more than others. If one elevation is used to define all tidal breaklines, digging along those breaklines must be acceptable. If the elevation of the tidal breaklines must be set to zero elevation, then there may still be digging portions of the breakline, but horizontal placement of the breakline may be affected as the breakline will need to be placed approximately along the zero contour line or slightly below so that the tidal breaklines will not be floating. Forcing tidal breaklines to zero elevation may artificially "flatten" or cover some ground if the exposed ground is at an elevation less than zero. Figure 15.48 shows two images where method #1 was used on tidal breaklines. In the left image, a single elevation was used while keeping as much ground as possible in the bare earth DEM. In the right image, the tidal breakline was set to 0 meters, which causes some low ground elevations to be hydro-flattened.

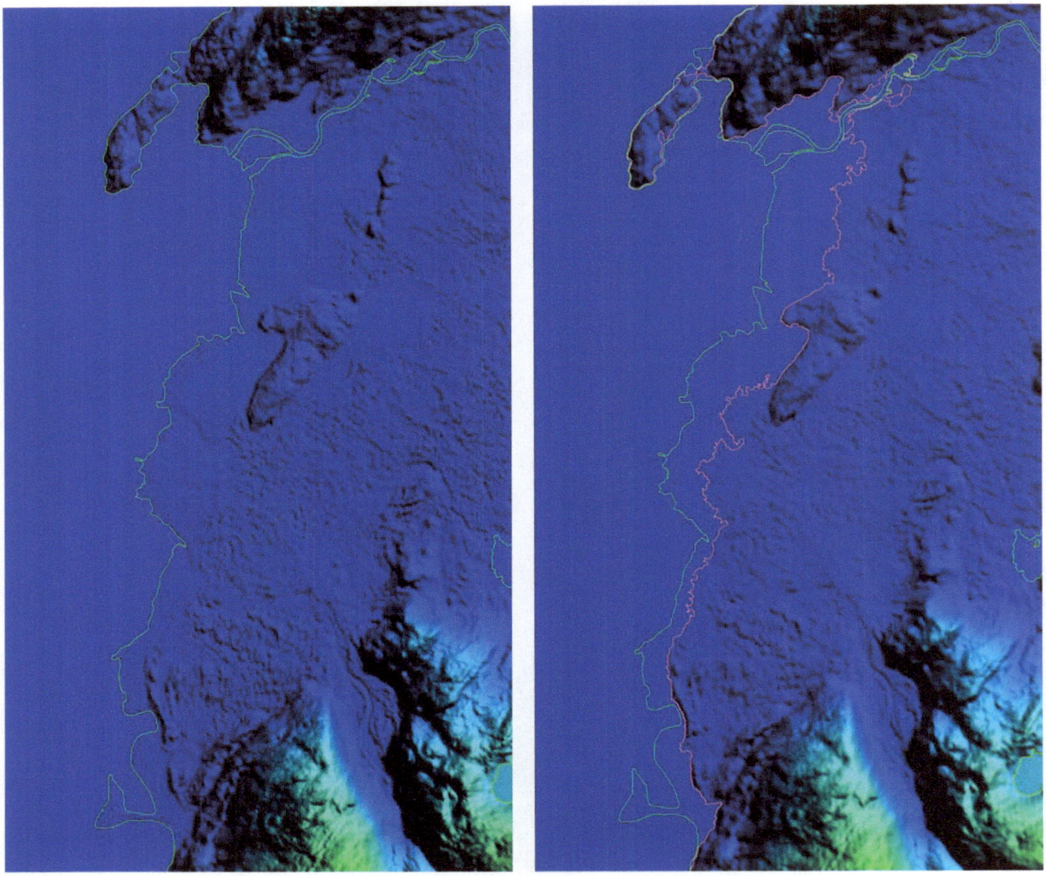

Figure 15.48. Both DEMs show the same coastal area. The left DEM has been hydro-flattened to include as much ground as possible. The elevations along these tidal breaklines (green) may vary between flight lines and may not be monotonic as they represent the ground and tidal conditions at the time of acquisition. The right DEM has been hydro-flattened with breaklines (pink) set to one constant elevation of 0 meters. The two breaklines overlaid on the right DEM show that setting the elevation to 0 meters causes some exposed ground to be hydro-flattened.

Method #2 for tidal breakline collection (specified by USGS in their Lidar Base Specification Version 1.0, 1.1, 1.2, and 1.3) is to retain all horizontal and vertical discontinuities resulting from tidal variations so that as much ground data as possible is retained and used in the DEMs. This method allows discontinuity in monotonicity between different flight lines. However, the breakline elevations should match the lidar elevations quite well with no floating or excessively digging breaklines. Some tidal breaklines may still dig if connected streams/rivers drain into the tidal breaklines and are at much lower elevations. Figure 15.49 shows an example where an island was only partially collected as it was partially submerged under higher tides during a portion of the acquisition. Using method #2 for tidal breakline collection, the portions of exposed island were maintained in the bare-earth DEM in order to retain as much ground data as possible, even if vertical and horizontal discontinuities are present.

Figure 15.49. This DEM shows a partial island collected offshore. Due to tidal variations, only a portion of this island was exposed during acquisition. Depending on client needs and requirements, all exposed ground may be kept 'as is' in the data resulting in some discontinuities, but preserving as much usable ground as possible. Alternatively, data can be artificially flattened or modified to only represent contiguous shorelines.

Method #3 allows tidal breaklines to be "feathered." Feathering will remove or mostly remove horizontal and vertical discontinuities between flight lines with tidal variations. Feathered tidal breaklines will usually be monotonic or mostly monotonic but the feathering requires artificial adjustment to both the horizontal location of tidal breaklines and the vertical elevations. The horizontal adjustments usually result in ground being artificially flattened. Vertical adjustments should minimize floating but there will be digging portions of the tidal breaklines where they must be carried through higher tidal conditions at lower elevation levels to maintain vertical consistency with the surrounding flight lines and areas that may have been acquired under different tidal conditions.

Non-hydrographic breaklines, such as 3-D road centerlines, must not be floating or excessively digging into the surface and generally should match the elevation surface models quite well. Non-hydrographic features that will not be used for hydro-enforcement or hydro-flattening, will not have higher ΔZ values between the breakline elevations and lidar elevations due to hydro-enforcement or hydro-flattening rules such as monotonicity or maintaining constant elevations. As non-hydrographic breaklines are not following hydro-flattening or hydro-enforcement rules, corrections to high ΔZ values on non-hydrographic breaklines are much more feasible and practical without causing additional issues. Additionally, non-hydrographic breaklines generally represent features that have less elevation change/undulation in the lidar elevation dataset compared to hydrographic features. While there are some exceptions, such as breaklines that are being used to enforce ridges, cliffs, or vertical features that may not have been completely modeled in the lidar dataset, most non-hydrographic breaklines should match the lidar elevations within a smaller tolerance than that allowed for hydrographic breaklines as these breaklines are not flattening features or enforcing monotonicity but generally represent features that may have a higher density of more reliable lidar points. Even though some non-hydrographic breaklines, such as road centerlines or transportation features, may be part of a connected network, the breakline elevations can always be adjusted as they are not bound by monotonicity or other hydro-enforcement and hydro-flattening requirements.

To summarize, the following guidelines should be used when producing and evaluating vertical breakline placement:

- All vertices on features of constant elevations, such as ponds or lakes, must be below the surrounding terrain. The feature elevation should be lowered if even a single vertex is identified as floating
- Continuous segments of streams or rivers (dual line or single line) that are floating above the surrounding terrain must always be corrected
- Isolated or even single floating vertices on streams, rivers, or other hydrographic features representing gradient flow that are a result of non-monotonic breaklines must be corrected
- Isolated or a minimal number of moderately floating vertices on a monotonic dual line stream may be acceptable if the elevations are consistent with the opposite bank which is not floating or if corrections would require severe or infeasible alterations to the stream network
- 3-D stream centerlines must match in elevation to the dual line stream banks. While some small differences are allowed near stream bends or in areas of complex geometry, differences in elevation that result in ridges or troughs along the centerline must be corrected
- All floating vertices on single line streams that result in the hydrographic feature appearing raised or elevated in the DEMs must be corrected
- Some high ΔZ values are acceptable if they result from normal or expected characteristics of the lidar dataset, such as increased interpolation along vegetated stream banks or elevated vertical features, i.e., seawalls and dams. If the horizontal breakline placement of the breakline is verified to be acceptable, the z-elevations are consistent and monotonic, and the high ΔZ values can be attributed to a characteristic of the lidar dataset, then no corrections should be applied in these locations
- Features can be considered excessively digging if they create anomalous sinks in the DEM or if they create an erroneous build-up of contours. The minimum appropriate contour interval suitable for each project dataset must be utilized. Features that would legitimately be represented by a stack of contours, such as quarries, features with lower than normal water levels, and streams/rivers with steep banks or features that only appear to be digging due to the surrounding terrain, should normally be acceptable. Water bodies should generally be corrected when identified as excessively digging. Streams/rivers, however, are generally acceptable if raising elevations would cause other portions of the stream network to float or fail monotonicity, connectivity, and consistent bank to bank elevation requirements
- Tidal breaklines may be captured and defined many different ways. Specific methods may be defined by clients for each project. When one of the defined three methods are used, tidal breaklines should not float, but may dig. Monotonicity requirements may also be relaxed along tidal breaklines
- Non-hydrographic breaklines generally should not be floating or excessively digging and should match the elevation surface models very closely. Non-hydrographic breaklines do not follow monotonicity, hydro-flattening, or hydro-enforcement rules which can cause high ΔZ values and in some instances make corrections to those high ΔZ values infeasible and impractical. While some non-hydrographic breaklines, such as transportation features, may be part of a network, the elevations can still be corrected to better match the source lidar

Breakline Topology

All breakline datasets should be verified for correct topology. Topology rules applicable to each project will depend on the collected breaklines and requirements for that specific project. But breaklines that will be used to hydro-flatten or hydro-enforce DEMs should be verified for the following:

- Monotonicity
- Vertical and Horizontal Connectivity
- Flatness or Consistency Bank to Bank
- Adjacent Vertex Elevation Changes (only on feature types that are supposed to be one constant elevation, such as water bodies)

Monotonicity should be checked on all stream features (single and dual banks) to ensure proper downhill flow. Connectivity ensures that features edge-match, snap, and intersect properly both horizontally and vertically (x,y, and z). All hydrographic breaklines should be verified for connectivity within each layer and between different layers. Connectivity may be required for other non-hydrographic breaklines as well, such as road breaklines. Dual line stream banks should be flat from bank to bank. If stream centerlines have been collected, these features should be verified for consistency between their z-elevations and the z-elevations of the dual line stream/river breaklines, as well as for proper connectivity with any connected network features. Pond and lake (or other water body features that should be a single elevation) breaklines that will be used for hydro-flattened and hydro-enforced DEMs should always be verified that all vertices on each feature have one, constant z-elevation.

In addition to topology rules specific for hydro-flattened or hydro-enforced DEMs all breaklines should always be verified for general integrity of the breaklines including dangle checks, duplicate breaklines, overlapping breaklines, and invalid geometry. These integrity checks will help ensure the breaklines were collected to project specifications and that the breaklines will only provide beneficial impacts to DEMs or other surfaces modeled from the breaklines or analyses performed with the breaklines. The following list includes common breakline topology checks. The types of breaklines collected for each project, as well as the intended use of the breaklines, will determine what topology checks may be applicable to each project. Additionally, some checks should be run on each breakline feature type and between breakline layers, meaning a single check could be run multiple times depending on the type of breaklines collected for each project.

- Dangles
- Duplicate Geometry
- Unnecessary Polygon Boundaries
- Overlapping Geometry (including polygons and incorrectly crossing polylines)
- Polygon Gaps
- Vertical and Horizontal Connectivity

Invalid dangles on line features are start/end points that exist due to incomplete or incorrect capture of a feature. Valid dangles are start/end points of line features at upstream origins or where breaklines end at project boundaries. All invalid dangles should be corrected. Duplicate geometry should be avoided as duplicate geometry increases the risks of versioning issues, processing issues, and can increase processing time for tools and analyses. Many features represented by polygons will be represented by a single polygon covering the entire feature. When numerous, smaller adjacent polygons are used to delineate a feature, these polygons are identified as segmented polygons. While segmented polygons may be acceptable or even desired in some circumstances, such as when used to define lift extents in tidal waters where each lift has a different elevation, unnecessary segmentation (unnecessary polygon boundaries) can decrease the aesthetics of breaklines and can increase the chances for other errors occurring between adjacent polygons.

Overlapping geometry (polygons and polylines) can lead to incorrect representations of features, incorrect versioning, and coding errors. Polygon gaps share similarity to invalid dangles that occur on polylines and can indicate incomplete or incorrect capture of a feature. However, as polygons are often used as boundaries for additional coding or identification of areas, polygon gaps may also lead to errors in coding, identification, and representation of features. While horizontal and vertical connectivity, or absolute agreement of Z-values for all points or vertices existing at the same X-Y coordinate, is usually associated first with features that will be used to hydro-flatten or hydro-enforce DEMs, connectivity is also important for any breakline that will be used to enforce a DEM as connectivity ensures all breaklines snap correctly (x,y, and z). If breaklines are 2-D, only horizontal connectivity should be verified.

Figures 15.50 through 15.55 show examples of topology errors along with some examples on how these topology errors can impact the final DEM surfaces.

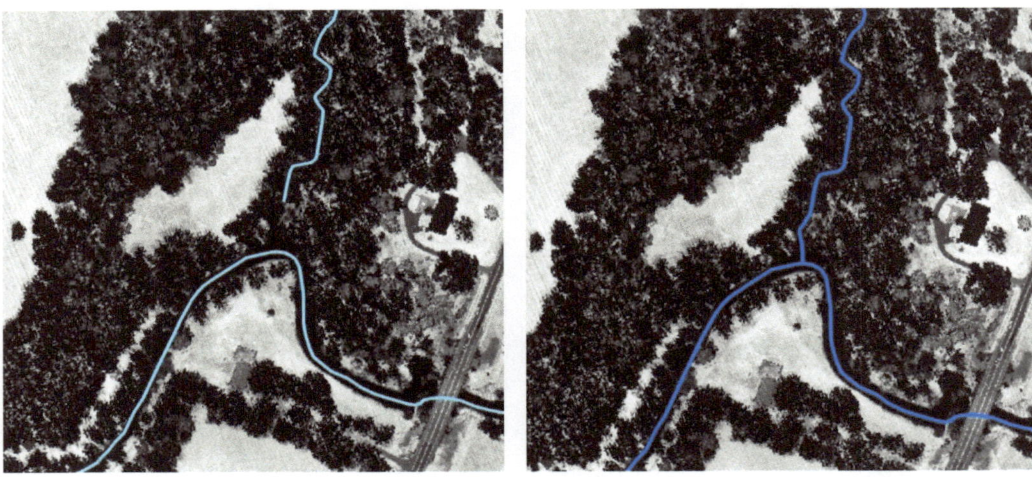

Figure 15.50. Both images show the same single line drain network overlaid on lidar full point cloud intensity imagery. The breaklines on the left illustrate a dangle where the northern reach does not connect to the southern reach. This dangle has been corrected in the breaklines shown on the right. Dangles on hydrographic features can cause hydro-enforcement or hydro-flattening issues and dangles on all types of breaklines can cause modeling and analysis issues because dangles create an incorrect disruption in the breakline network and connectivity.

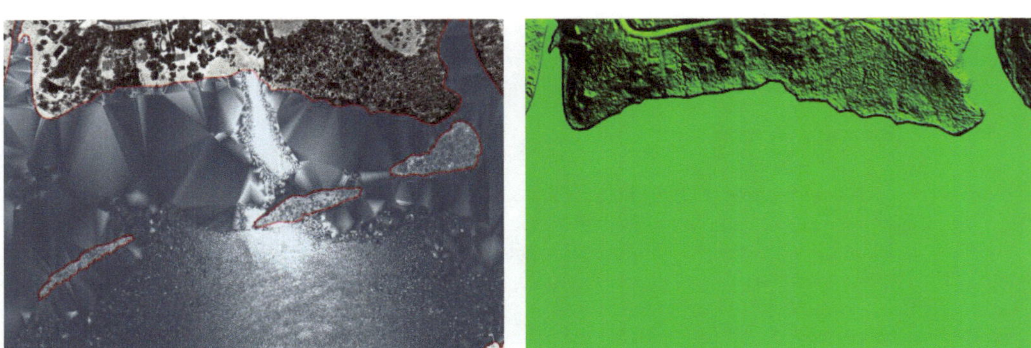

Figure 15.51. The image on the left shows hydrographic breaklines (partially transparent blue polygons outlined in red) overlaid on lidar intensity imagery. The three islands are erroneously covered by overlapping polygons. The result of the overlapping polygons is the incorrect flattening of these features in the DEM, shown in the image on the right. The three islands should instead be represented as "holes" or voids within the larger water polygon.

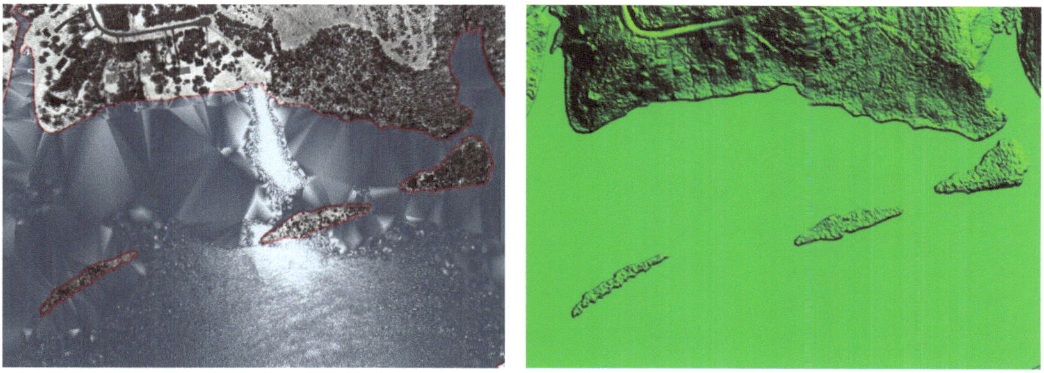

Figure 15.52. The image on the left shows hydrographic breaklines (partially transparent blue polygons outlined in red) overlaid on lidar intensity imagery. The incorrect overlapping polygons covering the three islands have been removed from the breakline dataset. These three islands are now modeled in the corrected DEM, shown in the image on the right.

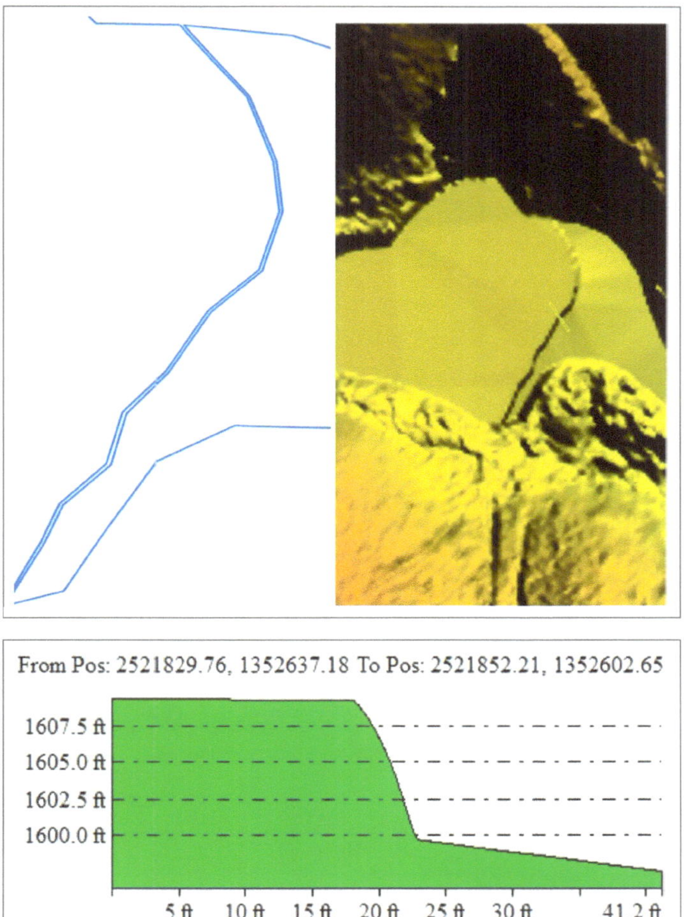

Figure 15.53. The image on the left shows a segmented hydrographic breakline (blue) at a very close scale. Not only is this segment unnecessary, but there are numerous errors associated with it including polygon gaps, horizontal and vertical connectivity, and monotonicity errors. The two adjacent segments do not snap horizontally (leaving a small gap between the segments), or vertically (creating a waterfall between the two segments). The profile graph shows the erroneous waterfall to be approximately 7.5 feet in height. This waterfall causes the dataset to fail monotonicity requirements as the dual line hydrographic feature does not flow downhill consistently or correctly. The right image shows that these errors are clearly visible in the resulting DEM (viewed at a smaller scale than the left image).

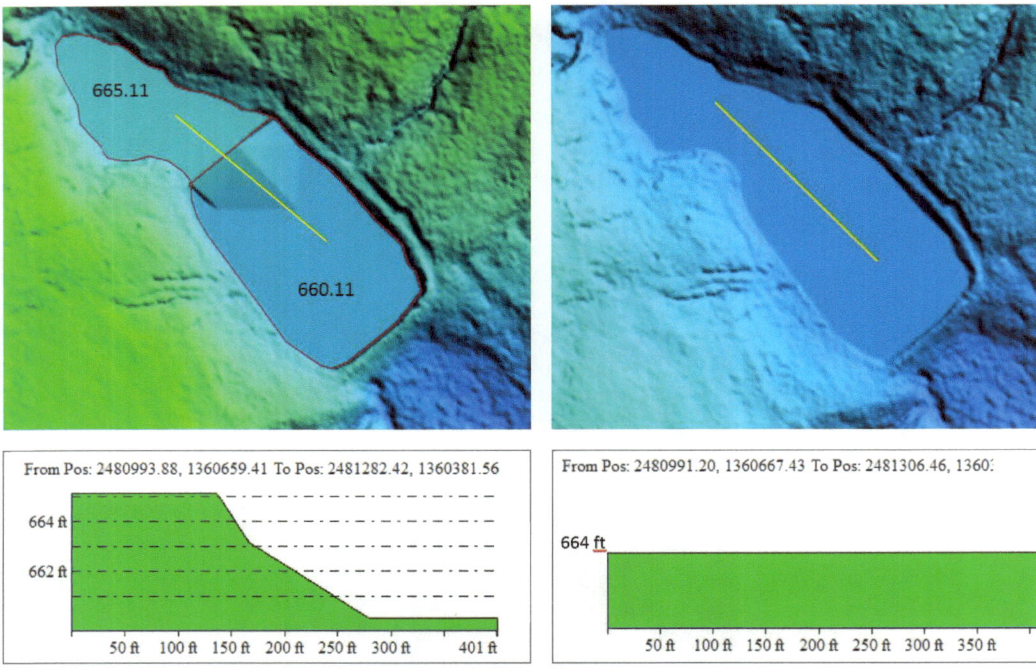

Figure 15.54. The top left image shows the result of a water body that should have been set to one constant elevation, but vertices on this water body breakline (red outline) actually varied in elevation (adjacent vertex elevation changes). The top portion of the breakline was set to an elevation of 665.11 feet while the bottom portion of the breakline was set to an elevation of 660.11 feet. The bottom left profile graph shows these multiple elevations in the final DEM and how the water body is not consistently flattened. The bottom right profile graph shows the corrected water body at one constant elevation and the top right image shows the corrected DEM with a consistently flattened water body feature.

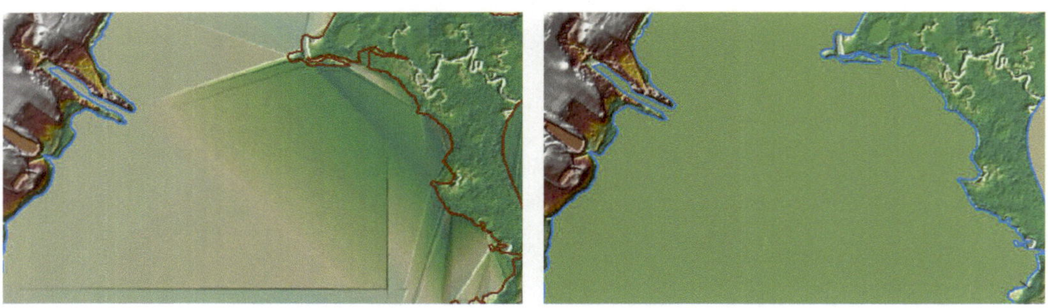

Figure 15.55. The image on the left shows a DEM overlaid with dual hydrographic breaklines (blue) and island shoreline breaklines (brown). This DEM is not flat bank to bank (or bank to island shoreline to bank). The numerous water artifacts and interpolation artifacts within the hydrographic feature result from inconsistent elevations on the hydrographic and island breaklines. Opposite or parallel elevations along shorelines should be consistent to create hydrographic features that are flat from bank to bank. The image on the right is the same feature overlaid with corrected hydrographic (and island shoreline) breaklines in blue. The corrected breaklines have consistent elevations from bank to bank, resulting in a DEM that is flat from bank to bank and does not contain the numerous water artifacts that are present in the DEM on the left.

Macro Level Review of DEM Data

Macro level reviews generally verify project requirements, such as formatting and tile size, and ensure all necessary files have been delivered and are readable. Macro level reviews should include checking the extent of coverage, formatting of the DEMs, and all pertinent DEM dataset properties. The DEM formatting of tiled versus larger mosaics and the number of DEMs should be verified against project requirements along with adherence to the required naming convention. Depending on the number of DEMs, the DEMs can be directly loaded into various software programs for an overview and verification of complete coverage. Alternatively, DEMs can be mosaicked into larger datasets that will load faster in some software programs due to the use of pyramid and overview files. This overview verification can generally occur at a small scale that would allow the reviewer to see the full project or large sections of the project at once, such as county or even multiple counties at once. The amount of data viewed at once will vary by computer, software, and data formatting. It is necessary to perform this overview verification rather than assuming that the correct number of tiles with the correct DEM properties translates into complete coverage. Tiles may be corrupt and may not load correctly even when the properties of the dataset are readable.

If project requirements specify that data should be organized in a certain folder structure, then the overall data organization and folder directory should be verified as part of the macro review as well.

All DEM properties specified as project requirements should be verified. DEM properties that should be verified are not limited to, but commonly include the following:

- DEM format type, i.e., TIFF, IMG, Esri GRID
- Spatial reference information
- Cell (or pixel) size
- Number of rows and columns of pixels
- Pixel type, i.e., floating or integer
- Pixel depth, i.e., 32 bit or 16 bit
- Type of compression used, if any
- Existence of pyramids
- Extents (top/bottom and left/right coordinates of the file)
- No Data value
- File names match tile names
- If filenames include coordinates, they match the file data's coordinates

Properties inherent to DEM creation, such as DEM format type, cell or pixel size, pixel type, and pixel depth, should always be verified. While DEM processing does not require the DEMs to have a defined spatial reference, all final and deliverable DEMs should have this property defined. Other DEM properties, such as compression types, pyramids, and file naming conventions, will be dependent on project requirements. These types of DEM properties should always be validated when applicable to a project. Most DEM properties can be verified quickly and thoroughly by using batch tools or customized scripts. Figure 15.56 shows an example of DEM properties.

Raster Information	
Columns and Rows	13836, 23794
Number of Bands	1
Cell Size (X, Y)	5, 5
Uncompressed Size	1.23 GB
Format	TIFF
Source Type	Generic
Pixel Type	floating point
Pixel Depth	32 Bit
NoData Value	-32767
Colormap	absent
Pyramids	level: 6, resampling: Nearest Neighbor
Compression	None
Mensuration Capabilities	Basic
Extent	
Top	1375575
Left	468195
Right	537375
Bottom	1256605
Spatial Reference	NAD83_Alaska_Albers
Linear Unit	Meter (1.000000)
Angular Unit	Degree (0.0174532925199433)
false_easting	0
false_northing	0
central_meridian	-154
standard_parallel_1	55
standard_parallel_2	65
latitude_of_origin	50
Datum	D_North_American_1983

Figure 15.56. Example of DEM properties that should be verified during the macro level reviews.

Micro Level Review of DEM Data

Whereas macro level reviews generally assess formatting requirements of data and ensure no data are corrupted, micro level reviews generally assess the quality of the data. While many projects may have statements requiring a certain level of cleanliness (freedom from artifacts), these requirements are usually much harder to quantify. As these requirements are harder to quantify, it is common to identify all issues during DEM QC that are either clearly outside of project requirements or that may negatively impact end-users of the data. Clients may choose to allow some minor issues to remain in the dataset, but by documenting all issues identified during the DEM QA/QC in reports and metadata, other end-users will be able to better determine if a dataset will meet their needs.

Micro level reviews are accomplished through manual visual reviews at larger scales than those used for macro level reviews. The scale used to perform micro level reviews should be dependent on the horizontal and vertical accuracies and resolution of the dataset. The manual visual reviews should be tailored to the specific type of DEM being reviewed in addition to overall project requirements. For example, a non-hydro-flattened bare-earth DEM will have different requirements than a hydro-flattened bare-earth DEM. And all types of bare-earth DEMs have different requirements than a Digital Surface Model (DSM). In addition to tailoring the QA/QC to the type of DEM being reviewed, the use of multiple visualization methods can achieve more thorough results.

DEM Visualization and Display

DEMs can be visually reviewed in many different software programs with many different methods of visualization. Some methods of visualization, such as hillshades, are much more common than others; imagery draped on terrain models, for example, require that imagery suitable for the project be available. Most DEM QA/QC reviews will benefit from using a variety of visualization methods as different methods will highlight different aspects of a DEM and allow for more thorough review. Using the visualization methods described below will help reviewers identify the issues described in the "DEM Qualitative Review" section below.

Hillshades

DEMs should always be reviewed with a hillshade. Hillshades are shaded relief surfaces created by setting an illumination source and angle for the DEM dataset. Hillshades allow users to view DEMs in a manner that illustrates depth and shows more detail and characteristics than strictly 2-D DEM images. Some software programs will create a default hillshade when viewing DEMs, while other software programs require a hillshade to be processed separately. In all cases, adjusting default settings on the hillshade may highlight or exaggerate certain aspects of the DEM more or less. Because many software programs use different methods and default settings for their hillshades, reviewing the same DEM dataset in different software programs can achieve the same results as manually adjusting hillshade settings. In many cases, it may be more efficient to use the default settings of another program versus manually adjusting the settings in one program to match another software program. Figure 15.57 shows an example of a DEM with and without a hillshade. Figure 15.58 shows an example of the differences between hillshades created from different software programs for the same geographic location. Users should be cognizant of the affect different hillshade settings will have on a DEM dataset. Changing the azimuth setting will change the angular direction of the illumination source and changing the altitude will change the slope or angle of the illumination source above the horizon. Changing these factors can affect how much depth and shadow is displayed and can affect where the shadows occur. And setting the hillshades to have high vertical exaggeration will cause minor differences to appear extreme. All of these settings affect how the DEM is displayed and interpreted. Because the settings, especially improperly used settings, can cause good data to appear flawed or flawed data to appear acceptable, it is beneficial to view DEMs with various hillshade settings or in different software programs that use different default settings. This allows the data to be thoroughly reviewed and helps to ensure that issues observed

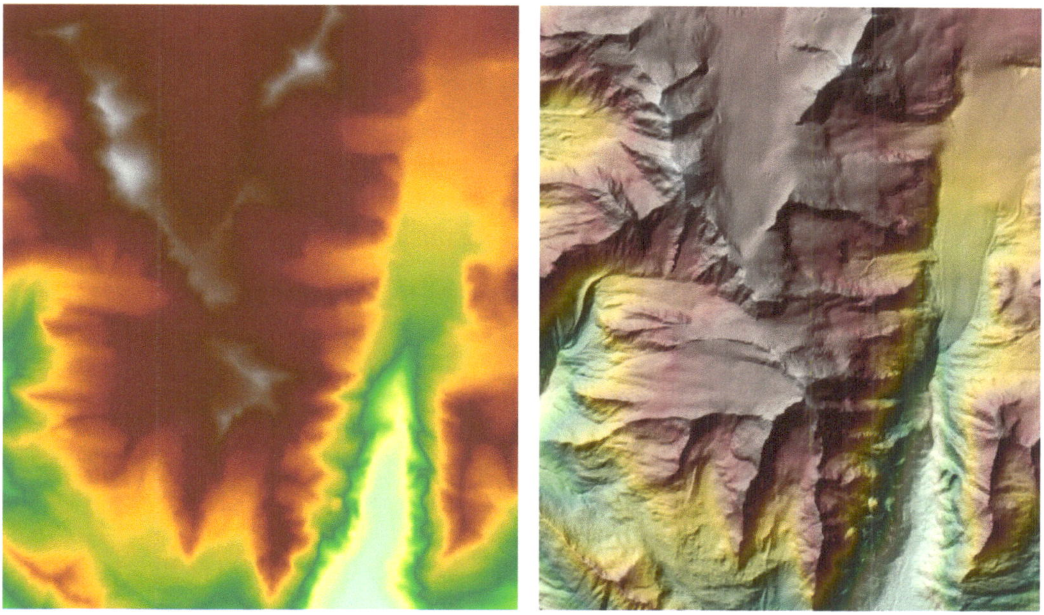

Figure 15.57. The image on the left shows a DEM, in 2-D, while the image on the right shows a DEM displayed with a hillshade, allowing the user to view depth and additional characteristics of the DEM.

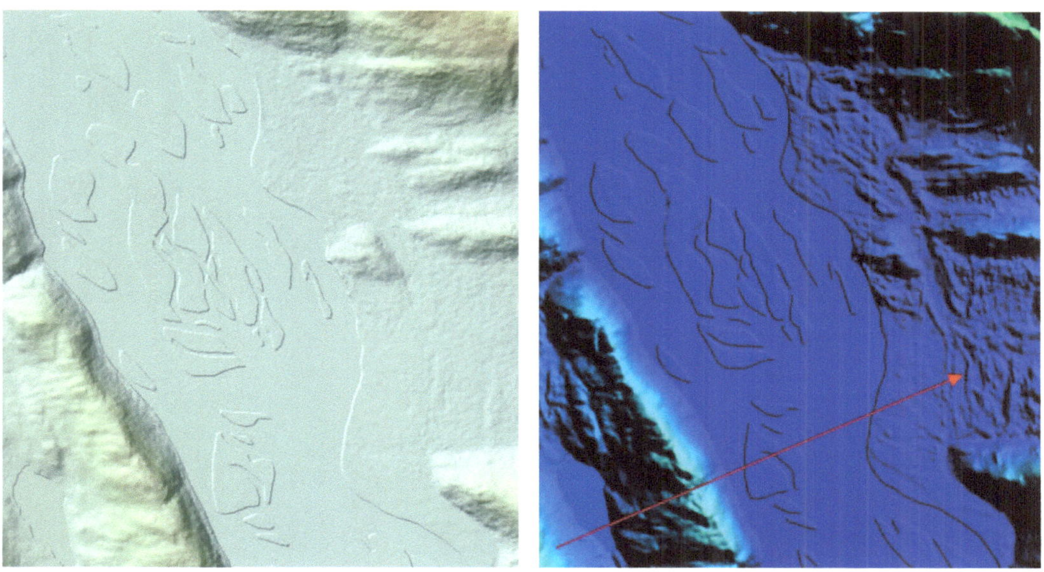

Figure 15.58. Both DEMs represent the same location, but are displayed with hillshades from different software programs. Reviewing DEMs in different software programs or with different hillshade settings allows users to more thoroughly review datasets as each software program or hillshade setting may highlight different aspects of DEMs better. The DEM on the right highlights elevation differences more and causes some additional surface details to stand out (red arrow) but has heavy shadowing. The DEM on the left may not highlight all surface details, but allows a more thorough review of all elevated features that may be too heavily shadowed in the DEM on the right.

during the hillshade review are in fact data issues, not simply display artifacts resulting from the hillshade display settings.

3-D Views

In addition to the familiar "top down" or "bird's eye view", many software programs offer other useful data visualization tools, including traditional 2-D profiles or 2-D cross sections, and 3-D perspective views that allow users to rotate and tilt data. Reviewing data from different perspectives allows the user to more thoroughly review DEMs and better evaluate all aspects of the dataset. While some issues may be exaggerated in 3-D views, other issues are more apparent from one perspective compared to others. Figure 15.59 shows spikes within a dataset that are visible when viewed in a 3-D view, but are not visible when viewed in a 3-D "top down" perspective. While adjusting the azimuth and/or altitude settings of the hillshade may also identify this spike in the top down perspective, the entire dataset would need to be reviewed with several hillshade settings to ensure all slopes were adequately reviewed. Spikes and divots are easily identified in 3-D views so using this additional perspective means the DEM dataset can be thoroughly reviewed with fewer hillshade variations.

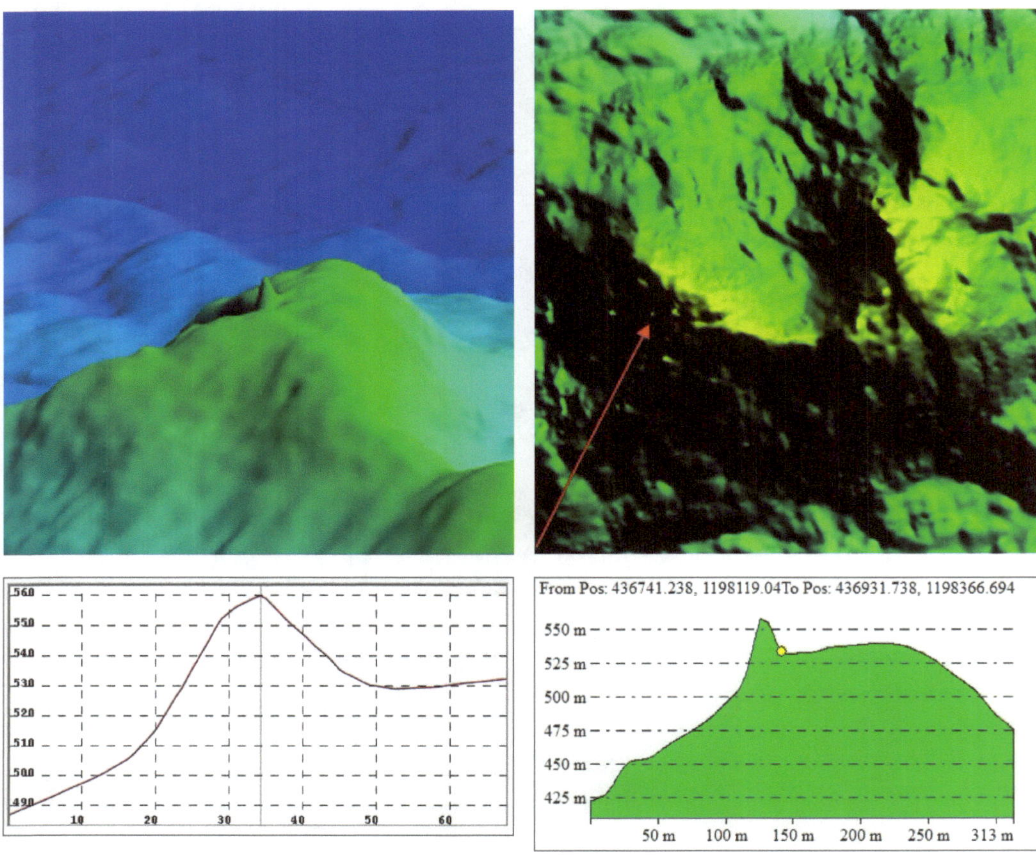

Figure 15.59. Both DEMs represent the same location but the DEM on the left is viewed in 3-D while the DEM on the right is viewed from the "top down". The 30 meter spike is measurable in both datasets, shown by the accompanying 2-D profiles. However, the spike is only visible in the 3-D view. Viewing data from multiple perspectives provides a much more thorough review of the dataset.

Imagery

If aerial or satellite orthoimagery is available for a project area, this imagery can be draped on the DEM data (usually bare-earth or DTM data). Draping available imagery over the DEM data can provide another method of review as anomalies in the DEM data may be more apparent when viewing the imagery in 3-D views. Consistent or realistic geometry and shape of imagery draped on DEM data can lend confidence to remotely sensed data. Inconsistent features, anomalies, smears, warps, and other unrealistic geometry that results from imagery draped over DEM data may indicate issues with the DEM data such as where additional breaklines and editing may be required. Figure 1.60 in Chapter 1 is an excellent example where RapidEye satellite imagery was draped on airborne IfSAR DTM data showing the top of Denali (Mt. McKinley) in Alaska and compared to a fly-by photo taken by a pilot looking out the window of an airplane (courtesy of E-Terra). The comparison of the two images is remarkably similar, lending confidence that the airborne IfSAR captured the full peak or summit of Denali when other verification methods are limited due to the remoteness and ruggedness of the terrain. Figure 15.60 shows an example of a topobathymetric DEM near the Oxbow Dam on Snake River along with imagery (courtesy of Watershed Sciences, Inc., a Quantum Spatial Company) draped on the DEM. The imagery was used to help evaluate acquired bathymetric data.

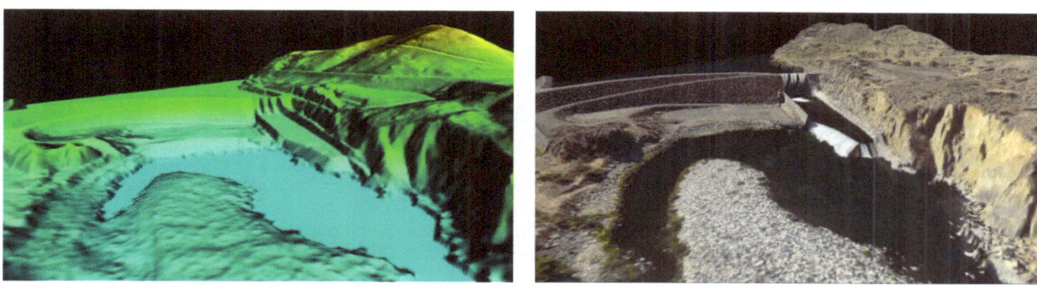

Figure 15.60. The image on the left shows a topobathymetric DEM in 3-D view while the image on the right shows aerial imagery. Image courtesy of Watershed Sciences, Inc., a Quantum Spatial Company, draped over the DEM.

Additional Raster Models

Sometimes reviewing different types of raster models created from the elevation data can provide additional insight into the quality of the elevation data. Common types of raster models used in DEM QA/QC include density models and difference rasters. Density models are generally used with lidar data and can identify areas in the surface where no, or few lidar returns exist. Density models can be used to visually represent the density of a specific surface, such as bare-earth density and where low confidence areas may exist. Density models, when created from first return, single swath lidar points, can also be used to assess if the data meets point density requirements. But when density models are coupled with visual review of the elevation data, density models can also be used to identify and evaluate specific features or issues, including voids in the data and areas of aggressive classification or over-smoothing in the ground. Figure 15.61 on page 516 shows an example of a density model.

Difference rasters are commonly used to evaluate how well overlapping DEMs match in elevation, which can also correlate to how well the overlapping DEMs match horizontally. These types of rasters can be used to help evaluate which available datasets may suit the needs and requirements for the desired final products. Difference rasters can also be used to determine which datasets should be given priority when performing analysis or help determine the best location of seamlines when merging or combining different datasets into one dataset. Difference rasters are created by subtracting one DEM from another DEM. These rasters can visually display where differences between overlapping DEMs exist, if there are patterns to the differences, and be used to quantify the differences. Difference rasters should be coupled with a visual review of the overlapping DEMs along with any vertical and/or horizontal accuracy statistics that exist for each dataset to fully evaluate the datasets. Figures 15.62 and 15.63 on page 517 show examples of difference rasters.

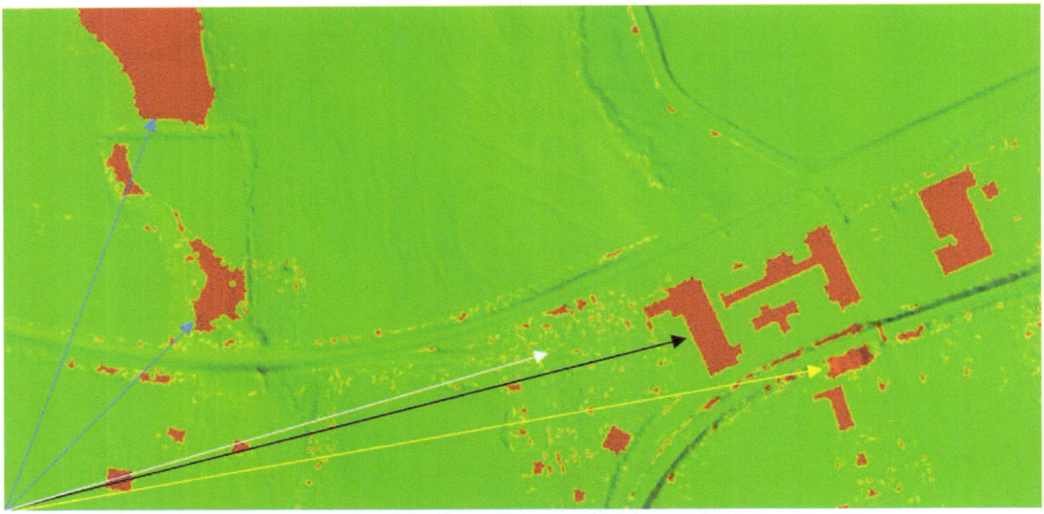

Figure 15.61. This figure shows a density model where areas meeting point density requirements are colored green and all other areas are colored red. As this density model is created from ground only points, red areas representing no or sparse ground points are expected in building footprints (black arrow) and hydrographic features (blue arrows). Areas of dense vegetation (white arrow) tend to have a mottled appearance or mixture of red and green as the number of points reaching actual ground in these areas is generally reduced. Areas of no ground points that represent elevated features, slopes, embankments, or other areas that are inconsistent with the surrounding areas (yellow arrow) should be reviewed for possible issues. The feature identified with the yellow arrow correlates to Figure 15.76 in the sections below.

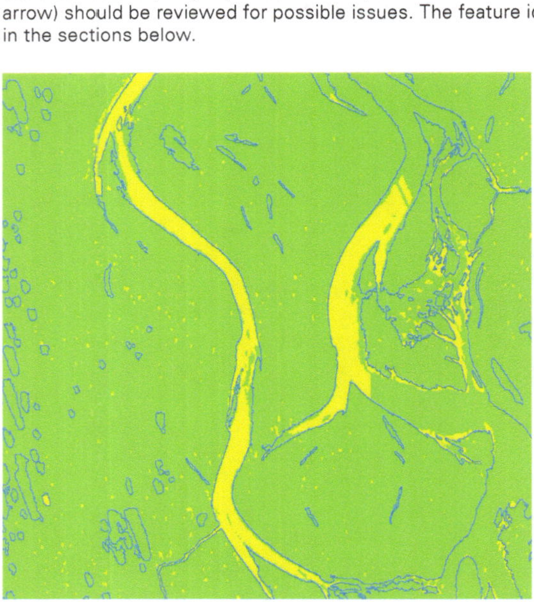

Figure 15.62. This example shows a difference raster created from two overlapping datasets acquired two years apart. All upland topographic features in the two overlapping datasets are within 5 cm of each other (green pixels). Hydrographic features (blue outlines) were at different water levels in the two acquisitions and the resulting hydro-flattened elevations are up to 30 cm different between the two datasets (yellow pixels within blue outlines). There are additional areas along hydrographic embankments and isolated areas within the floodplain that differ by up to 30 cm (yellow pixels outside of the blue outlines).

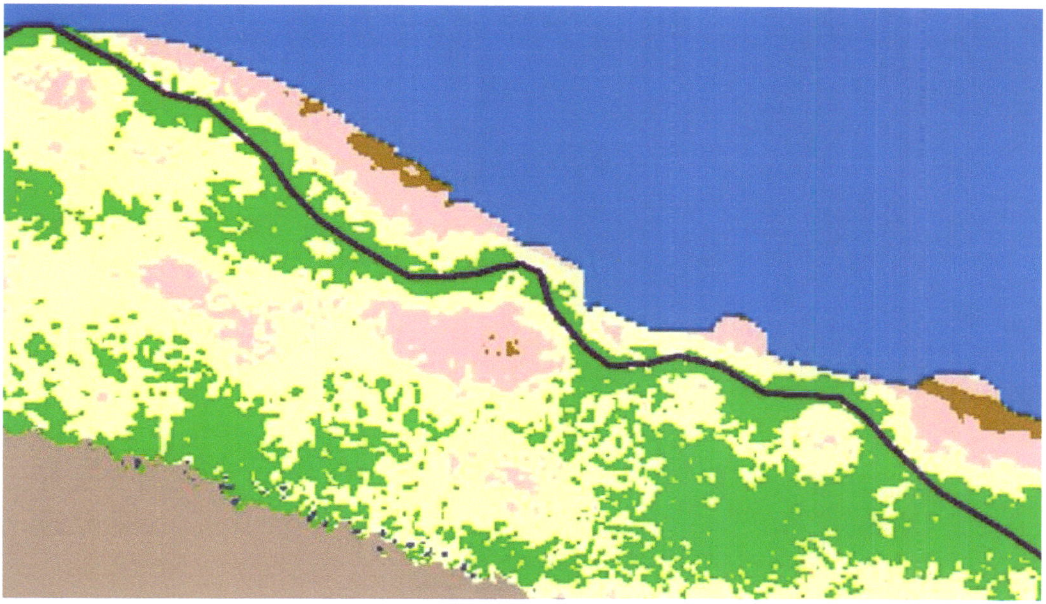

Figure 15.63. This difference raster was created between two different datasets that would be merged into a single dataset. Reviewing the difference raster allowed the seam line (black line) to be placed where the two datasets best matched (green). Choosing the placement of the seam line where the datasets best matched reduced the visible difference between the datasets in the merged product.

DEM Raster Types

The most common types of elevation rasters requested as final deliverable products are hydro-flattened bare-earth DEMs, hydro-enforced bare-earth DEMs, DSMs, and increasingly topobathymetric DEMs. While other types of DEMs exist, the sections below will focus on these most common types. The concepts and examples in the sections below may apply to other DEMs, but users will need to determine applicability based on DEM specifications and project requirements. See Chapter 1 for full descriptions and definitions of DEMs.

Digital Surface Models

Digital Surface Models (DSMs) are also known as reflective surface models or first return surface models because they represent the initial signal return from a surface, including all buildings and vegetation. DSMs are created from the "first return" data and when using lidar, all classes except for noise are usually used. As DSMs are created from first return data, some issues found in bare-earth DEMs, such as vegetation and building artifacts, would not be considered problems in the DSM. Since DSMs have fewer possible issues, less micro level QA/QC is generally required. However, DSMs can still have issues such as spikes, divots, noise, and voids. Spikes, divots, and noise issues are sometimes more difficult to identify in DSMs because of the presence of other first return elevated features. DSM issues are generally issues that are not always related to the final bare-earth surface. Examples in the sections below that relate to the cleanliness and acceptability of the final bare-earth surface are applicable to bare-earth DEMs and not to DSMs. Other examples below that are not ground specific, such as voids (Figure 15.68 on page 521), missing pixels (Figure 15.70 on page 523), spikes (Figure 15.71 on page 523), and divots (Figure 15.72 on page 523), do apply to DSMs. Figures 15.64 and 15.65 on page 518 represent the same geographic location and show the difference between a DSM and bare-earth DEM.

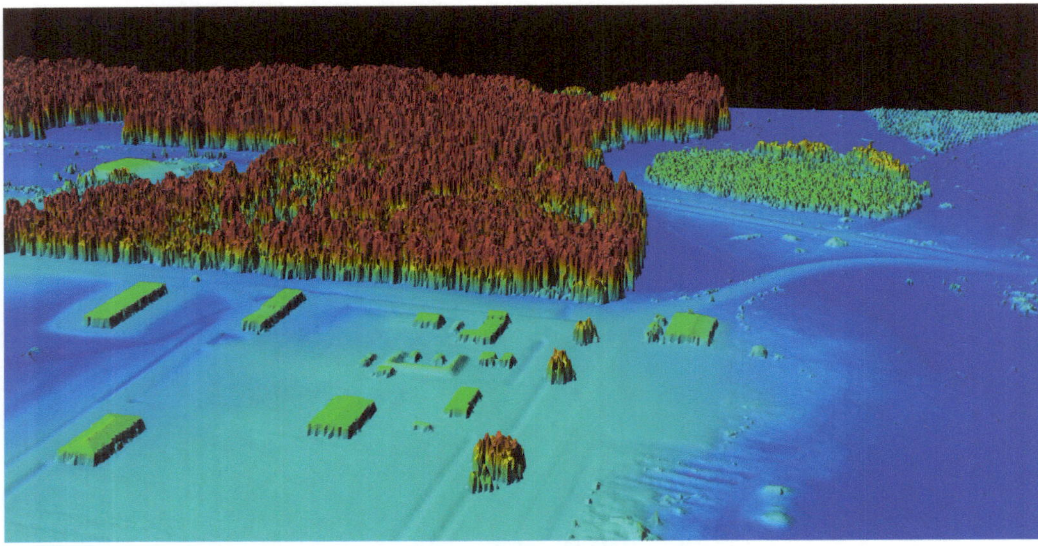

Figure 15.64. A Digital Surface Model (DSM) will model all above-ground features, including vegetation, buildings, and bridges. DSMs are reflective surfaces and represent a sensor's initial signal return or first return of all objects or obstructions encountered. If no above-ground features are present, DSMs will model the bare earth ground.

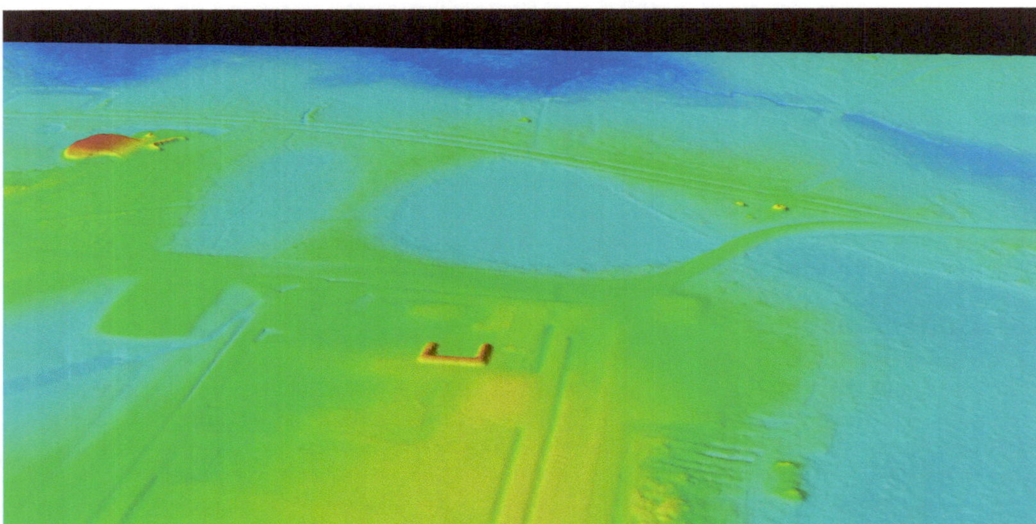

Figure 15.65. A bare-earth DEM only models the bald or bare earth, void of all above-ground features. All above-ground features, including vegetation, buildings, and bridges are filtered from the dataset to produce bare-earth DEMs.

Bare-Earth DEMs

Bare-earth DEMs should be void of all above-ground features, including vegetation, buildings, bridges, and overpasses (see Appendix B for definition of bridge and overpass). When compared to DSMs, bare-earth DEMs require more editing to produce and more QA/QC to ensure acceptability. Bare-earth DEM surfaces must be meticulously edited and reviewed in detail to ensure they are truly bare-earth. Most bare-earth DEMs have additional requirements for hydro-flattening or hydro-enforcement. If neither hydro-flattening nor hydro-enforcement is required, then the bare-earth DEM will only be reviewed for quality of the ground surface. When hydro-flattening or hydro-enforcement are required, hydrographic features must also be reviewed for flattening/enforcement requirements and for aesthetics.

It is important to understand the difference between hydro-flattened and hydro-enforced DEMs as it impacts how the DEMs are QA/QC'd. Certain types of features may or may not need to be identified as issues, depending on the type of DEM surface being produced. Please see the Hydrologic Surface Modeling section in Chapter 1 for more details on hydro-flattened and hydro-enforced DEMs. Hydro-flattened is a term coined by USGS with the release of their "National Geospatial Program Lidar Guidelines and Base Specification, Version 1.0." Hydro-flattened DEMs refer to DEMs that depict the ground surface from a bird's eye view, maintaining all apparent sinks and depressions naturally occurring on the landscape. Road fills over culverts remain in the ground surface as the culvert features are underground (as are tunnels) and are not visible from above. As in most topographic DEMs, bridges are removed from hydro-flattened DEMs as they are above-ground features inappropriate for a bare-earth DEM. Hydrographic features meeting minimum specifications should be flattened and at an elevation that is just below the surrounding terrain surface to produce aesthetically acceptable water surfaces for cartographic purposes. Linear streams and rivers will be monotonic and model downhill gradient flow in hydro-flattened or topographic DEMs. Hydro-enforced or hydrologic DEMs also have leveled hydrographic features and monotonic streams, but surface features that block or impede the actual flow of surface water are removed. Though there are others, the most common example of these flow interrupting features is the road fill over a culvert (see Appendix B for definition of culvert). On a topographic (hydro-flattened) DEM surface, the road fill creates a "digital dam" that stops water flow at the culvert location. On a hydrologic (hydro-enforced) DEM surface, a channel is cut across the road fill at the culvert location, allowing water flow to be modeled continuously as it actually flows. Whereas hydro-flattened DEMs are produced for cartographic purposes, hydro-enforced DEMs are produced for hydrologic and hydraulic (H&H) modeling purposes. It should be noted that breaklines are commonly used to both hydro-flatten and hydro-enforce DEMs. When performing QA/QC on DEMs, the biggest difference between hydro-flattened and hydro-enforced DEMs is how flow is modeled through drainage structures. Hydrographic breaklines should break on either side of in-ground drainage features, such as culverts, in hydro-flattened or topographic DEMs whereas hydrographic breaklines should be carried through drainage features in hydro-enforced or hydrologic DEMs. As above-ground features, bridges are removed in all bare-earth DEMs, and hydrographic breaklines should be carried through the bridge crossing location in both hydro-flattened and hydro-enforced DEMs. More information on these various types of DEM surfaces can be found in Chapter 1.

Figure 15.66 on page 520 shows the difference between a hydro-flattened DEM and hydro-enforced DEM near a drainage feature.

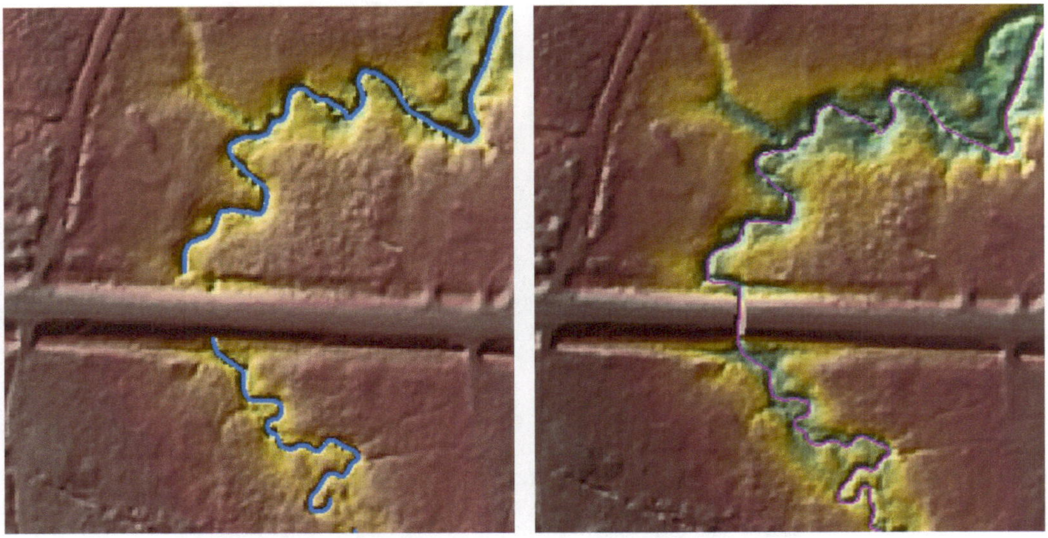

Figure 15.66. The DEM on the left is hydro-flattened while the DEM on the right is hydro-enforced. Hydro-flattened DEMs do not show flow through in-ground drainage features, such as culverts, and hydrographic breaklines should stop on either side of these features. Hydro-enforced DEMs model flow through all drainage features and hydrographic breaklines should continue through these drainage features.

Bathymetric and Topobathymetric DEMs

Topobathymetric sensors that will map shallow or near-shore waters along with traditional ground topography have become more common. See Chapter 10 on Airborne Bathymetric Lidar for more information. The increased use of topobathymetric sensors has resulted in an increased request for topobathymetric DEMs. Topobathymetric DEMs are a seamless model representing ground and as much bathymetry or submerged topography as was acquired by the sensor. It has also become more common to merge traditional acoustic or multibeam sonar data with topobathymetric data to create seamless DEMs covering ground, near-shore and deeper submerged topography. See Chapter 11 on Sonar for additional information regarding sonar. As these topobathymetric DEMs model as much submerged topography as possible, hydrographic breaklines are not used in these types of DEMs. Breakline variance and examples in the sections below that are related to the use of hydrographic breaklines are not applicable to topobathymetric DEMs. All examples in the section below related to the quality of the bare-earth are also applicable to topobathymetric DEMs. Some concepts, such as spikes and divots, apply to all surfaces (topographic and bathymetric), but may be harder to interpret within bathymetry data as an issue or a legitimate change in submerged topography. The examples relating to submerged topography are specific to topobathymetric and bathymetric DEMs and do not apply to other topographic-only DEMs. Figure 15.67 shows an example of a topobathymetric DEM compared to a hydro-flattened DEM.

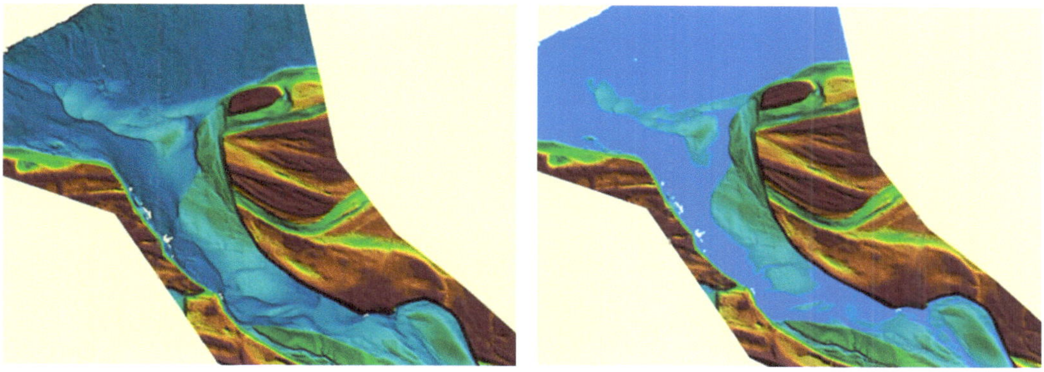

Figure 15.67. Both images are the same geographic location but the image on the left is a topobathymetric DEM while the image on the right is a hydro-flattened DEM. The topobathymetric DEM models bare-earth and submerged topography. The hydro-flattened DEM models bare-earth and flattens all water surfaces. While some ground exposed above the water surface is visible as island features in the hydro-flattened DEM, more of the channel geometry and submerged topography is modeled in the topobathymetric DEM.

DEM Qualitative Review

Using the visualization methods described in the "DEM Visualization and Display" section above, visually review all DEM data to identify issues, anomalies, and to assess the overall quality of the data. As described in the "DEM Raster Types" section, not all DEMs require the same review or will contain the same type of issues. The examples below identify the most common types of issues found in bare-earth DEMs (including hydro-flattened and hydro-enforced), DSMs, and topobathymetric DEMs.

Voids and Missing Pixels

Voids are areas of missing data and can occur in any type of DEM. When DEMs are created from specific classifications of lidar data (such as bare-earth), voids can occur due to misclassification of data. In these instances, the data can be re-classified properly to correct the apparent void. However, true voids occur when there is a data holiday and no data exists in that location. Data holiday voids can occur due to gaps between adjacent flight lines or due to sensor malfunctions. Figure 15.68 shows an example of a void within a bare-earth DEM along with the corrected DEM.

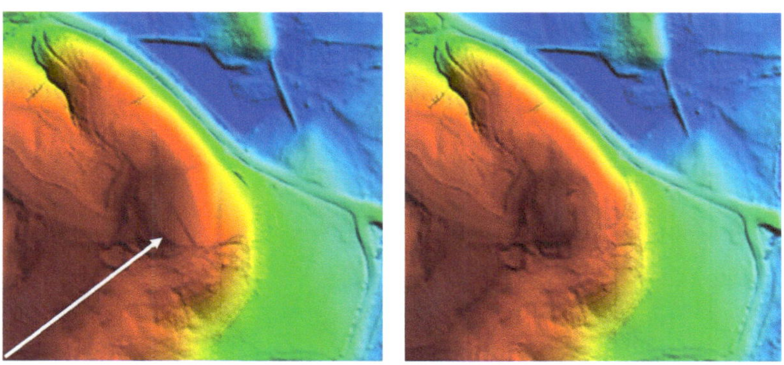

Figure 15.68. The image on the left shows a DEM where a gap between flight lines created a void (arrow) on the top of a hill. Please note that voids in the ground surface due to misclassification of lidar will appear the same as data holiday voids in bare-earth DEMs; the lidar full point cloud must be reviewed to determine what caused the void. The image on the right shows the corrected DEM after re-flights occurred to acquire data fully capturing this feature.

It is common to have voids within IFSAR data due to radar shadow and layover. Please see Chapter 7 on IFSAR for more details and explanations. While the amount and number of voids present can be reduced through additional flight lines in different directions, some allowance for voids is usually granted on IFSAR projects. Voids are generally filled in through proprietary interpolation methods and these methods may incorporate ancillary data, such as The National Map, 3DEP, or Shuttle Radar Topography Mission (SRTM) data, to help fill and shape the final surface. When voids are allowable within a project, the DEM data should be accompanied by a void layer which defines all void areas and identifies the source of all ancillary data used in the final DEM dataset. Project metadata should document the process used to incorporate the ancillary data into the final DEMs. The void layer should be reviewed as part of the overall QA/QC and all voids in the DEM data should be reviewed for adherence to project specifications, including the total amount of void areas allowable. Figure 15.69 shows an extreme example of a void area in IFSAR data that has been filled to create a realistic continuous surface. In this particular IFSAR dataset, no more than 5% of each 15-minute tile were allowed to be voids.

Figure 15.69. Void areas are expected in IfSAR data. Thresholds should be identified in project specifications and all DEMs should be delivered with a void polygon layer, shown in pink for this example. As shown in this example, void areas should be interpolated to create a realistic, continuous surface consistent with the surrounding terrain.

DEMs should also be reviewed for missing pixels within the surface. Missing pixels are actually pixels with void or bad values and are usually a result of DEM processing and not acquisition or classification errors. Missing pixels can act as a void within the DEMs and may cause erroneous breaks or interruptions in the DEM surface. Figure 15.70 shows two examples of missing pixels.

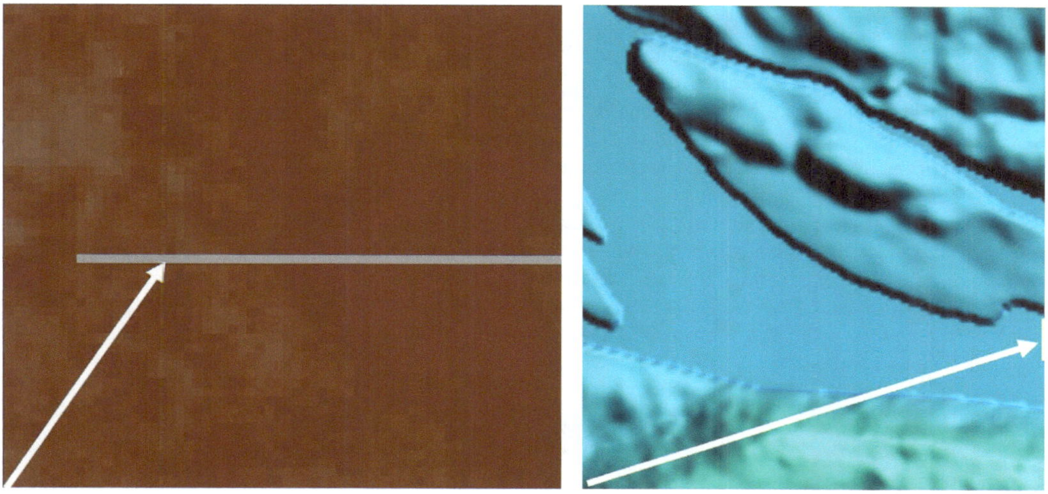

Figure 15.70. The left image shows a missing row of pixels within a DEM while the right image shows missing pixels at the eastern edge of the tile within a hydrographic feature.

Spikes/Divots

Spikes are erroneous high points and divots are erroneous low points. Spikes and divots can occur in any type of raster elevation model. Figures 15.71 and 15.72 show examples of spikes and divots, respectively.

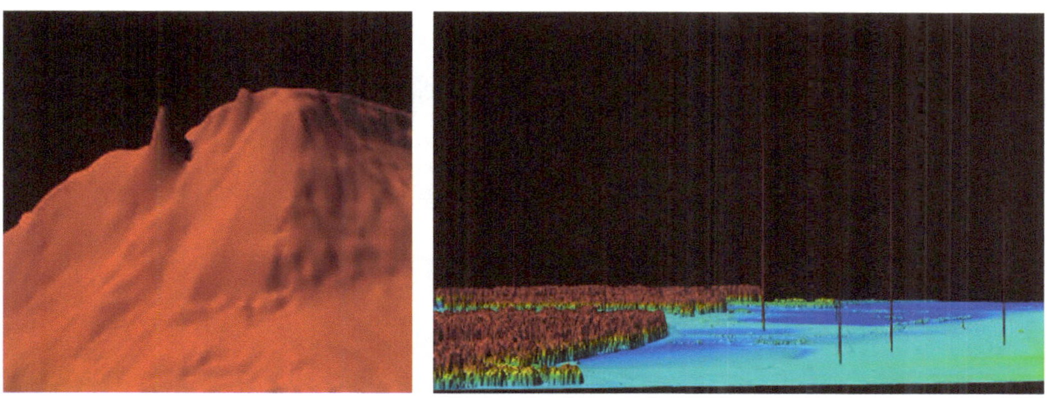

Figure 15.71. The left image illustrates a spike in a bare-earth surface while the right image illustrates multiple spikes in a DSM.

Figure 15.72. Divots in this bare-earth surface are several meters deep and visually look like deep holes within the surface. Legitimate holes and depressions within the bare-earth surface usually have better definition of the low feature compared to these depressions which have one or two random, very low points when compared to the surrounding terrain.

Artifacts

Artifacts are features that have been inappropriately included in a surface. Artifacts in bare-earth surfaces occur where any above-ground features have been incorrectly classified as ground, rather than their correct class. Common types of artifacts include vegetation, buildings, bridges, and other structures that have been included in a bare-earth DEM. Artifacts can also appear in bathymetric surfaces where features other than true bathymetric returns are modeled as the submerged topography. Surfaces with numerous artifacts are sometimes called noisy surfaces. Artifacts are less commonly identified in DSMs as these surfaces model all above-ground features. Spikes and divots are specific types of artifacts, resulting from error or noise points being included in the surface model. Many projects also have separate specifications for spike/divot tolerances compared to other artifact features. For these reasons, spikes and divots are generally considered their own category of issue. The examples below show vegetation (Figure 15.73) and building (Figure 15.74) artifacts within a bare-earth DEM. An example of artifacts within a topobathymetric DEM is provided in the Bathymetry Data section below.

Figure 15.73. The image on the left shows a bare-earth surface model that contains vegetation artifacts. The vegetation artifacts appear as random features with higher elevations within the surface. The image on the right is the corrected bare-earth surface after all vegetation artifacts have been reclassified from the ground classification to a more appropriate classification.

Figure 15.74. The image on the left shows a bare-earth DEM with a building artifact. The image on the right shows the corrected bare-earth DEM after the building has been reclassified from the ground class to a more appropriate classification.

As elevation information is not captured beneath buildings and other solid structures, bare-earth DEMs must interpolate across the area where the structures were present, leaving an irregularity in the surface. As with water bodies, breaklines can be used to smooth and flatten these areas (see Figure 1.36 on page 23 in Chapter 1), but the cost and effort is usually prohibitive. If the buildings or other structures removed were on a slope, then the interpolation created by their removal from the bare-earth surface may be more visible in the final DEM, as shown in Figure 15.75. These interpolated areas may also exist in areas of very dense vegetation where few reliable returns were acquired from the actual bare-earth ground. These areas of interpolation are expected in bare-earth DEMs due to the removal of above-ground features and are acceptable. Different interpolation methods span or fill voids differently; the results from some methods can be more noticeable, or less acceptable, than from others.

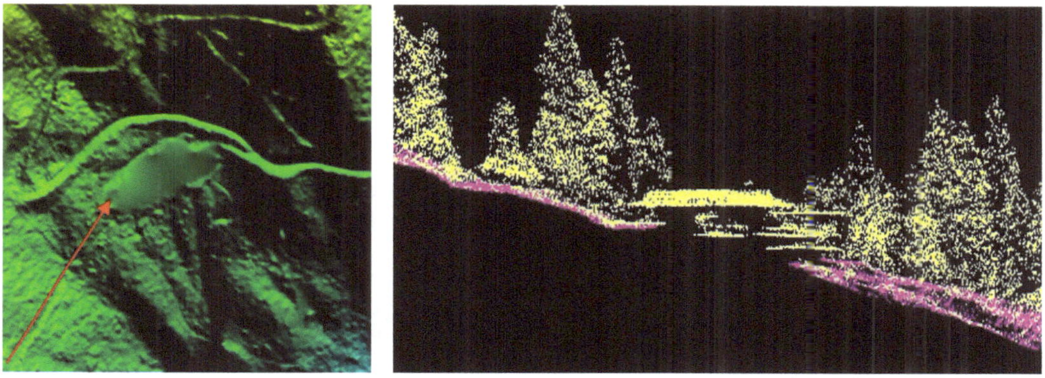

Figure 15.75. The image on the left shows a bare-earth DEM where interpolation resulting from the removal of a building on a slope is more visible than building removal interpolation that occurs on flat terrain. The image on the right is profile of the lidar data used to create the bare-earth DEM. The profile shows that all above-ground features (yellow), including vegetation and structures, have been removed from the bare-earth ground surface (pink).

Aggressive Ground Misclassification or Over-Smoothing

Aggressive ground misclassification or over-smoothing is the opposite of artifacts and occurs when too much information or too many features are incorrectly classified out of the surface. Aggressive ground misclassification generally does not occur with DSMs as all features, including the ground and above-ground features are modeled in the first return or reflective surfaces. Aggressive misclassification generally occurs with bare-earth and bathymetric surfaces. This type of misclassification usually results in the loss of detail in a surface by partially or completely incorrectly classifying ground features as another class and by over-smoothing the bare earth surface. Figure 15.76 on page 526 shows an example of aggressive misclassification of bare-earth terrain features.

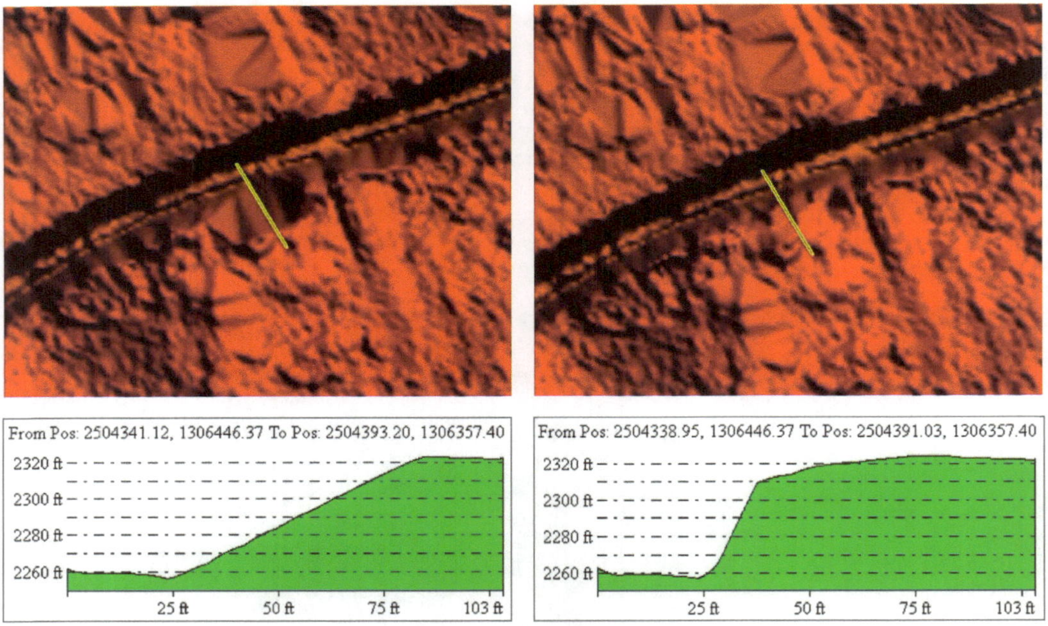

Figure 15.76. The bare-earth DEM on the left shows an embankment that has been aggressively classified out of the bare-earth surface to another class. This feature has been corrected in the bare-earth DEM on the right. As the accompanying profiles show, the slope and detail of the surface was completely altered by the aggressive misclassification.

Bridges/Culverts

As discussed in the DEM Raster Types section in this chapter and in Chapter 1, bridges should remain in DSM surfaces as these surfaces are first return or reflective surfaces, but bridges should be removed from all types of bare-earth surfaces. Figure 15.77 shows a DSM surface with a bridge compared to a bare-earth DEM surface where the bridge has been removed.

While bridges should be removed from all bare-earth surfaces, the treatment of culverts depends on whether the bare-earth surface is hydro-flattened (topographic) or hydro-enforced (hydrologic). Additionally, many projects may have specific requirements regarding culverts and bridges as these features are often very important to clients. If bridges are left in a bare-earth surface rather than being completely removed, they are usually considered an artifact, though some specialty applications may desire this (e.g., transportation models). If road fills over culverts are removed from bare-earth surfaces when they should be fully modeled, it is considered aggressive ground misclassification of features. While both artifacts and aggressive ground misclassification of features are described in the sections above, it is worth discussing bridges and culverts separately as these features are usually very important to clients. The importance of these features to clients usually results in more scrutiny of bridges and culverts by clients. It can be very beneficial for DEM QA/QC reviewers to have a QA/QC check specifically tailored to bridge/culvert requirements for each project, separate from other artifacts, aggressive ground misclassification, and other visual assessment QA/QC checks or steps. Figure 15.78 shows a bridge correctly removed from a bare-earth surface along with a culvert fully and correctly modeled in a bare-earth surface.

Bridges removed from the bare-earth surface can also leave interpolation artifacts, as demonstrated by buildings on sloped terrain in Figure 15.75. Bridge interpolation artifacts are also known as bridge saddles. Just as with buildings and other structures, there is often no valid elevation data beneath an elevated bridge.

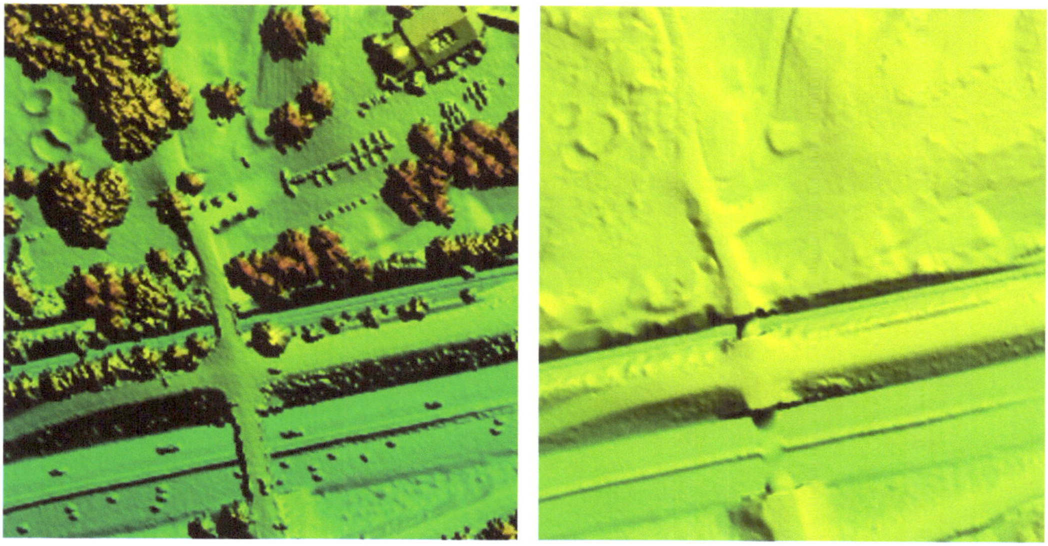

Figure 15.77. Both images show the same geographic location but the left image is a DSM where the bridge along with all other above-ground features is modeled in the surface. The right image is a bare-earth DEM where the bridge and all other above-ground features have been removed from the surface; unless project requirements specify the use of breaklines to flatten and smooth interpolated areas beneath bridges that have been removed, the remaining interpolation artifacts are regarded as acceptable.

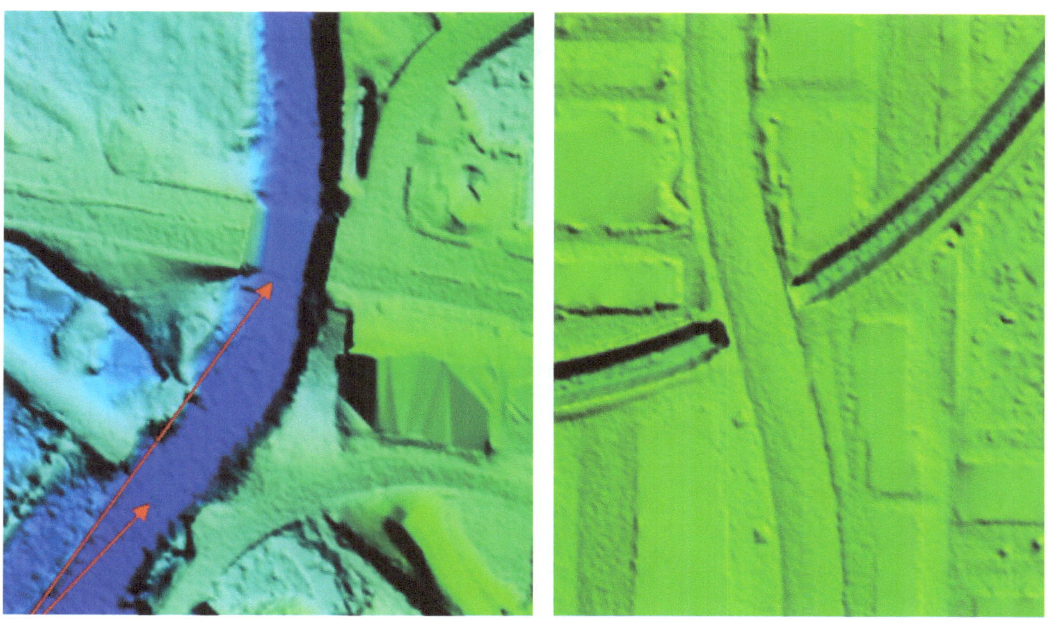

Figure 15.78. The left image shows a bare-earth DEM where two bridges have been correctly removed from the surface (red arrows). This is the correct treatment for both topographic and hydrologic DEM surfaces. The right image shows a bare-earth topographic DEM where the road fill over a culvert has been correctly modeled in the surface. In a hydrologic (hydro-enforced) DEM, the road fill would have been cut across to allow the surface to model water flow through the culvert beneath.

All DEMs are continuous surfaces (except for areas that may be masked out by the use of breaklines) so when bridges are removed from the surface, interpolation occurs where the bridge feature was located. Because bridge features are elevated above the surrounding terrain and generally given extra scrutiny by reviewers and clients, the interpolated areas where bridges existed can be quite visible. These interpolated areas, however, are expected. Just as with interpolated areas resulting from the removal of structures, the interpolated areas can only be modified with the use of breaklines. Unless project requirements specify the use of breaklines to flatten and smooth interpolated areas, then these interpolations must be regarded as acceptable. Figure 15.79 shows an interpolated area in a DEM resulting from the removal of a bridge. The accompanying lidar profile shows that all elevated points have correctly been classified out of the ground surface. The interpolated area due to bridge removal is acceptable as it does not result from incorrectly processed data. If clients or project requirements specify that bridge interpolation artifacts should be removed from the bare-earth DEMs to better represent the actual ground, then breaklines are commonly used to flatten the interpolation artifacts by enforcing the ground elevations beneath the bridge deck. USGS Lidar Base Specifications v1.3 require bridge interpolation artifacts, or bridge saddles, to be removed from the final bare earth DEM products. Removing bridge interpolation artifacts can be very beneficial if contours will be created from the DEMs. Interpolation artifacts can cause contours to incorrectly portray an elevated feature (the artifact itself) rather than correctly representing the ground elevations beneath the bridge deck. Figure 15.80 shows a bridge interpolation artifact before and after breaklines were used to remove the artifact.

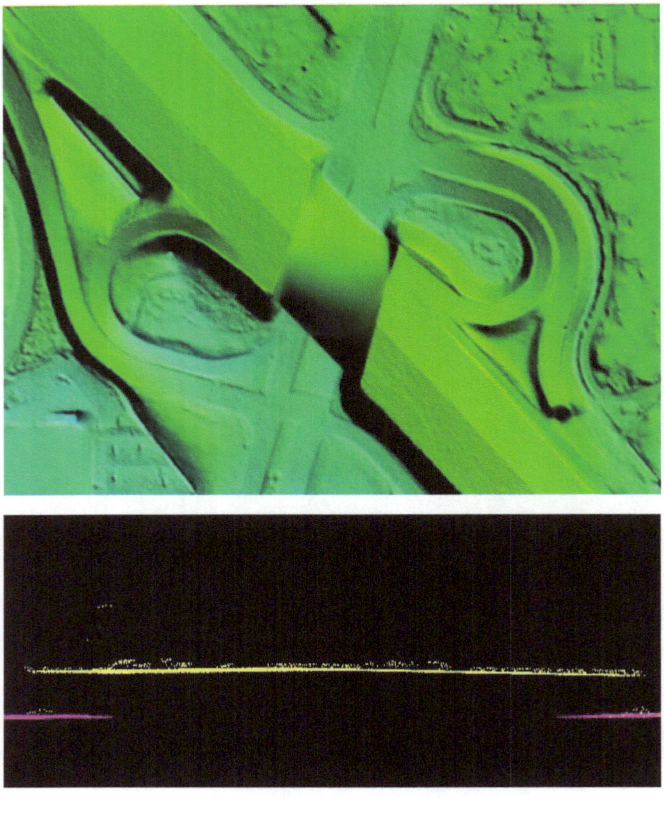

Figure 15.79. The top image shows an interpolated area or bridge saddle is present in the bare-earth DEM where a bridge was removed. The profile on the bottom shows that all elevated bridge points (yellow) have been correctly filtered from the ground surface (purple). The interpolated area in the bare-earth DEM is expected and acceptable if clients or project specifications do not require the removal of such interpolation artifacts. Lidar projects processed to USGS Lidar Base Specifications must have these artifacts corrected and removed from the final bare earth DEMs, as shown in Figure 15.80.

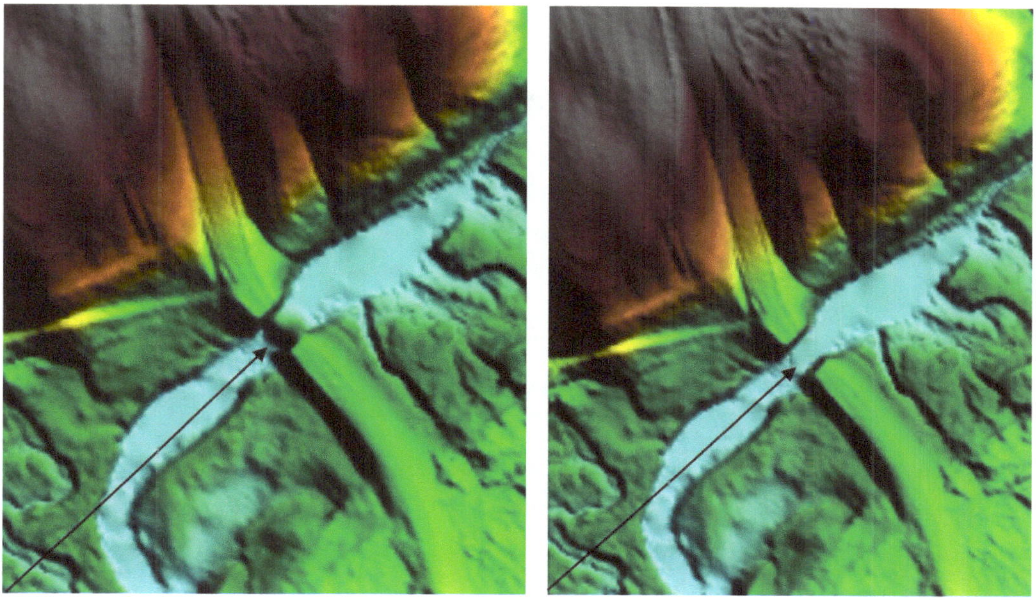

Figure 15.80. The left image shows an interpolation artifact (arrow) where a bridge has been removed from the bare-earth surface. The right image shows how applying breaklines to enforce the ground elevations beneath the bridge deck removes the bridge interpolation artifact. Removing bridge interpolation artifacts can be very beneficial, especially if contours will be created from the bare-earth DEMs. Lidar projects processed to USGS Lidar Base Specifications require these corrections to be applied to bridge interpolation artifacts or bridge saddles in the final bare-earth DEMs.

Inconsistent Editing

Inconsistent editing occurs when the surface or individual features are edited or processed differently across a single project, most commonly between tiles. The results of inconsistent editing are most visible along tile edges where one tile may have been processed or edited with different methods or by different editors. Inconsistent editing can occur on any type of DEM. Figure 15.81 on page 530 shows an example of inconsistent editing across tile boundaries that resulted in discrepancies of surface detail, elevation, and point density between the adjacent tiles.

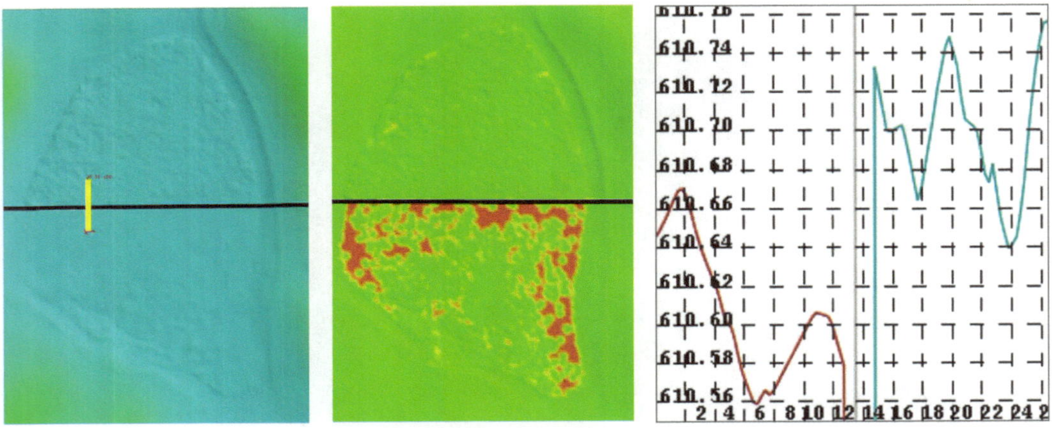

Figure 15.81. The left image shows two adjacent bare-earth DEMs colored by elevation. The bottom DEM is much smoother and lacks surface detail consistent with the top DEM. The middle image is a density model of the two DEMs and red areas identify locations where no ground points exist in the dataset. This density model shows that many more points were removed from the bottom DEM than the top DEM. The right image is a profile graph, taken at the location of the yellow cross section shown in the left image, where the red line represents the bottom DEM and the blue line represents the top DEM. The bottom DEM is approximately 12 cm lower than the adjacent top DEM.

Misclassification

When reviewing DEMs created from lidar data, most issues result from literal misclassification of point data according to the project classification schema. Because IFSAR does not generate point data, there is no classification, and the term "misclassification" is used to refer to ancillary features or data used incorrectly to create a surface, resulting in issues. In all cases, misclassification is usually used as a generic call when the issue does not fall in one of the specific categories mentioned in this section, such as artifacts or inconsistent editing.

Hydro-Flattening or Hydro-Enforcement

Hydro-flattening and hydro-enforcement rules will only apply to DEMs required to be hydro-flattened or hydro-enforced. As mentioned in the "DEM Raster Types" section above, hydro-flattening and hydro-enforcement is commonly accomplished through the use of breaklines. Breakline QA/QC should always be performed on the final breaklines, as described in the "Breakline QA/QC" section of this chapter. Some steps of the breakline QA/QC, such as breakline variance, will only be feasible when breaklines can be compared directly to the source elevation dataset, such as lidar point cloud data. The breakline QA/QC should always be performed to the fullest extent allowed by the delivered data. If steps such as breakline variance are not possible, then the DEMs must be reviewed for the issues that would have been identified by those checks, such as floating and digging hydrographic features. These issues can usually be identified during the visual review. Even if all breakline QA/QC steps were performed, hydro-flattened and hydro-enforced DEMs should always be reviewed to ensure flattening or enforcement requirements have been met. This review will verify and double-check that no breakline issues were missed during the breakline QA/QC.

Additionally, there are some issues and phenomena that may occur due to raster processing and not necessarily because of breakline issues. The most common issue created from raster processing is interpolation artifacts. Interpolation artifacts can be created when the raster model interpolates from one bank to the opposite bank. Elevation values along breaklines that are not consistent from bank to bank will cause interpolation artifacts to be much worse. Interpolation artifacts can usually be corrected post-processing or prevented by using a "stair-stepped" gradient rather than a smooth gradient (see Figures 1.28 and 1.29 in Chapter 1) as each step is assigned one elevation value. One elevation value per step forces the raster to be flat from bank to bank as the elevation value can be applied to all pixels within each step, but stair-steps do not look as natural as smooth gradients. With non-stair-stepped gradients, pixel elevations within hydrographic features that represent flow will be interpolated from the elevations along opposite banks. The interpolated pixel elevations may not always be flat from bank to bank, especially in areas of rapidly changing bank elevations and in areas of complex stream geometry. Each project may have specifications regarding tolerances for interpolation artifacts, bank to bank flatness thresholds, and the maximum amount of vertical change each stair-step may represent. Figure 15.82 shows an example of interpolation artifacts located within complex stream geometry where numerous islands are present within a broad stream channel.

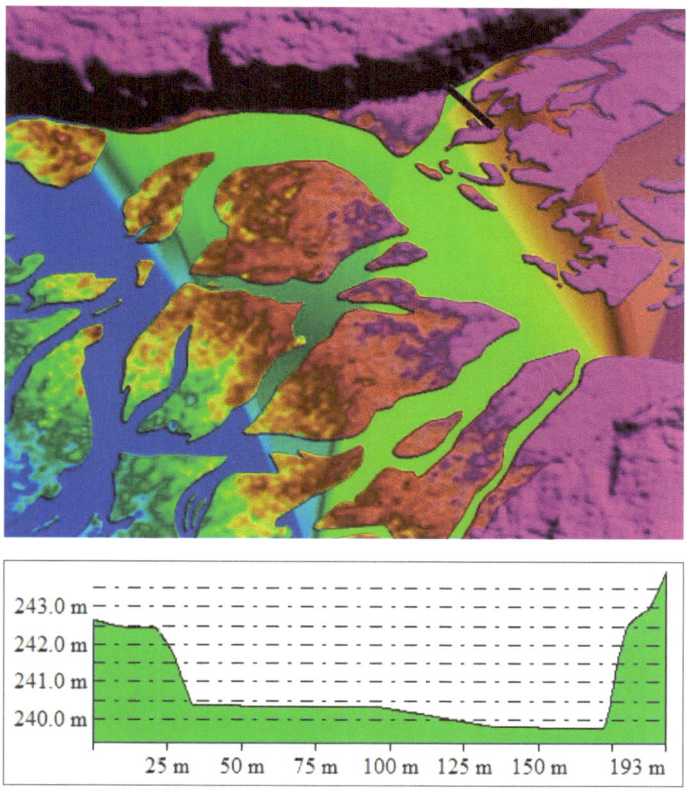

Figure 15.82. The top image shows a DEM symbolized to highlight changing elevations within the linear hydrographic feature (blue represents lower elevations while purple represents higher elevations and the green, yellow, orange, and red show the uneven elevation changes in the water between upstream and downstream). Due to the complex stream geometry, river elevations are not always consistent or flat from bank to bank and can even cause issues with monotonicity. The profile graph in the bottom image, taken at the location of the black line in the northeast corner of the DEM, shows that in some areas parallel banks vary in elevation by approximately 50 cm. Project specifications should help determine which interpolation artifacts will need to be corrected and which are within acceptable tolerances.

Surface Detail

The surface detail present in the final DEMs will depend on several factors, including pixel resolution of the DEMs, point density or resolution of the source elevation data, and the terrain type and land cover of the area being mapped. Generally areas of very dense vegetation will show less detail than open or sparsely vegetated terrain as the vegetation limits the number of reliable signals from the actual ground surface that are returned to lidar and IfSAR sensors. Similarly, heavily vegetated areas are usually obscured in imagery used for traditional photogrammetry, precluding the collection of dense mass points and breaklines in these areas. Regardless of the source data collection method, low detail and low confidence areas such as these should be identified and delineated in a separate file. Less surface detail should be expected in these heavily vegetated areas, as shown in Figure 15.83 on page 532.

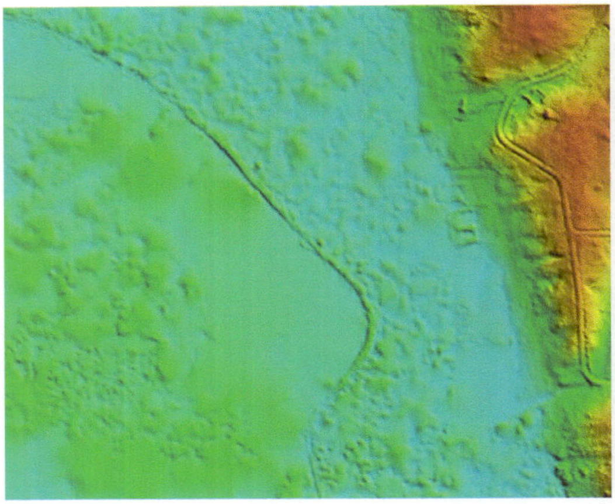

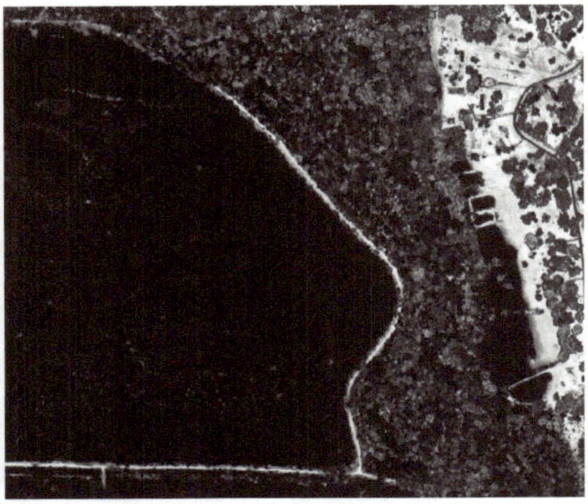

Figure 15.83. The left image shows a bare-earth DEM while the right image shows an intensity image for the same location. Areas of water and heavily vegetated forest (left and middle portions of images) visibly have less surface detail than areas of open or sparsely vegetated terrain (right side of images). This result is expected in these types of areas.

When reviewing DEMs, it is common to identify features that appear different or inconsistent with the surrounding terrain. These features should always be evaluated to determine if they are artifacts or legitimate terrain feature. Features that are well-modeled are more likely to be valid terrain features compared to features that are partial or inconsistent in their geometry and shape. Features that are determined to be valid terrain features, but may raise questions by clients or end-users should be discussed in metadata and project reports to show that the features were reviewed and evaluated. Figures 15.84 and 15.85 show examples of legitimate terrain features in DEM surfaces that are inconsistent with the surrounding terrain and required additional explanations in project metadata and documentation.

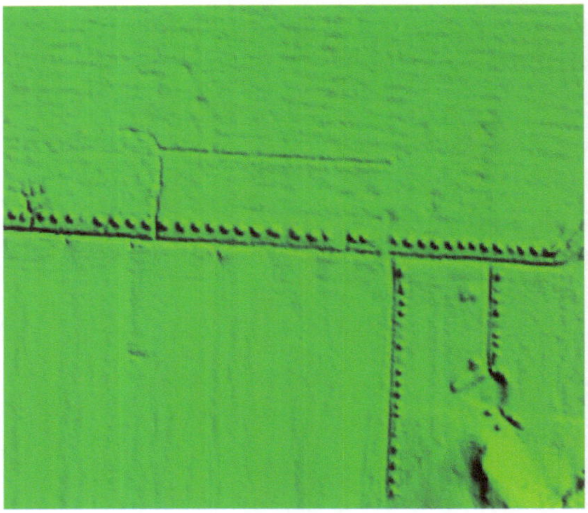

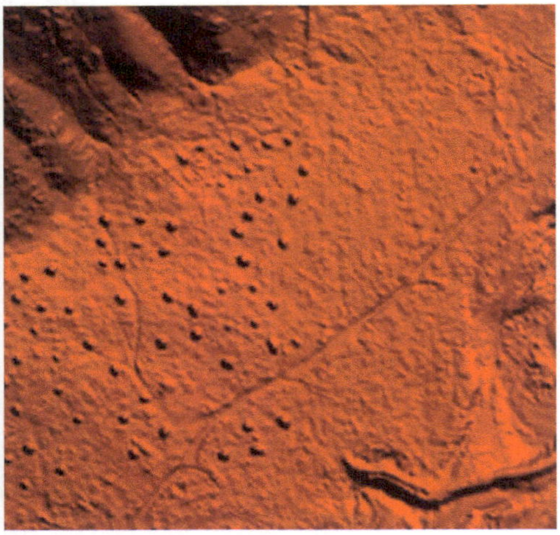

Figure 15.84. The left image shows a bare-earth DEM containing spoil mounds along an irrigation canal and the right image shows a bare-earth DEM containing small mounds in a valley. In both DEMs, the mounds are approximately 1-3 meters in height and appear slightly incongruent with the surrounding farm fields and valley floor. These features were verified in the full lidar point cloud and ancillary imagery as valid terrain features and discussed in final project documentation and metadata. Temporary features, such as large hay bales, would be removed from the bare-earth DEM.

Figure 15.85. The left image shows a bare-earth terrain modeling in-ground bunkers while the right image shows a lidar intensity image for the same location. These features are fully modeled in the DEM and clearly visible in the intensity imagery due to the strong reflectance of the NIR wavelength from the bare-earth soil or grass covering the bunkers compared to the surrounding vegetation. Using intensity imagery and ancillary color imagery layers when necessary can help identify questionable or uncertain features. Once features are identified, it can be determined if they are valid terrain features. As some end-users may mistake these features as above-ground features that should have been removed from the bare-earth surface, these features were discussed in final project documentation and metadata.

Bathymetry Data

Bathymetry data maps submerged topography. Bathymetry data from one sensor or data source can be merged with topographic data from another source to create a seamless topobathymetric DEM. Or, both topographic and near-shore bathymetry data can be collected with a topobathymetric sensor. Topobathymetric DEMs are becoming more common as new sensors are developed and used commercially. Topobathymetric DEMs can be a powerful tool for analysis and modeling as they provide a seamless surface along the shoreline or land/water interface. As this seamlessness is one of the main goals with topobathymetric DEMs, the transition between topographic and bathymetric data should be reviewed carefully. If multiple datasets from different sensors were merged together, the seams between the datasets should be reviewed in addition to the transition between ground and submerged topography. Figure 15.86 shows an example of issues along the land/water interface and Figure 15.87 shows an example of a seamless topobathymetric DEM created from multiple datasets from different sensors.

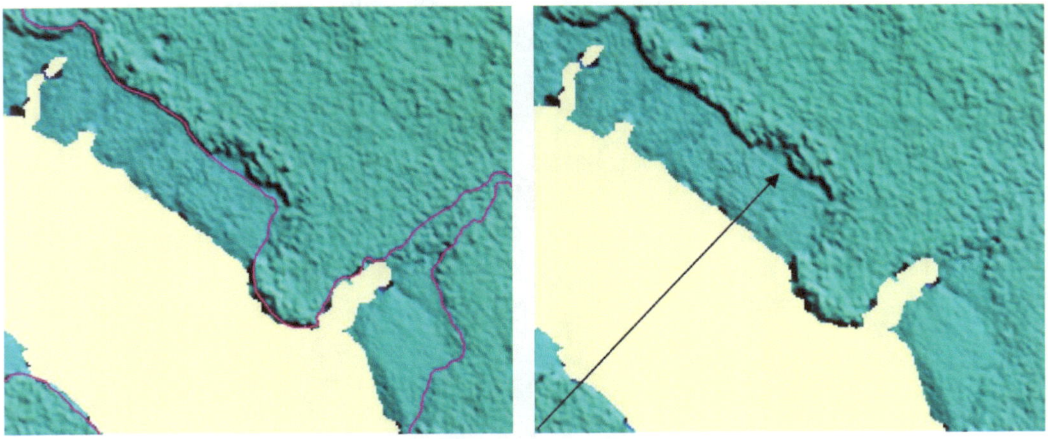

Figure 15.86. These two images show a close-up of the transition area between topographic and bathymetric data. The yellow areas are areas of no data. The purple outline in the left image shows the delineation between ground and submerged topography. In the right image, the arrow shows one area where the transition along this seam line creates a hard edge. There should be no hard edges along the land/water transition. This visible seam should be corrected to match the rest of the seamless transition.

Figure 15.87. This seamless topobathymetric DEM was created by merging topographic lidar, bathymetric lidar, and in the deeper water, multibeam acoustic sonar. Transition areas between the different datasets were smoothed and reviewed to ensure no hard edges in the final product. Image courtesy of Dewberry.

Many of the DEM issues above, such as artifacts, spikes and divots, aggressive ground misclassification or over-smoothing, and inconsistent editing, apply to bathymetric surfaces. Evaluating bathymetric surfaces can be more challenging than bare-earth surfaces. Whereas the "low" points or returns, with the exceptions of divots, are usually the ground surface in open terrain or sparsely vegetated topographic data, the low returns in bathymetric data are not necessarily submerged topography. The lowest acquired returns in bathymetric data may still represent the water column and not bathy bottom depending on how far the signal was able to penetrate the water. If bathymetric data was collected with a lidar sensor, reviewing the full point cloud rather than just the DEM is necessary to fully evaluate and review the dataset, just as with topographic lidar datasets. If full wave form data was collected, reviewing and analyzing the full wave form may provide additional information as well. It may not be plausible to fully review all collected waveform due to size, storage, processing, software constraints, and the required specialized skillset, but full wave form data should be reviewed in complex areas or areas where the interpretation and classification of submerged topography is questionable. Color imagery acquired with the bathymetric data, in addition to intensity imagery if acquired with lidar sensors, can be used as an additional tool to help determine areas where environmental conditions, including water clarity, are more likely to result in full penetration to the submerged bottom.

Bathymetric data may not always capture the submerged bottom due to environment variables and sensor limitations, resulting in voids. Whereas voids are unacceptable in topographic data, voids are acceptable and expected in bathymetric data. Generally, the contractor must show or demonstrate that a reasonable effort was made to collect as much bathymetry as possible, limiting the number of bathymetric voids. A maximum void threshold cannot be established as the size of the bathymetric voids will be entirely dependent on environmental variables and sensor limitations. Bathymetric voids may be small localized areas or may span an entire channel or bay. As bathymetric voids will exist, bathymetry projects should establish minimum void threshold where all bathymetric voids greater than or equal to the minimum threshold must be identified in a separate void layer and delivered with the final deliverables. Project specifications should state a minimum void size that incorporates accuracy of the data, user requirements and goals, and the pixel size of the final DEMs. Bathymetric voids smaller than the minimum threshold are generally interpolated in the final bathymetric DEMs and are not considered part of the final void layer. Depending on client requirements and project specifications, the voids meeting minimum threshold requirements may be left as no data in the final DEMs, or the void areas may still be interpolated. Some clients may require two sets of DEMs-one set with the final void layer interpolated and one set with the final void layer left as no data. Interpolating across void areas will ensure the final DEM is a continuous, uninterrupted surface but interpolation across areas void of bathymetry data can also result in incorrect depth values. For example, if bathymetry data are only collected in shallow waters and the deeper channel of a stream is a void area, interpolating across the void area will falsely raise the level of the deeper channel portion to be more consistent with the shallow areas closer to the banks. Figure 15.88 on page 536 shows an example of a void area in both the final topobathymetric DEM and the full point cloud lidar data. Figure 15.91 on page 537 shows an example where voids have been interpolated in the final DEM to create a continuous, uninterrupted surface.

As the lowest bathymetric returns may not truly represent submerged topography and there may be legitimate voids in the bathymetric bottom surface, QA/QC of bathymetry data should ensure the bathymetric surface is consistent and represents logical geometry. Figure 15.89 on page 537 illustrates how water column points misclassified as submerged topography have caused artifacts and inconsistent geometry in the bathymetric surface. Figure 15.90 on page 537 shows the same feature after corrections have been applied. Figure 15.91 shows consistent and logical geometry within a topobathymetric DEM where voids have been interpolated to create a continuous surface.

Supplemental rasters or DEMs specific to bathymetry can help further evaluate bathymetric DEMs. Depth rasters are a specific type of difference raster where the submerged bottom DEM is subtracted from a DEM representing the water surface. Depth rasters can be used to help evaluate depths penetrated by the bathymetric sensor as well as to quantify the percentage of the project area that reached specific depths or depth ranges. Statistics such as maximum and average depths reached can be calculated from the depth rasters.

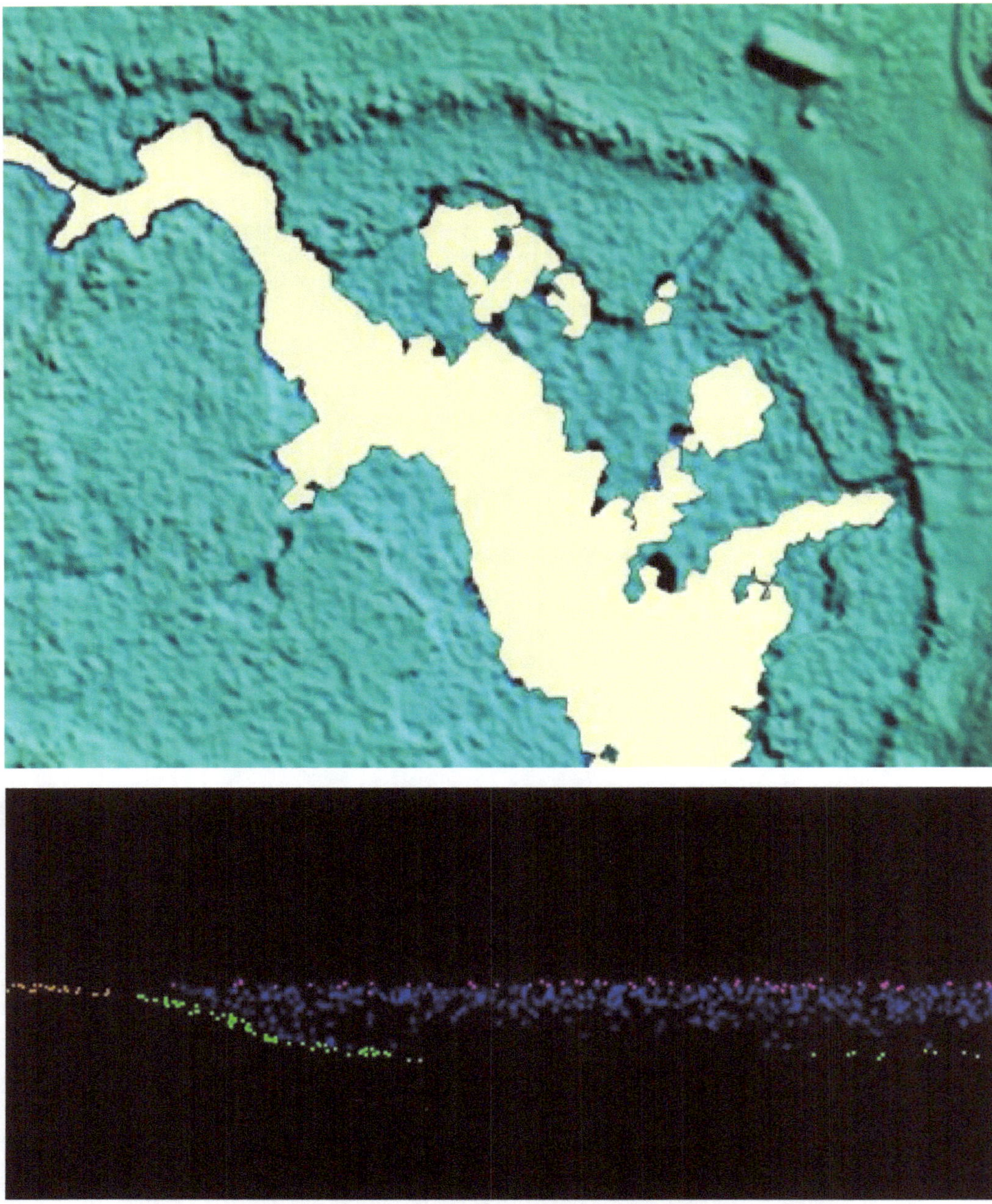

Figure 15.88. The top image shows a topobathymetric DEM where void areas are left as no data and are colored yellow. The bottom image shows a cross section of the lidar full point cloud colored by classification (orange-ground, pink-water surface, blue-water column, green-submerged topography). The profile shows a large gap or void in the green points where no returns were able to penetrate through the full water column to the submerged topography.

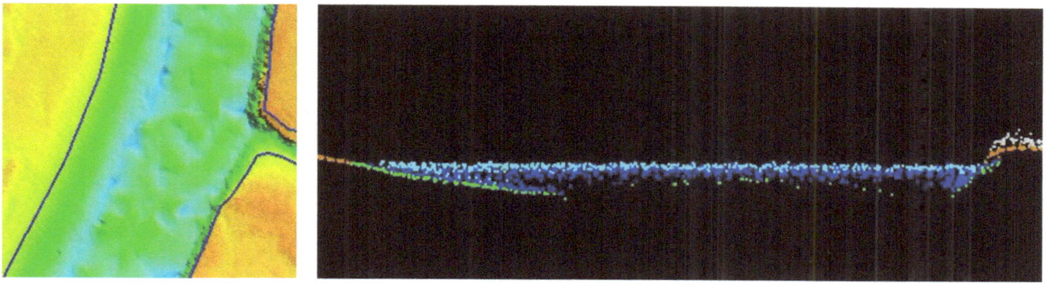

Figure 15.89. The left image shows a portion of a topobathymetric DEM with lines representing the land/water interface overlaid in blue. In this DEM, the hydrographic channel has an irregular bottom with undulating elevations which do not actually exist. The lidar full point cloud profile colored by classification (orange-ground, turquoise-water surface, blue-water column, green-submerged topography) is shown on the right. The profile shows that water column points have erroneously been classified as submerged topography. This has caused the channel geometry to be inconsistent in the middle of the channel compared to consistent and logical channel geometry along the outer banks. These misclassified water column points create artifacts in the bathymetric surface.

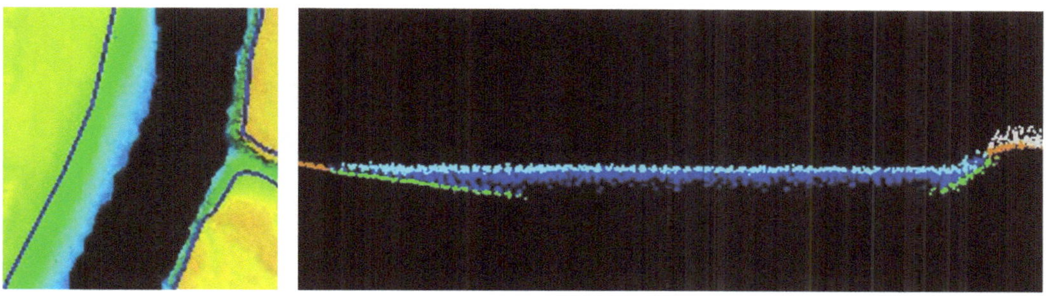

Figure 15.90. The left image shows the same topobathymetric DEM as above and the right image shows the lidar full point cloud profile, colored by classification. Erroneously classified water column points (dark blue) have been removed from the submerged bottom (green). A legitimate void now exists in the topobathymetric DEM, but artifacts are no longer present in the bathymetric surface.

Figure 15.91. Channel geometry and all submerged bottom were reviewed for consistency and logical geometry. Voids in this topobathymetric DEM have been interpolated rather than left as no data.

A second type of layer common to bathymetric DEMs is a confidence layer. A confidence layer represents the reliability of elevation values used in the final bathymetric DEM and is usually created as a raster where each cell value is the standard deviation of all bathymetric or all topobathymetric returns in that cell. Using the standard deviation, this layer represents the confidence of the elevation values, where low standard deviations correlate to higher confidence and higher standard deviations correlate to lower confidence. Areas of higher precision or lower standard deviations indicate areas where the elevation values are consistent and the sensor received repeatable returns from an object. Areas of high slope and other valid terrain changes will result in legitimate higher standard deviation values. Slope maps can be used to help identify areas where higher standard deviation values will be expected. It is beneficial to use the same cell size as the final bathymetric DEM so that the bathymetric DEM and confidence layer align. The raster layer can then be converted to a vector layer if the client or project specifications state a vector layer is preferred.

A third layer common to bathymetric data is a density layer. A density layer is generally created as a raster that records the number of bathymetric or topobathymetric returns per cell. Using the same cell size as the final bathymetric DEM can be very useful for later comparisons. Higher density also lends higher confidence as dense returns are more likely to model consistent bottom geometry compared to sparse returns that may be hard to differentiate between water column and submerged bottom. Using density layers in conjunction with confidence layers can be very effective tools to help evaluate bathymetric data and analyze the reliability of classified submerged topography. Figure 15.92 shows an example of a confidence layer and a density layer.

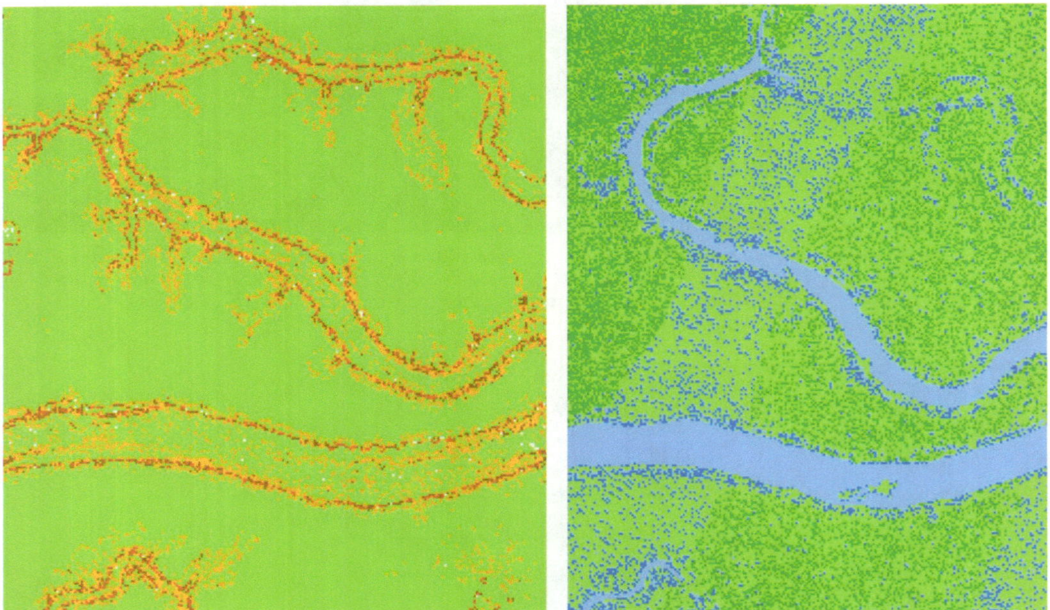

Figure 15.92. A raster confidence layer is shown on the left and a raster density layer of the same area is shown on the right. In the confidence layer, areas of green have low standard deviations less than 5 cm whereas orange and brown pixels have higher standard deviation values between 5 cm–20 cm. Higher standard deviation values are expected along the banks of the channels due to sloped terrain. In the density raster, blue areas represent pixels where 1-5 returns were used to derive that pixel's elevation in the bathymetric DEM while green areas represent pixels where 5-15 returns were used to derive an elevation value in the final bathymetric DEM. While not all low density areas correspond to high standard deviation values, most high standard deviation values correspond to lower density areas, indicating less confidence in those areas.

Edge-Matching

All DEMs within a single project area that were acquired under temporally consistent conditions should edge-match with no visible seams between individual tiles. Seams that exist between individual tiles within the same project AOI usually result from inconsistent editing (see section above) or calibration and/or relative accuracy issues (see Relative Vertical Accuracy section of this chapter) and should always be corrected. Edge-matching issues within a project due to calibration or relative accuracy issues are generally identified during the validation of relative accuracy whereas edge-matching issues within a project due to inconsistent editing are generally identified during the visual review of data.

Edge-matching should always be verified if existing DEM datasets relevant to the client are adjacent to the current project area. If there is overlap between the current project area and adjacent datasets, creating difference rasters is usually the best method to evaluate the edge-match as they allow the user to both visually review and quantify all differences. If adjoining datasets merely touch but do not overlap, then profiles in addition to visual review can be used to evaluate the edge-matching between the two datasets. If differences between adjacent, non-overlapping datasets must be quantified, a sample number of profile lines can be drawn across the edge-match so that one endpoint of each profile line is located in one dataset while the other endpoint is located in the adjacent dataset. The elevation of each dataset at the location of each endpoint can then be extracted and the elevations of the profile lines quantitatively compared. Slope changes along the profile lines must be considered when using this method to evaluate adjacent datasets. Temporal differences should always be considered when evaluating adjacent datasets as significant terrain changes can occur between datasets collected over different time periods. Because difference rasters will highlight changes between overlapping datasets, they are an excellent tool for use in change detection analyses. Figures 15.93 and 15.94 show difference rasters created for overlapping datasets. Figure 15.93 shows an area with significant temporal and ecologic change occurring between the acquisitions of each dataset while Figure 15.94 on page 540 represents datasets that edge-match within project specifications even though the datasets were collected years apart. Figure 15.95 on page 540 shows how profiles can be used to visually check edge-matching of non-overlapping datasets.

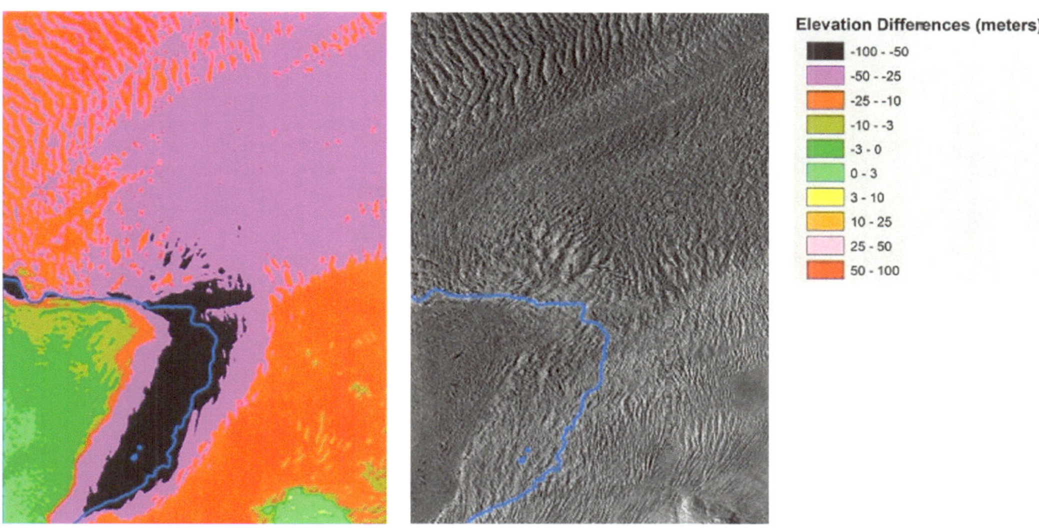

Figure 15.93. The left image is a difference raster created for two overlapping datasets acquired 2 years apart. Significant change, up to 100 meters, is identified by the difference raster. The ortho-rectified radar image (right image) shows that this significant change occurred over a glaciated area where the glacier has retreated and lost elevation in the more recent dataset compared to the previously acquired dataset. The blue line in both images represents the extents of a water body formed from the melted glacier in the more recent dataset. The radar image shown in the right image is from the older dataset, which shows this area as fully glaciated with no water body present. This difference raster is highlighting ecologic change and could be used in change detection analyses.

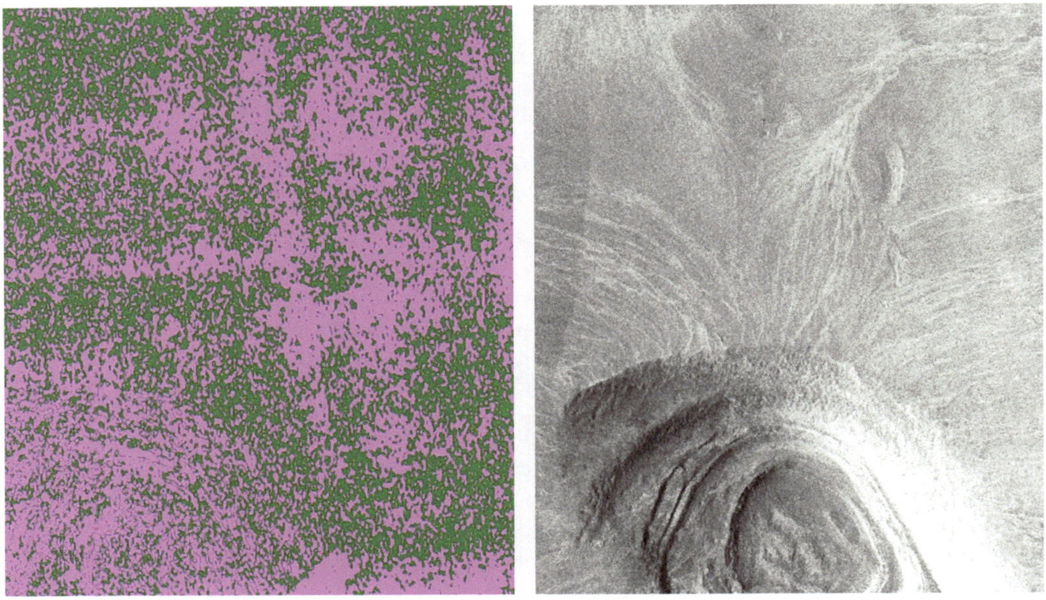

Figure 15.94. The left image is a difference raster created for two overlapping datasets acquired 2 years apart. Pixels colored green represent areas where elevations from the two datasets are within 10 cm of each other and the purple pixels represent areas where elevations from the two datasets are within 20 cm of each other. The ortho-rectified radar image (right image) shows that this area does not contain any highly dynamic features, such as hydrographic features or glaciers, and consistent elevations between the two datasets would be expected. As elevations from the two datasets are consistent, the edge-matching of these datasets is within project specifications.

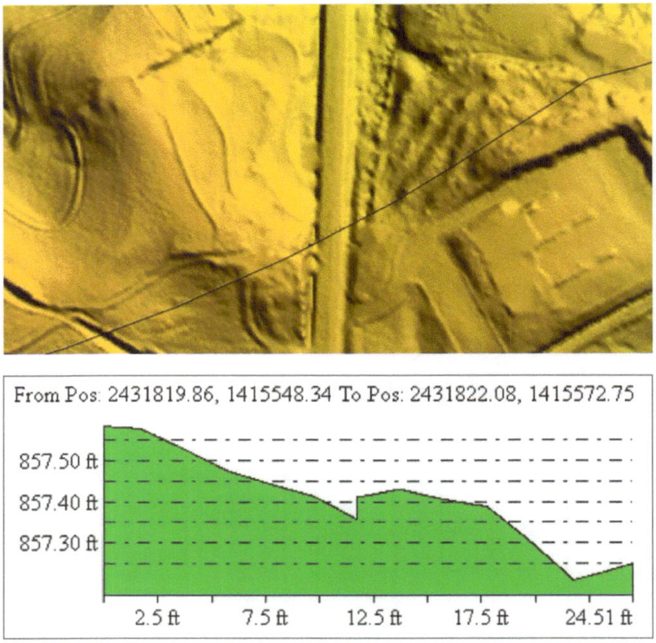

Figure 15.95. The top image shows two non-overlapping but adjacent DEM datasets. The datasets edge-match along the black boundary. A sample number of profiles were taken across the edge-match area, including along the planar road feature where both datasets should match reasonably well. There is only a small 0.05 foot difference where the two datasets edge-match; the datasets edge-match within project specifications.

Contours

Some clients require contours. Contours are usually delivered in addition to DEMs. As described in Chapter 1, contours are generally referred to as either engineering contours or cartographic contours. The level of QA/QC effort will depend on the type of contours requested. Engineering contours should be reviewed for:

- Contour interval (1 ft, 2 ft, 10 ft, etc.)
- File type (shapefile, GDB, DGN, etc.)
- Formatting (2-D or 3-D and tiled or non-tiled)
- Coding (Intermediate, Index, Depression)
- Topology (Intersections, self-intersections, dangles)
- Completeness

Cartographic contours should additionally be reviewed for:

- Appropriate smoothing to remove jagged angles
- Removal of small closed contours that do not represent valid peaks or depressions, commonly referred to contour noise or isolations
- Modifications to ensure contours point upstream in a valley, cross roads appropriately, represent building footprints accurately, and model overpasses and bridges correctly
- Spot heights or any other additional features required by project specifications

Reviews of cartographic contours generally require more large-scale visual review than for engineering contours. As true cartographic contours can be very expensive but basic engineering contours can be very hard to interpret and use, many clients request a middle approach to contours. Many clients request contours that start as engineering contours, but then have a moderate amount of automated smoothing along with automated and limited manual clean-up applied. The moderate amount of smoothing and clean-up will reduce the jaggedness of the engineering contours and remove most of the small closed contours that are solely noise in the dataset. Contours with a moderate amount of smoothing may not always cross roads perpendicularly or point upstream in a valley, but they should at least smoothly cross hydrographic channels and should not be fully contained within any water body feature. Moderately smoothed contours may intersect the edges of hydrographic features due to the applied smoothing algorithms. Moderately smoothed contours may not represent building footprints accurately or model overpasses and bridges with as much detail as cartographic contours, but they should generally show large bridges correctly. Figure 15.36 on page 542 shows an example of moderately smoothed contours overlaid on hydro-flattened bare-earth DEMs.

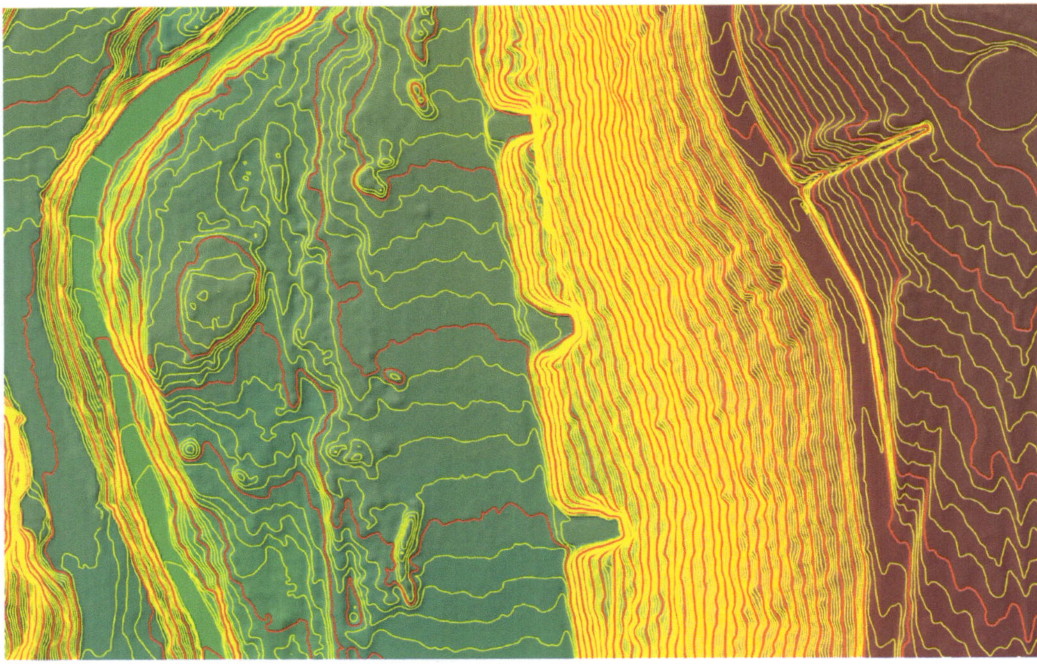

Figure 15.96. Moderately smoothed contours are overlaid on hydro-flattened bare-earth DEMs. These are 1-ft contours (yellow) with Index contours every 5 feet symbolized as the thicker red lines. No spot heights or depression contours were created for this dataset.

Metadata

Metadata are files that document all pertinent information relevant to a project or group of deliverables, such as source information, contact information, date of acquisition, data attributes, control points used, data processing methods, data accuracy, and the full lineage of data from acquisition to delivery. Metadata should always be delivered with geospatial data products. The Federal Geographic Data Committee (FGDC) adopted the Content Standard for Digital Geospatial Metadata (CSDGM) in 1994 and revised it in 1998 (FGDC-STD-001-1998). Since then, all federal agencies have been required to use this standard. Many other state and local agencies have adopted the same standard to increase compatibility and interoperability between datasets. These metadata files are commonly in eXtensible Markup Language (XML), but may also be requested as text (TXT) or HyperText Markup Language (HTML). In 2010, the FGDC endorsed several non-federally authored metadata standards, including ISO metadata (specifically the ISO 191** series). While the transition to ISO metadata has not been fully implemented, many organizations are starting to transition from the CSDGM to the ISO metadata standard. The type of metadata delivered for each project should follow the format and specifications outlined in the project requirements. Many clients will also provide templates to contractors to ensure consistency, thoroughness, and compatibility between all of their metadata records.

Metadata should always be reviewed to ensure it follows all formatting requirements outlined for each project. If the FGDC CSDGM is requested, all metadata files should be verified against the USGS Metadata Parser (MP). The freely available MP tool was developed by USGS and will validate the structure and format of a metadata record against the FGDC CSDGM. In addition to proper formatting, metadata files should always be reviewed for correct and sufficient content that fully describes all data processing methods, tested or expected accuracy values, and the full lineage of data from acquisition to delivery. Additional requirements stated within project specifications should always be verified as well.

Metadata may also include files or data that cannot be contained within XML files. Additional documents typically included as part of the metadata include project reports, acquisition reports, calibration reports, processing reports, QA/QC reports, survey reports, and accuracy reports or accuracy spreadsheets. These documents usually contain images, charts, graphs, maps, and tables that cannot be added to the XML files. Additional data files typically included as part of the metadata include control points, survey checkpoints, flight lines or swaths as flown, project boundary files, and tile grids. All additional metadata files should be reviewed for completeness, sufficient content, and adherence to project specifications.

AUTHOR BIOGRAPHY

Jennifer Novac is the Quality Manager at Dewberry for the remote sensing department and reviews numerous types of datasets, including topographic lidar, IfSAR, topobathymetric lidar, breaklines, contours, planimetrics, and ortho-imagery. She received her B.A. in anthropology and certificate in Geographic Information Systems from Indiana University-Purdue University-Indianapolis in Indianapolis, Indiana in 2006.

ACKNOWLEDGEMENTS

The author gratefully acknowledges the vertical accuracy background summaries and definitions provided by Dr. David Maune of Dewberry and the editorial comments and recommendations from Mr. H. Karl Heidemann of the USGS for the entire chapter.

REFERENCES

ASPRS (1990). ASPRS accuracy standards for large-scale maps. In *Photogrammetric Engineering and Remote Sensing,* Vol 56, No. 7, (pp 1068-1070). Bethesda, MD: American Society for Photogrammetry and Remote Sensing.

ASPRS (2014). ASPRS positional accuracy standards for digital geospatial data. In *Photogrammetric Engineering and Remote Sensing, 81(3) (pp. A1-A26).* Bethesda, MD. , American Society for Photogrammetry and Remote Sensing.

Bureau of the Budget (1947). *United States national map accuracy standards.* Washington, D.C.

FEMA (2015). *Guidance for flood risk analysis and mapping: elevation guidance, November 2015.* Federal Emergency Management Agency.

FGDC (1998). Geospatial positioning accuracy standards, Part 3: *National Standard for Spatial Data Accuracy (NSSDA),* Federal Geographic Data Committee, 1998.

FGDC (2014). *Geospatial metadata.* Federal Geographic Data Committee.

Heidemann, Hans Karl (2012). *Lidar base specification (ver. 1.0): U.S. Geological Survey Techniques and Methods,* book 11, chap. B4, 63 p.

Heidemann, Hans Karl (2014). *Lidar base specification (ver. 1.2): U.S. Geological SurveyTechniques and Methods,* book 11, chap. B4, 67 p.

Heidemann, Hans Karl (2018). *Lidar base specification (ver. 1.3): U.S. Geological Survey Techniques and Methods,* book 11, chap. B4, 101 p.

NDEP (2004). *Guidelines for digital elevation data (ver. 1.0): National Digital Elevation Program (NDEP).*

NOAA (1997). *Guidelines for establishing GPS-derived ellipsoid heights (standards: 2 cm and 5 cm),* Technical Memorandum NOS NGS-58 (ver. 4.3): National Geodetic Survey, Silver Spring, MD.

USGS (2012). *Final report of the national enhanced elevation assessment,* U.S. Geological Survey, March 29, 2012.

USGS (2017). *Geospatial metadata validation service.* Formal metadata: information and software.

Appendix A

ACRONYMS

A

AAA	American Automobile Association
AAGS	American Association for Geodetic Surveying
AASG	Association of American State Geologists
ABGPS	Airborne GPS
ABS	Acoustic Backscatter
A/D	Analog–Digital
AGC	Army Geospatial Center
AGC	Automatic Gain Control
AGL	Above Ground Level
AINS	Aided Inertial Navigation System
ALB	Airborne Lidar Bathymetry
ALH	Airborne Lidar Hydrography
ALPS	Airborne Lidar Processing System
ALS	Airborne Laser Scanning
ALTM	Airborne Laser Terrain Mapper (or Mapping)
AM	Amplitude Modulation
AMT	Above Mean Terrain
ANPD	Aggregate Nominal Pulse Density
ANPS	Aggregate Nominal Pulse Spacing

ANSI	American National Standards Institute
AOC	Airport Obstruction Chart
AOI	Area of Interest
AOL	Airborne Oceanographic Lidar
AOR	Area of Responsibility
APD	Avalanche Photodiode
ARC	Automatic Radar Controller
ARRA	American Recovery and Reinvestment Act
ASCII	American Standard Code for Information Interchange
ASFPM	Association of State Floodplain Managers
ASI	Italian Space Agency
ASL	Above Sea Level
ASPRS	American Society for Photogrammetry and Remote Sensing
ASVD	American Samoa Vertical Datum
ATC	Air Traffic Control
ATM	Airborne Topographic Mapper
ATON	Aids to Navigation
ATV	All-Terrain Vehicle
AUV	Autonomous Underwater Vehicle

B

BAG	Bathymetric Attributed Grid
BATS	Bathymetric and Topographic Survey
BCA	Benefit-Cost Analysis

BCR	Benefit/Cost Ratio
BFE	Base Flood Elevation
BIL	Band Interleaved by Line (File Format)
BIN	Binary
BIP	Band Interleaved by Pixel (File Format)
BIT	Built-In Test
BLM	Bureau of Land Management
BLS	Bureau of Labor Statistics
BMP	Best Management Practice
BPA	Buffered Project Area
BSQ	Band Sequential (File Format)

C

C3I	Command, Control, Communications and Intelliger.ce
C/A	Coarse/Acquisition
CAD	Computer-Aided Design
CADD	Computer-Aided Design and Drafting
CAES	Computer-Aided Earthmoving System
CASI	Compact Airborne Spectrographic Imager
C-CAP	Coastal Change Analysis Program
CCCL	Coastal Construction Control Line
CCD	Charge Coupled Device
CCRS	Canada Centre for Remote Sensing
CDC	Centers for Disease Control and Prevention

CDED	Canadian Digital Elevation Data
CE	Circular Error
CEP	Circular Error Probable
CFIT	Controlled Flight Into Terrain
CFR	Code of Federal Regulations
CGVD	Canadian Geodetic Vertical Datum
CHARTS	Compact Hydrographic Airborne Rapid Total Survey
CHS	Canadian Hydrographic Service
CI	Contour Interval
CIR	Color Infrared
cm	Centimeter
CMAS	Circular Map Accuracy Standard
COGO	Coalition of Geospatial Organizations
CoNED	Coastal National Elevation Dataset
CONUS	Conterminous United States, the 48 states
CO-OPS	Center for Operational Oceanographic Products and Services
CORPSCON	Corps Conversion
CORS	Continuously Operating Reference Station
COTS	Commercial Off-the-Shelf
CNMI	Commonwealth of the Northern Mariana Islands
CRAB	Coastal Research Amphibious Buggy
CRS	Coordinate Reference System
CSDGM	Content Standard for Digital Geospatial Metadata
CSH	Cross-Scan Height

CTOG	Contour to Grid
CVA	Consolidated Vertical Accuracy
CUBE	Combined Uncertainty and Bathymetry Estimator
CW	Continuous Wave
CZMIL	Coastal Zone Mapping and Imaging Lidar

D

DARPA	Defense Advanced Research Projects Agency
DAS	Data Acquisition System
DAT	DEM Analysis Tool
DCM	Direction Cosine Matrix
DDDMMSS	Degrees, Minutes, and Seconds
DED	Digital Elevation Data
DEM	Digital Elevation Model
DES	DEM Edit System
DFIRM	Digital Flood Insurance Rate Map
DG	Direct Georeferencing
DGN	MicroStation Design (File Format)
DGPS	Differential GPS
DHQ	Diurnal High Water Inequality
DHS	Department of Homeland Security
DLG	Digital Line Graph
DLQ	Diurnal Low Water Inequality
DLT	Digital Linear Tape

DLR	German (Deutsche) Aerospace Center
DO	DLG Optional (file format)
DOD	Department of Defense
DOE	Department of Energy
DOQ	Digital Orthophoto Quadrangle
DOQQ	Digital Orthophoto Quarter-Quadrangle
DORIS	Doppler Orbitography and Radiopositioning Integrated by Satellite
DOT	Department of Transportation
DPA	Defined Project Area
DPI	Dots per Inch
DRG	Digital Raster Graphic
DSM	Digital Surface Model
DTED	Digital Terrain Elevation Data
DTG	Dry Tuned Rotor Gyro
DTHM	Digital Tree Height Model
DTL	Diurnal Tide Level
DTM	Digital Terrain Model
DVD	Digital Video Disk
DVS	DEM Verification System
DWG	AutoCAD Drawing (File Format)
DXF	Drawing Exchange (File Format)
DZ	Difference in Elevation

E

E00	ArcInfo Export (File Format)
EAARL	Experimental Advanced Airborne Research Lidar
ECDIS	Electronic Chart Display and Information System
ECEF	Earth-Centered-Earth-Fixed
ECS	Electronic Chart System
ED	Existence Doubtful
EDME	Electronic Distance Measuring Equipment
EDNA	Elevation Derivatives for National Applications
EEC	European Economic Community
EEZ	Exclusive Economic Zone
EFTN	Elevation for the Nation (now the 3DEP)
EGI	Embedded GPS-Inertial
EGM	Earth Gravitational Model
EHA	Erosion Hazard Area
ENC	Electronic Navigational Chart
EPA	Environmental Protection Agency
ERDC	U.S. Army Engineer Research and Development Center
ERIM	Environmental Research Institute of Michigan
EROS	Earth Resources Observation and Science
ERS	Ellipsoidally-Referenced Survey
ERS	European Remote Sensing
ESA	European Space Agency

ESRI	Environmental Systems Research Institute
EU	European Union
EVLR	Extended Variable Length Record
EVS	Effective Vertical Spacing

F

FA	Functional Activity
FAA	Federal Aviation Administration
FAR	Federal Acquisition Regulation
FDC	Frequency-to-Distance Conversion
FEMA	Federal Emergency Management Agency
FERC	Federal Energy Regulatory Commission
FGCC	Federal Geodetic Control Committee
FGCS	Federal Geodetic Control Subcommittee
FGDC	Federal Geographic Data Committee
FID	Feature Identifier
FIRM	Flood Insurance Rate Map
FLPMA	Federal Land Policy Management Act
FM	Frequency Modulation
FMS	Flight Management System
FOG	Fiber Optic Gyro
FOV	Field of View
FR	Far-Range
ft	Feet

FTP	File Transfer Protocol
FVA	Fundamental Vertical Accuracy
FWD	Full Waveform Digitizing
FWHM	Full Width at Half Maximum
FWS	Fish and Wildlife Service

G

GB	Gigabyte
GDEM	Global Digital Elevation Model
GDOP	Geometric Dilution of Position
GEC	Geocoded Ellipsoid Corrected
GeoTIFF	Georeferenced Tagged Image File Format
gHz	Gigahertz
GIS	Geographic Information System
GLISTIN	Glacier and Land Ice Surface Topography Interferometer
GLONASS	Global Navigation Satellite System (Russian GNSS)
GLORIA	Geological Long Range Inclined Asdic
GMSL	Global Mean Sea Level
GMT	Greenwich Mean Time
GNSS	Global Navigation Satellite System
GOTS	Government Off-the-Shelf
GPRA	Government Performance and Results Act
GPS	Global Positioning System
GPSC	Geospatial Products and Services Contract

GRASS	Geographic Resources Analysis Support System
GRAV-D	Gravity for the Redefinition of the American Vertical Datum
GSD	Ground Sample Distance
Gt	Great Diurnal Range
GTC	Geocoded Terrain Corrected

H

ha	Hectare
HAG	Highest Adjacent Grade
HARN	High Accuracy Reference Network
H&H	Hydrologic and Hydraulic
HRBS	Hydrography Requirements and Benefits Study
HRTE	High Resolution Terrain Elevation
HTML	HyperText Markup Language

L

IA	Interagency Agreement
IAG	International Association of Geodesy
IBWC	International Boundary and Water Commission
ID	Identifier
IDW	Inverse Distance Weighted
IEC	International Electrotechnical Commission
IFOV	Instantaneous Field of View
IFR	Instrument Flight Rules
IfSAR	Interferometric Synthetic Aperture Radar (also InSAR)

IFSARE	IFSAR-Elevation
IGLD	International Great Lakes Datum
IGS	International GNSS Service
IHO	International Hydrographic Organization
IMG	ERDAS Image (file format)
IMO	International Maritime Organization
IMU	Inertial Measurement Unit
in	Inch
INS	Inertial Navigation System
InSAR	Interferometric Synthetic Aperture Radar (also IFSAR)
INU	Inertial Navigation Unit (also INS)
ION	Institute of Navigation
IOS	Institute of Oceanographic Sciences
IPCC	Intergovernmental Panel on Climate Change
IPP	Interpulse Period
IR	Infrared
IRU	Inertial Reference Unit (also IMU)
ISAR	Interferometric Synthetic Aperture Radar (also IFSAR and InSAR)
ISO	International Organization for Standardization
ISPRS	International Society for Photogrammetry and Remote Sensing
ITRF	International Earth Rotation Service Terrestrial Reference Frame
IV & V	Independent Validation and Verification
IWG-OCM	Interagency Working Group on Ocean and Coastal Mapping

J

JALBTCX	Joint Airborne Lidar Bathymetry Technical Center of Expertise
JFA	Joint Funding Agreement
JPL	Jet Propulsion Laboratory
JPSD-PO	Joint Precision Strike Demonstration Project Office

K

KAR	Kinematic Ambiguity Resolution (also see OTF)
KGPS	Kinematic GPS
kHz	Kilohertz
km	Kilometers
KW	Kilowatts

L

L1	Level 1 frequency of GPS satellites
L2	Level 2 frequency of GPS satellites
L3	Level 3 frequency of GPS satellites
L5	Level 5 frequency of Block IIF GPS satellites
LAAS	Local Area Augmentation System
LADAR	LAser Detection and Ranging
LADS	Laser Airborne Depth Sounder
LAG	Lowest Adjacent Grade
LAPF	Laser Altimeter Processing Facility
LARSEN	Laser Airborne Remote SENsor
LAS	LASer (File Format and Extension)

LASER	Light Amplification by Stimulated Emission of Radiation
LAT	Lowest Astronomical Tide
LBMS	Laser-Baseline Measurement System
LBS	Lidar Base Specification (USGS)
LCD	Liquid Crystal Display
LE	Linear Error
LESS	Lidar Enhanced Soil Survey
LF	Line Feed
LFE	Lowest Floor Elevation
LiDAR	Light Detection and Ranging
LIF	Laser Induced Fluorescence
LiMWA	Limit of Moderate Wave Action
LINZ	Land Information New Zealand
LIS	Land Information System
LLR	Lunar Laser Ranging
LMSL	Local Mean Sea Level
LPC	Lidar Point Cloud
LRF	Laser Rangefinder
LVIS	Laser Vegetation Imaging Sensor
LWD	Low Water Datum

M

m	Meter
m^2	Square Meter

MAPPS	Management Association for Private Photogrammetric Surveyors
MASER	Microwave Amplification by Stimulated Emission of Radiation
MB	Megabytes
MBES	Multi-Beam Echo Sounder
MDG	Multilook Detected Ground
MEIS	Multi-spectral Electro-optic Imaging Sensor
MHHW	Mean Higher High Water
MHW	Mean High Water
mHz	Megahertz
mi	Mile
mi^2	Square Mile
MIT	Massachusetts Institute of Technology
MLLW	Mean Lower Low Water
MLW	Mean Low Water
MMP	Modular Mission Payload
MODIS	Moderate Resolution Imaging Spectroradiometer
MP	Metadata Parser
MR	Mid-Range
MRR	Measurement Residual Ratio
MRS	Microwave Ranging System
MSL	Mean Sea Level
MTL	Mean Tide Level
MTS	Marine Transportation System

N

N/A	Not Available or Not Applicable
NAD 27	North American Datum of 1927 (horizontal datum)
NAD 83	North American Datum of 1983 (horizontal datum)
NADCON	North American Datum Conversion
NAI	No Adverse Impact
NALL	Navigable Area Limit Line
NAPP	National Aerial Photography Program
NASA	National Aeronautics and Space Administration
NAVD 88	North American Vertical Datum of 1988
NAVOCEANO	Naval Oceanographic Office
NAVSTAR	NAVigation Signal Timing and Ranging
NBA	No Bottom At
NBCD	National Biomass and Carbon Dataset
NCALM	National Center for Airborne Laser Mapping
NCITS	National Committee for Information Technology Standards
NCS	Navigation Computer System
NDEP	National Digital Elevation Program
NDGPS	Nationwide Differential GPS
NDOP	National Digital Orthophoto Program
nDSM	Normalized Digital Surface Model
NED	National Elevation Dataset
NEDF	National Elevation Data Framework (Australia)
NEEA	National Enhanced Elevation Assessment

NEOS	National Earth Orientation Service
NERC	North American Electric Reliability Corporation
NFIP	National Flood Insurance Program
NGA	National Geospatial-Intelligence Agency
NGDA	National Geospatial Data Asset
NGP	National Geospatial Program (USGS)
NGS	National Geodetic Survey
NGTOC	National Geospatial Technical Operations Center (USGS)
NGVD 29	National Geodetic Vertical Datum of 1929
NHD	National Hydrography Dataset
NHRBS	National Hydrography Requirements and Benefits Study
NIMA	National Imagery and Mapping Agency
NIR	Near Infra-red
NL	New Line
NMAS	National Map Accuracy Standard
NOAA	National Oceanic and Atmospheric Administration
NOS	National Ocean Service
NP	Navigation Processor
NPD	Nominal Pulse Density
NPS	Nominal Pulse Spacing
NPS	National Park Service
NR	Near-Range
NRC	National Research Council
NRCS	Natural Resources Conservation Service

NSDI National Spatial Data Infrastructure

NSGIC National States Geographic Information Council

NSRS National Spatial Reference System

NSSDA National Standard for Spatial Data Accuracy

NSTC National Science and Technology Council

NTBMS National Tidal Bench Mark System

NTDE National Tidal Datum Epoch

NTM National Technical Means

NVA Non-Vegetated Vertical Accuracy

NWI National Wetlands Inventory

NWLON National Water Level Observation Network

O

O&M Operations and Maintenance

OC Order/Class Code

OCM Office for Coastal Management

OCONUS Outside the Continental, Contiguous or Contermincus United States

OCRM Office of Ocean and Coastal Resource Management

OCS Office of Coast Survey

OEM Original Equipment Manufacturer

OGC Open Geospatial Consortium

OIM Orthorectified Image Mosaic

OIS Obstruction Identification Surface

OLOPS Off-Line Orthophoto Production System

ORI	Ortho-rectified Radar Image
OTF	On-the-fly (GPS ambiguity resolution), also see KAR
OTS	Off-the-shelf

P

PA	Position Approximate
PCT	Polarization Coherence Tomography
PD	Position Doubtful
PDOP	Position Dilution of Position
PDU	Power Distribution Unit
PE&RS	Photogrammetric Engineering & Remote Sensing
PGPS	P-code GPS
PID	Point Identifier or Point Identification (number)
pls/m^2	Pulses per Square Meter
PLU	Planning Land Unit
PMT	Photomultiplier Tube
PNEOS	Preliminary National Earth Orientation Service
POD	Print on Demand
POL	Polarization
PolInSAR	Polarimetric-Interferometric SAR
POS	Position and Orientation System
POS/AV	Position and Orientation System for Airborne Vehicles
POS/LV	Position and Orientation System for Land Vehicles
POS/MV	Position and Orientation System for Marine Vessels

POS/TG	Position and Orientation System for Railroad Track Geometry
PPA	Platform Precision Autopilot
PPM	Parts Per Million
PPP	Precise Point Positioning
PRF	Pulse Repetition Frequency
PRR	Pulse Repetition Rate
PRVD	Puerto Rico Vertical Datum
PSP	Pseudo Stereo Pair

Q

QA	Quality Assurance
QA/QC	Quality Assurance/Quality Control
QC	Quality Control
QL	Quality Level
QL0	Quality Level 0
QL1	Quality Level 1
QL2	Quality Level 2
QL3	Quality Level 3
QL4	Quality Level 4
QL5	Quality Level 5

R

R&D	Research & Development
RAMMS	Rapid Airborne Multibeam Mapping System
RAN	Royal Australian Navy

RASCAL	RAster SCanning Airborne Laser
REST	Representational State Transfer
RFI	Radio Frequency Interference
RGB	Red Green Blue
RLE	Run Length Encoded
RLG	Ring Laser Gyro
RMS	Root Mean Square
RMSD	Root Mean Square Difference
RMSDz	Root Mean Square Difference in the z Direction
RMSE	Root Mean Square Error
RMSEr	Radial RMSE (horizontal, as function of RMSEx and RMSEy)
RMSEx	RMSE in the x dimension (Easting or longitude)
RMSEy	RMSE in the y dimension (Northing or latitude)
RMSEz	RMSE in the z dimension (Elevation)
RNC	Raster Navigational Chart
ROI	Return On Investment
ROV	Remotely-Operated Vehicle
RPH	Roll-Pitch-Heave
RPI	Repeat Pass Interferometry
RSS	Root Sum Squaring
RTCM	Radio Technical Commission for Maritime Services
RTK	Real Time Kinematic
RTKGPS	Real Time Kinematic Global Positioning System
RVoG	Random Volume over Ground

S

SA	Selective Availability
SAR	Synthetic Aperture Radar
SASS	Sonar Array Sounding System
SAT	Single Antenna Transmit
SAV	Surface Autonomous Vehicle
SBCD	FGDC Subcommittee on Base Cartographic Data
SBES	Single-Beam Echosounder
SCS	Single-look Complex Slant
SD	Sounding Doubtful
SDDS	Seamless Data Distribution System
SDTS	Spatial Data Transfer Standard
SDWA	Safe Drinking Water Act
SE	Squared Errors
SENC	System Electronic Navigational Chart
SFHA	Special Flood Hazard Area
SfM	Structure from Motion
SGM	Semi-Global Matching
SHOALS	Scanning Hydrographic Operational Airborne Lidar Survey
SHP	ArcView Shape (File Format)
SIO	Scripps Institution of Oceanography
SIR	Space Imaging Radar
SLA	Shuttle Laser Altimeter
SLICER	Scanning LIDAR Imager of Canopies by Echo Recovery

SLR	Sea Level Rise
SLR	Satellite Laser Ranging
SOLAS	Safety of Life at Sea
SONAR	Sound Navigation and Ranging
SOVI	Social Vulnerability Index
SPCS	State Plane Coordinate System
SPI	Single Pass Interferometry
SQRT	Square Root
SRTM	Shuttle Radar Topography Mission
SSDM	Seabed Survey Data Model
SST	Sea Surface Topography
SSURGO	Soil Survey Geographic
SVA	Supplemental Vertical Accuracy
SVGA	Super Video Graphic Array
SWMB	Shallow-Water Multibeam

T

TDOP	Time Dilution of Position
TEC	(U.S. Army) Topographic Engineering Center
TEC	Total Electron Count
THED	Terrain Height Error Data
THU	Total Horizontal Uncertainty
TIFF	Tagged Image File Format
TIN	Triangulated Irregular Network

TLM	Topographic Line Map
TLS	Terrestrial Laser Scanning
TMS	Thermal Management System
TOF	Time-of-Flight
TPE	Total Propagated Error
TPU	Total Propagated Uncertainty
TVF	Tagged Vector File
TVG	Time Varied Gain
TVU	Total Vertical Uncertainty
TWF	Tagged World File

U

UAS	Unmanned Aerial System
UAV	Unmanned Aerial Vehicle
ULCC	Ultra Large Crude Carrier
URL	Uniform Resource Locator
USA	United States of America
USACE	U.S. Army Corps of Engineers
USC	University of South Carolina
USCG	U.S. Coast Guard
USD	U.S. Dollar
USDA	U.S. Department of Agriculture
USFS	U.S. Forest Service
USGS	U.S. Geological Survey

USIEI	U.S. Interagency Elevation Inventory
UTC	Universal Time, Coordinated, the same as Greenwich Mean Time
UTM	Universal Transverse Mercator
UUV	Unmanned Underwater Vehicle
UV	Ultraviolet

V

VERTCON	Vertical Conversion
VFR	Visual Flight Rules
VG	Vertical Gyro
VHF	Very High Frequency
VIVD	Virgin Islands Vertical Datum
VLBI	Very Long Baseline Interferometry
VLR	Variable Length Record
VMAS	Vertical Map Accuracy Standard
VPF	Vector Product Format
VPS	Vertical Point Spacing
VRS	Virtual Reference Station
VTS	Vessel Traffic System
VTUAV	Vertical takeoff and landing Tactical UAV
VVA	Vegetated Vertical Accuracy

W

WAAS	Wide Area Augmentation System (or Service)
WADGPS	Wide Area Differential GPS

WGS 84	World Geodetic System of 1984
WHP	Wellhead Protection
WHPA	Wellhead Protection Area
WiFi	Wireless Fidelity
WiMAX	Worldwide Interoperability for Microwave Access
WKT	Well Known Text
WMS	Web Mapping Service
WRELADS	Weapons Research Establishment LADS

X

XML	eXtensible Markup Language
XY	2-D Coordinates in X (Easting or Longitude) and Y (Northing or Latitude)
XYZ	3-D Coordinates in X (Easting/Longitude), Y (Northing/Latitude) and Z (Elevation)

Y

Z

OTHER

2-D	2-Dimensional (feature vertices defined in x and y with no elevation information)
2.5-D	2.5-Dimensional (feature vertices defined in x and y with a single elevation attribute for the entire feature)
3-D	3-Dimensional (feature vertices defined in x, y and z)
3DEP	3D Elevation Program
ΔX	Difference in X-coordinates (Eastings)
ΔY	Difference in Y-coordinates (Northings)
ΔZ	Difference in Z-coordinates (Elevation)

Appendix B

DEFINITIONS

3D Elevation Program (3DEP): A national program managed by the USGS to acquire high-resolution elevation data. Lidar data are being acquired for the conterminous U.S., Hawaii, and the U.S. territories. IFSAR data are being acquired for Alaska. The goal of the cooperatively funded program is to be fully operational in 2015 and to complete national coverage in 8 years. Existing USGS elevation, lidar and IFSAR data, and their management and delivery systems are being consolidated under 3DEP. These data serve government, public and private sector needs for a wide range of activities that include flood hazard mapping, precision agriculture, infrastructure planning and development, natural resource management, environmental assessment and a host of other applications.

A

Absorption: The process or action by which one thing absorbs or is absorbed by another, as when light (including lidar pulses) are absorbed by water, converting the energy of photons to internal energy. Once light enters the medium, the amount of absorption is a function of the distance traveled within the medium. Absorption is expressed as a ratio per unit distance (m^{-1}) and is a function of the thickness of the target object. Absorption also varies by wavelength. For example, the absorption of light by water varies significantly based on wavelength. At the green wavelength of 520-560 nm, the absorption is 0.14-0.04 m^{-1}. At 1,064 nm wavelength, the absorption is ~60 m^{-1}. At 1,540 nm, the absorption is over 5,000 m^{-1}.

Accelerometer: A device that measures the total specific force with respect to an inertial reference comprising gravity plus acceleration.

Accuracy: The closeness of an estimated value (e.g., measured or computed) to a standard or accepted (true) value of a particular quantity. Not to be confused with precision. See precision.

- **Absolute Accuracy:** A measure that accounts for all systematic and random errors in a data set. Absolute accuracy is stated with respect to a defined datum or reference system
- **Accuracy$_r$ (ACC$_r$):** The NSSDA reporting standard in the horizontal component that equals the radius of a circle of uncertainty, such that the true or theoretical horizontal location of the point falls within that circle 95-percent of the time. Accuracy$_r$ = 1.7308 x RMSE
- **Accuracy$_z$ (ACC$_z$):** The NSSDA reporting standard in the vertical component that equals the linear uncertainty value, such that the true or theoretical vertical location of the point falls within that linear uncertainty value 95-percent of the time. Accuracy$_z$ = 1.9600 x RMSE$_z$
- **Consolidated Vertical Accuracy (CVA):** An obsolete NDEP term for reporting lidar vertical accuracy at the 95th percentile for all land cover categories combined, assuming that errors do not necessarily follow a normal distribution. There is no equivalent in the 2014 ASPRS Positional Accuracy Standards for Digital Geospatial Data which instead uses a combination of NVA and VVA
- **Fundamental Vertical Accuracy (FVA):** An obsolete NDEP term for reporting lidar vertical accuracy at the 95th percent confidence level in open terrain only (e.g., dirt, sand, rocks, and short grass) based on RMSEz x 1.9600. In the 2014 ASPRS Positional Accuracy Standards for Digital Geospatial Data, the

prior FVA term was merged with the former SVA term for urban terrain (concrete and asphalt) to form the new term called Non-vegetated Vertical Accuracy (NVA) where all errors are assumed to follow a normal distribution

- **Horizontal Accuracy:** The horizontal (radial) component of the positional accuracy of a dataset with respect to a horizontal datum at a specified confidence level. See $Accuracy_r$
- **Local Accuracy:** The uncertainty in the coordinates of points with respect to coordinates of other directly connected, adjacent points at the 95-percent confidence level
- **Network Accuracy:** The uncertainty in the coordinates of mapped points with respect to the geodetic datum at the 95-percent confidence level. According to the National Standard for Spatial Data Accuracy (NSSDA): "For NSRS network accuracy classification, the datum is considered to be best expressed by the geodetic values at the Continuously Operating Reference Stations (CORS) supported by NGS. By this definition, the local and network accuracy values at CORS sites are considered to be infinitesimal, i.e., to approach zero."
- **Non-vegetated Vertical Accuracy (NVA):** The vertical accuracy at the 95-percent confidence level in non-vegetated open terrain where errors should approximate a normal distribution. Replaces Fundamental Vertical Accuracy (FVA) used in the past
- **Positional Accuracy:** The accuracy at the 95-percent confidence level of the position of features, including horizontal and vertical positions, with respect to horizontal and vertical datums
- **Relative Accuracy:** A measure of variation in point-to-point accuracy in a data set. In lidar, this term may also specifically mean the positional agreement between points within a swath, adjacent swaths within a lift, adjacent lifts within a project, or between adjacent projects
- **Supplemental Vertical Accuracy (SVA):** An obsolete NDEP term for reporting lidar vertical accuracy at the 95th percentile for individual land cover categories (e.g., weeds and crops, brush lands, and fully forested) where vertical errors may not follow a normal error distribution. In the 2014 ASPRS Positional Accuracy Standards for Digital Geospatial Data, the multiple SVAs were merged into the Vegetated Vertical Accuracy (VVA)
- **Vegetated Vertical Accuracy (VVA):** An estimate of the vertical accuracy, based on the 95th percentile, in vegetated terrain where errors do not necessarily approximate a normal distribution. Replaces Supplemental Vertical Accuracy (SVA) for individual land cover categories. See percentile
- **Vertical Accuracy:** The measure of the positional accuracy of a dataset with respect to a specified vertical datum, at a specified confidence level or percentile. See $Accuracy_z$, NVA, and VVA

Acoustic Wave: See Sound Wave.

Adjustment: The process of changing the values of a given set of quantities so that results calculated using the changed set will be better than those calculated using the original set. The concept of "better" is vague. The most common interpretation is that the sum of the squares of differences between results obtained by measurement and results obtained by calculation shall be a minimum. With this criterion, the method of least squares is the required process.

Aerial Triangulation (Aerotriangulation): The process of measuring a number of points on overlapping images and/or ground control points to determine the most probable values of the six exterior orientation elements of aerial photographs (x/y/z coordinates of the camera focal point plus the camera's roll, pitch and yaw). The output of this process includes ground space coordinates for all points measured on at least two images.

Aggregate Nominal Pulse Density (ANPD): A variant of nominal pulse density (NPD) that expresses the total expected or actual density of pulses occurring in a specified unit area resulting from multiple passes of the lidar instrument, or a single pass of a platform with multiple lidar instruments, over the same target area. In all other respects, ANPD is identical to NPD. In single coverage collection, ANPD and NPD will be equal. See aggregate nominal pulse spacing, nominal pulse density, and nominal pulse spacing.

Aggregate Nominal Pulse Spacing (ANPS): A variant of nominal pulse spacing (NPS) that expresses the typical or average lateral distance between pulses in a lidar dataset resulting from multiple passes of the lidar instrument, or a single pass of a platform with multiple lidar instruments, over the same target area. In all other respects, ANPS is identical to NPS. In single coverage collections, ANPS and NPS will be equal. See aggregate nominal pulse density, nominal pulse density, and nominal pulse spacing.

Airborne GPS: A method for collecting Global Positioning System (GPS) data continuously for an aerial camera, lidar or IFSAR sensor so as to: (1) Provide Direct Georeferencing of photo centers and/or the origin of individual laser or radar pulses used in mapping, (2) Strengthen the positioning accuracy or larger mapping projects while drastically reducing the number of ground control points necessary, and/or (3) Enable accurate execution of pre-planned flight lines in automated Flight Management Systems.

Ambiguity Resolution: Combining the phase data from two or more GPS receivers so that, after eliminating all other significant errors, the unknown number of integer wavelengths can be determined for signals coming from GPS satellites. Redundant L1 and L2 phase observations from two or more receivers each tracking five or more satellites, provide the information for rapid unambiguous resolution. Once the ambiguities are resolved, the corrected phases for each observed satellite become precise ranges that allow the computation of the baseline vector(s) between the receivers with a typical accuracy of 2-10 centimeters.

Analog Photogrammetry: Stereo photogrammetric procedures that utilize direct viewing of film imagery in analog stereoplotters that optically and/or mechanically replicate, at reduced scale, the physical geometry that existed when stereo photography was acquired. Analog stereoplotters are less accurate than analytical stereoplotters, and they are ill-suited for DEM production.

Analytical Photogrammetry: Stereo photogrammetric procedures that utilize direct viewing of film imagery in analytical stereoplotters that mathematically replicate, at reduced scale, the physical geometry that existed when stereo photography was acquired. Analytical stereoplotters can also mathematically apply camera calibration (interior orientation) parameters, and correct for atmospheric refraction and earth curvature.

Aqueduct: A structure designed to transport domestic or industrial water from a supply source to a distribution point, often by gravity.

Arc Second (Arc-Second or Second of Arc): In angular measurements, 1/3600th of a degree. Commonly used in descriptions of geographic coordinate reference systems, for example, the horizontal resolution of the National Elevation Dataset (NED), 1 arc-second, ⅓ arc-second, ⅑ arc-second. For most of the contiguous United States, 1 arc-second is approximately 30 meters, linear.

Artifacts: An inaccurate observation, effect or result, especially one resulting from the technology used in scientific investigation or from experimental error. In bare-earth elevation models, artifacts are detectable surface remnants of buildings, trees, towers, telephone poles or other elevated features; also, detectable artificial anomalies that are introduced to a surface model by way of system specific collection or processing techniques, e.g., corn-row effects of profile collection, star and ramp effects from multidirectional contour interpolation, or detectable triangular facets caused when vegetation canopies are removed from lidar data. Orthophotos and other geospatial datasets also have blemishes referred to as artifacts.

Aspect: The direction which a building or topographic surface faces with respect to points of a compass. The slope direction or steepest downslope across a surface, normally measured clockwise in degrees from due north. The value of each location in an aspect dataset indicates the direction the surface slope faces.

Atmospheric Refraction: The bending of light rays as they pass through the atmosphere.

Attitude: The position of a body defined by the angles between the axes of the coordinate system of the body and the axes of an external coordinate system. In photogrammetry, the attitude is the angular orientation of a camera (roll, pitch, yaw), or of the photograph taken with that camera, with respect to some external reference system. With lidar and IFSAR, the attitude is normally defined as the roll, pitch and heading of the instrument at the instant an active pulse is emitted from the sensor.

Azimuth: Horizontal direction reckoned clockwise from the meridian plane.

B

Bald Earth: Non-preferred term. See bare-earth.

Bare Earth (Bare-Earth): Digital elevation data of the terrain, free from vegetation, buildings and other man-made structures. Elevations of the ground.

Base Flood: A flood that has a one-percent chance of being equaled or exceeded in any given year. Often called the 100-year flood.

Base Flood Elevation (BFE): An elevation that has a one-percent chance of being equaled or exceeded in any given year by a base flood.

Baseline: The vector between two survey-grade GPS (or GNSS) receivers, one being typically the base station (see definition below), and the other roving or moving in a vehicle or aircraft, and both collecting data from the same GPS/GNSS satellite configuration at the same time.

Base Map: A map on which information can be placed for comparison or for georeferencing and graphical correlation, including digital orthophotos for horizontal georeferencing and DEMs for vertical georeferencing. A map from which other maps are prepared by adding information; in particular, a planimetric map used in preparing topographic maps.

Base Station: A survey station with known coordinates from which all unknown differentially measured positions are derived.

Bathymetric Chart: A topographic chart of the bed of a body of water. Generally, bathymetric charts show depths by contour lines and gradient tints. Regional bathymetric surveys are conducted to determine the location and orientation of depth contours in water depths and areas where safety of surface navigation is not a major concern. This type of surveying might be accomplished with an interferometric sonar, a vertical beam echo sounder, or a multibeam sonar. Shallow water bathymetry is also determined from airborne lidar bathymetry. Software used for bathymetric surveying is attuned to production of a digital elevation model, or bathymetric surface, from cleaned or smoothed soundings.

Bathymetric Surface: The underwater terrain model.

Bathymetry: The measurement and study of water depths. Traditionally bathymetry has been expressed with contours and hydrography with spot depths. Bathymetry may not meet hydrographic standards, as it may not show all of the bottom characteristics important to the mariner who is navigating. Bathymetric contours depicted on nautical charts may be displaced cartographically to accommodate sounding data which need to be displayed. NOAA's hydrographic depths and bathymetric contours represented on a nautical chart are all related to the national chart datum of MLLW. These depths are expressed as positive numbers below the chart datum. Minus signs are not depicted on nautical charts unless above MLLW, meaning the point is above water when the water level is at MLLW or below.

Beam Divergence: An angular measure that describes the increase in laser pulse diameter as a pulse gets further away from the laser emitter. The beam divergence and diameter of the beam waist characterize the quality of a laser beam. Beam divergence is typically defined as the half-angle beam divergence and expressed in milliradians. A wider beam divergence will create a bigger "spot size" on the target and thereby reduces precision and accuracy, especially when the target is sloping or changing rapidly. Laser pulses are intentionally expanded for the eye safety of casual observers in the survey area.

Bench Mark (Benchmark or BM): A relatively permanent, natural or artificial, material object bearing a marked point whose elevation above or below an adopted vertical datum is known.

- **Permanent Bench Mark:** A bench mark as nearly permanent in character as is practicable. The National Spatial Reference System classifies bench marks in terms of their "stability," with Stability A being the most stable, and Stability D being of questionable stability
- **Temporary Bench Mark:** A bench mark established to hold, temporarily, the end of a completed section of a line of levels and to serve as a starting point from which the next section is run. Spikes and screws in poles, bolts on bridges, and chiseled marks on masonry are typically used as temporary bench marks (TBMs)
- **Tidal Bench Mark:** A bench mark whose elevation has been determined with respect to mean sea level at a nearby tide gauge. The tidal bench mark is used as reference for that tide gauge

Bias: A systematic error inherent in measurements due to some deficiency in the measurement process or subsequent processing.

Blunder: A mistake resulting from carelessness or negligence.

Boresight: Calibration of a sensor system equipped with GPS and an inertial measurement unit (IMU) to determine the accurate position of the instrument (x/y/z) with respect to the GPS antenna, and orientation (roll, pitch, heading) of active sensor source (IFSAR, lidar, sonar) or the optical axis of a film camera or a digital imaging system.

Breakline: A linear feature that describes a change in the smoothness or continuity of a surface. Breaklines may be 2-D, 2.5-D or 3-D and collected using several different methodologies. 2-D breaklines only contain x/y coordinates and do not include elevation values; these breaklines only define edges of features but do not enforce or maintain z-elevations in the surface. 2.5-D breaklines contain x/y coordinates in the internal geometry but have an elevation assigned to the feature such as an attribute in an attribute table. 3-D breaklines include x/y and z coordinates in the internal geometry of the breakline, i.e., each vertex contains an x/y/z coordinate; these breaklines will maintain z-elevations in the surface and can be used to define breaks or interruptions in the surface as well. The two most common forms of breaklines are as follows:

- A **soft breakline** ensures that known z-values along a linear feature are maintained (e.g., elevations along a pipeline, road centerline or drainage ditch), and ensures that linear features and polygon edges are maintained in a TIN surface model, by enforcing the breaklines as TIN edges. They are generally synonymous with 3-D breaklines because they are depicted with series of x/y/z coordinates. Somewhat rounded ridges or the trough of a drain may be collected using soft breakline
- A **hard breakline** defines interruptions in surface smoothness, e.g., to define streams, rivers, shorelines, dams, ridges, building footprints, and other locations with abrupt surface changes. Although some hard breaklines are 3-D breaklines, they are typically depicted as 2-D breaklines because features such as shorelines and building footprints are normally depicted with series of x/y coordinates only, often digitized from digital orthophotos that include no elevation data

Bridge: A structure carrying a road, path, railroad, canal, aircraft taxiway, or any other transit between two locations of higher elevation over an area of lower elevation. A bridge may traverse a river, ravine, road, railroad, or other obstacle. "Bridge" also includes but is not limited to aqueduct, drawbridge, flyover, footbridge, overpass, span, trestle, and viaduct. In mapping, a bridge is distinguished from a roadway over a culvert in that a bridge is a man-made elevated deck that is not underlain with earth or soil. See culvert.

Building Footprint: The outline of a building, normally as viewed orthogonally from above.

C

C-Factor: An empirical measure of the inherent contouring accuracy of a photogrammetric plotter and its operator.

Calibration: The process of identifying and correcting for systematic errors in hardware, software, or data. Determining the systematic errors in a measuring device by comparing its measurements with the markings or measurements of a device that is considered correct.

- **Camera Calibration:** The geometric calibration of a conventional film mapping camera includes the determination of the following quantities: (a) The calibrated focal length, (b) The location of the principal point with respect to the fiducial marks, (c) The location of the point of symmetry, (d) The distortion effective in the focal plane of the camera and referred to the particular calibrated focal length, (e) The resolution of the lens system, (f) The degree of flatness of the focal plane, (g) The opening and closing cycle of the shutter as a function of time, and (h) The locations of fiducial marks—all of which help to ensure that µm measurements made on aerial film will translate correctly into accurate ground coordinates via photogrammetric calculations. Depending on camera design, digital cameras have different forms of geometric calibration and also include radiometric calibration, e.g., spectral response of CCD sensors over the spectral range of the sensor, and determination of the pixel-to-pixel uniformity
- **Lidar System Calibration:** (1) Instrument Calibration: Factory calibration includes both radiometric and geometric calibration unique to each manufacturer's hardware, and tuned to meet the performance specifications for the model being calibrated; instrument calibration can only be assessed and corrected by the factory. (2) Data Calibration: The lever-arm calibration determines the sensor-to-GPS-antenna offset vector (lever arm) components relative to the antenna phase center. The offset vector components are re-determined each time the sensor or aircraft GPS antenna is moved or repositioned. Because normal aircraft operations can induce slight variations in component mounting, the components are normally field calibrated for each project, or even daily, to determine corrections to the roll, pitch, yaw, and scale calibration parameters

Camera Station: The point in space where the forward node of the camera lens was located at the instant the photographic exposure was made.

Cartesian Coordinates: Coordinate values of two types:

- **2-D Cartesian Coordinates:** A pair of numbers (x/y coordinates) that locate a point by its distance from two intersecting, normally perpendicular lines in the same plane. Each distance is measured along a parallel to the other line. UTM and State Plane coordinates are examples of 2-D Cartesian coordinates
- **3-D Cartesian Coordinates:** A triad of numbers (x/y/z coordinates) that locate a point by its distance from three fixed planes that intersect one another at right angles. However, except for unique applications, 3-D Cartesian coordinates with z-coordinates are rarely used. Instead, z-values are more popularly understood as heights or elevations above a curved surface defined by the vertical datum, ellipsoid, or geoid. Thus, when 3-D coordinates are mentioned, the term normally pertains to x/y coordinates and z-values

Cartography: The science and art of making maps and charts. It is customary to distinguish between maps in general and maps used for navigation; the latter are generally called charts.

Cell (Pixel): A single element of a raster dataset. Each cell contains a single numeric value of information representative of the area covered by the cell. Although the terms "cell" and "pixel" are synonymous, "cell" is normally used in reference to non-image rasters such as DEMs whereas "pixel" is used in reference to image rasters such as orthoimagery, lidar intensity images, or orthorectified radar images.

Charts: Maps used for nautical or aeronautical navigation.

Check Point (Checkpoint): A surveyed point (x,y or x,y,z) used to estimate the positional accuracy of a geospatial dataset against an independent source of greater accuracy. Check points are independent from, and may never be used as, control points on the same project.

Classification (of lidar): The classification of lidar point cloud returns in accordance with a classification scheme to identify the type of target from which each lidar return is reflected. The process allows future differentiations between bare-earth terrain points, water, noise, vegetation, buildings, other manmade features , and objects of interest.

Cleanness: A subjective term to describe the degree to which artifacts have been removed from a DEM or TIN.

Compilation: The production of a new or revised map or chart, or portions thereof, from existing maps, aerial photographs, field surveys, new data and other sources. The process is called stereocompilation if stereoscopic plotting instruments (stereo plotters) are used.

Confidence Level: The percentage of points within a dataset that are estimated to meet the stated accuracy, e.g., accuracy reported at the 95-percent confidence level means that 95 percent of the positions in the data set will have an error with respect to true ground position that are equal to or smaller than the reported accuracy value.

Consolidated Vertical Accuracy (CVA): The term used by the National Digital Elevation Program (NDEP) guidelines for vertical accuracy at the 95th percentile in all land cover categories combined. This term is no longer used for the 3DEP which instead uses Vegetated Vertical Accuracy (VVA).

Contours: Lines of equal elevation on a surface. An imaginary line on the ground, all points of which are at the same elevation above or below a specified reference surface (vertical datum).

Contour Interval: The difference in elevation (z-values) between two adjacent contours.

Control: The coordinates of a control station. The geometric data associated with a collection of control stations, such as coordinates, distances, angles, or directions between control stations. The data associated with a set of control stations and used as the basis for detailed surveys.

Control Point (Calibration Point): A surveyed point used to geometrically adjust a geospatial dataset to establish its positional accuracy relative to the real world. When used for aerotriangulation, control points are either chosen as photo-identifiable points or are "paneled" prior to flying. Control points are independent from, and may never be used as, check points on the same project.

Control Tide Station: A primary water level station, located with open ocean exposure or within the mouth of a bay or estuary, that has operated for 19 or more years, is expected to continue to operate in the future, and is used to obtain a continuous record of the water levels in a locality. A control tide station is sited to provide datum control for national applications.

Controlled Flight Into Terrain (CFIT): An accident in which an airworthy aircraft, under pilot control, is unintentionally flown into the ground, a mountain, a body of water or an obstacle.

Coordinates: A set of N numbers designating the location of a point in N-Dimensional space. Horizontal coordinates are 2-Dimensional coordinates, normally expressed as x/y coordinates, Eastings and Northings, or Longitude and Latitude (geographic coordinates). A vertical coordinate may be 1-Dimensional, i.e., the vertical distance of a point above or below a reference surface (vertical datum) such as the elevation of a bench mark without known x/y coordinates. However, most vertical coordinates are specified as 3-Dimensional coordinates, i.e., x/y coordinates and z-values.

Correlation: The extent to which one randomly varying quantity can be expressed as a function of another, or to which both quantities can be expressed as function of a third, nonrandom quantity. With IFSAR, interferometric correlation is a measure of the similarity of the signal received at the two antennae. See Image Correlation.

Culvert: A tunnel carrying a stream or open drain under a road or railroad or through another type of obstruction to natural drainage. A culvert is typically constructed of formed concrete or corrugated metal and surrounded on all sides, top and bottom, by earth or soil.

D

Data Format: The specific arrangement of data in a file, including records, data types, and data labels. For example, floating point numbers as x/y pairs with one set per record.

Data Model: The conceptual view of what information is to be represented. For example, the surface of the Earth can be represented as a grid of posts of varying heights (i.e., raster data model), or as lines of equal elevation (i.e., vector data model).

Data Void: In lidar, a gap in the point data coverage, caused by surface non-reflectance of the lidar pulse, instrument or processing anomalies or failure, obstruction of the lidar pulse, or improper collection flight planning or execution. Any area greater than or equal to (4 * ANPS)^2 measured using first-returns only, is considered to be a data void.

Datum (plural Datums): A reference point, surface or coordinate system from which position measurements are made. For mapping purposes, datum examples include WGS-84, NAD83 and NAVD88.

- **Chart Datum:** The datum to which soundings on a chart are referred, usually taken to correspond to a low-water elevation
- **Diurnal Tide Level (DTL):** A tidal datum which is the average of Mean Higher High Water and Mean Lower Low Water
- **Geodetic Datum:** A coordinate system, and a set of reference points, used to locate places on the Earth or similar objects. Geodetic datums define the size and shape of the Earth and the origin and orientation of the coordinate systems used to map the Earth
- **Horizontal Datum:** A geodetic datum for any extensive measurement system of positions, usually expressed as latitude-longitude coordinates, on the Earth's surface
- **Hydrographic Datum:** The level of water that charted depths displayed on a nautical chart are measured from. A datum for depths (soundings), depth contours, and elevations of foreshore and offshore features. Also called chart datum
- **Local Datum:** A datum defining a coordinate system that is used only over a region of very limited extent
- **Low Water Datum (LWD):** An approximation to mean low water that has been adopted as a standard tidal datum for a specific region although it may differ slightly from a later determination
- **Lower Low Water Datum (LLWD):** An approximation of mean lower low water that has been adopted as a standard reference for a limited area and is retained for an indefinite period regardless of the fact that it may differ slightly from a better determination of mean lower low water from a subsequent series of observations. Used primarily for river and harbor engineering purposes. The Columbia River lower low water datum is an example
- **Mean Higher High Water (MHHW):** A tidal datum computed as the average of the higher high water height of each tidal day observed over a specific 19-year Metonic cycle denoted as the National Tidal Datum Epoch (NTDE). Only the higher high water of each pair of high waters of a tidal day is included in the mean. For stations with shorter series, a comparison of simultaneous observations is made with a primary control tide station in order to derive the equivalent of the 19-year value
- **Mean High Water (MHW):** A tidal datum computed as the arithmetic mean of the high water heights observed over a specific 19-year Metonic cycle. For stations with shorter series, a comparison of simultaneous observations is made with a primary control tide station in order to derive the equivalent of the 19-year value
- **Mean Lower Low Water (MLLW):** A tidal datum computed as the arithmetic mean of the lower low water heights of the tide observed over a specific 19-year Metonic cycle. Only the lower low water of each pair of low waters of a tidal day is included in the mean. For stations with shorter series, a comparison of simultaneous observations is made with a primary control tide station in order to derive the equivalent of the 19-year value. MLLW is the chart datum used in the U.S.
- **Mean Low Water (MLW):** A tidal datum computed as the arithmetic mean of the low water heights observed over a specific 19-year Metonic cycle. For stations with shorter series, a comparison of simultaneous observations is made with a primary control tide station in order to derive the equivalent of the 19-year value

- **Mean Sea Level (MSL):** A tidal datum which is computed as part of the National Tidal Datum Epoch (NTDE) based on data collected over a 19-year cycle. It pertains to Local Mean Sea Level (LMSL) at the tide station at which it was observed and should not be confused with any other vertical datum, including LMSL at other tide stations
- **Mean Tide Level (MTL):** A tidal datum which is the average of Mean High Water and Mean Low Water. Also known as the half tide level
- **Sea Level Datum (SLD):** An obsolete term. See National Geodetic Vertical Datum of 1929 and mean sea level
- **Tidal Datum:** A surface with a designed elevation from which heights or depths are reckoned, defined by a certain phase of the tide. A tidal datum is local, usually valid only for a restricted area about the tide gauge used in defining the datum. Examples include Mean Sea Level (MSL), Mean Low Water (MLW) and Mean Higher High Water (MHHW)
- **Vertical Datum:** A surface of zero elevation to which heights of various points are referred in order that those heights be in a consistent system. More broadly, a vertical datum is the entire system of the zero elevation surface and methods of determining heights relative to that surface. Over the years, many different types of vertical datums have been used; the most predominant types today are tidal datums and geodetic datums

Decorrelation (IFSAR): A measure of the dissimilarity of the signal received at the two antennae caused by geometry, thermal noise, and other factors.

Deflection of the Vertical: The angle between the direction of gravity defining astronomical latitude and longitude, and the normal to the reference ellipsoid defining geodetic latitude and longitude.

Delaunay Triangulation: A mesh of contiguous, non-overlapping, triangles established from a set of points in the plane such that each triangle's circumscribing circle is empty, containing no point from the set in its interior.

Diapositive: A positive photograph on a transparent medium. A positive film transparency as opposed to a positive paper print which is not transparent.

Differential GPS: A GPS enhancement procedure whereby GPS timing and atmospheric errors are determined at a GPS base station with precisely known coordinates so that corrections can be applied to nearby GPS rover units, improving location accuracy from the 15-meter nominal GPS accuracy to about 10-cm at GPS rover units where GPS timing and atmospheric errors are nearly identical to those at the base station. The corrected information can be applied to data from the roving receiver in real time in the field using radio signals or through post-processing after data capture using special processing software. See Real Time Kinematic (RTK).

Differential Leveling: Survey procedures used to measure the elevation differences between two or more points, to extend vertical control, and to monitor the stability of water level measurement gauges. The quality of leveling is a function of the procedures used, the sensitivity of the leveling instruments, the precision and accuracy of the leveling rod, the attention given by surveyors, and the refinement of the computations.

Digital Correlation (Digital Image Correlation): Image correlation performed automatically through the use of digital imagery and digital (softcopy) photogrammetry, using knowledge of sensor parameters and orientations and correlation algorithms that directly compare pixel patches on conjugate images or indirectly compare features derived from digital stereo images. See semi-global matching.

Digital Elevation Model (DEM): See three different definitions below:

- A popular acronym, as in the title of this book, used as a generic term for digital topographic and/or bathymetric data in all its various forms. Unless specifically referenced as a Digital Surface Model (DSM), the generic DEM normally implies x/y coordinates and z-values of the bare-earth terrain, void of vegetation and man-made features
- As used by the U.S. Geological Survey (USGS), a DEM is the digital cartographic representation of the elevation of the land at regularly spaced intervals in x and y directions, using z-values referenced to a common vertical datum
- As typically used in the U.S. and elsewhere, a DEM has bare-earth z-values at regularly spaced intervals in x and y, where Δx and Δy are normally measured in feet or meters in even units; however, grid spacing, datum, coordinate systems, data formats, and other characteristics may vary widely

Digital Elevation Model (DEM) Resolution: The linear size of each cell of a raster DEM. Features smaller than the cell size cannot be explicitly represented in a raster model. DEM resolution may also be referred to as cell size, grid spacing, or ground sample distance.

Digital Line Graph (DLG): Geospatial data, digitized as node, line and area features, using hundreds of different attribute codes to define basic cartographic data categories such as hypsography (contours), hydrography, transportation, manmade features, vegetation, boundaries, survey control, etc.

Digital Orthophoto: A digital photograph prepared from a perspective photograph by digitally removing displacements of points caused by tilt, relief, and perspective. A "true digital orthophoto" is defined as one in which the sides of vertical features are not visible, as though looking straight down on each natural and man-made feature from infinity; this feature is especially desired in urban areas with skyscrapers and tall buildings, the sides of which are normally photographed with aerial photography.

Digital Photogrammetry: See Softcopy Photogrammetry.

Digital Surface Model (DSM): Similar to DEMs or DTMs, except that they depict the elevations of the top surfaces of buildings, trees, towers, and other features elevated above the bare earth. DSMs are especially relevant for telecommunications management, air safety, forest management, and 3-D modeling and simulation

Digital Terrain Elevation Data (DTED): A uniform matrix of terrain elevation values produced by the National Geospatial-Intelligence Agency (NGA) which provides basic quantitative data for military systems that require terrain elevation, slope, and/or surface roughness information. Data density depends on the level produced. Similar to USGS DEMs.

- DTED Level 0 elevation post spacing is 30 arc seconds (nominally one kilometer)
- DTED Level 1 elevation post spacing is three arc seconds (approximately 100 meters)
- DTED Level 2 post spacing is one arc second (approximately 30 meters), corresponding

Digital Terrain Model (DTM): See two different definitions below.

- In some countries, DTMs are synonymous with DEMs, representing the bare-earth terrain with uniformly-spaced z-values, as in a raster
- As used herein, DTMs may be similar to bare-earth DEMs but they frequently incorporate the elevation of important topographic features on the land via irregularly spaced mass points, spot elevations, and breaklines. With DTMs, distinctive terrain features are more clearly defined and precisely located,

and contours generated from DTMs more closely approximate the real shape of the terrain and thus are technically superior to standard DEMs for many applications. DTMs are normally more expensive and time consuming to produce than uniformly-spaced DEMs because breaklines are less suitable for automated collection

Direct Georeferencing: The direct measurement of Exterior Orientation parameters, i.e., position (x/y/z coordinates) and attitude (roll/pitch/heading) at the instant an aerial photograph or digital image is taken, to aid or replace aerial triangulation. The term is also applicable to the position and orientation of airborne lidar or IFSAR sensors.

Direction of Gravity: The direction, toward the Earth's center, indicated by a plumb line. Gravity is a force and therefore has both magnitude and direction. The direction of gravity is independent of any coordinate system, but its components may be described with respect to two mutually perpendicular planes through the local normal, one being that of the local meridian. Both components are given in terms of the deflection of the vertical from those planes.

Discrete Return Lidar: Lidar system or data in which important peaks in the waveform are captured and stored as a discrete value in time and intensity. Each peak represents a return from a different target, discernible in vertical or horizontal domains. Most modern lidar systems are capable of capturing multiple discrete returns from each emitted laser pulse. See waveform lidar, linear mode lidar, Geiger mode lidar, single photon lidar, and bathymetric lidar.

Displacement: Any shift in the position of an image on a photograph due to tilt during photography or scale changes in the imagery normally caused by changing elevations (relief) of the area photographed.

Distortion: A lens aberration that causes a difference between the position of any part of the image and its theoretically correct position.

Diurnal: Having a period or cycle of approximately one tidal day. A tide is said to be diurnal when only one high water and one low water occur during a tidal day.

Diurnal High Water Inequality (DHQ): The difference between Mean Higher High Water (MHHW) and Mean High Water (MHW).

Diurnal Low Water Inequality (DLQ): The difference between Mean Low Water (MLW) and Mean Lower Low Water (MLLW).

Diurnal Range: The difference in height between mean higher high water (MHHW) and mean lower low water (MLLW).

Drainage-Enforced: See Hydro-Enforced.

Drape: The superimposition of 2-D features over a 3-D surface, normally for viewing of all features in 3-D perspective, for 3-D fly-throughs or walk-throughs in virtual reality.

Dual Frequency: A high performance GPS receiver for precision applications that tracks the L1 C/A pseudoranges and carrier phase, and either tracks L2 pseudoranges and carrier phase or computes derived L2 pseudoranges and carrier phase using the known cross correlation between the L1 and L2 code modulations.

E

Easting: In a plane coordinate system, the coordinate in a general east-west direction increasing to the east. The grid coordinate of a point eastward (positive) or westward (negative) from a reference meridian. The grid coordinate eastward (positive) or westward (negative) from the central (zero) meridian (the line of zero eastings or the y-axis) on a gridded map. The central meridian is frequently replaced by another called the false meridian, sufficiently far to the west of the central meridian so that all eastings on the map are then positive with respect to the false meridian. A false easting is the constant value added to all eastings so that only positive values of easting are recorded.

Ebb Tide: The period between high tide and low tide, during which water flows away from the shore. Compare flood tide.

Echogram: The graphic record produced by an echo sounder and showing, as a function of time, the strength of the echo signal and the time taken for the echo to return.

Echo Sounder: An instrument for determining the depth of water by measuring the time of travel of a sound pulse from the surface of a body of water to the bottom and back. An echo sounder consists of an oscillator for generating pulses, a hydrophone for detecting the echo, a clock for timing the pulses, and a recorder for converting the time of travel to a depth and plotting that depth on a continuous chart.

Echo Sounding: Determining the distance from the surface of water to the bottom by measuring the time interval required for sound waves to go from a sound generator to the bottom and back again. The principal source of error in echo sounding is uncorrected refraction; another source is schools of fish that reflect sound back before it reaches the bottom.

Electromagnetic Spectrum: The complete range of wavelengths of electromagnetic radiation, beginning with the longest radio waves and extending through visible light all the way to the extremely short gamma rays.

Elevation: The distance measured upward along a plumb line between a point and the geoid. The elevation of a point is normally the same as its orthometric height, defined as "H" in the equation: $H = h - N$, where h is the ellipsoid height and N is the geoid height.

Elevation Certificate: A form on which the lowest floor elevation, lowest adjacent grade, and highest adjacent grade of a building are certified relative to the base flood elevation for the location of the building. Other descriptive information is also provided to help identify the flood risk to the building surveyed.

Elevation Derivatives: Products derived from 3-D source data. With lidar, the point cloud data are the source from which all derivative products are produced, including Digital Surface Models (DSMs), Digital Terrain Models (DTMs), gridded Digital Elevation Models (DEMs), contour lines, hillshades, slope and aspect maps, curvature maps, height-above-ground maps, intensity imagery, cross sections, etc. Elevation derivatives should always begin from the same certified source, if possible, to ensure consistent results.

Elevation Post: The vertical component of a DEM lattice mesh point, having height above the vertical datum equal to the z-value of its mesh point.

Elevation Post Spacing: The constant sampling interval in x and y directions of a DEM lattice or grid. The horizontal resolution of a raster DEM. See raster.

Ellipsoid: A closed surface whose planar sections are either ellipses or circles. The Earth's ellipsoid is a biaxial ellipsoid of revolution, defined by an ellipse with major axis "a" and its minor axis "b," obtained by rotating the ellipse about its minor (shorter) axis.

Ellipsoid Height: See Height.

Ensonify: See Insonify.

Ephemeris: A tabulation of the locations and related data for a celestial body, including Earth-orbiting satellites, for given epochs (dates) at uniform intervals of time.

- **Broadcast Ephemeris:** The ephemeris broadcast from a satellite, from which Earth-fixed satellite positions can be computed
- **Precise Ephemeris:** The ephemeris of a satellite computed by adjustment of observations obtained from a worldwide tracking network in order to obtain maximum accuracy

Epoch: A particular instant of time from which an event or a series of events is calculated. A starting point in time to which events are referred.

Equator, Geodetic: The circle of the reference ellipsoid midway between its poles of rotation. The geodetic equator is the line on which geodetic latitude is 0° and from which geodetic latitudes are reckoned, north and south, to 90° at either pole.

Equipotential Surface: A surface with the same gravity potential at every point. An equipotential surface is also referred to as a level surface. The force of gravity is everywhere perpendicular to this surface. The surface of a body of still water is an equipotential surface.

Error: The difference between the observed value of a quantity and the theoretical or defined value of that quantity. In the computation of root-mean-square errors (RMSE), x-, y- and z-errors are the differences in x- or y-coordinates or z-values between a sample dataset and a dataset of higher accuracy for the same sample points.

- **Mean Error:** The average positional error in a set of values for one dimension (x, y, or z), obtained by adding all errors in a single dimension together and then dividing by the total number of errors for that dimension
- **Positional Error:** The difference between data set coordinate values and coordinate values from an independent source of higher accuracy for identical points
- **Random Error:** An error, produced by irregular causes whose effects upon individual observations are governed by no known law that connects them with circumstances and so cannot be corrected by use of standardized adjustments
- **Systematic Error:** An error whose algebraic sign and, to some extent, magnitude bears a fixed relation to some condition or set of conditions. Systematic errors follow some fixed pattern and are introduced by data collection procedures, processing or given datum

Estuary: An embayment of the coast in which fresh river water entering at its head mixes with the relatively saline ocean water. When tidal action is the dominant mixing agent it is usually termed a tidal estuary. Also, the lower reaches and mouth of a river emptying directly into the sea where tidal mixing takes place; the latter is sometimes called a river estuary.

Exterior Orientation: The position and attitude of an acquisition sensor with respect to a ground coordinate system. The six parameters that define the position (x/y/z coordinates) and attitude (roll, pitch and heading) of a remote sensing system, relative to a ground coordinate reference frame, at the instant that a pulse is emitted from an active sensor (e.g., IFSAR, lidar or sonar) or an image is acquired by a passive sensor (e.g., exposure station of an aerial photograph). With digital imaging systems, the exposure station differs with each image line; therefore, the exterior orientation elements vary over time and are often modeled as a non-linear function requiring some measure of time tied to the image lines, ABGPS and IMU data. See Direct Georeferencing.

F

Fathom: A measure of water depth equal to exactly 6 feet.

Feathering: Adjusting the relative weighting of the different data sets over a specified distance to smoothly transition from one data set or combination of data sets to another.

Field of View: The angular extent of the portion of object space surveyed by an aerial camera or lidar sensor, measured in degrees. To avoid confusion, a typical airborne lidar sensor with a field of view of 30 degrees is commonly depicted as ±15 degrees on either side of nadir. In bathymetric lidar, Field of View is often used as a term for Receiver Field of View, in which case, it is the receiver telescope field-of-view that is usually wider than the transmitted laser pulse to capture the return signal that spreads due to the presence of the water column. The receiver Field of View is usually measured in milliradians.

First Return: The first important measurable portion of a return lidar pulse.

Flash Lidar (3-D Flash Lidar): Similar to a digital camera with a flash illumination but with the flash provided by a pulsed laser and use of a flash-focal-plane-array (FFPA) detector with clocks in each readout integrated circuit pixel cell to determine the time needed for the flash to depart, reflect from a target, and return. The time-of-flight of the reflected laser pulse gives a range measurement along with intensity for each pixel in the image.

Flight Altitude: The vertical distance of the aircraft above mean sea level.

Flight Height: The vertical distance from average terrain elevation to the point from which aerial photography or other remote sensing data are collected.

Flightline (Flight Line): A single pass of the collection aircraft over the target area. Commonly misused to refer to the data resulting from a flightline of collection. See swath.

Flightline Overlap: Percent of redundant coverage in adjoining flightlines to ensure there are no data gaps between the usable portions of lidar swaths. Collections in high relief terrain are expected to require greater overlap so as to avoid data voids.

Flight Plan: All factors related to aircraft and aerial sensor operation contributing to producing suitable data. A flight plan includes flight altitude, flight lines, and additional information, e.g., preplanned camera stations for each frame image to be acquired.

Flood Insurance Rate Map (FIRM): An official map of a community on which FEMA has delineated both the special flood hazard areas and the risk premium zones applicable to the community.

Flood Mitigation: Activities that lessen the potential for future flood damages.

Floodplain: The low lying area along a river, stream, or coast that is subject to flooding. Any land area susceptible to being inundated by water from any source.

Flood Tide: The incoming or rising tide, occurring between the time when the tide is lowest and the time when the following tide is highest. Compare ebb tide.

Floodway: The channel and portion of the adjoining area required to discharge the base flood without increasing flood heights more than one foot.

Fly-Through: A terrain visualization technique, used in computer simulations, in which the 3-D perspective view of the terrain continuously changes relative to the location of the viewer who appears to be moving as though flying over the terrain in an aircraft. The terrain models used for fly-throughs may be overlaid with digital imagery, simulated clouds or smoke, or added or enhanced virtual features to depict landmarks or objects of unique importance to a specific fly-through.

Focal Length: The distance between the rear node of a lens (or the vertex of a mirror) and the point at which the image of an infinitely distant object comes into critical focus.

Footprint: The beam size or surface area measured by a single beam from an active sensor such as IFSAR, lidar or sonar. See Building Footprint.

Foreshore: That part of shore which lies between high and low water mark at ordinary tide.

Foreshortening: IFSAR relief displacement towards the direction of the radar, opposite to that of optical imagery where relief displacement is away from the direction of the camera.

Forward Overlap: Overlap of any two successive aerial photographs or images in the direction of the flight line.

Full Waveform Lidar: Lidar system or data in which the entire reflection of the laser pulse is fully digitized, captured, and stored. Discrete return point clouds can be extracted from the waveform data during post processing. See discrete return lidar.

Full Width at Half Maximum (FWHM): A measure of the extent of a function, given by the difference between the two extreme values of the independent variable at which the dependent variable is equal to half of its maximum value, commonly used to define the duration of pulse waveforms. For example, the duration of a laser pulse (pulse width) from a lidar sensor is usually defined as the time during which the laser output pulse power remains continuously above half its maximum value.

Fundamental Vertical Accuracy (FVA): Replaced by the term Nonvegetated Vertical Accuracy (NVA). FVA is the term used by the NDEP guidelines for vertical accuracy at the 95% confidence level in open terrain only where errors should approximate a normal error distribution. It is obtained utilizing standard tests for RMSE, where FVA = $Accuracy_z = RMSE_z \times 1.9600$.

G

Geiger-Mode Lidar: A photon-sensitive lidar sensor that uses the Geiger-mode Avalanche Photo-Diode (APD), also known as Single Photon Avalanche Diode (SPAD), for laser ranging. It is a solid-state photodetector in which a photon-generated carrier can trigger an avalanche current by exceeding the breakdown voltage of the p-n junction between two types of semiconductors. The device is able to detect low-intensity signals down to the single photon and measure the arrival time of the photons with a jitter of a few tens of picoseconds. The biggest advantage of using a Geiger-mode detector is the ability to detect range using very few photons; this allows ranging to be done using a low-power pulse with short pulse width from very high altitudes by using an array of Geiger-mode avalanche photo diodes.

Geocentric: Referred to the center of the Earth. The exact meaning of "center" must usually be inferred from the context. The following kinds of center are in common use: (a) The center of an ellipsoid representing the Earth's shape or figure, as in "geocentric longitude"; (b) The Earth's center of mass, as in "geocentric orbit"; (c) The point at which the Earth's axis of rotation intersects the plane of the celestial Equator.

Geodesy: The science that determines the size and shape of the Earth, locates positions on the Earth, and determines the Earth's gravity field. Geodesy also includes the large-scale extended surveys for determining positions and elevations of points, in which the size and shape of the Earth must be taken into account.

Geodetic Coordinates: The 3-D position of a point described by latitude, longitude, and height above the ellipsoid.

Geodetic Height: See Height.

Geodetic Survey: A survey that takes into account the size and shape of the Earth, as distinguished from a plane survey in which the surface of the Earth is considered a plane.

Geographic Coordinates: The 2-D position of a point described by latitude and longitude only.

Geographic Information System (GIS): A system of spatially referenced information, including computer programs that acquire, store, manipulate, analyze, and display spatial data.

Geoid: An equipotential surface coinciding with mean sea level for the oceans and extended in land areas so the surface is always perpendicular to the direction of gravity. The geoid undulates up and down with local variations in the mass and density of the Earth.

Geoid Height: See geoid undulation.

Geoid Undulation: The distance along the ellipsoid normal from the reference ellipsoid to the geoid. See Chapter 2, Figure 2.8 on page 51.

Geomorphology: The study of the physical features of the Earth, the arrangement and form of the Earth's crust and of the relation between the physical features and geological structures beneath.

Geop: A geopotential surface that approximates the Earth's potentials of gravitation. The geoid is a specific geop, selected to pass through the average sea surface for the Earth.

Geopotential Number: The geopotential number (C) of a bench mark is the difference in potential measured from the reference geopotential surface to the equipotential surface passing through the survey mark. It is the amount of work required to raise a unit mass of 1 kg against gravity through the orthometric height to the mark. Geopotential differences are differences in potential which indicate hydraulic head, i.e., the direction water flows.

Geospatial Data: Information that identifies the geographic location and characteristics of natural or constructed features and boundaries of the Earth. This information may be derived from, among other things, remote sensing, mapping, and surveying technologies. Geospatial data is generally considered to be synonymous with spatial data; however, the former is always associated with geographic or cartesian coordinates linked to a horizontal or vertical datum, whereas the latter (e.g., generic architectural house plans) may include dimensions and other spatial data not linked to any physical location.

Global Navigation Satellite System (GNSS): A constellation of global positioning satellites intended to be interoperable, including the U.S.'s Global Positioning System (GPS), the Russian Federation's Global Orbiting Navigation Satellite System (GLONASS), and Europe's Galileo when fully developed.

Global Positioning System (GPS): A system of satellites, computers and receivers that provide highly accurate position and velocity information in three dimensions of a GPS receiver on Earth by calculating the time difference for signals from different satellites to reach the receiver.

GPS Time: The GPS time stamp at which the laser point was emitted from the aircraft. The time is in GPS seconds of the week.

Gravity: The force which is the resultant of gravitation, the force exerted by the mass of the Earth, and the centrifugal force, caused by the rotation of the Earth. The acceleration is a function of the body's location but is independent of the mass, shape, size, or other properties of the body. The term gravity is used for both the force and the acceleration and for both the vector and its magnitude. Thus the word may have any one of several different meanings. At the Equator and close to the surface, the acceleration of gravity (g) is approximately 978.03 cm/s^2; at the poles, it is approximately 983.22 cm/s^2. The quantity "1 cm/s^2" is called the gal.

Great Diurnal Range (Gt): The difference between Mean Higher High Water and Mean Lower Low Water.

Greenwich Mean Time (GMT): Mean solar time at the Greenwich (England) meridian.

Grid: A rectangular array of cells, each of which stores an elevation value for the centroid of the cell, that forms the basic structure of the most common and uniformly-spaced DEMs. The resolution of the grid (the width and height of the cells) determines the precision of the grid representation.

Grid Post: See Elevation Post.

Ground Sample Distance (GSD): The linear dimension of a sample pixel's footprint on the ground. The GSD is used when referring to the collection GSD of the raw image, assuming near-vertical imagery. The actual GSD of each pixel is not uniform throughout the raw image and varies significantly with terrain height and other factors. In this manual, GSD is assumed to be the value computed using the calibrated camera focal length and camera height above average horizontal terrain.

Ground Truth: Verification of a situation, without errors introduced by sensors or human perception and judgment.

Gyro: A device that measures angular rate or angle increments with respect to an inertial reference.

H

Half Tide Level: A tidal datum. The arithmetic mean of mean high water (MHW) and mean low water (MLW). Same as Mean Tile Level (MTL).

Hardcopy Photogrammetry: Stereo photogrammetric procedures that utilize direct viewing of hardcopy film images in analog or analytical stereoplotters.

Heading: The azimuth of the longitudinal axis of a vehicle; the "straight-ahead" direction. True heading is the vehicle azimuth with respect to true North; magnetic heading is the azimuth with respect to the magnetic North pole.

Head of Tide: The inland or upstream limit of water affected by the tide. For practical application in the tabulation for computation of tidal datums, head of tide is the inland or upstream point where the mean range becomes less than 0.2 foot.

Height: The distance, measured along a perpendicular, between a point and a reference surface, e.g., the height of an airplane above the ground surface. The distance, measured upwards along a plumb line (line of force), between a point and a reference surface of constant geopotential. Elevation is preferred if the reference surface is the geoid. Height systems are called by different names depending on the geopotential number (C) and gravity (G) selected. When G is computed using the Helmert height reduction formula, which is what was used in NAVD 88, the heights are called Helmert orthometric heights. When G is computed using the International formula for normal gravity, the heights are called normal orthometric heights. When G is equal to normal gravity at 45 degrees latitude, the heights are called normal dynamic heights, which is what was used in IGLD 85.

- **Dynamic Height:** The value, assigned to a point, determined by dividing the geopotential number for that point by the value of gravity on the reference ellipsoid at 45° latitude, as calculated from a standard gravity formula. Dynamic heights are geopotential numbers scaled by a constant, using normal gravity at 45 degrees latitude equal to 980.6199 gals. Points that have the same geopotential number will have the same dynamic height. The primary users of dynamic heights use the heights for measuring water-level heights and/or height differences on large bodies of water, i.e., the Great Lakes, because they are interested in knowing the hydraulic head
- **Ellipsoid Height:** The height above or below the reference ellipsoid, i.e., the distance between a point on the Earth's surface and the ellipsoidal surface, as measured along the normal (perpendicular) to the ellipsoid at the point and taken positive upward from the ellipsoid. Defined as "h" in the equation: $h = H + N$. Same as ellipsoidal height and geodetic height
- **Geoid Height:** See geoid undulation
- **Helmert Orthometric Height:** An approximate value HNH found for the elevation of a point PN by dividing the geopotential number WN at that point by an approximation (gNH) to the average value of gravity along the vertical between PN and the geoid. The value gNH is calculated from the measured value gN of gravity at PN and two forms of gravity corrections, i.e., the Bouguer and free-air gravity corrections. The difference between adjusted Helmert orthometric heights is computed using geopotential differences based on observed gravity

- **Orthometric Height (Elevation):** The height above the geoid as measured along the plumbline between the geoid and a point on the Earth's surface, taken positive upward from the geoid. Defined as "H" in the equation: H = h - N. The difference between adjusted orthometric heights is computed using a normal gravity formula. The orthometric height (H) and the geopotential number (C) are related through the following equation: C = G x H, where G is the gravity value estimated for a particular system

Higher High Water (HHW): The higher of the two high waters of a tidal day where the tide is of the semidiurnal or mixed type. The single high water occurring daily during periods when the tide is diurnal.

Higher Low Water (HLW): The higher of the two low waters of a tidal day where the tide is of the semidiurnal or mixed type.

High Water (HW): The high tide or maximum height reached by a rising tide.

Hillshade: A function used to create an illuminated representation of the surface, using a hypothetical light source, to enhance terrain visualization effects.

Histogram: A graphic representation of a frequency or relative frequency distribution consisting of vertical rectangles whose widths correspond to a defined range of frequencies and whose heights correspond to the number of frequencies occurring within the range.

Holiday: An area of missing coverage, caused by missing or unresolvable data, data edits, or incorrectly positioned flightlines, normally identified for further investigation or reflying. An unintentionally unsurveyed part of a region that was to have been completely surveyed. See Data Void.

Horizontal Accuracy: Positional accuracy of a dataset with respect to a horizontal datum. According to the NSSDA, horizontal (radial) accuracy at the 95% confidence level is defined as "$Accuracy_r$" or "ACC_r."

Horizontal Error: Magnitude of the displacement of a feature's recorded horizontal position in a dataset from its true or more accurate position, as measured radially according to the NMAS and NSSDA standards, or as measured separately, for x and y, according to NSSDA and ASPRS standards. Horizontal errors comprise the errors that are squared in the computation of horizontal root-mean-square errors ($RMSE_s$), whether $RMSE_x$, $RMSE_y$, or $RMSE_r$. Note: $RMSE_r$ = square root of ($RMSE_x^2 + RMSE_y^2$).

Horizontal Post Spacing: The ground distance interval between grid posts in a uniformly gridded data model. It is important to note that features of a size equal to, or even greater than the post spacing, may not be detected or explicitly represented in a gridded model. For gridded elevation data, the horizontal post spacing may be referenced as the cell size, the grid spacing, the posting interval, or the ground sample distance.

Horses-for-Courses: The practice of choosing the best response to a situation or the best means to an end.

Hydraulic Modeling: The use of digital elevation data, rainfall runoff data from hydrologic models, surface roughness data, and information on hydraulic structures (e.g., bridges, culverts, dams, weirs, sewers) to predict flood levels and manage water resources. Hydraulic models are based on computations involving liquids under pressure, and many other definitions of hydraulic modeling exist that are not associated with terrain elevations, e.g., modeling of hydraulic lines in aircraft and automobiles.

Hydrographic Survey: A survey conducted to determine the configuration of the bottom of water bodies and to identify and locate all features, natural and man-made, that may affect navigation. Hydrographic surveys are conducted for the purpose of safety of navigation and require very accurate bathymetry and the detection of all hazards to surface navigation like wrecks and obstructions. This type of surveying is most often conducted with a multibeam sonar and/or a sweep sonar, which might be augmented with towed side scan sonar to ensure full coverage of the seabed. Hydrographic surveys may include the delineation of the submerged contours of channels, banks and shoals, and the collection of bottom specimens and samples of water. Also included are consideration of tides and currents, and information on temperature and salinity of the water and their effects on the accurate measurement of depths by echo sounding.

Hydrography: That branch of applied science which deals with the measurement and description of the physical features of the navigable portion of the Earth's surface and adjoining coastal areas, with special reference to their use for the purpose of navigation. Hydrography also includes the collection of other oceanography, geodesy and geologic data (e.g., bottom classification) in conjunction with the depth data, and it may be used for other coastal engineering purposes, e.g., maintenance dredging, construction, in addition to nautical charting.

Hydrologic Modeling: The computer modeling of rainfall and the effects of land cover, soil conditions, and terrain slope to estimate rainfall runoff into streams, rivers and lakes. Digital elevation data are used as part of hydrologic modeling.

Hydrologically-Conditioned (Hydro-Conditioned): Processing of a DEM or TIN so that the flow of water is continuous across the entire terrain surface, including the removal of all isolated sinks or pits. The only sinks that are retained are the real ones on the landscape. Whereas "hydrologically-enforced" is relevant to drainage features that are generally mapped, "hydrologically-conditioned" is relevant to the entire land surface and is done so that water flow is continuous across the surface, whether that flow is in a stream channel or not. The purpose for continuous flow is so that relations and/or links among basins and/or catchments can be known for large areas.

Hydrologically-Enforced (Hydro-Enforced): Processing of mapped water bodies so that lakes and reservoirs are level and so that streams flow downhill. For example, a DEM, TIN or topographic contour dataset with elevations removed from the tops of selected drainage structures (bridges and culverts) so as to depict the terrain under those structures. Hydro-enforcement enables hydrologic and hydraulic (H&H) models to depict water flowing under these structures, rather than appearing in the computer model to be dammed by them because of road deck elevations higher than the water levels. Hydro-enforced TINs also utilize breaklines along shorelines and stream centerlines, for example, where these breaklines form the edges of TIN triangles along the alignment of drainage features. Shore breaklines for streams and rivers would be 3-D breaklines with elevations that decrease as the stream flows downstream; however, shore breaklines for lakes or reservoirs would have the same elevation for the entire shoreline if the water surface is known or assumed to be level throughout.

Hydrologically-Flattened (Hydro-Flattened): Processing of a DEM or TIN so that mapped water bodies, streams, rivers, reservoirs, and other cartographically polygonal water surfaces are flat and, where appropriate, level from bank-to-bank. Additionally, surfaces of streams, rivers and long reservoirs demonstrate a gradient change in elevation along their length, consistent with their natural behavior and the surrounding topography. In traditional maps that are compiled photogrammetrically, this process is accomplished automatically through the inclusion of measured breaklines in the DTM. However, because lidar does not inherently include breaklines, a DEM or TIN derived solely from lidar points will depict water surfaces with unsightly and unnatural artifacts or triangulation. The process of hydro-flattening typically involves the addition of breaklines along the banks of specified water bodies, streams, rivers, and ponds. These breaklines establish elevations for the water surfaces that are consistent with the surrounding topography, and produce aesthetically acceptable water surfaces in the final DEM of TIN. Unlike hydro-conditioning and hydro-enforcement, hydro-flattening is not driven by any hydrologic and hydraulic modeling requirements, but solely by cartographic mapping needs.

Hydrology: The scientific study of the waters of the Earth, especially with relation to the effects of precipitation and evaporation upon the occurrence and character of water in streams, lakes, and on or below the land surfaces.

Hypsography: The description of elevations or heights of land surfaces with reference to a specified surface (usually the geoid). The configuration of land or underwater surfaces with respect to a horizontal and vertical datum. Hypsography includes topographic and bathymetric contours, spot heights, mass points, breaklines, and all forms of generic DEM data, except DSMs that depict surfaces above the ground.

I

Image Correlation: Directly comparing hardcopy or softcopy images, or patches of pixels on conjugate digital images, or indirectly comparing information derived from the stereo images, to determine that points on stereo images (viewed from different perspectives) represent the same points on the imaged surface. Automated image correlation is a computerized technique to match the similarities of pixels in one digital image with comparable pixels in its digital stereo image in order to automate or semi-automate photogrammetric compilation. Automated image correlation provides an efficient method for generating DSMs photogrammetrically, instead of DEMs, because such correlation generates elevations of rooftops, treetops and other surface features as imaged on the stereo photographs.

Image Resolution: The size of the smallest feature that can be resolved or interpreted in an image. Often confused with ground sample distance (GSD). With digital imagery, the GSD is the size of the ground footprint of a pixel and is reported as linear units/pixel, such as 1 meter/pixel, 1 foot/pixel, etc. The image resolution is normally superior to the GSD; for example, it is common to interpret highway paint stripes only 4"-6" wide on digital imagery with a GSD of 1 foot or 1 meter.

Independent Source of Higher Accuracy: Data of higher accuracy, acquired independently of the geospatial data set being tested, that shall be of the highest accuracy feasible and practicable to evaluate the accuracy of a dataset. The ASPRS Positional Accuracy Standards for Digital Geospatial Data states that the independent source of higher accuracy for checkpoints shall be at least three times more accurate than the required accuracy of the geospatial data set being tested.

Inertial Measurement Unit (IMU): The primary component of an inertial navigation system (INS). The IMU measures three components of acceleration and three components of rotation using orthogonal triads of accelerometers and gyros.

Inertial Navigation System (INS): A self-contained navigation system, comprised of several subsystems: IMU, navigation computer, power supply, interface, etc. Uses measured accelerations and rotations to estimate velocity, position and orientation. An unaided INS loses accuracy over time, due to gyro drift.

Infrared: The portion of the invisible spectrum consisting of electromagnetic radiation with wavelengths in the range from 750 nanometers to 1 millimeter.

Insonify (Ensonify): To expose an area, or portion of seabed, to sonar energy. Seabed that has been covered by imaging sonar, provided acceptable backscatter and interpretable sonar data, is said to have been insonified. Ping rates (numbers of output pulses per second) are often referred to as "insonification rates."

Intensity, Lidar: The return strength of the laser pulse that generated the lidar point. For discrete-return lidar instruments, intensity is the recorded amplitude of the reflected lidar pulse at the moment the reflection is captured as a return by the lidar instrument. Lidar intensity values can be influenced by numerous factors such as the instantaneous setting of the instrument's automatic gain control and angle of incidence and cannot be equated to a true measure of "energy." In full-waveform systems, the entire reflection is sampled and recorded, and true energy measurements can be made for each return or overall reflection. Intensity values for discrete returns derived from a full-waveform system may or may not be "calibrated" to represent true energy. Lidar intensity data make it possible to map variable textures in the form of a gray-scale image. Intensity return data enables automatic identification and extraction of objects such as buildings and impervious surfaces, and can aid in lidar point classification. In spite of their similar appearance, lidar intensity images differ from traditional panchromatic images in several important ways: (1) Lidar intensity is a measure of the reflection of an active laser energy source, not natural solar energy (2) Lidar intensity images are aggregations of values at point samples; the value of a pixel does not represent the composite value for the area of that pixel; (3) Lidar intensity images depict the surface reflectivity within an extremely narrow band of the infra-red spectrum, not the entire visible spectrum as in panchromatic images; (4) Lidar intensity images are strongly influenced by the angle of incidence of the laser to the target and are subject to "unnatural" shadowing artifacts; and (5) The values on which lidar intensity images are based may or may not be calibrated to any standard reference. Intensity images usually contain wide variation of values within swaths, between swaths, and between lifts. For these reasons, lidar intensity images must be interpreted and analyzed with unusually high care and skill.

Interferometer: An instrument that measures differences between the phases of two different electromagnetic signals originating from a common source, but which have traversed different paths. The phase differences are measured by combining the two signals. The amplitude of the combined signal is a function of the phase difference between the two signals. The phenomenon of fluctuations in the amplitude of the combined signals in response to phase changes in the input signals if sometimes referred to as interference.

Interferometric Synthetic Aperture Radar (IFSAR): An airborne or spaceborne interferometer radar system, flown aboard rotary or fixed wing aircraft or space-based platform, that is used to acquire 3-Dimensional coordinates of terrain and terrain features that are both manmade and naturally occurring. IFSAR systems form synthetic aperture images of terrain surfaces from two spatially separated antennae over an imaged swath that may be located to the left, right or both sides of the imaging platform.

Interior Orientation: Definition of the elements of the image and how they relate to the optical system used to acquire them, e.g., the camera's focal length, location of the "principal point" on the image, and any known distortions induced by the optics and/or film.

International Great Lakes Datum of 1985 (IGLD 85): The elevation reference system (datum) used to define water levels within the Great Lakes and St. Lawrence River system, based on a single reference point (Father Point, Rimouski, Quebec, Canada) considered as sea level on the Atlantic Ocean. See also the North American Vertical Datum of 1988 (NAVD 88).

Interpolation: As used in this manual, interpolation is the estimation of z-values at a point with x/y coordinates, based on the known z-values of surrounding points. There are many different forms of interpolation:

- **ANUDEM:** A variation of spline interpolation optimized for the creation of hydrologically correct (hydro-enforced) terrain models
- **Inverse Distance Weighted Interpolation:** A method for interpolating the elevations of new points by using a linearly weighted combination of values from nearby points, where the weight is a function of inverse distance
- **Kriging:** A method for interpolating the elevations of new points by using geostatistical techniques that weigh the surrounding measured values to derive a prediction for each location. However, the weights are based not only on the distance between the measured points and the prediction location but also on the strength of the overall correlation among the measured points. With kriging, the data are used to define the spatial correlation model to determine the weights for neighboring samples that are used to fit a surface to the points
- **Natural Neighbor Interpolation:** A method for interpolating the elevations of new points by using an area based weighting scheme to interpolate heights
- **Spline:** A method for interpolating the elevations of new points by using a mathematical formula to create a surface that minimizes the overall surface curvatures, resulting in a smooth surface that passes through the input points

Intertidal Zone: The zone between the mean higher high water (MHHW) and mean lower low water (MLLW) lines.

Intervisibility: The identification of all points that are visible from a designated input point. Generally synonymous with "viewshed" and "line-of-sight."

J

K

Kalman Filter: A recursive minimum variance estimation technique that uses statistical models to weight each new measurement relative to past information and estimate quantities of interest having statistical characteristics. It can be thought of as a recursive least squares adjustment.

Kriging: See Interpolation.

Kurtosis: The measure of relative "peakedness" or flatness of a distribution compared with a normally distributed data set. Positive kurtosis indicates a relatively peaked distribution near the mean whereas negative kurtosis indicates a flat distribution near the mean.

L

Laser: An acronym for light amplification by stimulated emission of radiation. When activated by an external energy source, a laser produces and emits a beam or pulse of monochromatic radiation in which all the waves are coherent and in phase. The emitted radiation is also highly collimated and directional, and is emitted as a narrow beam or pulse (constituting a multitude of photons) traveling in a specific direction.

LAS File Format: A public file format for the interchange of 3-D point cloud data between data users. The file extension is .las.

Last Return: The last important measurable part of a return lidar pulse.

Latitude: The angular distance, in degrees, minutes and seconds, of a point north or south of the equator, measured along a meridian of longitude.

Latitude, Geodetic: The angle that the normal to the ellipsoid at a point makes with the equatorial plane (the geodetic Equator) of the ellipsoid. Geodetic latitudes depend on the chosen geodetic datum, which determines the orientation and dimensions of the ellipsoid.

Lattice: A 3-D vector representation method created by a rectangular array of points spaced at a constant sampling interval in x and y directions relative to a common origin. A lattice differs from a grid in that it represents the value of the surface only at the lattice "mesh points" or "elevation posts" of the lattice, rather than the elevation of the cell area surrounding the centroid of a grid cell.

Layover: A limiting case of foreshortening where points arranged with increasing ground coordinates appear reversed in the radar imagery, precluding the determination of elevation in layover regions. Geometrically this happens when the slope of the terrain is greater than the angle the incident radiation makes with respect to vertical.

Lead Line: A long chain or line with a lead weight attached at one end. The line is used to measure water depths (usually up to 10 fathoms) by survey parties not equipped with electronic depth-measuring equipment. It also can be used to check such electronic equipment.

Least Squares Adjustment: An adjustment satisfying the condition that the sum of the squares of the differences between the given and changed quantities be a minimum.

Leveling: The process of finding differences of elevation.

- **Differential Leveling:** Determining the difference in elevation between two points by the sum of incremental vertical displacements of a graduated leveling rod
- **Spirit Leveling:** Leveling with a leveling instrument that depends on a spirit level for making its line of sight horizontal. A spirit level uses a closed glass container filled with a fluid with free space so that a bubble of air rises to the top to indicate when the instrument is level
- **Trigonometric Leveling:** Determining differences of elevation by observing vertical angles between points. Either the horizontal distance between the points must be known or the straight-line distance must be measured

Lever Arm: A relative position vector of one sensor with respect to another in a direct georeferencing system. For example, with aerial mapping cameras, there are lever arms between the inertial center of the IMU and the phase center of the GPS antenna, each with respect to the camera perspective center within the lens of the camera. Lidar sensors also have similar lever arm vectors.

Lidar: An instrument that measures distance to a reflecting object by emitting timed pulses of light and measuring the time difference between the emission of a laser pulse and the reception of the pulse's reflection(s). The measured time interval for each reflection is converted to distance, which when combined with position and attitude information from the GPS, IMU, and the instrument itself, allows the derivation of the 3-D coordinates of the reflecting target's location.

Lift: A single take-off and landing cycle of a collection platform (fixed or rotary wing) within an aerial data collection project.

Linear Mode Lidar: A lidar system in which the optical detector on the receiver detects the entire laser pulse as a single entity, by receiving thousands of reflected photons per pulse in order to exceed the detector noise floor and have sufficient signal to accurately determine the range to the illuminated target. The optical receiver is sensitive to and will trigger when several hundred or thousands of photons are detected. The detection could be based on a first return, last return, multiple return, or as a full waveform. Linear lidar generally has a high signal-to-noise ratio.

Line-of-Sight: See Intervisibility.

Littoral: Bordering on the sea.

Local Accuracy: See Accuracy.

Longitude: The angular distance, in degrees, minutes and seconds, of a point east or west of the Greenwich meridian.

Longitude, Geodetic: The angle between the plane of the local geodetic meridian and the plane of an arbitrarily chosen, geodetic meridian, normally the prime meridian in Greenwich, England. A geodetic longitude may be measured by the angle, at one of the poles of the ellipsoid, between the local and geodetic meridian, or by the arc of the ellipsoid's Equator intercepted by these meridians. In recording a geodetic location, it is essential that the geodetic datum on which it is based also be stated.

Lower High Water (LHW): The lower of the two high waters of any tidal day where the tide is of the semidiurnal or mixed type.

Lower Low Water (LLW): The lowest of the low waters (or single low water) of any specified tidal day due to the declination effects of the Moon and Sun.

Lower Low Water Datum (LLWD): An approximation of mean lower low water that has been adopted as a standard reference for a limited area and is retained for an indefinite period regardless of the fact that it may differ slightly from a better determination of mean lower low water from a subsequent series of observations; used primarily for river and harbor engineering purposes.

Low Water (LW): The low tide or minimum height reached by a falling tide.

Low Water Datum (LWD): (1) The geopotential elevation (geopotential difference) for each of the Great Lakes and Lake St. Clair and the corresponding sloping surfaces of the St. Marys, St. Clair, Detroit, Niagara and St. Lawrence Rivers to which are referred the depths shown on the navigational charts and the authorized depths for navigation improvement projects. (2) An approximation of mean low water that has been adopted as a standard reference for a limited area and is retained for an indefinite period regardless of the fact that it may differ slightly from a better determination of mean low water from a subsequent series of observations; used primarily for river and harbor engineering purposes.

M

Map: A representation, usually on a plane surface and at an established scale, of the physical features (natural, artificial, or both) or a part or the whole of the Earth's surface. Features may be identified by means of signs and symbols, and geographical orientation is indicated.

- **Orthophoto Map:** A map produced by assembling orthophotographs at a specified uniform scale in a map format
- **Planimetric Map:** A map that shows only the horizontal positions of area, line and point features. It does not show relief (elevations) in measurable form
- **Topographic Map:** A map showing the horizontal and vertical locations of natural and manmade features. It is distinguished from a planimetric map by the presence of numbered contour lines or comparable symbols to indicate elevations of mountains, valleys, and plains; in the case of hydrographic charts, symbols and numbers are used to show depths in bodies of water

Map Manuscript: The original drawing of a map as compiled or constructed from ground surveys, photogrammetry, etc.

Map Projection: A function relating 3-D coordinates of points on a curved surface (usually an ellipsoid or sphere) to 2-D coordinates of points on a plane map.

Mark: A dot, the intersection of a pair of crossed lines, or any other physical point corresponding to a point in a survey. It is the physical point to which distances, elevations, heights, or other coordinates refer.

Mass Points (Masspoints): Irregularly spaced points, each with x, y, z coordinates, typically (but not always) used to form a TIN. When generated manually, mass points are ideally chosen to depict the most significant variations in the slope or aspect of TIN triangles. However, when generated automatically, e.g., by lidar or IFSAR scanners, mass point spacing and pattern depend upon the characteristics of the technologies used to acquire the data Mass points are usually used in conjunction with breaklines.

Mean (Arithmetic Mean): The average of all numbers in a dataset.

Mean Diurnal Tide Level (MDTL): A tidal datum. The arithmetic mean of mean higher high water (MHHW) and mean lower low water (MLLW).

Mean High Water (MHW): A tidal datum. The average of all the high water heights observed over the National Tidal Datum Epoch.

Mean Higher High Water (MHHW): A tidal datum. The average of the higher high water heights of each tidal day observed over the National Tidal Datum Epoch. Three states (HI, LA, and TX) use MHHW to define the official shoreline boundary between privately-owned and state-owned lands. See Mean High Water Line (MHWL).

Mean High Water Line (MHWL): The line on a chart or map which represents the intersection of the land with the water surface at the elevation of MHW. Fifteen states (AL, AK, CA, CT, FL, GA, MD, MS, NJ, NY, NC, OR, RI, SC, and WA) use MHW to define the official shoreline boundary between privately-owned uplands and state-owned inland waters, tidelands, and submerged lands. Three states (HI, LA, and TX) use Mean Higher High Water (MHHW) and six states (DE, MA, ME, NH, PA, and VA) use Mean Lower Low Water (MLLW) for similar purposes.

Mean Lower Low Water (MLLW): A tidal datum. The average of the lower low water heights of each tidal day observed over the National Tidal Datum Epoch.

Mean Lower Low Water Line (MLLWL): The line on a chart of map which represents the intersection of the land with the water surface at the elevation of mean lower low water (MLLW). Six states (DE, MA, ME, NH, PA and VA) use MLLW to define the official shoreline boundary between privately-owned and state-owned lands. See Mean High Water Line (MHWL).

Mean Low Tide: The average of all the low tides (high low tides and low low tides) occurring over a National Tidal Datum Epoch.

Mean Range of a Tide (Mn): The difference between Mean High Water (MHW) and Mean Low Water (MLW).

Mean Sea Level (MSL): A tidal datum. The arithmetic mean of hourly heights observed over the National Tidal Datum Epoch. Shorter series are specified in the name, e.g., monthly mean sea level and yearly mean sea level. Also called Mean Tile Level.

Mean Tide Level (MTL): A tidal datum. The arithmetic mean of mean high water (MHW) and mean low water (MLW). Same as half tide level.

Median: The value in a dataset at which there are as many larger numbers as there are smaller numbers.

Meridian: A plane curve on the surface of the Earth passing through the axis of rotation and any given point on the Earth's surface. All points on a given meridian have the same longitude. A north-south line from which differences of longitude and azimuths are reckoned. At the International Meridian Conference held in Washington D.C. in 1884, the adoption of the Greenwich Meridian as the prime meridian for the Earth was approved by the representatives of 22 governments, including the U.S.

Metadata: Any information that is descriptive or supportive of a geospatial dataset, including formally structured and formatted metadata files, e.g., eXtensible Markup Language (XML)-formatted Federal Geographic Data Committee (FGDC) metadata; reports (collection, processing, and quality assurance/quality control (QA/QC) reports; and other supporting data, e.g., survey points, shapefiles.

Metonic Cycle: A period of 6,940 days (approximately 19 years) devised by the Athenian astronomer Meton (c. 450 B.C.) for obtaining a period in which new and full moon would recur on the same day of the year.

Micron: One-millionth of a meter. One micrometer.

Mode: The value at which the greatest number of values is concentrated in a dataset. In a histogram, the mode is the value or range of values which has the highest abscissa value at the peak of the histogram.

Model: The copy of a physical object such as the Earth, normally at reduced scale.

- **Mathematical Model:** The mathematical reconstruction of a physical object such as the Earth, normally for computer display and analyses
- **Photogrammetric Model:** The visual or optical image produced by optically combining light transmitted through a stereoscopic pair of photographs or digital images. A stereoscopic model is not a material object but an image or pair of images that give the same visual effect as would a material model

Monotonic: In mathematics, a function that varies such that it either increases or decreases, but never both. As used herein, it describes a hydrographic breaklines that continuously flows either level or downhill, but never uphill.

Monument (Surveying): A permanent physical structure marking the location of a survey point. Common types of monuments are inscribed metal tablets set in concrete posts, solid rock, or parts of buildings; distinctive stone posts; and metal rods driven in the ground.

Morphology: The form and structure of a surface. In TINs, the morphology of a surface is defined by mass points and breaklines used to build the TIN. Breaklines are very significant in defining surface morphology. In lattices, the morphology of a surface cannot be directly represented in the model by linear features (e.g., breaklines) and must be implied from the mesh point z-values.

Mosaic: An assembly of vertical aerial images and/or DEMs to form a continuous representation of the terrain with a common datum, map projection, and data format.

Mosaicking: The process whereby multiple images and/or DEMs are merged into a single continuous image and/or DEM to create a mosaic.

Multipath: Reflected signals that traverse different longer paths than the direct line-of-sight signal before arriving at the receiving device.

Multipath Error: Errors in GPS positioning caused by signals reflected off of fixed surfaces or refracted by foliage, both causing signals with multiple path lengths to impinge on the GPS antenna.

N

Nadir: The point or line directly beneath the collection platform, corrected for attitude variations.

National Elevation Dataset (NED): A single, nationwide seamless elevation database with product areas defined by the customers, available in several product format options. It includes the "best available" data that meet a common standard. It is one of the framework data themes within The National Map. Standard datasets have 1-arc-second, ⅓-arc-second, and ⅑-arc-second post spacing and typically comprise the best available, multi-resolution dataset. See Chapter 4.

National Geodetic Vertical Datum of 1929 (NGVD 29): A fixed reference previously adopted as a standard geodetic datum for elevations determined by leveling. The datum was derived for surveys from a general adjustment of the first-order leveling nets of both the U.S. and Canada. In the adjustment, mean sea level was held fixed as observed at 21 tide stations in the U.S. and five in Canada. Also called the Sea Level Datum of 1929, NGVD 29 has been superseded by NAVD88 as the official geodetic vertical datum in the U.S.

National Standard for Spatial Data Accuracy (NSSDA): A document published by the U.S. Federal Geographic Data Committee (FGDC) that implements a statistical and testing methodology for estimating the positional accuracy of points on maps, and in digital geospatial data, with respect to georeferenced ground positions of higher accuracy. It is not truly a "standard," in the sense of the National Map Accuracy Standard or the ASPRS Accuracy Standard for Digital Geospatial Data, because the NSSDA does not define threshold accuracy values.

National Tidal Datum Epoch (NTDE): A specific 19-year period (closest full year to the 18.6-year nodal cycle of the moon) used to average out local meteorological effects on sea level and to compute tidal datums.

National Water Level Observation Network (NWLON): A coastal observing network of hundreds of tide and water level stations nationwide that collect continuous long-term water level observations to a known vertical reference. This network is used to establish tidal datums and for coastal processes and tectonic studies, tsunami and storm surge warnings, and climate change. The network supports regulation, navigation and charting, river and harbor improvement, power generation, various scientific activities, and the adjustment for vertical movement of the Earth's crust in the Great Lakes Basin. It is the foundation of the comprehensive system for observing, communicating, and assessing the impact of changing water levels nationwide.

Natural Neighbor Interpolation: See Interpolation.

Nautical Chart: A chart specifically designed to meet the requirements of marine navigation, showing depths of water, nature of bottom, elevations, configuration and characteristics of coast, dangers, and aids to navigation.

Neap Tide: A tide that occurs approximately midway between the times of the new and full moon. The neap tidal range is usually 10 to 30 percent less than the mean tidal range.

Network Accuracy: See Accuracy.

No Bottom At: Historically, depths apparently deeper than the maximum length of a lead line. Now, used to describe areas that have been "cleared" of hazards to an assumed lidar extinction depth, or areas that may be surveyed by sonar without fear of danger to the vessel.

Nodes: The beginning and ending location of a vector or arc, topologically linked to all vectors or arcs that meet at the node. When specifically applied to elevation data, nodes represent the corner points of each triangle in a TIN. A triangle node is topologically linked to all triangles that meet at that node.

Nominal Pulse Density (NPD): A common measure of the density of a lidar dataset; NPD is the typical or average number of pulses within a specified areal unit. NPD is typically expressed as pulses per square meter. This value is predicted in mission planning and empirically calculated from the collected data, using only the first (or last) return points as surrogates for pulses. As used herein, NPD refers to single swath, single instrument data, whereas ANPD describes the overall pulse density resulting from multiple passes of the lidar instrument, or a single pass of a platform with multiple lidar instruments, over the same target area. NPD is more commonly used in high-density collections (greater than or equal to 1 pulse per square meter [pls/m^2]), with its inverse, NPS, being used in low-density collections (less than or equal to 1 pls/m^2). Assuming meters are being used in both expressions, NPD can be calculated from NPS using the formula $NPD=1/NPS^2$. See aggregate nominal pulse density, aggregate nominal pulse spacing, nominal pulse spacing.

Nominal Pulse Spacing (NPS): A common measure of the density of a lidar dataset, NPS is the typical or average lateral distance between pulses in a lidar dataset, typically expressed in meters and most simply calculated as the square root of the average area per first return point. This value is predicted in mission planning and empirically calculated from the collected data, using only the first (or last) return points as surrogates for pulses. As used herein, NPS refers to single swath, single instrument data, whereas ANPS describes the overall pulse spacing resulting from multiple passes of the lidar instrument, or a single pass of a platform with multiple lidar instruments, over the same target area. NPS is more commonly used in low-density collections (greater than or equal to 1 m NPS), with its inverse, NPD, being used in high-density collections (less than 1 m NPS). Assuming meters are being used in both expressions, NPS can be calculated from NPD using the formula NPS = 1/√NPD. See aggregate nominal pulse density, aggregate nominal pulse spacing, nominal pulse density.

Non-vegetated Vertical Accuracy (NVA): Replaces fundamental vertical accuracy (FVA). The vertical accuracy at the 95-percent confidence level in non-vegetated open terrain, where errors should approximate a normal distribution. See fundamental vertical accuracy (FVA).

Normal: A straight line perpendicular to a surface or to another line. Also, the condition of being perpendicular to a surface or line. In geodesy, a normal is a straight line perpendicular to the surface of a particular ellipsoid.

Normalized Digital Surface Model (nDSM): A height-above-ground model produced by subtracting the DTM from the DSM.

North American Vertical Datum of 1988 (NAVD 88): A fixed reference for elevation determined by geodetic leveling. The datum was derived from a general adjustment of the first-order terrestrial leveling nets of the U.S., Canada, and Mexico. In the adjustment, only the height of the primary tidal bench mark, referenced to the International Great Lakes Datum of 1985 (IGLD 85) local mean sea level height value, at Father Point, Rimouski, Quebec, Canada was held fixed, thus providing minimum constraint. NAVD 88 and IGLD 85 are nearly identical.

Northing: In a plane coordinate system, the coordinate in a general north-south direction, increasing to the north. A linear distance, in the coordinate system of a map grid, northwards from the east-west line through the origin or false origin. A false northing is a value assigned the east-west line through the origin on a map grid to avoid the inconvenience of negative northings at some points.

O

Optical Detector: A device that efficiently converts optical signals into electrical currents. The optical detector used in linear or photon-based lidar is typically either an Avalanche Photodiode Detector (APD) or a Photo Multiplier Tube (PMT). Detectors can be triggered at different amounts of returned laser energy, from being triggered after detection of a single photon, to recording the entire signal returned per pulse (e.g., waveforms).

Orientation: The rotation or set of rotations needed to make the axes of one rectangular Cartesian coordinate system parallel to the axes of another.

Orthometric Height: See Height.

Orthophoto: A photograph prepared from a perspective photograph by removing scale differences and image displacements caused by tilt, relief, and perspective.

Orthophoto DEMs: Digital Elevation Models used to correct relief displacement in the production of digital orthophotos.

Orthophotomap: An orthophotomap is an orthophoto or mosaic of multiple orthophotos to which other optional information may be added, such as a map grid, north arrow, scale bar, contour lines, names, text, marginalia, and/or other information characteristic of a map, but missing on the orthophoto or orthophotos from which produced.

Outlier: An error larger than justified by a dataset of errors having a normal distribution, often assumed to be errors larger than three times the standard deviation, i.e., "3-sigma" which is equivalent to the 99.75% confidence level for a normal distribution.

Overage: Those parts of a swath that are not necessary to form a complete single, non-overlapped, gap-free coverage with respect to the adjacent swaths. The non-tenderloin parts of a swath. In collections designed using multiple coverage, overage are the parts of the swath that are not necessary to form a complete non-overlapped coverage at the planned depth of coverage. In LAS specification version 1.4–R13 (ASPRS, 2011), these points are identified by using the incorrectly named "overlap" bit flag. See overlap, tenderloin.

Overlap: Any portion of a swath that is also covered by any portion of any other swath. The term overlap is incorrectly used in the LAS v1.4 specification to describe the flag intended to identify overage points. See overage, tenderloin. In photogrammetry, forward overlap (also called end lap) is the overlapping of successive images along a flight strip, and sidelap is the overlap of adjacent flight strips, both customarily expressed as a percentage.

P

Panchromatic: Black-and-white imagery that is sensitive to all colors of the visible spectrum.

Panel Points: Ground control points used for aerotriangulation and designed to present clearly visible points that can be accurately measured on aerial photographs. They may be removable plastic or wooden panels, for example, or painted over the reference point, normally as "X", "V", or "T" symbols, and surveyed at the center of the intersecting lines of these symbols. The symbols are sized to be accurately measured based on the image scale.

Parallax: Small changes in the direction from sensor to object as the sensor or object moves. The apparent displacement of the position of an object with respect to a reference system, because of a shift in the location of the observer. Because photogrammetry is based on the imaging of the terrain from two different perspectives, parallax is essential for determining 3-D measurements from stereo imagery.

Parallel: A line which has the same latitude at every point.

Pass Point: A photo-identifiable point with unknown ground coordinates whose horizontal and/or vertical position is determined from photogrammetric methods (aerotriangulation) and is intended for use as a control point in the relative orientation of adjoining photographs/images in the same flight line. An image point that appears in the overlap area of two or more photos in the same strip of aerial photography so as to "pass" control from one photograph to the next in a single flight line. Pass points typically fall within the triple-overlap areas when the aerial photography has a forward overlap of at least 60%, meaning that some points on the ground appear on three consecutive aerial images, adding strength to the aerotriangulation process. See tie point (similar to pass point but pertaining to an image point that appears in the overlap area of two or more photos in adjacent flight lines).

P-Code: Precision Code (or Private Code) in relationship with the usage of GPS for military applications. It has a 10.23 mHz rate, a period of 7 days and is the principal navigation ranging code.

Peak: A point around which all slopes are negative.

Percentile: A measure used in statistics indicating the value below which a given percentage of observations in a group of observations fall. For example, the 95th percentile is the value below which 95 percent of the observations may be found, and 5% will be larger. For accuracy testing, percentile calculations are based on the absolute values of the errors, as it is the magnitude of the errors, not the sign that is of concern. Although Microsoft Excel and other software packages use slightly different formulas to compute percentiles, most methods yield similar results, particularly with large sample sizes. For example, with the absolute value of 20 errors in cells A1 through A20 of an Excel spreadsheet, the 95th percentile would be computed using "=PERCENTILE(A1:A20,0.95)."

Photogrammetry: The art, science, and technology of obtaining reliable information about physical objects and the environment through the process of recording, measuring, and interpreting photographic images and patterns of electromagnetic radiant energy and other phenomena.

Photon-Sensitive Lidar: Lidar sensors that detect individual or groups of photons that bounce off the target and are received by the sensor. Photon counting detectors record the arrival of single photons. The detector is highly sensitive to individual photons and can be easily triggered by stray photons from the same laser source or photons from another light source such as sunlight. Photon-sensitive lidars generally have a very low signal-to-noise ratio.

Ping Rate: Number of sonar output pulses per second.

Pit: A point around which all slopes are positive.

Pitch: The angular deviation from the horizontal of a craft or sensor about the horizontal axis normal to its direction of movement. The pitch is positive if it causes the craft's front end to rise. In photogrammetry, the term is applied to a rotation of the camera, or of the coordinate system of the photograph, about either the y-axis of the photograph's coordinate system or about the y-axis of the ground coordinate system Pixel: A two-dimensional raster cell, normally used for computer display of image rasters such a digital orthoimage or lidar intensity image. See cell.

Pixel Resolution or Pixel Size: The ground size of a pixel in a digital orthoimage after all rectifications and resampling procedures.

Plane Survey: A survey that treats the surface of the Earth as though it were a plane.

Planimetrics: Data about non topographic features on the Earth's surface that are represented only by their horizontal position.

Playa: An undrained desert basin that periodically fills with water to form a temporary lake. Playas drain directly into the ground. They can often be identified by having apparent inlet channels but no outlets, and with flat and level floors.

Plumb Line: A line perpendicular to all equipotential surfaces of the Earth's gravity field that intersect with it. It is a line of force in the Earth's gravity field and thus is curved rather than straight.

p-n Junction: A boundary or interface between two types of semiconductor materials, p-type and n-type, inside a single crystal of semiconductor. The "p" (positive) side contains an excess of holes, while the "n" (negative) side contains an excess of electrons in the outer shells of the electrically neutral atoms there. This allows electrical current to pass through the junction only in one direction. Credit: Wikipedia.

Point Classification: The assignment of a target identity classification to a particular lidar point or group of points, allowing differentiation between bare-earth terrain points, water, noise, vegetation, buildings, other man-made features and objects of interest.

Point Cloud: One of the fundamental types of geospatial data (others being vector and raster), a point cloud is a large set of 3-D points, typically from a lidar collection. As a basic GIS data type, a point cloud is differentiated from a typical point dataset in several key ways: (1) Point clouds are almost always 3-D, though some consider them 4-D (x/y/z/i) because they typically also include intensity values for each point; (2) Point clouds have an order of magnitude more features than point datasets; and (3) Individual point features in point clouds do not typically possess individually meaningful attributes; the information value in a point cloud is derived from the relations among large numbers of features. See raster, vector.

Point Family: The complete set of multiple returns reflected from a single lidar pulse.

Position: The location of a point on the surface of the Earth, expressed in terms of one of several coordinate systems. Examples are geographic position (latitude, longitude and altitude); UTM northing, easting and height; or State Plane northing, easting and height.

Positional Dilution of Precision (PDOP): An indicator of a GPS receiver's positional accuracy that can be derived from the geometry of the visible GPS satellites in relation to the GPS receiver. The 3-D positional accuracy is approximated by the product of PDOP and the statistical pseudorange errors.

Postprocessing: In lidar, postprocessing refers to the processing steps applied to lidar data, including point classification, feature extraction (for example, building footprints, hydrographic features, and others), tiling, and generation of derivative products (DEMs, DSMs, intensity images, and others). For comparison, see preprocessing.

Precision (repeatability): The closeness with which measurements agree with each other, even though they may all contain a systematic bias. With geospatial data, standard deviation(s) is a common measure of precision whereas root-mean-square-error (RMSE) is a common measure of accuracy. See accuracy.

Preprocessing: In lidar, preprocessing of data most commonly refers to those steps employed in converting the collected GPS, IMU, instrument, and ranging information into an interpretable x/y/z point cloud. This includes generation of trajectory information, calibration of the dataset, and controlling the dataset to known ground references.

Primary Control: Survey stations surrounding a local network used to tie the local network to the National Spatial Reference System (NSRS).

Principal Point: The intersection of the optical axis of the camera with the image and origin of an image coordinate system. Film cameras use imprinted "fiducial points" at the corners and sides of each image that define the orientation of the image coordinate system and a context for locating the principal point. Lens distortion is generally modeled as being both radial and tangential to the principal point.

Profile: The side view of a cross section of a terrain surface.

Profilometer: A device to measure the vertical profiles of surfaces on the Earth.

Projection: A function relating points on one surface to points on another surface so that for every point on the first surface corresponds exactly to one point on the second surface. A map projection is a special case in requiring that one of the surfaces be a spheroid or ellipsoid and the other be a developable surface (normally a plane or a cylinder or cone that can be "cut" and flattened into a plane).

Pseudorange: A GPS receiver's computed pseudorange to each satellite is its estimate of the time of signal propagation from the satellite to the GPS antenna multiplied by the speed of light. The computed pseudoranges comprise the true ranges between the satellites and the GPS antenna plus several pseudorange errors, the most dominant being the receiver clock error.

Puddle: One or more DEM grid cells totally surrounded by cells of higher elevation.

Pulse Duration: Time, measured in nanoseconds, during which the optical power of a laser pulse is at Full Width at Half Maximum (FWHM). Also called "pulse width." The pulse duration determines the range resolution of the pulse in multiple return systems, or the minimum distance between consecutive returns from a pulse. Traditionally, pulse durations for topographic systems have been in the range of 10 ns, where the laser pulse is ~3 m long, affecting its ability to discriminate between targets that are close to each other as with low vegetation. Newer laser technology has enabled the use of much shorter pulse durations (1-2 ns) for topographic and topobathymetric applications. Photon-detecting sensors typically have very low laser power and can have pulse durations (pulse widths) in the 600-800 picosecond range.

Pulse Energy: The total energy content of a laser pulse measured in micro joules (μJ). The pulse energy of topographic lidar systems are typically low (10-100 μJ) to allow for a tightly focused beam with low beam divergence that is also eye safe. Bathymetric lidar systems have pulse energies up to [illegible] μJ, which are typically much higher than the near-infrared laser used in topographic applications; the higher power is needed for the laser pulse to penetrate through the water column to map the bottom. Bathymetric sensors with very high laser pulse power also have a large footprint so that the energy is spread across a larger area for eye-safety reasons.

Pulse Repetition Frequency (PRF): The number of emitted laser pulses per second also known as Pulse Repetition Rate. For example, a 200 kHz lidar system generates 200,000 pulses of laser energy in one second.

Pulse Width: See Pulse Duration

Q

Quadrangle (Quad): A map or plat of a rectangular or nearly rectangular area usually bounded by given meridians of longitude and parallels of latitude.

Quality: Conformance to defined specifications in terms of performance, price and delivery.

Quality Assurance (QA): Steps taken: (1) To ensure the government receives the quality products it pays for and/or (2) To ensure an organization's Quality Program works effectively. QA aims to prevent defects with a focus on the process used to make the product.

Quality Control (QC): Steps taken by data producers to ensure delivery of products that satisfy standards, guidelines and specifications identified in the Scope of Work. These steps typically include production flow charts with built-in procedures to ensure quality at each step of the work flow, in-process quality reviews, and/or final quality inspections prior to delivery of products to a client. QC is a set of activities for ensuring quality in products by identifying and correcting defects in the actual products produced.

Quality Plan: A document that identifies specific customer requirements, quality standards, specifications, practices, resources, and sequence of activities necessary to deliver conforming products or services.

Quality Program: A program that ensures an organization is adhering to quality standards by systematically evaluating the adequacy and appropriateness of products and/or services.

R

Radar: An instrument for determining the distance and direction to an object by measuring the time needed for radio signals to travel from the instrument to the object and back, and by measuring the angle through which the instrument's antenna has traveled.

Radar, Synthetic-Aperture: A radar containing a moving or scanning antenna; the signals received are combined to produce a signal equivalent to that which would have been received by a larger, stationary antenna.

Random Error: See Error.

Range (IFSAR): The distance in a direction perpendicular to the flight path (cross path direction) imaging the terrain below. Range or cross track resolution is achieved by finely gating the received echo in time.

Rangefinder: A device which uses laser energy for determining the distance from the device to an object.

Range Walk: An error in laser ranging due to un-corrected time measurement error or ambiguities in a target surface. In laser ranging, range walk occurs when highly reflective targets appear higher than the surrounding less-reflective targets.

Raster: One of the fundamental types of geospatial data (others being vector and point cloud), a raster is an array of cells (or pixels) that each contain a single piece of numeric information representative of the area covered by the cell. Raster datasets are spatially continuous; with respect to DEMs this quality creates a surface from which information can be extracted from any location. As spatial arrays, rasters are always rectangular; cells are most often square. Co-located rasters can be stored in a single file as layers, as with color digital images. See vector, point cloud.

Real Time Kinematic (RTK): A method of utilizing carrier phase differential GPS with on-the-fly ambiguity resolution transmitted from a base station with known coordinates to a rover receiver via a radio link. This technique will compute a position in real time, relative to the base station within an accuracy of a few centimeters.

Receiver: A device that collects and records laser pulse energy returning from the ground, water surface, water column, and seafloor.

Rectification: The process of producing, from a tilted or oblique photograph, a photograph from which displacement caused by tilt has been removed. Ortho-rectification, in addition to correction of tilt displacement, also corrects for perspective and relief displacement.

Reflectance: The measure of the proportion of light or other radiation striking a surface that is reflected off it. In remote sensing, an object's reflectance is the amount of energy that arrives back to a receiver and is directly proportional to the percentage of energy that reflects off the object. The reflectance of an object is wavelength-dependent; because lidar systems are monochromatic, the reflectance at that particular wavelength determines how detectable an object is given the laser power.

Relative Accuracy: See Accuracy.

Relief: Topography. The deviation of a surface, or portions thereof, from some surface such as a reference ellipsoid.

Relief Displacement: The displacement of an image, outward from the center of an aerial photograph, caused by the elevation (relief) of features above an established base elevation.

Resolution: The smallest unit a sensor can detect or the smallest unit a raster DEM depicts. The degree of fineness to which a measurement can be made. "Resolution" is also used to describe the linear size of an image pixel or raster cell.

Riparian: Relating to or situated or dwelling on the bank of a river, lake, tidewater, or other body of water.

Roll: The angular deviation from the horizontal of a craft about its longitudinal axis, i.e., its direction of movement. The roll is positive if it causes the craft's left wing to rise. In photogrammetry, the angular deviation of an aerial camera or coordinate system of a photograph about either the photograph's x-axis or about the x-axis of the ground coordinate system.

Root Mean Square Difference (RMSD): The square root of the average of the set of squared differences between two dataset coordinate values taken at identical locations. The term RMSD differentiates from RMSE because neither dataset is known to be more or less accurate than the other, and the differences cannot be regarded as errors. RMSD is used in lidar when assessing relative accuracy, both intraswath and interswath. See root mean square error (RMSE).

Root Mean Square Error (RMSE): The square root of the average of the set of squared differences between dataset coordinate values and coordinate values from an independent source of higher accuracy for identical points. The RMSE is used to estimate the absolute accuracy of both horizontal and vertical coordinates when standard or accepted values are known, as with GPS-surveyed check points of higher accuracy than the data being tested. In the U.S., the independent source of higher accuracy is expected to be at least three times more accurate than the dataset being tested. The absolute accuracy of elevation datasets are commonly evaluated in terms of $RMSE_r$, $RMSE_x$, $RMSE_y$ and $RMSE_z$:

- $RMSE_r$ The horizontal root mean square error in the radial direction that includes both x and y coordinate errors.

  $$\sqrt{(RMSEx^2 + RMSE_y{}^2)} \qquad (3)$$

  Where:
  $RMSE_x$ is the RMSE in the x direction, and
  $RMSE_y$ is the RMSE in the y direction.

- $RMSE_x$ The horizontal root mean square error in the x direction (easting).

  $$\sqrt{\sum \frac{(x_n - x^1{}_n)^2}{N}} \qquad (4)$$

  Where:
  x_n is the set of N x coordinates being evaluated,
  x_n is the corresponding set of check point x coordinates for the points being evaluated,
  N is the number of x coordinate check points, and
  n is the identification number of each check point from 1 through N.

- $RMSE_y$ The horizontal root mean square error in the y direction (northing).

  $$\sqrt{\sum \frac{(y_n - y^1{}_n)^2}{N}} \qquad (5)$$

  Where:
  y_n is the set of N y coordinates being evaluated,
  y_n is the corresponding set of check point y coordinates for the points being evaluated,
  N is the number of y coordinate check points, and
  n is the identification number of each check point from 1 through N.

- $RMSE_z$ The vertical root mean square error in the z direction (elevation).

  $$\sqrt{\sum \frac{(z_n - z^1{}_n)^2}{N}} \qquad (6)$$

  Where:
  z_n is the set of N z values (elevations) being evaluated,
  z_n is the corresponding set of check point elevations for the points being evaluated,
  N is the number of z check points, and
  n is the identification number of each check point from 1 through N.

S

Saddle: The lowest area between two opposing higher terrain features, usually connecting two areas of lower terrain, as in a mountain. Used herein to reference the area exposed by the removal or exclusion of a bridge deck from a DEM.

Scale: The ratio of the size of the image or representation of an object on a photograph or map to its true size. Scale may be expressed as a representative fraction (e.g., $^1/_{1,200}$) or ratio (e.g., 1:1,200), or it may be expressed as the number of feet to an inch (e.g., 100-scale). Because 1/2,000th of something is smaller than 1/1,200th of something, a 1:2,000-scale map is considered to be a smaller scale map than a 1:1200-scale map.

Scale Factor: (1) A number by which a distance obtained from a map by computation or measurement is multiplied to obtain the actual distance on the datum of the map. (2) The relation between the input and output of the different IMU sensor sets; scale factor effects are most apparent in times of high acceleration and rotation.

Scan Angle: A value in degrees between -90 and +90. At 0 degrees, the laser pulse is directly below the aircraft at nadir. Most lidar systems use scan angles less than ±30 degrees. See Field of View.

Scan Direction: The direction the laser scanning mirror is traveling at the time of the output laser pulse. A value of 1 is a positive scan direction (the scanner is moving from the left side to the right side of the in-track flight direction), and a value of 0 is a negative scan direction (the scanner is moving from the right side to the left side of the in-track flight direction).

Scanner: A device that directs laser pulses into a pattern on the ground or water surface below the aircraft.

Scan Period: The period of time, measured in nanoseconds, taken by a scanner to complete a scan pattern, and return to its starting point.

Scan Rate: The number of times per second a scanning device samples its field of view, measured in Hz.

Sea Level: In general, the reference elevation of the surface of the sea from which elevations are measured. This term is used as a curtailed form of "mean sea level."

Sea Level Datum: An obsolete term. See National Geodetic Vertical Datum of 1929 and Mean Sea Level.

Secchi Depth: An intuitive water clarity measure that is the depth at which a standard black and white disc, deployed over the side of a boat, is no longer visible to the human eye.

Secondary Tide Stations: Water level stations, located within the mouth of a bay or estuary, that have operated for more than 1 year but less than 19 years, and have a planned finite lifetime. When reduced by comparison with simultaneous observations at a suitable control tide station, they provide accurate control in bays and estuaries where localized tidal effects are not realized at the nearest control station. They also provide data for the reduction of soundings in connection with hydrographic surveys.

Semidiurnal: Having a period or cycle of approximately one-half of a tidal day. The predominant type of tide throughout the world is semidiurnal, with two high waters and two low waters each tidal day. The tidal current is said to be semidiurnal when there are two flood and two ebb periods each day.

Semi-Global Matching (SGM): A form of digital image correlation, developed by Heiko Hirschmüller, used for automated generation of DSMs in which a global cost function based on disparity is optimized at each pixel along eight separate paths, thereby striking a balance between surface continuity and precise disparity extraction.

Shadow: A dark area in a radar image where the radar beam cannot reach a portion of the terrain being imaged because it is occulted by other parts of the terrain or other objects in the scene. Shadowed regions have no useful interferometric signal and consequently no elevation values can be determined.

Shoreface: The narrow zone, seaward or lakeward from the shoreline at low water, that is permanently covered by water and over which the sands and gravel actively move with the action of the waves.

Shorelands: A general term including tidelands and navigable freshwater shores below the ordinary high water mark.

Shoreline: A boundary component in legal descriptions, with different tidal datum variations, as the point of origin for jurisdictional boundaries and as the boundary between public and private ownership. Official definitions vary by state in the U.S. See Mean High Water (MHW), Mean Higher High Water (MHHW), and Mean Lower Low Water (MLLW).

Sidelap: The overlap between adjoining swaths of lidar data or adjoining strips of aerial photography.

Signal-to-Noise Ratio (SNR): The quantitative ratio between the part of an event that is considered signal to the part that is considered noise.

Single Frequency: A GPS receiver that tracks only the C/A pseudoranges and carrier phases on the L1 carrier from each satellite.

Single Photon Counting Lidar: A lidar sensor that detects individual photons that bounce off the target and are received by the sensor. Photon counting detectors record the arrival of single photons. The detector is highly sensitive to individual photons and can be easily triggered by stray photons from the same laser source or photons from another light source such as sunlight. Photon-sensitive lidars generally have a very low signal-to-noise ratio.

Skew: A measure of symmetry or asymmetry within a dataset. Symmetric data will have skewness towards zero.

Slope: The measure of change in elevation over distance, expressed either in degrees or as a percent. For example, a rise of 4 meters over a distance of 100 meters describes a 2.3° or 4% slope. The maximum rate of change in elevation, either from cell to cell in a gridded surface or of a triangle in a TIN. Every triangle in a TIN has a slope value; the lower the slope value, the flatter the terrain; the higher the slope value, the steeper the terrain.

Softcopy Photogrammetry: Stereo photogrammetric procedures that utilize digital imagery in digital stereo photogrammetric workstations (DSPWs): Also called softcopy workstations: that have significant advantages compared with analytical stereoplotters. These advantages include: automatic digital image correlation, efficient production of DEMs and digital orthophotos, and superimposition of all types of geospatial data over digital imagery. For DEM generation, superimposition means that all elevation mass points, breaklines and contours can be reviewed in stereo against the actual ground form, and old 3-D data can be superimposed on new stereo models to see where DEMs, breaklines or contours need to be revised.

Sonar: An apparatus that detects the presence of, or determines the distance or direction of, an object underwater by receiving and interpreting sound from the object. The term is applied principally to an apparatus that itself generates the sound; the object then reflects or scatters the sound back to the apparatus.

Sounding: A measurement of depth in water: Usually a measurement of the distance of the bottom below the boat or ship from which the measurement is taken. The usual method is to transmit a sonic pulse downwards and to measure the time required for the pulse to travel to the bottom and back. Since the path of the pulse is considerably bent by refraction and the pulse itself also penetrates some distance into the bottom, both factors must be accounted for in depth calculations. In shallow water, alternative techniques are used, to include airborne lidar bathymetry or weighted graduated plumb lines.

Sound Wave: Normally used interchangeably with "acoustic wave" for frequencies within the range of human hearing (sound) plus ultrasonic and sub-sonic frequencies that are respectively above and below the range of human hearing. Acoustic waves are based on vibrations of the actual material of the medium and are manifested as periodic variations of pressure in the medium.

Spatial Distribution: In lidar, the regularity or consistency of the point density within the collection. The theoretical ideal spatial distribution for a lidar collection is a perfect regular lattice of points with equal spacing on X and Y axes. Various factors prevent this ideal from being achieved, including the following: (1) Instrument design (e.g., oscillating mirrors); (2) Mission planning (difference between along-track and cross-track point spacing); and (3) In-flight attitude variations (roll, pitch, and yaw).

Special Flood Hazard Area (SFHA): The land area in the floodplain within a community subject to a 1 percent or greater change of flooding in any given year.

Spot Elevations: Elevation points collected at identifiable points, to leave behind a record of points that can be checked, or at local maxima or minima not represented by contour lines. Spot elevations are most typically collected for hill tops, dam spillway elevations, tops of selected bridge decks and road intersections, and bottoms of volcano craters and local sink holes. Also called spot heights.

Spring Tide: A tide that occurs at the time of the new and full moon when tides have their greatest amplitude. The times when highest and lowest waters are recorded.

Standard (1): An agreed-upon procedure within a particular industry or profession, that is to be followed in producing a particular product or result.

Standard (2): A number, or set of numbers, established within an industry, a science, or a technology, setting limits on precision or accuracy with which operations, measurements, or products are to be made.

Standard Deviation: A measure of spread or dispersion of a sample of errors around the sample mean error. It is a measure of precision, rather than accuracy. The standard deviation does not account for uncorrected systematic errors.

State Plane Coordinate System: One of the plane rectangular coordinate systems, one for each State in the U.S., established by the U.S. Coast and Geodetic Survey in 1933 for use in defining locations of geodetic stations in terms of plane-rectangular Cartesian coordinates. Each State is mapped by a conformal map projection in one or more zones, over each of which is placed a rectangular grid. Zones of limited east-west extent and indefinite north-south extent are mapped by a Transverse Mercator map projection; zones of limited north-south extent and indefinite east-west extent are mapped by the Lambert Conformal Conic map projection with two standard parallels. Zone One of Alaska is on the Oblique Mercator map projection. The use of the projections assures that, for zones having a width of 250 km, the greatest departure from exact scale (scale error) is 1 part in 10,000.

Stereo Comparator: A photogrammetric instrument that permits stereoscopic viewing of two photographs and measurement of the coordinates of corresponding points, the coordinates of one point and the stereoscopic parallax of its conjugate point on the other image.

Stereo Compilation: Drafting of a map or chart manuscript from stereo aerial images and geodetic control data by means of photogrammetric procedures.

Stereo Model: The surface area of elevation and feature models visible in 3-D by viewing the overlapping areas of stereo imagery in an analog, analytical, or softcopy stereoplotter.

Stereoscopic: Pertaining to the use of binocular vision for observation of a pair of overlapping photographs or other perspective views, giving the impression of depth.

Stereoscopic Image: A three-dimensional impression given by viewing one of a pair of overlapping pictures of an object with the left eye and the other with the right eye, either simultaneously or in rapid succession.

Strip: A set of overlapping photographs that can be arranged in sequence so that, except for the last photograph, part of the object space shown in one photograph is also shown in the succeeding photograph. A set of overlapping photographs obtained sequentially from a moving aircraft or satellite.

Striping: A DEM anomaly caused by photogrammetric profiling where the floating mark is allowed to pass either above the ground or below the ground on a profiling pass.

Submerged Lands: Lands covered by water at any stage of the tide.

Subsidence: The loss of land surface elevation due to removal of subsurface support.

Superimposition: The imposing, placing, and/or registration of geospatial data over orthoimagery so all can be seen at once in their correct relative positions. See Softcopy Photogrammetry.

Supplemental Control: Surveys between primary control points to establish additional points necessary to control detailed mapping in a local network.

Supplemental Vertical Accuracy (SVA): Merged into the Vegetated Vertical Accuracy (VVA) in the ASPRS Positional Accuracy Standard for Digital Geospatial Data, SVA is the NDEP guideline's term for reporting the vertical accuracy at the 95th percentile in each separate land cover category where vertical errors may not follow a normal error distribution. See percentile, vegetated vertical accuracy.

Surface: A 3-D geographic feature represented by computer models built from uniformly- or irregularly-spaced points with x/y coordinates and z-values.

Swath: The data resulting from a single flightline of lidar collection.

Systematic Error: See Error.

T

Tenderloin: The central portion of a lidar swath that, when combined with adjacent swath tenderloins, forms a complete, single, non-overlapping, gap-free coverage. In collections designed using multiple coverage, tenderloins are the parts of the swath necessary to form a complete non-overlapped, gap-free coverage at the planned depth of coverage. See overage, overlap.

Terrain: An ESRI terrain is a multi-resolution, TIN-based surface that must reside in an ESRI Geodatabase feature dataset and is built from feature classes stored within the same feature dataset. It is a dataset that creates a continuous surface by connecting a set a vertices or xyz points which become the nodes of the underlying triangles. The edges between nodes are interpolated and build the triangular faces which create the continuous surface. Because terrains are TIN-based, terrains identify and build the same topological data structure between adjoining triangles as TINs. Terrain pyramid levels are defined for each terrain dataset, and each pyramid level is a different resolution associated with specified map scale ranges and relative accuracy that shows a different amount of surface detail. Once a terrain is built, these pyramid levels can be viewed on-the-fly in order to quickly visualize data at multiple resolutions. Terrains are different from TINs in that they are capable of storing much greater amounts of information. Terrain datasets can be used to produce TINs and gridded DEMs. While ESRI terrain datasets have become common and prevalent within the geospatial industry, other software may have their own versions of terrain surfaces. Often these surfaces are created through a TIN-based structure, but will not have associated multi-resolution pyramid levels and may not be compatible with geodatabase storage.

Terrain Visualization: Viewing the terrain in three dimensions, normally from differing perspectives.

The National Map: A seamless, continuously maintained set of public domain geographic base information, in digital form, that replaces the USGS (hardcopy) topographic map program. It includes orthoimagery, elevation, geographic names, hydrography, boundaries, transportation, structures, and land cover framework data themes. It is accessible via the Web, as products and services, and as downloadable data. It serves as a foundation for integrating, sharing, and using other geospatial data easily and consistently. It is part of the National Geospatial Program (NGP).

Three Dimensional (3-D): Having horizontal (x/y) coordinates plus elevations (z-values)

Tidal Day: The time of rotation of the Earth with respect to the moon. Its mean value is approximately equal to 24.84 hours.

Tide: The alternating rise and fall of the oceans with respect to the land, produced by differential variations in the gravitational attraction of the moon and sun. There are three types of tides: Diurnal, semidiurnal, and mixed. A diurnal tide has one high tide and one low tide per tidal day. A semidiurnal tide has two high tides and two low tides each tidal day. A mixed tide is similar to a semidiurnal tide except that the two high tides and two low tides of each tidal day have marked differences in their heights.

Tidelands: The zone between the mean high water and mean low water lines. It is identical with intertidal zone when the type of tide is semidiurnal or diurnal.

Tide Gauge: A device for measuring the rise and fall and the current height of the tide. The simplest form is a graduated staff placed vertically in the water; the height of the water is read visually from markings on the staff. A more common version has a float with pointer attached to the staff; the float is free to move up and down the staff with the tide. Readings may be made visually, or may be recorded on paper or electronically.

Tie Point: An image point that appears in the sidelap area of photos in overlapping strips of aerial photography, used to "tie" adjoining strips together. When the aerial photography is flown with 60% forward overlap and 30% sidelap, and when tie points are measured at the top and bottom corners of each image with adjoining strips, tie point images will appear on 4-6 images, adding extra strength to the aerotriangulation process. See pass point (similar to tie point but pertaining to an image point that appears in the overlap area of two or more photos in the same flight line).

Tilt Displacement: The displacement of features in an image caused by the roll and pitch of the aircraft at the instant a photograph is taken.

Topography: The 3-D shape of the terrain surface and the identification of specific landforms. Topography is displayed graphically by contour lines, elevation tints, and/or shaded relief. Topography is displayed digitally by mass points, breaklines, TINs, DEMs, DTMs, and/or hillshades. Also called relief. Topography is normally mapped in terms of elevations above a geodetic datum, whereas bathymetry, the underwater equivalent to topography, is normally mapped in terms of depths below a tidal datum.

Topology: The spatial relationships which exist between adjacent digital objects. The digitization of node, point, line and area features so that computers are able to determine relationships that humans recognize logically when they determine what feature is connected to, adjacent to, or in close proximity to a point, line, area or surface; what feature intersects a line, area or surface; or what features are contained within an area or surface. Examples include precise coincidence of ending and beginning vertices of lines in a continuous network (stream connectivity), and consistent line direction (from high to low) corresponding to stream flow direction.

Triangulated Irregular Network (TIN): A vector data structure that partitions geographic space into contiguous, non-overlapping triangles. In lidar, the vertices of each triangle are lidar points with x, y, and z values. The TIN data structure is based on irregularly-spaced point, line, and polygon data interpreted as mass points and breaklines and stores the topological relationship between triangles and their adjacent neighbors. In most geographic applications, TINs are based on Delaunay triangulation algorithms in which no point in any given triangle lies within the circumcircle of any other triangle. The TIN model may be preferable to a DEM when it is critical to preserve the precise location of narrow or small surface features, such as levees, ditch or stream centerlines, isolated peaks or pits in the data model.

Two Dimensional (2-D): Having horizontal (x/y) coordinates only, without elevations (z-values)

U

Uncertainty (of measurement): A parameter that characterizes the dispersion of measured values, or the range in which the "true" value most likely lies. It can also be defined as an estimate of the limits of the error in a measurement where "error" is defined as the difference between the theoretically-unknowable "true" value of a parameter and its measured value. Standard uncertainty refers to uncertainty expressed as a standard deviation.

Undulation of the Geoid: The rise and fall of the geoid, sometimes used synonymously with geoid height. Used with the geoid separation.

Universal Time (UT): Same as Greenwich mean time (GMT).

Universal Transverse Mercator (UTM) Coordinate System: A 2-Dimensional Cartesian coordinate system used to identify horizontal locations on the Earth independently of vertical positions. The system divides the Earth into 60 zones, each a six-degree band of longitude, and uses a secant transverse Mercator projection in each zone.

Upland: Land above the mean high water line (shoreline) and subject to private ownership, as distinguished from tidelands, the ownership of which is prima facie in the state but also subject to divestment under state statutes. See tidelands.

V

Vector: One of the fundamental types of geospatial data (others being raster and point cloud), vectors include a variety of data structures that are geometrically described by x and y coordinates, and potentially z values. Vector data subtypes include points, lines, and polygons. A DTM composed of mass points and breaklines is an example of a vector dataset; a TIN is a vector surface. See raster, point cloud.

Vegetated Vertical Accuracy (VVA): An estimate of the vertical accuracy, based on the 95th percentile, in vegetated terrain where errors do not necessarily approximate a normal distribution. The VVA is determined with well-distributed checkpoints located in all vegetated land cover categories combined, including tall weeds and crops, scrub forests, and fully forested areas, as well as other localized vegetation categories such as mangrove or sawgrass.

Vegetation Removal: Classification of reflective surface elevations on vegetation so as to depict the elevation of the bare-earth terrain beneath the vegetation. Lidar mass points on vegetation are not actually removed, but classified to a category other than bare earth.

Veronoi (Thiessen) Polygon: A polygon that contains areas which are nearer to the point around which it is constructed than to any other point.

Vertical: The direction in which the force of gravity acts. Whereas the vertical is the perpendicular to an equipotential surface of gravity (e.g., the geoid), a normal is the perpendicular to a given ellipsoid.

Vertical Accuracy: See Accuracy.

Vertical Datum: See Datum.

Vertical Error: The displacement of a feature's recorded elevation in a dataset from its true or more accurate elevation.

Vertical Photograph: A photograph (or image) taken with the camera axis directed downward along (or nearly along) a vertical line.

Viewshed: See Intervisibility.

Void: Portions of a digital elevation dataset where no elevation data are available. In USGS DEMs, each elevation post located within a void area is assigned a discrete false value representing the void. Treatment of void areas should be documented in the metadata file.

Voxel: A three-dimensional "cube" that serves as a counterpart to a traditional pixel, used for three-dimensicnal visualization of elevation data. A 3-D pixel.

W

Water: A synonym for tide, in the sense of the elevation of the surface of the hydrosphere at a particular longitude and latitude.

- **High Water:** The maximum height reached by a rising tide
- **Higher High Water:** The higher of the two high waters occurring on a particular tidal day
- **Higher Low Water:** The higher of the two low waters occurring on a particular tidal day
- **Low Water:** The lowest height reached by a falling tide
- **Lower High Water:** The lower of two high waters of any tidal day
- **Lower Low Water:** The lower of two low waters of any tidal day
- **Mean High Water (MHW):** A tidal datum. The average height of all high waters recorded at a particular point or station over the National Tidal Datum Epoch
- **Mean Higher High Water (MHHW):** A tidal datum. The average of the higher high water height of each today day observed over the National Tidal Datum epoch
- **Mean Low Water (MLW):** The average height of all low waters recorded at a particular point or station over the National Tidal Datum Epoch
- **Mean Lower Low Water (MLLW):** A tidal datum. The average of the lower low water heights of each tidal day observed over the National Tidal Datum Epoch
- **Mean Water Level:** A datum. The mean surface elevation as determined by averaging the heights of the water at equal intervals of time, usually hourly. Mean water level is used in areas of little or no range in tide

Water Clarity: A general term describing the combined scattering and absorption properties of the water column; the primary limitation for depth performance of bathymetric lidar systems.

Waveform Lidar: Lidar system or data in which the entire reflection of the laser pulse is fully digitized, captured, and stored. Discrete return point clouds can be extracted from the waveform data during post processing. See discrete return lidar.

Weight: A factor by which a quantity is multiplied to increase or decrease the effect of that quantity on the results of an adjustment.

Well-Defined Point: A photo-identifiable point, used in horizontal accuracy testing, that represents a feature for which the horizontal position is known to a high degree of accuracy and position with respect to the geodetic datum.

Well-Distributed: For a dataset covering a rectangular area that has uniform positional accuracy, checkpoints should be distributed so that points are spaced at intervals of at least 10 percent of the diagonal distance across the dataset and at least 20 percent of the points are located in each quadrant of the dataset adapted from the National Standards for Spatial Data Accuracy (NSSDA). As related to lidar accuracy assessment, this guideline may be applied to each land cover class for which checkpoints are being collected if individual land cover categories are well represented in each quadrant.

Wellhead Protection Area: The surface and subsurface area surrounding a water well, or well field, supplying a public water system, through which contaminants are reasonably likely to move toward and reach such water well or well field.

Withheld: Within the LAS file specification, a single bit flag indicating that the associated lidar point is geometrically anomalous or unreliable and should be ignored for all normal processes. These points are retained because of their value in specialized analysis. Withheld points typically are identified and tagged during preprocessing or through the use of automatic classification routines. Examples of points typically tagged as withheld include spatial outliers in either the horizontal or vertical domains and geometrically unreliable points near the edge of a swath.

World Geodetic System (WGS): A standard for use in cartography, geodesy, and navigation including the Global Positioning System (GPS). WGS 84 is the reference coordinate system used by the GPS.

X

X-coordinate: The distance along the x-axis from the origin of a 2-D or 3-D Cartesian coordinate system. An x-coordinate is the first half (the easting) of UTM coordinates or the Easting of State Plane coordinates.

Y

Yaw: The variation, or amount of variation, of the longitudinal axis of a craft from the direction in which the craft is moving, i.e., the difference between the direction in which a craft is pointed and the direction in which it is moving (when typically altered by winds).

Y-coordinate: The distance along the y-axis from the origin of a 2-D or 3-D Cartesian coordinate system. A y-coordinate is the second half (the northing) of UTM coordinates or the Northing of State Plane coordinates.

Z

Z-bump: The shift of an entire elevation dataset either up or down by the same offset value to correct a systematic error that impacted all prior elevations in a dataset. Z-bumps are performed prior to delivery of data to a client and should not be applied to previously published datasets already in the public domain.

Z-coordinate: The distance along the z-axis from the origin of a 3-D Cartesian coordinate system. Note, this is not the same as the elevation or height above the vertical datum.

Z-units: The units of measure used for the z-values in a geographic dataset.

Z-values: The elevations of the 3-D surface above the vertical datum at designated x/y locations.

SAMPLE DATASETS

INTRODUCTION

This appendix provides links for where to find geospatial data primarily related to elevation datasets. Many of the links provided were current as of 2018 and the links that were used were the ones that are likely to be stable for a number of years. As such the user may need to navigate the page to identify specific datasets that are available. This is especially true with federal sites like The National Map from USGS and The Digital Coast from NOAA. Both sites contain a significant amount of geospatial data including DEMs and LiDAR. The types of geospatial data listed are the ones that are most relevant to elevation processing and are not inclusive of all the data available from those sites. Many individual states and countries provide geospatial portals as well to provide this type of data to end users. These sites can be found by searching for the state/country and their geospatial program.

In addition to links where data downloads are available, the U.S. provides an Interagency Elevation Inventory website which allows users to spatially see where elevation data is available. This site is incredibly useful in identifying topographic LiDAR, bathymetric LiDAR, IfSAR, and Sonar data availability. The site attempts to provide an aggregate of elevation data available and provides download links to many of the other sites in this list. For each project shown on the map, information is provided about where to download the data if it is publicly available. The URL to that site is the first one listed in the table.

Data Source	URL		
Interagency Elevation Inventory	https://coast.noaa.gov/inventory/		
Types of Geospatial Data Available			
Topographic Lidar	Bathymetric Lidar	Lidar	IfSAR (DEM)
IfSAR (DSM)	IfSAR (ORI)	SONAR	
Data Source	**URL**		
The National Map	https://nationalmap.gov/		
Types of Geospatial Data Available			
Topographic Maps	DEMs	Lidar	IfSAR (DEM)
IfSAR (DSM)	IfSAR (ORI)	Hydrography	Imagery
Data Source	**URL**		
Digital Coast	https://coast.noaa.gov/digitalcoast/		
Types of Geospatial Data Available			
Topographic Lidar	Bathymetric Lidar	Lidar	IfSAR (DEM)
IfSAR (DSM)	IfSAR (ORI)	Hydrography	UAS

Data Source	URL		
National Centers for Environmental Information	https://www.ngdc.noaa.gov/		
Types of Geospatial Data Available			
Coastal and Marine Elevation Models		Bathymetry	SONAR
Data Source	URL		
Open Topography	https://opentopography.org/		
Types of Geospatial Data Available			
Lidar	DEMs		
Data Source	URL		
Earth Explorer	https://earthexplorer.usgs.gov/		
Types of Geospatial Data Available			
SAR (SRTM)	IfSAR	Lidar	SIR-C
UAS (LiDAR)	UAS(SfM)		

INDEX

All page numbers in bold font lead to term definitions used in this manual

A

All page numbers in bold font lead to term definitions used in this manual

All page numbers in bold font lead to term definitions used in this manual

C

All page numbers in bold font lead to term definitions used in this manual

E

www.ingramcontent.com/pod-product-compliance
Lightning Source LLC
LaVergne TN
LVHW070942210926
848482LV00002B/72
9798895690116